Nonlinear Spectroscopy of Solids

Advances and Applications

NATO ASI Series

Advanced Science Institutes Series

A series presenting the results of activities sponsored by the NATO Science Committee, which aims at the dissemination of advanced scientific and technological knowledge, with a view to strengthening links between scientific communities.

The series is published by an international board of publishers in conjunction with the NATO Scientific Affairs Division

A	**Life Sciences**	Plenum Publishing Corporation
B	**Physics**	New York and London
C	**Mathematical and Physical Sciences**	Kluwer Academic Publishers
D	**Behavioral and Social Sciences**	Dordrecht, Boston, and London
E	**Applied Sciences**	
F	**Computer and Systems Sciences**	Springer-Verlag
G	**Ecological Sciences**	Berlin, Heidelberg, New York, London,
H	**Cell Biology**	Paris, Tokyo, Hong Kong, and Barcelona
I	**Global Environmental Change**	

PARTNERSHIP SUB-SERIES

1. Disarmament Technologies	Kluwer Academic Publishers
2. Environment	Springer-Verlag
3. High Technology	Kluwer Academic Publishers
4. Science and Technology Policy	Kluwer Academic Publishers
5. Computer Networking	Kluwer Academic Publishers

The Partnership Sub-Series incorporates activities undertaken in collaboration with NATO's Cooperation Partners, the countries of the CIS and Central and Eastern Europe, in Priority Areas of concern to those countries.

Recent Volumes in this Series:

Volume 338—Electroweak Physics and the Early Universe
edited by Jorge C. Romão and Filipe Freire

Volume 339—Nonlinear Spectroscopy of Solids: Advances and Applications
edited by Baldassare Di Bartolo

Volume 340—Confined Electrons and Photons: New Physics and Applications
edited by Elias Burstein and Claude Weisbuch

Volume 341—Turbulence: A Tentative Dictionary
edited by P. Tabeling and O. Cardoso

Series B: Physics

Nonlinear Spectroscopy of Solids

Advances and Applications

Edited by

Baldassare Di Bartolo

Boston College
Chestnut Hill, Massachusetts

Assistant Editor

Brian Bowlby

Boston College
Chestnut Hill, Massachusetts

Plenum Press
New York and London
Published in cooperation with NATO Scientific Affairs Division

Proceedings of a NATO Advanced Study Institute on
Nonlinear Spectroscopy of Solids: Advances and Applications,
held June 15–30, 1993,
in Erice, Italy

NATO-PCO-DATA BASE

The electronic index to the NATO ASI Series provides full bibliographical references (with keywords and/or abstracts) to about 50,000 contributions from international scientists published in all sections of the NATO ASI Series. Access to the NATO-PCO-DATA BASE is possible in two ways:

—via online FILE 128 (NATO-PCO-DATA BASE) hosted by ESRIN, Via Galileo Galilei, I-00044 Frascati, Italy

—via CD-ROM "NATO Science and Technology Disk" with user-friendly retrieval software in English, French, and German (©WTV GmbH and DATAWARE Technologies, Inc. 1989). The CD-ROM also contains the AGARD Aerospace Database.

The CD-ROM can be ordered through any member of the Board of Publishers or through NATO-PCO, Overijse, Belgium.

Library of Congress Cataloging-in-Publication Data

On file

ISBN 0-306-44916-1

Printed in the United States of America

For the sake of persons of . . . . different types, scientific truth should be regarded as equally scientific, whether it appears in the robust form and the vivid coloring of a physical illustration, or in the tenuity and paleness of a symbolic expression.

JAMES CLERK MAXWELL

PREFACE

This report presents an account of the course "Nonlinear Spectroscopy of Solids: Advances and Applications" held in Erice, Italy, from June 16 to 30, 1993. This meeting was organized by the International School of Atomic and Molecular Spectroscopy of the "Ettore Majorana" Centre for Scientific Culture.

The purpose of this course was to present and discuss physical models, mathematical formalisms, experimental techniques, and applications relevant to the subject of nonlinear spectroscopy of solid state materials.

The universal availability and application of lasers in spectroscopy has led to the widespread observation of nonlinear effects in the spectroscopy of materials. Nonlinear spectroscopy encompasses many physical phenomena which have their origin in the monochromaticity, spectral brightness, coherence, power density and tunability of laser sources.

Conventional spectroscopy assumes a linear dependence between the applied electromagnetic field and the induced polarization of atoms and molecules. The validity of this assumption rests on the fact that even the most powerful conventional sources of light produce a light intensity which is not strong enough to equalize the rate of stimulated emission and that of the experimentally observed decay. A different situation may arise when laser light sources are used, particularly pulsed lasers. The use of such light sources can make the probability of induced emission comparable to, or even greater than, the probability of the observed decay; in such cases the nonlinearity of the response of the system is revealed by the experimental data and new properties, not detectable by conventional spectroscopy, will emerge.

The course focused on the additional knowledge that is acquired when the spectroscopic investigations are performed using electromagnetic energy strong enough to show the intrinsic nonlinearity of the atom-radiation interaction.

Each lecturer developed a coherent section of the program starting at a somewhat fundamental level and ultimately reaching the frontier of knowledge in the field. The sequence of the lectures was in accordance with the logical development of the subject of the meeting. The formal lectures were complemented by seminars and discussions on specific topics and applications and by problem sessions.

The course was addressed to workers in the field of solid-state physics from universities and industries. Senior scientists in related fields were encouraged to participate.

A total of 79 participants came from 56 laboratories and 20 different countries (Australia, Belgium, Brazil, Denmark, France, Germany, Greece, Ireland, Israel, Italy, Mexico, Poland, Portugal, Russia, Slovenia, Spain, Switzerland, Turkey, the United Kingdom and the United States).

The secretaries of the course were Brian Bowlby and Stamatios Kyrkos.

35 lectures divided in 10 series were given. In addition 12 (one-, two- or three-hour) "long seminars", 2 "special seminars", 27 "short seminars" and 6 posters were presented. The sequence of lectures was in accordance with the logical development of the subject of the meeting. Each lecturer started at a rather fundamental level and ultimately reached the frontier of knowledge in the field.

Two round-table discussions were held. The first round-table discussion took place after three and a half days of lectures in order to evaluate the work done in the first days of the course and consider suggestions and proposals regarding the organization, format and presentation of the lectures. The second one was held at the conclusion of the course, so that the participants could comment on the work done during the entire meeting and discuss various proposals for the next course of the International School of Atomic and Molecular Spectroscopy.

I wish to express my sincere gratitude to Ms. Zaini and Dr. Gabriele, to Ms. Savalli and Mr. Pilarski and to all the personnel of the "Ettore Majorana" Centre, who contributed so much to create a congenial atmosphere for our meeting. I also wish to acknowledge the sponsorship of the meeting by NATO Scientific Affairs Division, The ENEA Organization, Boston College, the European Physical Society, the Italian Ministry of Education, the Italian Ministry of University and Scientific Research and the Sicilian Regional Government.

I would like to thank the members of the Organizing Committee (Doctors Auzel, Baldacchini and MacFarlane and Professor Klingshirn), the secretaries of the course (Brian Bowlby and Stamatios Kyrkos), Angela Siraco, Daniel Di Bartolo and John Di Bartolo for their valuable help.

I would like to acknowledge the many relatives and friends from Trapani and Erice who were close to me during the time of the meeting, especially my brother Francesco and his children, Michele Strazzera, Pino Luppino, Salvatore and Crispino Coppola, Professore Adragna, Signori Piacentino, and my former schoolmates of the Liceo Scientifico of Trapani. I would also like to thank Dr. Elena Rutkowska Buscemi of the Azienda Provinciale del Turismo of Trapani for her very kind assistance during out memorable excursion in Trapani.

Every Erice meeting gives me the opportunity to return to Sicily and to my hometown Trapani. The memories of my early youth come back to my mind with renewed intensity and I savor them with the same relish with which I drink the strong Marsala wine. To direct a course at hte Majorana Centre adds to my feelings because in Erice I spent several pleasant summers with my parents and relatives.

During the preparation of this book I have been making plans for the next meeting of the International School of Spectrscopy in 1995 and I am looking forward to it.

To encounter in my native land so many fine people who participate in my courses, to be able to share the Erice experience with them and to call them friends gives me the opportunity to tie in my thoughts the present with the past and to project towards the future a serene hope.

B. Di Bartolo
Director of the International
School of Atomic and Molecular
Spectroscopy of the "Ettore
Majorana" Centre

Erice, June 30, 1993

CONTENTS

LONG SEMINARS

SPECIAL SEMINARS

SHORT SEMINARS

POSTERS

INTRINSIC NONLINEARITY OF THE INTERACTION BETWEEN RADIATION FIELDS AND ATOMS

B. Di Bartolo

Department of Physics
Boston College
Chestnut Hill, Massachusetts 02167, USA

ABSTRACT

The purpose of this article is to present the background material necessary to deal with the nonlinear effect of the interaction between light and atomic systems.

The first section of this article is an introduction to the treatment of nonlinear effects in solids. The experimental conditions in which they are observed are first considered. The usage of nonlinear spectroscopy is reviewed. This usage takes place at three levels: 1) It may deal with the phenomena themselves, 2) it may deal with information about the nonlinear coupling of light to different classes of material systems, and 3) it may be used to probe the material properties themselves.

The second section presents a treatment of the density matrix formalism, first in a general context, and then for two-level systems. The formalism is applied to the treatment of the interaction of radiation with atoms and molecules. The optical Bloch equations, the Rabi oscillations and the connections of steady state solutions with transition rates are examined. The two characteristic relaxation times, T_1 and T_2, and the mechanisms responsible for them, are introduced.

The third section deals with the quantum treatment of the radiation field. The energy levels and the eigenfunctions of this field are derived.

The fourth section treats the interaction of the quantized radiation field with atomic systems. The special effects of the quantization of the radiation field, such as spontaneous emission, are considered.

In the last two sections two applications are presented, one dealing with the spontaneous emission of two photons from the metastable 2s state of the hydrogen atom, and the other with the radiative (Lamb) shift of atomic levels.

I. INTRODUCTION

The availability and application of lasers have led to the widespread observation of nonlinear effects. Nonlinear spectroscopy encompasses many physical phenomena which have their origin in the monochromaticity, spectral brightness, coherence, power density and tunability of laser sources.

Conventional spectroscopy assumes a linear dependence between the applied electromagnetic fields and the induced polarization of atoms and molecules. The validity of this assumption rests on the fact that even the most powerful conventional sources of

Nonlinear Spectroscopy of Solids: Advances and Applications
Edited by B. Di Bartolo and B. Bowlby, Plenum Press, New York, 1994

light produce a light intensity which is not strong enough to equalize the rate of stimulated emission and that of the experimentally observed decay. A different situation may arise when lasers are used, particularly pulsed lasers. The use of such light sources can make the probability of induced emission comparable to or even greater than the probability of the observed decay; in such cases the nonlinearity of the response of the systems is revealed by the experimental data and new properties not detectable by conventional spectroscopy, will emerge.

Nonlinear spectroscopy can be used at three levels:

1) It may deal with the phenomena themselves:
coherent phenomena (like the optical or ac Stark effect), quantum beats, dressed states, homogeneous linewidth hidden by inhomogeneous broadening in the frequency domain (hole-burning) and in the time domain (photon echoes), etc.

2) It may deal with information about nonlinear coupling to different classes of material systems:
inorganic insulator, semiconductors, organics. For example in Two-Photon Spectroscopy (TPS) the selection rules for the transitions are different that those for One-Photon Spectroscopy (OPS): TPS can be used in exciton resonances which are opaque to OPS.

3) It may be used to probe the material properties themselves:
phonon dynamics, exciton dynamics, quantum size effects, excited state spectroscopy, many particle effect, etc.

II. DENSITY MATRIX: FORMALISM AND APPLICATIONS

II. A. The Density Matrix

The density matrix formalism of quantum mechanical problems may treat effects such as broadening of atomic transitions that cannot be handled by the simpler treatment based on atomic wavefunctions.

Consider a physical system with a wavefunction $\psi(\vec{r},t)$ which satisfies the equation

$$i\hbar \frac{\partial \psi}{\partial t} = H \psi \tag{2.1}$$

Let $\psi_n(\vec{r})$ be a complete set of orthonormal time-independent functions. In the Schroedinger representation we may write

$$\psi(\vec{r},t) = \sum_n a_n(t)\ \psi_n(\vec{r}) \tag{2.2}$$

Then (2.1) can be written as follows

$$i\hbar \sum_n \dot{a}_n\ \psi_n = \sum_m a_m\ H\ \psi_m \tag{2.3}$$

If we multiply by ψ_k^* on the left side of each term and integrate over space coordinates we obtain

$$i\hbar\, \dot{a}_k = \sum_n a_n H_{kn} \tag{2.4}$$

where

$$H_{kn} = \langle\, \psi_k | H | \psi_n \rangle = H_{nk}^* \tag{2.5}$$

Similarly

$$i\hbar\, \dot{a}_m = \sum_n a_n H_{mn} \tag{2.6}$$

Consider the following

$$\begin{aligned} \frac{d(a_k a_m^*)}{dt} &= \dot{a}_k a_m^* + a_k \dot{a}_m^* \\ &= \frac{1}{i\hbar} \sum_n a_n a_m^* H_{kn} + a_k \left(\frac{1}{i\hbar} \sum_n a_n H_{mn} \right)^* \\ &= \frac{1}{i\hbar} \sum_n [H_{kn} (a_n a_m^*) - (a_k a_n^*) H_{mn}] \end{aligned} \tag{2.7}$$

Let

$$\rho_{km} = a_k a_m^* \tag{2.8}$$

Then

$$\frac{d\rho_{km}}{dt} = \frac{i}{\hbar} \sum_n (\rho_{kn} H_{nm} - H_{kn} \rho_{nm}) \tag{2.9}$$

or

$$\dot{\rho} = \frac{i}{\hbar} [\rho, H] \tag{2.10}$$

If the ψ_n functions in (2.2) are N in number, this relation represents N^2 equations in the N^2 unknown matrix elements of ρ. The N^2 constants that appear in the solutions are fixed by the initial conditions which may be specified by giving the entire matrix ρ at time t=0.

Consider now an operator Ω. The expectation value of the observable it represents is given by

$$\langle \Omega \rangle = \langle \psi | \Omega | \psi \rangle = \left\langle \sum_m a_m \psi_m | \Omega | \sum_n a_n \psi_n \right\rangle$$
$$= \sum_m \sum_n a_m a_n^* \langle \psi_m | \Omega | \psi_n \rangle = \sum_m \sum_n \rho_{nm} \Omega_{mn} \tag{2.11}$$
$$= \mathrm{Tr}\,(\rho\, \Omega) = \mathrm{Tr}\,(\Omega\, \rho)$$

If we take the trace of the matrices in (2.10) we can write

$$\sum_m \dot{\rho}_{mm} = \frac{i}{\hbar} \mathrm{Tr}\,(\rho H - H\rho) = \frac{i}{\hbar} [\mathrm{Tr}\,(\rho H) - \mathrm{Tr}\,(H\rho)] = 0 \tag{2.12}$$

Hence

$$\sum_m \dot{\rho}_{mm} = 0 \rightarrow \sum_m \rho_{mm}(t) = \mathrm{const} \tag{2.13}$$

This constant is 1 if the wavefunction $\psi(\vec{r},t)$ is normalized. Once the initial state of ρ is specified, we may obtain the time dependence of the matrix elements of ρ. We may then calculate the product of ρ and any other operator Ω and the time dependence of the expectation value of the observable related to Ω.

II. B. The Effect of a Perturbation

Let H_0 be the unperturbed Hamiltonian of a conservative (t-independent) system, and let $H'(t)$ be an external time-dependent perturbation. In a representation in which H_0 is diagonal, H' will not in general be so. The total Hamiltonian is

$$H = H_0 + H'(t) \tag{2.14}$$

We may write

$$\dot{\rho} = \frac{i}{\hbar} [\rho, H] = \frac{i}{\hbar} [\rho, H_0] + \frac{i}{\hbar} [\rho, H'] \tag{2.15}$$

Then

$$\dot{\rho}_{mn} = \frac{i}{\hbar} \sum_k [\rho_{mk} (H_0)_{kn} - (H_0)_{mk} \rho_{kn}]$$
$$+ \frac{i}{\hbar} \sum_k [\rho_{mk} H'_{kn} - H'_{mk} \rho_{kn}] \tag{2.16}$$

But

$$\begin{aligned}(H_0)_{kn} &= E_n\,\delta_{kn}\\ (H_0)_{mk} &= E_m\,\delta_{mk}\end{aligned} \tag{2.17}$$

Then

$$\begin{aligned}\dot{\rho}_{mn} &= \frac{i}{\hbar}(\rho_{mn}E_n - \rho_{mn}E_m) + \frac{i}{\hbar}\sum_k(\rho_{mk}H'_{kn} - H'_{mk}\rho_{kn})\\ &= \frac{i}{\hbar}(E_m - E_n)\,\rho_{mn} + \frac{i}{\hbar}\sum_k(\rho_{mk}H'_{kn} - H'_{mk}\rho_{kn})\end{aligned} \tag{2.18}$$

and

$$\dot{\rho}_{mm} = \frac{i}{\hbar}\sum_k(\rho_{mk}H'_{km} - H'_{mk}\rho_{km}) = \frac{i}{\hbar}\sum_k(\rho_{mk}H'_{km} - \text{c.c.}) \tag{2.19}$$

We shall now apply the above treatment to a two-level system.

In the representation in which H_0 is diagonal

$$H_0 = \begin{pmatrix} E_a & 0\\ 0 & E_b\end{pmatrix} \tag{2.20}$$

The matrix of a perturbation - $exE(t)$ will look like

$$H'(t) = -\,exE(t) = -\,e\,E(t)\begin{pmatrix} 0 & P_{ab}\\ P_{ba} & 0\end{pmatrix} \tag{2.21}$$

where $P_{ab} = \langle a|x|b\rangle$ and where we have assumed that, because of the definite parity of atomic states:

$$P_{aa} = P_{bb} = 0 \tag{2.22}$$

Then

$$\dot{\rho}_{aa} = \frac{i}{\hbar}(\rho_{aa}H'_{aa} - \rho_{ab}H'_{ba} + \text{c.c.}) = \frac{i}{\hbar}(\rho_{ab}H'_{ba} + \text{c.c.}) \tag{2.23}$$

$$\dot{\rho}_{bb} = \frac{i}{\hbar}(\rho_{ba}H'_{ab} - \rho_{bb}H'_{bb} + \text{c.c.}) = \frac{i}{\hbar}(\rho_{ba}H'_{ab} + \text{c.c.}) \tag{2.24}$$

$$\begin{aligned}\dot{\rho}_{ab} &= -\frac{i}{\hbar}(E_a - E_b)\,\rho_{ab} + \frac{i}{\hbar}(\rho_{aa}H'_{ab} - H'_{ab}\rho_{bb})\\ &= i\,\omega_0\,\rho_{ab} + \frac{i}{\hbar}(\rho_{aa} - \rho_{bb})\,H'_{ab}\end{aligned} \tag{2.25}$$

where

$$\omega_0 = \frac{E_b - E_a}{\hbar} \tag{2.26}$$

Summarizing

$$\begin{aligned}
\dot{\rho}_{aa} &= \frac{i}{\hbar}(\rho_{ab} H'_{ba} - \rho_{ba} H'_{ab}) \\
\dot{\rho}_{bb} &= -\frac{i}{\hbar}(\rho_{ab} H'_{ba} - \rho_{ba} H'_{ab}) = -\dot{\rho}_{aa} \\
\dot{\rho}_{ab} &= i\,\omega_0\,\rho_{ab} + \frac{i}{\hbar}(\rho_{aa} - \rho_{bb})\,H'_{ab} \\
\dot{\rho}_{ba} &= \dot{\rho}^*_{ab} = -i\,\omega_0\,\rho_{ab} - \frac{i}{\hbar}(\rho_{aa} - \rho_{bb})\,H'_{ba}
\end{aligned} \tag{2.27}$$

The expectation value of the operator *ex* is

$$\langle ex \rangle = \mathrm{Tr}\,(ex\rho) \tag{2.28}$$

The relevant matrix is

$$ex\rho = \begin{pmatrix} 0 & P_{ab} \\ P_{ba} & 0 \end{pmatrix} \begin{pmatrix} \rho_{aa} & \rho_{ab} \\ \rho_{ba} & \rho_{bb} \end{pmatrix} = \begin{pmatrix} P_{ab}\,\rho_{ba} & P_{ab}\,\rho_{bb} \\ P_{ba}\,\rho_{aa} & P_{ba}\,\rho_{ab} \end{pmatrix} \tag{2.29}$$

Then

$$\mathrm{Tr}\,(ex\rho) = P_{ab}\,\rho_{ba} + P_{ba}\,\rho_{ab} \tag{2.30}$$

and

$$\langle ex \rangle = \mathrm{Re}\,(2\,P_{ab}\,\rho_{ba}) \tag{2.31}$$

II. C. Interaction of Monochromatic and Polarized Radiation with a Two-level System

Let a polarized electromagnetic wave interact with an atom. Let the nucleus of the atom be at the origin of a system of coordinated as in Fig. 1 and let the electrons have coordinates $\vec{r}_i$; let also x be the direction of polarization of the $\vec{E}$ field. The size of an atom is of the order of the Bohr radius

$$a_0 = \frac{4\pi\varepsilon_0\,\hbar^2}{me^2} = 5 \times 10^{-9}\ \mathrm{cm} << \lambda \text{ radiation} \tag{2.32}$$

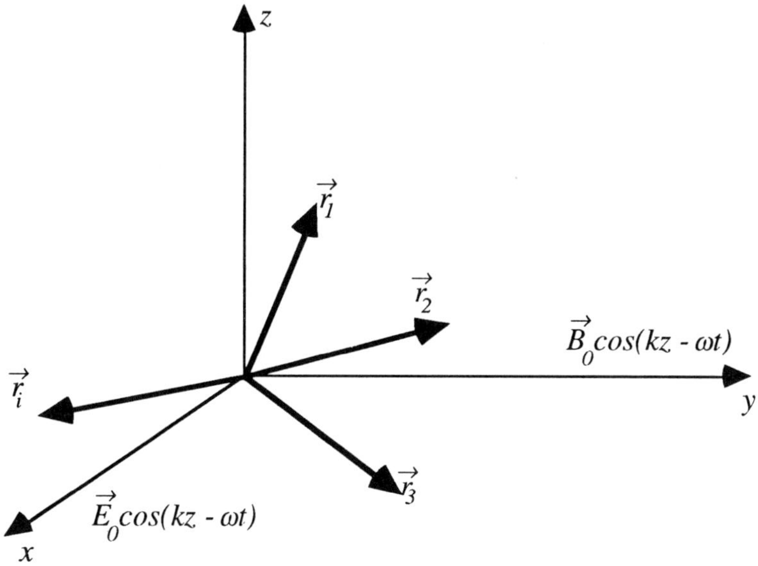

Figure 1. An atom interacting with radiation.

Then

$$k\, a_0 << 1 \tag{2.33}$$

and the field acting on the atom is given by

$$\vec{E} = \vec{E}_0 \cos \omega t \tag{2.34}$$

with no variation of $\vec{E}$ across the atom. The total electric dipole moment is given by

$$\vec{\mu} = -e \sum_{i=1}^{Z} \vec{r}_i \tag{2.35}$$

where Z = number of electrons in the atom. The interaction Hamiltonian can be written as follows

$$H = -e\, \vec{\mu} \cdot \vec{E}_0 \cos \omega t \tag{2.36}$$

If we set

$$\begin{aligned} V_{ij} &= \frac{1}{\hbar} \langle \psi_i(0) | H' | \psi_j(0) \rangle \\ &= -\frac{1}{\hbar} \langle \psi_i(0) | \vec{\mu} \cdot \vec{E}_0 \cos \omega t | \psi_j(0) \rangle \\ &= -\frac{E_0}{\hbar} \cos \omega t \langle \psi_i(0) | \mu^x | \psi_j(0) \rangle = -\frac{E_0 \, \mu_{ij}^x}{\hbar} \cos \omega t \end{aligned} \tag{2.37}$$

we obtain for a two-level system

$$V_{aa} = V_{bb} = 0 \tag{2.38}$$

because $\vec{\mu}$, and therefore H', are odd operators. But

$$V_{ab} = V \cos \omega t \tag{2.39}$$

where

$$V = -\frac{E_0 \, \mu_{12}^x}{\hbar} \tag{2.40}$$

and

$$\mu^x = -\sum_{i=1}^{Z} e x_i \tag{2.41}$$

<u>Example</u>: Atom of Hydrogen [1]

Let $|\psi_1\rangle = 1s$ state
$|\psi_2\rangle = 2p_x$ state

We can write:

$$\psi_1 = \frac{1}{\sqrt{\pi a_0^3}} e^{-r/a_0}$$

$$\psi_2 = \frac{1}{\sqrt{2^5 \pi a_0^5}} e^{-r/2a_0} x$$

The energies of the various levels are given by

$$E_n = -\frac{me^4}{32\pi^2 \varepsilon_0^2 \hbar^2 n^2}$$

Then

$$V = \frac{eE_0}{\hbar} \langle \psi_1 | x | \psi_0 \rangle = 5.93 \times 10^4 \, E_0 \, \text{sec}^{-1}$$

But

$$\hbar\omega_0 = E_2 - E_1 = -\frac{me^4}{32\pi^2 \varepsilon_0^2 \hbar^2 n^2} \left(\frac{1}{4} - 1\right) = \hbar \left(1.63 \times 10^{16}\right)$$

In order to make $V \approx \omega_0$ we need a field strength of $\sim 3 \times 10^{11}$ V/m. For light beams produced by conventional (non-laser) sources

$$V \ll \omega_0$$

The upper limit for conventional spectroscopic sources is represented by the field $E_0 = 10^3$ V/m produced by a mercury lamp with its emission line at 2537 A.

Returning now to our treatment of a two-level atom, we note that

$$H'_{aa} = H'_{ab} = 0 \tag{2.42}$$

$$H'_{ab} = \hbar V \cos \omega t \tag{2.43}$$

Then the equations of motion of the density matrix can be written

$$\begin{aligned} \dot{\rho}_{aa} &= -\frac{i}{\hbar} \left(\rho_{ab} \hbar \, V^* - \rho_{ab} \hbar \, V\right) \cos \omega t = -\dot{\rho}_{bb} \\ \dot{\rho}_{ab} &= i \, \omega_0 \, \rho_{ab} + \frac{i}{\hbar} \left(\rho_{aa} - \rho_{bb}\right) V \cos \omega t = \dot{\rho}_{ba}^* \end{aligned} \tag{2.44}$$

We shall now use a slightly different notation. We set

$$\begin{aligned} a_1 &= c_1 \, e^{-E_a t/\hbar} \\ a_2 &= c_2 \, e^{-E_b t/\hbar} \end{aligned} \tag{2.45}$$

and we use the following equations

$$\begin{aligned} \rho_{aa} &= a_1 \, a_1^* = c_1 \, c_1^* = \rho_{11} \\ \rho_{bb} &= a_2 \, a_2^* = c_2 \, c_2^* = \rho_{22} \\ \rho_{ab} &= a_1 \, a_2^* = c_1 \, c_2^* \, e^{i\omega_0 t} = \rho_{12} \, e^{i\omega_0 t} \\ \rho_{ba} &= a_1^* \, a_2 = c_1^* \, c_2 \, e^{-i\omega_0 t} = \rho_{21} \, e^{-i\omega_0 t} \end{aligned} \tag{2.46}$$

where $\hbar\omega_0 = E_b - E_a$, in order to define the new quantities ρ_{ij} $(i,j=1,2,3)$.

Using (2.46) and (2.44) we find

$$\begin{aligned} \dot{\rho}_{22} &= -\dot{\rho}_{11} = -i\cos\omega t\left(\rho_{12}\, e^{i\omega_0 t}\, V^* - \rho_{21}\, e^{-i\omega_0 t}\, V\right) \\ \dot{\rho}_{12} &= \dot{\rho}_{21}^* = i\, e^{-i\omega_0 t}\left(\rho_{11} - \rho_{22}\right) V \cos\omega t \end{aligned} \tag{2.47}$$

From now on we shall use for the density matrix the elements $\rho_{ij} = c_i\, c_j^*$.

II. D. Optical Bloch Equations without Losses

We consider now the equations (2.47) and we apply to them the so-called "rotating wave approximation", which consists in neglecting the terms oscillating with frequency $\omega_0+\omega$ with respect to those oscillating with frequency $\omega_0-\omega$. This is certainly reasonable if ω, the frequency of the monochromatic electromagnetic radiation interacting with the two level system is not far off resonance.

Then we find

$$\begin{aligned} \dot{\rho}_{22} &= -\dot{\rho}_{11} = -\frac{i}{2}\, V^*\, e^{i(\omega_0-\omega)t}\, \rho_{12} + \frac{i}{2}\, V\, e^{-i(\omega_0-\omega)t}\, \rho_{21} \\ \dot{\rho}_{12} &= \dot{\rho}_{21}^* = \frac{i}{2}\, V\, e^{-i(\omega_0-\omega)t}\left(\rho_{11} - \rho_{22}\right) \end{aligned} \tag{2.48}$$

These equations are called the Optical Bloch Equations [2].

II. E. Rabi Oscillations without Losses

The optical Bloch equations can be written as follows:

$$\begin{aligned} \dot{\rho}_{11} &= \frac{i}{2}\, V^*\, e^{i(\omega_0-\omega)t}\, \rho_{12} - \frac{i}{2}\, V\, e^{-i(\omega_0-\omega)t}\, \rho_{21} \\ \dot{\rho}_{22} &= -\frac{i}{2}\, V^*\, e^{i(\omega_0-\omega)t}\, \rho_{12} + \frac{i}{2}\, V\, e^{-i(\omega_0-\omega)t}\, \rho_{21} \\ \dot{\rho}_{12} &= \frac{i}{2}\, V\, e^{-i(\omega_0-\omega)t}\, \rho_{11} - \frac{i}{2}\, V\, e^{-i(\omega_0-\omega)t}\, \rho_{22} \\ \dot{\rho}_{21} &= -\frac{i}{2}\, V^*\, e^{i(\omega_0-\omega)t}\, \rho_{11} + \frac{i}{2}\, V^*\, e^{i(\omega_0-\omega)t}\, \rho_{22} \end{aligned} \tag{2.49}$$

We shall introduce the following trial solutions

$$
\begin{aligned}
\rho_{11} &= \rho_{11}^{0}\, e^{\alpha t} \\
\rho_{22} &= \rho_{22}^{0}\, e^{\alpha t} \\
\rho_{12} &= \rho_{12}^{0}\, e^{-i(\omega_0 - \omega)t}\, e^{\alpha t} \\
\rho_{21} &= \rho_{21}^{0}\, e^{i(\omega_0 - \omega)t}\, e^{\alpha t}
\end{aligned}
\tag{2.50}
$$

with the quantities ρ_{ij}^{0} and α independent of time.

Using the relations (2.50) in the equation (2.49) we obtain a system of homogeneous equations in the unknown quantities ρ_{ij}^{0}:

$$
\begin{pmatrix}
-\alpha & 0 & \frac{i}{2}V^* & -\frac{i}{2}V \\
0 & -\alpha & -\frac{i}{2}V^* & \frac{i}{2}V \\
\frac{i}{2}V & -\frac{i}{2}V & i(\omega_0 - \omega) - \alpha & 0 \\
-\frac{i}{2}V^* & \frac{i}{2}V^* & 0 & -i(\omega_0 - \omega) - \alpha
\end{pmatrix}
\begin{pmatrix}
\rho_{11}^{0} \\ \rho_{22}^{0} \\ \rho_{12}^{0} \\ \rho_{21}^{0}
\end{pmatrix}
= 0 \quad (2.51)
$$

These equations admit solutions if the determinant of the coefficients is equal to zero:

$$
\alpha^2 \left[\alpha^2 + (\omega_0 - \omega)^2 + |V|^2 \right] = 0 \tag{2.52}
$$

The possible values of α are

$$
\begin{aligned}
\alpha_1 &= 0 \\
\alpha_2 &= i\,\Omega \\
\alpha_3 &= -i\,\Omega
\end{aligned}
\tag{2.53}
$$

where

$$
\Omega = \sqrt{(\omega_0 - \omega)^2 + |V|^2} \tag{2.54}
$$

We note here that the dependence of Ω on $|V|^2$ indicates a change in the frequency of oscillations of the "coupled" system which consist of (atom and light beam): this effect is called the dynamic Stark effect.

The most general solution is then

$$
\rho_{ij} = \rho_{ij}^{(1)} + \rho_{ij}^{(2)}\, e^{i\Omega t} + \rho_{ij}^{(3)}\, e^{-i\Omega t} \tag{2.55}
$$

Additional oscillatory exponentials are present in the off-diagonal elements. By using the initial conditions and the optical Bloch equations we can obtain the constant coefficients.

Example

$$c_1(0) = 1 \rightarrow \rho_{22}(0) = 0$$
$$c_2(0) = 0 \rightarrow \rho_{12}(0) = 0 \tag{2.56}$$

The density matrix elements are

$$\rho_{22} = \frac{|V|^2}{\Omega^2} \sin^2 \frac{1}{2}\,\Omega\, t$$
$$\rho_{11} = 1 - \rho_{22} = 1 - \frac{|V|^2}{\Omega^2} \sin^2 \Omega\, t$$
$$\rho_{12} = e^{-i(\omega_0 - \omega)t} \frac{V}{\Omega^2} \sin \frac{1}{2}\,\Omega\, t\left[-(\omega_0 - \omega) \sin\left(\frac{1}{2}\,\Omega\, t\right) + i\,\Omega \cos\left(\frac{1}{2}\,\Omega\, t\right)\right] \tag{2.57}$$
$$\rho_{21} = e^{i(\omega_0 - \omega)t} \frac{V^*}{\Omega^2} \sin \frac{1}{2}\,\Omega\, t\left[-(\omega_0 - \omega) \sin\left(\frac{1}{2}\,\Omega\, t\right) - i\,\Omega \cos\left(\frac{1}{2}\,\Omega\, t\right)\right]$$

For zero detuning ($\omega = \omega_0$): $\Omega = |V|$ and

$$\rho_{22} = \sin^2 \frac{1}{2}|V|t$$
$$\rho_{12} = i\frac{V}{|V|} \sin\left(\frac{1}{2}|V|t\right) \cos\left(\frac{1}{2}|V|t\right) \tag{2.58}$$

The behavior of the quantity ρ_{22} for different values of the detuning

$$d = \frac{\omega_0 - \omega}{|V|} \tag{2.59}$$

is represented in Fig. 2. We shall make the following observations:

1) For zero detuning ($\omega = \omega_0$) the atom oscillates between ground and excited states: these oscillations are called <u>Rabi oscillations</u> and their frequency $|V|$ is called <u>Rabi frequency</u>. Solutions for the similar problem of a spin system in an oscillatory magnetic field were obtained by Rabi [3]. When

$$\frac{|V|}{2}t = \frac{E_0|\mu_{12}^x|}{2\hbar}t = \pi \tag{2.60}$$

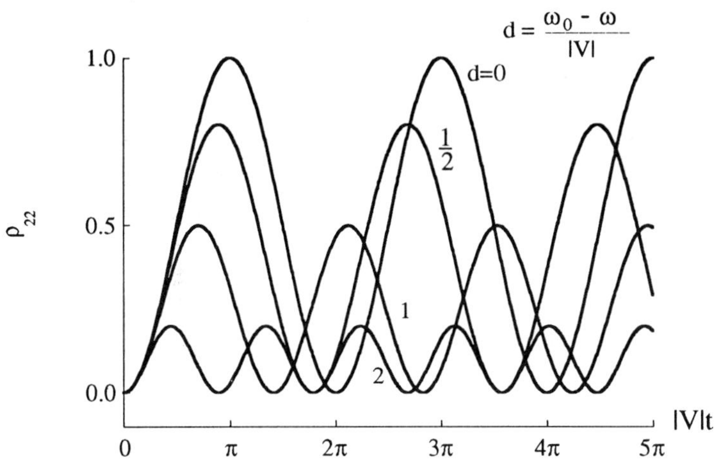

Figure 2. Probability of excitation as a function of time.

$\rho_{22} = 1$, all the atoms are in the upper state, i.e. the population is completely inverted. When

$$\frac{|V|}{2}t = \frac{E_0|\mu_{12}^x|}{2\hbar}t = 2\pi \tag{2.61}$$

the original situation, with all the atoms in the ground state, is restored.

2) Since the rotating wave approximation was originally used in deriving the optical Bloch equations the solutions (2.57) are valid only if $(\omega - \omega_0) \ll (\omega + \omega_0)$.

3) The solutions for ρ_{22} and ρ_{12} refer to monochromatic radiation. In effect the oscillations can be seen experimentally when the frequency spread of the electromagnetic radiation is much smaller than the linewidth of the transition.

4) We have to dedicate some attention to the fact that the processes that broaden the linewidth of the transition introduce modifications in the optical Bloch equations. We shall return to this last point.

II. F. Extension to an Ensemble

1. Ensemble of Quantum Systems. Measurements of physical parameters are usually performed on quantum systems in equilibrium whose states are controlled by a Boltzmann distribution over the available stationary energy levels. This distribution is characterized by temperature. An ensemble of systems which are so distributed is called a canonical ensemble.

Averages of physical parameters for such an ensemble must include:
1) quantum averaging, i.e. calculation of the expectation value: $\langle \Omega \rangle$, and
2) classical averaging, i.e. ensemble averaging: $\bar{\Omega}$.
A combination of the two averaging processes results in the notation $\langle \bar{\Omega} \rangle$. In general, $\langle \bar{\Omega} \rangle \neq \langle \Omega \rangle$.

2. Perturbations. The perturbations that may influence the systems of the ensemble are called coherent if they affect every system in the same way, and incoherent if they affect the systems differently.

Examples of coherent perturbations are the oscillating electric and magnetic fields produced at radio frequencies by oscillators, at microwave frequencies by klystrons and masers, and at optical and infrared frequencies by lasers.

As for the incoherent perturbations the oscillating fields may be produced by such sources as globars, heated filaments or electric discharges. Relaxation processes related to the so-called T_1 and T_2 time constants, to be introduced later, represent also incoherent perturbations always present in spectroscopic measurements. Spontaneous and nonradiative deexcitation processes almost always act as incoherent perturbations. On account of the often pervasive incoherent perturbations no two systems of the ensemble feel each other or the thermal bath with which they are in contact in the same way.

3. Ensemble-Quantum Averages. We shall focus our attention on quantum systems whose wavefunctions do not have much spatial overlap. These systems are coupled together for two reasons:
1) they may be affected by the presence of the same radiation field, and
2) they are all relaxing to the same thermal bath.

The ensemble-quantum average of the physical parameter represented by the operator Ω is given by the ensemble average of the possible expectation values of Ω (each one of these expectation values corresponding to a system of the ensemble):

$$\langle \bar{\Omega} \rangle = \frac{1}{N} \sum_{n=1}^{N} \langle \Omega \rangle_n \tag{2.62}$$

where N = number of systems in the ensemble,

$$\langle \Omega \rangle_n = \lim_{\nu \to \infty} \frac{1}{\nu} \sum_{i=1}^{\nu} \omega_{ni} \tag{2.63}$$

and ω_{ni} = result of a measurement on a system which is in state n.

4. The Density Matrix of the Ensemble. We shall assume that the systems of the ensemble have little wavefunction overlap and interact with one another weakly. In addition:

1) any perturbation applied to the systems is coherent,
2) at the time the perturbation is switched on, all the systems of the ensemble are in identical states, and
3) the experiments do not last longer than a time $\tau \ll T_1, T_2$, i.e. relaxation effects are not important.

Under these restrictive conditions we may define a matrix ρ for which two relations are of relevance:

$$\text{Equation of motion for } \rho: \quad \dot{\rho} = \frac{i}{\hbar}[\rho, H] \tag{2.64}$$

$$\text{Average value of } \Omega: \quad \langle \Omega \rangle = \mathrm{Tr}\,(\Omega\,\rho) = \langle \bar{\Omega} \rangle \tag{2.65}$$

If the systems of the ensemble have wavefunctions with little overlap, but do not follow the three additional requirements referred to above, then we have to describe the ensemble in different terms:

1) Let the ensemble consist of N identical systems and let the nth system be represented by a wavefunction

$$\psi^{(n)}(\vec{r},t) = c_1^{(n)}\, e^{-iE_1 t/\hbar}\, \psi_1(\vec{r}) + c_2^{(n)}\, e^{-iE_2 t/\hbar}\, \psi_2(\vec{r}) + \ldots \tag{2.66}$$

2) We define an operator $\rho^{(n)}$ as follows

$$\rho_{ij}^{(n)} = c_i^{(n)}\, c_j^{(n)*} \tag{2.67}$$

3) The expectation value of a physical parameter represented by the operator Ω is

$$\langle \Omega \rangle_n = \mathrm{Tr}\left[\rho^{(n)}\, \Omega\right] \tag{2.68}$$

4) The ensemble average of Ω is

$$\langle \bar{\Omega} \rangle = \frac{1}{N} \sum_{n=1}^{N} \langle \Omega \rangle_n = \frac{1}{N} \sum_{n=1}^{N} \left[\mathrm{Tr}\,(\rho^{(n)}\, \Omega)\right] \tag{2.69}$$

5) If we utilize the following properties of matrices

$$AB + AC = A(B + C) \tag{2.70}$$

$$\text{Tr}\,(AB) + \text{Tr}\,(AC) = \text{Tr}\,[A(B + C)] \tag{2.71}$$

we obtain

$$\langle\bar{\Omega}\rangle = \frac{1}{N}\sum_{n=1}^{N}[\text{Tr}\,(\rho^{(n)}\,\Omega)] = \text{Tr}\left[\left(\frac{1}{N}\sum_{n=1}^{N}\rho^{(n)}\right)\Omega\right] = \text{Tr}\,[\bar{\rho}\,\Omega] \tag{2.72}$$

where the matrix

$$\bar{\rho} = \frac{1}{N}\sum_{n=1}^{N}\rho^{(n)} \tag{2.73}$$

represents the density matrix of the ensemble.

6) The equation of motion of the matrix $\bar{\rho}$ is

$$i\hbar\frac{\partial\bar{\rho}}{\partial t} = [H,\bar{\rho}] \tag{2.74}$$

7) Prescriptions on how to calculate $\langle\bar{\Omega}\rangle$:

a) Choose an initial state for the ensemble by choosing the elements of $\bar{\rho}$ at time t=0. If the number of possible quantum states is M, M^2-1 constants are needed because $\bar{\rho}_{ij} = \bar{\rho}^*_{ji}$ and the diagonal elements are real.

b) Solve the equation of motion for $\bar{\rho}$ given by (2.74) and find $\bar{\rho}$ at the time t of the measurement.

c) Calculate the time-independent matrix of Ω using the same eigenfunctions.

d) Multiply the two matrices and sum the diagonal elements. The result is the experimental value of the parameter Ω, averaged over the systems of the ensemble.

5. Properties of the Ensemble Density Matrix. We can now investigate the properties of the ensemble density matrix $\bar{\rho}$:

1) Since the trace of each $\rho^{(n)}$ matrix is 1, the trace of $\bar{\rho}$ is also 1, if the ensemble is closed.

2) The total Hamiltonian of a system in the ensemble contains three terms:

H_0: time independent, very large, establishes the stationary states of a system,
H': time dependent, a perturbation, induces transitions among eigenstates of H_0, and
H_R: provides for the exchange of energy among the systems of the ensemble, and between the systems and the heat bath.

3) The most convenient choice of basis functions is the set of eigenfunctions of H_0.

4) Since all the $\rho^{(n)}$ are Hermitian, $\bar{\rho}$ is Hermitian.

5) The off-diagonal elements of a matrix $\rho^{(n)}$ can be expressed as follows:

$$\rho_{ij}^{(n)} = \left|\rho_{ij}^{(n)}\right| \exp\left(i\, \phi_{ij}^{(n)}\right) \tag{2.75}$$

Then

$$\bar{\rho}_{ij} = \frac{1}{N} \sum_{n=1}^{N} \left|\rho_{ij}^{(n)}\right| \exp\left(i\, \phi_{ij}^{(n)}\right) \tag{2.76}$$

6) We can state the <u>Principle of Equal a Priori Probabilities</u> as follows: "All configurations of any system, of ensemble of systems, which have the same energy are equally probable at equilibrium."

A consequence of this principle is the <u>Hypothesis of Random Phases</u>: "In an ensemble at equilibrium all the off-diagonal elements of the ensemble density matrix must be zero, because of the destructive interferences among the corresponding elements of the various individual $\rho^{(n)}$ matrices."

7) The diagonal elements of $\bar{\rho}$ do not vanish when the ensemble is in thermal equilibrium. They are given by

$$\bar{\rho}_{ii}^{e} = \frac{1}{N} \sum_{n=1}^{N} |\rho_{ii}^{(n)}| = \frac{1}{N} \sum_{n=1}^{N} |a_i^{(n)}|^2 = \frac{1}{N} \sum_{n=1}^{N} |c_i^{(n)}|^2 \tag{2.77}$$

where the superscript e indicates equilibrium and where

$$|c_i^{(n)}|^2 = \frac{e^{-E_i/kT}}{Z} = \begin{array}{c}\text{fraction of systems}\\ \text{with energy } E_i\end{array} \tag{2.78}$$

with

$$Z = \sum_{i=1}^{M} e^{-E_i/kT} = \text{partition sum} \tag{2.79}$$

Then

$$\bar{\rho}_{ii}^{e} = \frac{1}{Z}\begin{pmatrix} e^{-E_1/kT} & & & & 0 \\ & e^{-E_2/kT} & & & \\ & & e^{-E_3/kT} & & \\ & & & \ddots & \\ 0 & & & & e^{-E_M/kT} \end{pmatrix} \tag{2.80}$$

8) If a perturbation H' which, by acting on the systems, has driven the ensemble away from equilibrium, is removed, each element of $\bar{\rho}$ begins to decay back to its equilibrium value according to

$$\frac{\partial \bar{\rho}_{ij}}{\partial t} = -\frac{\bar{\rho}_{ij} - \bar{\rho}_{ij}^{e}}{T_{ij}} \tag{2.81}$$

The effect of the Hamiltonian H_R is felt through these relaxation mechanisms.

9) Not all the T_{ij}'s are independent of each other:

$$T_{ij} = T_{ji} \tag{2.82}$$

because $\bar{\rho}_{ij} = \bar{\rho}_{ji}^{*}$. Also there are conditions on the T's for the diagonal elements because, in any case,

$$\sum_{i=1}^{M} \bar{\rho}_{ii} = 1 \tag{2.83}$$

10) For two-level systems (M=2) there are two relaxation times: T_1 = relaxation time for the diagonal elements, called longitudinal relaxation time, and T_2 = relaxation time for the off-diagonal elements, called transversal relaxation time.

II. G. Optical Bloch Equations with Losses

1. The Four Equations. We shall now reconsider the optical Bloch equations and add to each of them the relaxation-related terms:

$$\dot{\rho}_{11} = \frac{i}{2} V^* \tilde{\rho}_{12} - \frac{i}{2} V \tilde{\rho}_{21} + 2\gamma\rho_{22}$$
$$\dot{\rho}_{22} = -\frac{i}{2} V^* \tilde{\rho}_{12} + \frac{i}{2} V \tilde{\rho}_{21} - 2\gamma\rho_{22}$$
$$\dot{\tilde{\rho}}_{12} = \frac{i}{2} V (\rho_{11} - \rho_{22}) + [i(\omega_0 - \omega) - \gamma'] \tilde{\rho}_{12} \tag{2.84}$$
$$\dot{\tilde{\rho}}_{21} = -\frac{i}{2} V^* (\rho_{11} - \rho_{22}) + [-i(\omega_0 - \omega) - \gamma'] \tilde{\rho}_{21}$$

where

$$\tilde{\rho}_{12} = e^{i(\omega_0 - \omega)t} \rho_{12}$$
$$\tilde{\rho}_{21} = e^{-i(\omega_0 - \omega)t} \rho_{21} \tag{2.85}$$

and

$$\gamma = \frac{1}{2T_1} \tag{2.86}$$

$$\gamma' = \frac{1}{2T_2} = \gamma + \gamma^{\text{coll}} \tag{2.87}$$

γ represents the decay rate of the excited state population. γ' is the dephasing rate of the dipoles induced by the radiation: it includes γ and γ^{coll}. The latter parameter γ^{coll} is the dipole dephasing rate due to processes not related to changes in the populations of the levels, such as elastic collisions.

2. Steady State Solutions and Transition Rates. In steady state the two independent equations of the set (2.84) become

$$\frac{i}{2} V^* \tilde{\rho}_{12} - \frac{i}{2} V \tilde{\rho}_{21} + 2\gamma\rho_{22} = 0$$
$$\frac{i}{2} V (\rho_{11} - \rho_{22}) + [i(\omega_0 - \omega) - \gamma'] \tilde{\rho}_{12} = 0 \tag{2.88}$$

with the solutions

$$\rho_{22} = \frac{\frac{1}{4}|V|^2 (\gamma'/\gamma)}{[(\omega_0 - \omega)^2 + \gamma'^2] + \frac{1}{2}|V|^2 (\gamma'/\gamma)}$$
$$\tilde{\rho}_{12} = \frac{\frac{1}{2} V (\omega_0 - \omega - i\gamma')\gamma}{[(\omega_0 - \omega)^2 + \gamma'^2] + \frac{1}{2}|V|^2 (\gamma'/\gamma)} \tag{2.89}$$

Let us consider now the dependence of ρ_{22} on γ and γ':

1) $\omega \approx \omega_0$

$$\rho_{22} \simeq \frac{\frac{1}{4}|V|^2 \gamma'}{\gamma \gamma'^2 + \frac{1}{2}|V| \gamma'} = \frac{\frac{1}{4}|V|^2}{\gamma \gamma' + \frac{1}{2}|V|} \tag{2.90}$$

The peak of ρ_{22} decreases with increasing γ and γ'.

2) $\omega \gg \omega_0$ or $\omega \ll \omega_0$

$$\rho_{22} \simeq \frac{\frac{1}{4}|V|^2}{(\Delta\omega)^2} \frac{\gamma'}{\gamma} \tag{2.91}$$

ρ_{22} increases with increasing γ' and decreases with increasingγ.

3) ρ_{22} decreases always with increasing γ.

4) $|V| \gg \Delta\omega, \gamma, \gamma'$

$$\rho_{22} = \frac{\frac{1}{4}|V|^2 (\gamma' / \gamma)}{\frac{1}{2}|V|^2 (\gamma' / \gamma)} = \frac{1}{2} \tag{2.92}$$

5) $|V| \ll \Delta\omega, \gamma, \gamma'$

$$\rho_{22} = \frac{\frac{1}{4}|V|^2 (\gamma' / \gamma)}{(\omega_0 - \omega)^2 + \gamma'^2} \tag{2.93}$$

We shall consider now the application of rate equations to the two-level system. With reference to Fig. 3, the equations of motion of the populations are

$$\dot{N}_1 = -\dot{N}_2 = N_2 A + (N_2 - N_1) B W \tag{2.94}$$

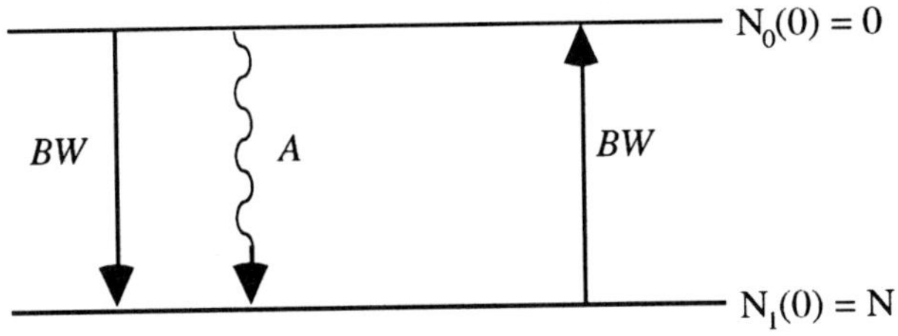

Figure 3. Transitions in a Two-Level System. A and B are the Einstein coefficients. BW is the probability per unit time of induced emission and of absorption.

which give the solutions

$$N_1(t) = \frac{NBW}{A + 2BW} e^{-(A + 2BW)t} + \frac{N(A + BW)}{A + 2BW} \underset{t \to \infty}{\longrightarrow} \frac{N(A + BW)}{A + 2BW} \tag{2.95}$$

$$N_2(t) = \frac{NBW}{A + 2BW} [1 - e^{-(A + 2BW)t}] \underset{t \to \infty}{\longrightarrow} \frac{NBW}{A + 2BW} \tag{2.96}$$

The steady-state value for N_2 is given by

$$N_2 = \frac{NBW}{A + 2BW} \tag{2.97}$$

Under the conditions of weak excitation ($A \gg BW$)

$$N_2 = \frac{NBW}{A} \tag{2.98}$$

On the other hand the formula for ρ_{22} with small $|V|$ is

$$\rho_{22} = \frac{(\gamma'/4\gamma)|V|^2}{(\omega_0 - \omega)^2 + \gamma'^2} \tag{2.99}$$

Let us now consider different conditions:

Broad Band Excitation

The exciting light is practically of constant intensity over the bandwidth $2\gamma'$ of the atomic transition. We note that

$$\int_{-\infty}^{+\infty} \frac{1}{(\omega_0 - \omega)^2 + \gamma'^2} \, d\omega = \frac{\pi}{\gamma'} \tag{2.100}$$

Then

$$\begin{aligned}\int_{-\infty}^{+\infty} \rho_{22} \, d\omega &= \frac{\gamma'}{4\gamma} |V|^2 \int_{-\infty}^{+\infty} \frac{1}{(\omega_0 - \omega)^2 + \gamma'^2} \, d\omega = \\ &= \frac{\gamma'}{4\gamma} |V|^2 \frac{\pi}{\gamma'} = \frac{\pi |V|^2}{4\gamma}\end{aligned} \tag{2.101}$$

independent of γ'. We can relate this quantity to the ratio N_2/N from (2.98):

$$B\,W = \frac{\pi |V|^2}{2} \tag{2.102}$$

We know from (2.40) that

$$|V|^2 = \frac{E_0^2 |\mu_{12}^x|^2}{\hbar^2} \tag{2.103}$$

Then

$$B\,W = \frac{\pi}{2} \frac{E_0^2 |\mu_{12}^x|^2}{\hbar^2} \simeq \frac{\pi}{2} \frac{E_0^2 |\mu_{12}|^2}{3\hbar^2} \tag{2.104}$$

On the other hand

$$\begin{aligned}&\tfrac{1}{2} \varepsilon_0 E_0^2 = W \\ &\text{or} \\ &E_0^2 = \frac{2W}{\varepsilon_0}\end{aligned} \tag{2.105}$$

Then we find

$$B = \frac{\pi |\mu_{12}|^2}{3\varepsilon_0 \hbar^2} \tag{2.106}$$

the known expression for the Einstein B coefficient.

Large Collision Broadening

We shall assume that $\gamma' \gg \gamma$ and the band of the exciting light is of arbitrary magnitude.

The steady state value of ρ_{22} at frequency ω is given, according to (2.89), by

$$\rho_{22} = \frac{r}{2\gamma} \tag{2.107}$$

where

$$r = \frac{1}{2} \frac{|V|^2 \gamma'}{(\omega_0 - \omega)^2 + \gamma'^2} \tag{2.108}$$

Since $\gamma' \gg \gamma$, $1/\gamma' \ll 1/\gamma$, then at time $1/\gamma$ the system is practically in steady state. ρ_{22} approaches its steady state value (2.107) as follows:

$$\dot{\rho}_{22} = \frac{r}{2\gamma}(1 - e^{-2\gamma t}) \tag{2.109}$$

Going back to the rate equations we find that (2.96) for the case of weak excitation gives us

$$\frac{N_2}{N} = \frac{BW}{A}(1 - e^{-At}) \tag{2.110}$$

We can then establish the following connection

$$\begin{aligned} B\,W &= \int_{-\infty}^{+\infty} r\, d\omega = \frac{1}{2} \frac{E_0^2 |\mu_{12}|^2}{3\hbar^2} \int_{-\infty}^{+\infty} \frac{\gamma'}{(\omega_0 - \omega)^2 + \gamma'^2} d\omega \\ &= \frac{1}{2} \frac{2W}{\varepsilon_0} \frac{|\mu_{12}|^2 \pi}{3\hbar^2} \end{aligned} \tag{2.111}$$

and we find again the result

$$B = \frac{\pi |\mu_{12}|^2}{3\varepsilon_0 \hbar^2} \tag{2.112}$$

We can state the following points:

1) In the case just treated the equivalence of the optical Bloch equations and of the rate equations holds in the following two circumstances:

   (*i*) broad band excitation, or
   (*ii*) large collision broadening, with any excitation band.

2) In the more general case when the Doppler broadening is present the equivalence holds in the two circumstances:

   (*i*) bandwidth of incident light > atomic transition bandwidth, or
   (*ii*) combined collision and Doppler linewidth » radiative linewidth of the transition.

3) We note that the rate equations depend on the intensity of the light interacting with the atomic system, whereas the Bloch optical equations depend on the phase and amplitude of the electric field, because they depend on V and V^*.

3. Damped Rabi Oscillations. The diagonal elements of the atomic density matrix give us, when multiplied by the total atomic population N, the populations of atoms in states ψ_1 and ψ_2

$$\begin{aligned} N_2 &= N\rho_{22} = \frac{N|V|^2/4}{(\omega_0-\omega)^2+\gamma^2+|V|^2/2} \\ N_1 &= N\rho_{11} = N - N_2 \end{aligned} \tag{2.113}$$

The expressions apply in the steady state. In the limit of weak intensity of incident light the value of N_2 is proportional to the intensity of light. We note also that the steady state value above is independent of the initial conditions.

For studying the transients we need, however, the initial conditions. The equations to consider are eqs. (2.84). Let us deal with the simple case in which $\omega_0=\omega$ (zero detuning), $\gamma=\gamma'$, and

$$\begin{aligned} &\rho_{22}(0) = 0 \quad (c_2 = 0) \\ &\rho_{12}(0) = 0 \quad (c_1 = 0) \end{aligned} \tag{2.114}$$

R. A. Smith [4] has given and C. Yang [5] has worked out in great detail the solutions for ρ_{22} in three different situations for this case of zero detuning:

$|V| > \frac{1}{2}\gamma$

$$\rho_{22} = \frac{|V|^2/2}{2\gamma^2 + |V|^2}\left[1 - \left(\cos at + \frac{3\gamma}{2a}\sin at\right)\exp\left(-\tfrac{3}{2}\gamma t\right)\right] \qquad (2.115)$$

where

$$a = \left(V^2 - \frac{\gamma^2}{4}\right)^{\frac{1}{2}} \qquad (2.116)$$

$|V| = \frac{1}{2}\gamma$

$$\rho_{22} = \frac{1}{18}\left[1 - \left(\tfrac{3}{2}\gamma t + 1\right)\exp\left(-\tfrac{3}{2}\gamma t\right)\right] \qquad (2.117)$$

$|V| < \frac{1}{2}\gamma$

$$\rho_{22} = \frac{|V|^2/2}{2\gamma^2 + |V|^2}\left[1 - \left(\cosh \tfrac{1}{2}a't + \frac{3\gamma}{a'}\sin \tfrac{1}{2}a't\right)\exp\left(-\tfrac{3}{2}\gamma t\right)\right] \qquad (2.118)$$

where

$$a' = \left(\gamma^2 - 4|V|^2\right)^{\frac{1}{2}} \qquad (2.119)$$

For $\gamma = 0, a = |V| = \Omega$, and we recover the expression (2.58)

$$\rho_{22} = \tfrac{1}{2}(1 - \cos|V|t) = \sin^2 \tfrac{1}{2}|V|t \qquad (2.120)$$

In Fig. 4 we represent the variations of ρ_{22} as given by (2.115) for the case of zero detuning. We note the following:

1) The greater is γ the more damped are the oscillations. For $\eta = \frac{\gamma}{|V|} = \frac{1}{3}$ a single maximum remains.

2) The incident light must be of large enough intensity in order to make

$$\eta \ll \frac{1}{3}, \text{ i.e. } |V| \gg 3\gamma$$

and generate significant oscillations in the populations of atoms in the two states.

3) The steady state value of ρ_{22}

$$\frac{|V|^2/2}{2\gamma^2 + |V|^2} = \frac{1}{2 + 4\eta^2} \tag{2.121}$$

decreasing with increasing η.

4) The oscillatory behavior of the populations is called optical nutation. The nutation frequency depends on $|V|$, detuning, and radiation damping.

5) The effect of a beam of light on atomic excitation can be summarized as follows:

a) The atoms are initially in the ground state. The beam is attenuated as energy is transferred from the beam to the atoms.

b) If the intensity of the beam is large enough that

$$|V| > (\omega_0 - \omega) \text{ and } |V| \gg \gamma$$

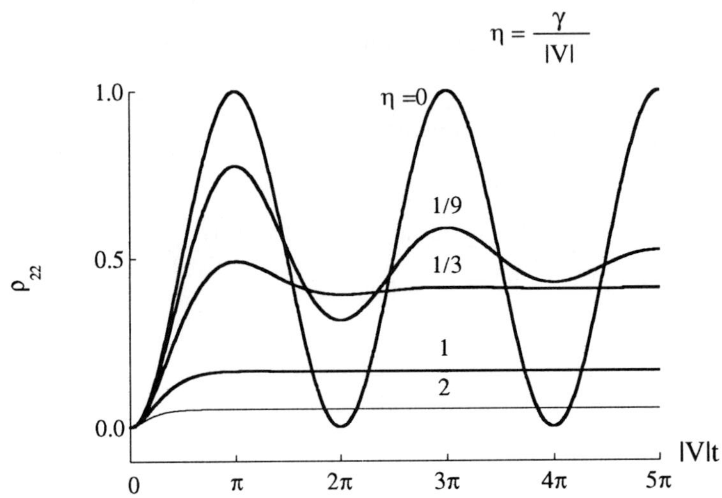

Figure 4. ρ_{22} for zero detuning and various values of $\eta = \frac{\gamma}{|V|}$.

the excited state component of the wave function will exceed the ground state component and energy is transferred back from the atoms into the beam, increasing its initial intensity.

c) This "cycle" repeats itself producing optical nutation. The oscillations in the atomic populations are accompanied by oscillations in the intensity of the transmitted light. Experimental observations of this type were made by MacGillivray and co-workers [6] who observed the effects of optical nutation in the D_2 transition of sodium atoms. A detailed theory of experiments of this type is given in [7].

III. THE QUANTIZED RADIATION FIELD

III. A. The Classical Radiation Field

The most general electromagnetic field in the presence of charges of density $\rho(\vec{r},t)$ and currents of density $\vec{j}(\vec{r},t)$ is given by the four Maxwell equations

$$\vec{\nabla}\cdot\vec{B}=0 \tag{3.1}$$

$$\vec{\nabla}\times\vec{E}+\frac{1}{c}\frac{\partial\vec{B}}{\partial t}=0 \tag{3.2}$$

$$\vec{\nabla}\cdot\vec{E}=4\pi\rho \tag{3.3}$$

$$\vec{\nabla}\times\vec{B}-\frac{1}{c}\frac{\partial\vec{E}}{\partial t}=\frac{4\pi}{c}\vec{j} \tag{3.4}$$

Since for a generic vector $\vec{u}$

$$\vec{\nabla}\cdot\vec{\nabla}\times\vec{u}=0 \tag{3.5}$$

we can set

$$\vec{B}=\vec{\nabla}\times\vec{A} \tag{3.6}$$

$\vec{A}$ is here called "vector potential". Using this expression of $\vec{B}$ in eq. (3.2) we obtain

$$\vec{\nabla} \times \left(\vec{E} + \frac{1}{c} \frac{\partial \vec{A}}{\partial t} \right) = 0 \tag{3.7}$$

and, since for a generic function ϕ

$$\vec{\nabla} \times \vec{\nabla} \phi = 0 \tag{3.8}$$

we can set

$$\vec{E} + \frac{1}{c} \frac{\partial \vec{A}}{\partial t} = - \vec{\nabla} \phi$$

or

$$\vec{E} = - \frac{1}{c} \frac{\partial \vec{A}}{\partial t} + \vec{\nabla} \phi \tag{3.9}$$

ϕ here is called "scalar potential".

Using now the expressions (3.6) of $\vec{B}$ and (3.9) of $\vec{E}$ we can write eq. (3.3) and eq. (3.4) as follows

$$\vec{\nabla} \cdot \left(- \vec{\nabla} \phi - \frac{1}{c} \frac{\partial \vec{A}}{\partial t} \right) = 4 \pi \rho \tag{3.10}$$

$$\vec{\nabla} \times \left(\vec{\nabla} \times \vec{A} \right) - \frac{1}{c} \frac{\partial}{\partial t} \left(- \vec{\nabla} \phi - \frac{1}{c} \frac{\partial \vec{A}}{\partial t} \right) = \frac{4\pi}{c} \vec{j} \tag{3.11}$$

In turn these two expressions can be reexpressed in the following fashion

$$\nabla^2 \phi + \frac{1}{c} \frac{\partial}{\partial t} \left(\vec{\nabla} \cdot \vec{A} \right) = - 4 \pi \rho \tag{3.12}$$

$$\left(\nabla^2 \vec{A} - \frac{1}{c^2} \frac{\partial^2 \vec{A}}{\partial t^2} \right) - \vec{\nabla} \left(\vec{\nabla} \cdot \vec{A} + \frac{1}{c} \frac{\partial \phi}{\partial t} \right) = - \frac{4\pi}{c} \vec{j} \tag{3.13}$$

For the passage from eq. (3.11) to eq. (3.13) we have taken advantage of the relation

$$\vec{\nabla} \times \left(\vec{\nabla} \times \vec{A} \right) = \vec{\nabla} \left(\vec{\nabla} \cdot \vec{A} \right) - \nabla^2 \vec{A} \tag{3.14}$$

The effect of our manipulations has been the replacement of the four Maxwell equations (3.1) to (3.4) with eqs. (3.6), (3.9), (3.12), and (3.13).

We note that the two potentials ϕ and $\vec{A}$ are not uniquely determined. In fact, if we set

$$\phi' = \phi - \frac{1}{c}\frac{\partial f}{\partial t} \tag{3.15}$$

$$\vec{A}' = \vec{A} + \vec{\nabla} f \tag{3.16}$$

where f is a generic function of position and time, the field $\vec{E}$ and $\vec{B}$ have the same expressions in terms of ϕ' and $\vec{A}'$ as they have in terms of ϕ and $\vec{A}$:

$$\begin{aligned} \vec{E} &= -\frac{1}{c}\frac{\partial \vec{A}}{\partial t} - \vec{\nabla}\phi = \\ &= -\frac{1}{c}\frac{\partial(\vec{A}' - \vec{\nabla} f)}{\partial t} - \vec{\nabla}\left(\phi' + \frac{1}{c}\frac{\partial f}{\partial t}\right) = \\ &= -\frac{1}{c}\frac{\partial \vec{A}'}{\partial t} + \frac{1}{c}\frac{\partial}{\partial t}\vec{\nabla} f - \vec{\nabla}\phi' - \frac{1}{c}\frac{\partial}{\partial t}\vec{\nabla} f = -\frac{1}{c}\frac{\partial \vec{A}'}{\partial t} - \vec{\nabla}\phi' \end{aligned} \tag{3.17}$$

$$\vec{B} = \vec{\nabla}\times\vec{A} = \vec{\nabla}\times(\vec{A}' - \vec{\nabla} f) = \vec{\nabla}\times\vec{A}' \tag{3.18}$$

It can be easily shown that eq. (3.12) and eq. (3.13) preserve the same form when we use with them the transformations (3.15) and (3.16).

The indeterminacy of $\vec{A}'$ and ϕ is further reduced by the use of an additional condition. We choose this condition to be the "Coulomb gauge"

$$\vec{\nabla}\cdot\vec{A} = 0 \tag{3.19}$$

When this gauge is used eqs. (3.12) and (3.13) become

$$\nabla^2\phi = -4\pi\rho \tag{3.20}$$

$$\nabla^2\vec{A} - \frac{1}{c^2}\frac{\partial^2\vec{A}}{\partial t^2} = -\frac{4\pi}{c}\vec{j} + \frac{1}{c}\frac{\partial}{\partial t}\vec{\nabla}\phi \tag{3.21}$$

The solution of eq. (3.20), called "Poisson's equation", is given by

$$\phi(\vec{r},t) = \int \frac{\rho(\vec{r}', t)}{|\vec{r} - \vec{r}'|} d^3\vec{r}' \tag{3.22}$$

The scalar potential is then determined by Coulomb's law from the charge distribution as in electrostatics (however, contrary to the case of electrostatics, the charge here may change with time). This is the origin of the expression "Coulomb gauge".

Let us consider now eq. (3.21). If we use the result (3.22) and the equation of continuity

$$\begin{aligned}\frac{\partial}{\partial t} \vec{\nabla}\phi &= \vec{\nabla}\dot{\phi} \\ &= \vec{\nabla} \int \frac{\dot{\rho}(\vec{r}', t)}{|\vec{r} - \vec{r}'|} d^3\vec{r}' = -\vec{\nabla} \int \frac{\vec{\nabla}' \cdot \vec{j}(\vec{r}', t)}{|\vec{r} - \vec{r}'|} d^3\vec{r}'\end{aligned} \tag{3.23}$$

Then

$$\nabla^2\vec{A} - \frac{1}{c^2}\frac{\partial^2\vec{A}}{\partial t^2} = -\frac{4\pi}{c}\vec{j} - \frac{1}{c}\vec{\nabla} \int \frac{\vec{\nabla}' \cdot \vec{j}(\vec{r}', t)}{|\vec{r} - \vec{r}'|} d^3\vec{r}' \tag{3.24}$$

We set

$$\vec{j}_t = \vec{j} + \frac{1}{4\pi}\vec{\nabla} \int \frac{\vec{\nabla}' \cdot \vec{j}(\vec{r}', t)}{|\vec{r} - \vec{r}'|} d^3\vec{r}' \tag{3.25}$$

and prove immediately the following:

$$\begin{aligned}\vec{\nabla} \cdot \vec{j}_t &= \vec{\nabla} \cdot \vec{j} + \frac{1}{4\pi}\nabla^2 \int \frac{\vec{\nabla}' \cdot \vec{j}(\vec{r}', t)}{|\vec{r} - \vec{r}'|} d^3\vec{r}' \\ &= \vec{\nabla} \cdot \vec{j} + \frac{1}{4\pi} \int -4\pi\,\delta(\vec{r} - \vec{r}')\,\vec{\nabla}' \cdot \vec{j}(\vec{r}', t)\, d^3\vec{r}' = 0\end{aligned} \tag{3.26}$$

since

$$\nabla^2\left[\frac{1}{|\vec{r} - \vec{r}'|}\right] = -4\pi\,\delta(\vec{r} - \vec{r}')$$

The vector $\vec{j}_t$ has zero divergence, or, as it is said, is "solenoidal". On the other hand, the remaining part of $\vec{j}$

$$\vec{j}_l = \vec{j} - \vec{j}_t = -\frac{1}{4\pi}\vec{\nabla} \int \frac{\vec{\nabla}' \cdot \vec{j}(\vec{r}', t)}{|\vec{r} - \vec{r}'|} d^3\vec{r}' \tag{3.27}$$

has zero curl, or, as it is said, is "irrotational". Indeed any vector function of position can be expressed as sum of a solenoidal part and of an irrotational part.

We note that

$$\vec{\nabla}\dot{\phi} = 4\pi \vec{j}_l \tag{3.28}$$

and

$$\nabla^2 \vec{A} - \frac{1}{c^2}\frac{\partial^2 \vec{A}}{\partial t^2} = \frac{4\pi}{c}\vec{j}_t \tag{3.29}$$

We can summarize the situation as follows. The electric field consists of two parts $\vec{E}_1$ and $\vec{E}_2$, one irrotational and the other solenoidal

$$\vec{E} = \vec{E}_1 + \vec{E}_2 \tag{3.30}$$

where

$$\vec{E}_1 = -\vec{\nabla}\phi \tag{3.31}$$

$$\vec{E}_2 = -\frac{1}{c}\frac{\partial \vec{A}}{\partial t} \tag{3.32}$$

The magnetic field $\vec{B}$ is entirely solenoidal

$$\vec{B} = \vec{\nabla} \times \vec{A} \tag{3.33}$$

$\vec{E}_2$ and $\vec{B}$ represent the "radiation field" determined only by the vector potential $\vec{A}$ and related to the solenoidal part of the current $\vec{j}_t$. The radiation field is determined by the four equations (3.32), (3.33), (3.19), and (3.29).

We shall now proceed with the treatment of this field by considering the case where the current is zero. We note that, contrary to the scalar potential ϕ, the vector potential $\vec{A}$ may exist even in the absence of currents or charges.

III. B. Solutions of the Field Equations

In the absence of currents, the field equation (3.29) becomes

$$\nabla^2 \vec{A} - \frac{1}{c^2}\frac{\partial^2 \vec{A}}{\partial t^2} = 0 \tag{3.34}$$

We try solutions of the type

$$\vec{A}(\vec{r}, t) = q(t)\, \vec{\mathcal{A}}(\vec{r}) \tag{3.35}$$

Replacing this expression in eq. (3.34) we obtain

$$q\, \nabla^2 \vec{\mathcal{A}} = \frac{1}{c^2}\vec{\mathcal{A}}\, \ddot{q} \tag{3.36}$$

or

$$\frac{\ddot{q}}{q} = \frac{c^2\, \nabla^2 \vec{\mathcal{A}}}{\vec{\mathcal{A}}} = -\,\omega^2 \tag{3.37}$$

where ω^2 is a constant. Therefore

$$\ddot{q} + \omega^2 q = 0 \tag{3.38}$$

$$\nabla^2 \vec{\mathcal{A}} + \frac{\omega^2}{c^2}\vec{\mathcal{A}} = 0 \tag{3.39}$$

with the solutions

$$q(t) = |q|\, e^{-i\omega t} \tag{3.40}$$

$$\vec{\mathcal{A}}(\vec{r}) = \vec{\pi}\left(\frac{4\pi c^2}{V}\right)^{\frac{1}{2}} e^{i\vec{k}\cdot\vec{r}} \tag{3.41}$$

where

$$|\vec{k}| = \frac{\omega}{c} \tag{3.42}$$

$\vec{\pi}$ is a unit (polarization) vector and the constant term in front of $e^{i\vec{k}\cdot\vec{r}}$ has been introduced for convenience. Using the expressions for q and $\vec{\mathcal{A}}$ in (3.35) we obtain

$$\vec{A}(\vec{r}, t) = \vec{\pi}\left(\frac{4\pi c^2}{V}\right)^{\frac{1}{2}} |q|\, e^{i(\vec{k}\cdot\vec{r} - \omega t)} \tag{3.43}$$

$\vec{A}(\vec{r}, t)$ represents a plane wave of wavelength $2\pi/|\vec{k}|$. Since $\vec{\nabla} \cdot \vec{A} = 0$

$$\pi_x k_x + \pi_y k_y + \pi_z k_z = \vec{\pi} \cdot \vec{k} = 0 \tag{3.44}$$

namely $\vec{A}$ is always prependicular to the direction of propagation.

Generalizing the expression of $\vec{A}$:

$$\vec{A}(\vec{r}, t) = \sum_{\alpha}\sum_{\sigma} \left[q_{\alpha}^{\sigma}(t)\, \vec{A}_{\alpha}^{\sigma}(\vec{r}) + q_{\alpha}^{\sigma}(t)^*\, \vec{A}_{\alpha}^{\sigma}(\vec{r})^* \right] \tag{3.45}$$

where the sum over α includes all the allowed values of $\vec{k}$ and the sum over σ includes, for each $\vec{k}$, the two allowed independent polarizations. The allowed values of $\vec{k}$ are determined by the boundary conditions.

III. C. Periodic Boundary Conditions and Density of States

We may assume the volume V to consist of a parallelepiped with sides of lengths L_x, L_y, and L_z. The periodic boundary conditions imply

$$\vec{A}(\vec{r} + L_x \hat{1}_x) = \vec{A}(\vec{r} + L_y \hat{1}_y) = \vec{A}(\vec{r} + L_z \hat{1}_z) = \vec{A}(\vec{r}) \tag{3.46}$$

where $\hat{1}_x$, $\hat{1}_y$, $\hat{1}_z$ are unit vectors in the x, y, z directions, respectively. Because of the form of $\vec{A}$, these relations give

$$e^{ik_xL_x} = e^{ik_yL_y} = e^{ik_zL_z} = 1 \tag{3.47}$$

or

$$\begin{aligned} n_x &= \frac{k_xL_x}{2\pi} \\ n_y &= \frac{k_yL_y}{2\pi} \\ n_z &= \frac{k_zL_z}{2\pi} \end{aligned} \tag{3.48}$$

where n_x, n_y, $n_z = 0, \pm 1, \pm 2, \pm 3....$ Clearly the index α in eq. (3.45) stands for a particular choice of these three numbers:

$$\alpha \equiv (n_x, n_y, n_z) \tag{3.49}$$

The number of allowed values of $\vec{k}$ in the interval $(\vec{k}, \vec{k} + \vec{dk})$ is given by

$$\frac{L_x dk_x}{2\pi} \frac{L_y dk_y}{2\pi} \frac{L_z dk_z}{2\pi} = \frac{V}{8\pi^3} d^3\vec{k} = \\ = \frac{V}{8\pi^3} k^2 \, dk \sin\theta \, d\theta \, d\phi = \frac{V}{8\pi^3} \frac{\omega^2}{c^3} d\omega \, d\Omega \tag{3.50}$$

where $v = L_x L_y L_z$, $\omega = kc$ and $d\Omega = \sin\theta d\theta d\phi$. We note again that two polarizations correspond to each allowed $\vec{k}$ value.

III. D. The Hamiltonian of the Radiation Field

The most general expression for $\vec{A}$ is now written

$$\vec{A}(\vec{r}, t) = \sum_{\alpha}\sum_{\sigma} \vec{\pi}_{\alpha}^{\sigma} \left(\frac{4\pi c^2}{V}\right)^{\frac{1}{2}} |q_{\alpha}^{\sigma}| \times \\ \times \left\{ \exp\left[i\left(\frac{2\pi}{L_x} n_{x\alpha} x + \frac{2\pi}{L_y} n_{y\alpha} y + \frac{2\pi}{L_z} n_{z\alpha} z - \omega_{\alpha} t \right)\right] + c.\,c. \right\} \tag{3.51}$$

Dropping for the moment the index σ for simplicity of notation, we note that

$$\vec{k}_{-\alpha} = -\vec{k}_{\alpha} \\ \vec{A}_{-\alpha} = \vec{A}_{\alpha}^{*} \tag{3.52}$$

and

$$\vec{\pi}_{-\alpha} = \vec{\pi}_{\alpha} \\ \omega_{-\alpha} = \omega_{\alpha} \\ q_{-\alpha} = q_{\alpha} \tag{3.53}$$

We set

$$\vec{A}_{\alpha}(\vec{r}) = \vec{\pi}_{\alpha} \left(\frac{4\pi c^2}{V}\right)^{\frac{1}{2}} e^{i\vec{k}_{\alpha}\cdot\vec{r}} \tag{3.54}$$

and the following orthogonality relations are derived

$$\int \vec{\mathcal{A}}_\alpha \cdot \vec{\mathcal{A}}_{\alpha'} \, d^3\vec{r} = 4\pi c^2 \, \delta_{\alpha,-\alpha'}$$

$$\int \vec{\mathcal{A}}_\alpha \cdot \vec{\mathcal{A}}_\alpha d^3\vec{r} = \int \vec{\mathcal{A}}^*_\alpha \cdot \vec{\mathcal{A}}^*_\alpha d^3\vec{r} = 0 \tag{3.55}$$

$$\int \vec{\mathcal{A}}^*_\alpha \cdot \vec{\mathcal{A}}_{\alpha'} \, d^3\vec{r} = 4\pi c^2 \, \delta_{\alpha,\alpha'}$$

The α-component of $\vec{A}(\vec{r}, t)$ is given by

$$\vec{A}_\alpha(\vec{r}, t) = q_\alpha(t)\, \vec{\mathcal{A}}_\alpha(\vec{r}) + q^*_\alpha(t)\, \vec{\mathcal{A}}^*_\alpha(\vec{r}) \tag{3.56}$$

and the α-component of the electric field by

$$\vec{E}_\alpha = -\frac{1}{c}\frac{\partial \vec{A}_\alpha}{\partial t} = \frac{i\omega_\alpha}{c}\left(q_\alpha \vec{\mathcal{A}}_\alpha - q^*_\alpha \vec{\mathcal{A}}^*_\alpha\right) \tag{3.57}$$

The total electric field can now be written

$$\vec{E} = \sum_\alpha \frac{i\omega_\alpha}{c}\left(q_\alpha \vec{\mathcal{A}}_\alpha - q^*_\alpha \vec{\mathcal{A}}^*_\alpha\right) \tag{3.58}$$

and its square

$$\begin{aligned}
(\vec{E})^2 &= \sum_\alpha \sum_{\alpha'} \left[\frac{i\omega_\alpha}{c}\left(q_\alpha \vec{\mathcal{A}}_\alpha - q^*_\alpha \vec{\mathcal{A}}^*_\alpha\right)\right] \\
&\times \left[\frac{i\omega_{\alpha'}}{c}\left(q_{\alpha'} \vec{\mathcal{A}}_{\alpha'} - q^*_{\alpha'} \vec{\mathcal{A}}^*_{\alpha'}\right)\right] \\
&= \sum_\alpha \sum_{\alpha'} \left(\frac{-\omega_\alpha \omega_{\alpha'}}{c^2}\right) \\
&\times \left[q_\alpha q_{\alpha'} \vec{\mathcal{A}}_\alpha \cdot \vec{\mathcal{A}}_{\alpha'} + q^*_\alpha q^*_{\alpha'} \vec{\mathcal{A}}^*_\alpha \cdot \vec{\mathcal{A}}^*_{\alpha'} \right. \\
&\left. - q^*_\alpha q_{\alpha'} \vec{\mathcal{A}}^*_\alpha \cdot \vec{\mathcal{A}}_{\alpha'} - q_\alpha q^*_{\alpha'} \vec{\mathcal{A}}_\alpha \cdot \vec{\mathcal{A}}^*_{\alpha'}\right]
\end{aligned} \tag{3.59}$$

Integrating over space

$$\begin{aligned}
&\int (\vec{E})^2 \, d^3\vec{r} \\
&= 4\pi \sum_\alpha \left[\omega^2_\alpha \left(q_\alpha q^*_\alpha + q^*_\alpha q_\alpha\right) - \omega^2_\alpha \left(q_\alpha q_{-\alpha} + q^*_\alpha q^*_{-\alpha}\right)\right]
\end{aligned} \tag{3.60}$$

The magnetic field is given by

$$\vec{B} = \vec{\nabla} \times \vec{A} = \vec{\nabla} \times \left[\sum_{\alpha} \left(q_\alpha \vec{A}_\alpha + q_\alpha^* \vec{A}_\alpha^*\right)\right]$$
$$= \sum_{\alpha} \left[q_\alpha \left(\nabla \times \vec{A}_\alpha\right) + q_\alpha^* \left(\nabla \times \vec{A}_\alpha^*\right)\right] \tag{3.61}$$

It is easy to show that

$$\int (\vec{B})^2 \, d^3\vec{r}$$
$$= 4\pi \sum_{\alpha} \left[\omega_\alpha^2 \left(q_\alpha q_\alpha^* + q_\alpha^* q_\alpha\right) + \omega_\alpha^2 \left(q_\alpha q_{-\alpha} + q_\alpha^* q_{-\alpha}^*\right)\right] \tag{3.62}$$

The energy of the radiation field is given by

$$\frac{1}{8\pi} \int \left[(\vec{E})^2 + (\vec{B})^2\right] d^3\vec{r} = \sum_{\alpha} \omega_\alpha^2 \left(q_\alpha q_\alpha^* + q_\alpha^* q_\alpha\right) \tag{3.63}$$

The Hamiltonian of the radiative field is then given by

$$H = \sum_{\alpha} \sum_{\sigma} \omega_\alpha^2 \left(q_\alpha^\sigma q_\alpha^{\sigma\,*} + q_\alpha^{\sigma\,*} q_\alpha^\sigma\right) = \sum_{\alpha} \sum_{\sigma} H_\alpha^\sigma \tag{3.64}$$

where

$$H_\alpha^\sigma = \omega_\alpha^2 \left(q_\alpha^\sigma q_\alpha^{\sigma\,*} + q_\alpha^{\sigma\,*} q_\alpha^\sigma\right) \tag{3.65}$$

We may note at this point that the Hamiltonian is the sum of independent terms and, therefore, the coordinates q_α^σ represent the normal coordinates of the radiation field. We note also that, we have arrived at the result (3.64) without making any approximation. The form (3.65) for the Hamiltonian, rather than the form $2\omega_\alpha^2 q_\alpha^\sigma q_\alpha^{\sigma\,*}$ has been preferred in preparation for our move into the quantum-mechanical treatment: we shall see later that q_α^σ and $q_\alpha^{\sigma\,*}$ are two operators that do not commute.

We may want at this point to introduce two new real variables for each α

$$Q_\alpha = q_\alpha + q_\alpha^*$$
$$P_\alpha = -i\,\omega_0 \left(q_\alpha - q_\alpha^*\right) = \dot{Q}_\alpha \tag{3.66}$$

Inversely

$$q_\alpha = \frac{1}{2}\left(Q_\alpha - \frac{1}{i\omega_\alpha} P_\alpha\right)$$
$$q_\alpha^* = \frac{1}{2}\left(Q_\alpha + \frac{1}{i\omega_\alpha} P_\alpha\right) \tag{3.67}$$

The Hamiltonian H_α, when written in terms of the new real coordinates, has the form

$$H_\alpha = \frac{1}{2}\,\omega_\alpha^2 Q_\alpha^2 + \frac{1}{2}\,P_\alpha^2 \tag{3.68}$$

The Hamiltonian's equations for Q_α and P_α are

$$\frac{\partial H_\alpha}{\partial P_\alpha} = P_\alpha = \dot{Q}_\alpha \tag{3.69}$$

$$\frac{\partial H_\alpha}{\partial Q_\alpha} = \omega_\alpha^2 Q_\alpha = -\dot{P}_\alpha \tag{3.70}$$

Equation (3.69) above gives us back a result already known from the second (3.66) equation. Equation (3.70) gives

$$\dot{P}_\alpha + \omega_\alpha^2 Q_\alpha = 0 \tag{3.71}$$

or

$$\ddot{Q}_\alpha + \omega_\alpha^2 Q_\alpha = 0 \tag{3.72}$$

which is true because

$$\ddot{q}_\alpha + \omega_\alpha^2 q_\alpha = 0 \tag{3.73}$$

a result already known as (3.38). Therefore Q_α and P_α satisfy Hamilton's equations; we can then write the following Poisson's brackets:

$$\{Q_\alpha, P_\alpha\} = \sum_i \left(\frac{\partial Q_\alpha}{\partial Q_i}\frac{\partial P_\alpha}{\partial P_i} - \frac{\partial Q_\alpha}{\partial P_i}\frac{\partial P_\alpha}{\partial Q_i} \right) = 1 \tag{3.74}$$

$$\{Q_\alpha, P_{\alpha'}\} = \delta_{\alpha\alpha'} \tag{3.75}$$

$$\{Q_\alpha, Q_{\alpha'}\} = \{P_\alpha, P_{\alpha'}\} = 0 \tag{3.76}$$

III. E. Quantization of the Radiation Field

The prescription for moving over from a classical to a quantum mechanical treatment is simple. Q_α and P_α become Hermitian operators and their commutator is obtained by replacing the Poisson brackets as follows

$$\{Q_\alpha, P_{\alpha'}\} \rightarrow \frac{1}{i\hbar}[Q_\alpha, P_{\alpha'}] \tag{3.77}$$

then

$$\begin{aligned} &[Q_\alpha, P_{\alpha'}] = i\hbar\, \delta_{\alpha\alpha'} \\ &[Q_\alpha, Q_{\alpha'}] = [P_\alpha, P_{\alpha'}] = 0 \end{aligned} \tag{3.78}$$

q_α and q^*_α, which are related to Q_α and P_α by the relations (3.67), become two (non-Hermitian) operators which we shall call q^*_α and q^+_α, respectively. The commutation relations of these two operators are easily derived

$$\begin{aligned} &[q_\alpha, q^+_{\alpha'}] = \frac{\hbar}{2\omega_\alpha}\, \delta_{\alpha\alpha'} \\ &[q_\alpha, q_{\alpha'}] = [q^+_\alpha, q^+_{\alpha'}] = 0 \end{aligned} \tag{3.79}$$

We may replace q_α and q^+_α with the dimensionless operators

$$a_\alpha = \left(\frac{2\omega_\alpha}{\hbar}\right)^{\frac{1}{2}} q_\alpha \qquad\qquad a^+_\alpha = \left(\frac{2\omega_\alpha}{\hbar}\right)^{\frac{1}{2}} q^+_\alpha \tag{3.80}$$

We find

$$\begin{aligned} &[a_\alpha, a^+_{\alpha'}] = \delta_{\alpha\alpha'} \\ &[a_\alpha, a_{\alpha'}] = [a^+_\alpha, a^+_{\alpha'}] = 0 \end{aligned} \tag{3.81}$$

The Hamiltonian of the radiation field can now be written

$$H = \sum_\alpha \omega^2_\alpha (q_\alpha q^+_\alpha + q^+_\alpha q_\alpha) = \sum_\alpha \hbar\omega_\alpha \left(a^+_\alpha a_\alpha + \frac{1}{2}\right) \tag{3.82}$$

Reintroducing the polarization index σ

$$H = \sum_\alpha \sum_\sigma \hbar\omega_\alpha \left(a^{\sigma +}_\alpha a^\sigma_\alpha + \frac{1}{2}\right) \tag{3.83}$$

III. F. Energy Levels and Eigenfunctions of a Radiation Field

The Hamiltonian

$$H^\sigma_\alpha = \hbar\omega_\alpha \left(a^{\sigma +}_\alpha a^\sigma_\alpha + \frac{1}{2}\right) \tag{3.84}$$

has the energy eigenvalues

$$E_{\alpha}^{\sigma} = \hbar\omega_{\alpha}\left(n_{\alpha}^{\sigma} + \tfrac{1}{2}\right) \tag{3.85}$$

where $n_{\alpha}^{\sigma} = 0, 1, 2, ...$ The eigenfunctions of H_{α}^{σ} are simply given by the kets $| n_{\alpha}^{\sigma} \rangle$.

The Hamiltonian, eigenvalues and eigenfuctions of the radiation field are now listed:

$$H = \sum_{\alpha}\sum_{\sigma} H_{\alpha}^{\sigma} = \sum_{\alpha}\sum_{\sigma} \hbar\omega_{\alpha}\left(a_{\alpha}^{\sigma +} a_{\alpha}^{\sigma} + \tfrac{1}{2}\right) \tag{3.86}$$

$$E_{n_1^{\sigma_1} n_1^{\sigma_2} n_2^{\sigma_1} \ldots} = \sum_{\alpha}\sum_{\sigma} \hbar\omega_{\alpha}\left(n_{\alpha}^{\sigma} + \tfrac{1}{2}\right) \tag{3.87}$$

$$\psi_{n_1^{\sigma_1} n_1^{\sigma_2} n_2^{\sigma_1} \ldots} = \prod_{\alpha\sigma} |n_{\alpha}^{\sigma}\rangle \tag{3.88}$$

One can see from the above relations how the radiation field may be thought of as a collection of an infinite number of harmonic oscillators, one for each (α,σ) component, with different degrees of excitation n_{α}^{σ}. Alternatively the radiation field may be thought of as an ensemble of photons: n_{α}^{σ} is the number of photons present for each direction $\vec{k}$ and polarization σ.

III. G. The Operator Vector Potential

In the quantum-mechanical treatment the vector potential represents an operator which can be expressed as follows:

$$\begin{aligned}\vec{A} &= \sum_{\alpha}\sum_{\sigma}\left[\vec{A}_{\alpha}^{\sigma} q_{\alpha}^{\sigma} + \vec{A}_{\alpha}^{\sigma *} q_{\alpha}^{\sigma +}\right] \\ &= \sum_{\alpha}\sum_{\sigma}\left(\frac{4\pi c^2}{V}\right)^{\frac{1}{2}}\left(\frac{\hbar}{2\omega_{\alpha}}\right)^{\frac{1}{2}} \vec{\pi}_{\alpha}^{\sigma}\left(e^{i\vec{k}_{\alpha}\cdot\vec{r}} a_{\alpha}^{\sigma} + e^{-i\vec{k}_{\alpha}\cdot\vec{r}} a_{\alpha}^{\sigma +}\right) \\ &= \sum_{\alpha}\sum_{\sigma}\left(\frac{hc^2}{\omega_{\alpha} V}\right)^{\frac{1}{2}} \vec{\pi}_{\alpha}^{\sigma}\left(e^{i\vec{k}_{\alpha}\cdot\vec{r}} a_{\alpha}^{\sigma} + e^{-i\vec{k}_{\alpha}\cdot\vec{r}} a_{\alpha}^{\sigma +}\right)\end{aligned} \tag{3.89}$$

We note that the operators a_{α}^{σ} and $a_{\alpha}^{\sigma +}$ operate as follows

$$\begin{aligned} a_{\alpha}^{\sigma}|n_{\alpha}^{\sigma}\rangle &= \sqrt{n_{\alpha}^{\sigma}}\,|n_{\alpha}^{\sigma} - 1\rangle \\ a_{\alpha}^{\sigma +}|n_{\alpha}^{\sigma}\rangle &= \sqrt{n_{\alpha}^{\sigma} + 1}\,|n_{\alpha}^{\sigma} + 1\rangle \end{aligned} \tag{3.90}$$

IV. INTERACTION OF RADIATION WITH CHARGED PARTICLES

IV. A. The Hamiltonian of a Charged Particle in an Electromagnetic Field

Assume that a particle of mass m and charge q is under the action of a radiation field $\vec{A}(\vec{r}, t)$ and of a potential field $\phi(\vec{r}, t)$.

The equation of motion of such a particle is given by the Hamiltonian

$$H = \frac{\left(\vec{p} - \frac{q}{c}\vec{A}\right)^2}{2m} + q\,\phi \tag{4.1}$$

where $\vec{p}$ = linear momentum of the particle. We may justify this claim by considering the Hamilton's equations. In the x direction

$$\frac{\partial H}{\partial p_x} = \frac{p_x - \frac{q}{c}A_x}{m} \tag{4.2}$$

and

$$\dot{x} = \frac{\partial H}{\partial p_x} = \frac{p_x - \frac{q}{c}A_x}{m} \tag{4.3}$$

In three dimensions

$$\vec{p} = m\,\dot{\vec{r}} + \frac{q}{c}\vec{A} \tag{4.4}$$

Also

$$\begin{aligned} \frac{\partial H}{\partial x} &= \frac{2\left(p_x - \frac{q}{c}A_x\right)\left(-\frac{\partial A_x}{\partial x}\frac{q}{c}\right)}{2\,m} + \frac{2\left(p_y - \frac{q}{c}A_y\right)\left(-\frac{\partial A_y}{\partial x}\frac{q}{c}\right)}{2\,m} \\ &+ \frac{2\left(p_z - \frac{q}{c}A_z\right)\left(-\frac{\partial A_z}{\partial x}\frac{q}{c}\right)}{2\,m} + q\,\frac{\partial \phi}{\partial x} \\ &= -\frac{q}{c}\left(\dot{x}\,\frac{\partial A_x}{\partial x} + \dot{y}\,\frac{\partial A_y}{\partial x} + \dot{z}\,\frac{\partial A_z}{\partial x}\right) + q\,\frac{\partial \phi}{\partial x} \\ &= -\frac{q}{c}\frac{\partial}{\partial x}(\vec{v} \cdot \vec{A}) + q\,\frac{\partial \phi}{\partial x} \end{aligned} \tag{4.5}$$

and from eq. (4.4)

$$\frac{dp_x}{dt} = m\,\ddot{x} + \frac{q}{c}\frac{dA_x}{dt} \tag{4.6}$$

Therefore the relation

$$\frac{dp_x}{dt} = -\frac{\partial H}{\partial x} \tag{4.7}$$

gives

$$m\,\ddot{x} + \frac{q}{c}\frac{dA_x}{dt} = \frac{q}{c}\frac{\partial}{\partial x}\left(\vec{v}\cdot\vec{A}\right) - q\frac{\partial \phi}{\partial x} \tag{4.8}$$

In three dimensions

$$m\,\ddot{\vec{r}} = q\left[-\vec{\nabla}\phi + \frac{1}{c}\vec{\nabla}\left(\vec{v}\cdot\vec{A}\right) - \frac{1}{c}\frac{d\vec{A}}{dt}\right] \tag{4.9}$$

But

$$\vec{v}\times\left(\vec{\nabla}\times\vec{A}\right) = \vec{\nabla}\left(\vec{v}\cdot\vec{A}\right) - \left(\vec{v}\cdot\vec{\nabla}\right)\vec{A} \tag{4.10}$$

$$\frac{d\vec{A}}{dt} = \frac{\partial\vec{A}}{\partial t} + \left(\vec{v}\cdot\vec{\nabla}\right)\vec{A} \tag{4.11}$$

Then

$$\begin{aligned} m\,\ddot{\vec{r}} &= \\ &= q\left\{-\vec{\nabla}\phi + \frac{1}{c}\left[\vec{v}\times\left(\vec{\nabla}\times\vec{A}\right) + \left(\vec{v}\cdot\vec{\nabla}\right)\vec{A}\right] - \frac{1}{c}\frac{\partial\vec{A}}{\partial t} - \frac{1}{c}\left(\vec{v}\cdot\vec{\nabla}\right)\vec{A}\right\} \\ &= q\left\{\left[-\vec{\nabla}\phi - \frac{1}{c}\frac{\partial\vec{A}}{\partial t}\right] + \frac{1}{c}\vec{v}\times\left(\vec{\nabla}\times\vec{A}\right)\right\} = q\left\{\vec{E} + \frac{1}{c}\vec{v}\times\vec{B}\right\} \end{aligned} \tag{4.12}$$

where we have used the relations (3.9) and (3.6) for $\vec{E}$ and $\vec{B}$, respectively.

IV. B. The Interaction of a Charged Particle with a Radiation Field

Let us consider now the situation in which a charged particle is under the action of a potential ϕ and of a radiation field $\vec{A}$. The Hamiltonian of the system which consists of the particle and the radiation field is given by

$$H = \frac{1}{2m}\left(\vec{p} - \frac{q}{c}\vec{A}\right)^2 + q\,\phi$$
$$+ \frac{1}{8\pi}\int \vec{E}^2 + \vec{B}^2\, d^3\vec{r}$$
$$= \left(\frac{p^2}{2m} + q\,\phi\right) + \frac{1}{8\pi}\int \left[\vec{E}^2 + \vec{B}^2\right] d^3\vec{r} \tag{4.13}$$
$$- \frac{q}{2mc}\left(\vec{p}\cdot\vec{A} + \vec{A}\cdot\vec{p}\right) + \frac{q^2}{2mc^2}\left(\vec{A}\right)^2$$

The condition for the existence of stationary states is generally valid:

$$\text{radiative lifetime } (\sim 10^{-18}\text{ sec}) \gg \text{period of oscillations } (\sim 10^{-15}\text{ sec}) \tag{4.14}$$

Since

$$\vec{p}\cdot\vec{A} = \vec{A}\cdot\vec{p} + \left[\vec{p}\,,\vec{A}\right] \tag{4.15}$$

and, on the other hand,

$$\left[\vec{p}\,,\vec{A}\right]\psi = -i\hbar\left[\vec{\nabla}\,,\vec{A}\right]\psi =$$
$$= -i\hbar\,\vec{\nabla}\cdot\left(\vec{A}\,\psi\right) + i\hbar\,\vec{A}\cdot\vec{\nabla}\psi = -i\hbar\left(\vec{\nabla}\cdot\vec{A}\right)\psi = 0 \tag{4.16}$$

because of the Coulomb gauge, then

$$\left[\vec{p}\,,\vec{A}\right] = 0 \tag{4.17}$$

or

$$\vec{p}\cdot\vec{A} = \vec{A}\cdot\vec{p} \tag{4.18}$$

and we can write

$$H = H_0 + H_1 + H_2 \tag{4.19}$$

where

$$H_0 = \left(\frac{p^2}{2m} + q\,\phi\right) + \frac{1}{8\pi}\int \vec{E}^2 + \vec{B}^2\, d^3\vec{r}$$
$$= -\frac{\hbar^2}{2m}\nabla^2 + q\,\phi + \sum_{\alpha}\sum_{\sigma}\hbar\omega_\alpha\left(a_\alpha^{\sigma +}a_\alpha^{\sigma} + \frac{1}{2}\right) \tag{4.20}$$

$$H_1 = -\frac{q}{mc}\vec{A}\cdot\vec{p}$$
$$= -\frac{q}{m}\sum_{\alpha}\sum_{\sigma}\left(\frac{h}{\omega_\alpha V}\right)^{\frac{1}{2}}\left(a_\alpha^\sigma e^{i\vec{k}_\alpha\cdot\vec{r}} + a_\alpha^{\sigma+} e^{-i\vec{k}_\alpha\cdot\vec{r}}\right)\vec{\pi}_\alpha^\sigma\cdot\vec{p} \tag{4.21}$$

$$H_2 = \frac{q^2}{2mc^2}(\vec{A})^2$$
$$= \frac{q^2}{2m}\frac{h}{V}\sum_{\alpha}\sum_{\sigma}\sum_{\alpha'}\sum_{\sigma'}\frac{1}{(\omega_\alpha\,\omega_{\alpha'})^{\frac{1}{2}}}\left(\vec{\pi}_\alpha^\sigma\cdot\vec{\pi}_{\alpha'}^{\sigma'}\right) \tag{4.22}$$
$$\times\left(a_\alpha^\sigma e^{i\vec{k}_\alpha\cdot\vec{r}} + a_\alpha^{\sigma+} e^{-i\vec{k}_\alpha\cdot\vec{r}}\right)\times\left(a_{\alpha'}^{\sigma'} e^{i\vec{k}_{\alpha'}\cdot\vec{r}} + a_{\alpha'}^{\sigma'+} e^{-i\vec{k}_{\alpha'}\cdot\vec{r}}\right)$$

IV. C. Radiative Processes

The method to be applied here consists in considering H_0 as the Hamiltonian of the "unperturbed" system, given simply by the sum of the Hamiltonian of the particle and the Hamiltonian of the radiation field and taking H_1 and H_2 as time dependent perturbations of the system which may induce transitions between the different eigenstates of H_0. These eigenstates are given by

$$\psi_{e\,;\,n_1\,,\,n_2\,,\,\ldots} = \psi^e \prod_{\alpha\,\sigma} |n_\alpha^\sigma\rangle \tag{4.23}$$

where ψ^e = eigenfunction of the particle and $|n_\alpha^\sigma\rangle$ = eigenfunction of the (α, σ) radiation oscillator. The energies of the states (4.23) are given by

$$E_{e\,;\,n_1\,,\,n_2\,,\,\ldots} = E^e + \sum_{\alpha}\sum_{\sigma}\hbar\omega_\alpha\left(n_\alpha^\sigma + \frac{1}{2}\right) \tag{4.24}$$

where E^e = energy of the particle and the sum over α and σ gives the energy of the radiation field.

We note at this point that H_1 contains terms with either a or a^+, whereas H_2 contains only terms with products of the type aa^+ , a^+a , aa , a^+a^+ . We can then distinguish the following.

1) First order Processes. These processes are produced when H_1 is used in first order. They can be subdivided in

1a. Annihilation of a photon of frequency ω_α or absorption.
1b. Creation of a photon of frequency ω_α or emission.

2) Second order Processes. These processes are produced when H_1 is used in second order and/or H_2 is used in first order. They can be subdivided in

2a. Photon scattering, consisting in the simultaneous absorption of a photon of frequency ω_α and emission of a photon of frequency ω_α' ($\omega_\alpha \lesseqgtr \omega_\alpha'$).

2b. Two photon absorption.

2c. Two photon emission.

Higher order processes are possible but are much less probable.

IV. D. First Order Processes

First order processes are obtained by the use of the Hamiltonian H_1 to first order. We can expand the exponentials appearing in H_1 as follows:

$$e^{\pm i\vec{k}_\alpha\cdot\vec{r}} = 1 \pm i\,\vec{k}_\alpha\cdot\vec{r} + \ldots \tag{4.25}$$

Then

$$\begin{aligned}
H_1 = &-\frac{q}{m}\sum_\alpha\sum_\sigma\left(\frac{h}{\omega_\alpha V}\right)^{\frac{1}{2}}(\vec{\pi}_\alpha^\sigma\cdot\vec{p})\,a_\alpha^\sigma \\
&-\frac{q}{m}\sum_\alpha\sum_\sigma\left(\frac{h}{\omega_\alpha V}\right)^{\frac{1}{2}}(\vec{\pi}_\alpha^\sigma\cdot\vec{p})\,a_\alpha^{\sigma+} \\
&-i\frac{q}{m}\sum_\alpha\sum_\sigma\left(\frac{h}{\omega_\alpha V}\right)^{\frac{1}{2}}(\vec{\pi}_\alpha^\sigma\cdot\vec{p})(\vec{k}_\alpha\cdot\vec{r})\,a_\alpha^\sigma \\
&+i\frac{q}{m}\sum_\alpha\sum_\sigma\left(\frac{h}{\omega_\alpha V}\right)^{\frac{1}{2}}(\vec{\pi}_\alpha^\sigma\cdot\vec{p})(\vec{k}_\alpha\cdot\vec{r})\,a_\alpha^{\sigma+} + \ldots
\end{aligned} \tag{4.26}$$

Given a function $F(x_i, p_i, t)$, we have classically

$$\begin{aligned}
\frac{dF(x_i, p_i, t)}{dt} &= \frac{\partial F}{\partial t} + \sum_i\left(\frac{\partial F}{\partial x_i}\frac{\partial x_i}{\partial t} + \frac{\partial F}{\partial p_i}\frac{\partial p_i}{\partial t}\right) \\
&= \frac{\partial F}{\partial t} + \sum_i\left(\frac{\partial F}{\partial x_i}\frac{\partial H}{\partial p_i} + \frac{\partial F}{\partial p_i}\frac{\partial H}{\partial x_i}\right) \\
&= \frac{\partial F}{\partial t} + \{F, H\}
\end{aligned} \tag{4.27}$$

where the { } brackets indicate a Poisson bracket. If we want to express the above relation quantum mechanically we replace the Poisson bracket by a commutator as follows

$$\{F, H\} \underset{Q.M.}{\longrightarrow} \frac{1}{i\hbar}[F, H] = \frac{i}{\hbar}[H, F] \tag{4.28}$$

When this is done (4.27) becomes

$$\frac{dF}{dt} = \frac{i}{\hbar}[H, F] + \frac{\partial F}{\partial t} \tag{4.29}$$

If we set $F = x$,

$$\dot{x} = \frac{i}{\hbar}[H, x] \tag{4.30}$$

and

$$p_x = m\,\dot{x} = i\,\frac{m}{\hbar}[H, x] \tag{4.31}$$

Taking a matrix element of $[H, x]$ between two eigenfunctions of the particle we find

$$\begin{aligned} \langle\, \psi_f^e | [H, x] |\, \psi_i^e \,\rangle &= \langle\, \psi_f^e | H\, x - x\, H |\, \psi_i^e \,\rangle \\ &= \left(E_f^e - E_i^e\right) \langle\, \psi_f^e | x |\, \psi_i^e \,\rangle \end{aligned} \tag{4.32}$$

and

$$\begin{aligned} \langle\, \psi_f^e | \vec{p} |\, \psi_i^e \,\rangle &= \frac{im}{\hbar} \left(E_f^e - E_i^e\right) \langle\, \psi_f^e | \vec{r} |\, \psi_i^e \,\rangle \\ &= im\omega_{fi} \langle\, \psi_f^e | \vec{r} |\, \psi_i^e \,\rangle \end{aligned} \tag{4.33}$$

where

$$\omega_{fi} = \frac{E_f^e - E_i^e}{\hbar} \tag{4.34}$$

On the basis of this result the first two terms in the expression of H_1 of eq. (4.26) can be written

$$\begin{aligned} &-\frac{q}{m} \sum_\alpha \sum_\sigma \left(\frac{h}{\omega_\alpha V}\right)^{\frac{1}{2}} a_\alpha^\sigma\, \vec{\pi}_\alpha^\sigma \cdot \vec{p} - \frac{q}{m} \sum_\alpha \sum_\sigma \left(\frac{h}{\omega_\alpha V}\right)^{\frac{1}{2}} a_\alpha^{\sigma +}\, \vec{\pi}_\alpha^\sigma \cdot \vec{p} \\ &= -\,i \sum_\alpha \sum_\sigma \left(\frac{h\omega_\alpha}{V}\right)^{\frac{1}{2}} a_\alpha^\sigma\, \vec{\pi}_\alpha^\sigma \cdot (q\,\vec{r}) + i \sum_\alpha \sum_\sigma \left(\frac{h\omega_\alpha}{V}\right)^{\frac{1}{2}} a_\alpha^{\sigma +}\, \vec{\pi}_\alpha^\sigma \cdot (q\,\vec{r}) \end{aligned} \tag{4.35}$$

Consider now the quantity $\left(\vec{k} \cdot \vec{r}\right) \vec{p}$. We can write

$$(\vec{k}\cdot\vec{r})\,\vec{p} = \tfrac{1}{2}\{(\vec{k}\cdot\vec{r})\,\vec{p} - (\vec{k}\cdot\vec{p})\,\vec{r}\} + \tfrac{1}{2}\{(\vec{k}\cdot\vec{r})\,\vec{p} + (\vec{k}\cdot\vec{p})\,\vec{r}\} \tag{4.36}$$

Also

$$\tfrac{1}{2}\{(\vec{k}\cdot\vec{r})\,\vec{p} - (\vec{k}\cdot\vec{p})\,\vec{r}\} = -\tfrac{1}{2}\,\vec{k}\times(\vec{r}\times\vec{p}) = -\tfrac{\omega}{2c}\{\vec{1}_{\vec{k}}\times\vec{L}\} \tag{4.37}$$

where $\vec{L}$ = angular momentum of the particle, and $\vec{1}_{\vec{k}}$ = unit vector in the $\vec{k}$ direction. The second term in (4.36) can be expressed as follows

$$\begin{aligned} \tfrac{1}{2}\{(\vec{k}\cdot\vec{r})\,\vec{p} + (\vec{k}\cdot\vec{p})\,\vec{r}\} &= \tfrac{m}{2}\{(\vec{k}\cdot\vec{r})\,\dot{\vec{r}} + (\vec{k}\cdot\dot{\vec{r}})\,\vec{r}\} \\ &= \frac{m}{2}\frac{d}{dt}[(\vec{k}\cdot\vec{r})\,\vec{r}] = \frac{im\omega}{2}[(\vec{k}\cdot\vec{r})\,\vec{r}] \end{aligned} \tag{4.38}$$

On the basis of the above result we can write the third and fourth terms in (4.26) as follows

$$\begin{aligned} &-i\frac{q}{m}\sum_{\alpha}\sum_{\sigma}\left(\frac{h}{\omega_\alpha V}\right)^{\frac{1}{2}} a_\alpha^\sigma\,(\vec{k}\cdot\vec{r})\,\vec{p}\cdot\vec{\pi}_\alpha^\sigma \\ &+i\frac{q}{m}\sum_{\alpha}\sum_{\sigma}\left(\frac{h}{\omega_\alpha V}\right)^{\frac{1}{2}} a_\alpha^{\sigma+}\,(\vec{k}\cdot\vec{r})\,\vec{p}\cdot\vec{\pi}_\alpha^\sigma \\ &= i\sum_{\alpha}\sum_{\sigma}\left(\frac{h\omega_\alpha}{V}\right)^{\frac{1}{2}}\left[\left(\vec{1}_{\vec{k}}\times\frac{q}{2mc}\vec{L}\right)\cdot\vec{\pi}_\alpha^\sigma\right] a_\alpha^\sigma \\ &+i\sum_{\alpha}\sum_{\sigma}\left(\frac{h\omega_\alpha}{V}\right)^{\frac{1}{2}}\left[\left(\vec{1}_{\vec{k}}\times\frac{q}{2mc}\vec{L}\right)\cdot\vec{\pi}_\alpha^\sigma\right] a_\alpha^{\sigma+} \\ &+\sum_{\alpha}\sum_{\sigma}\left(\frac{h\omega_\alpha}{V}\right)^{\frac{1}{2}}\frac{1}{2}\,(\vec{k}_\alpha\cdot\vec{r})\,(q\,\vec{r}\cdot\vec{\pi}_\alpha^\sigma)\,a_\alpha^\sigma \\ &+\sum_{\alpha}\sum_{\sigma}\left(\frac{h\omega_\alpha}{V}\right)^{\frac{1}{2}}\frac{1}{2}\,(\vec{k}_\alpha\cdot\vec{r})\,(q\,\vec{r}\cdot\vec{\pi}_\alpha^\sigma)\,a_\alpha^{\sigma+} \end{aligned} \tag{4.39}$$

The interaction Hamitonian H_1 can then be written

$$H_1 = H_1(E1) + H_1(M1) + H_1(E2) \tag{4.40}$$

where

$$\begin{aligned} H_1(E1) &= \underline{\textit{electric}}\ \underline{\textit{dipole}}\ \textit{interaction} \\ &= -i\sum_{\alpha}\sum_{\sigma}\left(\frac{h\omega_\alpha}{V}\right)^{\frac{1}{2}}(q\,\vec{r}\cdot\vec{\pi}_\alpha^\sigma)\,a_\alpha^\sigma + i\sum_{\alpha}\sum_{\sigma}\left(\frac{h\omega_\alpha}{V}\right)^{\frac{1}{2}}(q\,\vec{r}\cdot\vec{\pi}_\alpha^\sigma)\,a_\alpha^{\sigma+} \end{aligned} \tag{4.41}$$

$$H_1(M1) = \underline{magnetic\ dipole}\ interaction$$

$$= i \sum_\alpha \sum_\sigma \left(\frac{h\omega_\alpha}{V}\right)^{\frac{1}{2}} \left[\left(\vec{1}_{\vec{k}} \times \frac{q}{2mc}\vec{L}\right) \cdot \vec{\pi}_\alpha^\sigma\right] a_\alpha^\sigma$$

$$+ i \sum_\alpha \sum_\sigma \left(\frac{h\omega_\alpha}{V}\right)^{\frac{1}{2}} \left[\left(\vec{1}_{\vec{k}} \times \frac{q}{2mc}\vec{L}\right) \cdot \vec{\pi}_\alpha^\sigma\right] a_\alpha^{\sigma +} \quad (4.42)$$

and

$$H_1(E2) = \underline{electric\ quadrupole}\ interaction$$

$$= \sum_\alpha \sum_\sigma \left(\frac{h\omega_\alpha}{V}\right)^{\frac{1}{2}} \frac{1}{2}\left(\vec{k}_\alpha \cdot \vec{r}\right)\left(q\,\vec{r} \cdot \vec{\pi}_\alpha^\sigma\right) a_\alpha^\sigma$$

$$+ \sum_\alpha \sum_\sigma \left(\frac{h\omega_\alpha}{V}\right)^{\frac{1}{2}} \frac{1}{2}\left(\vec{k}_\alpha \cdot \vec{r}\right)\left(q\,\vec{r} \cdot \vec{\pi}_\alpha^\sigma\right) a_\alpha^{\sigma +} \quad (4.43)$$

The three Hamiltonians in (4.40) produce electric dipole, magnetic dipole, and electric quadrupole transitions, respectively.

We note at this point that the electric dipole operator is an odd operator and that the magnetic dipole and the electric quadrupole operators are even operators. This leads to the well known "Laporte rule" which states that electric dipole transitions can take place only between quantum states of opposite parity, whereas magnetic dipole and electric quadrupole transitions can take place only between states of the same parity. A system has states of definite parity when it presents a center of symmetry: this is the case of free atoms and ions, of centrosymmetric molecules and of impurity ions in centrosymmetric solids.

The dimension d of free atoms and molecules is in general such that

$$|kd| \ll 1 \quad (4.44)$$

or

$$d \ll \lambda \ (= \text{wavelength of absorbed or emitted light}).$$

In this case it is possible to use the approximation

$$e^{i\vec{k}\cdot\vec{r}} \approx 1 \quad (4.45)$$

in H_1 and the electric dipole transitions are predominant. On the other hand, impurity ions in non-centrosymmetric solids may sometimes present electric dipole and magnetic dipole transitions with probabilities of the same order of magnitude.

IV. E. Absorption and Emission Rates

In the model set up in the previous section the "unperturbed" system consists of the charged particle and the radiative field and is related to the Hamiltonian H_0 of (4.20). The interaction Hamiltonian H_1 of (4.21) is considered to be a small perturbation which may cause transitions of the unperturbed system from an initial to a final state.

In the case of one photon absorption the initial and the final states are given by

$$\psi_i = |\psi_i^e\rangle |n_\alpha^\sigma\rangle |n_{\alpha'}^{\sigma'}\rangle \ldots \tag{4.46}$$

$$\psi_f = |\psi_f^e\rangle |n_\alpha^\sigma - 1\rangle |n_{\alpha'}^{\sigma'}\rangle \ldots \tag{4.47}$$

respectively. Since

$$\langle n_\alpha^\sigma - 1 |\, a_\alpha^\sigma \,| n_\alpha^\sigma\rangle = \sqrt{n_\alpha^\sigma} \tag{4.48}$$

the relevant matrix element for the process of absorption of one photon is

$$\begin{aligned} &\langle \psi_f^e ; n_\alpha^\sigma - 1 |\, H_1 \,| \psi_i^e ; n_\alpha^\sigma \rangle = \\ &= -\frac{q}{m}\left(\frac{h}{\omega_\alpha V}\right)^{\frac{1}{2}} \langle \psi_f^e |\, e^{i\vec{k}_\alpha \cdot \vec{r}}\, \vec{\pi}_\alpha^\sigma \cdot \vec{p} \,| \psi_i^e \rangle \sqrt{n_\alpha^\sigma} \end{aligned} \tag{4.49}$$

In the case of one photon emission, the initial and the final states are given by

$$\psi_i = |\psi_i^e\rangle |n_\alpha^\sigma\rangle |n_{\alpha'}^{\sigma'}\rangle \ldots \tag{4.50}$$

$$\psi_f = |\psi_f^e\rangle |n_\alpha^\sigma + 1\rangle |n_{\alpha'}^{\sigma'}\rangle \ldots \tag{4.51}$$

respectively. But

$$\langle n_\alpha^\sigma + 1 |\, a_\alpha^{\sigma +} \,| n_\alpha^\sigma\rangle = \sqrt{n_\alpha^\sigma + 1} \tag{4.52}$$

then the relevant matrix element for the process of emission of one photon is

$$\begin{aligned} &\langle \psi_f^e ; n_\alpha^\sigma + 1 |\, H_1 \,| \psi_i^e ; n_\alpha^\sigma \rangle \\ &= -\frac{q}{m}\left(\frac{h}{\omega_\alpha V}\right)^{\frac{1}{2}} \langle \psi_f^e |\, e^{i\vec{k}_\alpha \cdot \vec{r}}\, \vec{\pi}_\alpha^\sigma \cdot \vec{p} \,| \psi_i^e \rangle \sqrt{n_\alpha^\sigma + 1} \end{aligned} \tag{4.53}$$

Since the radiative field has a continuous density of states (see for this section III.C), both absorption and emission processes are associated with a probability per unit time. By applying the Fermi Golden Rule we derive that the probability per unit time of finding the system (particle + radiation field) with one less or one more photon of energy $\hbar\omega_\alpha$ and polarization $\vec{\pi}_\alpha^\sigma$ in the solid angle $(\Omega_\alpha, \Omega_\alpha + d\Omega_\alpha)$ is given by

$$P_\alpha^\sigma \, d\Omega_\alpha = \frac{2\pi}{\hbar^2} |M_\alpha^\sigma|^2 \, g(\omega_\alpha) \tag{4.54}$$

where

$$g(\omega_\alpha) = \frac{V^2 \omega_\alpha}{8\pi^3 c^3} \, d\Omega_\alpha \tag{4.55}$$

and

$$|M_\alpha^\sigma|^2 = \left\langle \begin{array}{l} \dfrac{q^2}{m^2} \dfrac{h}{\omega_\alpha V} \left| \langle \psi_f^e | \, e^{i\vec{k}_\alpha \cdot \vec{r}} \, \vec{\pi}_\alpha^\sigma \cdot \vec{p} \, | \psi_i^e \rangle \right|^2 n_\alpha^\sigma \\ \dfrac{q^2}{m^2} \dfrac{h}{\omega_\alpha V} \left| \langle \psi_f^e | \, e^{-i\vec{k}_\alpha \cdot \vec{r}} \, \vec{\pi}_\alpha^\sigma \cdot \vec{p} \, | \psi_i^e \rangle \right|^2 (n_\alpha^\sigma + 1) \end{array} \right. \tag{4.56}$$

where the upper (lower) row corresponds to the process of absorption (emission) of one photon. Replacing (4.56) in (4.54) and taking (4.55) into account we find:

$$P_\alpha^\sigma \, d\Omega_\alpha = \frac{\omega_\alpha q^2}{hc^3 m^2} \left| \left\langle \psi_f^e \left| \begin{array}{c} e^{i\vec{k}_\alpha \cdot \vec{r}} \, \vec{\pi}_\alpha^\sigma \cdot \vec{p} \\ e^{-i\vec{k}_\alpha \cdot \vec{r}} \, \vec{\pi}_\alpha^\sigma \cdot \vec{p} \end{array} \right| \psi_i^e \right\rangle \right|^2 \left\{ \begin{array}{c} (n_\alpha^\sigma) \\ (n_\alpha^\sigma + 1) \end{array} \right\} d\Omega_\alpha \tag{4.57}$$

Let us consider two quantum states of the particle ψ_l^e (l stands for lower) and ψ_u^e (u stands for upper) with energies E_l and E_u, respectively and $E_l < E_u$. The process of emission of the photon is related to the squared matrix element

$$\left| \langle \psi_l^e | \, e^{-i\vec{k} \cdot \vec{r}} \, \vec{\pi} \cdot \vec{p} \, | \psi_u^e \rangle \right|^2 \tag{4.58}$$

Since $\vec{\pi} \cdot \vec{k} = 0$ we can interchange $\vec{\pi} \cdot \vec{p}$ with $e^{-i\vec{k} \cdot \vec{r}}$:

$$\begin{aligned} &\left| \langle \psi_l^e | \, e^{-i\vec{k} \cdot \vec{r}} (\vec{\pi} \cdot \vec{p}) \, | \psi_u^e \rangle \right|^2 = \left| \langle \psi_l^e | \, (\vec{\pi} \cdot \vec{p}) \, e^{-i\vec{k} \cdot \vec{r}} \, | \psi_u^e \rangle \right|^2 \\ &= \left| \langle (\vec{\pi} \cdot \vec{p}) \, \psi_l^e | \, e^{-i\vec{k} \cdot \vec{r}} \, | \psi_u^e \rangle \right|^2 = \left| \langle \psi_u^e | \, e^{-i\vec{k} \cdot \vec{r}} (\vec{\pi} \cdot \vec{p}) \, | \psi_l^e \rangle \right|^2 \end{aligned} \tag{4.59}$$

The last squared matrix element is the one that would enter the transition probability for a $l \rightarrow u$ (absorption) process.

On the basis of the above result, (4.57) becomes

$$P_\alpha^\sigma \, d\Omega_\alpha = \frac{\omega_\alpha q^2}{hc^3 m^2} \left| \langle \psi_u^e | \, e^{i\vec{k} \cdot \vec{r}} \, \vec{\pi}_\alpha \cdot \vec{p} \, | \psi_l^e \rangle \right|^2 \left(\begin{array}{c} (n_\alpha^\sigma) \\ (n_\alpha^\sigma + 1) \end{array} \right) d\Omega_\alpha \tag{4.60}$$

The transition probability for absorption is always proportional to the number of photons n_α^σ present; the transition probability for emission consists of one part, called induced emission, which is proportional to n_α^σ and of another part, called spontaneous emission, which is present even when $n_\alpha^\sigma = 0$. We note here that the transition probability of absorption and the transition probability for induced emission between two states are equal. If two or more charged particles are present

$$P_\alpha^\sigma \, d\Omega_\alpha = \frac{\omega_\alpha}{hc^3} \left| \left\langle \sum_i \frac{q_i}{m_i} (\vec{\pi}_\alpha^\sigma \cdot \vec{p}_i) \, e^{i\vec{k}_\alpha \cdot \vec{r}_i} \right\rangle_{ul} \right|^2 \begin{pmatrix} (n_\alpha^\sigma) \\ (n_\alpha^\sigma + 1) \end{pmatrix} d\Omega_\alpha \qquad (4.61)$$

The probability per unit time of spontaneous emission of a photon of frequency ω_α and polarization $\vec{\pi}_\alpha^\sigma$ in the solid angle $d\Omega_\alpha$ is given by

$$P_\alpha^\sigma(sp) \, d\Omega_\alpha = \frac{\omega_\alpha}{hc^3} \left| \left\langle \sum_i \frac{q_i}{m_i} (\vec{\pi}_\alpha^\sigma \cdot \vec{p}_i) \, e^{i\vec{k}_\alpha \cdot \vec{r}_i} \right\rangle_{ul} \right|^2 d\Omega_\alpha \qquad (4.62)$$

The probability per unit time of absorption or induced emission of a photon of frequency ω_α and polarization $\vec{\pi}_\alpha^\sigma$ in the solid angle $d\Omega_\alpha$ is given by

$$P_\alpha^\sigma(abs;\, emi) \, d\Omega_\alpha = [P_\alpha^\sigma(sp) \, d\Omega_\alpha] \, n_\alpha^\sigma \qquad (4.63)$$

IV. F. Absorption and Emission in the Electric Dipole Approximation

In the electric dipole approximation the expression (4.57) becomes

$$\begin{aligned} P_\alpha^\sigma \, d\Omega_\alpha &= \frac{\omega_\alpha}{hc^3} \left| \left\langle \sum_l \frac{q_l}{m_l} (\vec{\pi}_\alpha^\sigma \cdot \vec{p}_l) \right\rangle_{fi} \right|^2 \begin{pmatrix} (n_\alpha^\sigma) \\ (n_\alpha^\sigma + 1) \end{pmatrix} d\Omega_\alpha \\ &= \frac{\omega_\alpha}{hc^3} \left| \langle \vec{\pi}_\alpha^\sigma \cdot \vec{M} \rangle_{fi} \right|^2 \begin{pmatrix} (n_\alpha^\sigma) \\ (n_\alpha^\sigma + 1) \end{pmatrix} d\Omega_\alpha \end{aligned} \qquad (4.64)$$

where

$$\vec{M} = \sum_l q_l \, \vec{r}_l = \textit{electric dipole operator} \qquad (4.65)$$

The radiation density per unit frequency range at $\omega = \omega_\alpha$, unit solid angle and polarization $\vec{\pi}_\alpha^\sigma$ is given by

$$\frac{g(\omega_\alpha) \, \hbar\omega_\alpha n_\alpha^\sigma}{V \, d\Omega_\alpha} = \frac{\omega_\alpha^2}{8\pi^3 c^3} \hbar\omega_\alpha n_\alpha^\sigma = \frac{\hbar\omega_\alpha^3}{8\pi^3 c^3} n_\alpha^\sigma \qquad (4.66)$$

and the radiation intensity per unit frequency range at $\omega = \omega_\alpha$, unit solid angle and polarization $\vec{\pi}_\alpha^\sigma$ is given by

$$I(\omega_\alpha, \sigma) = \frac{\hbar\omega_\alpha^3}{8\pi^3 c^2} n_\alpha^\sigma \tag{4.67}$$

We can also write

$$\frac{4\pi^2 I}{\hbar^2 c} = \frac{4\pi^2}{\hbar^2 c} \frac{\hbar\omega_\alpha^3}{8\pi^3 c^2} n_\alpha^\sigma = \frac{\omega_\alpha^3}{hc^3} n_\alpha^\sigma \tag{4.68}$$

Therefore we can express the formula (4.64) as follows

$$P_\alpha^\sigma = \left\langle \begin{array}{l} \dfrac{4\pi^2 I(\omega_\alpha; \sigma)}{\hbar^2 c} |M_{fi}|^2 \\ \left\{ \dfrac{4\pi^2 I(\omega_\alpha; \sigma)}{\hbar^2 c} + \dfrac{\omega_\alpha^3}{hc^3} \right\} |M_{fi}|^2 \end{array} \right. \tag{4.69}$$

where the upper (lower) row corresponds to absorption (emission) and where

$$|M_{fi}|^2 = |\langle \vec{\pi}_\alpha^\sigma \cdot \vec{M} \rangle_{fi}|^2 \tag{4.70}$$

IV. G. The Radiative Broadening and Shift of Atomic Levels

The interaction of an atom with a radiation field is essentially twofold:

1) The exchange of real photons through the processes of emission and absorption leads to level-broadening.
2) The exchange of virtual photons leads to level-shift.

We consider now a system with a Hamiltonian H_0 under the action of a time dependent perturbation H'. The Hamiltonian of the perturbed system is

$$H = H_0 + H'(t) \tag{4.71}$$

and the wavefunction is given by

$$\psi = \sum_i c_i(t)\, \psi_i(t) \tag{4.72}$$

where the $\psi_i(t)$'s are eigenfunctions of H_0 and where the time dependent coefficients $c_i(t)$ are given by

$$\begin{aligned} i\hbar\, \dot{c}_k(t) &= \\ &= \sum_i c_i(t) \langle \psi_k(t) | H' | \psi_i(t) \rangle \\ &= \sum_i c_i(t) \langle \psi_k(0) | H' | \psi_i(0) \rangle e^{i\omega_{ki} t} \\ &= \sum_i c_i(t) H'_{ki} e^{i\omega_{ki} t} \end{aligned} \tag{4.73}$$

where

$$\omega_{ki} = \frac{E_k - E_i}{\hbar} \tag{4.74}$$

and

$$H'_{ki} = \langle \psi_k(0) | H' | \psi_i(0) \rangle \tag{4.75}$$

Let us assume now that at time $t = 0$ the system is in a definite eigenstate of H_0;

$$c_n(0) = 1; \quad c_i(0) = 0 \quad \text{for } i \neq n \tag{4.76}$$

Let us assume also that there is only one state l to which the system can go:

$$\begin{aligned} &\langle \psi_n(0) | H' | \psi_l(0) \rangle \neq 0, \\ &\langle \psi_m(0) | H' | \psi_k(0) \rangle = 0 \text{ for } m, k \neq n \end{aligned} \tag{4.77}$$

The equations (4.73) will then reduce to

$$i\hbar\, \dot{c}_l(t) = c_n \langle \psi_l(0) | H' | \psi_n(0) \rangle e^{i\omega_{ln} t} \tag{4.78}$$

$$i\hbar\, \dot{c}_n(t) = c_l \langle \psi_n(0) | H' | \psi_l(0) \rangle e^{-i\omega_{ln} t} \tag{4.79}$$

where

$$\omega_{ln} = \frac{E_l - E_n}{\hbar}$$

We shall assume that the probability of finding the system in the state n decreases esponentially

$$c_n = e^{-(\gamma/2)t} \tag{4.80}$$

with γ independent of time. We have then from (4.78)

$$i\hbar\, \dot{c}_l(t) = H'_{ln}\, e^{[i\omega_{ln} - (\gamma/2)]\, t} \tag{4.81}$$

where $H'_{ln} = \langle \psi_l(0) |\, H' \,| \psi_n(0)\rangle$. If H'_{ln} is independent of time, integrating from 0 to t we get

$$c_l = -\frac{i}{\hbar} H'_{ln} \frac{e^{[i\omega_{ln} - (\gamma/2)]\, t} - 1}{i\omega_{ln} - (\gamma/2)} \tag{4.82}$$

Using the expression found for c_l and taking (4.80) into account, (4.79) gives

$$\gamma = \frac{2}{\hbar^2} |H'_{ln}|^2 \frac{1 - e^{[i\omega_{ln} - (\gamma/2)]\, t}}{i\omega_{ln} - (\gamma/2)} \tag{4.83}$$

γ is not independent of time as we assumed. We seem to be on the wrong track, but we may proceed in a somewhat different way. We assume that the quantum state l is in a continuum of energies. Equations (4.78) and (4.79), changing the subscript l to λ, and integrating over the continuum of energies, become

$$i\hbar\, \dot{c}_\lambda(t) = c_n H'_{\lambda n}\, e^{i\omega_{\lambda n} t} \tag{4.84}$$

$$i\hbar\, \dot{c}_n(t) = \int c_\lambda H'_{n\lambda}\, e^{-i\omega_{\lambda n} t}\, d\lambda \tag{4.85}$$

Using the assumption (4.80) and integrating (4.84) from 0 to t, we find

$$c_\lambda = -\frac{i}{\hbar} H'_{\lambda n} \frac{e^{[i\omega_{\lambda n} - (\gamma/2)]\, t} - 1}{i\omega_{\lambda n} - (\gamma/2)} \tag{4.86}$$

From (4.85), using (4.86),

$$i\hbar \left(-\frac{\gamma}{2}\right) e^{-(\gamma/2)\, t} = \frac{i}{\hbar} \int |H'_{\lambda n}|^2\, e^{-i\omega_{\lambda n} t}\, \frac{1 - e^{[i\omega_{\lambda n} - (\gamma/2)]\, t}}{i\omega_{\lambda n} - (\gamma/2)}\, d\lambda$$

or

$$\gamma = \int \frac{2i}{\hbar^2} |H'_{\lambda n}|^2 \frac{1 - e^{[i\omega_{\lambda n} + (\gamma/2)]\, t}}{\omega_{n\lambda} - i(\gamma/2)}\, d\lambda \tag{4.87}$$

But

$$d\lambda = \frac{d\lambda}{d\omega_{n\lambda}} d\omega_{n\lambda} = \rho(\omega_{n\lambda})\, d\omega_{n\lambda} \tag{4.88}$$

and then,

$$\gamma = \frac{2i}{\hbar^2} \int |H'_{\lambda n}|^2 \rho(\omega_{n\lambda}) \frac{1 - e^{[i\omega_{\lambda n} + (\gamma/2)]t}}{\omega_{n\lambda} - i(\gamma/2)} d\omega_{n\lambda} \tag{4.89}$$

In evaluating the integral above we shall assume γ very small in comparison to $\omega_{n\lambda}$. We may then write

$$\gamma = \frac{2i}{\hbar^2} \int |H'_{\lambda n}|^2 \rho(\omega_{n\lambda}) \frac{1 - e^{i\omega_{n\lambda} t}}{\omega_{n\lambda}} d\omega_{n\lambda} \tag{4.90}$$

But

$$\frac{1 - e^{i\omega_{\lambda n} t}}{\omega_{\lambda n}} = \frac{1 - \cos \omega_{n\lambda} t}{\omega_{n\lambda}} - i \frac{\sin \omega_{n\lambda} t}{\omega_{n\lambda}} \tag{4.91}$$

For very large t the first term may be replaced by $1/\omega_{n\lambda}$ for $\omega_n \neq \omega_\lambda$ because the rapidly oscillating $\cos \omega_{n\lambda} t$ does not give any contribution to the integral; at $\omega_{n\lambda} = 0$ this term goes to zero. When it is multiplied by the rest of the integrand and integrated over $\omega_{n\lambda}$, the result is the principal value of the integral. For very large t the second term in (4.91) can be expressed as $i\pi\delta(\omega_{n\lambda})$. In fact, by definition, the function $\delta(x)$ can be expressed in the following way:

$$\delta(x) = \frac{1}{\pi} \lim_{\alpha \to \infty} \frac{\sin \alpha x}{x} \tag{4.92}$$

Therefore we may write

$$\lim_{t \to \infty} \frac{1 - e^{i\omega_{\lambda n} t}}{\omega_{\lambda n}} = \frac{\mathcal{P}}{\omega_{n\lambda}} - i\pi\delta(\omega_{n\lambda}) \tag{4.93}$$

where P indicates the principal value. Now γ can be expressed as

$$\begin{aligned} \gamma &= \frac{2i}{\hbar^2} \int |H'_{\lambda n}|^2 \rho(\omega_{n\lambda}) \left\{ \frac{\mathcal{P}}{\omega_{n\lambda}} - i\pi\delta(\omega_{n\lambda}) \right\} d\omega_{n\lambda} \\ &= \frac{2\pi}{\hbar^2} |H'_{\lambda n}|^2 \rho(\omega_n = \omega_\lambda) + \frac{2i}{\hbar^2} \mathcal{P} \int \frac{|H'_{\lambda n}|^2 \rho(\omega_{n\lambda})\, d\omega_{n\lambda}}{\omega_{n\lambda}} \\ &= W_{n\lambda} + i\, \mathcal{J} \end{aligned} \tag{4.94}$$

where

$$\mathrm{Re}\,(\gamma) = W_{n\lambda} = \frac{2\pi}{\hbar^2} |H'_{\lambda n}|^2 \rho\,(\omega_n = \omega_\lambda) \tag{4.95}$$

is in effect the transition probability per unit time from state n to state λ. The imaginary part of γ is

$$\mathrm{Im}\,(\gamma) = \mathcal{I} = \frac{2}{\hbar}\,\mathcal{P}\int \frac{|H'_{\lambda n}|^2 \rho\,(\omega_{n\lambda})}{\hbar\omega_{n\lambda}}\, d\omega_{n\lambda} \tag{4.96}$$

Let us consider now the expression for c_λ after a time $t \gg 1/W_{n\lambda}$:

$$\begin{aligned} c_\lambda &= -\frac{i}{\hbar} H'_{\lambda n} \frac{e^{[i\omega_{\lambda n} + (\gamma/2)]\,t} - 1}{i\,\omega_{n\lambda} - \frac{1}{2}(W + i\,\mathcal{I})} \\ &\xrightarrow[t \gg 1/W]{} \frac{i}{\hbar} H'_{\lambda n} \frac{1}{i\left(\omega_\lambda - \omega_n - \frac{1}{2}\mathcal{I}\right) - \frac{1}{2}W} \end{aligned} \tag{4.97}$$

and

$$|c_\lambda(\infty)|^2 = \frac{1}{\hbar^2} |H'_{\lambda n}|^2 \frac{1}{\left[\left(\omega_n + \frac{1}{2}\mathcal{I}\right) - \omega_\lambda\right]^2 - W^2/4} \tag{4.98}$$

It is interesting to apply the above results to the case of an atom in an excited state interacting with the radiation field. In this case the unperturbed system consists of the atom and the radiation field and the perturbing Hamiltonian H' is given by the interaction between them. The following observations can be made:

1. The real part of γ (equal to the transition probability per unit time) produces a finite width of the transition. In the case of an atom interacting with radiation this width is called the natural linewidth. Spectral lines with the shape (4.98) are called Lorentzian lines. The appearance of a line breadth derives from the fact that we have allowed the probability $|c_n|^2$ of the atom being in the initial state to decay for a time which is greater than the lifetime of the initial state. For times much shorter than this lifetime, the probability $|c_n|^2$ maintains practically the initial value and the line is very sharp.

2. The imaginary part of γ produces a shift in the energy of the initial level of the transition. This shift is called the Lamb shift.

3. By the use of the pricipal value P, (4.96) takes care of the resonances that may occur in correspondence to the emission and reabsorption of virtual photons with energy equal to the energy difference between the initial level and a different level connected to it by H'.

From (4.98) we can see that *the width at half maximum intensity is related to the total transition probability as follows*:

$$\Delta E = \hbar W \tag{4.99}$$

We may call W^{-1} the lifetime τ of the state n,

$$(\Delta E)\ \tau = \hbar \tag{4.100}$$

This result is in agreement with the uncertainty relation between energy and time which expresses the fact that we may know the energy of a system with an accuracy $\Delta E = \frac{\hbar}{\tau}$ if only the time τ is available to measure it.

Such effects as linewidth and energy shift are a consequence of the interaction between the atomic system and the radiation; similar effects are however found when an atom in a crystal interacts with the thermal vibrations of the lattice. The formalism developed in this section applies whenever an isolated atom interacts with a large number of other degrees of freedom whose density is practically continuous.

Several lifetime shortening mechanisms may be present in a system; for example, thermal vibrations and radiation damping may be operating at the same time; also several transitions may originate from the same level. In this case the breadth of the level is given by the sum of the breadths due to all these processes.

V. TWO-PHOTON SPONTANEOUS EMISSION

V.A. System and Hamiltonians

We shall examine the spontaneous decay of a hydrogen atom in the 2s state. The Hamiltonian of the unperturbed system which consists of the atom and of the radiative field is

$$H_0 = \frac{p^2}{2m} - e\,\phi + \int \frac{E^2 + B^2}{8\pi}\, d\tau$$

The interaction Hamiltonian is

$$\begin{aligned} H' &= \frac{e}{2mc}\left[\vec{p}\,,\vec{A}\right]_+ + \frac{e^2}{2mc^2}\left(\vec{A}\right)^2 \\ &= \frac{e}{mc}\,\vec{p}\cdot\vec{A} + \frac{e^2}{2mc^2}\left(\vec{A}\right)^2 = H_1 + H_2 \end{aligned} \tag{5.1}$$

where

$$\vec{A} = \sum_{\alpha}\sum_{\sigma}\left(\frac{h\,c^2}{\omega_\alpha V}\right)^{\frac{1}{2}} \vec{\pi}_\alpha^\sigma \left(e^{i\vec{k}_\alpha\cdot\vec{r}}\, a_\alpha^\sigma + e^{-i\vec{k}_\alpha\cdot\vec{r}}\, a_\alpha^{\sigma\,+}\right) \tag{5.2}$$

$$H_1 = \frac{e}{m}\sum_{\alpha\sigma}\left(\frac{h}{\omega_\alpha V}\right)^{\frac{1}{2}} \left(a_\alpha^\sigma\, e^{i\vec{k}_\alpha\cdot\vec{r}} + a_\alpha^{\sigma\,+}\, e^{-i\vec{k}_\alpha\cdot\vec{r}}\right) \vec{\pi}_\alpha^\sigma \cdot \vec{p} \tag{5.3}$$

$$\begin{aligned} H_2 = \frac{e^2}{2m}\,\frac{h}{V}\sum\sum_{\alpha\sigma\alpha'\sigma'} \frac{1}{(\omega_\alpha\,\omega_{\alpha'})^{\frac{1}{2}}}\left(\vec{\pi}_\alpha^\sigma \cdot \vec{\pi}_{\alpha'}^{\sigma'}\right) \\ \times\left(a_\alpha^\sigma\, e^{i\vec{k}_\alpha\cdot\vec{r}} + a_\alpha^{\sigma\,+}\, e^{-i\vec{k}_\alpha\cdot\vec{r}}\right)\left(a_{\alpha'}^{\sigma'}\, e^{i\vec{k}_{\alpha'}\cdot\vec{r}} + a_{\alpha'}^{\sigma'\,+}\, e^{-i\vec{k}_{\alpha'}\cdot\vec{r}}\right) \end{aligned} \tag{5.4}$$

We shall now show that one-photon $l = 0 \rightarrow l = 0$ transitions are strictly forbidden. Consider two atomic states spherically symmetric, say 2s and 1s. The integral

$$\int \psi_{2s}^{*}\, \vec{p}\, \psi_{1s}\, d\tau = -\,i\hbar \int \psi_{2s}^{*}\, \vec{\nabla}\, \psi_{1s}\, d\tau = 0$$

This can be seen by taking the system of coordinates for the integration with the x axis in the direction of polarization $\vec{\pi}$. Then

$$e^{i\vec{k}\cdot\vec{r}} = e^{i\,(k_y\,y\,+\,k_z\,z)}$$

because $\vec{k}\cdot\vec{\pi} = 0$ and $\vec{\nabla}\psi_{1s}$ is an odd function of x. Thus the integrand is an odd function of x and the integral vanishes.

V.B. Two-Photon Decay

A hydrogen atom in the 2s state can decay to the ground state 1s via the spontaneous emission of two photons. The Feynman diagrams of this process are shown in Fig. 5 where a) and b) refer to the intervention of the perturbing Hamiltonian H_1 and c) refers to the intervention of H_2. We shall now examine these cases individually.

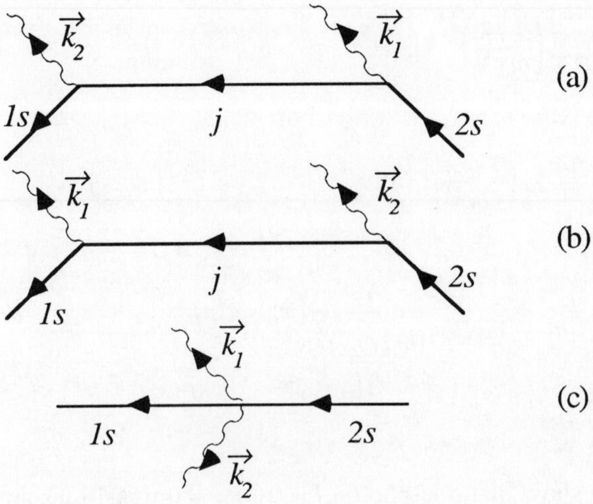

Figure 5. Diagrams of three types of interactions which contribute to the two-photon decay.

Case a

$|\, i \,\rangle$ = atom in state 2s .... $\left(E_i = E_i^{el}\right)$
$|\, j \,\rangle$ = atom in state $j + (\hbar\omega_1)$ photon .... $\left(E_j = E_j^{el} + \hbar\omega_1\right)$

$|\, f \,\rangle$ = atom in state 1s + $(\hbar\omega_1) + (\hbar\omega_2)$ photons
where the superscript "el" indicates a pure electronic state.

Emission of photon $\hbar\omega_1$: $\langle\, j \,|\, H_1 \,|\, i \,\rangle = \frac{e}{m}\left(\frac{h}{\omega_1 V}\right)^{\frac{1}{2}} \left\langle\, \psi_j^{el} \,\middle|\, e^{-i\vec{k}_1 \cdot \vec{r}}\, \vec{p} \cdot \vec{\pi}_{k_1}^{\sigma_1} \,\middle|\, \psi_i^{el} \right\rangle$ (5.5)

Emission of photon $\hbar\omega_2$: $\langle\, f \,|\, H_1 \,|\, j \,\rangle = \frac{e}{m}\left(\frac{h}{\omega_2 V}\right)^{\frac{1}{2}} \left\langle\, \psi_f^{el} \,\middle|\, e^{-i\vec{k}_2 \cdot \vec{r}}\, \vec{p} \cdot \vec{\pi}_{k_2}^{\sigma_2} \,\middle|\, \psi_j^{el} \right\rangle$ (5.6)

Energy denominator: $\hbar\omega_{ij} + \hbar\omega_1$ (5.7)

where $\hbar\omega_{ij} = E_i^{el} - E_j^{el}$

Case b

$|\, i \,\rangle$ = atom in state 2s .... $\left(E_i = E_i^{el}\right)$
$|\, j \,\rangle$ = atom in state $j + (\hbar\omega_2)$ photon .... $\left(E_j = E_j^{el} + \hbar\omega_2\right)$

$| f \rangle$ = atom in state 1s + $(\hbar\omega_2) + (\hbar\omega_1)$ photons

Emission of photon $\hbar\omega_2$: $\langle H_1 \rangle_{ji} = \frac{e}{m}\left(\frac{h}{\omega_2 V}\right)^{\frac{1}{2}} \left\langle \psi_j^{el} \right| e^{-i\vec{k}_2\cdot\vec{r}}\, \vec{p}\cdot\vec{\pi}_{k_2}^{\sigma_2} \left| \psi_i^{el} \right\rangle$ (5.8)

Emission of photon $\hbar\omega_1$: $\langle H_1 \rangle_{fj} = \frac{e}{m}\left(\frac{h}{\omega_1 V}\right)^{\frac{1}{2}} \left\langle \psi_f^{el} \right| e^{-i\vec{k}_1\cdot\vec{r}}\, \vec{p}\cdot\vec{\pi}_{k_1}^{\sigma_1} \left| \psi_j^{el} \right\rangle$ (5.9)

Energy denominator: $\hbar\omega_{ij} + \hbar\omega_2$ (5.10)

where $\hbar\omega_{ij} = E_i^{el} - E_j^{el}$

Case c

Of all the terms contained in H_2 the only one that gives a contribution to the decay probability is the term that contains $a_{k_1}^{+}, a_{k_2}^{+}$

$| i \rangle$ = atom in state 2s
$| f \rangle$ = atom in state 1s + $(\hbar\omega_2) + (\hbar\omega_1)$ photons

$$\langle f | H_2 | i \rangle = \frac{e^2}{2m}\frac{h}{V}\frac{\vec{\pi}_{k_1}^{\sigma_1}\cdot\vec{\pi}_{k_2}^{\sigma_2}}{(\omega_1\,\omega_2)^{1/2}} \left\langle f \right| e^{-i(\vec{k}_1+\vec{k}_2)\cdot\vec{r}} \left| i \right\rangle \tag{5.11}$$

Using now the formulae (5.5) to (5.11) we can write down the effective matrix element of the two-photon transition

$$\begin{aligned} \langle f | H' | i \rangle &= \frac{\langle f | H_1 | j \rangle \langle j | H_1 | i \rangle}{E_i - E_j} + \langle f | H_2 | i \rangle \\ &= \frac{e^2}{m}\frac{h}{V}\frac{1}{(\omega_1\,\omega_2)^{\frac{1}{2}}} \\ &\times \frac{1}{m}\left[\sum_j \left(\frac{\left\langle f \right| e^{-i\vec{k}_2\cdot\vec{r}}\, \vec{p}\cdot\vec{\pi}_{k_2}^{\sigma_2} \left| j \right\rangle \left\langle j \right| e^{-i\vec{k}_1\cdot\vec{r}}\, \vec{p}\cdot\vec{\pi}_{k_1}^{\sigma_1} \left| i \right\rangle}{\hbar\omega_{ij} - \hbar\omega_1} \right.\right. \\ &+ \left.\frac{\left\langle f \right| e^{-i\vec{k}_1\cdot\vec{r}}\, \vec{p}\cdot\vec{\pi}_{k_1}^{\sigma_1} \left| j \right\rangle \left\langle j \right| e^{-i\vec{k}_2\cdot\vec{r}}\, \vec{p}\cdot\vec{\pi}_{k_2}^{\sigma_2} \left| i \right\rangle}{\hbar\omega_{ij} - \hbar\omega_2}\right) \\ &+ \left.\frac{1}{2}\left(\vec{\pi}_{k_1}^{\sigma_1}\cdot\vec{\pi}_{k_2}^{\sigma_2}\right) \left\langle f \right| e^{-i(\vec{k}_1+\vec{k}_2)\cdot\vec{r}} \left| i \right\rangle\right] \end{aligned} \tag{5.12}$$

We want now to compare the magnitude of the (H_1-based) first two terms with that of the (H_2-based) third term. Considering the latter

$$\langle f | e^{-i(\vec{k}_1+\vec{k}_2)\cdot\vec{r}} | i \rangle \approx \left\langle f \left| 1 - i(\vec{k}_1+\vec{k}_2)\cdot\vec{r} - \frac{[(\vec{k}_1+\vec{k}_2)\cdot\vec{r}]^2}{2} \right| i \right\rangle \approx$$
$$\approx (\vec{k}_1+\vec{k}_2)^2 a_0^2$$

$$\lambda \approx 1{,}215 \times 10^{-8} \text{ cm} \approx 10^{-3} \text{ (Lyman series)}$$

$$k \approx 10^3 \qquad k^2 \approx 10^6$$

$$(\vec{k}_1+\vec{k}_2)^2 a_0^2 \approx 10^6 \times 10^{-16} \approx 10^{-10} \tag{5.13}$$

As for the first two terms, we consider the quantity

$$\frac{p^2}{m}\frac{1}{\hbar\omega} = \frac{m\text{v}^2}{\hbar\omega}$$

$$\text{v} = \frac{2\pi r}{T} = 2\pi a_0 \, \nu = 6 \times 10^{-8} \times 3 \times 10^{15} \approx 2 \times 10^8 \text{ cm/s}$$

because

$$\nu = \frac{c}{\lambda} = \frac{3 \times 10^{10}}{1.2 \times 10^{-5}} = 3 \times 10^{15}$$

$$m\text{v}^2 = 9 \times 10^{-28} \times 4 \times 10^{16} = 4 \times 10^{-11}$$

$$h\nu = 6.6 \times 10^{-27} \times 3 \times 10^{15} = 20 \times 10^{-12}$$

Then

$$\frac{m\text{v}^2}{h\nu} \approx \frac{4 \times 10^{-11}}{2 \times 10^{-11}} \approx 1 \tag{5.14}$$

Since the quantity (5.14) far exceeds the quantity (5.13), the H_2-based third term is negligible with respect to the two H_1-based terms. We are left with

$$\langle f | H' | i \rangle =$$
$$= \frac{e^2}{m^2}\frac{h}{V}\frac{1}{(\omega_1\,\omega_2)^{\frac{1}{2}}}$$
$$\times \left[\sum_j \left(\frac{\langle f | e^{-i\vec{k}_2\cdot\vec{r}}\, \vec{p}\cdot\vec{\pi}_{k_2}^{\sigma_2} | j \rangle \langle j | e^{-i\vec{k}_1\cdot\vec{r}}\, \vec{p}\cdot\vec{\pi}_{k_1}^{\sigma_1} | i \rangle}{\hbar\omega_{ij} - \hbar\omega_1} \right.\right. \tag{5.15}$$
$$\left.\left. + \frac{\langle f | e^{-i\vec{k}_1\cdot\vec{r}}\, \vec{p}\cdot\vec{\pi}_{k_1}^{\sigma_1} | j \rangle \langle j | e^{-i\vec{k}_2\cdot\vec{r}}\, \vec{p}\cdot\vec{\pi}_{k_2}^{\sigma_2} | i \rangle}{\hbar\omega_{ij} - \hbar\omega_2} \right) \right]$$

In the electric dipole approximation we can write:

$$\begin{aligned}
&\left\langle f\right| e^{-i\vec{k}_2\cdot\vec{r}}\,\vec{p}\cdot\vec{\pi}_{k_2}^{\sigma_2}\left|j\right\rangle \approx \langle f|\,\vec{p}\,|j\rangle\cdot\vec{\pi}_{k_2}^{\sigma_2} = -\,im\omega_2\,\langle\vec{r}\rangle_{fj}\cdot\vec{\pi}_{k_2}^{\sigma_2}\\
&\left\langle j\right| e^{-i\vec{k}_1\cdot\vec{r}}\,\vec{p}\cdot\vec{\pi}_{k_1}^{\sigma_1}\left|i\right\rangle \approx \langle j|\,\vec{p}\,|i\rangle\cdot\vec{\pi}_{k_1}^{\sigma_1} = -\,im\omega_1\,\langle\vec{r}\rangle_{ji}\cdot\vec{\pi}_{k_1}^{\sigma_1}\\
&\left\langle f\right| e^{-i\vec{k}_1\cdot\vec{r}}\,\vec{p}\cdot\vec{\pi}_{k_1}^{\sigma_1}\left|j\right\rangle \approx \langle f|\,\vec{p}\,|j\rangle\cdot\vec{\pi}_{k_1}^{\sigma_1} = -\,im\omega_1\,\langle\vec{r}\rangle_{fj}\cdot\vec{\pi}_{k_1}^{\sigma_1}\\
&\left\langle j\right| e^{-i\vec{k}_2\cdot\vec{r}}\,\vec{p}\cdot\vec{\pi}_{k_2}^{\sigma_2}\left|i\right\rangle \approx \langle j|\,\vec{p}\,|i\rangle\cdot\vec{\pi}_{k_2}^{\sigma_2} = -\,im\omega_2\,\langle\vec{r}\rangle_{ji}\cdot\vec{\pi}_{k_2}^{\sigma_2}
\end{aligned} \tag{5.16}$$

We have then

$$\begin{aligned}
&\langle f|\,H'\,|i\rangle =\\
&= \frac{e^2 h}{V}(\omega_1\,\omega_2)^{\frac{1}{2}}\\
&\times\left[\sum_j\left(\frac{\langle\vec{r}\cdot\vec{\pi}_{k_2}^{\sigma_2}\rangle_{fj}\langle\vec{r}\cdot\vec{\pi}_{k_1}^{\sigma_1}\rangle_{ji}}{\hbar\omega_{ij}-\hbar\omega_1}+\frac{\langle\vec{r}\cdot\vec{\pi}_{k_1}^{\sigma_1}\rangle_{fj}\langle\vec{r}\cdot\vec{\pi}_{k_2}^{\sigma_2}\rangle_{ji}}{\hbar\omega_{ij}-\hbar\omega_2}\right)\right]\\
&= \frac{e^2}{V}(\omega_1\,\omega_2)^{\frac{1}{2}}\\
&\times\left[\sum_j\left(\frac{\langle\vec{r}\cdot\vec{\pi}_{k_2}^{\sigma_2}\rangle_{fj}\langle\vec{r}\cdot\vec{\pi}_{k_1}^{\sigma_1}\rangle_{ji}}{\nu_{ij}-\nu_1}+\frac{\langle\vec{r}\cdot\vec{\pi}_{k_1}^{\sigma_1}\rangle_{fj}\langle\vec{r}\cdot\vec{\pi}_{k_2}^{\sigma_2}\rangle_{ji}}{\nu_{ij}-\nu_2}\right)\right]
\end{aligned} \tag{5.17}$$

The quantity in square brackets will be referred to later as [ ].

V.C. Density of Final States

The number of photon states with $\vec{k}$ in the interval $(\vec{k}_1, \vec{k}_1 + d\vec{k}_1)$ is given by

$$\begin{aligned}
d^3n &= \frac{V}{8\pi^3}\,d^3\vec{k}_1 = \frac{V}{8\pi^3}\,k_1^2\,dk_1\,d\Omega_1\\
&= \frac{V}{8\pi^3}\,\frac{\omega_1^2}{c^2}\,\frac{1}{c}\,d\omega_1\,d\Omega_1 = \frac{V}{c^3}\,\nu_1^2\,d\nu_1\,d\Omega_1
\end{aligned} \tag{5.18}$$

The number of states of photon $\hbar\omega_2$ per unit energy range is

$$\frac{V}{8\pi^3}\,\frac{\omega_2^2}{c^3}\,\frac{1}{\hbar}\,4\pi = \frac{V}{2\pi^2\hbar}\,\frac{\omega_2^2}{c^3} = \frac{2V\nu_2^2}{\hbar c^3} \tag{5.19}$$

The transition rate into a state in which photon 1 has frequency in $(\nu_1, \nu_1 + d\nu_1)$ given direction and given polarization is then given by

$$d\lambda = \frac{2\pi}{\hbar}\left|H'_{fi}(\nu_1)\right|^2\frac{2V^2}{\hbar c^6}\,\nu_1^2\,\nu_2^2\,d\nu_1\,d\Omega_1 \tag{5.20}$$

V.D. Transition Probability and Spectrum

The transition probability per unit time that the hydrogen atom decays via a two-photon emission, with one of the two photons having its frequency in the interval $(\nu_1, \nu_1 + d\nu_1)$, with given direction and given polarization is

$$\begin{aligned} d\lambda &= \frac{4\pi e^4}{\hbar^2 c^6} 4\pi^2 \nu_1^3 \nu_2^3 [\,]^2 d\nu_1 \, d\Omega_1 \\ &= \frac{16\pi^3 e^4}{\hbar^2 c^6} \nu_1^3 \nu_2^3 [\,]^2 d\nu_1 \, d\Omega_1 \end{aligned} \tag{5.21}$$

where we have used (5.17) and (5.20) and the symbol [ ] indicates the quantity in square brackets in (5.17). Integrating over all space (4π extra factor) and summing over directions of polarization (4 extra factor) we get the following expression for the probability per second of the simultaneous emission of two quanta with one quantum in the frequency range $(\nu_1, \nu_1 + d\nu_1)$:

$$\begin{aligned} W\, d\nu_1 &= \frac{16\,(2\pi)^4\, e^4}{\hbar^2 c^6} \nu_1^3 \nu_2^3 \\ &\times \left\langle \left| \sum_j \frac{\left\langle \vec{r} \cdot \vec{\pi}_{k_2}^{\sigma_2} \right\rangle_{fj} \left\langle \vec{r} \cdot \vec{\pi}_{k_1}^{\sigma_1} \right\rangle_{ji}}{\nu_{ij} - \nu_1} + \frac{\left\langle \vec{r} \cdot \vec{\pi}_{k_1}^{\sigma_1} \right\rangle_{fj} \left\langle \vec{r} \cdot \vec{\pi}_{k_2}^{\sigma_2} \right\rangle_{ji}}{\nu_{ij} - \nu_2} \right|^2 \right\rangle \end{aligned} \tag{5.22}$$

where the average is made over all the directions of propagation and over the directions of polarization.

Since no distinction is made in regard to the order in which photons are emitted, the probability of emission is

$$W = \frac{1}{2} \int_0^{\nu_0} w \, d\nu_1 \tag{5.23}$$

where $\nu_0 = \dfrac{E_{2s} - E_{1s}}{h}$.

We do not consider the effect of the fine and hyperfine structures due to the electronic and nuclear spins respectively because the operator $\vec{r}$ is diagonal in both spins: then the s levels are nondegenerate and the p levels are triply degenerate. It is convenient to use for the p levels three linearly independent states which transform like the cartesian coordinates x, y, z. We call these new states α, β, γ. The sum $\sum_j$ for each energy E_j contains a sum over α, β, γ (see Fig. 6):

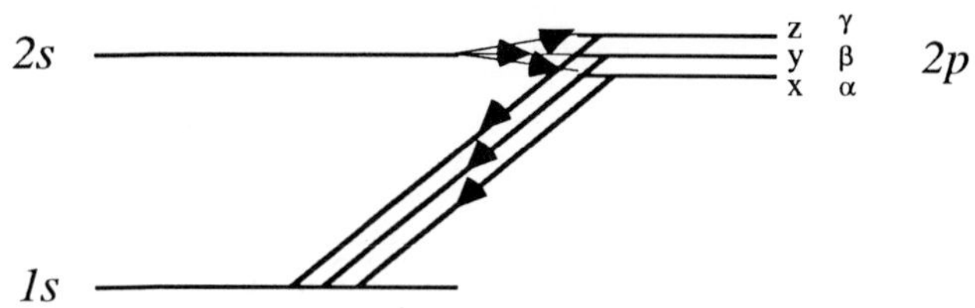

Figure 6. Details of levels of the hydrogen atoms.

$$
\begin{aligned}
&\sum_{\alpha\beta\gamma}^{\infty} \left\langle \vec{r}\cdot\vec{\pi}_{k_1}^{\sigma_1}\right\rangle_{fj} \left\langle \vec{r}\cdot\vec{\pi}_{k_2}^{\sigma_2}\right\rangle_{ji} \\
&= \left(x\,\pi_x'\right)_{f\alpha}\left(x\,\pi_x''\right)_{\alpha i} + \left(y\,\pi_y'\right)_{f\beta}\left(y\,\pi_y''\right)_{\beta i} + \left(z\,\pi_z'\right)_{f\gamma}\left(z\,\pi_z''\right)_{\gamma i} \\
&= z_{f\gamma} z_{\gamma i}\left(\pi_x'\,\pi_x'' + \pi_y'\,\pi_y'' + \pi_z'\,\pi_z''\right) \\
&= z_{f\gamma} z_{\gamma i}\left(\pi'\cdot\pi''\right)
\end{aligned}
\tag{5.24}
$$

because

$$
x_{f\alpha}x_{\alpha i} = y_{f\beta}y_{\beta i} = z_{f\gamma}z_{\gamma i} \tag{5.25}
$$

and

$$
y_{f\alpha} = y_{f\gamma} = x_{f\beta} = x_{f\gamma} = \ldots = 0 \tag{5.26}
$$

Then the quantity in ⟨ ⟩ brackets in (5.22) becomes

$$
\begin{aligned}
&\left\langle \left| \sum_j \text{———} + \text{———} \right|^2 \right\rangle = \\
&= \left\langle \left(\vec{\pi}_{k_1}^{\sigma_1}\cdot\vec{\pi}_{k_2}^{\sigma_2}\right)\right\rangle \sum_{m=2}^{\infty}\left(\frac{1}{\nu_{im}-\nu_1} + \frac{1}{\nu_{im}-\nu_2}\right)^2 \left(z_{fm}\,z_{mi}\right)^2
\end{aligned}
\tag{5.27}
$$

where

m = principal quantum number of p state

z_{fm} = matrix element between the state 1s and a state of type γ with the principal quantum number m

z_{mi} = matrix element between a state of type γ and principal quantum number m and the state 2s. Also

$$\langle \pi' \cdot \pi' \rangle = \frac{1}{4\pi}\frac{4\pi}{3} = \frac{1}{3} \tag{5.28}$$

Then

$$W\,d\nu_1 = = \frac{16\,(2\pi)^4\,e^4}{3\hbar^2 c^6}\,\nu_1^3\,\nu_2^3 \left| \sum_{m=2}^{\infty} z_{fm}\,z_{mi}\left(\frac{1}{\nu_{mi}+\nu_1}+\frac{1}{\nu_{mi}+\nu_2}\right)\right|^2 \tag{5.29}$$

The most significant contribution to the sum comes from $m = 2$. Let us then consider only this contribution

$$W\,d\nu_1 = \frac{16\,(2\pi)^4\,e^4}{3\hbar^2 c^6}\,\nu_1^3\,\nu_2^3\,(z_{1s,\,2p}\,z_{2p,\,2s})\left(\frac{1}{\nu_1}+\frac{1}{\nu_2}\right)^2 d\nu_1 \tag{5.30}$$

But

$$\begin{aligned} \nu_1^3\,\nu_2^3\left(\frac{1}{\nu_1}+\frac{1}{\nu_2}\right)^2 &= \nu_1^3\,\nu_2^3\left(\frac{\nu_1+\nu_2}{\nu_1\,\nu_2}\right)^2 = \nu_1\,\nu_2\,\nu_0^2 \\ &= \nu_0^2\,\nu_1\,(\nu_0-\nu_1) \end{aligned} \tag{5.31}$$

where

$$\nu_0 = \nu_1 + \nu_2 = \frac{E_{2s}-E_{1s}}{h} \tag{5.32}$$

The formula (5.31) gives us the spectrum of the two-photon spontaneous emission. The emission is zero at $\nu_1 = 0$ or ν_0 and it is maximum at $\nu_1 = \frac{\nu_0}{2}$.

We want now to calculate the radiative lifetime of the 2s state. We find

$$\begin{aligned} \int_0^{\nu_0} \nu_1^3\,\nu_2^3\left(\frac{1}{\nu_1}+\frac{1}{\nu_2}\right)^2 d\nu_1 &= \nu_0^2\int_0^{\nu_0}\nu_1\,(\nu_0-\nu_1)\,d\nu_1 = \\ &= \nu_0^2\int_0^{\nu_0}(\nu_0\,\nu_1-\nu_1^2)\,d\nu_1 = \nu_0^2\left(\frac{\nu_0^3}{2}-\frac{\nu_0^3}{3}\right) = \frac{\nu_0^5}{6} \end{aligned} \tag{5.33}$$

Then

$$\frac{1}{2}\int_0^{\nu_0} W\,d\nu_1 = \frac{16\,(2\pi)^4\,e^4}{6\hbar^2 c^6}\,\frac{\nu_0^5}{6}\,[z_{1s,\,2p}\,z_{2p,\,2s}]^2 \tag{5.34}$$

We need at this point to calculate some quantities:

$$z = r\cos\theta$$

$$|\,1s\,\rangle = \frac{1}{\sqrt{\pi a_0^3}}\, e^{-r/a_0}$$

$$|\,2s\,\rangle = \frac{1}{\sqrt{128\pi a_0^3}}\, e^{-r/2a_0}\left[2\left(\frac{r}{a_0}\right) - 4\right]$$

$$|\,2p\ (m{=}0)\,\rangle = \frac{1}{\sqrt{32\pi a_0^3}}\, e^{-r/2a_0}\left(\frac{r}{a_0}\right)\cos\theta$$

Then

$$\langle\,1s|\,z\,|2p_0\,\rangle = \frac{1}{\sqrt{32\ \pi a_0^3}}\int_0^\infty e^{-3r/2a_0}\,\frac{r^4}{a_0}\,dr\int_0^\pi \cos^2\theta\,\sin^2\theta\,d\theta\,2\pi = 0.44\,a_0$$

Similarly we find

$$\langle\,2p|\,z\,|2s\,\rangle = 1.78\,a_0$$

Therefore

$$\left[z_{1s,\,2p}\,z_{2p,\,2s}\right]^2 = (0.44\times 1.78)^2\,a_0^4 = 0.61\times 0.53^4\times 10^{-32} = 0.05\times 10^{-32}$$

The wavelength corresponding to the energy gap between 2s and 1s state is

$$\lambda = 1215\ \text{Å} = 1.215\times 10^{-5}\ \text{cm}$$

Then

$$\nu_0 = \frac{3\times 10^{10}}{1.215\times 10^{-5}} = 2.469\times 10^{15}\ \text{Hz}$$

$$\nu_0^5 = 9.18\times 10^{76}$$

Therefore

$$\frac{1}{2}\int_0^{\nu_0} W\,d\nu_1 = \frac{16\,(2\pi)^4\,e^4}{6\hbar^2c^6}\,\frac{\nu_0^5}{6}\left[z_{1s,\,2p}\,z_{2p,\,2s}\right]^2 =$$

$$= \frac{16\times 6.28^4\times 9.2\times 10^{76}\times 0.05\times 10^{-32}}{36\times 81\times 10^{40}\times 137^2} = 2.1$$

and

$$\tau = \frac{1}{\frac{1}{2}\int_0^{\nu_0} W \, d\nu_1} \approx 0.5 \text{ sec} \tag{5.35}$$

More detailed calculations [9] yield the value 0.125 sec.

V.E. Experimental Conditions

The lifetime of the 2s state is quite long and an atom in such a state collides many times before it can radiate spontaneously by emitting two photons. Proper conditions seem to be present in planetary nebulae. The emission from these celestial bodies includes a continuous spectrum in the long-wavelength side of the Balmer continuum.

A possible source of such continuous radiation is based on the simultaneous emission of two photons from a hydrogen atom in the metastable state 2s:

1) an appreciable fraction of electrons captured by photons will reach the 2s level on their way down to the ground state.

2) Lα radiation may be converted into two photons of this visual continuum in the following way. The absorption of Lα radiation will excite an H atom to the 2p level; during the brief interval before the Lα photon is reemitted, a collision with a free electron may induce a transition from the 2p to the 2s level.

All one-photon transitions from the 2s $\rightarrow$ 1s level are forbidden, and unless a collision with another free electron induces a transition back to the 2p level, the excited electron will jump down to the 1s level, emitting two photons.

VI. THE RADIATIVE SHIFT OF ATOMIC LEVELS

VI.A. The Data [10]

The radiative (Lamb) shift arises from the fact that both the $2\,^2S_{1/2}$ and the $2\,^2P_{1/2}$ states of hydrogen are displaced, the latter state by a negligible amount, and the former by ~1040 MHz (Fig. 7). The cause of the shift is the exchange of virtual quanta between the atom and the radiation field.

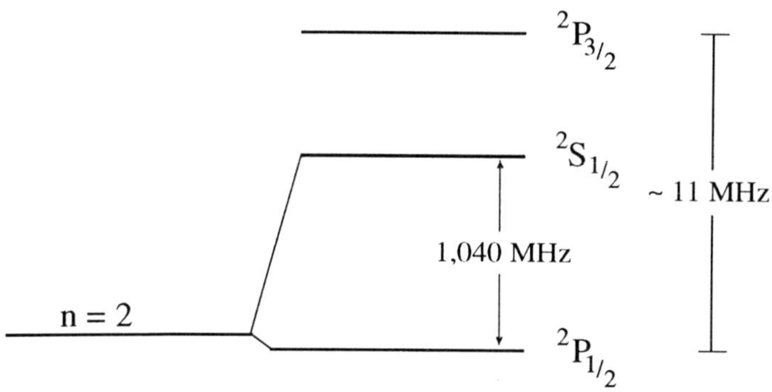

Figure 7. The n=2 states of the hydrogen atom.

VI.B. Setting of the Problem

The second order perturbation using H_1 and the first order using H_2 give us the following corrections to the energy of an atom in quantum state m

$$\Delta E = \langle m | H_2 | m \rangle + \sum_n \frac{|\langle m | H_1 | n \rangle|^2}{E_m^0 - E_n^0} \tag{6.1}$$

where

$$H_1 = \frac{e}{mc} \vec{p} \cdot \vec{A}, \; H_2 = \frac{e^2}{2mc^2} (\vec{A})^2 \tag{6.2}$$

We shall neglect the contribution $\langle m | H_2 | m \rangle$ to the energy shift, because it depends only on $\vec{A}$, namely on the electromagnetic field and is common to all the electronic systems, bound or not, and is also independent of the particular state under consideration.

The electron may also interact with states: electron plus 2 photons, electron plus 3 photons, etc.; in these cases higher order perturbation theory should be used. The probability of finding a bare electron decreases as we consider more divergent terms. The second-order perturbation theory is, however, sufficient in general to account for the Lamb shift [11].

VI.C. Processes and Calculations

We consider the second order process which consists of two parts.

Emission of a virtual photon:

$$\langle n|\, H_1\, |m\rangle = \frac{e}{m}\left(\frac{h}{\omega V}\right)^{\frac{1}{2}} \langle n|\, e^{-i\vec{k}\cdot\vec{r}}\, \vec{p}\cdot\vec{\pi}_k^{\sigma}\, |m\rangle \tag{6.3}$$

Absorption of a virtual photon:

$$\langle m|\, H_1\, |n\rangle = \frac{e}{m}\left(\frac{h}{\omega V}\right)^{\frac{1}{2}} \langle m|\, e^{-i\vec{k}\cdot\vec{r}}\, \vec{p}\cdot\vec{\pi}_k^{\sigma}\, |n\rangle \tag{6.4}$$

Denominator:

$$E_m^0 - E_n^0 = E_m - E_n - \hbar\omega \tag{6.5}$$

$$\begin{aligned} &\sum_n \frac{|\langle H_1\rangle_{mn}|^2}{E_m^0 - E_n^0} \\ &= \int \sum_n \frac{|\langle H_1\rangle_{mn}|^2}{E_m - E_n - \hbar\omega}\, \frac{V}{8\pi^3}\, k^2\, dk\, d\Omega \\ &= \sum_n \int \frac{e^2}{m^2}\, \frac{h}{\omega V}\, \frac{\langle m|\, e^{i\vec{k}\cdot\vec{r}}\, \vec{p}\cdot\vec{\pi}_k^{\sigma}\, |n\rangle\langle n|\, e^{-i\vec{k}\cdot\vec{r}}\, \vec{p}\cdot\vec{\pi}_k^{\sigma}\, |m\rangle}{E_m - E_n - \hbar\omega} \\ &\times \frac{V}{8\pi^3}\, k^2\, dk\, d\Omega \end{aligned} \tag{6.6}$$

Using the dipole approximation:

$$\begin{aligned} &\sum_n \frac{|\langle H_1\rangle_{mn}|^2}{E_m^0 - E_n^0} = \\ &= -\frac{e^2 h}{c\, 8\pi^3} \sum_n \int \frac{|\langle \vec{p}\cdot\vec{\pi}_k^{\sigma}\rangle_{mn}|^2}{m^2 (E_n - E_m + \hbar\omega)}\, k\, dk\, d\Omega \\ &= -\frac{e^2 h}{c\, 8\pi^3} \sum_n \int \frac{|\langle \vec{v}\rangle_{mn}\cdot \vec{\pi}_k^{\sigma}|^2}{(E_n - E_m + \hbar\omega)}\, k\, dk\, d\Omega \end{aligned} \tag{6.7}$$

Summing over polarization directions and integrating over all space we get an extra factor $8\pi/3$.

$$W_m = -\frac{8\pi}{3}\frac{e^2 h}{8\pi^3 c}\sum_n \int \frac{|\langle \vec{v}_{mn}\rangle|^2}{(E_n - E_m + \hbar\omega)} k\, dk$$

$$= -\frac{e^2 h}{3\pi^2 c}\int_0^K k\, dk \sum_n \frac{|\vec{v}_{mn}|^2}{(E_n - E_m + \hbar\omega)} \tag{6.8}$$

Again

$$W_m = -\frac{e^2 h}{3\pi^2 c}\int_0^K k\, dk \sum_n \frac{|\vec{v}_{mn}|^2}{(E_n - E_m + \hbar\omega)} \tag{6.9}$$

where

$$\vec{v} = \frac{\vec{p}}{m} = -i\frac{\hbar}{m}\vec{\nabla} = \text{velocity of the electron} \tag{6.10}$$

and

$\sum_n$ extends to all atomic states $|n\rangle$ for which $\vec{v}_{mn} \neq 0$

and

K = maximum value of k.

VI.D. Mass Renormalization

For a free electron the velocity is a constant of the motion, namely it has only diagonal elements. The sum $\sum_n$ reduces to one term and we can write

$$W_0 = -\frac{e^2 h}{3\pi^2 c}\int_0^K \frac{k\, dk\, v^2}{\hbar\omega} \tag{6.11}$$

This quantity can be interpreted as a kinematical correction to the mass of the electron. This correction is already contained in the experimental electron mass, and, therefore, we have to subtract it from the value of W_m.

Since the electron is bound v^2 has to be replaced by its expectation value

$$(v^2)_{mm} = \sum_n |\vec{v}_{mn}|^2 \tag{6.12}$$

$$W_0' = -\frac{e^2 h}{3\pi^2 c}\int_0^K \frac{k\,dk}{\hbar\omega}\sum_n |\vec{v}_{mn}|^2 \tag{6.13}$$

Then

$$\begin{aligned}
W_k' &= W_m - W_0' = \\
&= -\frac{e^2 h}{3\pi^2 c}\int_0^K k\,dk \sum_n |\vec{v}_{mn}|^2\left[\frac{1}{E_n - E_m + \hbar\omega} - \frac{1}{\hbar\omega}\right] \\
&= \frac{e^2 h}{3\pi^2 c}\int_0^K k\,dk \sum_n |\vec{v}_{mn}|^2 \frac{E_n - E_m}{E_n - E_m + \hbar\omega}\,\frac{1}{\hbar\omega} \\
&= \frac{2e^2}{3\pi c^3 \hbar}\sum_n |\vec{v}_{mn}|^2 (E_n - E_m)\int_0^{\Omega = Kc} \frac{\hbar\,d\omega}{E_n - E_m + \hbar\omega} \\
&= \frac{2e^2}{3\pi c^3 \hbar}\sum_n |\vec{v}_{mn}|^2 (E_n - E_m)\ln\frac{E_n - E_m + \hbar\Omega}{E_n - E_m} \\
&\approx \frac{2e^2}{3\pi c^3 \hbar}\sum_n |\vec{v}_{mn}|^2 (E_n - E_m)\ln\frac{\hbar\Omega}{E_n - E_m}
\end{aligned} \tag{6.14}$$

Relativity theory provides a natural cutoff Ω for the frequency ω of the virtual photons. The argument of the ln is very large and in first approximation we shall consider the ln independent of n. We then calculate

$$\begin{aligned}
\sum_n |\vec{v}_{mn}|^2 (E_n - E_m) &= \frac{1}{m^2}\sum_n |\vec{p}_{mn}|^2 (E_n - E_m) \\
&= \frac{1}{m^2}\sum_n \vec{p}_{nm}(E_n - E_m)\,\vec{p}_{mn} = -\frac{1}{m^2}\sum_n \vec{p}_{mn}(E_m - E_n)\,\vec{p}_{nm} \\
&= -\frac{1}{m^2}\sum_n [H, \vec{p}]_{mn}\,\vec{p}_{nm} = -\frac{1}{m^2}\sum_n \frac{\hbar}{i}\,\dot{\vec{p}}_{mn}\,\vec{p}_{nm} \\
&= \frac{1}{m^2}\sum_n \frac{\hbar}{i}(\vec{\nabla}V)_{mn}\left(\frac{\hbar}{i}\vec{\nabla}\right)_{nm} \\
&= -\frac{\hbar^2}{m^2}(\vec{\nabla}V\cdot\vec{\nabla})_{mm}
\end{aligned} \tag{6.15}$$

Also

$$\begin{aligned}
\sum_n |\vec{v}_{mn}|^2 (E_n - E_m) &= \frac{1}{m^2}\sum_n |\vec{p}_{mn}|^2 (E_n - E_m) \\
&= \frac{1}{m^2}\sum_n \vec{p}_{nm}(E_n - E_m)\,\vec{p}_{mn} = \frac{1}{m^2}\sum_n [H, \vec{p}]_{nm}\,\vec{p}_{mn} \\
&= \frac{1}{m^2}\frac{\hbar}{i}\sum_n \vec{p}_{mn}\,\dot{\vec{p}}_{nm} = \frac{1}{m^2}\frac{\hbar}{i}\sum_n \left(\frac{\hbar}{i}\vec{\nabla}\right)_{mn}(-\vec{\nabla}V)_{nm} \\
&= \frac{\hbar^2}{m^2}(\vec{\nabla}\cdot\vec{\nabla}V)_{mm} = \frac{\hbar^2}{m^2}(\nabla^2 V)_{mm} + \frac{\hbar^2}{m^2}(\vec{\nabla}V\cdot\vec{\nabla})_{mm}
\end{aligned} \tag{6.16}$$

because

$$\vec{\nabla} \cdot \vec{\nabla} V \psi = \psi \nabla^2 V + \vec{\nabla} V \cdot \vec{\nabla} \psi \tag{6.17}$$

Then, equating the two results

$$-(\vec{\nabla} V \cdot \vec{\nabla})_{mm} = (\nabla^2 V)_{mm} + (\vec{\nabla} V \cdot \vec{\nabla})_{mm} \tag{6.18}$$

and

$$(\vec{\nabla} V \cdot \vec{\nabla})_{mm} = -\frac{1}{2} (\nabla^2 V)_{mm} \tag{6.19}$$

We can write then

$$\sum_n |\vec{v}_{mn}|^2 (E_n - E_m) = \frac{1}{2} \frac{\hbar^2}{m^2} \int \nabla^2 V \, \psi_m^2 \, d\tau \tag{6.20}$$

But V is a Coulomb potential:

$$V = -\frac{Ze^2}{r} \tag{6.21}$$

and

$$\nabla^2 V = Ze^2 \, 4\pi \delta(\vec{r}) \tag{6.22}$$

Then

$$\begin{aligned} \sum_n |\vec{v}_{mn}|^2 (E_n - E_m) &= \frac{1}{2} \frac{\hbar^2}{m^2} 4\pi \int Ze^2 \delta(\vec{r}) \, \psi_m^2 \, d\tau \\ &= \frac{2\pi\hbar^2 e^2}{m^2} Z \, \psi_m^2(0) \end{aligned} \tag{6.23}$$

The term $\sum_n |\vec{v}_{mn}|^2 (E_n - E_m)$ is different from zero only for eigenfunctions of s states (l=0), because the other eigenfunctions are equal to zero at r=0.

For an s state we find that the square of the eigenfunction at r=0 is given by

$$\psi_n^2(0) = \frac{Z^3}{(na_0)^3 \, \pi} \qquad \left(a_0 = \text{Bohr radius} = \frac{\hbar^2}{me^2}\right) \tag{6.24}$$

We now insert this expression in (6.23) to get

$$\sum_n |\vec{v}_{mn}|^2 (E_n - E_m) = \frac{2\pi\hbar^2 e^2}{m^2} Z \frac{Z^3}{(na_0)^3 \pi}$$
$$= \frac{2\hbar^2 e^2}{m^2} \frac{Z^4}{n^3 \frac{\hbar^2}{me^2} a_0^2} = \frac{2e^4 Z^4}{n^3 a_0^2 m} \tag{6.25}$$

and

$$W' = \frac{2e^2}{3\pi c^3 \hbar} \frac{2e^4 Z^4}{n^3 a_0^2 m} \ln \frac{\hbar\Omega}{\langle E_n - E_m \rangle_{av}} =$$
$$= \frac{4e^8 Z^4}{3\pi c^3 \hbar^3 a_0} \ln \frac{\hbar\Omega}{\langle E_n - E_m \rangle_{av}} = \frac{8}{3\pi} \left(\frac{e^2}{\hbar c}\right)^3 \frac{e^2}{2a_0} \ln \frac{\hbar\Omega}{\langle E_n - E_m \rangle_{av}} \tag{6.26}$$

VI.E. Results

$$W_m' = \frac{8}{3\pi} \left(\frac{e^2}{\hbar c}\right)^3 \frac{e^2}{2a_0} \frac{Z^4}{n^3} \ln \frac{\hbar\Omega}{\langle E_n - E_m \rangle_{av}} =$$
$$= \frac{8}{3\pi} \left(\frac{e^2}{\hbar c}\right)^3 \mathrm{Ry} \frac{Z^4}{n^3} \ln \frac{\hbar\Omega}{\langle E_n - E_m \rangle_{av}} \tag{6.27}$$

where the average is made over all the n states.

The state m is the metastable state 2s and the states n are all the possible p states, plus all the states in the continuum range of the spectrum.

Since we pulled $\ln \frac{\hbar\Omega}{E_n - E_m}$ out of the sum $\sum_n$ (which is really in part a sum and in part an integral) we have to calculate the average of the ln... . The quantity $\langle E_n - E_m \rangle_{av}$ has been calculated numerically and found equal to 17.8 Ry. If we introduce this quantity in ln... and set $\hbar\Omega = mc^2$ we find

$$\ln \frac{mc^2}{17.8\ \mathrm{Ry}} = \ln \frac{mc^2\, 2a_0}{17.8\, e^2} = \ln \frac{mc^2\, 2}{17.8\, e^2} \frac{\hbar^2}{me^2} =$$
$$= \ln \frac{2}{17.8} \left(\frac{\hbar c}{e^2}\right)^2 = \ln \frac{2 \times 137^2}{17.8} = 7.63$$

We have then for the 2s state of H (Z=1, n=2)

$$\frac{8}{3\pi} \left(\frac{e^2}{\hbar c}\right)^3 \mathrm{Ry} \frac{1}{n^3} = \frac{8}{3\pi} \frac{1}{137^3} \frac{2.176 \times 10^{-11}}{8} = 8.98 \times 10^{-19}\ \text{ergs}$$

$$\frac{8.98 \times 10^{-19}\ \text{ergs}}{h = 6.625 \times 10^{-27}} \approx 136 \times 10^6\ \text{cycles}$$

$$136 \times 10^6 \times 7.63 \approx \underline{1040\ \text{Megacycles}} \tag{6.28}$$

We can make the following observations:

1) The self energy diverges linearly with the frequency. But when we subtract the self energy of the free electron from the self energy of the electron, the energy diverges logarithmically. It is to be expected that when we use "hole theory" in which self energy diverges logarithmically we shall have a convergence of the energy difference. This fact provides a cutoff to the frequencies of the virtual photons which contribute to the shift of the level.

2) The effective limit to these frequencies is of the order $\frac{mc^2}{\hbar}$. For an exact knowledge of the true limit relativistic calculations should be made.

3) The value 1040 Mc/sec is in good agreement with the value 1000 Mc/sec found experimentally by Retherford and Lamb [10].

We note that in the Lamb shift the assumption is made that all the radiation modes are in their ground states. The excitation of the radiation modes increases the probability of the exchange of virtual photons between the atom and the radiative field. The effect of such an excitation is a "light shift" which depends on the intensity of the radiation present at the atomic site. This effect was predicted by the author of this article [12] and was observed later by Liberman, Pinard, and Taleb [13] who measured the light shift of a Rydberg level of a Rb atom, in the presence of a strong nonresonant laser radiation. More recently Collins and Di Bartolo [14] have examined the possibility of observing similar shifts in the 4f levels of trivalent rare earths in ionic solids. For the case of Pr-doped LaF_3 they showed that shifts on the order of tens of MHz can be easily generated and proposed and experiment to observe them.

ACKNOWLEDGEMENTS

The author wishes to thank the graduate students Stamatios Kyrkos, Brian Bowlby, Gang Lei, and Chunlai Yang for helpful criticism of this paper.

He wishes also to acknowledge the benefit of reading or consulting the following books which he would like to recommend to the readers of this article for further readings:

1. R. Loudon, The Quantum Theory of Light, Clarendon Press, Oxford, 1983.
2. J. D. Macomber, The Dynamics of Spectroscopic Transitions, John Wiley & Sons, New York, 1976.

3. L. Allen and J. Eberly, Optical Resonance and Two-Level Atoms, John Wiley & Sons, New York, 1975.
4. R. W. Boyd, Nonlinear Optics, Academic Press, San Diego, CA, 1992.

REFERENCES

1. L. Pauling and E. B. Wilson, Introduction to Quantum Mechanics, McGraw Hill, New York, 1935, pp. 133-136.
2. F. Bloch, Phys. Rev. 70, 460 (1946).
3. I. I. Rabi, Phys. Rev. 51, 652 (1937).
4. R. A. Smith, Proc. Roy. Soc. A362, 1 (1978).
5. C. Yang, Boston College, Department of Physics, Spectroscopy Laboratory Internal Report, 1992.
6. W. R. MacGillvray, D. T. Pegg, and M. C. Standage, Optics Commun. 25, 355 (1978).
7. R. G. Brewer in Nonlinear Spectroscopy, N. Bloembergen, editor, North-Holland, Amsterdam (1977), p. 87.
8. R. A. Smith, Proc. Roy. Soc. A362, 13 (1978); *ibid.* A368, 163 (1979).
9. L. Spritzer and J. L. Greenstein, Astrophys. Journal 114, 407 (1951).
10. W. E. Lamb and R. C. Retherford, Phys. Rev. 72, 241 (1947).
11. H. A. Bethe, Phys. Rev. 72, 739 (1947).
12. B. Di Bartolo, Optical Interaction in Solids, Wiley, New York, 1968, p. 411.
13. S. Liberman, J. Pinard, and A. Taleb, Phys. Rev. Lett. 50, 888 (1983).
14. J. M. Collins and B. Di Bartolo, J. of Luminescence 58, 290 (1994).

THE SPIN-PHOTON ECHO REVISITED

E. L. Hahn

Physics Department
University of California
Berkeley, California 94720

ABSTRACT

Pulse echo phenomena in general are briefly reviewed in relation to the spin echo as the first observed case. Two and three pulse NMR and photon echo amplitudes are calculated on the basis of a geometric model which emphasizes the role of non-linear coupling between the macroscopic polarization components of the ensemble and applied pulses. Coherent radiation damping in NMR is analyzed taking inhomogeneous broadening into account. Interpretations of entropy and thermodynamics are presented in light of the apparent contradiction of reversibility represented by the formation of the echo.

I. INTRODUCTION

Here in Sicily, a region very rich in Greek and Roman history, it is most appropriate at this scientific school to speak about "Echo" who was a mischievous female nymph described in Greek and Roman mythology. Because of her misbehavior she was punished by Hera who limited her speech to those words she had just heard (in NMR those words represent first and subsequent rf pulses). In her grief at this torture her flesh shrank away (denoted as the "free induction decay" (FID) following the pulses) until her voice replied later (the spin echoes) in response to all (the initial rf pulses) who called to her in the mountains.

II. ECHOES AND NONLINEARITY

A review now follows which emphasizes the theme that resonance pulse echoes can only occur in systems which are nonlinearly interactive. The role of non-linear coupling with pulses is outlined in an elementary calculation of two and three pulse spin echo formation using the vector model. The cavity response of spin magnetization to phase shift of the applied rf field in the rotating frame of reference will be shown to correspond directly to the changes in directions of laser pulse propagation vectors which apply to photon echo behavior. The free precession signal is analyzed in terms of the coherent radiation damping mechanism in a tuned cavity in the presence of inhomogeneous broadening. Portions of the non-linearity discussion, relevant only to NMR, are based on a review which will appear in the Educational Quarterly Journal Concepts in Magnetic Resonance, reproduced here with the kind permission of the Editors of NMR Concepts, Kingston, RI.

Nonlinear Spectroscopy of Solids: Advances and Applications
Edited by B. Di Bartolo and B. Bowlby, Plenum Press, New York, 1994

Spontaneous magnetic and electric dipole echo signals which follow the application of two or more radio-frequency, microwave, laser optical, or ultrasonic pulses at resonance are known to occur in a variety of systems if one of two types of non-linear coupling is present [1]. Non-linear coupling between the spin ensemble and applied rf or microwave pulses accounts for the nuclear or electron spin echo. Similarly the photon echo occurs because of non-linear coupling between applied laser pulses and optically induced atomic electric dipole moments.

In systems which respond linearly to pulses at resonance, a second class of echoes occurs because of non-linear coupling or damping during the long time periods between pulses. The clearest example of this type of echo is the cyclotron echo in plasmas [2]. Cyclotron transitions induced by microwave pulse power between equally spaced Landau levels of an ideal dilute electron gas will not occur because the response of the system to the pulses is linear. In principle the manifold of Landau levels cannot be "saturated". The induced electric dipole moment becomes larger and larger without limit during the driven response of the electrons because there is no terminal top level. Induced absorption is not blocked by a top "roof" level as in the case of a simple two level system. Cyclotron echoes however are observed in spite of the lack of a "roof" because each Landau level may acquire a different broadening in the actual experiment, depending upon the orbital velocity which characterizes that level. Consequently a degree of saturation or blocking takes place because of the disparity of cyclotron transition probabilities due to damping over the Landau spectrum. This variation of damping or of level widths is a consequence of velocity dependent elastic electron-buffer gas collision cross-sections. In the real experiment the buffer gas itself is ionized in order to provide a neutral plasma.

Another non-linearity of the second type is more subtle and difficult to analyze. It comes about when the particles which emit the echo interact non-linearly among themselves if they possess modes of excitation which can convert coherently from one internal mode to another. This non-linearity is shown by the electron spin microwave pulse echo effect in truncated ferrite spheres [3]. Although the usual spin echo of the first kind may occur in this experiment, it is negligible in magnitude compared to the echo which results because of the non-linear coupling between the primary spin precession mode and spin wave modes excited in the ferrite sample. A remarkable amplification effect takes place. A very weak first pulse, followed by a delayed strong second pulse, results in a spin echo which is amplified in power of order 50 dB relative to the input weak pulse. The echo power gain derives from the strong second pulse power which launches spin waves in the sample. After an optimum period of time following the second pulse, the spin wave modes couple optimum power into the precession mode, thus enhancing the size of the two pulse echo amplitude.

The usual nuclear or electron spin echo starts out at a maximum amplitude for short pulse pair separations and falls off monotonically as the pulse pair separation is increased. In the case of the second type of non-linearity, the echo has essentially zero amplitude for short pulse separations. As the pulse pair separation is increased to an optimum value the echo builds up to a maximum amplitude. After non-linear collisions or mode coupling events build up the echo to a maximum, the echo then begins to die off in amplitude for further pulse separations as irreversible inelastic damping mechanisms take over and destroy the echo.

Echoes also result from pulse excited intrinsic electric dipole moments in glasses, crystals, powders and piezoelectric semiconductors, and from trapped charges in these materials as well [4]. The holographic spectral memory storage principle of the stimulated echo is displayed particularly by trapped charges when they are excited by combinations of static fields and oscillating electric field pulses [5]. The trapped charges are first produced by optical excitations across band gaps in certain types of semiconductors. In some of these cases electric field pulse oscillations couple to anharmonic lattice oscillators which in turn produce ultrasonic phonon traveling wave echoes. Pulse ultrasonic pulse power may also be applied to such systems [6], and echoes are produced and detected ultrasonically . In type II superconductors the application of magnetic rf pulses induces the anharmonic oscillation of pinned magnetic field vortices. These vortices in turn rephase and yield a magnetic echo [7] which looks the same as a two pulse nuclear spin echo.

III. RADIATION DAMPING OF AN INHOMOGENEOUSLY BROADENED SPIN ENSEMBLE

Essential to the observation of free precession signals is the action of radiation damping, which accounts for coherent dipole radiation signal power. The following treatment analyzes radiation damping in the presence of inhomogeneous broadening, followed by the next section which will present a non-linearity echo analysis. The sharp line case of coherent radiation damping due to resistive loss in NMR is well known [8] for a magnetization $M_x(t) = M_0 \sin\theta(t)$ freely precessing at Larmor frequency $\omega_0 = \gamma H_0$, where $\theta(t) = \theta_0 + \Delta\theta(t)$ is the angle at which M_0 is tipped away from the direction of H_0. At time t=0 a sharp pulse orients M_0 at an initial angle θ_0. In the frame of reference rotating at frequency ω_0 the magnetization M_0 precesses about a reaction field $H_1(t)$ through an angle

$$\Delta\theta(t) = \gamma \int_0^t H_1(t)dt$$

where $H_1(t)$ is proportional to the current in the tuned NMR circuit. The initial angle θ_0 is diminished by $\Delta\theta(t)$ in time t as the spins loose energy to the circuit resistance. The magnetization M_0 is distributed over a normalized inhomogeneous broadening distribution $g(\delta)$ of off-set frequencies δ in the rotating frame .

The equation for the time dependence of the reaction field $H_1(t) = \omega_1 / \gamma$ expresses the coupling between a tuned circuit and the precessing magnetization. Neglecting relaxation, conservation of energy shared between spins and circuit requires that

$$LI^2/2 + \int_0^t RI^2 dt = -V_s H_0 \int_{-\infty}^{+\infty} M_z(\delta,t) g(\delta) d\delta. \tag{1}$$

The circuit current is $I(t) = i(t)\cos\omega t$; V_s is the sample volume, and the slowly varyng current modulus is given by

$$i(t) = [V_c / \pi L]^{1/2} \omega_1(t), \tag{2}$$

where V_c is the effective volume of the inductance L. Also $\tau = 2Q/\omega \rangle\rangle \omega^{-1}$ where τ is the circuit decay time constant and $Q = \omega L / R$ is the circuit figure of merit. After Eq. (1) is differentiated with respect to time and written first in terms of the slowly varying current, the Bloch equation $\dot{M}_z(\delta,t) = \omega_1(t) v(\delta,t)$ is substituted into the right side to gives

$$Li\,di/dt + Ri^2/2 = -V_s H_0 \int_{-\infty}^{+\infty} \omega_1(t) v(\delta,t) d\delta \tag{3}$$

The relation $\dot{\theta} = \omega_1$ combined with (2) allows (3) to be written in the convenient form

$$\ddot{\theta} + \omega\dot{\theta}/2Q = -\pi\gamma\omega\xi \int_{-\infty}^{+\infty} v(\delta,t) g(\delta) d\delta, \tag{4}$$

where $\xi = V_s / V_c$ is the sample filling factor. In the so-called "low Q" case, typical of standard NMR tuned circuits, the first term on the left side of (4) may be neglected compared to the second term, which implies the inequality $\tau \langle\langle 2\pi / \omega_1$. In this limit we have

$$d\theta / dt = -(1 / M_0 \tau_r) \int_{-\infty}^{+\infty} v(\delta,t) g(\delta) d\delta, \tag{5}$$

where $\tau_r = 1/2\pi M_0 \gamma Q \xi$ is the radiation damping time constant. In the sharp line case, $v(0,t) = \sin\theta$, and (5) reduces to

$$d\theta / dt = -(1/\tau_r)\sin\theta. \tag{6}$$

A small angle change $\Delta\theta(t \to \infty) = \theta(t \to \infty) - \theta_0 \langle\langle 1$ is now evaluated from a special solution of (5), based upon a manipulation in the theory of self-induced transparency [9]. At $\delta = 0$ in the inhomogeneous distribution the angle $\theta(t)$ pertains exactly to the orientation of the total magnetization $[M_z^2(0,t) + v^2(o,t)]^{1/2}$ with respect to the z axis, and $v(0,t)/M_z(0,t) = \tan\theta(t)$. The time integration of both sides of (5) yields

$$\Delta\theta(t \to \infty) = -(1/M_0\tau_r) \int_0^{t\to\infty} \int_{-\infty}^{\infty} v(\delta,t) g(\delta) d\delta dt. \tag{7}$$

The Bloch equation $v(\delta,t) = \dot{u}(\delta,t)/\delta$ is now substituted into (7) which is integrated to give

$$\Delta\theta = -(1/M_0\tau_r) \int_{-\infty}^{+\infty} (1/\delta)[u(\delta,t \to \infty) - u(\delta,0)] g(\delta) d\delta . \tag{8}$$

At $t = 0$ after the pulse the component $u(\delta,0) = 0$. Let $t = T + t'$ approach infinity with T defined as a very long time after which the reaction field H_1 decays to zero. Arbitrarily T serves as an initial time for the "hidden" evolution of the term

$$u(\delta, t = T + t' \to \infty) = u(\delta,T)\cos\delta t' + v(\delta,T)\sin\delta t' .$$

By "hidden " evolution it is meant that although the FID signal has disappeared and the spins are all out of phase, the individual isochromats defined at δ are still alive and precessing. Therefore (8) becomes

$$\Delta\theta = -(1/M_0\tau_r) \int_{-\infty}^{+\infty} (1/\delta)[u(\delta,T)\cos\delta t' + v(\delta,T)\sin\delta t'] g(\delta) d\delta. \tag{9}$$

In (9) the first term in the integrand averages to zero since both δ and $u(\delta,t)$ are odd functions; and the second term remains to be evaluated as a delta function at $\delta t'$=0. For large t' and finite δ the sine term averages to zero accept in the limit $\delta = 0$. Consequently

$$\begin{aligned}\Delta\theta &= -(1/M_0\tau_r) \int_{-\infty}^{+\infty} g(0) v(0,T) [\sin\delta t' / \delta t']_{\delta=0} d(\delta t') \\ &= -(T_2^* / \tau_r)\sin(\theta_0 + \Delta\theta).\end{aligned} \tag{10}$$

The maximum $g(0) = T_2^* / \pi$ of a symmetric Lorentzian distribution is assumed, and $v(0,T) = M_0 \sin(\theta_0 + \Delta\theta)$ is the solution to Bloch's equation. This solution is valid only for the isochromat at $\delta = 0$ and does not depend upon the time behavior of $H_1(t)$, but only upon its integral over time expressed by $\Delta\theta = \int_0^{t\to\infty} \gamma H_1(t) dt$.

Let $\theta_0 = \pi / 2$ pulse. With the condition $\Delta\theta \langle\langle 1$, it follows from Eq.(10) that

$$\Delta\theta \approx -T_2^* / \tau_r. \tag{11}$$

Therefore $T_2^* / \tau_r \langle\langle 1$ and the spins loose very little energy due to radiation damping. Typical decay times of $T_2^* \approx 0.01 - 0.001$ seconds are short compared to τ_r, where τ_r may be in the range of seconds or more. In approximately a time T_2^* the spin energy

$$\Delta W = -M_0 H_0[\cos(\theta_0 + \Delta\theta_0) - \cos\theta_0] \approx -M_0 H_0 \theta_0 \Delta\theta_0 \tag{12}$$

is consumed by the tuned circuit. Assuming $\theta_0 = \pi / 2$ and inserting (11) into (12), the average precession signal power is approximately

$$dW / dt \approx P_{ave} \approx -M_0 H_0 \theta_0 \Delta\theta / T_2^* \approx M_0 H_0 / \tau_r.$$

This estimate does not express any dependence upon inhomogeneous broadening, which one would expect. A rigorous expression for the signal power which does show an indirect dependence is given by

$$dW / dt = -V_s H_0 \int_{-\infty}^{+\infty} \dot{M}_z g(\delta) d\delta = -V_s H_0 \omega_1(t) \int_{-\infty}^{+\infty} v(\delta,t) g(\delta) d\delta. \tag{13}$$

After replacing the integral in (13) by the relation for $\omega_1(t) = d\theta / dt$ from (5) , the signal power at any time becomes

$$dW / dt = -V_s M_0 H_0 \omega_1(t)^2 \tau_r. \tag{14}$$

Here the factor $\omega_1(t)^2$ does depend upon $g(\delta)$, but $\omega_1(t)^2$ cannot be evaluated directly except by computation.

Eq. 14 is essential in reconciling a curious property. Eq. 10 gives the result that $\Delta\theta = 0$ if the inhomogeneous spectrum $g(\delta)$ is symmetric about $\delta = 0$ and there is a hole in the spectrum at $\delta = 0$ ($g(0) = 0$). The θ_0 pulse must excite the spectrum symmetrically so that $v(0,t) = 0$. However this does not mean that $H_1(t) = 0$ but rather that $H_1(t)$ is oscillatory. The average of $H_1(t)$ is zero, whereas the square of $H_1(t)$ in (14) does not average to zero. Of course (14) reduces to $P_{ave} = -V_s M_0 H_0 / \tau_r$ by estimating that $\omega_1^2(t)_{ave} \approx 1 / \tau_r^2$ from (6).

The $\ddot{\theta}$ term in (4) can be retained as a perturbation to the "low Q" problem solved above . With this term included the steps which lead to (10) will add to (10) an extra term on the left side as follows:

$$-\omega_1(0) L / 2R + \Delta\theta \approx -(T_2^* / \tau_r)\sin(\theta_0 + \Delta\theta). \tag{15}$$

The initial reaction field $H_1(0) = \omega_1(0) / \gamma$ is not to be confused with the field responsible for the external rf θ_0 pulse. From (6) we assume $\omega_1(0) \approx -(1 / \tau_r)\sin\theta_0$ to be a good approximation in (15), and therefore

$$\Delta\theta \approx -[(\tau_d / \tau_r)\sin\theta_0 + (T_2^* / \tau_r)\sin(\theta_0 + \Delta\theta)], \tag{16}$$

where the conditions $\tau_d \langle\langle \tau_r, T_2^*$ apply. Again, for $\theta_0 = \pi / 2$ and $\Delta\theta \langle\langle 1$,

$$\Delta\theta \approx -(\tau_d + T_2^*) / \tau_r. \tag{17}$$

which is a slight modification of (11). A very small amount of spin energy is now stored in the circuit before it dissipates into the resistance. The $\Delta\theta$ in (17) is thus enhanced by an amount τ_d / τ_r over that expressed by (11) because the reaction field $H_1(t)$ is correspondingly enhanced.

A generalization from (10) cannot be made to include the sharp line limit when $T_2^* \rangle\rangle \tau_r$ because the delta function argument becomes invalid. In this limit the inhomogeneous broadening is insufficient to dephase the transverse magnetization completely. Before complete dephasing takes place the magnetization tilts much more rapidly toward the equilibrium ground state because of coherent radiation damping.

IV. THE EXPLICIT NON-LINEARITY OF ECHO GENERATION

The discussion here is not intended to present a rigorous or exhaustive review of non-linearity concerning all kinds of echoes. The non-linearity concerning production of two and three pulse echoes will be outlined. The reader is referred to a general discussion by Bagguley [10] and to the stated literature concerning various types of echoes.

Routine predictions of nuclear spin echo signals may be carried out by the application of Bloch's equations directly, by sequences of unitary transformations of the initial density matrix, or by classical rotation transformations of the macroscopic u (dispersion), v (absorption) and longitudinal M_z magnetization components. One does not have to inject the linearity mechanism into any of these methods of calculation or even be conscious of it, because the mechanism is automatically built into the calculation. In order to expose the non-linearity it is of interest to calculate the nuclear spin echo signal in terms of the explicit non-linearity required to obtain the echo. A modified vector model is formulated to predict echo formation, all based on the dynamics of the macroscopic Bloch vector model. It is assumed that the reader is familiar with the physical application of Bloch's equations and the interpretation of macroscopic spin magnetization in the rotating frame of reference in order to describe the nuclear spin echo [11].

The echo is often referred to as a reversal phenomena [12, 13,14]. In the language of classical statistical mechanics the rephasing of spin vectors appears to be equivalent to the arbitrary reversal of particle momenta **P** from -**P** to +**P**, independent of **P**. All the components of the particle momentum change sign together, and their amplitudes remain the same as they were before the transformation. This argument applies as well to the dye experiment [12], in which a streak of dye in a viscous syrup between two cylinders is made to refocus after spreading it out by reverse rotation of one of the cylinders. However, these are analogues of reversibility which cannot show the essential non-linearity imposed upon oscillating dipole moments in the real echo experiment. The interpretation of entropy in light of the reversibility property will be discussed in the last section of this paper.

At thermal equilibrium a first impulse transformation, due to a $\theta_1 = \gamma H_y t_w$ pulse of width $t_w \langle\langle \tau$,creates a x component of magnetization $M_x(0,\delta) = g(\delta)\sin\theta_1$ in the x-y plane of reference rotating at angular frequency ω. The rotating rf field H_y is defined along the y axis, the off-resonance parameter is given by δ, and the inequality $1/t_w \rangle\rangle \delta$ applies for all δ in the normalized inhomogeneous spectral distribution $g(\delta)$. Relaxation times are taken to be infinite. The free precession equation of motion following the pulse is

$$\dot{M}_{xy}(t,\delta) - i\delta M_{xy}(t,\delta) = 0 \tag{18}$$

with the solution $M_{xy}(t,\delta) = g(\delta)e^{i\delta t}$. So far linearity prevails, with the FID expressed by

$$F(t) = \sin\theta_1 \int_{-\infty}^{+\infty} g(\delta)e^{i\delta t} d\delta. \tag{19}$$

In a two pulse echo experiment the non-linearity which produces the echo is imposed by a second θ_2 pulse applied at time τ. At the same time the pre-existing longitudinal magnetization $M_z = \cos\theta_1$ is projected by the θ_2 pulse onto the x axis by an amount $M_x = \sin\theta_2 \cos\theta_1$ .This represents the maximum of another FID signal

$$F(t \geq \tau) = \sin\theta_2 \cos\theta_1 \int_{-\infty}^{\infty} g(\delta) e^{i\delta(t-\tau)} d\delta \tag{20}$$

at $t = \tau$ which is linearly additive to the FID of the first θ_1 pulse. These linear signals are assumed to decay to zero over short enough times because of inhomogeneous broadening so that they do not run into subsequent pulses and echoes. The precession initiated by (20) is omitted for the present since it does not influence the two pulse $\theta_1 - \theta_2$ echo to be evaluated next.

The solution of (18) may be written as

$$M_{xy}(t,\tau) = \sin\theta_1 e^{i\delta\tau} e^{i\delta(t-\tau)} = \sin\theta_1 (\cos\delta\tau + i\sin\delta\tau) e^{i\delta(t-\tau)} \tag{21}$$

with τ defined at the time the θ_2 pulse is to be applied. The torque imposed on M_{xy} by the rf field H_y during the θ_2 pulse switches the component $M_x = \sin\theta_1 \cos\delta\tau$ in (4) to the value $M_x = \sin\theta_1 \cos\theta_2 \cos\delta\tau$, while the component $M_y = \sin\theta_1 \sin\delta\tau$ remains unchanged (see Fig. 1). Therefore the θ_2 pulse transforms (21) to

$$M_{xy} = \sin\theta_1 (\cos\theta_2 \cos\delta\tau + i\sin\delta\tau) e^{i\delta(t-\tau)} \tag{22a}$$

$$= \sin\theta_1 [-\sin^2(\theta_2/2) e^{-i\delta\tau} + \cos^2(\theta_2/2) e^{i\delta\tau}] e^{i\delta(t-\tau)}, \tag{22b}$$

valid for $t \geq \tau$. The product of two sinusoidal functions $e^{i\delta\tau} e^{i\delta(t-\tau)}$ in (22b) expresses the non-linearity introduced by the θ_2 pulse. Therefore the full fledged echo, including its shape, is obtained from

$$E(t \geq \tau) = \int_{-\infty}^{\infty} M_{xy}(t \geq \tau)\ g(\delta) d\delta, \tag{23}$$

where $M_{xy}(t \geq \tau)$ is given by (22). Independent of the integration of (23) over δ, from (22) the echo maximum at $t = 2\tau$ is

$$E_{\max} = -\sin\theta_1 \sin^2(\theta_2/2). \tag{24}$$

The $\cos^2(\theta_2/2)$ term in (22b) accounts for incomplete focusing of the echo. Complete focusing is achieved only for $\theta_2 = \pi$. For simplicity let $\theta_1 = \pi/2$ in order to make the following argument. A circle of unit vectors with phase angles $\delta\tau$ (see Eq. 21) is tilted by the θ_2 pulse through an angle θ_2 about a diameter along the y axis . The tilted circle projects back on to the x-y plane as an ellipse which has x minor axis A= $\cos\theta_2$ and y major axis B=1, as shown in Fig. 1. These axes are respectively the coefficients of $\cos\delta\tau$ and $\sin\delta\tau$ in (22a). Each isochromat which had phase angle $\delta\tau$ prior to the θ_2 pulse has its phase angle switched to

$$\eta = \tan^{-1}[(\tan\delta\tau)/\cos\theta_2] \tag{25}$$

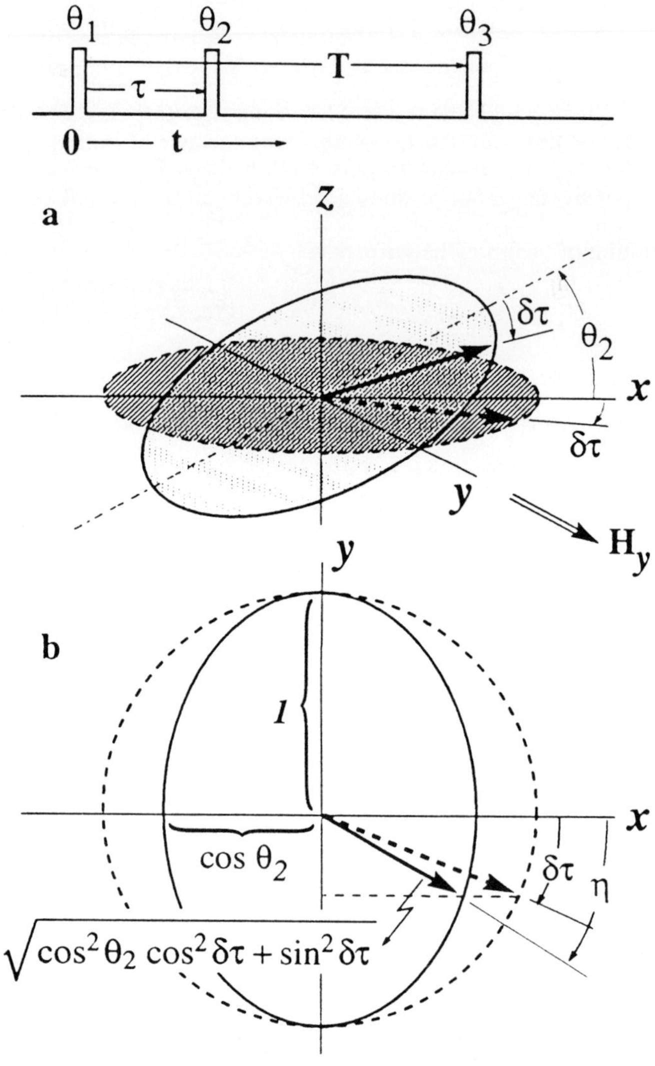

Fig. 1. — Transformation of precessing magnetization in the rotating frame immediately after the θ_2 pulse of a $\theta_1 - \theta_2$ pulse echo sequence. The unit radius circle of precession following the θ_1 pulse projects (Fig. 1-a) as an ellipse onto the $x-y$ plane after the θ_2 pulse (Fig. 1-b). The transformations of phase ($\delta\tau \rightarrow \eta$) and magnitude ($1 \rightarrow \sqrt{\cos^2\theta_2\cos^2\delta\tau + \sin^2\delta\tau}$) are shown.

after the pulse. The unit radius magnitude of M_{xy} in the circle of precession switches to the value

$$|M_{xy}| = [\cos^2\theta_2 \cos^2\delta\tau + \sin^2\delta\tau]^{1/2} \tag{26}$$

for $t \geq \tau$, and continues to precess in the x-y plane at frequency δ.

The δ dependence of (26) signifies a "hologram" of transverse spectral memory contained in $|M_{xy}|$ which stores the primary echo. At the same time $|M_{xy}|$ is created, a longitudinal spectral memory magnetization

$$M_z(\tau,\delta) = \sin\theta_1 \sin\theta_2 \cos\delta\tau \tag{27}$$

appears along the z axis. It is this component which is responsible for the stimulated echo at $t = T + \tau$ following the application of a third θ_3 pulse at time T. It was pointed out that the magnetization $M_z(0) = \cos\theta_1$ which remains after the first pulse does not participate in the $\theta_1 - \theta_2$ pulse echo memory. However a $\theta_2 - \theta_3$ pulse echo is formed at $t = 2T - \tau$ in the same way the $\theta_1 - \theta_2$ echo is formed . With respect to the θ_3 pulse the θ_2 pulse acts as a first pulse which projects an initial magnetization $M_x = \sin\theta_2 \cos\theta_1$ at $t = \tau$.

The concept of the two-pulse echo as a reversal phenomenon is one more of appearance than of fact. However the gradient echo, which is not a two-pulse echo, does show literally a reversal in phase. The second pulse is replaced by a sudden switch in sign of a completely controlled field gradient. The two-pulse 90-180 degree echo occurs because of a reflection transformation $M_x(\delta,\tau) \rightarrow -M_x(\delta,\tau)$, and not because of phase reversal. Partial transformations of this sort occur for any $\theta_1 - \theta_2$ pulse sequence. The sudden advance in phase of an isochromat after the θ_2 pulse is expressed in general by $\Delta\phi = \eta - \delta\tau$, where η is given by (25). It is not correct to say that the echo is merely a consequence of "constructive interference" among spins with different Larmor frequencies. It is more realistic to say that the echo from each isochromat will not stick its head out as well if it were not for "destructive interference" among all the isochromats that make up the distribution $g(\delta)$. Each isochromat possesses an intrinsic echo response and contributes its own constant at the time of the echo. If one could measure the signal from an isolated isochromat it would produce a continuous type sine wave at the Larmor frequency in the laboratory frame with a characteristic amplitude and phase at time $t = 2\tau$, evident from (22b). Put back the interference signals from the rest of the isochromats (they contribute the same constant $\sin\theta_1 \sin^2(\theta_2/2)$ at $t = 2\tau$, weighted by $g(\delta)$), and the echo appears out of nowhere.

The logic developed above may be applied to calculate echo maxima which occur following the application of a third θ_3 pulse at $t = T$. An expression analogous to Eq. 21 must be formulated in which the components of magnetization $M_x(T,\delta)$ and $M_y(T,\delta)$ are identified explicitly at time T just before the application of the third θ_3 pulse. The θ_3 pulse imposes the transformation $M_x(T,\delta) \rightarrow M_x(T,\delta)\cos\theta_3$ while $M_y(T,\delta)$ remains unchanged, in analogy to (22a). At the same time the stimulated echo memory magnetization $M_z(\tau,\delta)$ given by (27) projects as $M_z(\tau,\delta)\sin\theta_3$ onto the x axis, which must be added to $M_x(T,\delta)\cos\theta_3$ in Eq. 31 below. At time $t = T_-$ just before the θ_3 pulse is applied, the initial magnetization in the xy plane is

$$M_{xy}(T_-,\delta) = M_x(T,\delta) + iM_y(T,\delta), \tag{28}$$

where

$$M_x(T,\delta) = \sin\theta_1 \sqrt{\cos^2\theta_2 \cos^2\delta\tau + \sin^2\delta\tau} \cos[\eta + \delta(T-\tau)] \tag{29}$$

and

$$M_y(T,\delta) = \sin\theta_1 \sqrt{\cos^2\theta_2 \cos^2\delta\tau + \sin^2\delta\tau} \sin[\eta + \delta(T-\tau)]. \tag{30}$$

Following the θ_3 pulse for all times $t \geq T$, Eq. 28 is transformed to

$$= [M_x(T,\delta)\cos\theta_3 + \sin\theta_1 \sin\theta_2 \sin\theta_3 \cos\delta\tau + iM_y(T,\delta)]e^{i(t-T)}. \tag{31}$$

Eq. 31 is now expanded to show the imaginary exponential terms, analogous to (22b).

$$M_{xy}(t \geq \tau,\delta) = \sin\theta_1 \sin^2(\theta_2/2)\sin^2(\theta_3/2)e^{i\delta[t-(2T-2\tau)]}$$

$$+(1/2)\sin\theta_1 \sin\theta_2 \sin\theta_3 e^{i\delta[t-(T+\tau)]}$$

$$-\sin\theta \sin^2(\theta_2/2)\cos^2(\theta_3/2)e^{i\delta(t-2\tau)}$$

$$-\sin\theta \cos^2(\theta_2/2)\sin^2(\theta_3/2)e^{i\delta(t-2T)}. \tag{32}$$

The terms in (32) represent maximum M_x polarization amplitudes which occur for times at which the corresponding exponents are equal to zero. The first term at $t = 2T - 2\tau$ indicates the echo formed from the alignment of the previous $t = 2\tau$ echo and the θ_3 pulse. This echo will not occur for $\tau \leq T \leq 2\tau$, in which case the θ_3 pulse alters the $\theta_1 - \theta_2$ echo amplitude, given by the third term in (32) as $-\sin\theta_1 \sin^2(\theta_2/2)\cos^2(\theta_3/2)$, . The second term represents the stimulated echo at $t = T + \tau$; and the last term is the echo formed at $t = 2T$ from the $\theta_1 - \theta_3$ pulse pair. Eq. 32 does not include the $\theta_2 - \theta_3$ pulse pair echo because the initial FID magnetization expressed by (20) was dropped in the analysis. This echo occurs at $t = 2T - \tau$ and has a maximum of $-\cos\theta_1 \sin\theta_2 \sin^2(\theta_3/2)$.

This short review of echo nonlinearity omits important points about the connection between various types of photon echo production in the pulse time domain and steady state non-linear optical effects, such as four wave mixing and phase conjugation, which involve propagation. As laser pulse sequences are conceived to merge with one another into the steady state, considerations of non-linearity persist but in a different form [15].

V. THE PHOTON ECHO

It is well known [16] that the description of the coherent superposition of states induced in a quantum two level system is the same whether magnetic dipole or electric dipole moments are involved. Therefore the time dependent dynamics of echo formation depicted in NMR will apply directly to the description of photon echo production, with the spatial dependence handled in addition by use of the semi-classical Bloch-Maxwell equations. The dimensions of a radiating sample in the optical electric dipole case is much larger than the optical wave length and wave vector propagation must be taken into account. For NMR or EPR single mode excitation in a cavity, the sample dimension is much smaller or comparable to the wavelength of the applied radiation. Also spontaneous emission of electric dipole radiation is a new factor. An elementary description now follows for the formation of the photon FID after a single laser θ_1 pulse, and for the photon echo which results after a second laser θ_2 pulse is applied to a two level atomic system. A sample of length $z = L$ is assumed to consist of a continuum of radiating macroscopic electric dipole moments p_0 at most per unit length which may be induced by a 90 degree pulse from equilibrium. The sample cross-sectional area is given by A. Diffraction will be neglected, and ideal plane waves are assumed throughout the analysis. The NMR notation $\theta = \omega_1 t_w$ for the laser pulse tipping angle is retained with the new definition $\omega_1 = \kappa\varepsilon$, where $\kappa = \mu_\varepsilon / \hbar$, the laser pulse electric field amplitude is ε, and μ_ε is the magnitude of the radiating electric dipole matrix element. Fig. 1 indicates again the pulse sequence, and propagating wave vectors k_1 and k_2 are assigned to the first and second pulses respectively. The propagation vector k_F is assigned to the FID radiated signal, and the vector k_E` is assigned to the echo signal. The accumulated phase shift $\vartheta = (k_1 - k_2)z$ at distance z occurs when the two pulses are not applied in the same direction.

V. A. The Photon Free Induction Decay

Within a volume $A\Delta z$, the time dependent dynamics for the formation of radiating electric dipole polarization in the $x-y$ plane is identical to the previous description for nuclear magnetization in pulsed NMR. In addition the θ_1 pulse sets up a spatially phased polarization $p_0 \sin\theta_1 \exp(-ik_1 z)$ along the z direction from $z=0$ to $z=L$. Immediately after the pulse the volume elements between z and $z+\Delta z$ radiate small FID signals in the forward z direction, all of which add constructively to provide a total electric field signal at $z=L$, given by $E(L,t) = \sqrt{I_0}$:

$$E_{fid}(L,t) = Cp_o A \sin\theta_1 e^{-t^2/T_2^2} \int_0^L e^{-ik_1 z} e^{-ik_F(L-z)} dz$$

$$= CLp_o A \sin\theta_1 e^{-t^2/T_2^2} e^{-ik_F L} \frac{\sin[(k_F - k_1)L/2]}{(k_F - k_1)L/2}. \tag{33}$$

A phenomenological damping term e^{-t^2/T_2^2} is included to account for signal decay due primarily to inhomogeneous broadening. In most experimental cases this term is dominant over spontaneous emission because of Stark electric field strain in crystals or because of Doppler broadening in gases. The signal intensity is I_0 and C is a constant of proportionality C. Since the dipoles are automatically phase matched by the first pulse, therefore $k_F = k_1$, and the photon FID signal is emitted in the same direction as the applied pulse. If the refractive index of the sample could be suddenly altered significantly by some sort of electro-optic effect immediately after the θ_1 pulse, a phase mismatch would then cause a reduced signal amplitude. This reduction would be imposed by the fractional function in (33), which otherwise reduces to unity at the matching condition .

V. B. Two Pulse Photon Echo

After the first θ_1, k_1 pulse, a second θ_2, k_2 pulse is applied in a slightly different direction. It will be seen that the photon echo signal [17] is then emitted in a third k_E direction , which conveniently avoids saturation of the optical detector by the applied laser pulses along the k_1 and k_2 directions. In a given volume element $A\Delta z$ at z the rotating electric laser modulus ε is shifted in phase angle by $\vartheta = (k_1 - k_2)z$ relative to its alignment along the y axis during the first pulse. This phase shift specifies another frame of reference x', y' with axes rotated by angle φ relative to the x,y axes, and ε is directed along y'. Analogous to Eq. 22-a, the echo signal in the prime frame is then determined by the time dependence of

$$S(t)_{x',y'} \propto \sin\theta_1[\cos\theta_2 \cos(\delta\tau - \vartheta) + i\sin(\delta\tau - \vartheta)]e^{i\delta(t-\tau)}. \quad (34)$$

For example, in the NMR case if the θ_2 pulse is applied at a phase $\vartheta = \pi/2$ with respect to the θ_1 pulse, it is easily seen from the rotation of coordinates that the echo in the x',y' frame appears along the y' axis, equivalent in this case to the positive x axis of the original frame of reference. In order to evaluate the phase shift in general for the echo relative to the original x,y frame, the x' and y' components of polarizations $P_{x'}$ and $P_{y'}$ must be transformed into the original frame by the transformations

$$P_x = P_{x'} \cos\vartheta - P_{y'} \sin\varphi \text{ and } P_y = P_{x'} \sin\vartheta + P_{y'} \cos\varphi. \quad (35)$$

Eq. 34 then transforms to the signal corresponding to (22b), now including the phase shift 2φ:

$$S(t)_{x,y} \propto \sin\theta_1[-\sin^2(\theta_2/2)e^{-i\delta\tau} + \cos^2(\theta_2/2)e^{i\delta\tau}]e^{i\delta(t-\tau)+2i\vartheta}. \quad (36)$$

All of these steps lead to the simple result that the phase shift of the echo signal is 2ϑ, which in fact can be visualized just from sketching out the geometry of the polarization vectors undergoing precession.

The photon echo is now described in its time dependence from a given volume element $A\Delta z$ by Eq. 36. Integration over all the signals from each volume element extending over distance $z=0$ to $z=L$ must be carried out regarding the phase dependencies $k_1 z$, and over the phase shift $2\varphi = 2(k_1 - k_2)z$. Including again the effect of inhomogeneous broadening, the result for $\xi \langle\langle 1$ is

$$E_{echo}(L,t\rangle\tau) = Cp_o[\sin\theta_1 \sin^2\theta_2/2]e^{-(t-2\tau)^2/T_2^2}\int_0^L e^{2i(k_1-k_2)z - ik_1 z - ik_{echo}(L-z)}dz,$$

$$E_{echo}(L,t=2\tau) = CLp_o[\sin\theta_1 \sin^2\theta_2/2]\sin\frac{[k_{echo} - (2k_2 - k_1)]L/2}{[k_{eccho} - (2k_2 - k_1)]}. \quad (37)$$

Since the magnitudes of the two pulse k vectors ($|k_1|=|k_2|$) are the same in the usual experiment the phase matching condition above cannot be satisfied exactly, given that the direction of k_2 is chosen at a small angle ξ with respect to the z direction of k_1. If ξ were not small, the phase factor in (37) should be written as $\exp i[\underline{k}_e - (2\underline{k}_2 - \underline{k}_1)]\cdot\underline{r}$, which in principle would require integration over all dipoles at points over $\underline{r}$ in the sample. Since we ignore diffraction from a strip of dipoles along $\underline{r} = \underline{z}$, then it is seen for $\xi\langle\langle 1$ that $i[\underline{k}_e - (2\underline{k}_2 - \underline{k}_1)]\cdot\underline{r} \approx [k_1 - |2\underline{k}_2 - \underline{k}_1|]z\cos 2\xi$, where the maximum photon echo amplitude is emitted at an angle $\approx 2\xi$ with respect to the $\underline{r} = \underline{z}$ direction when $\underline{k}_e$ is parallel to $2\underline{k}_2 - \underline{k}_1$. However, because $|\underline{k}_1| = |\underline{k}_2|$, the propagation vector triangle cannot form an exact closed triangle, and the intensity of the photon echo is therefore reduced by a very small factor. The important point is that the photon echo is emitted in the k_E direction at an angle $\approx 2\xi$ with respect to the k_1 direction along z in order to avoid the blanking effect of the applied laser pulses.

V. C. The Three Pulse Stimulated Photon Echo

The procedure for evaluating the stimulated photon echo maximum at $t = \tau + T$ in the direction of $\underline{k}_{SE}$ must now take into account the phase shifts of the second θ_2, k_2 and third θ_3, k_3 pulses relative to the initial phase of the first θ_1, k_1 pulse applied along the z direction. Following the NMR case for the stimulated echo stored magnetization component (Eq. 32), the phase shift of $\vartheta_2 = (k_1 - k_2)z$ appears in the stored longitudinal polarization

$$p_z \propto \sin\theta_1 \sin\theta_2 \sin\theta_3 \cos(\delta\tau - \vartheta_2) \tag{38}$$

immediately after the second pulse at $t = \tau$. The frame of reference in which this takes place is defined such that the applied rotating laser electric field vector of the θ_2 laser pulse is defined in a new $x' - y'$ frame of reference along a the y' axis. This axis is at an angle ϑ_2 with respect to the y axis along which the rotating electric field from the first pulse was directed. Directly after the third pulse applied at time $t = T$, a radiating signal component in the $x' - y'$ frame results , proportional to

$$S(t = T) \propto \sin\theta_1 \sin\theta_2 \sin\theta_3 \cos(\delta\tau - \varphi_2) e^{i\vartheta_3}, \tag{39}$$

where $\vartheta_3 = (k_1 - k_3)z$. Now, according to the transformation into the original $x - y$ frame using (35), including phase factors as shown for the primary echo in (37), the stimulated echo for $t \geq T$ becomes proportional to

$$S(t \geq T) \propto (1/2)\sin\theta_1 \sin\theta_2 \sin\theta_3 e^{i\vartheta_3 + i\vartheta_2} e^{-ik_1 z} e^{-ik_{SE}(L-z)} e^{i\delta[t-(T+\tau)]}$$

$$\propto e^{i(k_1 - k_2 - k_3 + k_{SE})z}. \tag{40}$$

The stimulated echo signal at $t = T + \tau$ is a maximum in the direction $\underline{k}_{SE} = -\underline{k}_1 + \underline{k}_2 + \underline{k}_3$.

VI. ECHOES AND STATISTICAL MECHANICS

The spin echo as a reversal phenomenon looks very much like a realization of Loschmidt's paradox. This paradox was introduced by Loschmidt in 1872 as a criticism of Boltzmann's H theorem for entropy increase toward thermal equilibrium. Loschmidt counter argued that by reversing the momenta of particles in a statistical ensemble, the reverse paths which they would take are no less probable than their forward paths toward equilibrium, since the energy of the system is the same for collisions in either direction.

Interestingly, there are differences and well as general agreement to this day regarding the interpretation of the spin echo in relation to entropy and the second law of thermodynamics. Various viewpoints [18] have been posed by Rothstein, Mayer and Mayer, Blatt, Denbigh, Ma, and by Waugh [14], Hahn [12], and Waugh, Pines and Rhim [13]. Waugh [14] particularly argues against certain conclusions by others.

The spin echo forces one to reconcile concepts of mechanical (or quantum mechanical) collision dynamics with the notions of thermodynamics, where the former are directly related to microscopic initial/final conditions, and the latter washes them out. Blatt and Denbigh [18] argue that a spin ensemble, though it may have a very long relaxation time, is never a truly isolated system because fluctuations eventually take over from the outside "world degrees of freedom." Disorder reigns supreme and entropy is necessarily an inviolable parameter of nature according to Denbigh, like susceptibility or density. However this begs the question as pointed out by Waugh [14], because in principle there is no fundamental limit that one can establish in defining an isolated system. Ensembles of particles could in principle be prepared in correlated states which exceed ones lifetime.

The observer and producer of the spin echo can argue that entropy is a man made concept and is not an intrinsic parameter, but rather a matter of ignorance versus information. After a 90 degree pulse he knows that if he applies a second 180 degree pulse at a later time t= τ he will see the spin ensemble refocus at t= 2τ. If the ensemble after the 90 degree pulse is presented to a thermodynamicist who doesn't know that the pulse was applied, and the system has completely dephased, he would determine by some sort of thermodynamic measurement that the spin system is in equilibrium and has a definite spin temperature and entropy. The observer of the echo interprets that the entropy of the system also has not changed after the pulses or during the time even after the echo. However, to him the system cannot be assigned a temperature; it is in a single coherent superposition of microscopic states [12], comprising a single grand state which adds zero entropy to the initial entropy.

Returning now to the thermodynamicist, if he is now shown the echo after he made his measurement, he is absolutely startled at first. What is he to think? He is told that the system is in a coherent superposition of accessible states, and that he made ensemble measurements of diagonal states, not knowing in doing so that he destroyed the superposition which is invisible to him. Whether the destruction took place or not, his measurement would be the same. Therefore he reaches the conclusion that there is a spin temperature, an infinite temperature in this case. The second law is valid for the thermodynamicist. Entropy for him is intrinsic in nature as long as dynamical paths in phase space are impossible to follow or calculate. Both observers have reached valid conclusions based on what knowledge they have of the situation. Shortly we will return to the thermodynamicist and find out how he wiggles out of his dilemma according to an argument by Waugh.

Mayer and Mayer [18, pg. 137] made the observation that the spin echo was just a rare, curious phenomenon, stating "similar echo effects will remain pretty well limited to nuclear spin systems". We know now this is not true because many different systems other than spins exhibit pulse echoes. Their interpretation of the echo did not meet the challenge of the real dilemma. The existence and exception of the echo must be faced as a generality, and must be reconciled with the second law. Our bewildered thermodynamicist is firmly convinced at first that the second law is contradicted by the echo. Then he begins to think about the granularity of matter and reconciliation of fluctuations with the interpretation of thermodynamic equilibrium. Although the system ensemble can be traced through many paths in hyper-phase space which characterize equilibrium over time, temporary non-equilibrium echoes could behave as precursors of pulse disturbances imposed on the ensemble, and yet equilibrium is defined. This is the argument of Waugh [14] who suggests that the contradiction can be eliminated for the thermodynamicist by a restatement of macroscopic thermodynamic principles as follows: "When a system is in such a state that after any slight temporary disturbance of external conditions it returns rapidly or slowly to the initial state, this state is said to one of *equilibrium."* However, if equilibrium exists, then a temperature must exist, but if the spins are to be conceived as one correlated coherent superposition state which exhibits the echo, then a temperature cannot be defined. The Waugh definition may satisfy the adherence to a traditional interpretation by the thermodynamicist, but that definition is not required for those who know the initial conditions imposed by outside pulses and what follows thereafter.

It should be emphasized that the echo can form in the presence of collisions in a closed system, as demonstrated [13] by the "magic sandwich" echo displayed fluorine dipolar coupled spins in calcium fluoride crystal. In principle it may be demonstrated [10] also in some future experiment by emission of electron charges from a source, where the electrons collide for a time in two dimensions. The electrons are then refocused back to the source in equal time following the reversal effect of short cyclotron precession magnetic field pulses applied normal to the surface.

Ma [18] makes the statement that the entropy remains constant during the formation of spin or plasma echoes only because these situations involve inhomogeneous broadening (Landau damping in the case of plasma echoes), and no collisions are involved. This is contradicted because echoes do form in the presence of collisions, pointed out above. Ma does not acknowledge the importance of correlations which account for constant entropy in echo experiments, including the echo demonstration with the dye streak in the rotating cylinder demonstration, and the "magic sandwich " echo. He states that only the absence of fluctuations accounts for constant entropy.

What is remarkable is the fact that pulses that produce echoes can simultaneously prepare superposition states of entire ensembles without requiring a knowledge of each particle response in a system of an enormous number of particles. Because nature has provided that interactions of nuclear moments with the environment is weak, this makes possible observable and controllable pulse transient free induction signals in time domains that are technically feasible. Consider instead what effect the slightest perturbation of gas kinetic collisions would have on the particle path memory in a gas at NTP. Borel [19] calculated that if a one gram mass is displaced through a distance of one cm. on a star at the distance of Sirius, it would change the magnitude of the gravitational field at the Earth's surface by a factor 10^{-100}. The magnification effect by multiple collisions of the slightest deviation in molecular path due to this gravity change on earth is sufficient to alter the molecular trajectories within a time as short as one microsecond. Another example by Berry [20] shows that the presence or absence of an electron at a distance of 10^{10} light years alters the gravitational force on earth sufficient to change the angles of molecular trajectories by as much as one radian after only 60 collisions. It is often lamented that NMR signals are inconveniently weak, particularly for purposes of magnetic resonance imaging. On the other hand this weakness is a virtue, making possible sharp NMR spectra and the utility of controllable pulse echo signal processing .

Acknowledgment

The author thanks the National Science Foundation for partial support.

REFERENCES

1. R. W. Gould, Phys. Lett. **19**, 477(1965).
2. G. F. Herrman, R. M. Hill and D. E. Kaplan, Phys. Rev. **156**, 118 (1967).
3. G. F. Herrman, D. E. Kaplan and R. M. Hill, Phys. Rev. **181**, 829 (1969).; 2, 2587 (1970).
4. K. Forsheim, K. Kajimura, T. G. Kazyaka, R. L. Mekcher and N. S. Shiren, Phys. Rev. **B17**, 964 (1978). An extensive list of references is given in this paper.
5. N. S. Shiren, R. L. Melcher, D. G. Garrod and T. G. Kazyaka, Phys. Rev Lett. **24**, 542 (1973).
6. B. Golding and J. E. Graebner, Phys. Rev. Lett. **37**, 852 (1976); N. S. Shiren, W. Arnold and and T. G. Kazyaki, Phys. Rev. Lett. **39**, 239 (1977); B. Golding, M. V. Schickfus, S. Hunklinger and K. Dransfeld, Phys. Rev. Lett. **43**, 1817 (1979).
7. H. Alloul and C. Froidevaux, Phys. Rev, Lett. **20**, 1235 (1968): B. Goldberg, E. Ehrenfreund and M. Weger, Phys. Rev. Lett. **20**, 539 (1968) and J. Appl. Phys. **39**, 5941 (1968).
8. N. Bloembergen and R. V. Pound Phys, **95**, 8 (1954).
9. S. L. McCall and E. L. Hahn, Phys. Rev. **183**, 457 (1969).
10. **Pulsed Magnetic Resonance: NMR, ESR, and Optics**, Chapter 1, Edited by D. M. S. Bagguley, Oxford University Press, 1992.
11. E. L. Hahn, Phys. Rev. **80**, 580 (1950).
12. R. G. Brewer and E. L. Hahn, Sci. Am., **251**,(6), 50 (1984); E. L. Hahn, Proc. of the Royal Inst. G. B. **44**, 26 (1970) and Proc. A. von Humboldt Fdn. Symposium Stanford, Ca. (1983), p. 135.

13. W. Rhim, A. Pines and J. S. Waugh, Phys. Rev. **B3**, 684 (1971).

14. J. S. Waugh, ref (10), Chapter 6.

15. **The Principles of Nonlinear Optics**, Y.R. Shen, Ch. 21, John Wiley and Sons Inc., 1984.

16. R. P. Feynman, F. L. Vernon Jr., and R. W. Hellwarth, J. Appl. Phys. **28**, 49 (1957).

17. I. D. Abella, N. A. Kurnit, and S. R. Hartmann, Phys. Rev. **141**, 391 (1966).

18. J. Rothstein, Am. J. Phys. **25**, 510 (1957); **Statistical Mechanics**, 2nd Ed., J. E. Mayer and M. G. Mayer, Chapter 6, John Wiley and Sons,1977; J. M. Blatt, Progr. Theo. Phys. **22**, 745 (1959); **Entropy in Relation to Incomplete Knowledge**, K. G. Denbigh and JK. S. Denbigh, p. 141, Cambridge University Press, 1985; **Statistical Mechanics**, Shang-Keng Ma, Ch. 24, World Scientific Pub. Co, 1985.

19. **Le Hasard**, E. Borel, p.174, Alcan, Paris,1928.

20. M. V. Berry, Am. Inst. of Phys. Conf. Proc., No. 46, p.16, 1978.

NONLINEAR SPECTROSCOPY OF INORGANIC MATERIALS: EXCITONS IN SEMICONDUCTORS

Jørn M. Hvam

Mikroelektronik Centret
The Technical University of Denmark
DK-2800 Lyngby, Denmark

ABSTRACT

Transient nonlinear optical spectroscopies for studying the dynamical properties of elementary excitations in semiconductors are being described and discussed. Particularly, the nonlinear optical properties of excitons in bulk semiconductors as well as in lower-dimensional semiconductor quantum structures are investigated by means of transient four-wave mixing, photon-echo and nonlinear quantum beat experiments.

I. INTRODUCTION

Nonlinear optical spectroscopy of inorganic materials has been a very active and expanding field of research ever since the introduction of high intensity laser sources made it possible to observe nonlinear optical responses with available detection techniques.[1],[2] Due to the discretenes of atomic levels and transitions, most of the nonlinear spectroscopies on solids were first performed on isolated ions in solid matrices[3] or on strongly localized defects and impurities such as F-centers in Alkalide Halides.[4] The strong interaction and scattering mechanisms of the electronic states in e.g. semiconductors made it more difficult to implement, and interprete, the same spectroscopies on these extended states. However, in the last decade there has been a development in experimental techniques as well as in the theoretical models of electronic interactions in semiconductors, so that nonliear spectroscopy has become a major tool to study the fundamental electronic and opto-electronic properties of semiconducting materials and structures.[5],[6],[7]

The linear optical response of direct gap semiconductors is governed by the large oscillator strength for the dipole-allowed direct electron transitions across the band gap giving rise to the creation of electron-hole (e-h) pairs, strongly correlated by the Coulomb interaction (exciton effect).[8] The nonlinearity is brought about by the final state (exciton) interactions, that occur even at fairly moderate exciton densities ($n_x = 10^{14}$ cm^{-3}). The observed response times are due to scattering and/or recombination, governed by impurity

Nonlinear Spectroscopy of Solids: Advances and Applications
Edited by B. Di Bartolo and B. Bowlby, Plenum Press, New York, 1994

scatteering, phonon scattering or carrier-carrier scattering depending on purity, temperature and excitation density.[8],[9],[10],[11],[12]

The strong interest in the nonlinear optical properties of semiconductors is twofold. In the first place, nonlinear optical spectroscopies have turned out to be extremely powerful for the investigation of carrier interaction and relaxation phenomena [8-12]. Secondly, the detailed knowledge of the nonlinear optical properties of semiconductor materials and structures may lead to the development of new electro-optical or all-optical devices.[13] The present paper is mainly concerned with the first aspect, but shall also try to point out application aspects of some of the observed nonlinear optical phenomena.

In the following section, will be presented the nonlinear optics formalisms employed in this article, and the microscopic origin of the nonlinear susceptibilities will be discussed briefly. Second and third order nonlinearities will be introduced, and transient degenerate four-wave mixing will be described in a simple medium, consisting of non-interacting two-level atoms. Quantum beat spectroscopy will also be introduced in simple three-level systems. In Sec. III, relevant electronic and optical properties of bulk semiconductors are briefly reviewed, followed by a discussion of the changes in these properties in lower dimensional semiconductor quantum structures. Section IV is devoted to the presentation of some nonlinear optical experiments on bulk semiconductors, semiconductor surfaces and semiconductor quantum well systems. In Sec. V, nonlinear quantum beat spectroscopy will be applied to bulk semiconductors as well as to quantum well structures. Summary and conclusions will be given in Sec. VI.

II. NONLINEAR OPTICS

The linear, as well as the nonlinear, optical response of materials is phenomenologically described by expressing the relation between the electric field **E** of the electromagnetic radiation and the resulting polarization **P** of the medium. In the nonlinear case, the nonlinear polarization $\mathbf{P}_{NL}$ is traditionally expanded in powers of **E** [2]

$$\begin{aligned} \boldsymbol{P}_{NL}(\boldsymbol{r},t) &= \chi^{(1)}\cdot\boldsymbol{E} + \chi^{(2)}\cdot\boldsymbol{E}\boldsymbol{E} + \chi^{(3)}\cdot\boldsymbol{E}\boldsymbol{E}\boldsymbol{E} + \cdots \\ &= \boldsymbol{P}^{(1)}(\boldsymbol{r},t) + \boldsymbol{P}^{(2)}(\boldsymbol{r},t) + \boldsymbol{P}^{(3)}(\boldsymbol{r},t) + \cdots \end{aligned} \tag{1}$$

where the terms on the right hand side are in a short-hand vector notation and in reality are multiple space and time integrals of the form, e.g.

$$\boldsymbol{P}^{(2)}(\boldsymbol{r},t) = \int \chi^{(2)}(\boldsymbol{r}-\boldsymbol{r}_1,t-t_1;\boldsymbol{r}-\boldsymbol{r}_2,t-t_2)\boldsymbol{E}(\boldsymbol{r}_1,t_1)\boldsymbol{E}(\boldsymbol{r}_2,t_2)d\boldsymbol{r}_1 d\boldsymbol{r}_2 dt_1 dt_2 \tag{2}$$

If the electromagnetic field is composed of a group of monochromatic plane waves

$$\boldsymbol{E}(\boldsymbol{r},t) = \sum_i \boldsymbol{E}_i(\boldsymbol{k}_i,\omega_i) = \sum_i (\boldsymbol{E}_{0i} e^{i(\boldsymbol{k}_i\boldsymbol{r}-\omega_i t)} + c.c.) \tag{3}$$

then the polarization can likewise be expanded in plane waves

$$P^{(n)}(r,t) = \sum_j P_j^{(n)}(k_j,\omega_j) = \sum_j (P_{0j}^{(n)} e^{i(k_j \cdot r - \omega_j t)} + c.c.) \tag{4}$$

In steady-state, the different components can be expressed by

$$P^{(n)}(k,\omega) = \chi^{(n)}(k,\omega) \cdot E_1(k_1,\omega_1) E_2(k_2,\omega_2) \cdots E_n(k_n,\omega_n) \tag{5}$$

where the n'th order nonlinear susceptibility $\chi^{(n)}(\mathbf{k},\omega)$ is the Fourier transform of the corresponding response function $\chi^{(n)}(\mathbf{r}-\mathbf{r}_1,t-t_1,\ldots\mathbf{r}-\mathbf{r}_n,t-t_n)$. Furthermore, the only components that survive the space and time integrations are those that conserve energy (for times $t-t_i \gg 2\pi/\omega_i$) and momentum:

$$\omega = \sum_i \pm\omega_i \quad \wedge \quad k = \sum_i \pm k_i \tag{6}$$

where +/- are entered for absorbed/emitted waves, respectively. Due to the fixed, and generally nonlinear, dispersion relation $\omega = \omega(\mathbf{k})$ in most media, the simultaneous fulfillment of the two Eqs. (6) is very restrictive. It is only met under special conditions, called phase-matching, and then usually only for one particular polarization wave. Phase-matching insures that the interacting waves and the nonlinear polarization propagate with the same velocity through the medium, so that an appreciable energy transfer between waves can take place over a certain length.[1],[2]

For steady-state conditions, the form in Eq. (5) is very convenient. In the usual dipole approximation, or local response theory, the susceptibility is, furthermore, independent of wave vector, so that the medium is fully described by the frequency dependent nonlinear susceptibilities

$$\chi^{(n)}(k,\omega) = \chi^{(n)}(\omega) = \chi^{(n)}(\omega;\pm\omega_1,\pm\omega_2,\ldots,\omega_n)) \tag{7}$$

where in the last expression we have adopted the usual notation for nonlinear susceptibilities [1],[2], indicating the applied frequencies and the resulting frequency from Eq. (6).

In the dynamic situation of ultrashort light pulses, one might expect the time dependent response functions in Eq. (2) to be the most appropriate representation. However, except for the most extreme cases, the time variations in the field amplitudes are slow compared to the optical periods $2\pi/\omega_i$, so that the energy conservation in Eq. (6) is still valid. In this case it is convenient to work in a mixed frequency and time representation of the form:

$$P^{(n)}(k,\omega,t) = \chi^{(n)}(\omega,t) \cdot E_1(k_1,\omega_1) \ldots\ldots E_n(k_n,\omega_n) \tag{8}$$

where the normalized time dependencies of the field amplitudes $f_i(t_i)$ are included in a time dependent susceptibility:

$$\chi^{(n)}(\omega,t) = \int_{-\infty}^{0} \chi^{(n)}(\tau_1,\tau_2,\ldots,\tau_n) f_1(\tau_1)\ldots f_n(\tau_n) \cdot e^{i(\omega_1\tau_1+\ldots+\omega_n\tau_n)} d\tau_1 \ldots d\tau_n \tag{9}$$

with $\tau_i = t - t_i$.

Physically, the response function $\chi^{(n)}(\tau_1,\tau_2,\ldots,\tau_n)$ expresses the polarization in the medium at time t following a series of δ-function pulses at times $t_1, t_2, ..., t_n$. They have to be calculated quantum mechanically, knowing the electronic structure of the medium, e.g. a semiconductor. They will thus contain the resonance enhancements as well as the coherence and dephasing properties of the optical transitions in the semiconductor in question.

The nonlinear polarization induced by the driving fields (Eqs. (1) and (8)) will in turn serve as source for an electromagnetic wave, as described by the wave equation [2]

$$(\nabla^2 + \frac{\epsilon}{c^2}\frac{\partial^2}{\partial t^2})\, \boldsymbol{E}(\boldsymbol{r},t) = -\frac{4\pi}{c^2}\frac{\partial^2}{\partial t^2}\boldsymbol{P}_{NL}(\boldsymbol{r},t) \qquad (10)$$

where ε is the dielectric function and c is the velocity of light. The wave equation has to be solved self-consistently with Eq. (1) to find the total nonlinear field $\mathbf{E}_{NL}(r,t)$. For waves propagating in the z-direction and with slowly varying amplitude as discussed above, the wave equation takes the form

$$(\frac{\partial}{\partial z} \pm \frac{1}{v_g}\frac{\partial}{\partial t})E(z,t) = \mp \frac{2\pi\omega^2}{ikc^2} P_{NL}(z,t)\, e^{i(kz\mp\omega t)} \qquad (11)$$

where $v_g = \partial\omega/\partial k$ is the group velocity of the wave and the upper (lower) signs correspond to forward (backward) propagation.

II.A. Microscopic Theory of Nonlinear Optical Susceptibilities

The (nonlinear) interation between light and a material system described by a set of atomic levels, $|g\rangle, |e_1\rangle, |e_2\rangle$,....., is found by solving the Schrödinger equation, or the rate equation for the density matrix ρ, as desribed by Di Bartolo in this volume[14]

$$\frac{\partial \rho}{\partial t} = \frac{i}{\hbar}[\rho, H] - \gamma\rho \qquad (12)$$

with the Hamiltonian

$$\boldsymbol{H} = \boldsymbol{H_0} + \boldsymbol{\mu}\cdot\boldsymbol{E} \qquad (13)$$

where $\mathbf{H}_0$ describes the atomic system and $\boldsymbol{\mu}$ is the dipole operator. Damping is phenomenologically introduced via the coefficient γ which is in general a tensor, where the diagonal elements express the longitudinal relaxation rates ($1/T_1$) and the off-diagonal elements express the transverse relaxation rates ($1/T_2$).

To find the macroscopic average expectation value of polarization **P** one has to calculate the trace of the operator product $\mu\rho$, i.e.

$$\boldsymbol{P} = N\, Tr[\boldsymbol{\mu\rho}] \qquad (14)$$

where N is the number of atoms in the system.

If the system can be reduced to a two-level system with ground state $|g\rangle$ and one excited state $|e\rangle$, Eq. (12) reduces to the well known two-level Bloch equations

$$\frac{\partial \rho_D}{\partial t} = -\frac{2i\mu E}{\hbar}(\rho_{eg} - \rho_{ge}) - \frac{\rho_D - \rho_{D0}}{T_1}$$

$$\frac{\partial \rho_{eg}}{\partial t} = \frac{\partial \rho^*_{ge}}{\partial t} = \frac{i\mu E}{\hbar}\rho_D - (\frac{1}{T_2} + i\omega_0)\rho_{eg} \tag{15}$$

where $\rho_D = \rho_{gg} - \rho_{ee}$ is the population difference between ground state and excited state with thermal equilibrium value ρ_{D0}. The longitudinal relaxation time T_1 describes the population relaxation towards thermal equilibrium, and the transverse relaxation time, or dephasing time $T_2 \le 2T_1$, describes the decay of the polarization, given by

$$P_{NL}(t) = N\,Tr[\mu\rho] = N\mu(\rho_{ge}(t) + \rho_{eg}(t)) = N\mu(\rho_{ge}(t) + c.c.) \tag{16}$$

To solve Eq. (12) or Eq. (15), one often uses a perturbational approach by expanding the density matrix operator in series of the interaction Hamiltonian $\boldsymbol{\mu}\cdot\mathbf{E}$

$$\rho = \rho^{(0)} + \rho^{(1)} + \rho^{(2)} + \cdots \tag{17}$$

where the density matrix is found to increasing order by the equations

$$\frac{\partial \boldsymbol{\rho}^{(1)}}{\partial t} - \frac{i}{\hbar}[\boldsymbol{\rho}^{(1)},\boldsymbol{H_0}] + \boldsymbol{\gamma\rho}^{(1)} = [\boldsymbol{\mu}\cdot\boldsymbol{E},\boldsymbol{\rho}^{(0)}]$$

$$\frac{\partial \boldsymbol{\rho}^{(2)}}{\partial t} - \frac{i}{\hbar}[\boldsymbol{\rho}^{(2)},\boldsymbol{H_0}] + \boldsymbol{\gamma\rho}^{(1)} = [\boldsymbol{\mu}\cdot\boldsymbol{E},\boldsymbol{\rho}^{(1)}] \tag{18}$$

$$\vdots$$

Solving the perturbational Eqs. (18) to n'th order gives an expression for the n'th order nonlinear susceptibility that contains the essential features of the nonlinear optical properties to be expected in a situation where the light intensity is not too high and one is still far away from saturation, e.g. in the small area pulse limit [2]

$$\chi^{(n)}(\omega;\pm\omega_1,\pm\omega_2,\cdots\pm\omega_n) = \frac{N}{n!}\boldsymbol{P}_{12\cdots n}\sum_{i,j,k,\cdots,l}\frac{\rho_{ii}^{(0)}}{\hbar^n}$$

$$\times\frac{\mu_{ij}(\omega)\mu_{jk}(\omega_1)\cdots\mu_{li}(\omega_n)}{(\pm\omega_1\pm\omega_2\cdots\pm\omega_n-\omega_{ij})(\pm\omega_2\pm\omega_3\cdots\pm\omega_n-\omega_{jk})\cdots(\pm\omega_n-\omega_{li})} \tag{19}$$

which is in general a tensor of order n+1. In the expression in Eq. (19) damping has been neglected, and $\mathbf{P}_{12\ldots n}$ means additional summing over all permutations of the input electric fields with frequencies ω_1, ω_2, ...,ω_n. The sum over $|i\rangle$ only contains terms with an initial population, $\rho_{ii}^{(0)} \neq 0$, i.e. usually the ground state or lowest excited state. All the other indices are summed over all the states where the corresponding dipole matrix elements are nonzero ($\mu_{ij} \neq 0$). The most important features in Eq. (19) are the resonance denominators. In a resonant situation usually only the fully resonant terms are of importance, but it should be remembered that multiple resonances may occur and that in $\chi^{(n)}$ up to n-photon resonances may occur. Energy conservation is still required as in Eq. (6), and +/- are entered for absorbed/emitted waves. In the next sections we shall look a little closer on the different orders of nonlinearity.

II.B. Second Order Optical Nonlinearities

It is easily seen that in systems with inversion symmetry there can be no second order nonlinearity, i.e. $\chi^{(2)} \equiv 0$. However, in systems without inversion symmetry such as $LiNbO_3$ KDP, BBO, GaAs or any semiconductor surface, the lowest order optical nonlinearities will be of second order as expressed by

$$\boldsymbol{P}^{(2)}(\boldsymbol{k},\omega) = \chi^{(2)}(\omega;\omega_1,\omega_2)\cdot \boldsymbol{E}_1(\boldsymbol{k}_1,\omega_1)\boldsymbol{E}_2(\boldsymbol{k}_2,\omega_2) \tag{20}$$

where the input fields may come from the same beam or from two different coherent beams. From the energy conservation in Eq. (6), this term is responsible for sum and frequency generations as well as for parametric generation, where one intense pump wave with frequency ω creates signal (ω_1) and idler (ω_2). Parametric generation and amplification is treated by Abram in this volume [15], so I shall only briefly discuss sum and difference generation which is used very conveniently to transform a signal into another wavelength range where one may have more sensitive or faster detectors, up- or down-conversion. The usual problem in these situations is phase-matching in order to obtain a reasonably efficient conversion. In the completely degenerate case where $\omega_1 = \omega_2 = \omega'$ and $\mathbf{k}_1 = \mathbf{k}_2 = \mathbf{k}'$, the second order processes become second harmonic generation, $\omega = 2\omega'$ and optical rectification $\omega = 0$. For second harmonic generation phase-matching is again an issue that has to be addressed. It is not obtained in general because in disperive media usually $\mathbf{k}(2\omega') \neq 2\mathbf{k}'(\omega')$. The standard way to overcome this problem is to use birefringent crystals for second harmonic generation and let the fundamental and second harmonic waves propagate as ordinary and extraordinary waves, respectively.[2] Accurate phase-matching is then achieved by varying the direction of propagation with respect to the optical axis of the crystal, e.g. KDP. Harmonic generation is not usually done in semiconductors because one wants the crystal to be transparent for the second harmonic wave.

Second harmonic generation has become a standard technique for measuring the duration of ultrafast light pulses for which no standard electro-optic detector is fast enough. The technique is called optical autocorrelation and is sketched in Fig. 1. The laser beam with ultrafast pulses is split into two beams in a Michelson interferometer, where a variable optical delay is introduced in one branch. The two beams are focussed into the same spot on the nonlinear crystal and only the second harmonic generation is detected after appropriate filtering. This signal will show a maximum when the two pulses have maximum overlap in space and time. By scanning the delay, one obtains an autocorrelation trace, from which the pulse length can be determined if the pulse shape is known.[16]

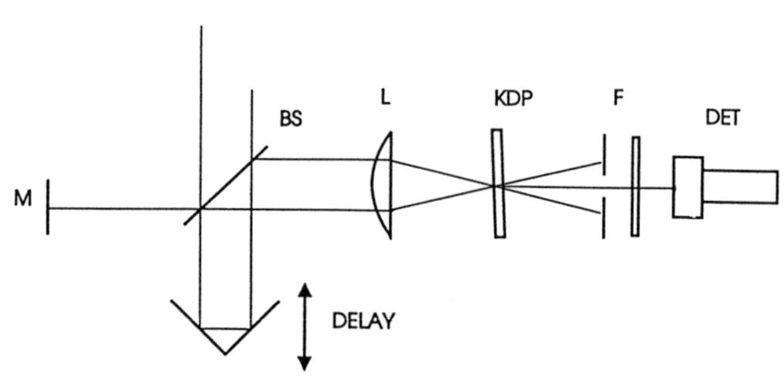

Fig. 1 Background free optical autocorrelation.

In Fig. 1 is shown background-free autocorrelation, where the two beams are incident on the nonlinear crystal under a mutual angle and the second harmonic is phase-matched and detected along the direction bisecting this angle. Hence, there is no signal when there is no temporal overlap between the two pulses. In this geometry, phase-matching can only be obtained over a short distance, but this is no problem for intense ultrashort laser pulses where too much material with dispersion is unwanted anyway, since it will inevitably broaden the pulse.

Optical rectification has also gained increased interest resently because of the possibility of generating very far infrared, i.e. megahertz and terahertz, radiation. Optical rectification follows from the DC polarization

$$\boldsymbol{P}^{(2)}(\omega \approx 0) = \chi^{(2)}(0;\omega,-\omega)\ |\boldsymbol{E}(\boldsymbol{k},\omega)|^2 \tag{21}$$

In an ultrafast laser with pulse length τ_L, the frequency components are spread over the range τ_L^{-1}. This will then also be the range of difference frequencies in the nonlinear polarization in Eq. (21) which may serve as an antenne for very far infrared radiation (FIR) according to the expression [2]

$$\boldsymbol{E}(\boldsymbol{r}) = \frac{1}{4\pi\epsilon c^2 r^3}\ \boldsymbol{r} \times \boldsymbol{r} \times \frac{\partial^2 \boldsymbol{P}}{\partial t^2} \tag{22}$$

where $\varepsilon = 1 + \chi$ is the dielectric function and the other symbols have their usual meaning.

There are three factors determining the efficiency of FIR emission from such a process: a) The spectral width of the laser pulse, b) phase-matching conditions around $\omega \approx 0$, and c) the dipole radiation efficiency. These factors are illustrated in Fig. 2 for pulses of 2 ps duration from a Nd^+-YAG laser propagating through a 1 mm thick crystal of $LiNbO_3$. The power spectrum is determined by the laser pulse, the phase-matching by the crystal orientation and the quadratic dependence of the radiation efficiency follows directly from Eq. (22). With the development of femtosecond lasers, terahertz emission from special antenna structures, or directly from semiconductor surfaces, have become possible, and terahertz spectroscopy is rapidly developing into a field of its own. I shall return to this subject in Sec. IV.

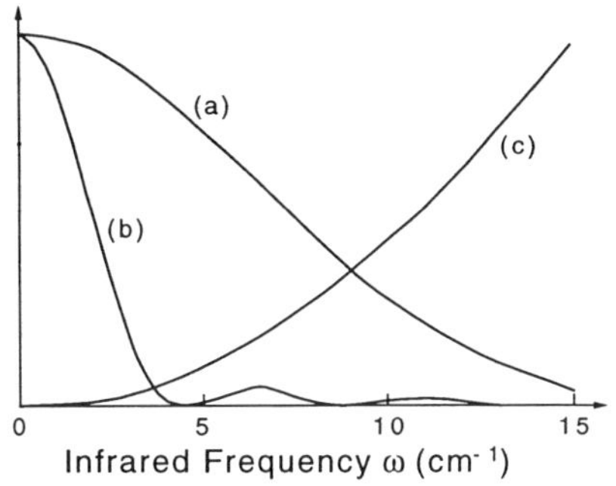

Fig. 2 Efficiency of FIR emission from optical rectification in $LiNbO_3$ (Y.R. Shen [2]).

II.C. Third Order Optical Nonlinearities

In materials with inversion symmetry ($\chi^{(2)} \equiv 0$) the lowest order nonlinearities are due to third order effects of the form

$$P^{(3)}(k,\omega) = \chi^{(3)}(\omega;\omega_1,\omega_2,\omega_3)\cdot E_1(k_1,\omega_1)E_2(k_2,\omega_2)E_3(k_3,\omega_3) \tag{23}$$

1. One Light Beam. The simplest case to consider is the one where only one monochromatic light beam is present in the medium. Then in Eq. (23), $\omega_1=\omega_2=\omega_3=\omega'$ and $\mathbf{k}_1=\mathbf{k}_2=\mathbf{k}_3=\mathbf{k}'$, which from Eq. (6) offers two possibilities: 1) Third harmonic generation with $\omega=3\omega'$ and $\mathbf{k}=3\mathbf{k}'$. In most media, however, phase-matching is difficult to achieve, since normally $\mathbf{k}(3\omega') \neq 3\mathbf{k}'$. Third harmonic generation will not be considered further in the present article. 2) Degenerate "four-wave-mixing" (DFWM) within the same beam with $\omega=\omega'$ and $\mathbf{k}=\mathbf{k}'$. This situation is always perfectly phase-matched and the nonlinear polarization

$$P^{(3)}(k,\omega) = \chi^{(3)}(\omega;\omega,-\omega,\omega)\ |E(k,\omega)|^2E(k,\omega) \tag{24}$$

expresses the situation of intensity dependent refractive index n(I) and absorption coefficient $\alpha(I)$. With $n(I) = n_0+ n_2I$ and $\alpha(I) = \alpha_0+ \alpha_2I$, we find n_2 and α_2 from the real and imaginary part of $\chi^{(3)}$

$$n_2 = \frac{Re\chi^{(3)}}{\epsilon_0 n_0^2 c} \quad \wedge \quad \alpha_2 = \frac{\omega Im\chi^{(3)}}{\epsilon_0 n_0^2 c^2} \tag{25}$$

The intensity dependent refractive index is responsible for self-focussing (or defocussing) and self-phase modulation. These phenomena are very important for solid state ultrafast lasers and for the propagation of intense light pulses (solitons) in optical fibers. The intensity dependent absorption coefficient describes saturable absorption or induced absorption, depending on the sign of α_2.

2. Two Light Beams. We shall limit ourselves to the degenerate case with $\omega_1=\omega_2=\omega$, $k_1=k_2=k$, and $\cos\theta = \mathbf{k}_1\cdot \mathbf{k}_2 / k^2 < 1$, as in Fig. 3. There are two possibilities with perfect phase-match: $\mathbf{k}=\mathbf{k}_1$ and $\mathbf{k}=\mathbf{k}_2$. As in the case with only one beam, the nonlinear signals are generated in a direction with a large linear signal and may therefore be difficult to detect at moderate intensities. There are also two possibilities with near phase-match: $\mathbf{k} = 2\mathbf{k}_2-\mathbf{k}_1-\Delta\mathbf{k}$ and $\mathbf{k}'= 2\mathbf{k}_1-\mathbf{k}_2-\Delta\mathbf{k}$, where $\Delta\mathbf{k}$ is the wave vector mismatch perpendicular to the sample plane. This geometry (see Fig. 3) has the advantage that the nonlinear signal is a collimated beam generated in a direction where there is no linear signal. It is therefore very well suited for the detection of even very small nonlinear signals. The linear background can to a high degree be eliminated by simple spatial filtering.

This DFWM can also be viewed upon as a case of light induced gratings. [11], [17],[18] The two incident beams set up stationary polarization gratings, in the nonlinear medium, with scattering vectors $\pm(\mathbf{k}_2-\mathbf{k}_1)$ and a grating constant $\Lambda=2\pi/|\mathbf{k}_2-\mathbf{k}_1|$. These gratings in turn scatter the incident beams $\mathbf{k}_1$ and $\mathbf{k}_2$ into the directions $\mathbf{k}'$ and $\mathbf{k}$, respectively, as also discussed above (see Fig. 3).

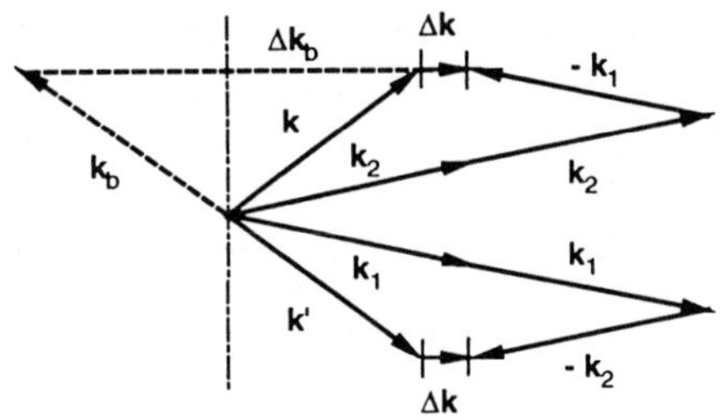

Fig. 3 Wave vector conservation for DFWM with three beams $\mathbf{k}_1$, $\mathbf{k}_2$, $\mathbf{k}_3$.

The signal intensity in the scattered direction is [11],[18]

$$I_s \propto d^2 \frac{\sin^2(\frac{\Delta k \cdot d}{2})}{(\frac{\Delta k \cdot d}{2})^2} \tag{26}$$

where d is the sample thickness, or nonlinear interaction length in general. Near-phase-match requires $\Delta kd \ll 1$, i.e. thin sample geometry. If the sample is very thin ($kd \ll 1$), the grating is essentially two-dimensional and back scattering (with $\Delta k_b \approx 2k$) will occur with about the same intensity as in the forward direction (see Fig. 3). Note, however, that in both directions the signal will be weak, because $I_s \propto d^2$. Therefore, strong resonance enhancement will normally be necessary to observe a back scattered signal. On the other hand, it does open up the possibility to observe the nonlinear interaction in a resonance where strong linear absorption prevents transmission even through a thin sample [11],[19].

3. Three light beams. With three incident light beams, more freedom is at hand to perform different types of four-wave mixing experiments. One advantage is, that two coherent light beams at an angle is sufficient to set up a light-induced grating in the medium, which can then be probed independently by the third beam that need no coherent relation to the first two beams. One can thus set up a nonlinear grating by exciting

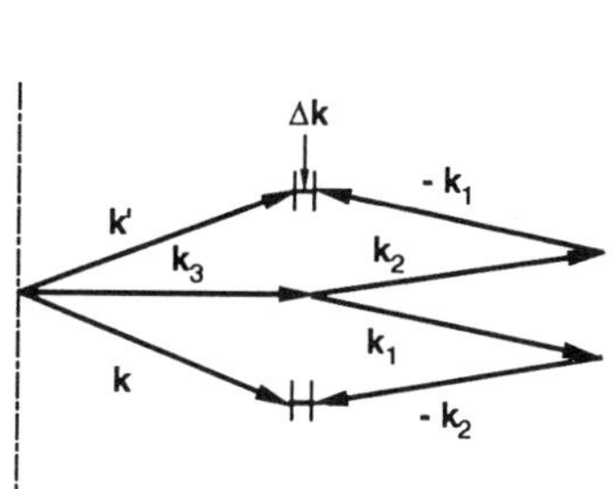

Fig. 4 Wave vector conservation for DFWM with two light beams, $\mathbf{k}_1$,$\mathbf{k}_2$.

resonantly with the two beams and then probe the grating in the transparent region. In the degenerate case $\omega_1=\omega_2=\omega_3$, and with three different beam directions, $\mathbf{k}_1$, $\mathbf{k}_2$ and $\mathbf{k}_3$, a certain phase mismatch $\Delta\mathbf{k}$ again requires thin samples, $d \ll 1/\Delta k$, and gives rise to first order scattering ($\chi^{(3)}$) in the three directions $\mathbf{k}=\mathbf{k}_1-\mathbf{k}_2+\mathbf{k}_3-\Delta\mathbf{k}$, $\mathbf{k}'=-\mathbf{k}_1+\mathbf{k}_2+\mathbf{k}_3-\Delta\mathbf{k}$ and $\mathbf{k}''=\mathbf{k}_1+\mathbf{k}_2+\mathbf{k}_3-\Delta\mathbf{k}$. These scattering directions are shown in Fig. 4, where accidentally $\mathbf{k}''=\mathbf{k}_3$, because $\mathbf{k}_3$ is bisecting the angle between $\mathbf{k}_1$ and $\mathbf{k}_2$. Even in the degenerate situation, there are some differences between four-wave mixing with two beams and with three beams, particularly considering dynamic effects as shall be discussed in the next section.

4. Measuring Nonlinear Coefficients. It is of course only the coherent contribution to the nonlinear signal that is well described by a nonlinear susceptibility as in Eq. (5). When using DFWM experiments to measure the magnitude (in resonance) of the nonlinearity, it is therefore essential to identify the coherent contribution as for example in a transient experiment with a time resolution better than the dephasing time of the resonance. In a resonant c.w. experiment, the incoherent contribution integrates up over the grating lifetime, and may thus exceed by several orders of magnitude the coherent contribution.

The intensity I_s of the coherent self-diffraction in a two-beam DFWM experiment, with the input intensities I_1 and I_2, can be calculated in steady-state as [18]

$$I_s = \frac{1}{4} I_1 I_2^2 e^{-\alpha d} \frac{|K|^2 l_a^2}{(1 + K_2 I_2 l_a)^3} \tag{27}$$

where $K = K_1 + iK_2 = \omega\chi^{(3)}/\varepsilon_0 n^2 c^2$ is the third order nonlinear coupling coefficient and $l_a=(1-e^{-\alpha d})/\alpha$ is the nonlinear interaction length, as determined by the linear absorption length $1/\alpha$ or the sample thickness, whichever is the smaller. It is here assumed that $l_a \ll 1/\Delta k$, where Δk is the wave vector mismatch. If nonlinear absorption of the pump beam is negligible compared to linear absorption ($K_2 I_2 l_a \ll 1$), a very simple expression for the nonlinear susceptibility is obtained in terms of the ratio between the transmitted signal and pump intensities $R_s \equiv I_s/I_2 e^{-\alpha d}$:

$$|\chi^{(3)}(\hbar\omega)|^2 = \frac{4\epsilon_0^2 n^2 c^4}{\omega^2 I_1 I_2 l_a^2} R_s(\hbar\omega) \tag{28}$$

Thus, knowing the input intensities and the linear absorption coefficient, a good estimate of $\chi^{(3)}$ can be obtained from measuring the ratio R_s [7],[18],[20].

II.D. Pulsed Degenerate Four-Wave Mixing

Performing degenerate four-wave mixing (DFWM) with ultrashort laser pulses, one can obtain not only spectral information about the nonlinear coefficients, but also dynamical information about the optical excitations. The type of information obtained depends on the actual experimental configuration, as well as character of the samples investigated.

1. Two-Beam DFWM. This situation is sketched in Fig. 5a. The two incident laser pulses are split off the same laser pulse (pulse length τ_L), and are impinging on the sample with a variable optical delay between them. In order for the two laser pulses to interact coherently in the nonlinear medium, for example by setting up a polarization grating, the delay between them should not exceed the dephasing time of the nonlinear polarization in the medium, caused by the first laser pulse [11]. The nonlinear DFWM signal is then self-diffraction of the second pulse in the grating set up by the coherent overlap between the polarizations from the first and the second pulse. For pulse #1 arriving first ($\tau_{12}>0$) as in Fig. 5a, a signal will thus be emitted in the direction $2\mathbf{k}_2-\mathbf{k}_1$ as indicated. For pulse #2 arriving first ($\tau_{12}<0$), the signal is emitted in the direction $2\mathbf{k}_1-\mathbf{k}_2$ (dashed in Fig.5a).

This result is obtained by solving the two-level Bloch equations in Eq. (15) to third order, which for δ-pulses yields [21],[22]

$$P^{(3)}(r,t,\tau_{12}>0) \propto \Theta(t)\mu_{eg}^4\, e^{i[(2k_2-k_1)r-\Omega_{eg}(t-\tau_{12})]} \tag{29}$$

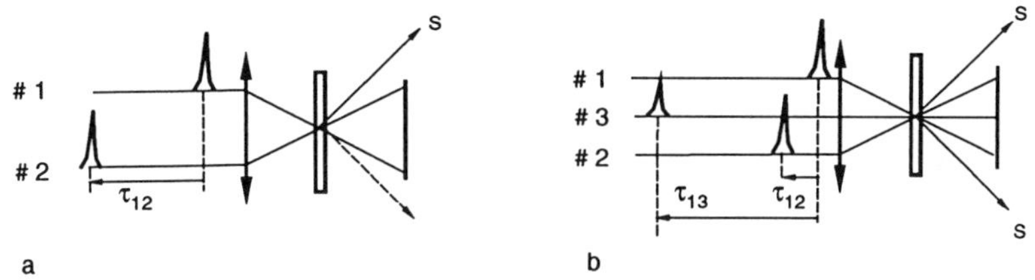

Fig. 5 Transient DFWM with two beams (a) and with three beams (b), respectively.

where t is the time after the arrival of the second pulse, Θ(t) is the Heaviside step function and $\Omega_{eg}= \omega_{eg}- \gamma_{eg}$ is the complex transition frequency with the damping $\gamma_{eg}= 1/T_2 + 2/T_1$. The intensity of the DFWM signal in the direction $2\mathbf{k}_2-\mathbf{k}_1$ is then

$$I_{DFWM} \propto |P^{(3)}(t,\tau_{12}>0)|^2 \propto \Theta(t)\mu_{eg}^8\, e^{-2\gamma_{eg}(t-\tau_{12})} \tag{30}$$

If the signal is spectrally resolved and integrated with a slow detector, it corresponds to detecting the absolute square of the Fourier transform of Eq. (29), which is

$$I_{DFWM} \propto |P^{(3)}(\omega,\tau_{12}>0)|^2 \propto \frac{\mu_{eg}^8}{|\Omega_{eg} - \omega|^2}\, e^{-2\gamma_{eg}\tau_{12}} \tag{31}$$

where ω is the optical frequency of the signal detected. These correlation traces of the time-integrated DFWM signal contain information about the dephasing rate $\gamma_{eg}= 1/T_2+ 2/T_1$, as well as the resonance enhancements of the nonlinear signal. Fig. 6 shows a calculation of the so-called free polarization decays in Eq. (30) for different delays τ_{12} and of the corresponding Fourier transform in Eq. (31) for a relevant case with picosecond dephasing times (excitons in semiconductors) and a finite pulse durations. [11]

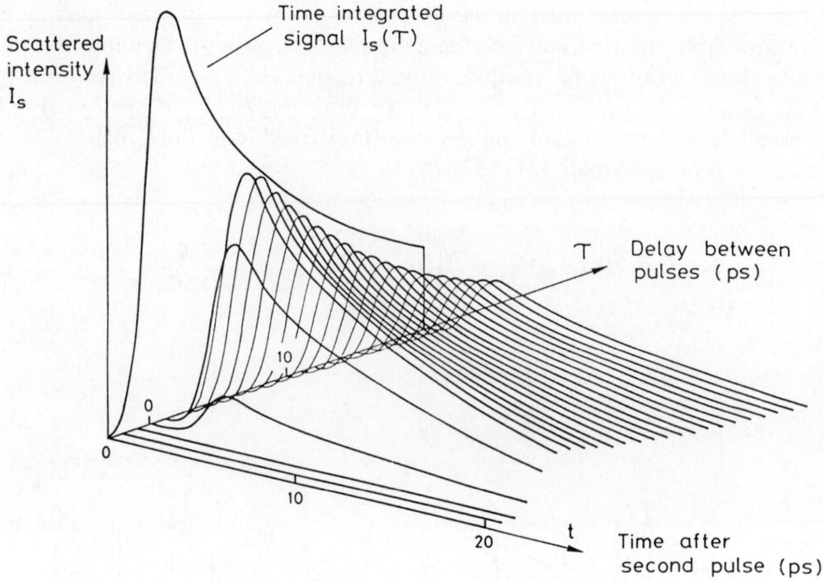

Fig. 6 Free polarization decays for different delays.

If the two-level system is inhomogeneously broadened, an integration is done over the distribution of transition frequencies, e.g. $g(\omega_{eg}) \propto \exp\{-(\omega_{eg}-\omega_0)^2/(\Delta\omega_{eg})^2\}$

$$P^{(3)}(t) = N\cdot\int_0^\infty \mu_{eg}\rho_{eg}^{(3)}(t)g(\omega_{eg})d\omega_{eg} \tag{32}$$

In this case, destructive and constructive interferences between the different transition frequencies play a significant role, and the signal in the $2\mathbf{k}_2$-$\mathbf{k}_1$ direction appears as a photon echo at the time $t = \tau$ as shown in Fig. 7. The time integrated Fourier transformed photon echo then takes the form [11]

$$I_{PE} \propto |P^{(3)}(\omega,\tau_{12}>0)|^2 \propto \mu_{eg}^8\, e^{-\frac{2|\Omega_{eg}-\omega|^2}{(\Delta\omega_{eg})^2}}\, e^{-4\gamma_{eg}\tau_{12}} \tag{33}$$

which for the same homogeneous damping decays twice as fast as the free polarization decay in Eq. (31). From a comparison of the observed DFWM decay with the linewidth, as observed in absorption or emission, it can be decided whether the transition is homogeneously or inhomogeneously broadened. Knowing that, the dephasing time, and thereby the homogeneous linewidth, can be determined from either Eq. (31) or Eq. (33).

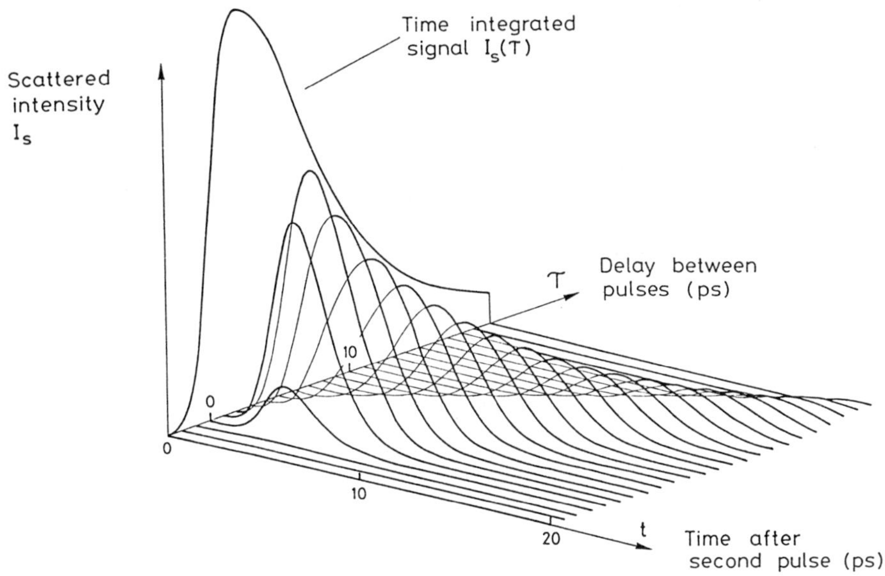

Fig. 7 Photon echoes for an inhomogeneously broadened 2-level system.

2. Three-Beam DFWM. This situation is sketched in Fig. 6b. The two first pulses arrive simultaneously, or well within the dephasing time of the material ($\tau_{12} \ll T_2$), and interfere coherently to set up a nonlinear grating in the medium. This grating can then be detected at variable time delays, τ_{13}, by diffraction of the third pulse. If also $\tau_{13} \ll T_2$, then pulse #3 will diffract off a coherent polarization grating set up by pulse #1 and pulse #2, as in the self-diffraction case above. If, however, ω is in resonance with an electronic excitation in the material a real excitation density grating may persist in the material long after the coherent polarization grating has disappeared by dephasing. This grating, however still bears the fingerprint of the coherent overlap between the two first pulses, and can even give rise to a photon echo signal, stimulated by the third pulse, but dependent on the delay τ_{12}. If, on the other hand, the scattered signal is recorded as a function of delay τ_{13} of the third pulse, the decay of the incoherent excitation population is being monitored. This type of experiments is therefore well suited to separate the purely coherent contribution to the optical nonlinearities from the more long-lived incoherent contributions from a high density of excited carriers in the medium.

The lifetime τ_G of the incoherent excitation density grating is determined by the lifetime τ_R of the excited carriers as well as by carrier diffusion since the latter will wash out the spatial modulation of the carrier density.[17],[23]

$$\tau_G^{-1} = \tau_R^{-1} + \frac{4\pi^2 D}{\Lambda} \qquad (34)$$

where D is the carrier diffusion coefficient and Λ is the grating constant, as determined by the wavelength λ of the exciting light and the angle θ between the two interfering beams ($\mathbf{k}_1$ and $\mathbf{k}_2$)

$$\Lambda = \frac{2\pi}{|k_1 - k_2|} = \frac{\lambda}{2\sin(\frac{\theta}{2})} \tag{35}$$

The grating lifetime is determined from the decay of the integrated intensity of the scattered test signal as a function of the delay τ_{13} of the test pulse (#3). Hence, the carrier lifetime and the diffusion coefficient can be determined separately from Eq. (34) by performing such transient grating experiments at different angles θ.

II.E. Quantum Beat Spectroscopy

Linear quantum beat spectroscopy (QBS) was developed already in the sixties as a high resolution spectroscopy operating in the time domain.[24] When two or more close-lying atomic levels are resonantly excited by the same coherent light pulse, the atom is prepared in a mixed state that is a coherent superposition of pure states. This will give rise to an interference term in the wave function that will manifest itself, e.g. by an oscillation of the fluorescence decay observed after the short-pulse excitation.

The situation is illustrated by the nearly degenerate two-level system in Fig. 8, where a simultaneous excitation from the ground state $|g\rangle$ to the close-lying excited levels $|e_1\rangle$ and $|e_2\rangle$ is provided by the same coherent laser pulse. The total Hamiltonian for the system is

$$H = \hbar\omega_0 |g\rangle\langle g| + \sum_i \hbar\omega_i |e_i\rangle\langle e_i| + \sum_k \hbar\omega(k) a_k^+ a_k$$
$$+ \hbar \sum_{k,\, i} [g_{ik} a_k^+ |g\rangle\langle e_i| + adj.] \tag{36}$$

where ħ is Planck's constant, $\hbar\omega_j$ is the energy of the atomic state $|j\rangle$, $\hbar\omega(k)$ is the energy of the electromagnetic mode k, and a_k^+ and a_k are the corresponding creation and annihilation operators, respectively. The light-atom dipole interaction is specified by the coupling coefficients [25]

$$g_{ik} = -\frac{\mu_{i0}\mathcal{E}_k}{2\hbar} U_k \tag{37}$$

where μ_{i0} is the transition dipole moment from the ground state $|g\rangle = 0$ to the atomic level $|e_i\rangle$ and $\mathcal{E}_k$ is the electric field 'per photon' of the surrounding electromagnetic mode k, characterized by the mode function U_k.

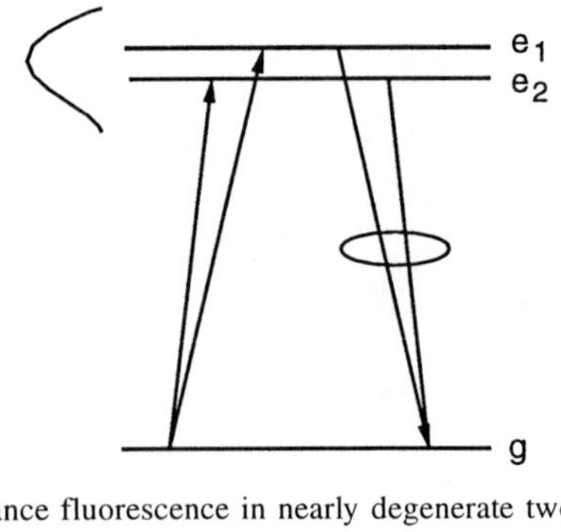

Fig.8 Resonance fluorescence in nearly degenerate two-level system.

The electric field operator for a single mode is thus

$$E_k(r, t) = E_k^+(r, t) + E_k^-(r, t) = \mathscr{E}_k\, [a(t) + a^+(t)]\, U_k(r) \tag{38}$$

where E^+ and E^- refer to the positive and negative frequency components of the field operator, respectively. After excitation with a short pulse laser, the system is left in a coherent superposition of excited states $|e_i\{0\}\rangle$ with dipole-allowed transitions from the ground state and no photons in any of the electromagnetic modes, i.e. with an initial wavefunction $|\psi(t)\rangle$

$$|\psi(0)\rangle = \sum_i c_{io}(0)\, |e_i\{0\}\rangle \tag{39}$$

where the coefficients c_{i0} contain the dipole moments as well as the integrated field strength of the laser pulse. At a later time t, one of the excited atomic states may have returned to the ground state and a photon emitted in mode k, as described by the wavefunction

$$|\psi(t)\rangle = \sum_i c_{i0}(t)\, e^{-i\omega_i t}\, |e_i\{0\}\rangle + \sum_k c_{gk}(t)\, e^{-i\omega_k t}\, |g\, \{1_k\}\rangle \tag{40}$$

with $\omega_k = \omega_0 + \omega(k)$.

The rate equations for the time dependent coefficients in Eq. (40) are obtained from the Schrödinger equation, and following the standard Wigner-Weisskopf approximations for the theory of spontaneous emision, one finds the solutions[25]

$$c_{i0}(t) = c_{i0}(0)\, e^{-\gamma_{i0} t} \tag{41}$$

and

$$c_{gk}(t) = \sum_{i=1}^{2} c_{i0}(0) \frac{i g_{ik}}{\gamma_{i0} + i\delta_{ik}} [e^{-(\gamma_{i0} + i\delta_{ik})t} - 1] \tag{42}$$

where $\delta_{ik} = \omega_i - \omega_0 - \omega(k) \equiv \omega_{i0} - \omega(k)$, and each level is charaterized by a single decay rate γ specifying the dephasing time $T_2 = 1/\gamma$ as well as the lifetime $T_1 = 1/2\gamma$.

The intensity of spontaneous emission, on the detector at position r is expressed by $I(t) = \langle\psi(t)|E^-(r)E^+(r)|\psi(t)\rangle$ which, with the state vector of Eq.(40) transforms into[25]

$$I(t) = |\sum_k c_{gk}{*}(t)\, e^{i\, k\cdot r}|^2 = \tag{43}$$

$$\frac{\Theta(t')}{r^2c^2}[\sum_{i=1}^{2} |g_{ik}\omega_{i0}c_{i0}(0)|^2 e^{i2\gamma_{i0}t'} + (g_{1k}g_{2k}^*\omega_{10}\omega_{20}c_{10}^*(0)c_{20}(0)e^{-(\gamma_{10}+\gamma_{20})t'}e^{-i\omega_{12}t'} + c.c.)]$$

where in the last expression only two excited states i=1,2 are considered and the retarded time $t' = t - r/c$ at the detector position has been introduced. $\Theta(t')$ is the Heaviside step function. The first two terms in the square brackets of Eq. (43) represent the exponential decays of the two excited states with decay constants γ_{10} and γ_{20} and the two last terms reveal the interference between the two excited states, the so-called quantum beat.

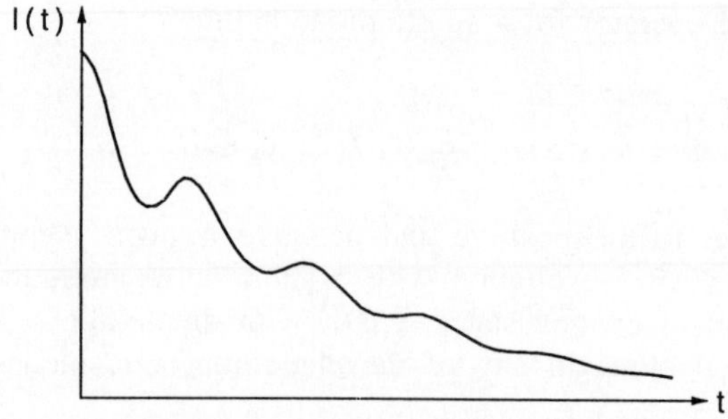

Fig. 9 Resonance fluorescence decay with quantum beats as in Eq. (44).

With coupling and expansion coefficients that are real, the fluorescence decay in Eq. (43) is expressed by

$$I(t) = \Theta(t)\, I_0\, [\mu_{10}^4 e^{-2\gamma_{10}t} + \mu_{20}^4 e^{-2\gamma_{20}t} + 2\mu_{10}^2\mu_{20}^2 e^{-(\gamma_{10}+\gamma_{20})t}\cos\omega_{21}t\,] \tag{44}$$

where I_0 is proportional to the integrated intensity of the exciting laser pulse arriving at t = 0. This decay is illustrated in Fig. 9.

From Eq. (44), it is seen that the overall resonance fluorescence decay is dominated by the level with the strongest oscillator strength and the smallest decay rate or damping. The interference term, however, is strongest when both oscillator strengths are large, and the damping of the modulations is governed by the fastest dephasing rate. The modulation frequency directly reveals the energy splitting between the excited states

$$\omega_{21} = \frac{E_2 - E_1}{\hbar} \tag{45}$$

Thus, in systems where the dephasing time is long enough, so that several oscillation periods can be observed in the coherent fluorescence decay, this is a very precise way to measure energy splittings of nearly degenerate systems in the time domain. At the same time, the coherence and dephasing time of the system can be measured in a simple linear fluorescence experiment.

The quantum beat modulation, as described here, is a very general quantum mechanical interference phenomenon, that is related to the fact, that in the present experiment there is no detailed knowledge of the path of the excitation-decay cycle. The system starts in the ground state and ends there again via two or more possible excited states, but we do not know which way it went. If the laser is only exciting one state resonantly, or if the emission is beeing spectrally resolved to better than the energy splitting between the excited states, we know which way, and the quantum beats dissapear. This is just like in any other interference phenomenon, as e.g. the double slit experiment.

It is interesting to note the detailed conditions for observing quantum beats as described above. The final state for the fluorescence decay may be an excited state without destroying the quantum beats, i.e. both elastic and inelastic scattering (fluorescence) is possible. However, quantum beats between different final states cannot be observed, because is this case it can be checked in which state the system is left after the decay. Again, only indistinguisable pathways give rise to interference, as illustrated in Fig. 10.

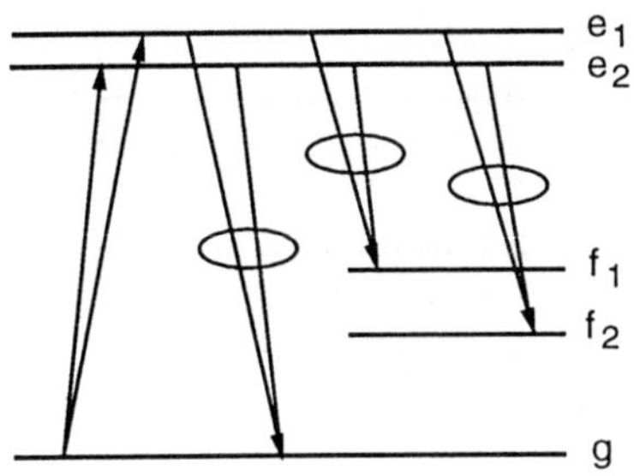

Fig. 10 Quantum beats from elastic and inelastic scattering. No beats from split final levels.

As already mentioned, the quantum beat spectroscopy was first used on independent single atom systems where the dephasing times are in the microsecond or even millisecond range allowing for coherent transient measurements without the need of ultrafast laser sources and detectors. Correspondingly, the energy splittings that can be investigated are in the megahertz to gigahertz range, i.e. ultrahigh spectral resolution from an optical point of view. Quantum beat spectroscopy is typically used to measure hyperfine splittings and Zeeman effects in atomic excitations in various environments. Since such nearly degenerate atomic levels can be rather complex with a number of optically active levels, it may be necessary to Fourier transform the fluorescence decay before an identification of the various beat levels can take place. Figure 11 shows an example of such a complex quantum beat signal and the corresponding Fourier transform for the hyperfine structure of the $7P_{3/2}$ level of Cs vapor excited from the F = 3 hyperfine ground state by a pulsed dye laser.[26] Figure 11a shows the fluorescence decays, $I_{\parallel}$ and $I_{\perp}$, with polarization vector parallel and perpendicular to the polarization direction of the incident laser, respectively. The two signals contain the same beat structure, however, with opposite phase. The Fourier transform of the difference signal $I_{\parallel} - I_{\perp}$ is shown in Fig. 11b and reveals the quantum beats between the F = 2, 3 and 4 hyperfine excited states, ν_{32}, ν_{43} and $\nu_{42} = \nu_{43} + \nu_{32}$.

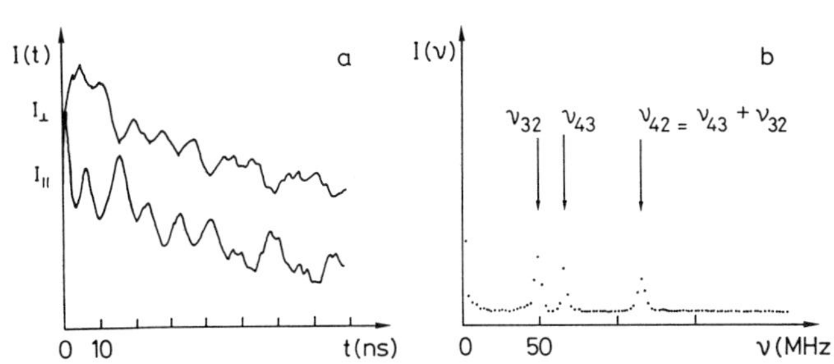

Fig. 11 Quantum beat signals (a) and Fourier spectrum (b) in fluorescence from the $7P_{3/2}$ level of Cs, excited from the F=3 hyperfine ground state (Ref. [26]).

1. Nonlinear Quantum Beats. Interference and beat phenomena may also appear in nonlinear optical signals and spectroscopies. The DFWM decays, or correlation traces, of Eqs. (31) and (33) assumed the sytems to be two-level like. If, on the other hand, there is a multiplicity of nearly degenerate levels, the decay of the nonlinear signal will be affected much the same way as in resonance fluorescence, described above. However, the quantum interferences, or beats, will in this case appear in the correlation traces, as indicated in Fig. 12. The first pulse (E_1) arrives at the sample at t = 0 and sets up a (linear) polarization that will decay according to the dephasing time and show interference modulations in the case of nearly degenerate coherently excited states (Fig. 9). This free polarization decay is responsible for the fluorescence decay as well as for the nonlinear interaction with the second pulse (E_2) arriving at $t = \tau$. It is intuitively obvious that the integrated intensity of

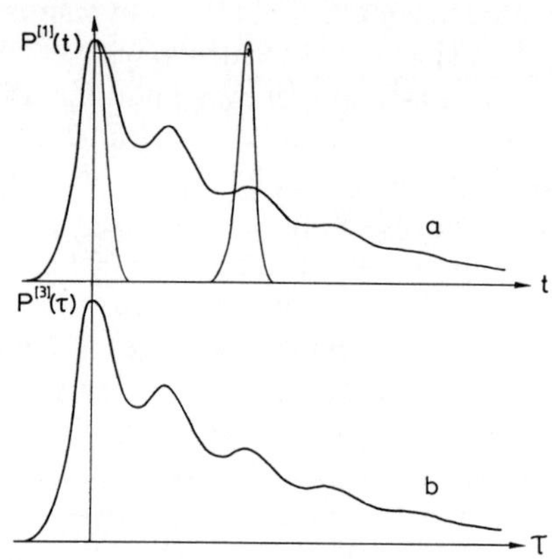

Fig. 12 Quantum beats in the linear polarization decay (a) and in the correlation trace of the nonlinear signal.

the nonlinear signal will carry over the beat modulation into the correlation trace of the nlinear signal as shown in Fig. 12. It can, however, be calculated by solving to third order, as in Sec. II.D., the Bloch equations for the three-level system shown in Fig. 8.

Assuming the complex frequencies of the three-level system

$$\begin{aligned} \Omega_{10} &= \omega_{10} - i\gamma_{10} \\ \Omega_{20} &= \omega_{20} - i\gamma_{20} \\ \Omega_{21} &= \omega_{21} \end{aligned} \tag{46}$$

the Bloch equations take the form

$$\frac{\partial \rho_{10}}{\partial t} + i\Omega_{10}\rho_{10} = \frac{iE}{\hbar}[\mu_{10}(\rho_{00}-\rho_{11}) - \mu_{20}\rho_{21}^{*}] \qquad (47)$$

$$\frac{\partial \rho_{20}}{\partial t} + i\Omega_{20}\rho_{20} = \frac{iE}{\hbar}[\mu_{20}(\rho_{00}-\rho_{22}) + \mu_{10}\rho_{21}] \qquad (48)$$

$$\frac{\partial \rho_{21}}{\partial t} + i\Omega_{21}\rho_{21} = \frac{iE}{\hbar}[\mu_{10}\rho_{20} - \mu_{20}\rho_{10}^{*}] \qquad (49)$$

$$\frac{\partial \rho_{11}}{\partial t} = -\frac{iE}{\hbar}\mu_{10}[\rho_{10} - \rho_{10}^{*}] \qquad (50)$$

$$\frac{\partial \rho_{22}}{\partial t} = -\frac{iE}{\hbar}\mu_{20}[\rho_{20} - \rho_{20}^{*}] \qquad (51)$$

$$\rho_{00} = N - \rho_{11} - \rho_{22} \qquad (52)$$

where **E** is the electric field, N is the density of three-level atoms and the other symbols have their usual meaning.

Solving to third order using a Greens functions technique, one obtains an expression for the third order nonlinear polarization, propagating in the $2\mathbf{k}_2 - \mathbf{k}_1$ direction[22]

$$P^{(3)}(t,\tau>0) \propto \Theta(t)[2\mu_{10}^4 e^{-i\Omega_{10}(t-\tau)} + 2\mu_{20}^4 e^{-i\Omega_{20}(t-\tau)} + \mu_{10}^2\mu_{20}^2(e^{-i\Omega_{10}(t)+i\Omega_{20}\tau} + e^{-i\Omega_{20}(t)+i\Omega_{10}\tau})] \qquad (53)$$

where again t is the time elapsed after the arrival of the second pulse.

The intensity of the DFWM signal is proportional to the absolute square of the nonlinear polarization. Hence, assuming equal dipole matrix elements of the two transitions, the following expression is obtained

$$I_{FWM} \propto 5(e^{-2\gamma_{10}t} + e^{-2\gamma_{20}t}) + 2e^{-(\gamma_{10}+\gamma_{20})t}\cos(\omega_{21}(t+\tau)) + 4(e^{-2\gamma_{10}t} + e^{-2\gamma_{20}t})\cos(\omega_{21}\tau) + 8e^{-(\gamma_{10}+\gamma_{20})t}\cos(\omega_{21}t) + 8e^{-(\gamma_{10}+\gamma_{20})t}\cos(\omega_{21}(t-\tau)) \qquad (54)$$

The oscillatory behaviour of the signal as a function of real time t, as well as a function of the delay time τ, is clearly revealed as shown in Fig. 13. Notice that also as a function of $t-\tau$, well defined oscillations with the beat frequency ω_{21} can be observed.

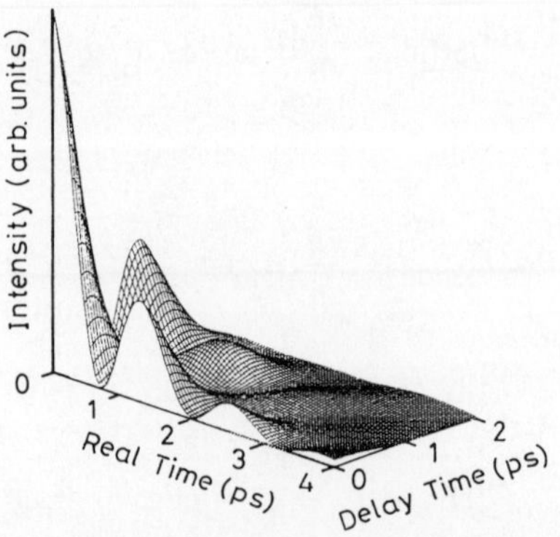

Fig. 13 DFWM signal from three level system featuring quantum beats.

In an experiment where real time resolution is not achieved, the maximum information is obtained by spectrally resolving the time integrated nonlinear signal. In this case the measured polarization is the Fourier transform of Eq. (53)

$$\tilde{P}^{(3)}(\omega,\tau>0) \propto \left[\frac{2\mu_{10}^4 e^{i\Omega_{10}^*\tau} + \mu_{10}^2\mu_{20}^2 e^{i\Omega_{20}^*\tau}}{\Omega_{10} - \omega} + \frac{2\mu_{20}^4 e^{i\Omega_{20}^*\tau} + \mu_{10}^2\mu_{20}^2 e^{i\Omega_{10}^*\tau}}{\Omega_{20} - \omega}\right] \tag{55}$$

The two terms illustrate the resonance enhancement of the signal near either of the involved resonances, and the quantum beats appear in each of the terms. Detecting the signal near the first resonance, the second term can be neglected, and in this case the nonlinear signal can be expressed

$$I_{FWM} \propto \frac{4\mu_{10}^8 e^{-2\gamma_{10}\tau} + \mu_{10}^4\mu_{20}^4 e^{-\gamma_{20}\tau} + 4\mu_{10}^6\mu_{20}^2 e^{-(\gamma_{10}+\gamma_{20})\tau}\cos(\omega_{21}\tau)}{(\omega_{10} - \omega)^2 + \gamma_{10}^2} \tag{56}$$

revealing the modulation intuitively argued above and shown in Fig. 12. Assuming again equal dipole matrix elements of the transitions and distinctely different dampings of the two levels involved in the beating, e.g. $\gamma_{20} \gg \gamma_{10}$, a simpler expression is obtained

$$I_{FWM} \propto \frac{e^{-2\gamma_{10}\tau}[1 + e^{-\gamma_{20}\tau}\cos(\omega_{21}\tau)]}{(\omega_{10} - \omega)^2 + \gamma_{10}^2} \tag{57}$$

showing that the average signal decays with slowest damping rate, whereas the decay of the beat modulation decays with the fastest damping rate. Both damping rates can thus be determined from such a nonlinear quantum beat experiment.

In the above is assumed a homogeneously broadened system with a free polarization decay, but a similar modulation of the correlation trace is obtained with inhomogeneous broadening, i.e. in a photon echo experiment, as first shown by Lampert, Compaan and Abella.[27] In the early days, linear as well as nonlinear quantum beat spectroscopy was applied to atomic systems in various environments, e.g. to determine the hyperfine splitting of different atomic states.[28] Figure 14 shows an example of the photon-echo modulation pattern of the $^3P_0$-$^3H_4$ transition of a crystal of $LaF_3:Pr^{3+}$ (0.03 at.%).[29] The upper trace is the experiment and the lower trace shows how well a theoretical calculation reproduces the complicated beat pattern. The actual level splittings and dampings are revealed by Fourier transforming the correlation trace, as shown in Fig. 15. The observed beat frequencies correspond to the level splitting of the $^3H_4$ ground state in the Pr^{3+} ions. The deduced linewidths of the nuclear transitions are of the order of 200 kHz

If, however, the observed modulations are due to classical polarization interferences occuring in the detector, then they will be smeared out by inhomogeneous broadening. In a homogeneously broadened system it may be difficult to distinguish between true quantum beats, e.g. from a three-level system, and polarization interferences from independent two-level systems.

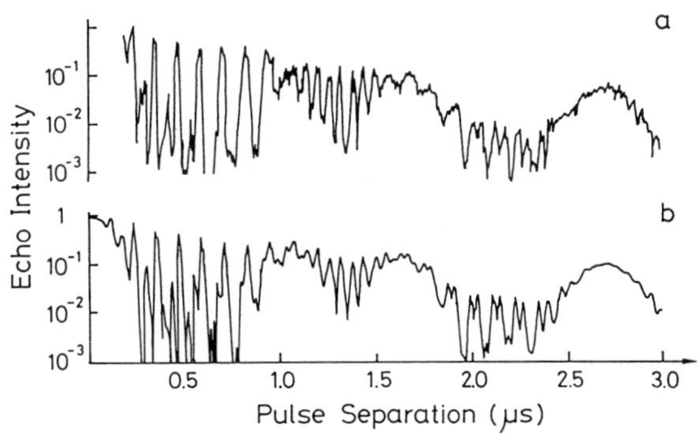

Fig. 14 Experimental (a) and theoretical (b) photon echo modulation pattern in $LaF_3:Pr^{3+}$ (Ref. [29]).

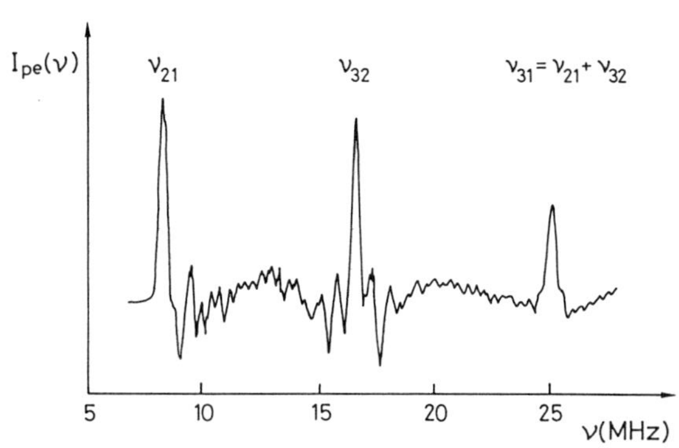

Fig. 15 Fourier spectrum of the echo modulation in Fig. 14 (Ref. [29]).

For two independent two-level systems with the complex transition frequencies $\Omega_{j0} = \omega_{j0} - i\gamma_{j0}$, the solution is straight forward. For $\tau>0$ (no signal for $\tau<0$), the total third order polarization has the form [22]

$$P^{(3)}(t,\tau>0) = \Theta(t)[M_{10}^4 N_1 e^{-i\Omega_{10}(t-\tau)} + M_{20}^4 N_2 e^{-i\Omega_{20}(t-\tau)}] \tag{58}$$

where t is the time after the arrival of pulse #2, M_{j0} is the dipole matrix elements and $\Theta(t)$ is the step function. N_j is a complicated function of the time-integrated pulse amplitude, its spectral overlap with the resonance number j and the concentration of two-level centers of

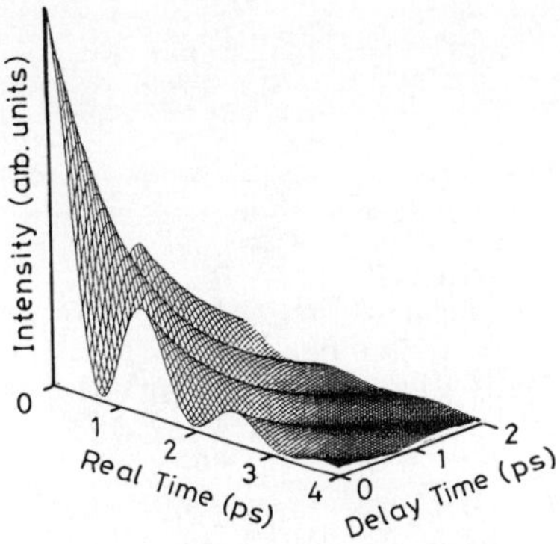

Fig. 16 Polarization interferences from two independent two-level systems

this type. The FWM signal intensity is proportional to the absolute square of this polarization, as shown in Fig. 16. The temporal interference results from the cross term which has a phase that varies as $t-\tau$.[30] With $\omega_{21} = \omega_{20} - \omega_{10}$

$$I_{FWM} \propto |P^{(3)}(t,\tau>0)|^2 \propto \Theta(t)[e^{-2\gamma_{10}t} + e^{-2\gamma_{20}t} + 2e^{-(\gamma_{10}+\gamma_{20})t}\cos(\omega_{21}(t-\tau))] \tag{59}$$

The spectral dependence is again obtained as the Fourier transform of Eq. (58)

$$\tilde{P}^{(3)}(\omega,\tau>0) \propto \left[\frac{N_1 M_{10}^4 e^{i\Omega_{10}^*\tau}}{\Omega_{10} - \omega} + \frac{N_1 M_{20}^4 e^{i\Omega_{20}^*\tau}}{\Omega_{20} - \omega}\right] \tag{60}$$

and the intensity of the signal is proportional to the absolute square of Eq. (60). Now the modulation is due to the interference from the cross term, that contains both resonance denominators. Thus, if we detect the signal near the resonance ω_{10} the following expression is valid

$$I_{FWM} \propto \frac{N_1^2 e^{-2\gamma_{10}\tau}}{(\omega_{10} - \omega)^2 + \gamma_{10}^2} + \frac{2N_1N_2 e^{-\gamma_{20}\tau}\cos(\omega_{21}\tau)}{(\omega_{10} - \omega)(\omega_{20} - \omega) + \gamma_{10}\gamma_{20}} \tag{61}$$

The important thing to notice here is that the modulation of the nonlinear signal as a function of τ wil undergo a phase shift of π, when passing through either one of the resonances. This is in contrast to the true quantum beat in Eq. (57) and therefore offers an experimental possibility of distinguishing the two phenomena, as we shall see.[31]

The reason why NQBS was restricted to noninteracting atomic systems lies in the fundamental conditions for observing quantum beats

$$\gamma < \omega_{21} < \Delta f \tag{62}$$

where $\Delta f = 1/\tau_L$ is the band width of the exciting laser. In isolated atoms, dampings are in the megaherz range, so that only moderately short laser pulses are required. In liquids or solids, e.g. semiconductors, electronic damping is in the terahertz range, so that picosecond or subpicosecond laser pulses are required to study possible quantum beats.

Another aspect of strong interatomic interactions in solids is the formation of electronic energy bands that will impede the observation of interferences between nearly degenerate discrete energy levels. In semiconductors, however, Coulomb interaction between photoexcited electron-hole pairs form hydrogen-like bound states, excitons, that can be either freely moving in the crystal or bound to local impurities (free or bound excitons). These excitonic states are the lowest elementary electronic excitations of semiconductors and introduce some narrow resonant optical transitions just below the fundamental band gap, that are very important for the optical properties, and in particular the nonlinear optical properties, of semiconductors in this spectral range. Thus, recently a number of observations of quantum beats have been made on excitons in semiconductors, particularly in GaAs quantum well structures. In the following section, I shall discuss the general optical properties of semiconductors, excitonic features in particular.

III. SEMICONDUCTORS

Semiconductors cannot in general be described by a two-level, but rather by a two-band model including the electronic states near the valence and conduction band edges, as we shall see. Concentrating on crystalline semiconductors, the translational symmetry, defined by the crystalline lattice vectors $\mathbf{R}_L = a_1\hat{\mathbf{e}}_1 + a_2\hat{\mathbf{e}}_2 + a_3\hat{\mathbf{e}}_3$ where $\hat{\mathbf{e}}_i$ are the lattice unit vectors, implies that the electronic wave functions take the form of Bloch functions[32]

$$\psi_{n,k}(r) = u_{n,k}(r)e^{i[k r - \omega_n(k)]t} \tag{63}$$

where $\mathbf{k} = k_1\hat{\mathbf{e}}_1 + k_2\hat{\mathbf{e}}_2 + k_3\hat{\mathbf{e}}_3$ is the wave vector, n is a band index and $u_{n,k}(\mathbf{r}+\mathbf{R}_L) = u_{n,k}(\mathbf{r})$

is the lattice periodic part of the wavefunction. The plane wave part of the wave function contains the energy band structure $E_n(\mathbf{k}) = \hbar\omega_n(\mathbf{k})$ as described in a one-electron Hartree-Fock approximation. The band structure can be fully specified within the 1. Brillouin zone

$$-\frac{\pi}{a_i} \leq k_i \leq \frac{\pi}{a_i} \tag{64}$$

Band structures of the common semiconductors silicon and germanium are shown in Fig.17 for major crystalline directions.[33] In both materials the maximum of the valence band, which is the highest (almost) filled band, is at the center of the Brillouin zone (BZ), $\mathbf{k} = 0$ (Γ-point). The minimum of the conduction band, which is the lowest (almost) empty band, on the other hand, is far away from the Γ-point. In such indirect gap materials, the optical interactions near the fundamental band gap are weak, because in pure crystalline materials, optical transitions conserve wavevector, and photon wavevectors $\mathbf{q}$ are negligibly small compared with the extent of the BZ.

Semiconductors with interesting optical properties should thus have a direct band gap, e.g. with the minimum of the conduction band also at the Γ-point as for GaAs, shown in Fig. 18. [33] The optically interesting region near the band edges is encircled in Fig. 18 and shown again in Fig. 19. The conduction band is s-type and only spin degenerate, whereas the valence band is p-type and dobly degenerate at the Γ-point with a band split off by spin-orbit coupling. An optical excitation across the gap gives an electron in the conduction band and a hole in the valence band. The hole can be treated as an electron-like particle with a change of sign of charge, mass, energy and wavevector (see Fig. 19). Effective masses of electrons (m_e) and holes (m_h) in their respective bands can in general be defined by the tensor

$$\left[\frac{1}{m_{e,h}}\right]_{i,j} = \frac{1}{\hbar^2}\frac{\partial^2 \mathscr{E}_{e,h}}{\partial k_i \partial k_j} \tag{65}$$

where $\mathscr{E}_{e,h}$ are electron and holes energies, respectively, calculated from the band extrema (see Fig. 19).
For isotropic bands, we have to second order (parabolic bands)

$$\mathscr{E}_e(\boldsymbol{k}) = \frac{\hbar^2 k_e^2}{2m_e}, \quad \mathscr{E}_h(\boldsymbol{k}) = \frac{\hbar^2 k_h^2}{2m_h} \tag{66}$$

where the electron and holes are characterized by simple scalar effective masses. As mentioned, optical transitions conserve energy and wavevector

$$\hbar\omega = \mathscr{E}_e + \mathscr{E}_h + E_g \quad \wedge \quad \boldsymbol{q} = \boldsymbol{k}_e + \boldsymbol{k}_h \simeq 0 \tag{67}$$

favouring the direct optical transitions across the energy gap E_g between filled valence band

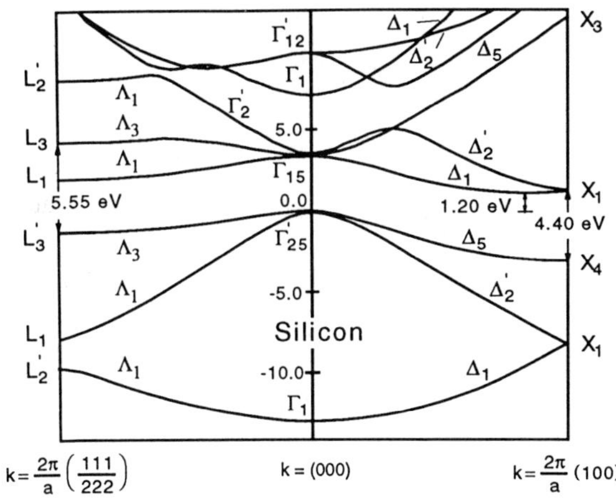

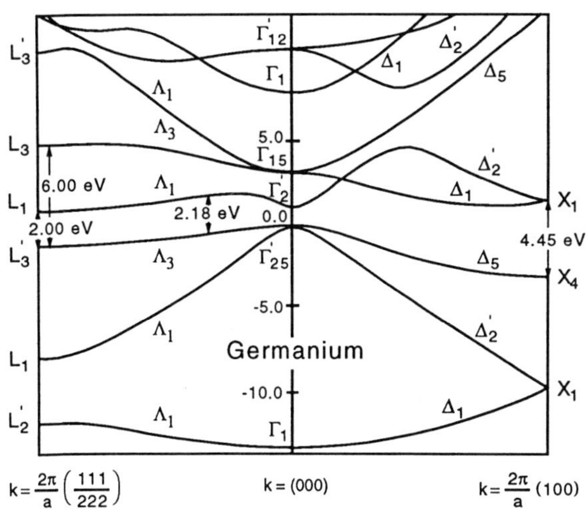

Fig. 17 Band structures of silicon and germanium, after Ref. [33].

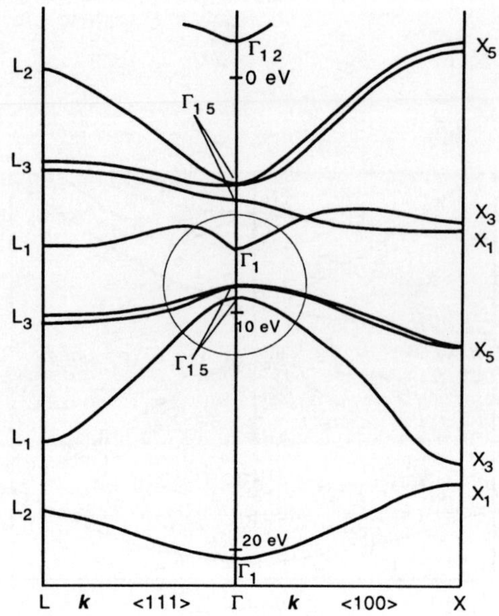

Fig. 18 Band structure of GaAs, after Ref. [33].

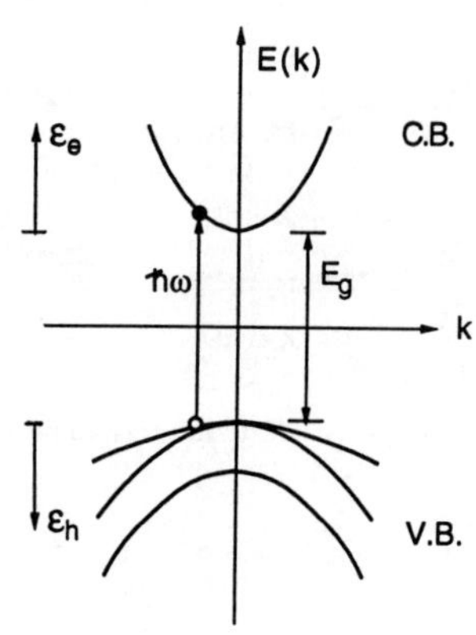

Fig. 19 Band edges of GaAs, schematic.

states and empty conduction band states. The resulting optical susceptibility can in a dipole approximation be expressed by [32]

$$\chi(\omega) = \frac{8\pi e^2/V}{(\omega + i\gamma)^2} \sum_k \{\frac{f_e(k)}{m_e(k)} + \frac{f_h(k)}{m_h(k)}\} + \frac{8\pi e^2}{V} \sum_k \frac{|\mu_{vc}|^2[1 - f_e(k) - f_h(k)]}{\hbar\omega - E_g - \mathscr{E}_e(k) - \mathscr{E}_h(k) + i\gamma} \tag{68}$$

where V is the crystal volume, γ is a phenomenological damping constant, and it is indicated that the masses may depend on wave vector (non parabolicity). The occupations of the electron and hole states are given by the Fermi functions

$$f_i = \frac{1}{e^{\beta(\mathscr{E}_i - \mathscr{F}_i)} + 1} \tag{69}$$

where $\beta = 1/kT$ and $\mathscr{F}_{e,h}$ are the chemical potentials, or Fermi energies, of the electrons and holes, respectively.

The first sum in Eq. (69) represents the free carrier contribution to the optical properties, whereas the last sum is the contribution from interband transitions. For intrinsic semiconductors at low excitation, $f_e \approx f_h \approx 0$. Hence, the linear optical properties are governed by interband transitions, where the sum over the joint density of states for parabolic bands yield the well-known square-root dependence of the optical absorption coefficient [32],[33]

$$\alpha(\omega) \propto Im\chi(\omega) \propto \Theta(\hbar\omega - E_g)\,(\hbar\omega - E_g)^{\frac{1}{2}} \tag{70}$$

where $\Theta(\omega)$ is the Heaviside step function.

Nonlinear optical properties can now have contributions from several sources: Intensity dependent broadening $\gamma = \gamma(I)$ can result from increased carier-carrier scattering, band-filling effects will appear when the occupation probabilities in Eq. (68) approach one, $f_e(I), f_h(I) \to 1$, and many-body effects may change the band energies, $\mathscr{E}_e + \mathscr{E}_h \to \mathscr{E}_e(I) + \mathscr{E}_h(I)$, so-called band renormalization.[34]

III.A. Exciton Effects

In the above, we have neglected the Coulomb interaction between the e-h pair involved in the optical transition. Including this exciton[35] effect leads to a hydrogen-like series of bound states just below the band gap as sketched in Fig. 20a with the exciton energies given by:

$$E_{x,n}(K) = E_g - \frac{E_x^b}{n^2} + \frac{\hbar^2 K^2}{2m_x} \quad ; \quad n = 1,2,3,\cdots \tag{71}$$

where $E_x^b = e^2/2\varepsilon a_x$ and $a_x = \hbar^2\varepsilon/\mu e^2$ are the exciton binding energy and Bohr radius, respectively. $m_x = m_e + m_h$ and $\mu = m_e m_h/m_x$ are the translational and reduced masses of the exciton and $\mathbf{K} = \mathbf{k}_e + \mathbf{k}_h$. Due to the large dielectric constant, $\varepsilon \approx 10\varepsilon_0$, and small reduced mass, $\mu \approx 0.1m_0$, in typical semiconductors, the exciton binding energy is small, $E_x \approx 10$meV, and the exciton Bohr radius correspondingly large, $a_x \approx 50$Å. This has two implications: Sharp exciton resonances are normally only observed at low temperatures and exciton interactions start to occur at relatively low exciton densities.

Excitons interact very strongly with light, because the electron and hole wave functions are correlated in space. The resulting exciton wavefunction is expressed by[35]

$$\Psi(\boldsymbol{r},\boldsymbol{R}) = u_{c0}u_{v0}\Phi_{n\ell}(\boldsymbol{r})e^{i\boldsymbol{K}\cdot\boldsymbol{R}} \tag{72}$$

where $\mathbf{R} = \mathbf{r}_e + \mathbf{r}_h$ is the center of mass coordinate and $\mathbf{r} = \mathbf{r}_e - \mathbf{r}_h$ is the relative coordinate of the electron-hole pair, or exciton. The functions u_{c0} and u_{v0} in Eq. (72) retain the atomic orbital character of the conduction and valence bands, respectively, entering in the strength and the selection rules of the optical transitions together with the exciton envelope function $\Phi_{n\ell}(\mathbf{r})$ which is hydrogen-like. The last part gives the plane wave character of the free exciton. The exciton contribution to the optical susceptibility can be expressed by[35]

$$\chi(\omega) = \frac{8\pi e^2}{V}\sum_n \frac{|\Phi_{n\ell}(0)|^2|\mu_{cv}|^2}{\hbar\omega - E_{x,n} + i\gamma_{x,n}} \tag{73}$$

Thus a series of sharp absorption resonances are observed at low temperatures, as indicated in Fig. 20b, where it is also shown that even in the continuum states (above the band gap) there is a strong enhancement of the absorption from the exciton effect.[35]

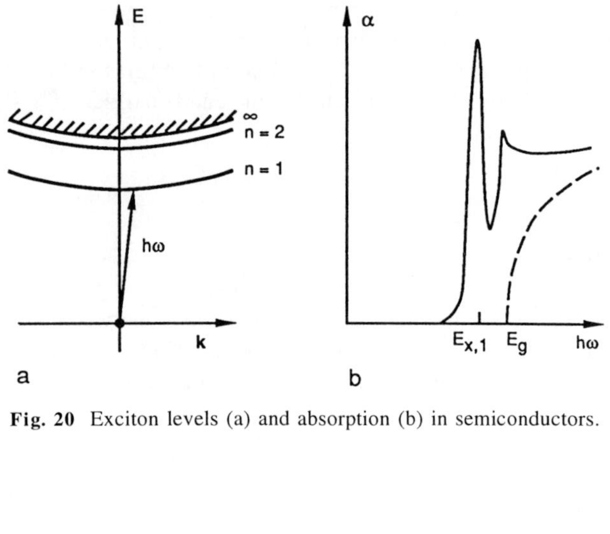

Fig. 20 Exciton levels (a) and absorption (b) in semiconductors.

1. Biexcitons. In many semiconductors, exciton interactions may lead to the formation of new bound complexes like excitonic molecules, or biexcitons, the ground state of which form a new energy band in the semiconductor [6],[8]:

$$E_m(\boldsymbol{K}) = 2E_{x,1} - E_m^b + \frac{\hbar^2 K^2}{4m_x} \tag{74}$$

where the biexciton binding energy, E_m^b, is typically 10-20% of the corresponding exciton binding energy E_x.

As sketched in Fig. 21, the formation of biexcitons open up new optical transitions. [6],[18] Besides the fundamental transitions from the ground state to the exciton state (g-x) for $\hbar\omega \geq E_x$, direct two-photon transitions (TPA) between the ground state and the biexciton state (g-m) are allowed for $\hbar\omega \geq E_x - E_m/2$, and induced transitions (IA) between the exciton and the biexciton states (x-m) for $\hbar\omega \geq E_x - E_m$. The two latter transitions are inherently nonlinear processes and therefore play a significant role when discussing the nonlinear optical properties of semiconductors near the band gap, as we shall see in the following. These transitions are expressed by the fully resonant biexciton contributions to the third order nonlinear optical susceptibility, taking into account population effects [36]

$$\chi^{(3)} \propto - \frac{\mu_{gx}^2 \mu_{xm}^2}{(\omega - \omega_x)(2\omega - \omega_{xx})} \left[\frac{1}{\omega - \omega_x} - \frac{1}{\omega + \omega_x - \omega_{xx}} \right] \tag{75}$$

A proper theoretical model for the exciton system, sketched in Fig. 22, should thus be a three-band model. For the one-photon transitions (g-x) and (x-m), however, a two-band model could be appropriate. As we shall see later for some purposes a two-band system can even be approximated with a simple two-level model, including inhomogeneous broadening to take into account phenomenologically the exciton and biexciton dispersions as well as any potential fluctuations due to real crystal inhomogeneities.[11],[21]

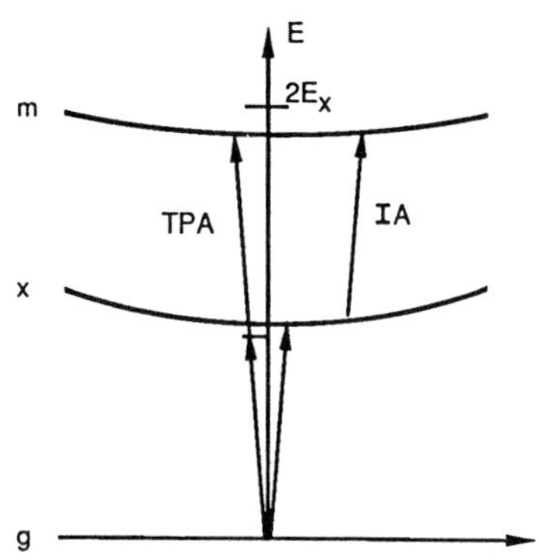

Fig. 21 Exciton and biexciton levels in semiconductors.

2. Optical Nonlinearities. Near the band edge of a semiconductor there are thus a number of optical nonlinearities that are resonantly enhanced due to real electronic, or excitonic, excitations. These nonlinearities are basically due to many-body effects in the excited semiconductor. Near the fundamental absorption edge, state filling due to the Fermi nature of the excited carriers will shift the absorption edge towards higher energies and give rise to absorption saturation, or bleaching, at a given wavelength near the edge (see Eq. (68)). A counteracting effect, giving rise to induced absorption, is a rigid shrinkage of the band gap with increasing carrier density due to many-body exchange and correlation effects (gap renormalization).[8] At low temperatures, the density of states rises very quickly with $E-E_g$, so that the latter effect dominates over state filling resulting in a net red-shift of the absorption edge (induced absorption). At higher temperatures, on the other hand, the density of states rises more slowly and state filling will dominate over band gap renormalization. This leads to a net blue shift of the absorption edge (bleaching).

In the above discussion exciton effects were excluded. This may be valid at high temperatures and/or very high excitation densities. At low temperatures and low to medium excitation densities, however, excitons dominate the linear, and even more so the nonlinear, optical properties.[6],[8] As mentioned above in Sec. III.A.1, two inherently nonlinear absorption resonances appear in the energy range $E_x - E_m < E < E_x$ with corresponding nonlinear refractive indices. In the exciton resonance itself, nonlinearities in the form of broadening (bleaching) and shift of the absorption peak appear due to exciton interactions (collisions and phase-space filling) and at higher densities by screening of the excitons by free carriers (see Sec. IV). In the high density e-h plasma limit, the band gap renormalization discussed above takes over. This is discussed by Klingshirn in the present volume.[37]

III.B. Low Dimensional Structures

Modern semiconductor crystal growth and processing techniques allow semiconductor structures of lower dimensionality than three to be produced. In these structures, the free electronic motion is restricted in one (quantum wells), two (quantum wires) or three dimensions (quantum dots or artificial atoms). This has important consequences for the electronic energies, wavefunctions and densities of states and thereby for the optical properties.[32]

1. Two-Dimesional Structures. By molecular beam epitaxy (MBE), crystalline semiconductors can be grown in a very controlled way atom layer by atom layer. By varying the semicondutor composition with monolayer precision, layered structures of semiconductors with different bandgap (but matching lattice constant) can be produced.[32],[38] In such semiconductors, the electrons (and holes) will be confined in the layers with low bandgap sandwiched between barrier layers with the higher bandgap. In such layered hetero-structures, two-dimensional, or quasi two-dimensional, electron gasses and excitons can be formed. In the direction perpendicular to the layers, the free motion of carriers is restricted, i.e. the wavefunctions become standing waves, and the energy becomes quantized, as known from introductory textbooks in quantum mechanics.

One can grow single quantum wells, double quantum wells or many (multiple) quantum wells (MQW's). If the barriers between adjacent wells are sufficiently thin for the wavefunctions in neighbouring wells to overlap, one can create a new "superlattice" with a lattice constant exceeding the lattice constant of the semiconductors themselves in a controlled fashion. In such a superlattice new minibands appear, and the bandgap can to some extent be egineered to match desired values. I shall here only be concerned with isolated wells forming two-dimensional structures.

If we for simplicity consider free electrons in such a quantum well with confinement in one direction (z), the elctron energies will be given by[38]

$$E^{2D} = E_{n_z} + \frac{\hbar^2(k_x^2 + k_y^2)}{2m} \quad ; \quad E_{n_z} = \frac{\hbar^2 n_z^2 \pi^2}{2m L_z^2} \tag{76}$$

where L_z is the well width, and $n_z = 1, 2, 3, ...$ is a quantum number. The wave functions take the form

$$\Psi_{n_z}^{2D} = (\frac{2}{L}) e^{i(k_x x + k_y y)} (\frac{2}{L_z})^{\frac{1}{2}} \sin(k_{nz} z) \tag{77}$$

where L is the crystal length along the layers. The density of states function then no longer has the well-known square-root energy dependence, but becomes independent of energy except for a step at each of the quantized energy levels E_{nz}

$$D^{2D}(E) = \frac{m}{\pi \hbar^2} \sum_{n_z} \Theta(E - E_{n_z}) \tag{78}$$

as shown in Fig. 22. This step function can be measured in a simple absorption experiment. A more peaked density of states function is often desirable for optical applications, and this can be obtained in structures where the dimensionality is lowered further.

2. One-Dimensional Structures. By even more elaborate growth and processing techniques, than mentioned above, one- dimentional structures, quantum wires, can be formed.[ ] In quantum wires, electron energies are confined and quantized in two directions (y and z) and only free to move in one direction (x). Hence, the energies are given by

$$E^{1D} = E_{n_z} + E_{n_y} + \frac{\hbar^2 k_x^2}{2m} \tag{79}$$

and the corresponding wavefunctions take the form

$$\Psi_{n_z, n_y}^{1D} = (\frac{2}{L})^{\frac{1}{2}} e^{ik_x x} (\frac{2}{L_y})^{\frac{1}{2}} \sin(k_{ny} y) (\frac{2}{L_z})^{\frac{1}{2}} \sin(k_{nz} z) \tag{80}$$

Now the density of states shows square-root singularities at the quantized energy levels

$$D^{1D}(E) = [\frac{2}{\pi \hbar^2}]^{\frac{1}{2}} \sum_{n_y, n_z} (E - E_{n_y} - E_{n_z})^{-\frac{1}{2}} \tag{81}$$

with a strong enhancement of the density of states around the resonances (see Fig. 22). This is favourable for optically active devices like lasers and modulators, and there is presently an intense research going on with prepartion and analysis of quantum wire structures.

3. Zero-Dimensional Structures. The ultimate goal of controlled growth is to form nanometer size structures, as artificial atoms, where the electron motion is restricted in all three dimensions with the energy levels fully quantized as in natural atoms [38]

$$E^{0D} = E_{n_z} + E_{n_y} + E_{n_x} \tag{82}$$

The wave function is a standing wave

$$\Psi^{0D}_{n_z,n_y,n_x} = \left(\frac{2}{L_z}\right)^{\frac{1}{2}} \left(\frac{2}{L_y}\right)^{\frac{1}{2}} \left(\frac{2}{L_x}\right)^{\frac{1}{2}} \sin(k_{nz}z)\,\sin(k_{ny}y)\,\sin(k_{nx}x) \tag{83}$$

and the density of states function is fully discretized

$$D^{0D}(E) = 2 \sum_{n_z,n_y,n_x} \delta(E - E^{0D}) \tag{84}$$

as shown also in Fig. 22. It is a fascinating thought to be able to create such artificial solid state atoms with fully engineered (optical) properties, but the technology still has a way to go before this is realized in all its aspects.

III.C. Low-dimensional excitons

The main effects of excitons on the optical properties of bulk semiconductor is, as already mentioned, to sharpen up the density of states and produce descrete optical transitions. Similarly, do excitons in low-dimensional structures enhance the linear and nonlinear optical properties of low-dimensional semiconductors. There is one extra important aspect, however, of the low-dimensional structures. The binding energy of the excitons is enhanced, so that excitonic properties prevail even at room temperature.[39] This is very important from an applications point of view. Significant confinement effects on excitons start as soon as a confining dimension becomes of the same order of magnitude as the exciton Bohr radius. The enhancement of the exciton binding energy is simply due to a forced increased overlap of the electron and hole wavefunctions, increasing the attractive Coulomb energy of the bound electron-hole state (exciton).[39]

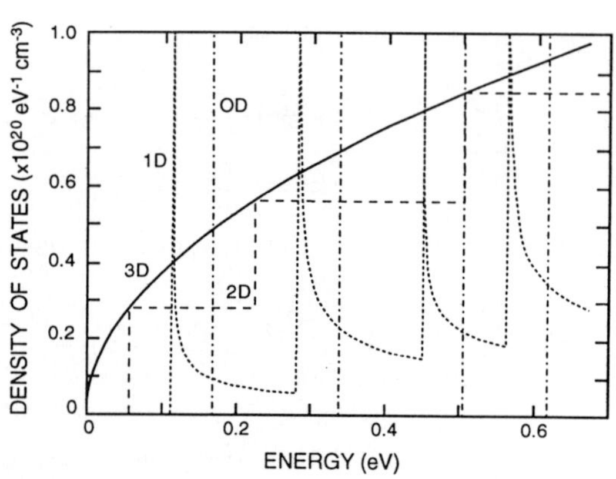

Fig. 22. Density of states for bulk semiconductors (3D), quantum wells (2D), wires (1D) and dots (0D), after Ref.[38].

The two-dimensional case can be solved analogously to the three-dimensional case and the results are summarized in Table 1 listing the important parameters in terms of bulk values of the exiton Bohr radius a_B, the exciton binding energy (in exciton Rydbergs $E_x{}^b$), the oscillator strength $f_0 = (2/m_0\hbar\alpha)|\mu_{vc}|^2$, the volume of the unit cell Ω, and the kinetic energy $\mathcal{E} = E - E_g$. The Sommerfeld factor $F(\mathcal{E})$ expresses the enhancement of the absorption in the continuum states due to Coulomb correlation.[38]

Table I Exciton parameters in 3-D and in 2-D, after Ref. [38].

	3D	2D
Bohr radius	a_B	$\frac{a_B}{2}$
Binding energy	$E_n = \frac{E_x^b}{n^2}$	$E_n = \frac{E_x^b}{(n - \frac{1}{2})^2}$
Density of states	$\frac{V}{2\pi^2 a_B^3 E_x^b}[\frac{\mathcal{E}}{E_x^b}]^{\frac{1}{2}}$	$\frac{V}{2\pi^2 a_B^2 E_x^b}\,\Theta(\mathcal{E})$
Oscillator strength	$f_0 \frac{\Omega}{\pi a_B^3}$	$f_0\, 4\frac{\Omega}{\pi a_B^2}$
Sommerfeld factor	$F(\mathcal{E}) = \frac{\pi}{\sqrt{\mathcal{E}}} \frac{e^{\frac{\pi}{\sqrt{\mathcal{E}}}}}{\sinh(\frac{\pi}{\sqrt{\mathcal{E}}})}$	$F(\mathcal{E}) = \frac{e^{\frac{\pi}{\sqrt{\mathcal{E}}}}}{\cosh(\frac{\pi}{\sqrt{\mathcal{E}}})}$

In many realistic situations strict two-dimensionality is not obtained. However, even in quasi two-dimesional structures, there is a significant enhancement of the exciton binding energy and also the biexciton binding energy. Figure 23 shows calculations of excitonic binding energies as a function of well widths in GaAs/AlGaAs quantum wells.[40] Similar calculations have been performed in one-dimensional quantum wires and zero-dimensional quantum dots. In quantum wires there is again an enhancement of the exciton binding energy, whereas in quantum dots the exciton binding energies are small compared to the large confinement energies (strong confinement). Optical transitions are, however, still affected by the Coulomb correlation between electron-hole pairs in the dots.

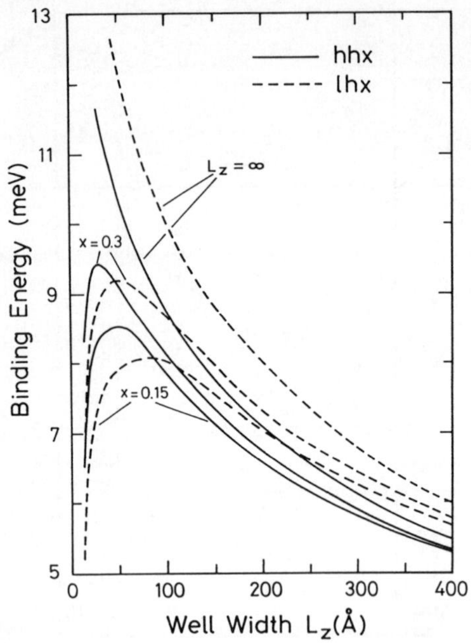

Fig. 23 Exciton binding energy as a function of quantum well width (Ref. [40]).

The nonlinear optical properties of the low-dimensional structures are of the same nature as discussed in section III.A.2 for bulk semiconductors except that the excitonic enhancements of the nonlinear optical transitions are even more pronounced in the low-dimensional structures, particularly at higher temperatures (room temperature). I shall return in the next chapters to several examples of nonlinear spectroscopy of low-dimensional semiconducting structures.

III.D. Semiconductor Dynamics

In direct gap semiconductors, the band edges are at the same point (k = 0) in crystal momentum space, facilitating strong optical interactions. For photon energies larger than the band gap, $\hbar\omega \geq E_g$, electron-hole (e-h) pairs are created by direct absorption of photons, resulting in a strong linear absorption.

Initially, the electrons and the holes are created in well-defined momentum states $\mathbf{k}_e$, $\mathbf{k}_h$ and the resulting polarization is in phase with the electromagnetic field. Scattering (even elastic) out of the initial momentum states will cause dephasing of the polarization, as characterized by the dephasing, or transverse relaxation, time T_2. Further relaxation in the form of inelastic intra-band or inter-band scattering gives rise to several longitudinal relaxation and recombination times, T_1, τ_r. The predominant scattering mechanisms are impurity scattering, phonon scattering and carrier-carrier scattering depending on purity, temperature and excitation density of the semiconductor, and the typical scattering times cover the range from femtoseconds to microseconds.

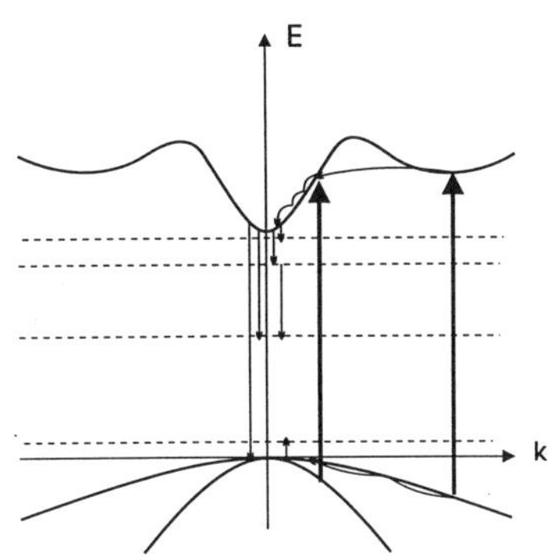

Fig. 24 Electronic relaxations in semiconductors.

If the carriers are created with kinetic energy exceeding the LO-phonon energy in the material ($\hbar\omega - E_g > \hbar\omega_{LO}$), LO-phonon scattering dominates with scattering times $\tau_{LO}\approx 150$fs,[41] except at very high carrier densities, where carrier-carrier scattering may give rise to even faster scattering. This is the hot carrier regime. In the band edge region, $0 < \hbar\omega - E_g < \hbar\omega_{LO}$, scattering times are in the sub-picosecond range, caused by acoustic phonon and impurity scattering. Interband recombination lifetimes are typically $\tau_r\approx 1$ns, [42] so that a thermalized population of carriers, or excitons, prevail. Alternatively, the carriers are being trapped in shallow or deeper localized gap states (traps), where they can live for microseconds or longer depending on localization depth.[43],[44] The diversity of electronic relaxations in semiconductors is illustrated in Fig. 24.

Linear and nonlinear optical spectroscopy has over the last decade or two proved to be an excellent tool to probe electron dynamics, as well as phonon dynamics, in semiconductors. In the next two chapters, I shall give a number of examples of such spectroscopies.

IV. NONLINEAR SPECTROSCOPIES

The most direct and simple way to use optical spectroscopy to probe the dynamics of elctronic states in semiconductors is to use a short laser pulse to photoexcite carriers and then observe the photoluminescence decay in direct time resolution either by using a streak-camera or by time correlated photon counting. In both cases time resolutions of about 20 picoseconds can be obtained. Better time resolution can only be obtained by optical gating techniques, where the emitted signal is mixed in a nonlinear crystal, e.g. by optical cross-correlation as in Fig. 1, with an ultrashort laser pulse with controlled and variable delay with respect to the laser pulse exciting the semiconductor. If the kinetics of the photoluminescence is already nonlinear

$$I_{lum}(\hbar\omega) = a(\hbar\omega)\cdot I_{exc} + b(\hbar\omega)\cdot I_{exc}^2 + \cdots \qquad (85)$$

the external optical gating technique can be avoided and replaced by e.g. an excitation correlation technique.[42] In this technique, the time-integrated luminescence (slow detector) excited by two pulse trains from the same laser with a variable delay between them is detected for different delays τ. By chopping the two beams with different frequencies f_1 and f_2 and using lock-in detection one can detect, at f_1,the total luminescence or, at $f_1 - f_2$, the leading nonlinear contribution in Eq. (95). Experimental luminescence spectra at $\tau = 0$ from excitons in CdSe at low temperature are shown in Fig. 25. The spectra contain luminescence from free excitons (E_x), possibly with simultaneous emission of one LO-phonon, from excitons bound to neutral acceptors (I_1), from excitons bound to neutral donors (I_2), from exciton-exciton collisions (P), and from biexcitons (M). The I_2-line shows sublinear kinetics (negative correlated intensity) whereas the remaining lines show superlinear kinetics (positive correlated intensity). From the correlated signal intensities as a function of the delay τ, recombination lifetimes of free and bound excitons are estimated.[42]

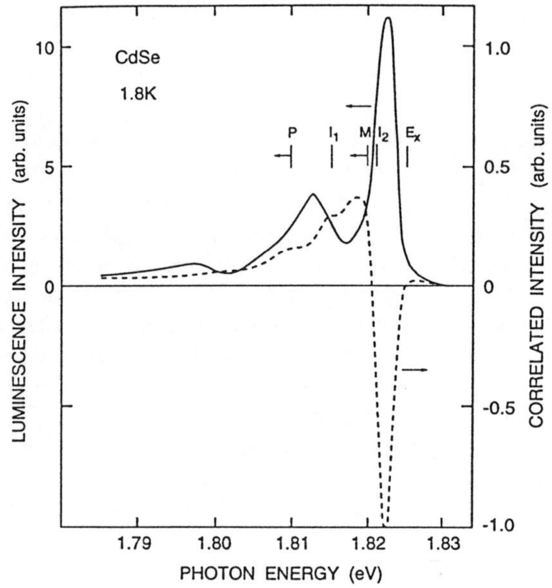

Fig. 25 Nonlinear luminescence in CdSe at 1.8K with picosecond excitation (full curve) and correlation excitation (dashed curve).

If the pulsed laser excitation is sufficiently intense to change the optical parameters of the semiconductor (nonlinear optical response), the general experimental technique to probe electron dynamics is the excite-and-probe technique. Again the pulsed laser beam is split in two beams. One beam is used to excite resonantly the semiconductor, and the other beam is used to probe the optical nonlinearity as a function of delay τ after the pump pulse. With a tunable transform limited pump pulse and a broadband (continuum) probe pulse, simple differential transmission experiments can be performed.

Early experiments of this type were performed with sub-picosecond time resolution near the band edge of GaAs. It was deduced that the screening of free excitons occurs on a subpicosecond time scale, whereas the cooling of the free carriers occurs with a time constant of about 4 ps.[45] Similar differential transmission experiments with even better time resolution have been performed later in CdSe by Fluegel et al.[46],[47]. In Fig. 26 is shown the situation with the pump pulse (150 fs FWHM) high in the band, burning a spectral "hole" in the continuum absorption, which is seen to broaden quickly with delay. It may seem mysterious that there is an effect for negative delays well before the arrival of the pump pulse. This is a typical coherent effect. The probe pulse sets up a coherent polarization in the semiconductor that will emit coherent radiation in the probe direction for as long as the dephasing time of the transitions excited by the probe. Hence, effectively the probe pulse is beeing significantly stretched and will be modified by the pump pulse arriving well after the arrival of the probe pulse. A calculations, based on the solution of the semiconductor counterpart of the two-level Bloch equations discussed in section II.A., simulates the observed behaviour quite well.[47]

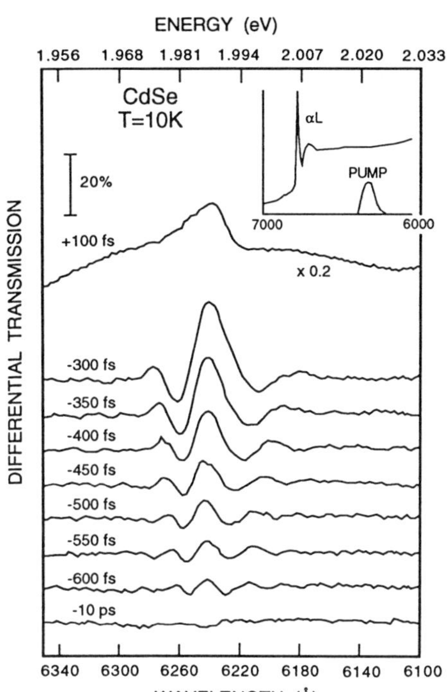

Fig. 26 Spectral hole burning in CdSe with femtosecond time resolution (Ref. [47]).

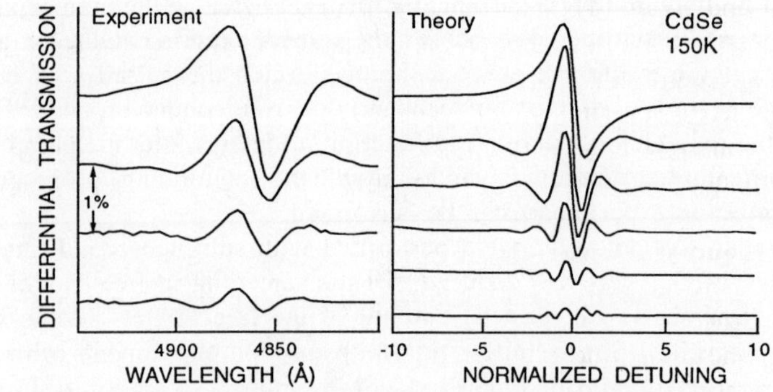

Fig. 27 Femtosecond differential transmission measuring Stark shift in CdS (Ref. [47]).

Even when semiconductors are excited well below the band edge by an intense laser pulse, can strong nonlinearities be observed in a differential transmission experiment. In Fig. 27, are shown results from CdS at 150K.[46] In the vicinity of the free exciton resonance, a significant shift, the so-called A.C. Stark shift, is observed for positive delays of the probe. For negative delays, are again observed ringings that are coherence effect of the same type as described above. The coherence effects will be treated in more detail in section IV.B. dealing with transient four-wave mixing, and in chapter V discussing nonlinear quantum beat spectroscopy. Before proceeding with these third order non-linearities, it is in place to mention also ultrafast time-resolved Raman spectroscopy as a tool to study phonon dynamics and phonon related electron dynamics.

IV.A. Time Resolved Raman Spectroscopy

If, in an excite-and-probe measurement, the probe is performing Raman spectroscopy, one can get dynamical information about the non-equilibrium phonon population after a short-pulse excitation of the electronic system. Von der Linde, Kuhl and Klingenberg[48] first measured the Anti-Stokes Raman signal of the LO-phonons in GaAs as a function of delay after a short-pulse excitation as shown in Fig. 28. From that they measured the lifetime of the non-equilibrium LO-phonons to be 7 ps at 77K.

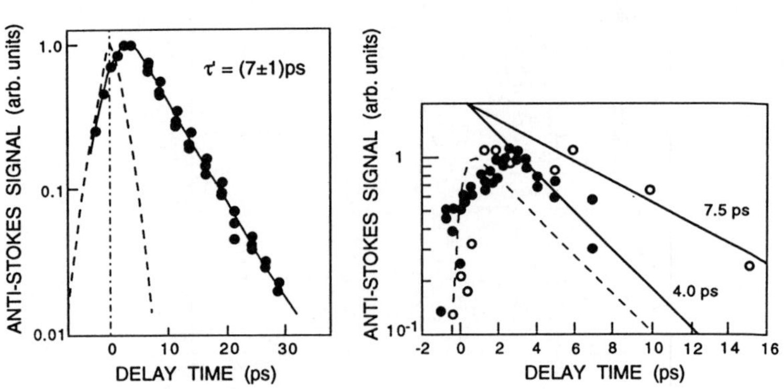

Fig. 28 Anti-Stokes Raman signal as a function of delay after picosecond (left - Ref. [48]) and sub-picosecond (right - Ref. [41]) excitation.

Kash, Tsang and Hvam [41] later repeated the experiment at low temperature as well as at room temperature. With an improved time resolution, we could also measure the risetime of the non-equilibrium LO-phonon population, as shown in Fig. 28. Knowing the excess energy of the carriers created by the pump, and assuming an energy relaxation by cascaded LO-phonon emission, the characteristic time for the LO-phonon emission could be estimated to about 150 fs. [41]

IV.B. Transient Four-Wave Mixing

Transient degenerate four-wave mixing (DFWM) as described in section II.D. is again a modification of the excite-and-probe technique, where the pump consists of two beams setting up a coherent polarization grating in the medium, and the probe is either one of the pumps self-diffracting in the grating (two-beam configuration) or a third delayed beam diffracting in the grating set up by the two first pulses (three-beam configuration). These experiments were first performed in semiconductors by Schultheis et al. [9],[10] to measure the coherence and dephasing of free excitons in GaAs. By applying a prepulse, setting up an incoherent exciton, or electron, density in the sample before performing the coherent scattering experiment, they obtained information about exciton-exciton and exciton-electron scattering rates.[10]

We have performed measurements of coherent DFWM as well as of incoherent laser induced gratings (LIG's) in the wide-gap II-VI semiconductors and in III-V low-dimensional heterostructures. We take examples from the binary compound CdSe [11], in which we have studied free excitons and biexcitons as well as impurity bound excitons [49], from mixed crystals of $CdSe_xS_{1-x}$,[50] where in particular the localization phenomena due to alloy disorder are important [51],[52], and from GaAs/AlGaAs quantum wells (MQW's).[53],[54]

A typical experimental setup to perform transient DFWM is shown in Fig. 29. The output from a synchronously pumped dye-laser system, or a self mode-locked Ti:sapphire laser, is divided into two or three beams with variable, computer controlled, delays in two of the beams. The samples are placed in a variable temperature liquid-helium cryostat and the output signal beam is selected, dispersed in a spectrometer, and measured time integrated by an OMA system. The simultaneous spectral resolution of the output signal and temporal resolution of the input signals yield, together with the energy conservation and the resonance enhancements of the nonlinear mixing process, the maximum information allowed by the uncertainty principle about the resonances involved in the FWM process.

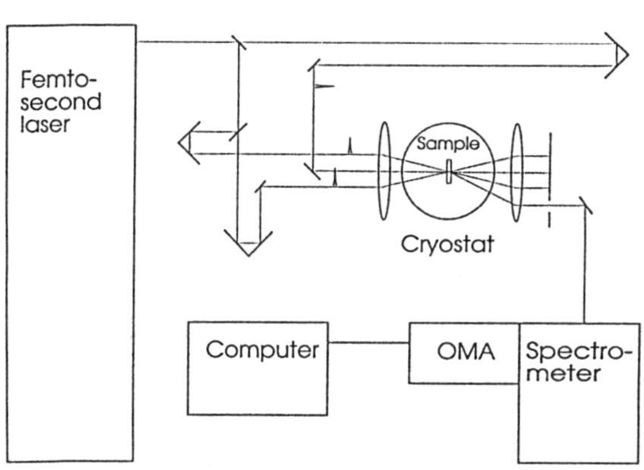

Fig. 29 Experimental setup for transient nonlinear spectroscopy.

1. Coherent Exciton Dynamics. For the coherent studies, we performed DFWM in a two-beam configuration. Figure 30 shows an example of the spectral dependence of the nonlinear signal for different delays τ between the two input pulses in a thin sample of CdSe at 4.2K.[55] Near $\tau = 0$, we observe several resonances due to free excitons and biexcitons that tend to have dephasing times T_2 of the order of picoseconds or tens of picoseconds, depending on temperature and exciton density. Bound excitons have 10-100 times longer dephasing times and the peak of their correlation trace also appears with a considerable delay. This is because the impurity bound excitons lines are inhomogeneously broadened and the nonlinear signal, in fact, occurs as a photon echo.[49] This is shown in Fig. 31 with streak-camera recordings of the nonlinear signal from the I_1 bound exciton line in a somewhat thicker CdSe sample at 4.2K.

The integrated intensity of the nonlinear signal decays exponentially with τ for delays that are long compared to the laser pulse duration: $I \propto \exp\{-a\tau/T_2\}$, where a = (4)2 for (in)homogeneously broadened systems. From these decays, we can thus measure the dephasing times for different experimental conditions and thereby determine the dominant scattering mechanisms. In Fig. 32 are shown the dephasing rates $1/T_2$ as a function of lattice temperature for free and bound exciton transitions in CdSe and for localized excitons in $CdSe_xS_{1-x}$.[12] In CdSe, the temperature dependence is well fitted with the expression

$$\frac{1}{T_2} = \gamma + \alpha T + \nu_0 e^{-\frac{E_a}{kT}} \tag{86}$$

where T is the temperature and k is the Boltzmann's constant. Thus, γ represents the background scattering, αT represents phonon scattering, and the last term expresses dephasing due to thermal dissociation of a bound complex with an activation energy E_a. For free excitons in CdSe, the background is impurity scattering with a rate $\gamma = 24$ GHz, and the phonon scattering contribution given by $\alpha = 1$ GHz/K. For transitions involving biexcitons, we find a dissociation rate governed by an activation energy $E_a = 5$ meV and with an attempt frequency $\nu_0 = 1$ THz. For excitons bound to neutral acceptors (I_1), the background scattering is much lower ($\gamma = 1$ GHz) and is in fact due to radiative recombination.[49] Resonantly excited I_1 bound excitons thus stay coherent with the exciting field for their entire lifetime. The resulting very narrow homogeneous linewidth ($<10^{-3}$meV) also implies a negligible interaction with phonons until thermal release of the bound exciton sets in (E_a= 5.6meV).

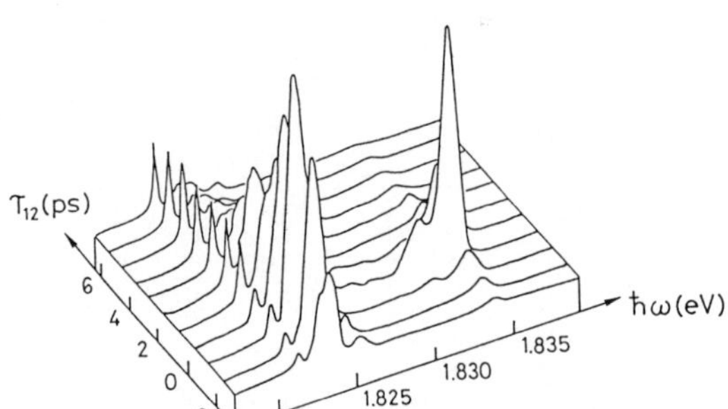

Fig. 30 Spectral dependencies of the DFWM signal in CdSe at 4.2K for different delays τ between the incoming pulses.

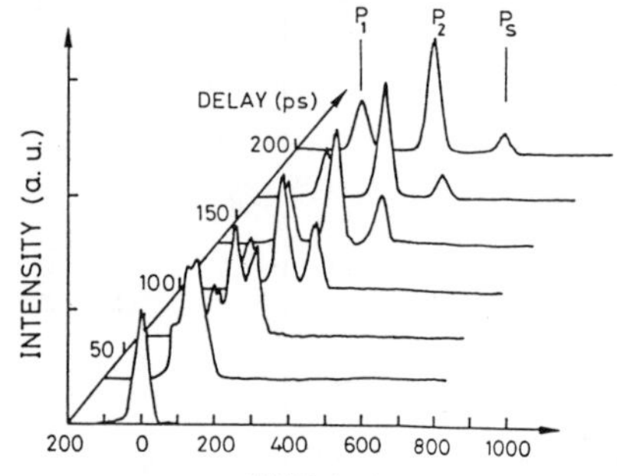

Fig. 31 Streak-camera recordings of the photon echoes (P_s) from the I_1 bound exciton in CdSe at 4.2K.

Localized excitons in $CdSe_xS_{1-x}$ have a more complex behavior depending on the localization depth. Very shallow states scatter by spectral relaxation into deeper states, whereas deeper localized states dephase rather efficiently by phonon interaction. The latter case is displayed in Fig. 32, where the low temperature limit again is determined by radiative recombination of the localized excitons ($\gamma = 0.14$ GHz).

We have also examined the exciton density dependence of the dephasing rate and found a linear dependence at low temperatures (see Fig. 32):

$$\frac{1}{T_2} = \gamma_0 + v_x \sigma_x N_x \tag{87}$$

where $v_x = 10^6$ cm s^{-1} is the thermal velocity and N_x is the density of free exciton. The collision cross section σ_x can then be found from the slopes in Fig. 32. For free excitons we find $\sigma_x = 6.25\times10^{-12}cm^2 \approx 6.8\pi a_x^2$, where $a_x = 54$ Å is the exciton Bohr radius in CdSe. For bound excitons, we find a substantially lower collision cross section $\sigma_x = 5.8\times10^{-13}cm^2 \approx 0.6\pi a_x^2$. Again, the localized excitons in $CdSe_xS_{1-x}$ constitute an intermediate case depending on the localization depth.

2. Incoherent Exciton Dynamics. Two input pulses arriving well within the dephasing time of a real resonance will create an exciton density grating that will last beyond the dephasing time, but eventually die out because of recombination and diffusion with a grating lifetime τ_G given by Eq. (34) in Sec II.D.2. as expressed by the exciton recombination lifetime and diffusion coefficient, τ_R and D, respectively. The grating constant Λ is determined by the variable angle θ between the input beams (Eq. 35). By measuring the diffracted intensity of a third delayed pulse $I_{diff} \propto \exp\{-2\tau/\tau_G\}$ for different grating constants, τ_R and D can be determined separately (see Eq. (34)).

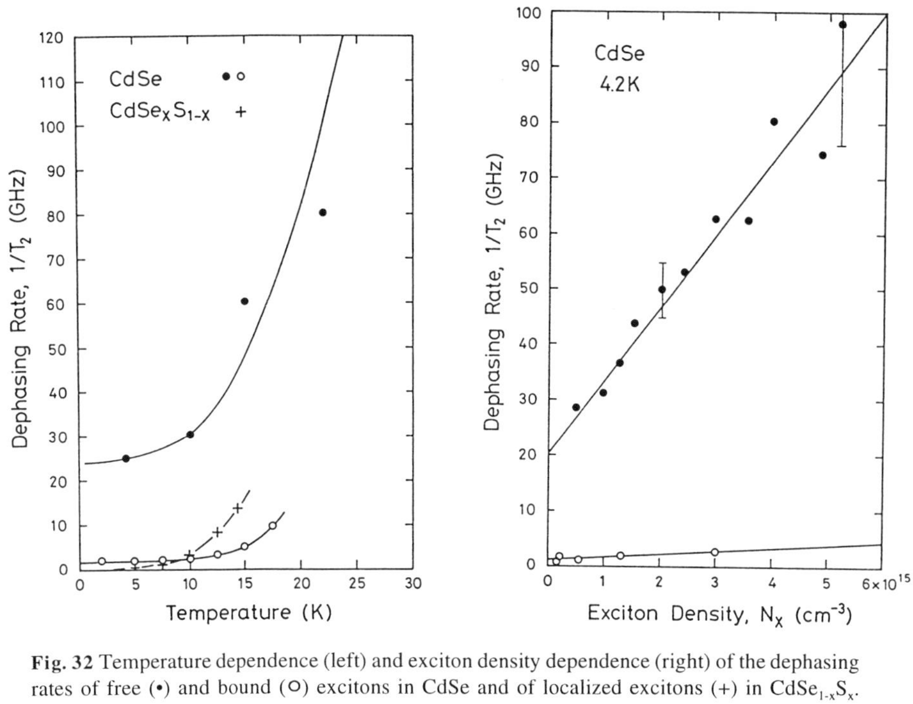

Fig. 32 Temperature dependence (left) and exciton density dependence (right) of the dephasing rates of free (•) and bound (O) excitons in CdSe and of localized excitons (+) in $CdSe_{1-x}S_x$.

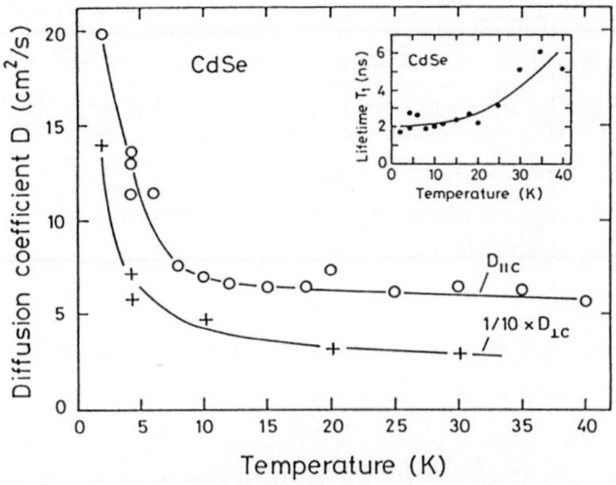

Fig. 33 Temperature dependence of the diffusion coefficient and the lifetime (inset) of excitons in CdSe.

Figure 33 shows the result of such a determination for free excitons in CdSe in the temperature range 2K-40K. It is seen that the free exciton recombination lifetime increases with temperature by a factor of about three (inset in Fig. 33). This is ascribed to an increase in the radiative recombination lifetime due to scattering out of the coherence volume of the exciton with rising temperature. The diffusion coefficient drops quickly when the sample is heated from 2K to about 10K, but then levels off. From the classical expression

$$D = \frac{kT}{m_x}\tau_p \tag{88}$$

where m_x and τ_p are the effective mass and the momentum relaxation time for the exciton. Hence, we find that the momentum relaxation rate $1/\tau_p$ increases linearly with temperature in most of the present temperature range indicating acoustic phonon scattering.[56]

The exciton mass is strongly anisotropic in the hexagonal CdSe and this anisotropy is reflected in the diffusion coefficients, as shown in Fig. 33. The ratio $D_\perp/D_\parallel$ varies from about 7 to about 5. The mass dependence in Eq. (88) account for the ratio $m_{x\parallel}/m_{x\perp}= 2.6$, while the rest comes from the exciton mass dependence of the momentum relaxation time. From the detailed evaluation of the ratio $\tau_{p\perp}/\tau_{p\parallel}$ as a function of temperature, one can therefore get information about the acoustic phonon scattering mechanisms.[56]

Similarly, we have also measured the diffusion coefficient of the localized excitons in $CdSe_xS_{1-x}$ just below the mobility edge (E_μ).[57] In Fig. 34 is shown the diffusion coefficient for localized excitons at E = 2.1982 eV as a function of temperature. The solid curve is a fit by the expression

$$D = D_0\, e^{\frac{E - E_\mu}{k_B T}} \tag{89}$$

where $D_0 = 0.3$ cm^2/s is the diffusion coefficient of the extended excitons, $E_\mu = 2.20$ eV and k_B is the Boltzmann's constant. For a fixed temperature, Eq. (99) also describes the energy dependence of D. This is shown in Fig. 34 at 20K, and with the same values for D_0 and E_μ, as the dashed curve which is in good agreement with the experimental values. From these fits, we conclude that the mobility of localized excitons is governed by thermal excitation to the mobility edge, where the mobility is still limited compared to the binary compound due to the fluctuating potential scattering.

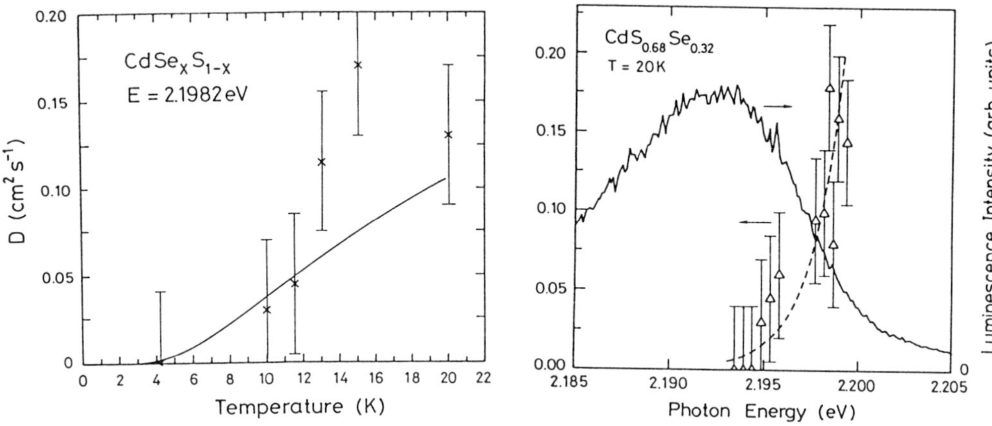

Fig. 34 Temperature dependence (left) and energy dependence (right) of the exciton diffusion coefficient in $CdSe_xS_{1-x}$ (x = 0.35) below the mobility edge. The fits are calculated from Eq. (89) with E_μ = 2.20 eV and D_0 = 0.3 cm^2/s. The solid curve (right) is a luminescence spectrum.

3. Low-Dimensional Semiconductors. It is interesting to use the powerful techniques of transient DFWM to investigate the scattering mechanisms of excitons in low-dimensional quantum structures. The influence of low dimensionality and of interface quality is particularly at focus. We have investigated the coherent as well as the incoherent dynamics of heavy-hole excitons (hhx) in GaAs/AlGaAs MQW's in the low temperature regime. [53],[54].

The observed dephasing times T_2 are slightly larger than in bulk GaAs, but the temperature dependence is relatively weak. However, it turns out that there is a significant difference in the scattering rates, that are measured in a coherent DFWM experiment, and those that are deduced from measuring exciton diffusion in the two-dimensional GaAs layers of about 100 Å width. Figure 35 shows the measured temperature dependence of the diffusion coefficient D and the recombination lifetime τ_R. The solid curve is a fit, to be discussed below. The dashed line is a linear fit to the exciton lifetime as a function of temperature with a slope of 100 ps/K. The lifetime increases due to increasing thermal population of nonradiative states. Only excitons with an in-plane wavevector smaller than the wave vector of a photon with the transition energy may decay radiatively. In addition, localized states enhancing radiative recombination are ionized with increasing temperature.

The momentum relaxation rate τ_p can be deduced from the diffucion constant applying Eq. (88). Figure 36 shows the temperature dependence of the momentum relaxation rate $1/\tau_p$ (circles) and the dephasing rates $1/T_2$ (squares). At low temperature, both rates are about 50 ns^{-1}. The dashed line is a fit obtained from linear regression to the dephasing results. The slope is $(2.6\pm0.7)\times10^9$ s^{-1}K^{-1}. The low-temperature limit is 40 ns^{-1} representing the residual interface-roughness scattering. The momentum relaxation rate increases much faster with temperature and shows nonlinear dependence. The solid curve is a fit by the formula

$$\frac{1}{\tau} = C + \frac{1}{\tau_{ir}} + \frac{1}{\tau_{ph}} = C + \frac{R\sqrt{T}}{L_z^2} + \frac{AT}{L_z} \qquad (90)$$

where τ_{ir} and τ_{ph} are the times for interface-roughness and phonon scattering, respectively, T is the temperature and L_z is the well width. The fitting parameters are A = 54 ms^{-1}K^{-1}, R = 9×10^{-6}m^2(s$\sqrt{T}$)$^{-1}$, and C = -6.6×10^{10}s^{-1}.

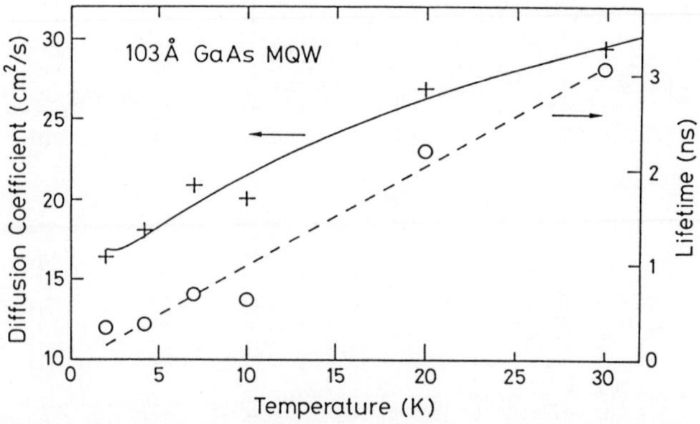

Fig. 35 Diffusion coefficient of excitons in in GaAs quantum wells (+) versus sample temperature. The solid curve represents a fit to the scattering rates (Eq. (90)) in Fig. 36. Also shown is the temperature dependence of the exciton lifetime (o) with a linear fit (dashed line).

The expression for the phonon scattering is intuitively easily understood. It is proportional to the temperature due to the classical phonon population, and inversely proportional to the well width, because of the density of exciton states to be scattered into. The expression for the interface-roughness scattering is also explained classically. The free excitons are considered to scatter off islands of slightly different well width (about one monolayer) The mean free path of excitons is assumed to be independent of temperature. Hence, the rate of scattering will be proportional to the exciton velocity ($\sqrt{T}$) and the height of the scattering potential ($1/L_z^2$). From the experimental results, it seems that this type of interface-roughness scattering is not effective for dephasing. This, however, is not too surprising, because the coherent excitons are formed with negligible vawe vector along the two-dimensional layers.

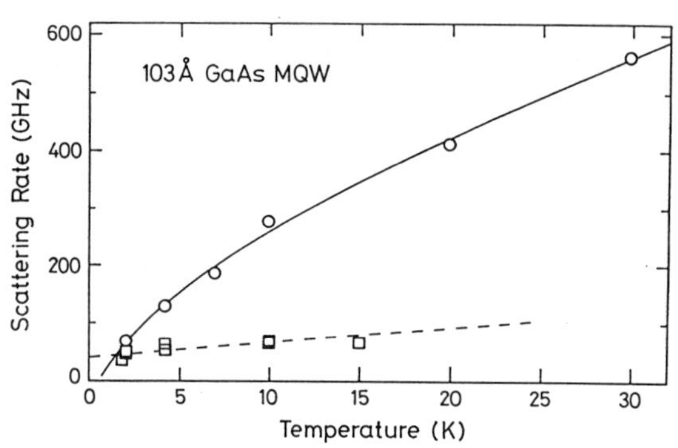

Fig. 36 Scattering rate of the thermalized excitons (o) with second order polynomial fit in $\sqrt{T}$ (solid curve) according to Eq. (90). Scattering rates of coherent excitons from dephasing measurements (□) with a linear fit (dashed curve).

IV.C. Terahertz Emission

In Sec. II.B, terahertz emission from optical rectification of an ultrashort laser pulse in a medium with a non-vanishing second order nonlinear susceptibilty was described. In material with inversion symmetry, the second order nonlinear susceptibility $\chi^{(2)} = 0$, unless there is a symmetry-breaking element e.g. an electric field. In semiconductors with an applied electric field there is an alternative strong source for terahertz emission, namely the creation of real photoexcited carriers being accelerated in an electric field.[58] With a bandwidth extending beyond 5 THz [59] and excellent signal-to-noise ratios, THz pulses are a promising tool for far infrared spectroscopy of solids liquids and gases.[60],[61],[62] Terahertz pulses can also be used to obtain information concerning the semiconductor materials they originate from. Recently this technique has been applied to study charge oscillations in double well quantum structures,[63] virtual carrier polarization,[64] and polarized excitons in quantum wells.[65]

We have applied an ultrafast pump-probe technique to study the carrier transport dynamics in the GaAs substrate of a terahertz antenna. [66] We observed ultrafast screening of the applied bias field resulting from the space charge field set up by the separating carriers. Two laser pulses are focused on the antenna; one with the purpose of generating free carriers, while the other is used to generate a terahertz pulse. The amplitude of the terahertz pulse is then used as a probe to monitor changes in the local field. Apart from measuring the internal field dynamics, the results can be used to interprete the shape of the generated terahertz pulses. Screening of the applied bias field, by optically injected carriers, strongly influences both the amplitude and the shape (phase) of the radiated terahertz pulse. For high carrier densities, the local field is screened in less than 1 ps, giving rise to strong carrier deceleration and broadening of the spectral bandwidth of the terahertz pulse.

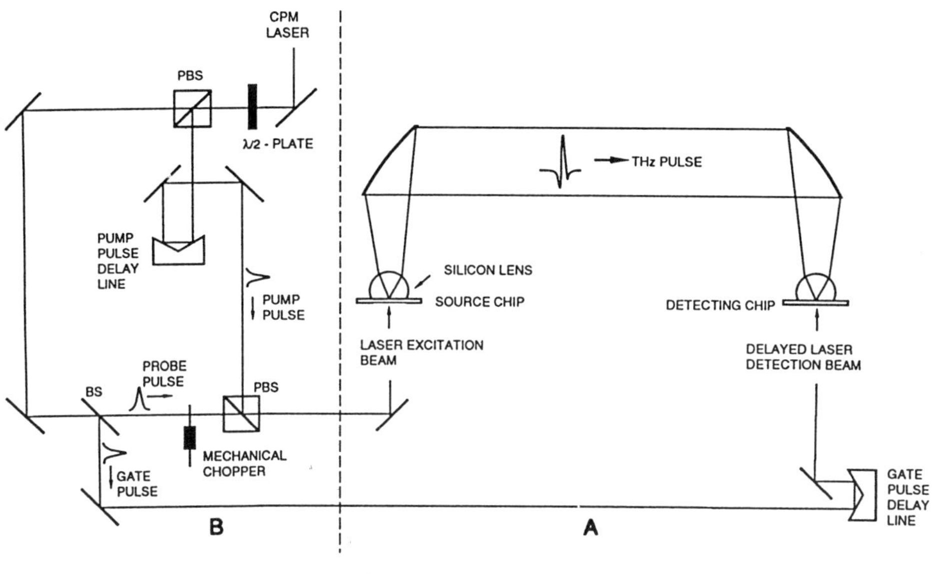

Fig. 37 Standard terahertz setup with emitter and detector (section A) with arrangements to generate pump, probe and gating pulses (section B).

V. NONLINEAR QUANTUM BEAT SPECTROSCOPY

As mentioned already in Sec. II.E, quantum beat spectroscopy was first introduced in atomic and molecular physics as a high-resolution linear spectroscopy working in the time domain. With modest time resolution, only extremely small splittings in the megahertz to gigahertz range can be observed with this technique. As faster and faster lasers have been developed, splittings in the terahertz range can be measured in media where dephasing times are short, e.g. in liquids [67] and in solids [68].

Recently, quantum beats have been observed in a number of solids, particularly involving excitons in semiconductors.[69],[70] Under high excitation densities, excitons may interact attractively to produce excitonic molecules or biexcitons. In Sec.V.A, will be reviewed quantum beats and polarization interferences from free excitons, impurity-bound excitons and bound biexcitons in bulk semiconductors observed by linear and nonlinear spectroscopy in CdS,[71] CdSe,[70] AgBr,[72] and Cu_2O,[73] stressing the importance of the spectral resolution of the nonlinear signal emitted from a FWM-process. In Sec. V.B, I shall review nonlinear quantum beats observed in GaAs MQW's.

V.A. Bulk Semiconductors

1. Free and Bound Excitons. Quantum beats from an extended exciton state in a crystalline solid were observed for the first time by V. Langer et al. in AgBr.[72] The authors time-resolved the polarized intensity of the two-phonon-assisted resonance fluorescence decay of the indirect free exciton states split in a magnetic field. The Raman-like absorption coherently excites the Zeeman split levels of excitons and quantum beats can be observed in the emitted signal. From the oscillation period, the authors determined the electron-hole exchange energy (Δ=0.13 meV) and the g-values for valence and conduction bands (g_v=2.64, g_c=1.41). Lineshape fits of the measured quantum beats gave dephasing times of several hundred picoseconds, comparable to the energy relaxation time, indicating that the coherence of the indirect exciton state is mainly destroyed by energy relaxation.

Similar measurements have been performed for the 1S yellow exciton-polariton in Cu_2O by Langer et al.[73] who time-resolved the resonant light scattering from the quadrupole polariton, split in a magnetic field. For T=2K, a dephasing time of 800 ps, decreasing with increasing temperatures due to acoustic phonon scattering, was observed. The g-values with $|g_c-g_v|$=1.78 were determined from the beat frequencies.

Quantum beats from impurity-bound excitons have been observed by Stolz et al., in linear spectroscopy.[71] They time-resolved the LO-phonon assisted fluorescence of the I_3-exciton complex in CdS, an exciton bound to an ionized donor, split in a magnetic field. From the damping of the beats, they determined dephasing times of several hundred picoseconds, and from the beat frequencies they determined magneto-optical parameters. A small splitting of the I_3-line for zero magnetic field could be explained by strain effects. For the I_1-line (excitons bound to neutral acceptors, A^0X) and the I_2-line (excitons bound to neutral donors, D^0X) beats could not be observed due to the rapid dephasing washing out the signal modulation. However, dephasing times of the order of 20 ps were found, measuring the depolarization of the luminescence after excitation with circularly polarized light.

With nonlinear quantum beat spectroscopy, excitonic transitions in bulk CdSe crystals were investigated by us.[70],[74] By applying a synchronously pumped dye laser with a coherence time of less than 500 fs, and a corresponding spectral width of about 4 meV, we coherently excited several bound and free exciton states and observed the beats in the resonantly enhanced four-wave mixing signal. There are two ways to select the beats to be studied. By tuning the broad laser line to cover the lines in question and by further spectrally selecting the resonantly enhanced nonlinear signal within this spectral range.

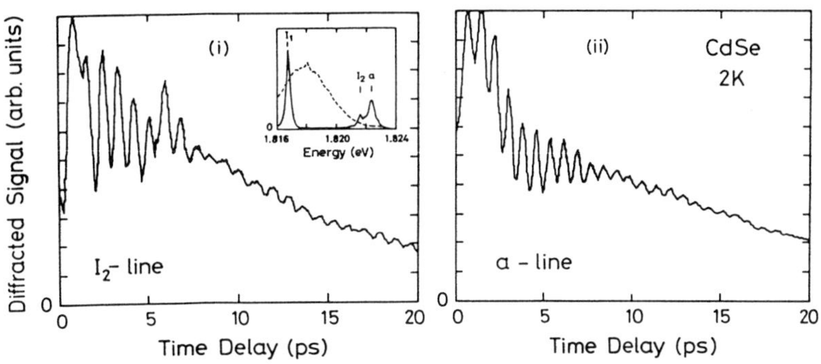

Fig. 38 Correlation traces with beats in bound exciton region of CdSe. The laser is centered near the I_1-line and detection is near the I_2-line (i) and the I_1^M-line (ii). The inset shows the laser profile (dashed) and the spectrum of the FWM signal.

Under excitation near the I_1-line the four-wave mixing signal is strongly enhanced at the positons of the I_1-line, the I_2-line and the I_1^M-line (biexciton bound to neutral acceptor, see Sec. V.A.2), as shown in the inset of Fig. 38. The correlation trace to the left in Fig. 38 is detected at the spectral position of the I_2-line, and the trace to the right is detected near the I_1^M-line. It is easily seen that they are distinctly different. Hence, one can conclude that spectral resolution of the nonlinear signal is an important key for the identification of the levels involved in the beats. In order to get information about the involved frequencies these data have been Fourier-analyzed. The Fourier spectrum of the beat signal detected near the I_2 line (Fig. 39, left) is dominated by the I_2-I_1 peak showing a fine structure due to a splitting of the I_2-line. The I_2'-I_2'' beat frequency can further be observed as a small shoulder near the zero frequency peak. Additionally, three pronounced beats I_2-I_1^M, I_2-b and I_2-c can be observed, where again the I_2-I_1^M beat clearly shows a doublet structure due to the splitting of the I_2-line. Comparing the intensities of the doublet structure in the I_2-I_1 peak and the I_2-I_1^M peak, showing a mirror image, we concluded that the energy of the I_1^M-level is higher than the energy of the I_2-level, which is confirmed by the spectral distribution of the emitted signal (see inset Fig. 39).

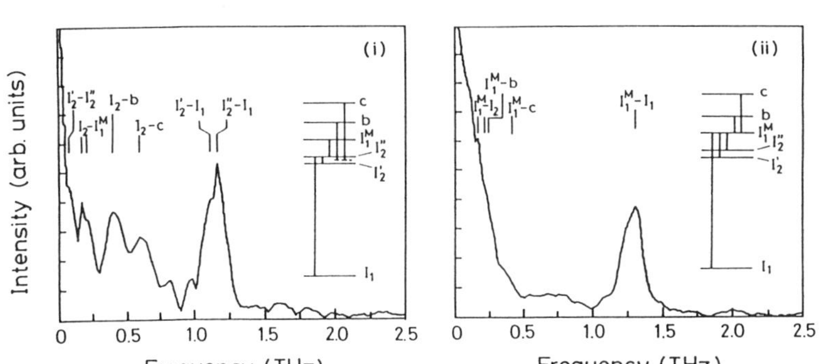

Fig. 39 Fourier transforms of the signals in Fig. 38. The insets show the infered levels (see text).

In order to establish a complete level diagram of the states, we analysed the signal detected near the I_1^M line as well. The corresponding Fourier spectrum (Fig. 39, right) shows a large peak at the I_1^M-I_1 beat frequency. Making use of the I_2-b and I_2-c frequencies in the left part, we may identify two shoulders at lower frequencies as I_1^M-b and I_1^M-c beats and thereby we have established a complete (tentative) level diagram, as shown in the insets of Fig. 39. It should be mentioned that the origins of the lines b and c are not known, the I_1^M-line is discussed in Sec. V.A.2. The energetic positions of the lines were calculated using the I_1-line as a reference. They are compared with literature values[75] in Table II.

The Fourier spectra contain further information about coherence times, spectral widths and intensity broadening of the resonances involved. We have determined the dephasing time of the I_1-bound exciton to be as long as 1 ns for low intensities and temperatures.[49] From the full width at half maximum in the Fourier spectra and assuming inhomogeneous broadening, we inferred a dephasing time of 16 ps for the I_2-bound exciton and somewhat lower values for the shallower lines, as shown also in Table II. These dephasing times were shortened with increasing excitation density, due to exciton collisions. Our results on exciton complexes in CdSe demonstrated that spectral resolution of the nonlinear signal is necessary for the identification of levels involved in the beats. Furthermore, as already discussed in the previous section, spectral resolution of a time-integated signal makes it possible to distinguish between quantum beats and polarization interference.

Figure 40(a) shows spectrally resolved correlation traces at the spectral positon of the I_2-resonance. At the I_2-resonance, $\hbar\omega = 1.8218$ eV, a phase shift of about π and a minimum of the amplitude of the modulation were observed. In Fig. 40(b) is shown the corresponding traces calculated for polarization interference according to Eq. (61). Hence, the I_1-I_2 beats are due to polarization interference, and excitons bound to different impurities can be considered as non-interacting.

Table II Position of bound exciton lines in CdSe and their dephasing times.

Line	Position (eV) this work	Position (eV) Ref. [75]	T_2 (ps)
I_1	1.8169	1.8169	1000 (Ref. [49])
I_2'	1.8217		
I_2''	1.8218	1.8218	16
I_1^M	1.8223		10
b	1.8234	1.8236	10
c	1.8241	1.8247	8

2. Impurity Bound Biexcitons. In an intensely excited semiconductor, biexcitons may form and possibly bind to impurity centers to form multi-exciton complexes.[76],[77] We observed a multi-exciton impurity complex, namely a biexciton bound to a neutral acceptor (A^0M), by nonlinear quantum beat spectroscopy [78], (see also Sec. V.B). Biexcitons can be created via direct two-photon absorption and via an induced transition between the exciton and the biexciton state at $\hbar\omega = E_x - E^b_{xm}/2$ and at $\hbar\omega = E_x - E^b_{xm}$, respectively.

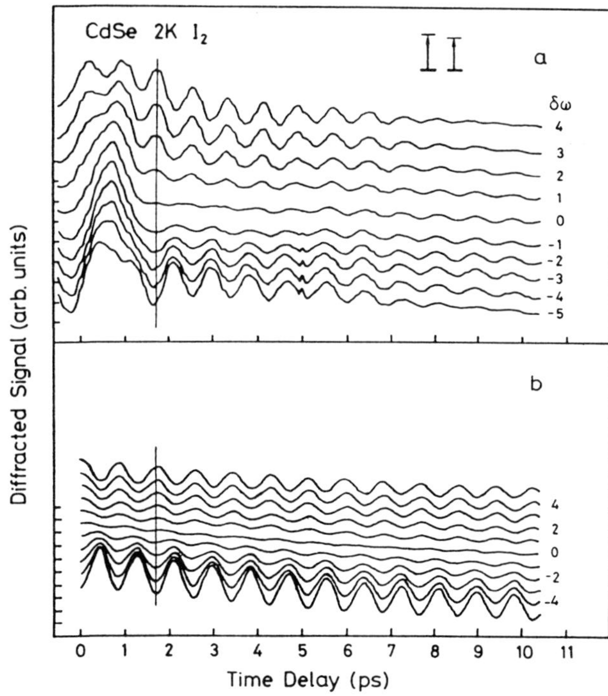

Fig. 40 Polarization interference detected near the I_2-resonance (a) and the corresponding calculation (b) after Eq. (61).

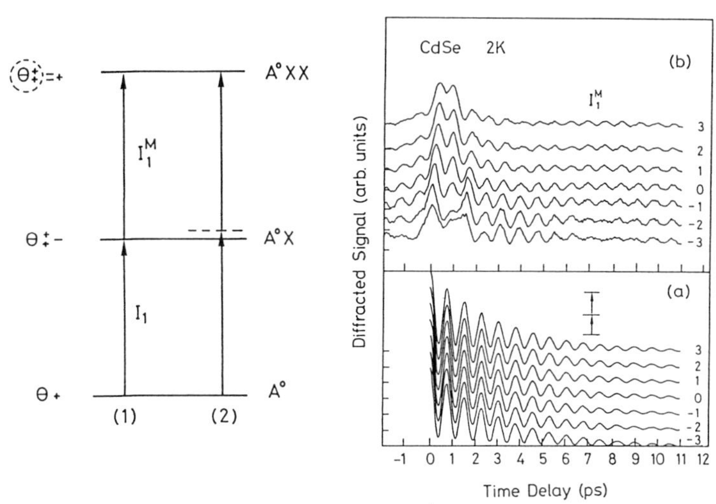

Fig. 41 Level diagram for bound exciton-biexciton system in CdSe (left) and quantum beats detected in the I_1^M-line experimentally (b) and as calculated from Eq. (56).

The excitation of an impurity bound biexciton complex A^0M via a neutal acceptor A^0 follows one of the optical transitions

$$A^0X + \hbar\omega \rightarrow A^0M \qquad or \qquad 2\hbar\omega \rightarrow A^0M \tag{91}$$

where $\hbar\omega$ is the energy of the absorbed photon as sketched in Fig. 41(left). The correlation trace of the self-diffracted signal, detected in the spectral region of the I_1^M-band and excited near the I_1-line, revealed pronounced oscillations with a period of 766 fs, as shown in Fig. 41(right). In contrast to the I_1-I_2 bound exciton polarization interference (see Fig. 40), no phase shift was observed near the resonance frequency ω_0 (I_1^M) which proves that the involved transitions are within the same quantum system (see Eq. (56)). The spectrally broad laser system prepares a coherent mixture of A^0X impurity bound excitons and A^0M impurity bound biexciton complexes by two-photon transitions transitions, either directly or via the A^0X resonance. The two FWM processes that are beating are the lower and upper transitions sketched in Fig. 41. From the beat period and from the spectral positions of the I_1 and I_1^M-lines (see Table II), we determined a complex binding energy E^b_{xm} = 5.4 meV per exciton (average). In this section, it was clearly demonstrated that the spectral resolution of the nonlinear signal reveals whether the transitions are interacting or not. This knowledge is of great importance for the identification of the electronic structure of the materials considered.

V.B. Quantum Well Structures

Quantum beats have been observed in type I and type II MQW structures. In type I structures the recombination takes place between electron states and hole states from the same well. In type II quantum well structures, the recombination takes place between the spatially separated electrons in the barriers and holes in the wells. These transitions reveal long lifetimes and dephasing times in the microsecond range.[79],[80] In this sections, I shall review quantum beats and polarization interferences observed by linear and nonlinear spectroscopy of free and bound excitons, exciton transitions split in different confinements, Zeeman split excitons and biexcitons, all in MQW structures.

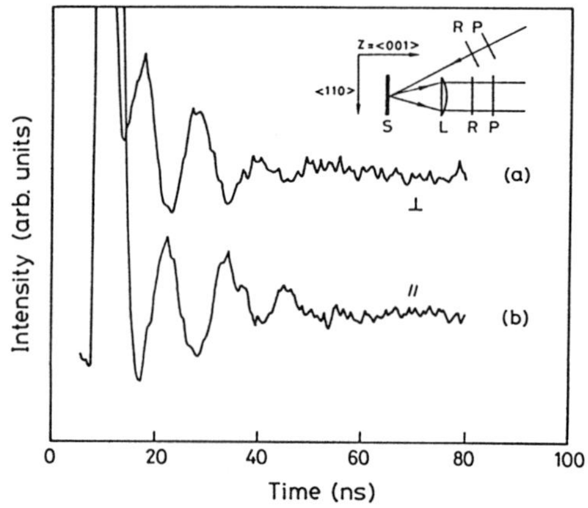

Fig. 42 Fluoresecence decay with quantum beats from type II GaAs/AlAs quantum well structure for perpendicular (a) and parallel (b) detection (Ref. [82]).

1. Free and Bound Excitons. Quantum beats in the exciton emission of type II GaAs/AlAs quantum wells have been measured by van der Poel et al.[81]. After resonantly exciting into the weakly allowed type II transition, with lifetimes in the μs region, the authors observed quantum beats in the time-resolved luminescence. From the beat frequency of 82 MHz observed in Fig. 42, a level-splitting of 0.3 μeV has been determined. The resolved level-splitting is far below the inhomogeneous linewidth of about 10 meV. Similar measurements are reported by Lavallard et al. [82], who related the origin of the observed splitting to localisation of excitons near the interface.

The light hole (lh) and heavy hole (hh) bands in bulk GaAs are degenerate at the Γ point. In a quantum well structure the degeneracy is lifted due to the confinement, and consequently, two distinct transitions of free excitons can be observed in the optical spectra. Quantum beats between lhx and hhx have been observed by different authors.[53],[83],[84],[85],[86],[87],[88]

In Fig. 43 are shown results of quantum beats in the four-wave mixing signal from hhx and lhx transitions excited coherently by laser pulses with a coherence length of about 200 fs (see inset). Pronounced oscillations could be observed in the time-integrated FWM-signal, (Fig. 43(a)). A beat period T = 525 fs is observed corresponding to an energy splitting of 7.9 meV, which coincides with the hhx-lhx splitting as measured in the absorption spectrum. From the line width in the Fourier spectrum $\Delta\nu$ = 500 GHz (Fig. 43(b)), a dephasing time of (4) 2 ps was determined assuming (in)homogeneous broadening of the resonances. This dephasing time is smaller than determined for the hhx [54],[84] and is probably to be ascribed to the lhx resonance.

Quantum beats within the hhx band are reported by Feldmann et al. [89] from a strained InGaAs/GaAs quantum well structure. In a transient four-wave mixing experiment the authors observed oscillations with a period of 600 fs corresponding to the splitting of 6.7 meV between the 1s and 2s hhx transitions. The fast initial decay of the FWM-signal at τ = 0 was explained by the simultaneous excitation of unbound continuum states. Calculations in a simple multilevel system support their explanation.

In MQW structures not only transitions due to hhx and lhx excitons can be observed, but also transitions of excitons bound to impurities. Oscillations in the decay of the self-diffracted signal in a DFM experiment due to the interference between the free heavy-hole exciton state and exciton bound to impurities states have been observed by Leo et al..[90] After excitation in the region below the hhx transition, the diffracted signal showed a modulation with a period of about 2.8 ps. The authors used a sample grown on n-type doped [100] GaAs substrate consisting of 10 GaAs wells of 170 Å nominal width. Increasing the lattice temperature or the exciton density, the oscillations disappeared very quickly due to a decrease of the phase relaxation time. The corresponding splitting of 1.47 meV agreed with the splitting between free and bound exciton transition observed in a luminescence experiment.

With DFWM measurements energy levels can be resolved, which are due to inhomogeneous broadening not visible in linear absorption or luminescence. This was done by Koch et al.[91] using a sample consisting of 100 periods 41 Å GaAs and 30 Å AlAs with a carbon concentration of about 10^{16} cm^{-3}. The heavy hole luminescence is red-shifted with respect to the absorption and shows an inhomogeneous broadening of about 12 meV. Under excitation in the spectral region of the luminescence band oscillations with a period of T = 0.8 ps corresponding to an energy splitting of 5.2 meV have been observed in the nonlinear signal of the DFWM experiment. The authors attributed the two transitions, which are not separated in the luminescence spectrum, to excitons localized by potential fluctuations and carbon acceptor bound excitons. A drastic increase in the dephasing time from about 10 ps for energies at the high energy band of the luminescence spectrum to about 130 ps in the low energy wing was measured in additional experiments and explained by spectral diffusion within the energy distribution of the localized excitons.

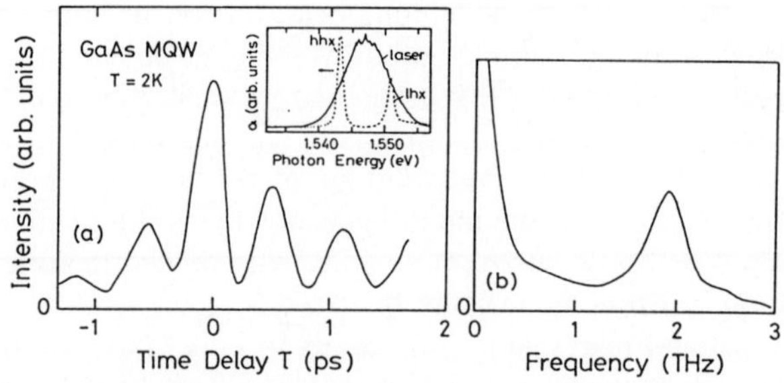

Fig. 43 Time integrated correlation trace with quantum beats from hhx an lhx transitions in GaAs MQW structure (a). The inset shows the spectral profile of the exciting laser (solid line) and the spectrum of the absorption coefficient (dotted). The beat spectrum is also shown (b).

From time-resolved measurements of the nonlinear signal via up-conversion, the authors concluded that true quantum beats had been observed. The optical excitation of localized excitons and impurity bound exciton states represents a three-level system, where the common state is the non-excited crystal ground state, and for sufficiently high impurity concentration, the mean distance between impurities is smaller than the diameter of the excitonic coherence volume. Then the bound exciton site lies within the coherence volume of the exciton which leads to the observed quantum beats in a three-level system as expressed by Eqs. (54) and (56).

2. Heavy Hole Excitons Split in Different Confinements. The confinement energy of excitons in a MQW structure depends on the thickness of the wells. Regions of different thicknesses within one well can be created if the molecular-beam epitaxy (MBE) growth is interrupted at the interfaces. The different confinement of the excitons then leads to a splitting of the exciton lines which can be observed in the optical spectra.

Polarization interference from regions with slightly different confinement energies were observed by Göbel et al.[92] in GaAs/AlGaAs quantum wells with a well thickness of 70 Å grown by MBE with growth interruption. In the photoluminescence excitation spectrum of this sample, three distinct exciton peaks with a spacing of 2.7 meV and a width of about 1.3 meV were observed corresponding to emission of excitons from regions with different thicknesses. The time-integrated signal from a FWM process reveals an exponentional decay superimposed oscillations with a period of 1.33 ps, corresponding to an splitting of 3.1 meV, which roughly coincides with the energy separation measured in the photoluminescence excitation. For comparison, the spectra of a sample grown without growth interruption were measured, where the photoluminescence spectrum shows one broad exciton peak and the transient-grating signal decays nearly exponentially with a time constant corresponding to a dephasing time of 3.2 ps, assuming inhomogeneous broadening. Excitons in samples containing separated wells with different widths represent a system of independent two-level transitions giving rise to polarization interferences, rather than quantum beats.

3. Magneto Excitons. The optical properties of a MQW structure is substantially modified when a magnetic field B is applied. The field lifts the spin degeneracy of the electronic states in the conduction and valence bands and creates two non-degenerate exciton states, which can couple to circularly polarized light, σ^+ and σ^-. In the inset of Fig. 44, the two allowed transitions for an orientation of the field normal to the surface of the sample and for light propagating parallel to the field are marked.

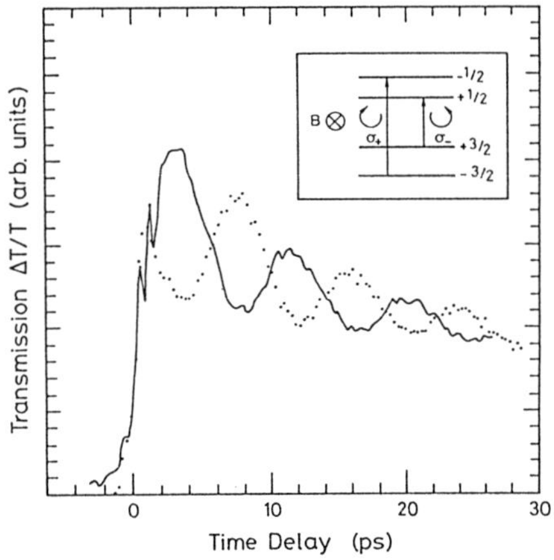

Fig. 44 Temporal evolution of the differential transmission at the hhx in GaAs quantum well for B = 4T. The solid curve was measured with orthogonal polarizations in pump and probe and the dotted curve with parallel polarization (Ref. [93]).

The Zeeman splitting $\Delta E = 2g\mu_B B\Delta m_j$, where g is the Landé factor, μ_B the Bohr magneton and m_j the quantum number, is smaller than the inhomogeneous line width of the exciton states for moderate field strength (B<10T), and can not easily be resolved using linear techniques like transmission or luminescense.

Simultaneously exciting σ^+ and σ^- exiton states by linearly polarized light, Bar-Ad and Bar-Joseph [93] observed induced absorption quantum beats in picosecond pump and probe experiments. The spectra in Fig. 44 for orthogonal and parallel pump and probe polarizations, respectively, are recorded slightly below the heavy-hole exciton absorption peak and at a magnetic field of 4T. From the period of the oscillations the authors estimated a Zeeman splitting of 0.5 meV and a Landé factor g_h= 0.6 for the hhx. The hhx-resonance is strongly inhomogeneously broadened to a width of about 6 meV, indicating that excitons are localized by potential disorder due to well width and alloy fluctuations. The authors assigned the decay time of the signal, of the order of 20 ps, to spectral diffusion of low-energy localized excitons. No oscillations of the induced absorption can be observed whenever the pump or probe beams are circularly polarized, because a circulary polarized beam only interacts with one of the eigenstates.

The same group investigated the Zeeman splittings of the excitons by performing four-wave mixing experiments.[94] The correlation traces of the time-integrated FWM-signal measured for orthogonal and parallel polarizations of the two linearly polarized beams revealed distinct oscillations due to σ^+ and σ^- exciton states coherently driven by the laser pulse. Detailed investigations of the observed Zeeman splittings resulted in a magnetic field dependence and well width dependence of the electron and hole g-values.

4. Biexcitons. Under high excitation density, excitons may interact attractively to produce excitonic molecules (biexcitons). In contrast to bulk GaAs with a calculated binding energy of biexcitons of about 0.13 meV,[95],[96] binding energies around 1 meV are expected in MQW structures[97] and make an experimental observation possible at temperatures below 10 K. A considerable effort has therefore been invested in the search for biexcitons in GaAs/AlGaAs multiple quantum well structures.[98],[99],[100],[101],[102],[103],[104],[105]

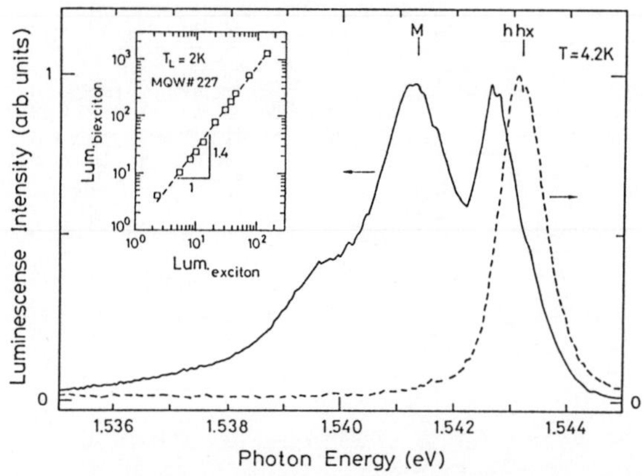

Fig. 45 Luminescence (solid curve) and absorption (dashed curve) spectra in GaAs quantum wells. In the inset is shown the ratio between the intensities of exciton and biexciton luminescence for excitation intensities increasing from 0.13 to 7.0 μJ/cm^2.

We have investigated biexcitons in MQW structures using samples grown by molecular-beam epitaxy on a {100} oriented GaAs substrate which was removed by selective chemical etching.[106] The linear absorption shows a narrow (≈1meV) hhx peak at 1.54315 eV (Fig. 45, dashed curve) indicating the high quality of the MQW's. The hhx luminescence, after band-to-band excitation with a power of 10 kW/cm^2 (Fig. 45, solid curve), is Stokes shifted about 0.5 meV relative to the absorption peak. A simple division by the transmission spectrum suggests that the observed shift is mainly due to reabsorption of the luminescence. In the low energy band of the hhx emission, a pronounced peak, which grows superlinearly with excitation power, and a small shoulder can be detected at 1.5412 eV and 1.5396 eV, respectively.

The spectrally broad laser system with a pulse coherence time of ≤200 fs cannot resolve any fine structure in the region of the hhx resonance. However, strong resonance enhancements of the signal, representing the third-order susceptibility $\chi^{(3)}$, are expected near the exciton resonance ω_x. Taking into account biexciton formation, two new resonances appear due to two-photon transitions (TPT) to the biexciton state, at $\omega_m/2$, and due to induced exciton-biexciton transitions (M-band), (see Fig. 21).

We performed DFWM experiments in the spectral range of the exciton and biexciton transitons. Time-integrated spectra of the emitted signal are shown in Fig. 46 for different delays τ between the input pulses. The dominant peak is due to the hhx transition. The signal is redshifted with respect to the absorption peak due to linear reabsorption just as for the hhx luminescence peak in Fig. 45. All the spectra in Fig. 46 are normalized to the maximum intensity near the hhx resonance that is strongly peaked around $\tau = 0$. For negative delays we observed, on the low energy side of the hhx-resonance, a small shoulder (TPT) at 1.54225 eV and a broader band (M-band) at lower energies around 1.5413 eV. By its reproducability, The shoulder at 1.54225 eV is well separated from the noise level (see inset in Fig. 46), and we assigned it to resonant enhancement of the FWM signal due to TPT directly to the biexciton state. From that we estimated a binding energy of 1.8 meV for the biexciton which is somewhat larger than theoretically calculated. The broader band for negative delays corresponds then to the resonant enhancement due to exciton-biexciton transitions and coincides with the M-band observed in luminescence. The shoulder at 1.5396 eV, that was observed in luminescence, could be due to bound excitons or possibly to biexciton-biexciton collisions.

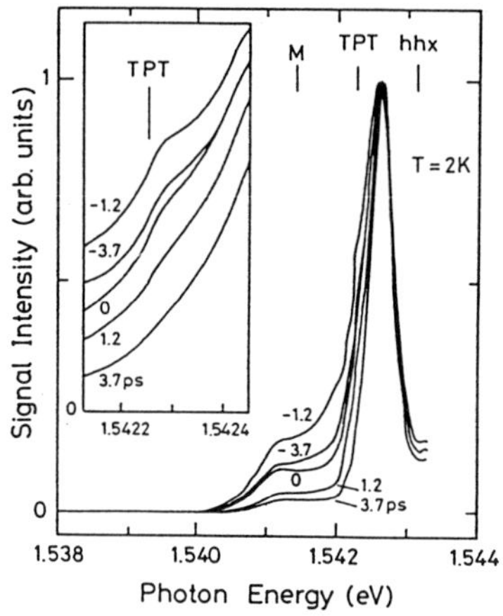

Fig. 46 Intensity of the self-diffracted signal for different delay times τ. All spectra are normalized to the maximum intensities.

The correlation trace of the self-diffracted signal, detected in the spectral region of the M-band, revealed pronounced oscillations with a period of 2.3 ps, as shown in Fig. 47. These oscillations were strongest when detecting in the M-band, and we interpreted them as quantum beats between exciton and biexciton transitions in the FWM process. The modulations of the correlation trace are clearly revealed as a single oscillation by the Fourier transform shown in the inset of Fig. 47, and the corresponding energy separation ΔE=1.8 meV agrees well with the biexciton binding energy as determined above. The modulation of the signal diminishes with increasing temperature, and above 20K no beats can be observed. From the line width in the Fourier spectrum, Δv = 100 GHz, a dephasing time of 20 ps was determined assuming inhomogeneous broadening of the resonances at T=2K. This is similar to the dephasing time of heavy hole excitons. Quantum beats in a FWM experiment performed on GaAs quantum wells with a well width of 102 Å were also observed by Lovering et al.[107] resulting in a somewhat lower biexciton binding energy.

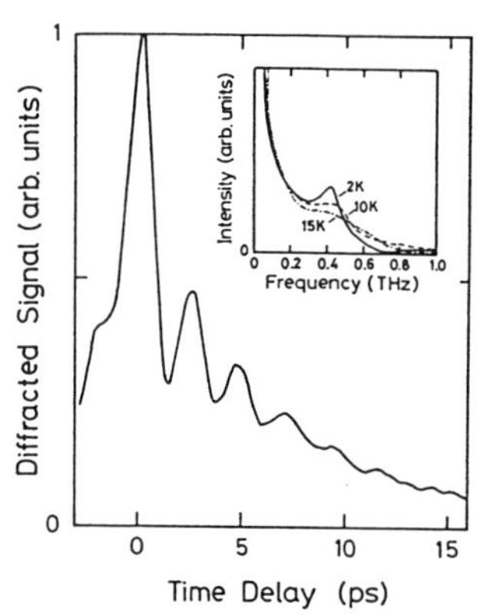

Fig. 47 FWM signal in the M-band of GaAs quantum well. Inset shows Fourier transform.

VI. CONCLUDING REMARKS

In this article I have tried to cover some of the important aspects of nonlinear spectroscopy of inorganic materials. Recapitulating basic theory and modelling and giving examples of experimental results to illustrate the various nonlinear spectroscopic techniques and demonstrate their capabilities of probing the elementary excitations in solids and their dynamics. The examples are almost exclusively taken from excitons in semiconductors, because this is the field with which I am most familiar. Even so, the article does not give credit to the vast amount of work that has been done in nonlinear optical spectroscopy of semiconductors over the last decade or more.

My aim with this work has not been to review the field, but merely to illustrate the power and the potential of nonlinear spectroscopy for high resolution spectroscopy as well as for ultrafast dynamical studies. The topics that are considered are strongly biased towards my own interests and I have in most cases taken examples from the work of my own group. I hope to have demonstrated that a number of the nonlinear optical processes and techniques that were studied in the sixties and the seventies on atomic and molecular systems have been succesfully applied to semiconductors in recent years. One major reason for this development is the progress in laser technology,[108],[109] particularly with respect to ultrafast laser systems working in the spectral range of the band gap of important semiconductors.

Another driving force behind this development is of course also the increasing technological importance of nonlinear optical processes in telecommunication[110] and in opto-electronics in general. It is great when interesting new physics goes hand in hand with exciting new technological developments.

ACKNOWLEDGEMENTS

I am deeply grateful to the large number of collegues, students and guests in my group that have made it possible to obtain new results in this strongly competitive field. Particularly, I would like to thank C. Dörnfeld, H. Schwab, D. Oberhauser, K.-H. Pantke, V. G. Lyssenko, J. Erland and I. Balslev for their significant contributions to this work, which was supported by the Danish Natural Science Research Council.

REFERENCES

1. N. Bloembergen, *Nonlinear Optics* (Benjamin Inc., New York, Amsterdam, 1965).
2. Y.R. Shen, *The Principles of Nonlinear Optics* (Wiley & Sons, New York, 1984).
3. R. M. Macfarlane, this volume.
4. G. Baldacchini, this volume.
5. A. Miller, D.A.B. Miller and S.D. Smith, Adv. in Phys. **30**, 697 (1981).
6. B. Hönerlage, R. Levy, J.B. Grun, C. Klingshirn, and K. Bohnert, Phys. Rep. **124**, 161 (1985).
7. J.M. Hvam in *Nonlinear Optics in Solids*, O. Keller (ed.), Springer Series in Wave Phenomena, Vol. **9** (Springer Verlag Berlin, Heidelberg, 1990) p. 213.
8. C. Klingshirn and H. Haug, Phys. Rep. **40**, 315 (1981).
9. L. Schultheis, J. Kuhl, A. Honold, and C.W. Tu, Phys. Rev. Lett **57**, 1797 (1986).
10. L. Schultheis, J. Kuhl, A. Honold, and C.W. Tu, Phys. Rev. Lett. **57**, 1635 (1986).
11. C. Dörnfeld and J.M. Hvam, IEEE J. Quantum Electron. **25**, 904 (1989).
12. K.-H. Pantke, V.G. Lyssenko, B.S. Razbirin, H. Schwab, J. Erland, and J.M. Hvam, phys. stat. sol. (b) **173**, 69 (1992).
13. S.D. Smith and R.F. Neale (eds.), *Optical Information Technology*, ESPRIT Basic Research Series (Springer Verlag, Berlin, Heidelberg,1993).
14. B. Di Bartolo, this volume.

15. I. Abram, this volume.
16. K.L. Sala, G.A. Kenney-Wallace, and G.E. Hall, IEEE J. Quantum Electron. **16**, 990 (1980).
17. H.J. Eichler, P. Günter, and D.W. Pohl, *Laser Induced Dynamical Gratings,* Springer Ser. Opt. Sci. **50** (Springer, Berlin, 1986).
18. A. Maruani and D.S. Chemla, Phys. Rev. B **23**, 841 (1981).
19. J.M. Hvam and C. Dörnfeld, in *Optical Switching in Low-Dimensional Systems*, H. Haug and L. Banyai, eds. (Plenum Publishing Corp., New York, 1989) pp. 233-241.
20. C. Dörnfeld and J.M. Hvam, J. de Physique **49 suppl. 6**, C2-205 (1988).
21. T. Yajima and Y. Taira, J. Phys. Soc. Japan **47**, 1620 (1979).
22. J. Erland and I. Balslev, Phys. Rev. B **48**, 1765 (1994).
23. S.C. Moss, J.R. Lindle, H.J. Mackey, and A. Smirl, Appl. Phys. Lett. **39**, 227 (1981).
24. S. Haroche, in *High Resolution Laser Spectroscopy*, K. Shimoda, ed. (Springer, Berlin, Heidelberg, 1976) pp. 253-313.
25. P. Meystre and M. Sargent III, *Elements of Quantum Optics* (Springer Verlag, Berlin, Heidelberg, 1990).
26. S. Haroche, J.A. Paisner, and A.L. Schawlow, Phys. Rev. Lett. **30**, 948 (1973).
27. L.Q. Lampert, A. Compaan, and I.D. Abella, Phys. Rev. A **4**, 2022 (1971).
28. L.Q. Lampert, Phys. Rev. B **7**, 1834 (1973).
29. Y.C. Chen, K. Chiang, and S.R. Hartmann, Phys. Rev. B **21**, 40 (1980).
30. M. Koch, J. Feldmann, G. von Plessen, E.O. Göbel, P. Thomas, and K. Köhler, Phys. Rev. Lett. **69**, 3633 (1992).
31. V.G. Lyssenko, J. Erland, I. Balslev, K.-H. Pantke, B.S. Razbirin, and J.M. Hvam, Phys. Rev. B **48**, 5720 (1993).
32. N. Peyghambarian, S.W. Koch, and A. Mysyrowicz, *Introduction to Semiconductor Optics* (Prentice-Hall, Inc., Englewood Cliffs, New Jersey, 1993).
33. D.L. Greenaway and G. Harbeke, *Optical Properties and Band Structure of Semiconductors* (Pergamon Press, Oxford, London, 1968).
34. H. Haug, ed., *Optical Nonlinearities and Instabilties in Semiconductors* (Academic Press, Boston, San Diego, 1988).
35. R.S. Knox, *Theory of Excitons*, Solid State Physics, Suppl. 5 (Academic Press, New York, London, 1963).
36. I. Abram, A. Maruani, and S. Schmitt-Rink, J. Phys. C **17**, 5163 (1988).
37. C. Klingshirn, this volume.
38. E.O. Göbel and K. Ploog, Prog. Quant. Electr. **14**, 289 (1991).
39. D.S. Chemla and D.A.B. Miller, J. Opt. Soc. Am. B **2**, 1155 (1985).
40. R.L. Greene, K.K. Bajaj, and D.E. Phelps, Phys. Rev. B **29**, 1807 (1984).
41. J.A. Kash, J.C. Tsang, and J.M. Hvam, Phys. Rev. Lett. **54**, 2151 (1985).
42. M. Jørgensen and J.M. Hvam, Appl. Phys. Lett. **43**, 460 (1983).
43. J.M. Hvam and M.H. Brodsky, Phys. Rev. Lett. **46**, 371 (1981).
44. H. Schwab, c. Dörnfeld, E.O. Göbel, J.M. Hvam, C. Klingshirn, J. Kuhl, V.G. Lyssenko, F.A. Majumder, G. Noll, J. Nunnenkamp, K.-H. Pantke, R. Renner, A. Reznitsky, U. Siegner, H.E. Swoboda, and Ch. Weber, phys. stat. sol. (b) **172**, 479 (1992).
45. Shank, Fork, Leheny, and J. Shah, Phys. Rev. Lett. **42**, 112 (1979).
46. B. Fluegel, N. Peyghambarian, M. Lindberg, S.W. Koch, M. Joffre, D. Hulin, A. Migus, and A. Antonetti, Phys. Rev. Lett. **59**, 2588 (1987).
47. S.W. Koch, N. Peyghambarian, and M. Lindberg, J. Phys. C **21**, 5229 (1988).
48. D. von der Linde, J. Kuhl and Klingenberg, Phys. Rev. Lett. **44**, 1506 (1980).
49. H. Schwab, V.G. Lyssenko, and J.M. Hvam, Phys. Rev. B **44**, 3999 (1991).
50. H. Schwab, V.G. Lyssenko, J.M. Hvam, and C. Klingshirn, Phys. Rev. B. **44**, 3413 (1991).
51. E. Cohen and M.D. Sturge, Phys. Rev. B **25**, 3828 (1982).
52. S. Permogorov, A. Reznitsky, S. Verbin, G.O. Müller, P. Flögel, and M. Nikifirova, phys. stat. sol. (b) **113**, 589 (1982).
53. D. Oberhauser, K.-H. Pantke, J.M. Hvam, K.-H. Schlaad, G. Weimann, and C. Klingshirn, Proc. Int. Conf. Laser '91, San Diego, (1991) p. 368.
54. D. Oberhauser, K.-H. Pantke, J.M. Hvam, G. Weimann, and C. Klingshirn, Phys. Rev. B **47**, 6827 (1993).
55. J.M. Hvam, C. Dörnfeld, and C.R. Paton, SPIE Proc. **1216**, 176 (1990).
56. J. Erland, B.S. Razbirin, K.-H. Pantke, V.G. Lyssenko, and J.M. Hvam, Phys. Rev. B **47**, 3582 (1993).
57. H. Schwab, K.-H. Pantke, J.M. Hvam, and C. Klingshirn, Phys Rev. B **46**, 7528 (1992).

58. P.R. Smith, D.H. Auston, and M.C. Nuss, IEEE J. Quantum Electron. **QE-24**, 255 (1988).
59. S.E. Ralph and D. Grischowsky, Appl. Phys. Lett. **60**, 1070 (1992).
60. H. Harde, S.R. Keiding, and D. Grischowsky, Phys. Rev. Lett. **66**, 1834 (1991).
61. M.C. Nuss. P.M. Mankiewich, M.L. O'Malley, E.H. Westerwick, and P.B. Littlewood, Phys. Rev. Lett. **66**, 3305 (1991).
62. J.E. Pedersen and S.R. Keiding, IEEE J. Quantum Electron. **QE-28**, 2518 (1992).
63. H.G. Roskos, M.C. Nuss, J. Shah, K. Leo, D.A.B. Miller, A.M. Fox, S. Schmitt-Rink, and K. Köhler, Phys. Rev. Lett. **68**, 2216 (1992).
64. B.B. Hu, X.-C. Zhang, and D. Auston, Phys. Rev. Lett. **67**, 2709 (1991).
65. P.C.M. Planken and M.C. Nuss, OSA Annual Meeting Alberquerque 1992, Paper TUTT2.
66. J.E. Pedersen, V.G. Lyssenko, J.M. Hvam, P. Uhd Jepsen, S.R. Keiding, C.B. Sørensen, and P.E. Lindelof, Appl. Phys. Lett. **62**, 1265 (1993).
67. P.C. Becker, H.L. Fragnito, J.Y. Bigot, C.H. Brito Cruz, R.L. Fork, and C.V. Shank, Phys. Rev. Lett. **63**, 505 (1989).
68. Y. C. Chen, K. Chiang, and S.R. Hartmann, Phys. Rev. B **21**, 40 (1980).
69. H. Stolz, in *Festkörperprobleme/Adv. Solid State Physics Vol. 30*, ed. U. Rössler (Vieweg Verlag, Braunschweig, 1991) p. 219.
70. K.-H. Pantke, V.G. Lyssenko, B.S. Razbirin, J. Erland, and J.M. Hvam, *Proc. 21. Int. Conf. Phys. Semicond.*, Beijing 1992 (World Scientific; Singapore, 1992) p. 129.
71. H. Stolz, V. Langer, E. Schreiber, S. Permogorov, and W. von der Osten, Phys. Rev. Lett. **67**, 679 (1991).
72. V. Langer, H. Stolz, and W. von der Osten, Phys. Rev. Lett. **64**, 854 (1990).
73. V. Langer, H. Stolz, W. von der Osten, D. Fröhlich, A. Kulik, and B. Uebbing, Europhys. Lett. **18**, 723 (1992).
74. K.-H. Pantke, P. Schillak, J. Erland, V.G. Lyssenko, B.S. Razbirin, and J.M. Hvam, phys. stat. sol. (b) **173**, 91 (1992).
75. E.F. Gross, B.S. Razbirin, V.P. Fedorov, and Yu P. Naumov, phys. stat. sol. **30**, 485 (1968).
76. V.D. Kulakovskii, G.E. Pikus, and V.B. Timofeev, Sov. Phys. Usp. **24**, 815 (1981).
77. M.L.W. Thewalt, in *Excitons*, eds. E.I. Rashba and M.D. Sturge (North-Holland Publishing Company, Amsterdam, New York, Oxford, 1982) p. 393.
78. J. Erland, B.S. Razbirin, V.G. Lyssenko, K.-H. Pantke, and J.M. Hvam, J. Crystal Growth, in press.
79. P. Dawson, B.A. Wilson, C.W. Tu, and R.C. Miller, Appl. Phys. Lett. **48**, 541 (1986).
80. E. Finkman, M.D. Sturge, and M.C. Tamargo, Appl. Phys. Lett. **49**, 1299 (1986).
81. W.A.J.A. van der Poel, A.L.G.J. Severens, and C.T. Foxon, Opt. Commun. **76**, 116 (1990).
82. P. Lavallard, C. Gourdon, and R. Planel, J. Lumin. **53**, 367 (1992).
83. K. Leo, T.C. Damen, J. Shah, E.O. Göbel, and K. Köhler, Appl. Phys. Lett. **57**, 19 (1990).
84. B.F. Feuerbacher, J. Kuhl, R. Eccleston, and K. Ploog, Solid State Commun. **74**, 1279 (1990).
85. K. Leo, G.O. Göbel, T.C. Damen, J. Shah, S. Schmitt-Rink, W. Schäfer, J.F. Müler, K. Köhler, and P. Ganser, Phys. Rev. B **44**, 5726 (1991).
86. D. Oberhauser, K.-H. Pantke, W. Langbein, V.G. Lyssenko, H. Kalt, J.M. Hvam, G. Weiman, and C. Klingshirn, phys. stat. sol. (b) **173**, 53 (1992).
87. S. Schmitt-Rink, D. Bennhardt, V. Heuckeroth, P. Thomas, P. Haring, G. Maidorn, H. Bakker, K. Leo, D.-S. Kim, J. Shah, and K. Köhler, Phys. Rev. B **46**, 10460 (1992).
88. R. Eccleston, J. Kuhl, D. Bennhardt, and P. Thomas, Solid State Commun. **86**, 93 (1993).
89. J. Feldmann, T. Meier, G. von Plessen, M. Koch, E.O. Göbel, P. Thomas, G. Bacher, C. Hartmann, H. Schweizer, W. Schäfer, and N. Nickel, Phys. Rev. Lett. **70**, 3027 (1993).
90. K. Leo, T.C. Damen, J. Shah, and K. Köhler, Phys. Rev. B **42**, 11359 (1990).
91. M. Koch, D. Weber, J. Feldmann, E.O. Göbel, T. Meier, A. Schulze, P. Thomas, S. Schmitt-Rink, and K. Ploog, Phys. Rev. B **47**, 1532 (1993).
92. E.O. Göbel, K. Leo, T.C. Damen, J. Shah, S. Schmitt-Rink, W. Schäfer, J.F. Müller, and K. Köhler, Phys. Rev. Lett. **64**, 1801 (1990).
93. S. Bar-Ad and I. Bar-Joseph, Phys. Rev. Lett. **66**, 2491 (1991).
94. O. Camel and I. Bar-Joseph, Phys. Rev. B **47**, 7606 (1993).
95. O. Akimoto and E. Hanamura, J. Phys. Soc. Japan **33**, 1537 (1972).
96. W.F. Brinkman, T.M. Rice, and B. Bell, Phys. Rev. B **8**, 1570 (1973).
97. D.A. Kleinman, Phys. Rev. B **28**, 871 (1983).
98. R.C. Miller, D.A. Kleinmann, A.C. Gossard, and O. Munteanu, Phys. Rev. B **25**, 6545 (1982).
99. R. Cingolani, Y. Chen, and K. Ploog, Phys. Rev. B **38**, 13478 (1988).
100. S. Charbonneau, T. Steiner, M.L.W. Thewalt, E.S. Koteles, J.Y. Chi, and B. Elman, Phys. Rev. B **38**, 3583 (1988).

101. D.C. Reynolds, K.K. Bajaj, C.E. Stutz, R.L. Jones, W.M. Theis, P.W. Yu, and K.R. Evans, Phys. Rev. B **40**, 3340 (1989).
102. R. Cingolani, K. Ploog, G. Peter, R. Hahn, E.O. Göbel, C. Moro, and A. Cingolani, Phys. Rev. B **41**, 3272 (1990).
103. R.T. Phillips, D.J. Lovering, G.J. Denton, and G.W. Smith, Phys. Rev. B **45**, 4308, (1992).
104. B.F. Feuerbacher, J. Kuhl, and K. Ploog, Phys. Rev. B **43**, 2439 (1991).
105. S. Bar-Ad and I. Bar-Joseph, Phys. Rev. Lett. **68**, 349 (1992).
106. K.-H. Pantke, D. Oberhauser, V. G. Lyssenko, J.M. Hvam, and G. Weimann, Phys. Rev. B **47**, 2413 (1993).
107. D.J. Lovering, R.T. Phillips, G.J. Denton, and G.W. Smith, Phys. Rev. Let. **68**, 1880 (1992).
108. A. Ferrario, this volume
109. A.I. Ferguson, this volume
110. L. Mollenauer, this volume.

HIGH RESOLUTION NONLINEAR SPECTROSCOPY OF RARE EARTH IONS IN SOLIDS

Roger M. Macfarlane

IBM Research Division
Almaden Research Center
650 Harry Road
San Jose, California 95120-6099

ABSTRACT

Spectroscopy is one of the most useful techniques for obtaining a fundamental understanding of the electronic structure and dynamical behavior of solids. From it, one can obtain a microscopic understanding of phenomena such as energy migration, trapping and upconversion, nonlinear absorption and bleaching, photorefractive effects and numerous other optical, magnetic and transport phenomena. The information available from spectroscopic studies depends to a great extent on the achievable resolution. At low temperatures this resolution is often limited by inhomogeneous strain broadening, which is intrinsically large in glassy systems but always present to some degree in real crystals, however perfect. We discuss a number of nonlinear spectroscopic techniques appropriate to both the frequency domain and the time domain and illustrate how these can overcome the effects of inhomogeneous broadening. Examples will be taken from a wide range of rare earth doped crystals and glasses, where it is possible to achieve a spectral resolution of better than 1 kHz corresponding to a resonance Q greater than 10^{12} for the optical transition. We describe two-step absorption and avalanche absorption leading to frequency upconversion, spectral holeburning, and photon-gated holeburning, as well as photon echoes and nonlinear optical free-induction decays which measure directly the coherent interaction of light with the materials. The effect of external perturbations such as electric and magnetic fields can be measured with great precision using several different nonlinear spectroscopy techniques. Examples of such measurements will be given. Time and frequency domain techniques can provide different windows on relaxation phenomena by probing different parts of the fluctuation spectrum of a relaxation process. This is manifest by time resolved holeburning measurements of spectral diffusion and non-exponential photon echo decays. Some of the limitations and possibilities for application to other materials will be presented.

Nonlinear Spectroscopy of Solids: Advances and Applications
Edited by B. Di Bartolo and B. Bowlby, Plenum Press, New York, 1994

I. INTRODUCTION

The high brightness of laser sources and the interaction of these with material systems have given birth to two related fields: nonlinear optics and nonlinear laser spectroscopy for which N. Bloembergen and A. L. Schawlow were awarded the Nobel Prize in physics in 1981. In nonlinear optics one is normally concerned with nonresonant or possibly near resonant interaction of light and matter and the phenomena are described by macroscopic expressions for the nonlinear susceptibilities $\chi^{(n)}$ [1]. In nonlinear spectroscopy, on the other hand, the principal concern is a study of the numerous optical resonances associated with electronically or vibrationally excited states and the response of the system to resonant excitation [2,3].

Many classes of solids have been studied by laser spectroscopy and one of the richest of these has been rare earth doped crystals and glasses [3] on which we concentrate here. Rare earth ions have an unfilled shell of 4f electrons which couple to give a very complex set of energy levels between which one observes optical transitions from the infrared to ultraviolet spectral regions. Many of these levels are metastable, i.e., have lifetimes of μsecs to msecs and numerous solid state laser transitions are known, [4] the most celebrated being the 1.06μ transition of Nd^{3+} in YAG ($^4F_{3/2} \rightarrow {}^4I_{11/2}$) [5]. We will begin by reviewing the energy level structure of the rare earth ions to set the stage for a discussion of their spectroscopy.

II. ENERGY LEVELS OF THE RARE EARTH IONS

Rare earth ions are most stable in the trivalent state where their ground state has the noble gas Xe electron core $1s^2, 2s^2, 2p^6, 3s^2, 3p^6, 3d^{10}, 4s^2, 4p^6, 4d^{10}, 5s^2, 5d^{10}$, plus a partly filled shell of $4f^n$ electrons. Hund's rule states that the lowest level has the maximum spin multiplicity and the maximum orbital angular momentum consistent with this spin. This fact gives most rare earth ions a large magnetic moment in their ground state. Several of the rare earth ions are stable in the divalent state where many still have a $4f^n$ ground state, but the levels of the $4f^{n-1}5d$ configuration lie lower than for the trivalent ions, often in the visible or infrared regions of the spectrum [6]. The most important divalent rare earth ions are $Sm^{2+}(4f^6)$ and $Eu^{2+}(4f^7)$.

Much effort has been expended in calculating the energy levels of both the free ions and the ions in many crystalline environments using parameterized models describing the Coulomb and exchange interaction between the electrons, the spin-orbit interaction and the effect of a crystal potential on the levels of the free ion. [7-11] Only some of the salient results will be mentioned here. The Coulomb and exchange interactions are the largest and split the $4f^n$ configuration into terms labelled by total spin S and total orbital angular momentum L. The separation between different ^{2S+1}L terms is $1-10{,}000\ cm^{-1}$. Spin-orbit interaction couples L and S resulting in a coupled angular momentum J and splittings of the ^{2S+1}L terms by about $1000\ cm^{-1}$. The crystal-field in turn splits these levels by several hundred cm^{-1} resulting in a level scheme shown in Fig. 1 which was prepared many years ago by Dieke and Crosswhite [6] and Carnall et al. [12] based on calculations and spectroscopy of rare earths in the $LaCl_3$ host. Since the crystal-field interaction is smaller than the Coulomb, exchange and spin-orbit terms in the free-ion Hamiltonian, this diagram serves approximately for rare earth doped solids in general. The thickness of the energy level bars signifies the approximate magnitude of the crystal field splittings.

An important distinction must be made between ions having odd and even numbers of f-electrons. For an odd number of f-electrons (Ce^{3+}, Nd^{3+}, Sm^{3+}, Gd^{3+}, Dy^{3+}, Er^{3+}, Yb^{3+}), Kramers' theorem, which is an expression of time-reversal symmetry, requires that levels be degenerate in pairs, even in the presence of the crystal field. A magnetic field can remove this degeneracy so these ions are paramagnetic, and because of the large angular momentum of the ground state, the paramagnetism can be strong--in some cases more than 10 Bohr magnetons/ion. For the even electron ions (Pr^{3+}, Pm^{3+}, Eu^{3+}, Tb^{3+}, Ho^{3+}, Tm^{3+}), the crystal field removes all of the electronic degeneracy for symmetries lower than axial resulting in non-magnetic levels. In axial symmetry, some levels can retain orbital degeneracy (so called non-Kramers' doublets), and these have an interesting magnetic character resulting from a highly anisotropic magnetic moment.

III. HYPERFINE INTERACTIONS

III.A The Hyperfine Hamiltonian

Hyperfine interactions in rare-earth systems produce splittings which typically lie in the range of 1 MHz to 1 GHz and therefore play an important role in the high resolution spectroscopy of rare-earth ions. In addition, fluctuating nuclear and electron spins couple to rare-earths through hyperfine and super-hyperfine interactions to produce what is often the dominant source of line broadening at low temperatures. This subject was treated in my previous Erice lectures but will be included here for convenience and completeness. For the most part, hyperfine interactions are

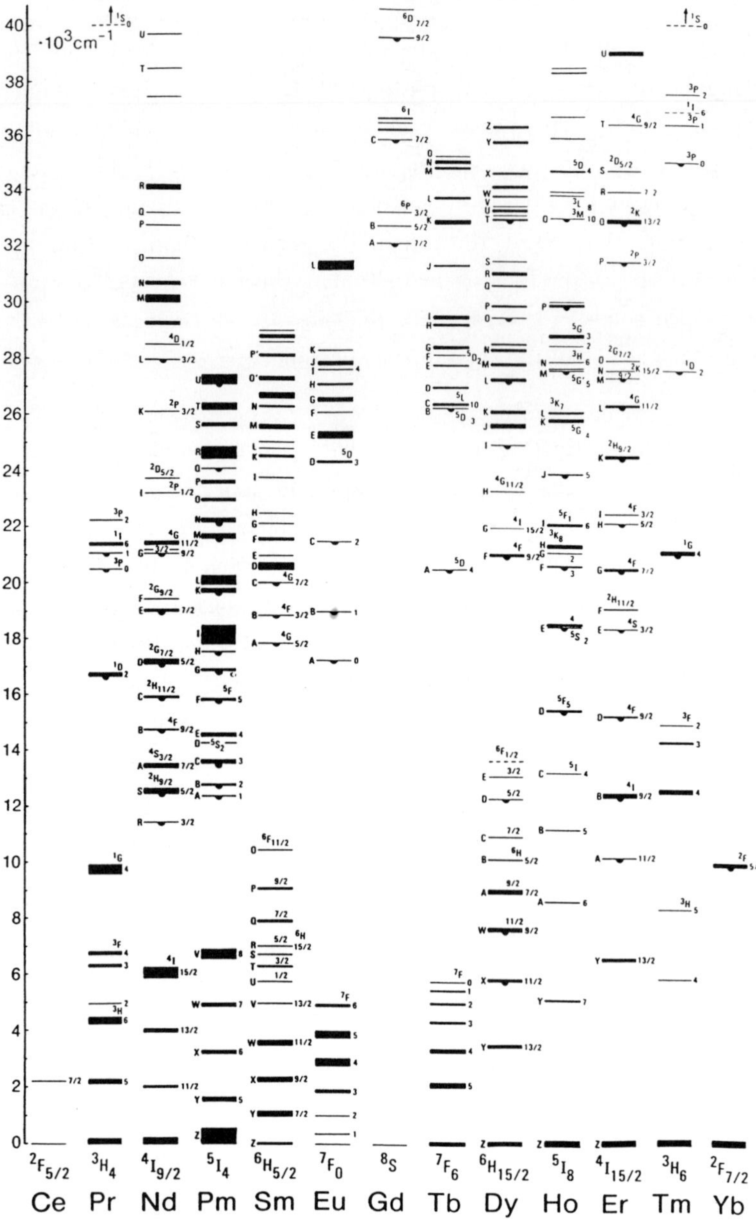

Figure 1. Energy levels of the $4f^n$ electrons of the trivalent rare earth ions.

adequately handled by effective, or spin-Hamiltonians starting from a zeroth order basis of the crystal field Hamiltonian. In most of the experiments to be reviewed below, the results are well accounted for by the Hamiltonian:

$$\mathcal{H} = [\mathcal{H}_{FI} + \mathcal{H}_{CF}] + [\mathcal{H}_{HF} + \mathcal{H}_{Q} + \mathcal{H}_{Z} + \mathcal{H}_{z}]. \tag{1}$$

The first two terms are the free ion Hamiltonian including spin-orbit coupling, and the crystal-field Hamiltonian discussed in the last section. The remaining terms have much smaller contributions of the order of optical inhomogeneous broadening and are, in order of appearance, the hyperfine coupling between the 4f electrons and the rare earth nucleus, the nuclear electric quadrupole interaction, the electronic Zeeman interaction, and the nuclear Zeeman interaction.

In the calculation of crystal field energy levels and electronic g-values, the first two terms of (1) are often diagonalized within the $4f^n$ configuration, thus taking into account intermediate spin-orbit coupling, and crystal field J-mixing. A simpler treatment of the zero-order Hamiltonian is, however, almost always used for a discussion of hyperfine interactions, i.e., the use of L-S coupling and a treatment of the crystal field interaction diagonal in J. This is generally a good approximation since the separation between $^{2S+1}L_J$ manifolds is larger than crystal field splittings.

The crystal-field wavefunctions, i.e., the eigenstates of the first two terms in (1), determine the hyperfine structure and magnetic response due to the last four terms. The results are most simply expressed in terms of an effective Hamiltonian for the electronic level of interest. The explicit forms of these four terms beginning with the magnetic hyperfine interaction are:

$$\mathcal{H}_{HF} = 2\mu_B h\gamma_N < r^{-3} > \vec{N} \cdot \vec{I} \tag{2}$$

where μ_B is the Bohr magneton, γ_N is the nuclear gyromagnetic ratio, $2\mu_B < r^{-3} > \vec{N}$ represents the field at the nucleus due to the f electrons and $< r^{-3} >$ is the expectation value of the inverse cube electron-nuclear distance. Within one LSJ state,

$$\mathcal{H}_{HF} = A_J \vec{I} \bullet \vec{J} \tag{3}$$

where

$$A_J = 2\mu_B \gamma_N \hbar < r^{-3} > < J\|N\|J > . \tag{4}$$

The quadrupole Hamiltonian has the form:

$$\mathcal{H}_Q = P[(I_{z'}^2 - I(I+1)/3) + (\eta/3)(I_{x'}^2 - I_{y'}^2)], \tag{5}$$

where P is the quadrupole coupling constant, η the electric field gradient (EFG) asymmetry parameter and x'y'z' are the principal axes of the EFG tensor. $\mathcal{H}_Q$ has contributions from the lattice, from the f electrons and from coupling to closed shells whose charge distribution is distorted by the electric field of the lattice. This latter, the so-called Sternheimer anti-shielding effect [13], often dominates.

The interaction of the electrons with an external magnetic field is given by:

$$\mathcal{H}_Z = \mu_B \vec{H} \bullet (\vec{L} + 2\vec{S}) \tag{6}$$

where $\vec{H}$ is the external magnetic field, $\vec{L}$ and $\vec{S}$ are the electronic orbital and spin angular momenta, and μ_B is the Bohr magneton. Within a single LSJ state, this can be written as

$$\mathcal{H}_Z = g_J \mu_B \vec{H} \cdot \vec{J} \tag{7}$$

where g_J is the Lande g-value, $g_J = (3/2) - [L(L+1) - S(S+1)]/[2J(J+1)]$. The nuclear Zeeman interaction is

$$\mathcal{H}_Z = -h\gamma_N \vec{H} \cdot \vec{I} \tag{8}$$

with γ_N and I being the nuclear gyromagnetic ratio and nuclear spin, respectively.

1. Electronic Singlets. For non-Kramers' ions in sites of low symmetry (e.g., the C_2 site in LaF_3 or the C_s site in $YAlO_3$) the electronic degeneracy of a given J-state is completely lifted, and the electronic angular momentum is quenched to first order. Hyperfine and electronic magnetic effects appear in second order of perturbation theory, and the effective Hamiltonian including nuclear Zeeman and nuclear quadrupole interactions, is [14]:

$$\begin{aligned}\mathcal{H}_{eff} &= -g_J^2\mu_B^2\vec{H} \cdot \tilde{\Lambda} \cdot \vec{H} - (2A_J g_J \mu_B \vec{H} \cdot \tilde{\Lambda} \cdot \vec{I} - \mathcal{H}_Z) - (A_J^2 \vec{I} \cdot \tilde{\Lambda} \cdot \vec{I} + \mathcal{H}_Q) \\ &\equiv -g_J^2\mu_B^2\vec{H} \cdot \tilde{\Lambda} \cdot \vec{H} + \mathcal{H}_Z' + \mathcal{H}_Q'\end{aligned} \tag{9}$$

where the tensor $\tilde{\Lambda}$ is given by

$$\Lambda_{\alpha\beta} = \sum_{n=1}^{2J+1}{}' \frac{<0|J_\alpha|n><n|J_\beta|0>}{\Delta E_{n,0}} \tag{10}$$

with $|0>$ being the level to which $\mathcal{H}_{eff}$ applies, $|n>$ the other crystal field levels of the LSJ term involved, and $\Delta E_{n,0}$ is the energy difference $E_n - E_0$. The first term in (9) is the quadratic electronic Zeeman shift, $\mathcal{H}_Z'$ is the enhanced nuclear Zeeman Hamiltonian [15], i.e.,

$$\mathcal{H}_Z' = -h[\gamma_x H_x I_x + \gamma_y H_y I_y + \gamma_z H_z I_z] \tag{11}$$

where x, y and z are the principal axes of the Λ-tensor and the effective enhanced nuclear gyromagnetic ratios are given by

$$\gamma_\alpha = \gamma_N + \frac{2g_J\mu_B A_J \Lambda_{\alpha\alpha}}{h}. \tag{12}$$

The term $A_J^2 \vec{I} \cdot \tilde{\Lambda} \cdot \vec{I}$ in (9) is the second order magnetic hyperfine interaction, or pseudo-quadrupole interaction [16], which can be written in the same form as $\mathcal{H}_Q$, i.e.,

$$\mathcal{H}_{pq} = D_{pq}\left[I_z^2 - I(I+1)/3\right] + E_{pq}\left[I_x^2 - I_y^2\right]. \tag{13}$$

In terms of the Λ tensor, the zero field parameters D_{pq} and E_{pq} can be written:

$$D_{pq} = A_J^2 \cdot \left[\frac{1}{2}(\Lambda_{xx} + \Lambda_{yy}) - \Lambda_{zz}\right] \tag{14}$$

$$E_{pq} = A_J^2 \cdot \frac{1}{2}\left[\Lambda_{yy} - \Lambda_{xx}\right]. \tag{15}$$

An accurate calculation of the enhanced nuclear Zeeman and pseudo-quadrupole contributions requires accurate crystal-field wavefunctions. These are rarely available for ions in low symmetry sites but in some cases it is possible to obtain good agreement with measured values. When combining the pseudo-quadrupole and pure quadrupole terms into one effective quadrupole Hamiltonian, it must be kept in mind that the principal axes of the Λ-tensor (x,y,z), and of $\mathcal{H}_Q(x', y', z')$ are in general different, although they coincide in cases of axial or higher symmetry. When the sum of these two terms is diagonalized, a third set of principal axes, x″, y″ and z″ will result for which:

$$\mathcal{H}_Q' = D\left[I_{z''}^2 - I(I+1)/3\right] + E\left[I_{x''}^2 - I_{y''}^2\right]. \tag{16}$$

D and E in this equation correspond to the combined quadrupole and second order hyperfine coupling constants in the double primed axis system. For certain rare earth ions, the enhanced nuclear Zeeman and pseudo-quadrupole coupling are much larger than the nuclear Zeeman and electric quadrupole interactions. In this case the axes x, y, z and x″, y″, z″ may nearly coincide.

2. Non-Kramers' Doublets. In axial or higher symmetry, not all of the levels have their electronic angular momentum along the symmetry axis quenched by the crystal field. As a consequence of time reversal symmetry, the angular momentum perpendicular to the symmetry axis is zero to first order. The resulting effective Hamiltonian is dominated by the parallel electronic Zeeman effect and a hyperfine term containing only $S_Z I_Z$. The second order terms discussed in the preceding section are present as well and can be treated in the form of enhanced nuclear Zeeman and pseudo-quadrupole spin Hamiltonians. Thus, a suitable effective Hamiltonian for a non-Kramers' doublet is

$$\mathscr{H} = g_Z \mu_B H_Z S_Z + A S_Z I_Z + \mathscr{H}'_Q + \mathscr{H}'_Z \tag{17}$$

where S_Z is an effective spin quantum number which can take the values $\pm 1/2$. The quantities g_Z and A are given by $g_Z = 2g_J < + | J_Z | + >$ and $A = 2A_J < + | J_Z | + >$ with $| + >$ and $| - >$ being the two components of the doublet. $\mathscr{H}'_Z$ and $\mathscr{H}'_Q$ are defined as in (9) and (16), respectively.

3. Kramers' Doublets. For ions with an odd number of electrons, Kramers' theorem dictates that all levels have electronic degeneracy which can only be lifted by time-odd interactions such as the Zeeman interaction with an external magnetic field. This electronic degeneracy and associated magnetic moment leads to large first order Zeeman and hyperfine interactions. Thus, an appropriate spin Hamiltonian is similar to (17) except that the g-values and hyperfine interactions are anisotropic and nonzero in all three principal directions:

$$\mathscr{H} = \mu_B \vec{H} \cdot \tilde{g} \cdot \vec{S} + \vec{I} \cdot \tilde{A} \cdot \vec{S} + \mathscr{H}'_Q + \mathscr{H}'_Z. \tag{18}$$

Since Kramers' degeneracy arises from time reversal symmetry, random strains cannot lift this degeneracy but they may contribute to inhomogeneous broadening of transitions between different Kramers' doublets.

IV. RELAXATION PROCESSES

Relaxation processes and spectroscopy are intimately related; spectroscopy is concerned with the frequency response of a material system and emphasizes a study of the quantum states of matter and their resonances whereas relaxation phenomena focus on the widths of these resonances and the time response of the system following excitation. A quantity of fundamental importance is the homogeneous linewidth of a transition between two quantum states. This has its origins in the coupling of the resonant transition to a fluctuating bath. The zero-point fluctuations such as those of the radiation field or the phonon field produce spontaneous emission and associated lifetimes, and the non zero-point fluctuations produce stimulated emission. The latter also introduce phase interruptions of a coherently prepared superposition state, i.e., they modulate the frequency of a transition producing a broadening but no population decay. The total homogeneous linewidth Γ_h is conveniently divided into population decay terms contributing a lifetime T_1 and phase fluctuation terms which give a dephasing time T_2:

$$\Gamma_h = \frac{1}{2\pi T_1} + \frac{1}{\pi T_2} \tag{19}$$

In nonlinear spectroscopy we are often concerned with the coherent interaction of light and matter which is usually described by a set of equations analogous to those describing 2-level spin systems in magnetic resonance first proposed by Bloch. [17] These so-called optical Bloch equations [18,19] describe the dynamics of coherent light -matter interaction by the motion of a Bloch vector $R = (u,v,w)$ in a three-space $(\vec{1}, \vec{2}, \vec{3})$.

The relaxation time T_2 corresponds to motion of this vector in the 1-2 plane and T_1 to a reduction in the length of this vector. Neglecting relaxation, the motion of R is described by $\frac{dR}{dt} = \Omega \times R$ where the driving vector $\Omega = [-\chi, 0, \Delta]$ has components of the Rabi frequency $\chi = \mu_{ab}E/\hbar$ (μ_{ab} is the transition dipole moment connecting two states a and b, E is the electric vector of the light), and Δ which is the frequency difference between the driving field and the 2-level resonance frequency ω_{ab}[20].

The optical Bloch equations in their usual form include the relaxation or damping terms T_1 and T_2 which are added phenomenologically:

$$\begin{aligned} \dot{u} + \Delta v + \frac{u}{T_2} &= 0 , \\ \dot{v} = \Delta u - \chi w + \frac{v}{T_2} &= 0 , \\ \dot{w} + \chi v + \frac{(w - w_{eq})}{T_1} &= 0 , \end{aligned} \tag{20}$$

where

$$\begin{aligned} u &= \rho_{ab} + \rho_{ba} , \\ v &= i(\rho_{ba} - \rho_{ab}) , \\ w &= \rho_{bb} - \rho_{aa} . \end{aligned} \tag{21}$$

with u being the component of the polarization in phase with the driving field of the light, v the component 90 degrees out of phase and w the population difference between levels a and b. These equations give a very useful description of the coherent light matter interaction although they are approximate and care is needed in applying them to individual cases. The most obvious complication arises if there are three or more levels coupled together so that multi-level coherences may be seen. This can occur in the presence of hyperfine or superhyperfine splittings, for example, when an exciting light pulse induces both optical and "sublevel" coherence and a variety of new phenomena appear [21-31]. Another limitation of the Bloch equations is that they assume rapid, uncorrelated motion of the bath responsible for relaxation. If the bath fluctuations become comparable to T_2, for example, the 2-level system can become decoupled from the bath at high values of χ. This effect was well known in magnetic resonance [32] and has been demonstrated in optical experiments [33-35] and analyzed theoretically [36-39].

An inspection of Fig. 1 shows that the energy levels of trivalent rare earth ions cluster into groups with sometimes relatively large gaps between them. Since the separation between energy levels within a group is of the order of one or two lattice phonon frequencies, spontaneous phonon emission broadens the optical transitions involving the upper components and they are homogeneously broadened even at low temperatures. The extent of this broadening depends strongly on the energy separation $\delta\omega$ to the levels below, since the density of states of acoustic phonon modes varies as $(\delta\omega)^2$. On the other hand, transitions to the lowest level of a group which has a large gap below it can have long lifetimes (μsec – msec). Here the decay is a multiphonon process which becomes exponentially less probable with increasing gap [40,41]. These metastable levels occupy a special position in optical spectroscopy because, at low temperatures (<2K), phonon induced dephasing is essentially absent. The homogeneous linewidth of transitions to these levels can be as sharp as ~ 0.1 kHz and is limited either by spontaneous phonon or photon emission (T_1) or by contributions to T_2 from the fluctuating nuclear spin fields which are present in almost all materials [3]. They are often six to nine orders of magnitude sharper than transitions to levels a few 10's or 100's of cm^{-1} above them!

The situation in glassy hosts is very different because of the presence of very low frequency tunnelling modes (so called two-level systems) which can only be frozen out in the mK temperature region and which contribute to anomalously large homogeneous linewidths for optical transitions of ions in glasses [42,43].

At higher temperature ($\gtrsim 5$K) phonon broadening sets in and one-phonon and phonon scattering processes increase the optical linewidths of transitions to metastable levels from $\sim$ MHz at 2K to $\sim$ 100 GHz at room temperature. A discussion of these processes can be found in [44-48].

V. INHOMOGENEOUS BROADENING

Inhomogeneous broadening in solid state systems normally arises from crystal strains or, in the case of glassy systems, from the presence of disorder which results in a different local environment, and hence a shifted resonance frequency, for different ions. It is usually the dominant source of spectral line broadening at low temperatures for energy levels whose lifetimes due to spontaneous phonon emission are longer than ~ 10 psec corresponding to a width of ~ 10 GHz [49]. The distinction between homogeneous broadening due to dynamical processes such as phonon coupling , local spin fluctuations or tunneling in glasses and inhomogeneous broadening due to a static distribution of environments is a convenient classification that works quite well. However, as we will see later, more and more examples are known where the distinction breaks down because of the time scale of the measurement process. For the moment, however, we retain the simple picture. Often, from a spectroscopist's point of view, inhomogeneous broadening is undesirable and several nonlinear spectroscopic techniques described below have been devised to eliminate its effects either by selectively exciting a narrow spectral packet with a monochromatic laser or using a "trick" such as the photon echo [2,3]. There are some cases, however, where it is desired to increase inhomogeneous broadening by intentional doping to induce disorder. Examples of this are materials in which one wishes to store frequency domain information in the form of spectral holes such as $Ba(Cl, Br)F{:}Sm^{2+}$ [50]. Provided the homogeneous linewidth is not broadened, the wider the line, the more information that can be stored in it.

A quantitative understanding of the origins of inhomogeneous broadening is elusive although, in some cases, the mechanisms are at least semi-quantitatively understood. One problem is that many microscopic mechanisms contribute together, such as point defects, dislocations, charge defects and isotope effects and they are difficult to separate. The crystalline environment is described by many parameters so there is no single variable such as the velocity of atoms or molecules in a gas. Some theoretical analyses have been carried out [51,52] and some specific systems are amenable to study. One particularly dramatic example is the case of $Er^{3+}{:}YLiF_4$ [53]. Even at moderate concentrations of the Er^{3+} dopant e.g. 0.1%, the width of the ${}^4I_{15/2}(1) \rightarrow {}^4I_{9/2}(1)$ line is less than 1 GHz and hyperfine structure is observed (Fig. 2). Crystal strains induced by the Er^{3+}-ion doping are felt by other Er^{3+} ions. As the concentration of Er^{3+} decreases to ~ 10 ppm, the line narrows to ~ 300 MHz and acquires a satellite structure with a splitting of 1.3 GHz. This structure originates from the presence of the two naturally abundant isotopes ${}^6Li$ and ${}^7Li$ which create slightly different crystal fields at the Er site. For the nearest neighbor lithium ions this causes structure, and for the more distant ones it contributes to inhomogeneous broadening. This was shown by growing a crystal with just the ${}^7Li$ isotope present, and measuring the Er line profile which showed remarkably narrow widths; as low as ~ 50 MHz and new structure never before directly resolved in rare earth spectroscopy: the isotope shift of the f-f transitions, together with the optically resolved hyperfine structure of the ${}^{167}Er$ isotope which has a non-zero nuclear spin of 7/2. (Fig. 2). The narrowest of these hyperfine lines corresponds to transitions between non-magnetic levels showing that some of the inhomogeneous broadening in the other transitions is magnetic in origin. It comes from the nearby fluorine ions which have a magnetic moment of 2.6 nuclear magnetons and they are "frozen" by the presence of the large magnetic moment of the Er^{3+} ion which is several Bohr magnetons. Thus starting at a concentration of Er^{3+} of about 0.1%, inhomogeneous broadening comes from the disturbance of the $YLiF_4$ lattice of the dopant ions, and then at a few ppm of Er^{3+} the distribution of lithium isotopes becomes limiting. For a crystal with a single lithium isotope the line splits into several components due to the Er isotope shift, so this isotope shiftfile was contributing to the width in other samples containing ^6Li and ^7Li. The remaining inhomogeneous width arises from "frozen" spin fields of the fluorine nuclei ($\lesssim 40$ MHz) and residual crystal strains ($\lesssim 30$ MHz), and this combines with ~ 10 MHz of homogeneous linewidth to produce the total width of ~ 50 MHz.

What is not yet understood is why $YLiF_4$ exhibits such a high degree of crystal perfection. One can speculate that other materials will be found with even narrower resonances that may find applications as secondary frequency standards.

An important thing to note is that while nonlinear spectroscopic techniques can be used to eliminate the effects of inhomogeneous broadening and extract otherwise hidden structure such as hyperfine, superhyperfine or tunneling splittings, only by reducing the overall inhomogeneous linewidth can the isotope shifts be revealed.

VI. NONLINEAR ABSORPTION

Nonlinear absorption involves the absorption of two or more photons and may proceed either via virtual intermediate states where it is essentially an instantaneous process [54], or via real intermediate states where the dynamics of the intermediate states confers a complicated dynamical behavior on the nonlinear absorption. Here we will be concerned only with the latter case, noting that rare earth doped solids have many metastable levels which make two- or multiple-step absorption relatively easy to observe, and since the final states or levels nearby are themselves metastable, nonlinear absorption leads to upconverted fluorescence and has been used to pump upconversion lasers, particularly in optical fibers. Three kinds of absorption will be discussed here. The first is two-step absorption in which two sequential one-photon absorptions take place. The second is one-photon absorption followed by internal cross-relaxation

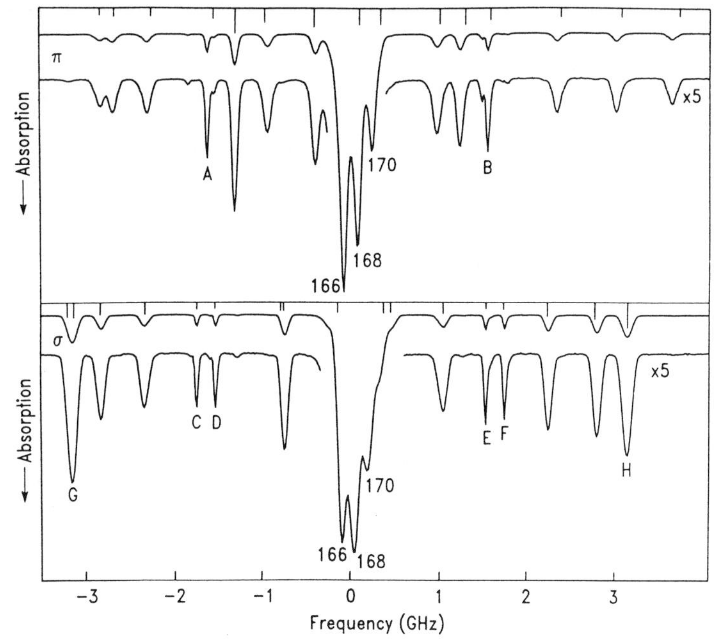

Figure 2. Excitation spectrum of the $^4I_{15/2}(1) \rightarrow {}^4F_{9/2}(1)$ transition of Er^{3+}:Y^7LiF_4 showing structure of the erbium isotopes of mass number 164-170 and hyperfine structure for mass 167.

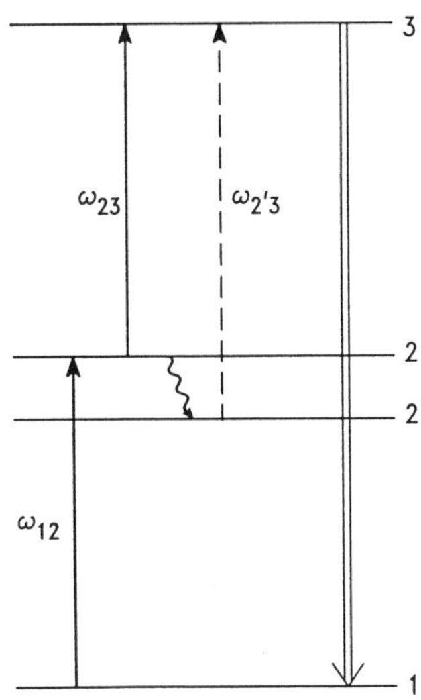

Figure 3. General energy level diagram for two-step nonlinear absorption. Level 2 may decay nonradiatively into level 2′. A subsequent emission step (thick arrow), is upconverted fluorescence.

between excited ions and the third is avalanche absorption which to some extent combines the first two; that is an initial two-step absorption followed by a cross-relaxation process.

VI.A. Two-Step Absorption

In this process (Fig. 3) an intermediate metastable level (2), with a lifetime of typically 0.1 – 10 msec, is populated by the first absorption step and a second photon further excites ions from (2) to (3). A more general scheme is one in which (2) relaxes to (2′) by radiative or non-radiative decay and the second step is (2′) → (3). (Fig. 3). For pulsed laser excitation, the time of the first and second absorption steps can be controlled and information obtained about internal relaxation processes. This is the basis for pump-probe techniques which are often used in the nsec-fsec time scales. In the case of cw excitation, steady-state populations favor the long lived metastable levels, for example, (2′) in Fig. 3 and processes such as (1) → (2) followed by (2′) → (3) dominate. If emission follows from level (3), this gives upconverted fluorescence, which is the basis of the infrared quantum counter idea of Bloembergen [55]. The spectrum of upconverted fluorescence can be very different following pulsed or cw excitation. In a crystal doped with ~ 1% of a rare-earth ion, a focused cw laser intensity of 10 kW/cm^2 can excite essentially all of the ions to a level which stores population for ~ 1 msec. This is a modest excitation density corresponding to 30 mW focused to a beam waist of 10μ. Under these conditions, the second step absorption can be comparable to the first step, or even higher if it is associated with a higher oscillator strength. These considerations led to the use of non-linear absorption processes to pump upconversion lasers [56].

The first cw upconversion laser was operated in 1% Er^{3+}:$YAlO_3$ using the pumping scheme shown in Fig. 4 [57]. The first step pumps Er^{3+} ions from the ground state to one of the crystal field components of $^4I_{9/2}$ from where it relaxes to $^4I_{11/2}$ from which the second step further excites the ions to $^4F_{5/2}$. Rapid relaxation then populates the $^4S_{3/2}$ upper level leading to laser emission at 549.8 nm. The choice of pump wavelengths was dictated to a large extent by available lasers, but evidently a number of other pumping schemes could lead to the same lasing transition, e.g., pumping the first step directly into $^4I_{11/2}$ around 970 nm. The original work used two dye lasers for pumping but diode lasers are now available for this purpose. The output of the upconversion laser is shown in Fig. 5 as a function of the power of the two

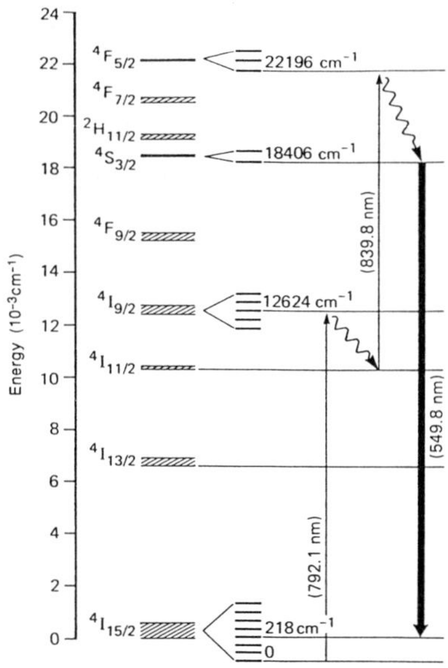

Figure 4. Nonlinear absorption pumping scheme for an upconversion laser in Er^{3+}:YAlO$_3$ showing the pumping and lasing (549.8nm) transitions.

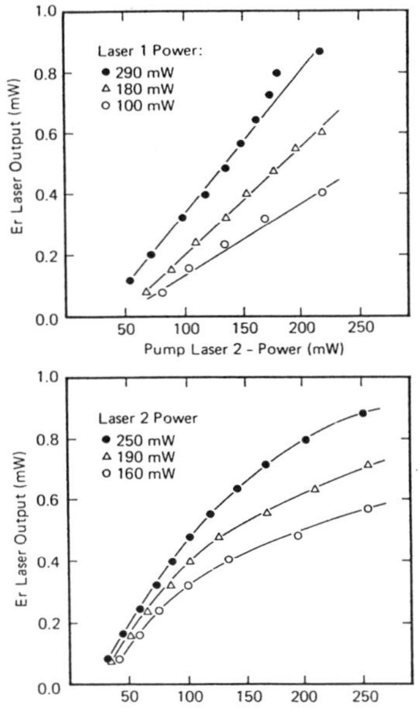

Figure 5. Output power of Er^{3+}:YAlO$_3$ laser as a function of the two pump-laser powers. Note the saturation due to ground state depletion in the lower curves.

pump lasers. When the first step is fixed, the output is linear in the second step, whereas under the conditions used of a 15μ pump waist and a 22μ cavity waist, the output obtained varying the first step shows saturation at higher first-step powers.

From a spectroscopic point of view, nonlinear absorption with two or even three laser sources can provide a wealth of information about the electronic structure of ions in solids notably about the coherence behavior of transitions between 2 or more excited states. This work is still in its infancy, but with the increasing availability of tunable diode laser sources, much more will be done. From a practical point of view, such as the pumping of upconversion lasers, it is desirable to use a single laser source to produce nonlinear absorption. Direct 2-photon absorption via a virtual intermediate state is too weak even with most pulsed sources. In the case of rare-earth doped optical fibers, nonlinear absorption using a single pump frequency is often rather efficient because of the broad absorption bands which result from the disordered structure of the glass. This provides an accidental resonance for two or more transitions at the same pump wavelength. A nice example of this is provided by the Ho^{3+} fiber laser which was the first room temperature upconversion laser reported [58]. The relevant energy levels and the nonlinear absorption steps used to pump this laser with a Kr^+ laser at 647.1 nm are shown in Fig. 6. The first step excites the $^5F_5$ level, which relaxes to $^5I_6$ in about 10 μsec, and a second step further excites from $^5I_6 \rightarrow {}^5G_6$ which quickly relaxes to the $^5S_2$ upper laser level. Another two-step pathway comes from excitation out of the longest lived excited state $^5I_7$, ($T_1 \simeq 15$ msec) to $^5F_3$ which in turn populates the $^5S_2$ upper laser level. Although one might expect a quadratic dependence of output power on pump power in such a laser, the high optical intensities achieved in the fiber easily leads to ground state depletion and a linear dependence as shown in Fig. 7 [59]. Depletion of the ground state population also lowers the threshold of this quasi three-level laser.

VI.B. Cross Relaxation Excitation

The nonlinear two-step, or in general multi-step, absorption discussed above is particularly applicable to the case of low dopant concentrations where the interaction between the optical centers is smaller than their radiative emission rates. At higher concentrations, typically greater than a few tenths of a percent for rare earth ions, a new nonlinear process of optical excitation occurs which relies on a cross relaxation between a pair of excited ions. This process was first described by Auzel [60] and

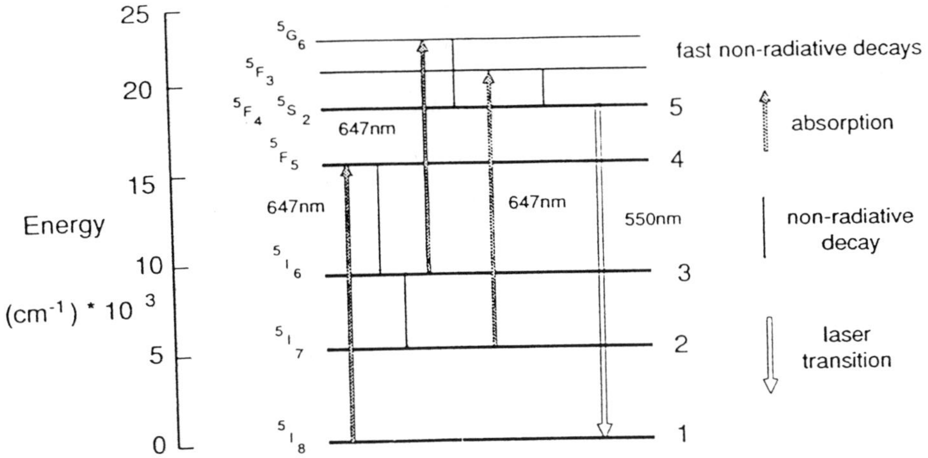

Figure 6. Energy levels for the Ho^{3+}:ZBLAN fiber laser showing multiple excitation steps of the 647.1 nm Kr^{+} pump laser (after ref. 59).

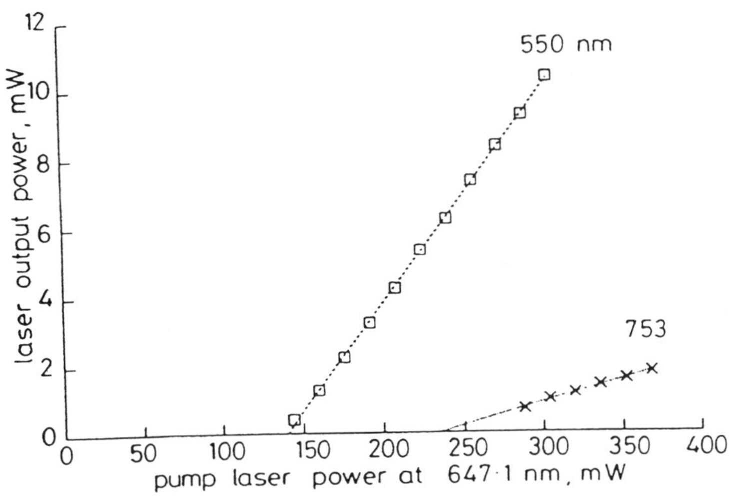

Figure 7. Output of the Ho^{3+}:ZBLAN fiber laser as a function of the pumping power (from ref 58).

termed APTE (Addition de Photons par Transfer d'Energie) (see Fig. 8). Two photons are absorbed creating two excited ions in a metastable state. This may be the state initially excited or a lower one to which they relax. The system then pumps itself internally by a cross-relaxation step in which one ion is de-excited and the other is promoted to a higher excited state. Subsequent relaxation populates an upper metastable level from which laser action can be observed. It is not necessary that the cross-relaxation process be exactly resonant, that is, that the final energy of the promoted ion be equal to the sum of the excited state energies of the two ions involved in the transfer. Phonon-assisted transfer can take place involving either the creation or (at elevated temperatures) annihilation of phonons. For the same reason, it is not necessary that the two ions are the same; a donor-acceptor transfer between different ions can also occur via APTE. An early example of the use of this nonlinear pumping process for an upconversion laser is provided by 1% Er^{3+}:$YLiF_4$ [61,62]. The energy level scheme is shown in Fig. 9. Excitation into the $^4I_{9/2}$ level results in a high concentration of ions in the metastable $^4I_{11/2}$ level which has a lifetime of $\sim$ 4 msec at room temperature. Two ions which are sufficiently close that their cross-relaxation rate exceeds the population decay rate, undergo an APTE process exciting one of them to the $^4F_{5/2}$ level which relaxes rapidly to the $^4S_{3/2}$ upper laser level. The advantage of this pumping scheme is that only one excitation source is used and the non-linearity comes from the cross-relaxation step. The efficiency of this laser was found to be higher than in the case of two-step excitation of Er^{3+}:$YAlO_3$ and pumping with a single diode laser was also demonstrated [63]. Due to ground state depletion, the output power was again linear with the pump despite the two-ion excitation process. An interesting characteristic of this excitation scheme is that laser emission continues after the pump is turned off because the reservoir of ions in the intermediate state (here $^4I_{11/2}$) continues to feed the upper laser level until the system drops below threshold. Depending on the pump wavelength and the possibilities of accidental two-step resonances, it is possible to observe intermediate behavior, i.e., a sudden drop in laser output after cessation of the pump followed by a slow decay [63]. All of these Er^{3+} lasers operated only below 100K.

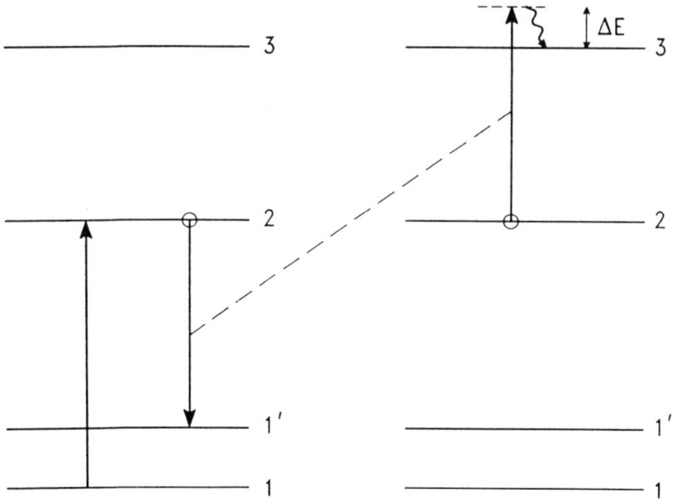

Figure 8. Nonlinear upconversion due to cross-relaxation between two excited ions in level 2. In general, the process is phonon assisted but resonant transfer ($\Delta E = 0$) is possible as a special case.

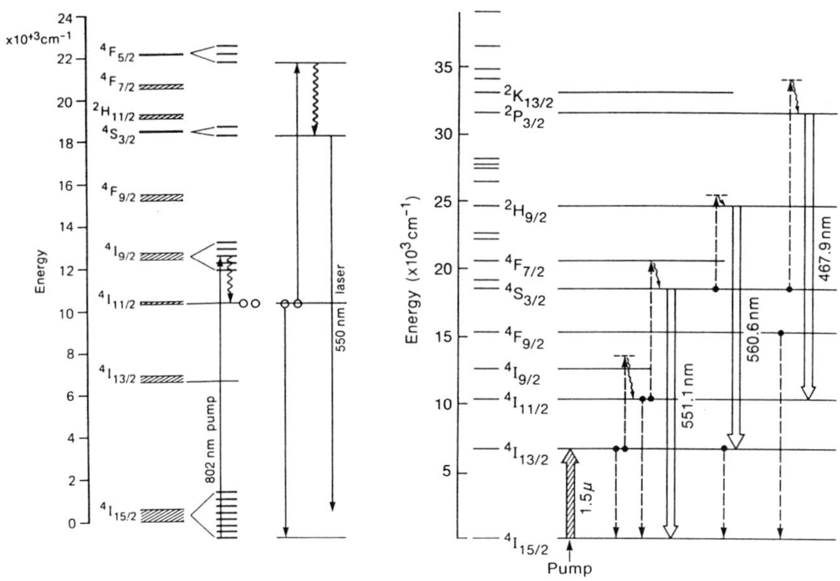

Figure 9. Energy level scheme for an Er^{3+}:$YLiF_4$ upconversion laser pumped by cross-relaxation a) 2-step upconversion b) multistep upconversion.

VI.C. Avalanche Absorption

The third nonlinear absorption process is the so-called avalanche-absorption (or "photon avalanche" [64]). Here a single excitation frequency is used, but it is resonant not with a transition from the ground state as in the last case but with a transition between two excited states. The choice of which excited state resonance produces an avalanche depends on the details of the situation. A simplified avalanche absorption model is illustrated in Fig. 10 and requires the participation of two ions. The pump frequency ω_{23} is resonant with the excited state transition (2) → (3) and excites above level (2) in which the population build-up will occur. Initially there is no population in (2) and the pump is only very weakly absorbed into a phonon sideband of (2) resulting in almost 100% transmission. This seeds a small population in (2) followed by absorption from the resonant pump (2) → (3). The next step is a cross relaxation from level 3 which effectively pumps the second ion from its ground state to level (2), at the same time relaxing the first ion to (2). Thus a single ion in (3) creates two ions in (2) which in turn absorb pump light to reach level (3) and the process continues. This is the essential step in the nonlinearity. The cross relaxation step does not have to be exactly resonant but can be phonon assisted as in the APTE process. In addition, the excited state resonance could involve a level above (3) which non-radiatively decays to (3). Such relaxation of the constraints on the level structure means that avalanche absorption can arise in a relatively large number of cases and for rare earths it has so far been demonstrated in Pr^{3+}, [64-66] Nd^{3+}, [56,67-69] Sm^{3+} [70] and Tm^{3+}[71,72] systems and for transition metals in Ni^{2+} [73].

A rate equation model which contains the elements outlined above was proposed [56] to describe the excitation of the first avalanche-pumped upconversion laser which was based on Nd:$YLiF_4$ [74]. Since the 2′ → 2 relaxation is fast, the dynamics can be described by three rate equations for the three populations, n_1, n_2 and n_3 in levels (1), (2) and (3) respectively.

$$
\begin{aligned}
\dot{n}_1 &= -R_1 n_1 + \frac{n_2}{\tau_2} + \frac{n_3 b}{\tau_3} - \sigma n_1 n_3 , \\
\dot{n}_2 &= R_1 n_1 - R_2 n_2 - \frac{n_2}{\tau_2} + \frac{n_3(1-b)}{\tau_3} + 2\sigma n_1 n_3 , \\
\dot{n}_3 &= n_2 R_2 - \frac{n_3}{\tau_3} - \sigma n_1 n_3 , \\
1 &= n_1 + n_2 + n_3 .
\end{aligned}
\tag{22}
$$

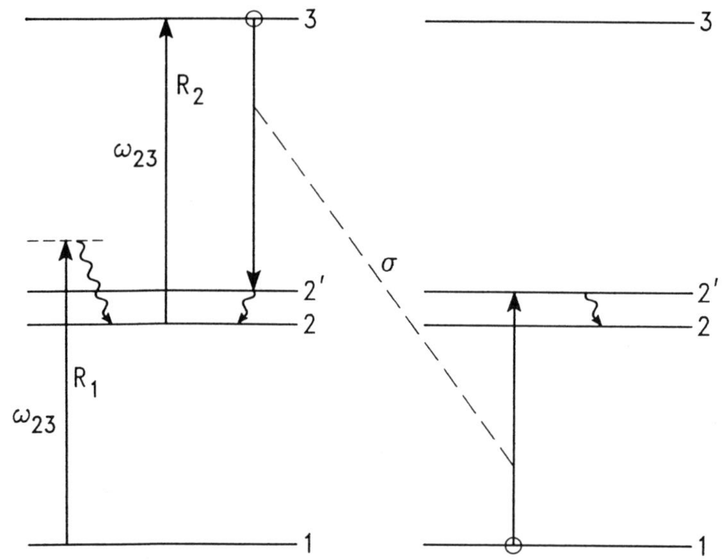

Figure 10. Energy level diagram illustrating the absorption avalanche process. The two diagrams refer to separate ions.

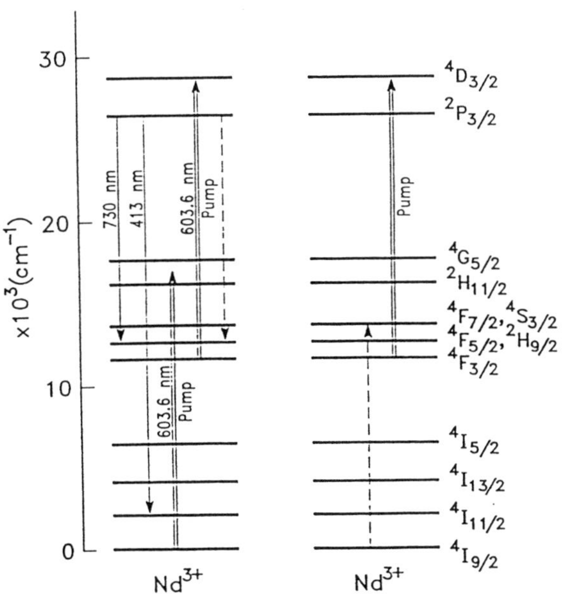

Figure 11. Energy levels of Nd^{3+}:$YLiF_4$, the first avalanche pumped upconversion laser, showing avalanche absorption steps and laser transitions at 730nm and 413nm pumped by this absorption.

The rates R_1 and R_2 are the pumping rates for the first non-resonant step (which is weak) and the second resonant step, both at frequency ω_{23}. The branching of level $(3) \rightarrow (1)$ is b and $(3) \rightarrow (2)$ is $(1 - b)$. The cross relaxation is described by the parameter σ and the rate depends on the product $\sigma n_1 n_3$. These equations must be solved numerically in the general case [56,69,73] although analytical expressions for limiting cases have been derived [69,73].

The energy level scheme for Nd:YLiF$_4$ is shown in Fig. 11. The key levels are the ground state $(1) \equiv {}^4I_{9/2}$, the metastable intermediate state $(2) \equiv {}^4F_{3/2}$ and the upper laser level $(3) \equiv {}^2P_{3/2}$. Note that in this case the resonant pump excites to the ${}^4D_{3/2}$ level from which the system relaxes to ${}^2P_{3/2}$ in $\sim 1\ \mu$ sec. Figure 12a shows the result of a numerical solution of Eq. 22 for the three populations and the transmitted pump light. These solutions describe well the observed pump transmission and rise of ${}^2P_{3/2}$ emission shown in Fig. 12b. A more detailed analysis of the avalanche absorption process, again treating the case of Nd^{3+}:YLiF$_4$ has been given by Joubert et al. [69]. An interesting property of the avalanche absorption process is that the time delay for the onset of absorption is often longer than any of the lifetimes of the excited states.

VII. SPECTRAL HOLEBURNING

In spectral holeburning, a laser is used to selectively excite a narrow portion of an inhomogeneously broadened spectral line or band. This produces bleaching of the absorption (Fig. 13) due to the removal of the ions from the ground state and their storage in a metastable level or the phototransformation of the excited species by photoionization, tautomerization etc. [75]. The concept goes back to selective saturation in nmr [76] where narrow band coherent sources have been available for a long time. Its application to optical spectroscopy came with the development of the laser. Early work was done on intra-cavity saturation of Ne in a He-Ne laser [77] and later applied to solids [78-80] where holeburning has developed as a branch of optical spectroscopy fueled by the availability of single frequency tunable lasers which allow the study of a wide range of inhomogeneously broadened resonances [3,75,81,82]. Most studies of solid state systems are made at low temperatures ($\leq$10K) where holewidths are typically 0.1 - 100 MHz, although holeburning has been observed in

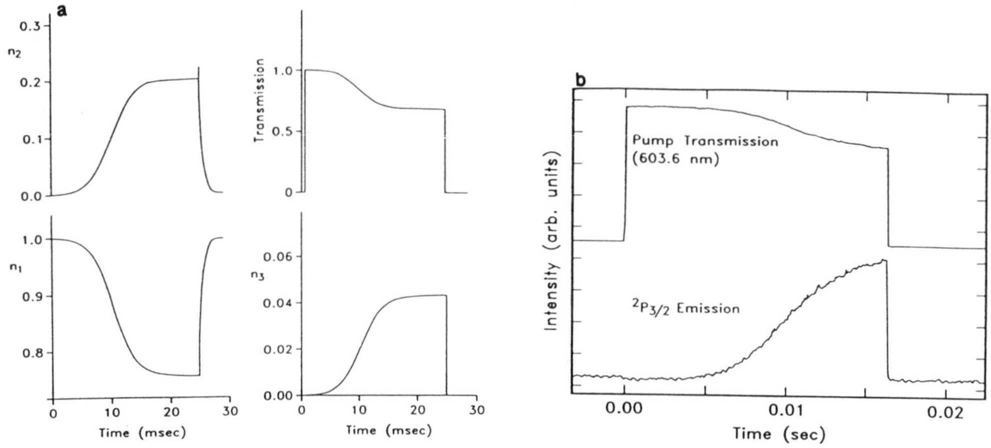

Figure 12. Effects of avalanche absorption on the time evolution of level populations and transmission of Nd^{3+}:$YLiF_4$ a) model calculations based on Eq. 22, b) experimental values of pump transmission and upper state population.

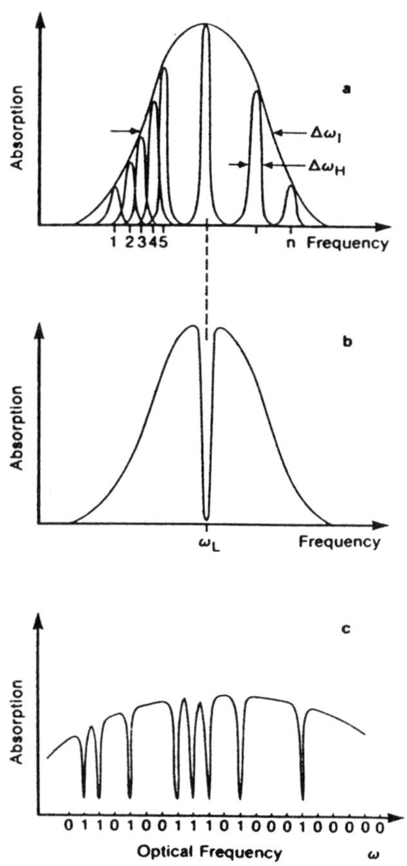

Figure 13 a) Inhomogeneous line profile as the envelope of a distribution of much narrower homogeneous linewidths. b) Spectral hole burned by the phototransformation of a subset of ions. c) Scheme for the storage of information in the frequency domain by burning a sequence of holes in the inhomogeneous profile.

several materials up to room temperature [83-85]. A unique feature of holeburning as a high resolution spectroscopic technique is the frequency memory provided by long lived holes. This enables frequency shifts to be measured over long time-scales and it also forms the basis of frequency domain [86] (Fig.13c) and time domain [87] information storage.

The observation of the spectral hole is made by monitoring the absorption of the bleaching laser, or possibly a second probe laser, while scanning it through the frequency originally excited. Thus, the instrumental resolution is determined by the frequency stability of the laser rather than that of a monitoring etalon or monochromator as is the case in fluorescence line narrowing for example. For commercially available, single frequency, cw dye or Ti:sapphire lasers this jitter width is ~ 1-2 MHz. Single-mode diode lasers have widths of around 50 MHz but in an external cavity linewidths of ~100 kHz or even much less have been obtained [88]. In a few cases highly frequency stabilized tunable lasers have been used for holeburning yielding a resolution of several kHz [89]. It is of course necessary that the optical resonance be very narrow to utilize such high resolution but this is often true of transitions between non-magnetic electronic singlet states of rare earth ions in crystals. Very few examples have been realized so far and much work remains to be done in this area.

The method of scanning spectral holes depends a great deal on their lifetime. Some general observations can be made but there is considerable overlap in the applicability of the techniques. For very short lived holes it is often convenient to use two laser sources and maintain saturation with one while probing with the other. For hole lifetimes of ~0.1 - 10 msec, a single laser can be used and its frequency scanned with an rf-chirped acousto-optic modulator [90] or the transition frequency of the material can be scanned with an electric [91] or magnetic [92] field. For longer lived holes, the laser can burn a hole at a fixed frequency and then probe the change in absorption by scanning under external control. Diode lasers have the advantage that they can be scanned quickly, which allows them to measure over a wide range of hole lifetimes [93]. To detect the holes one can either measure directly the change in transmission or monitor the light output in a fluorescence excitation spectrum. The latter can be very sensitive and it is well suited to cases where the sample absorption is very small but the response time of the measurement is limited by the fluorescence lifetime. If the material is not luminescent or a calibration of a small absorption is required, FM spectroscopy [94] provides a sensitive technique for the detection of holes [95].

VII.A Holeburning Mechanisms

Holeburning experiments can be classified according to the population reservoir in which the ions are stored following their removal from the ground state by the laser. (fig. 14). The simplest, two-level saturation holeburning, uses the excited state itself as the reservoir [80]. The usefulness of holeburning is dramatically enhanced, however, if a long-lived reservoir can be used, such as a metastable excited electronic state [96] or hyperfine [97] or superhyperfine [98] levels. The most important population reservoirs for rare earth ions are provided by hyperfine splittings of the ground electronic state, yielding holes in the inhomogeneous line which persist for a time (usually of order secs to mins) determined by spin relaxation among the hyperfine levels. If the coupling between the magnetic moment of the rare earth ion and those of nuclear spins in the host crystal is sufficiently strong, near-neighbor nuclear spins have well defined super-hyperfine levels. These levels can also be optically pumped by a laser, and holeburning results from population redistribution in this super-hyperfine reservoir [98]. Another low-lying population reservoir that has been used is a Zeeman component of a ground state electron-spin doublet [99], since the spin-lattice relaxation times in these systems can be even longer than 1 sec at low temperatures [100].

In addition to population holeburning which leads to hole lifetimes of msec→mins, there are mechanisms which produce "persistent" holeburning with lifetimes of hours or longer [75]. These processes can be described by a generalized energy diagram in which a barrier is crossed in the excited state, followed by relaxation to another stable ground-state configuration, (see Fig 15). In glasses these are the almost equivalent configurations of ions or molecules surrounding an optical center which have been modelled as "two-level systems"(TLS) [101]. Holeburning is thought of as originating from an optical absorption accompanied by a photo-induced flip of a nearby TLS which shifts the frequency of the resonant ions. This has been observed in rare-earth doped glasses [102] as well as in many examples of organic molecules in glasses and polymers [81,82]. Other cases of persistent holeburning include photoionization of divalent rare-earth ions (see below) or color centers [103], phototautomerism of porphyrins [104] and photoinduced tunneling of off-center ions [105] or molecular rotational states [106].

In general, one might expect to see antiholes due to absorption from levels which receive the population burned out of the hole. However the transition giving rise to such an anti-hole is often inhomogeneously broadened and cannot be observed. This is

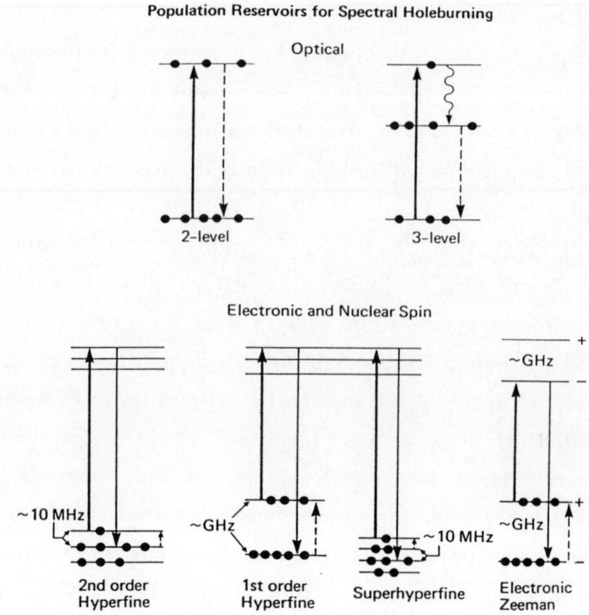

Figure 14 Population storage reservoirs for spectral holeburning. The hole lifetime is limited by the relaxation shown by the dotted arrow.

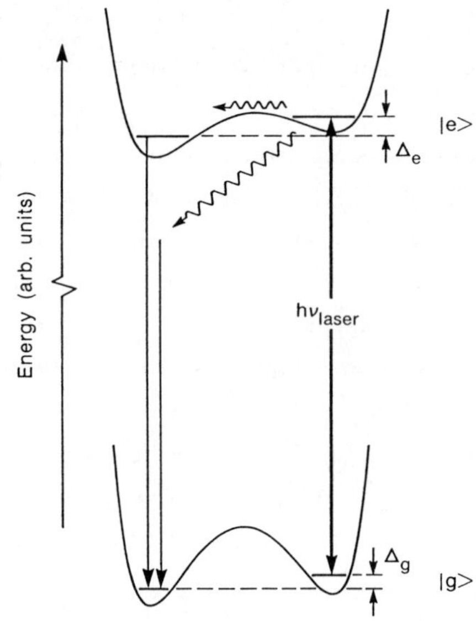

Figure 15 Schematic diagram of spectral holeburning mechanism involving coupled electron-double well potential system. In a glass the double well potentials are called "two-level-systems" or TLS.

because the particular set of ions or molecules which suffered holeburning are selected on the basis of their frequency for the transition excited and this corresponds to a range of environments. For another transition this range of environments will, in general, correspond to different transition frequencies and hence be inhomogeneously broadened. Notable exceptions are when the population is stored in different hyperfine [97,3] or magnetic [99] sublevels since the change in transition frequency associated with inhomogeneous broadening of these sublevel splittings is small.

We will now discuss three examples of spectral holeburning to illustrate the variety of mechanisms and detection schemes and conclude this section on holeburning with a description of some time resolved holeburning experiments. The three examples are the Tm^{3+} ion where population is stored in a metastable excited state, holeburning by light induced deuterium ion motion in $SrF_2:Pr^{3+}:D^-$ and photon-gated holeburning due to ionization of divalent samarium. The important case of holeburning due to optical pumping of ground state hyperfine components was discussed in the last Erice lectures [107].

VII.B Storage in a Metastable Optical Level

The trivalent thulium ion provides a good example of holeburning by population storage in a metastable optical level, here $^3F_4$, which has a lifetime of ~ 10msec in many materials. Fig. 16 shows the relevant energy levels and population storage mechanism. This mechanism is one of the most general, being applicable to almost all of the rare-earths as well as to organic molecules, transition metal ions and defect states in semiconductors. Holeburning in $LaF_3:Tm^{3+}$ was demonstrated [108] by irradiating the $^3H_6(1)\rightarrow{}^3H_4(1)$ transition at 12554.6 $cm^{-1}$with a Ti:sapphire laser, but a GaAlAs diode laser would serve equally well. In LaF_3, the Tm ions substitute for La in sites of C_2 symmetry with three local C_2 axes perpendicular to the trigonal axis of the crystal. The transition is σ polarized and connects non-degenerate (i.e. non-magnetic) levels having the same symmetry in the C_2 site group. Because of the short hole-lifetime the hole was probed by applying an electric field ramp to scan the transition frequency as shown in Fig. 17. The Stark effect is non-zero only for E-fields parallel to the C_2 axes, so applying this field in the plane perpendicular to the the crystal c-axis as shown in Fig 17a, equally shifts the levels of two of the inequivalent ions and leaves the third unshifted. A hole burned and scanned in this way is shown in Fig 17c. In general, of course, the Stark coefficients are not known so it is necessary to provide a frequency calibration. This can be done by burning multiple holes a fixed distance apart by frequency modulating the laser to put on sidebands several MHz apart. The width of the hole was only 1.7MHz and was limited by the laser frequency stability. This is a rather common ocurrence for transitions between levels which do not have an electronic magnetic moment since the homogeneous broadening in this case comes from nuclear spins coupling to the nuclear moment of Tm. In separate photon echo experiments the homogeneous linewidth was measured to be 45kHz.

This technique of Stark scanning a transient hole originating from population storage in a metastable optical level has wide applicability. If the levels are magnetic, a magnetic field ramp can be used. One limitation is that the width of scan is restricted by the Stark or Zeeman coefficients. If this is not sufficient, an acousto-optic modulator driven by a chirped rf source can be used to scan the laser frequency over ~100 MHz.

VII.C Holeburning by Deuteron Motion

Crystals of CaF_2 and SrF_2 doped with trivalent rare-earths into which H^-, D^- or T^- ions have been diffused, exhibit a novel form of spectral holeburning [109] caused by photoinduced H^- (or D^-, T^-) motion in the lattice [110,111]. This occurs following resonant excitation of the f-f transitions of the rare-earth ions. The resulting spectral holes are persistent at low temperatures due to the large barrier for thermal reorientation of the hyrogenic ions. Following optical excitation, relaxation of the excited rare-earth ions is dominated by non-radiative processes involving local mode vibrations of nearby hydrogenic ions. This shows itself as a substantially reduced quantum efficiency for light emission. The vibrational energy imparted to the hydrogenic ions provides the energy to produce ionic motion. There is a strong isotope effect since the lightest ion is more effective at quenching and hence gains more energy and it is also easier for it to move in the lattice. For this reason holeburning efficiencies are highest for the H^- doped materials. This persistent holeburning mechanism often competes with optical pumping of hyperfine levels and in some

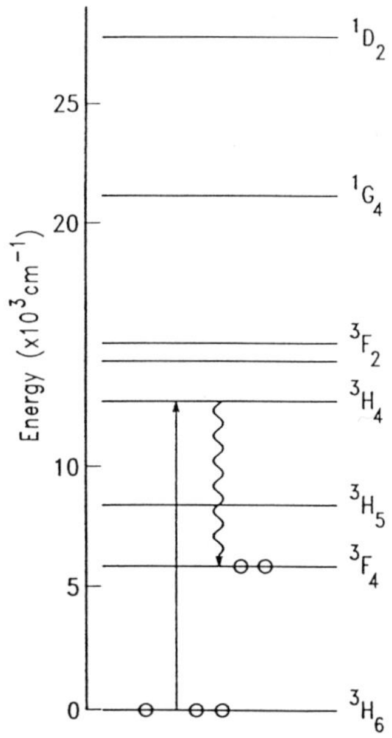

Figure 16 Energy level diagram for the Tm^{3+} ion showing the holeburning mechanism by population storage in the metastable $^3F_4$ level which typically has a lifetime of ~10 msec.

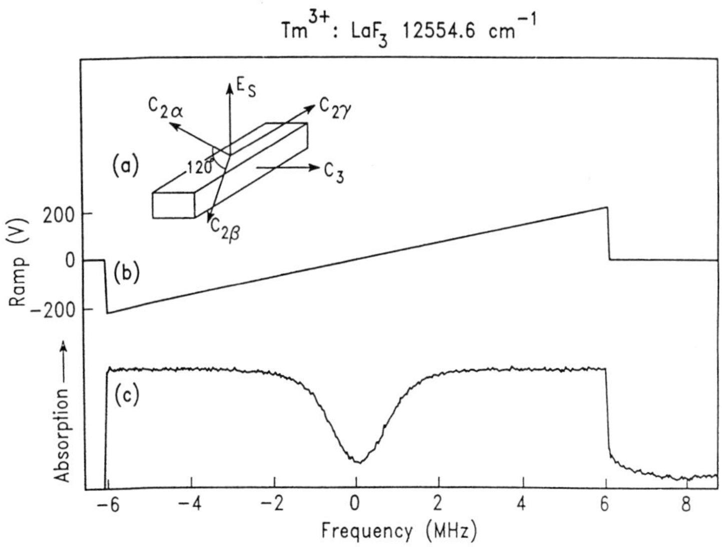

Figure 17 a) Geometry for the Stark field used to scan a hole in LaF_3:Tm^{3+} b)Stark ramp c) Spectral hole in the $^3H_6(1)\leftarrow\rightarrow{}^3H_4(1)$ transition.

materials both mechanisms can be observed. We illustrate this mechanism with a discussion of CaF_2 and SrF_2 doped with Pr^{3+} and D^- which is the system which has been most studied to date [109,111,112].

When hydrogenic ions are diffused into the alkaline earth fluorides, they replace F^- ions at substitutional sites and also occupy interstitial positions that charge compensate the trivalent rare earths. The predominant hydrogenic center in these crystals is the tetragonal (C_{4v}) center in which Pr^{3+} ions are charge compensated by D^- ions in vacant, body center positions along ⟨100⟩ directions (the "$C_{4v}H^-$" sites). The centers of interest to us here, since they exhibit persistent holeburning have rhombic symmetry and are modified from the C_{4v} center by the addition of extra hydrogenic ions in the near neighbor coordination cube [111]. Fig. 18 shows a schematic diagram of these centers containing two or three D^- ions. Four such centers have been identified in both CaF_2 and SrF_2 and they are labelled $C_s(1)$....(4). Persistent spectral holeburning occurs for these rhombic sites but does not occur in single hydrogenic ion C_{4v} centers.

Rhombic centers have singlet ground states because of the removal of the two-fold degeneracy of the tetragonal E level that is present in the parent C_{4v} site [111]. The separation Δ, between the two lowest levels (Fig. 19) depends on the strength of the rhombic distortion and differs greatly among different members of the family. This has important consequences for the magnetic properties of the ground state of ^{141}Pr [112] because the pseudo-quadrupole splittings and enhanced nuclear moments depend inversely on Δ.

Fig 20 shows the holeburning spectra for the four C_s sites in CaF_2 and SrF_2. For the $C_s(1)$ and $C_s(2)$ sites the holeburning mechanism is D^- motion and no antiholes appear since the photoproduct appears at a different wavelength or polarization and in any case the antiholes due to this mechanism would be very broad. For the $C_s(3)$ and $C_s(4)$ sites, the most efficient mechanism at 1.6K is optical pumping of the pseudoquadrupole splittings and antiholes appear at the sums and differences of ground and excited state splittings. The hole lifetimes are several seconds and as temperature increases the spin relaxation time between the pseudoquadrupole levels gets shorter and the antiholes decay more rapidly so that above about 5K, the persistent mechanism due to deuteron motion takes over.

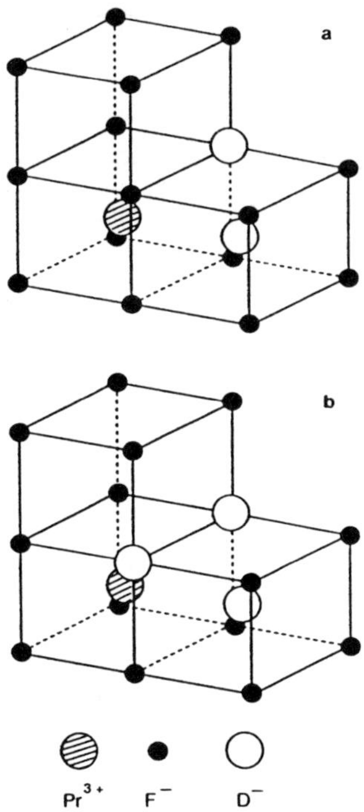

Figure 18 Model for (a) two and (b) three D^- centers in $CaF_2:Pr^{3+}$ which exhibit holeburning by photoinduced D^- ion motion.

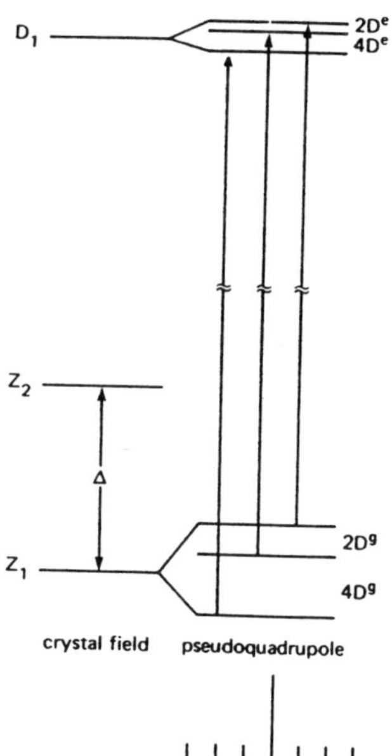

Figure 19 Energy level diagram for the transition from the groundstate to the lowest component of $^1D_2$ of Pr^{3+} in a bleachable D^- center of CaF_2. Such sites are almost axial with values of Δ ranging from 0.5 to 20 cm^{-1}. Holeburning by D^- motion produces a hole and six side holes as shown.

The first step in proposing a mechanism for the persistent holeburning is a model for the bleachable sites. One technique which has proven very useful here is the analysis of the D^- ion local mode contribution to the vibronic sidebands of the Pr^{3+} ion which was used to determine a model for the $C_s(2)$ center shown in Fig. 21 [111]. Three D^- ions were determined to be close to the Pr^{3+} center, two of them on fluorine lattice sites. Two kinds of deuteron motion can occur. The first (Fig. 21a) produces a center equivalent to the original but rotated by 90 degrees. This is consistent with bleaching that is recovered by rotating the polarization of the bleaching light. The second (Fig. 21b) produces an inequivalent center that absorbs at a different wavelength. This eventually leads to the bleaching of the entire line that cannot be recovered by rotation of the polarization.

The selective bleaching of rare-earth centers associated with hydrogenic ions appears to be a general property of the alkaline earth fluorides because of their open structure and the relative ease of cation motion. It is not known how general this phenonenon might be in other materials but undoubtedly new examples will be found. The long term memory of the persistent holes can be used to measure the effects of perturbations that can only be applied slowly such as high magnetic fields or changes in temperature.

VII.D Photon-Gated Holeburning

In photon-gated spectral holeburning [113], two photons are used as shown in Fig. 22. The first, or frequency selective step at ω_s, excites a narrow bandwidth within an inhomogeneously broadened line. The second or gating step at frequency ω_g, induces the photo-transformation of this excited packet. Excitation may be directly from level 2, or from one to which the system subsequently relaxes. Because holeburning ideally occurs only in the presence of both light sources, bleaching of the hole does not occur while reading at ω_s with a single source.

For $\omega_s < \omega_g$, the frequency selective photon is not sufficiently energetic to also act as a gating photon. This gives the highest gating ratio (*i.e.*, the ratio of the hole areas with and without gating light). For $\omega_s > \omega_g$ on the other hand, 'self-gating' can occur using two photons of frequency ω_s. However, since the holeburning is nonlinear, sufficient attenuation of the reading light still provides rather good hole stability.

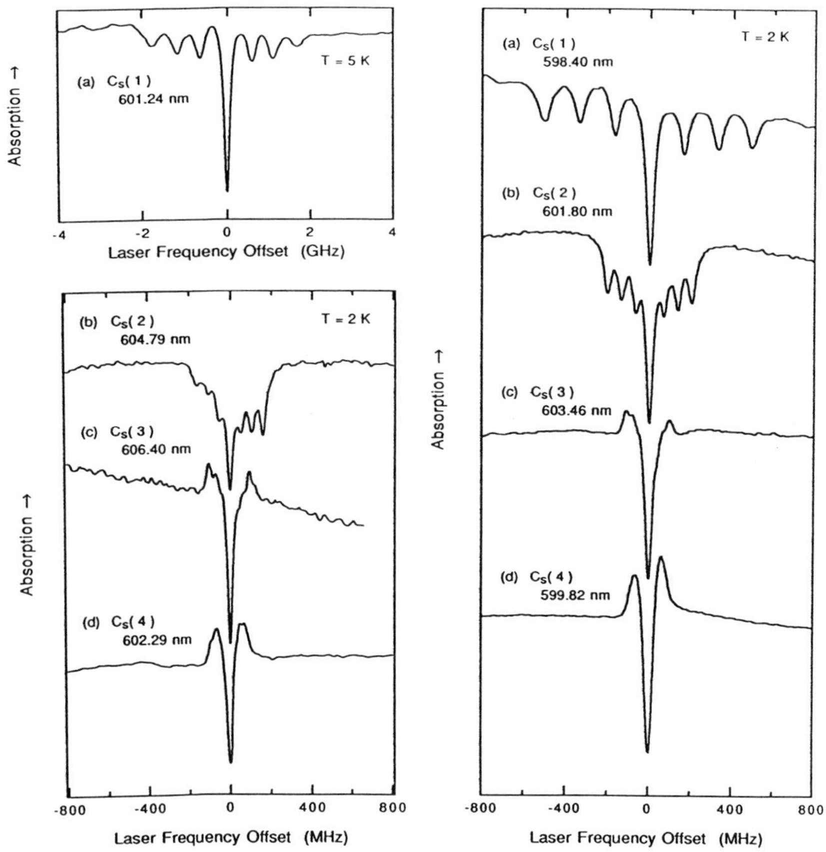

Figure 20 Holeburning spectra for four C_s Pr^{3+} centers in CaF_2 (left) and SrF_2(right) measured at 2K (except as shown). In the upper two traces the holeburning was persistent due to D^- motion and in the lower two are transient due to optical pumping of the hyperfine levels.

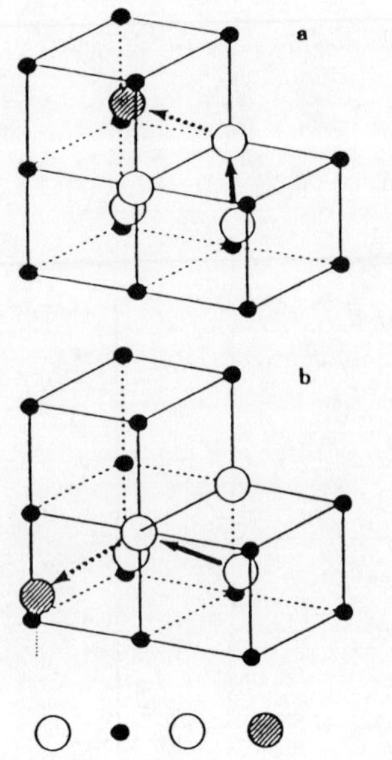

Figure 21 Proposed model for photoinduced motion of the D^- ion producing in a) an equivalent center rotated by 90° and in b) an inequivalent photoproduct absorbing at a different wavelength.

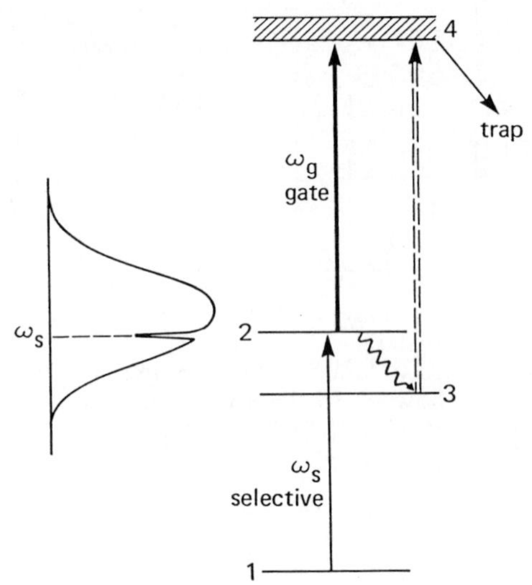

Figure 22 Schematic of photon-gated holeburning, in which the gating photon induces a phototransformation of the population selectively excited to level 2.

Photon-gated holeburning was first observed in $BaClF:Sm^{2+}$ [113], the mechanism being two-step photoionization of Sm^{2+}. Since then it has been demonstrated in a number of other inorganic [114] and organic [115-118] systems. In almost all of these cases photoionization has been responsible for the holeburning, but a wide variety of mechanisms can be envisaged.

1. The Case of Divalent Samarium. Rare earth ions in solids are most stable in their trivalent state, but samarium and europium can often be produced as divalent ions which can then be photoionized to the more stable trivalent state. If this photoionization occurs as a two-step process, photon-gated holeburning can result. Divalent samarium is the most studied ion and presents a number of interesting aspects, so we will review the nonlinear holeburning measurements which have been made so far.

The first example was provided by the $^7F_0 \leftarrow\rightarrow {}^5D_J$ transitions of Sm^{2+} substituted at ~0.01% in the C_{4v} barium sites of BaClF [113]. The relevant energy levels are shown in Fig. 23a. The groundstate and lowest metastable levels belong to the $4f^6$ configuration. Fig. 24 shows typical gated holeburning results for transitions to the metastable $^5D_0$ level (T_1 = 1.5 msec) where gating ratios of $> 10^5$ could be obtained. Measurements were made at 1.8K with approximately 2 W/cm^2 of frequency selective light (bandwidth 2 MHz) and 20 – 100 W/cm^2 of gating light at 514.5 nm. For the higher lying $^5D_1$ and $^5D_2$ levels which are also metastable with lifetimes of 0.8 msec and 0.75 msec respectively, the ionization threshold is lower. Thus the inequality $\omega_s < \omega_g$ is not as well satisfied and some holeburning occurs with light at ω_s. In most cases the width of the gated holes varied between 20 MHz and 100 MHz depending on the concentration of Sm^{2+} and the burning conditions. The origin of these holewidths is not understood in any detail but it is thought that time varying internal fields created by electrons released in the photoionization process contribute to spectral diffusion. $BaClF:Sm^{2+}$ is very unusual in that holes burned at 2K can be thermally cycled to room temperature (Fig. 25), held there for a day, and recovered at 2K with an increase in width of less than 200 MHz [119]. This shows that very little annealing of the microenvironment takes place on thermal cycling and that electron traps are sufficiently deep that detrapping at room temperature has a low probability.

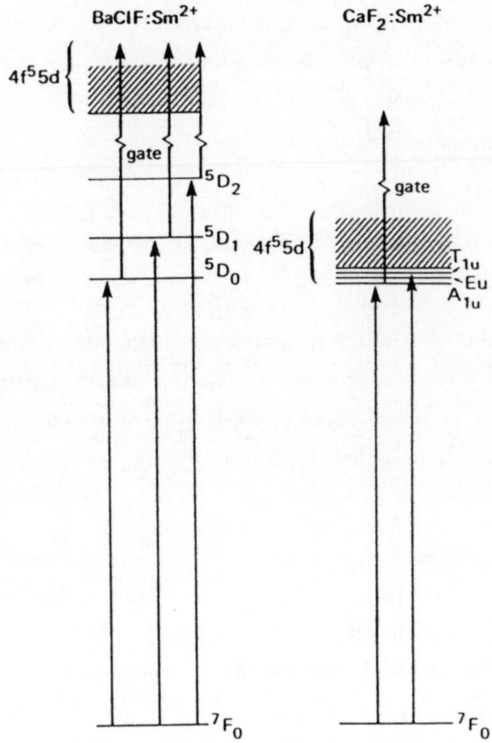

Figure 23 Energy level diagram for a) $BaClF:Sm^{2+}$ and b)$CaF_2:Sm^{2+}$. Holeburning occurs by photoionization of Sm^{2+}

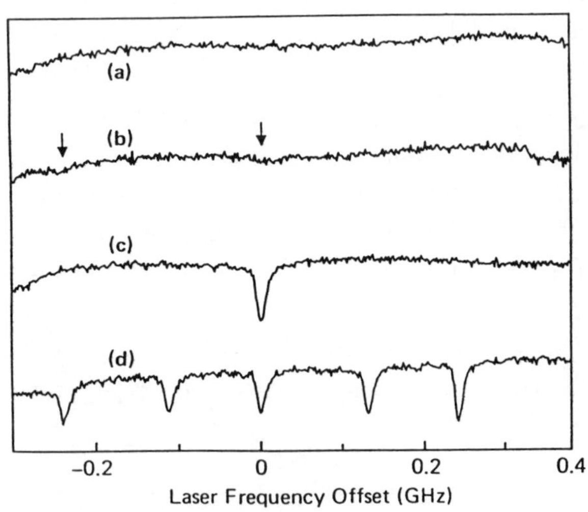

Figure 24 Gated holes in $BaClF:Sm^{2+}$ at 2K. a) part of the inhomogeneous line profile. b) hole burned by selective photons at 6879Å for 2000s at $2W/cm^2$. c)Gated hole burned for 3 sec by the addition of $20W/cm^2$ of green gating light. d) sequence of gated holes burned from left to right.

The isostructural compounds $SrClF:Sm^{2+}$ [120,121] and $CaClF:Sm^{2+}$ [122] also show holeburning by photoionization of the Sm^{2+} although some self-gating occurs in these materials which reduces the 'gating ratios'. This is particularly true of transitions to the $^5D_1$ and $^5D_2$ transitions where the photon frequencies are higher and photoionization can be achieved in two-steps of the frequency selective light. Holewidths are again generally larger than the laser jitter limit, but it is interesting to note that optical free induction decay measurements made in the $^7F_0 \leftrightarrow {}^5D_0$ transition, [120] without producing photoionization gave narrow linewidths of ~1MHz. (see Fig.26).

This family of materials is also interesting in that they exhibit high temperature holeburning. This was a natural extension of the fact that holes burned at low temperature could be cycled to room temperature, but the single crystals used in early experiments had such narrow inhomogeneous linewidths (15 - 20 GHz) that holes could only be observed at low temperatures. In $SrClF:Sm^{2+}$ powders with inhomogeneous linewidths of ~8 cm^{-1} holeburning was reported at 77K with holewidths of 1.2cm^{-1} [121]. More interesting is the case of mixed crystals such as $BaCl_{0.5}Br_{0.5}F:Sm^{2+}$ where the inhomogeneous linewidth increases by two orders of magnitude, to about 50 cm^{-1} with little change in the homogeneous linewidth. This is very desirable from the point of view of information storage and it also allowed holes to be burned at elevated temperatures [50]. Room temperature holeburning was first reported in $SrCl_{0.5}Br_{0.5}F:Sm^{2+}$ [83] where the homogeneous linewidths are remarkably narrow at 1 - 2 cm^{-1} and it was later shown [84] in a compound in which the cations were also disordered *viz* $Sr_{0.5}Mg_{0.5}Cl_{0.5}Br_{0.5}F:Sm^{2+}$.

Photon-gated holeburning has also been observed in $CaF_2:Sm^{2+}$ [123]. The energy level structure of this system is very different from that in $BaClF:Sm^{2+}$ since the lowest optical levels belong to the $4f^5 5d$ configuration (Fig. 23b). Because of the parity allowed transition to $^7F_1$, the lowest A_{1u} metastable level has a lifetime of 2 μsecs. Transitions between A_{1u} and the groundstate, however, are symmetry forbidden but can be made weakly allowed by the application of a magnetic field. Holes were burned in both the $^7F_0 \rightarrow A_{1u}$ and $^7F_0 \rightarrow T_{1u}$ absorptions the latter being broadened to 3.8 GHz by rapid nonradiative relaxation between T_{1u} and the E_u and A_{1u} levels below it. In this system $\omega_s > \omega_g$, so some holeburning occurs at ω_s, but this can be increased by more than three orders of magnitude by gating with green light. It was found by using a 2 μsec pulse at ω_s and a 10 nsec frequency doubled Nd:YAG pulse at 532 nm

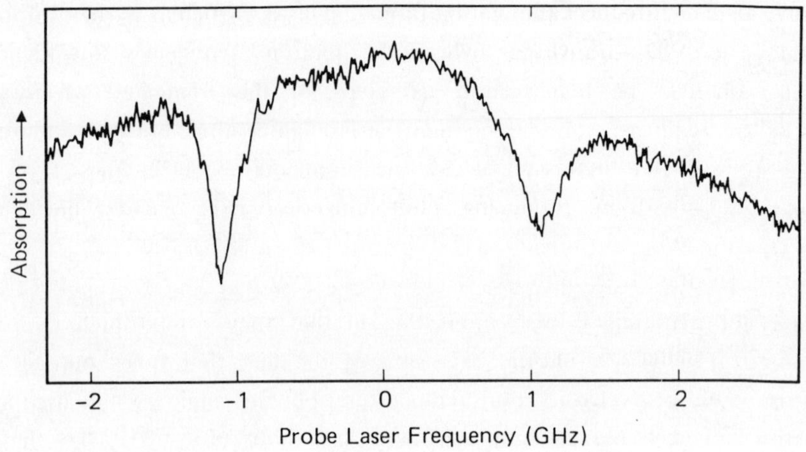

Figure 25 Temperature cycling of a spectral hole in BaClF:Sm^{2+} a) hole burned at 1.6K and cycled to room temperature for 1 day. b) new hole burned at 1.6K.

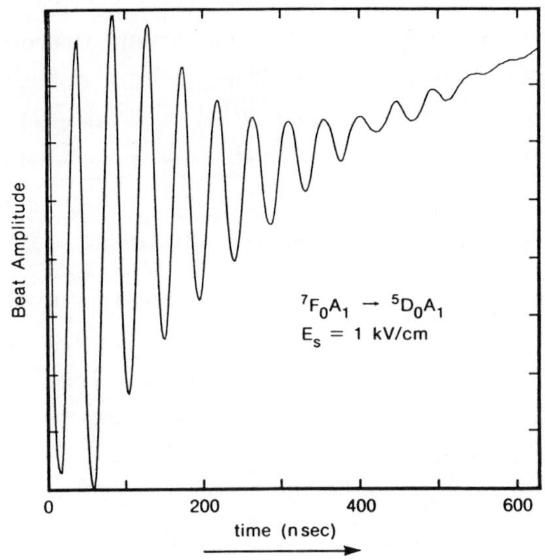

Figure 26 Heterodyne detected optical free induction decay in SrClF:Sm^{2+} at 1.6K, using Stark switching. The homogeneous linewidth measured was 1.2 MHz.

delayed from the first pulse, that gating takes place out of the A_{1u} level. The forbiddeness of transitions to A_{1u} is a consequence of the cubic symmetry of the Sm^{2+} site. Unfortunately there are extremely few known examples of zero phonon lines to metastable $4f^5 5d$ levels in other Sm^{2+} systems. These are potentially of great interest in combining photon-gated holeburning with relatively high oscillator strengths.

A more recent development has been the observation of spectral holeburning in samarium doped glass. As noted above holeburning in glasses is a rather general phenomenon involving photo-transformation of the tunneling levels of the glass. In the case of divalent samarium doped fluorohafnate glass however, photoionization of Sm^{2+} was again invoked as the mechanism. The first reports showed holeburning up to 180K [126] and later up to room temperature [85].

VII.E Time Resolved Holeburning and Spectral Diffusion

So far we have considered spectral holeburning by cw lasers in essentially steady state, where the population distribution in the reservoirs is achieved by long pulse excitation or by a number of optical pumping cycles on a time scale of msecs to secs.

The basic concept of homogeneous broadening leads one to expect Lorentzian hole shapes yielding the homogeneous width (Γ_h) in the frequency domain and for these to be equivalent to time domain measurements of the dephasing time T_2 using coherent transient techniques (see below). This simple picture often works rather well, but there are more and more cases where it must be modified. The reason for this is that the time scale of the fluctuations in the bath which causes line broadening (*e.g.* of nuclear or electron spins), can be comparable to the measurement times involved. It is therefore important to keep in mind that different experiments may be probing a different part of the fluctuation spectrum. We will explore these ideas in the context of time resolved holeburning in the absorptions of rare earth impurities coupled to a fluctuating spin bath [125,126]. Analogous results can be observed in disordered or glassy materials where the fluctuating bath is provided by tunneling of so-called two level systems (TLS) or by the motion of polymer chains for example [127]. In all cases a wide distribution of fluctuation rates occurs and it is this property which gives rise to spectral diffusion *i.e.* a linewidth which varies in time. Such phenomena are well established in magnetic resonance but only recently have they become the focus of optical experiments. In the case of broadening due to phonons, the fluctuations of the bath are sufficiently rapid that the relevant time scale lies in the psec or fsec regime.

At low temperature ($\lesssim$2K) the sharpest optical transitions of ions in solids are broadened by coupling to the bath of nuclear spin fluctuations in the host. The presence of the ionic impurity and its associated magnetic moment can strongly perturb the host nuclear spin dynamics by detuning the nuclear spins from each other and inhibiting their mutual spin-flips. This "frozen-core" of nuclear spins is a central concept which was first developed in nuclear resonance [128] and electron spin resonance [129] where it gives rise to spin diffusion barriers. It was later extended to the optical regime by Shelby *et al* [130] and it has proved to be of almost universal application to the field of high resolution spectroscopy of ions in crystals.

For the case of transitions between electronic singlet states which have no electronic magnetic moment, the resonances are sharp ($\sim$ 1 - 100 kHz) and the frozen core extends over only a few lattice constants. In this regime, time resolved holeburning using frequency scans is difficult because it requires that the laser frequency stability be of the order of several kHz over the duration of the measurement. In these cases it can be an advantage to make coherent transient measurements in the time domain which are less sensitive to laser frequency jitter. This is discussed below.

For the case of degenerate electronic levels which have a magnetic moment of the order of a Bohr magneton or more, the coupling to the spin bath is much stronger and the corresponding homogeneous linewidths are typically tens of MHz. These materials are more amenable to study with commercially available lasers which have a frequency stability of $\sim$ 1 MHz. The example of Er^{3+}:$YLiF_4$ will be used to illustrate this case [126].

$YLiF_4$ is a crystal with the scheelite structure (C^6_{4h}) and Er^{3+} ions can replace Y^{3+} on axially symmetric sites of S_4 symmetry. The magnetic moment is therefore anisotropic with $g_{\parallel} = 3.137$, $g_{\perp} = 8.105$ in the ground state and $g_{\parallel} = 9.9$, $g_{\perp} = 0.37$ in $^4F_{9/2}$ (1), the lowest component of the $^6F_{9/2}$ manifold which lies 15298.6 cm^{-1} above the ground state. These large magnetic moments create a large frozen core of ^{19}F nuclear spins, the size of which changes to some extent upon optical excitation. The fluorine spins near the Er^{3+} ion suffer the greatest detuning and flip slowly, while the spins further away which are weakly coupled have a faster flip rate eventually approaching that of the bulk if the Er^{3+} concentration is sufficiently low, that the frozen cores do not overlap. Because of this correlation with distance from the optical center, it is expected that there will be small frequency shifts at short times and larger frequency shifts at long times. This is what is observed.

Time-resolved holeburning experiments have been carried out on $Er^{3+}:YLiF_4$ using an external magnetic field, produced in a small coil around the sample, to rapidly scan the transition frequency following a burn pulse of 10μsec [126] (Fig. 27). Various time delays were used between burning and probing the hole. A large dc magnetic field of 26.5 kG was applied parallel to the c-axis to reduce the rate at which mutual spin-flips occurred between Er^{3+} ions (doped at 0.02%) by depopulating their upper Zeeman components. The results of the measurements are shown in Fig 28. For probe delays after holeburning of 100μsec to 380μsec the holewidth increases from 4MHz to 25MHz. At shorter delays the laser frequency jitter was a significant part of the measured width, since photon echo measurements on a 5-10 μsec timescale gave a width of 0.1 MHz. Note also the broad underlying component corresponding to spectral shifts of tens of MHz.

In order to understand the spectral diffusion better, it can be modelled by a computer simulation in which two elements are included. The first is the distribution of mutual spin flips of rate R_{ij} between ^{19}F nuclei at i and j separated by $\vec{r}_{ij}$ and detuned by $\Delta E_{ij} = \gamma_F(H_i - H_j)$ where γ is the gyromagnetic ratio of fluorine (4 kHz/G) and H_i, H_j the local magnetic field at sites i,j:

$$R_{ij} = R_o \left(\frac{\Gamma^2}{\Gamma^2 + \Delta E_{ij}^2} \right) \left(\frac{r_{nn}}{r_{ij}} \right)^6 (3 \cos^2 \theta_{ij} - 1)^2 \tag{23}$$

Here r_{nn} is the nearest neighbor F-F distance, Γ the bulk ^{19}F nmr linewidth, R_o the nearest neighbor mutual flip-rate of fluorines in the absence of Er^{3+} and θ_{ij} the angle between $\vec{r}_{ij}$ and the c-axis.

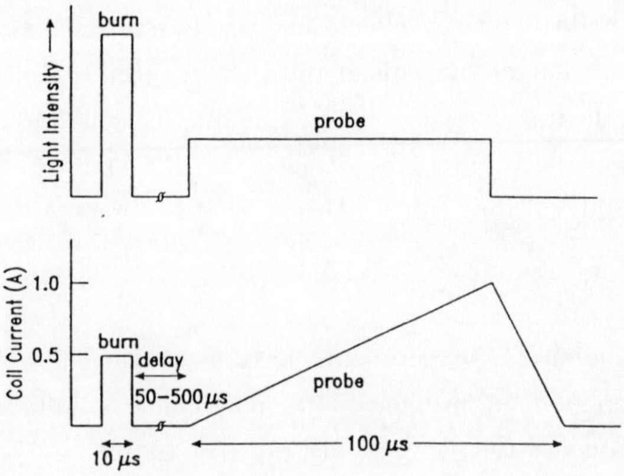

Figure 27 Time sequence of burning and Zeeman scanning of holes to measure spectral diffusion. (See also Fig. 28).

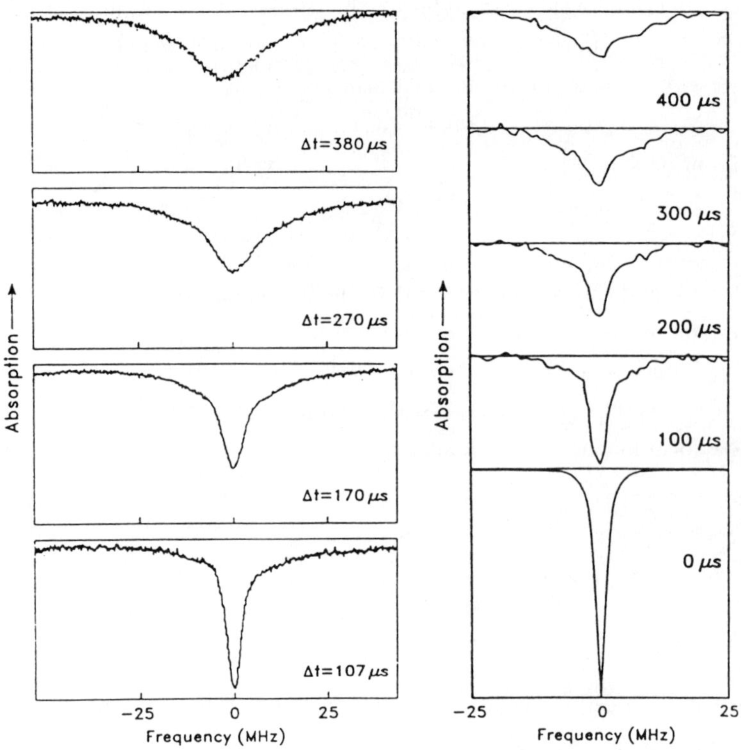

Figure 28 (left) Sequence of Zeeman scans of a hole in the 15298.6 cm^{-1} transition of $YLiF_4:Er^{3+}$, probed as a function of delay time from burning, showing broadening due to spectral diffusion. (right) Numerical modelling of the data at left for spectral diffusion due to a distribution of ^{19}F nuclear spin flips. (ref. 126)

The second component of the spectral diffusion comes from ^{19}F spin-flips produced by Er^{3+} electronic spin-flips. The main source of these flips are thermally induced spin-lattice relaxation transitions whose rate can be estimated by extrapolating EPR measurements to a field of 26.5 kG ($\sim H^5$). This gives $2.1 \times 10^5\ sec^{-1}$ for the upper to lower component relaxation or, at 1.6K, $160\ sec^{-1}$ for the upward spin-flip which is the dominant term since the upper level population is only 3%. The local magnetic field change at a fluorine nucleus at r_i produced by these Er^{3+} flips, varies as r_i^{-3} so that the probability of producing a ^{19}F spin-flip is

$$R'_i = R'_o \left(\frac{r_{nn}}{r_i} \right)^6 \tag{24}$$

where R'_o is the rate induced on a ^{19}F which is a nearest neighbor of an Er^{3+} ion.

The result of a numerical calculation over a lattice of 1000 fluorine spins of the local field fluctuations produced by these two contributions is shown in Fig. 28. The parameters in the model were obtained by reasonable extrapolation from other experiments. The agreement with the experimental hole-widths as a function of time shows that the essential features of the model are correct.

The spectral diffusion discussed above is a very general phenomenon in the case where homogeneous linewidths are determined by nuclear spin-flips, for both odd electron Kramer's doublet, and even-electron singlet systems. At higher concentrations of paramagnetic species or at lower magnetic fields, the mutual spin-flips of the paramagnetic ions will dominate the broadening.

VIII. COHERENT TRANSIENT TECHNIQUES

Coherent transient techniques investigate the coherent interaction of light and matter on a time scale of the order of the optical dephasing time T_2. This time varies dramatically with temperature and the material system under study. In isolated rare-earth centers it can be as long as ~ 1 msec at low temperatures and ~psec to fsec at ambient temperatures. From this it can be seen that the techniques and laser sources needed to study a wide range of materials vary considerably. Here we concentrate on low temperature measurements on rare-earth doped solids using narrow-band gated cw laser sources which can be tuned to specific resonances. Most of the optical coherent transient effects have been adapted from the field of nmr where they have long been of fundamental importance for determining dynamical and structural properties of materials.

Coherent transient experiments generally aim to do two things. The first is to determine optical dephasing times T_2 or an equivalent description of the time-scale of the dephasing in the case of non-exponential decays. Along with this goes a determination of the dynamics of the relaxation bath responsible for dephasing. The second goal is to use the coherent interaction to measure a spectrum and hence obtain structural or analytical information or information about interactions between the atoms and molecules which make up the material being studied. In addition, applications to information storage and processing and communications are currently under investigation.

VIII.A. Optical Free Induction Decay

This is, in many ways, the simplest coherent transient. As normally practised, a cw laser at frequency ω_1 irradiates a sample for a time long compared to T_2 and thereby induces a coherent polarization. This partially saturates a frequency packet in the inhomogeneous line of the order of the homogeneous linewidth and when the laser is switched off, the sample continues to emit coherently (the free induction decay by analogy with nmr) with its characteristic dephasing time T_2 which is related to the width of the saturation hole by eqn. (19). In terms of the vector model [20], the preparation pulse tips the Bloch vector and gives a component in the x – y plane which corresponds to a coherent sample polarization. The polarization so produced is proportional to the cube of the driving field χ, and has the form [19]:

$$P_3(t) \propto \chi^3 \exp\{(-t/T_2)[1 + (1 + \chi^2 T_1 T_2)^{1/2}]\} \qquad (26)$$

The term containing χ^2 is the power broadening term, emphasizing that this third order FID should be measured in the low power regime. It has been found that at high light intensities the solutions based on the Bloch equations can break down [33-35]. This depends on the correlation times in the fluctuating relaxation bath. At high Rabi frequencies, the optical two-level system becomes partially decoupled from the bath, and power broadening is smaller than predicted by eq. (25) (Fig. 29). This deviation from the behavior predicted by the Bloch equations has received much theoretical attention in recent years [36-39]. If the pulse used to prepare the FID is short, it may prepare a packet larger than the homogeneous linewidth. The resulting coherence decays as the different oscillators in the inhomogeneous ensemble get out of step and the coherent emission has a shorter decay given by the inverse of the frequency width prepared. In nmr, the preparation pulse is usually made short enough to prepare the entire inhomogeneous line and the decay of the FID then gives the inhomogeneous linewidth. In optical spectroscopy this is rarely used as it is generally simpler to scan the line profile with a tunable laser.

A convenient way to detect the optical free decay is to switch the frequency of the laser from ω_1 to ω_2 at the end of the preparation time [131,132], and use the laser as a local oscillator for the detection at a beat frequency of $(\omega_1 - \omega_2)$. The frequency-switching of the laser can be made with an extra-cavity acousto-optic modulator [133], or with an intra-cavity phase modulator [132]. This involves a relatively simple experimental set-up with a cw laser transmitted through a sample and detected with a p-i-n diode for example. An alternative approach is to switch the frequency of the sample to detune the group of atoms or ions in the frequency packet from the incident laser. A Stark field [134] is very convenient for this purpose and for systems with a large electronic magnetic moment, Zeeman switching of an external magnetic field has been used [135]. An example of an FID produced by Stark switching was given in Fig. 26. In practice a limitation on the applicability of the free-induction decay is the frequency jitter of the laser which contributes to the width of the saturation hole and shortens the decay. For commercial cw dye lasers this

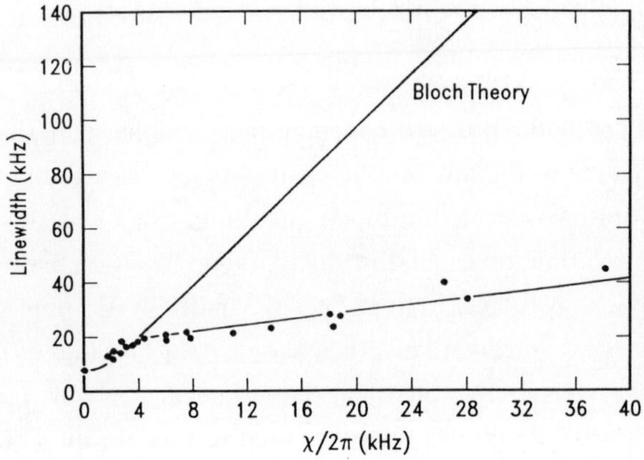

Figure 29 Power dependence of the optical FID in the 16873 cm^{-1} transition of $LaF_3:Pr^{3+}$ showing deviations from the optical Bloch equations (ref. 33)

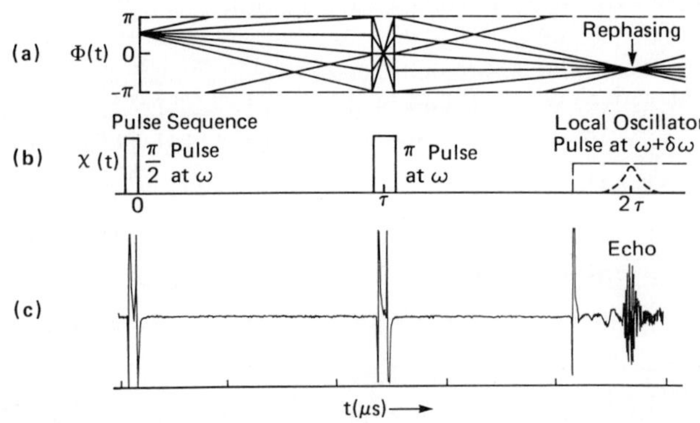

Figure 30 a) The inhomogeneous dephasing and rehasing which occurs during the photon-echo pulse sequence shown in b). The echo can be directly detected at 2τ or a heterodyne scheme can be used with a frequency-switched local oscillator.

contribution is typically several hundred kHz, but stability of better than 1kHz can be obtained and in expert hands less than 1Hz has been achieved [136]. Note that it is the short term frequency jitter which is important since the experiment is completed in a time of the order of T_2 which is rarely longer than 10μsec. Dye lasers suffer from high-frequency noise because of the free surface of the liquid jet, whereas solid-state lasers such as Ti:sapphire are quieter and better suited to the measurement of long coherence times.

A related technique is the delayed free induction decay [137,138]. Here there is a waiting time T between the preparation pulse which burns a narrow hole in the inhomogeneous line profile and the heterodyne probe pulse, such that $T_2 < T < T_1$. The two-level optical coherence has decayed by the time the probe pulse is turned on but the saturation hole remains. A short weak pulse is therefore applied just before the probe to induce some coherence and allow the observation of the FID. This is followed by a frequency switch and the heterodyne probe. The decay of the delayed FID is a measure of the width of the original hole and hence can be used to measure spectral diffusion [138,139].

VIII.B. Photon Echoes

Photon echoes, the optical analogue of spin echoes ('Hahn echoes') long known in nmr [140], were first observed by Kurnit et al. [141] in ruby using a Q-switched ruby laser. Since that time photon echoes have found wide application in the study of coherence decay of optical transitions in atoms, in molecules, and in solids [3,142].

The photon-echo technique is quite different from that described above for the FID in that the sample is excited by short pulses whose spectral widths are large compared to the homogeneous width of the optical transition. In contrast to nmr, the optical pulses do not usually excite the entire inhomogeneously broadened line, but rather a small segment of the line. The effect of inhomogeneous broadening is removed by the pulse sequence itself. The two-pulse echo experiment is carried out by exciting the sample with two laser pulses, separated by a delay τ (Fig. 30b). The first pulse creates a coherent superposition of the ground and excited state wave functions. This coherence is detectable because it leads to a macroscopic oscillating dipole which is capable of emitting coherent radiation. There is a rapid *inhomogeneous* dephasing on a timescale given by the inverse of the broad portion of the inhomogeneous line that was excited. The second pulse has the effect of exchanging, to some degree, the amplitudes of the ground and excited states in the coherent superposition, i.e the sign of this accumulated phase is reversed for each given ion. Following an additional evolution period of time τ, the net phase shift is cancelled for every ion, leading to a rephasing of the coherence and emission of a burst of coherent radiation (the photon echo) at time 2τ. The intensity or amplitude of the echo reflects the decay of coherence due to *homogeneous* relaxation processes during the time 2τ, and a plot of echo amplitude versus 2τ, yields the dephasing time, T_2. Ideally, one wants to excite the sample to a coherent superposition state with equal amplitudes for being in the ground and excited states, i.e., the first pulse should have pulse area, $\Theta = \int \mu_{12}E(t)dt = \pi/2$, a so-called $\pi/2$ pulse, which rotates the Bloch vector from -z, up into the $x-y$ plane. The second pulse should be a $\Theta = \pi$ pulse, to perfectly interchange the amplitudes and phase factors for the ground and excited states. This reverses the signs of the Bloch vectors which are arrayed in the x-y plane. However, echoes can be observed with non-optimal pulse areas. An important case is that of small area pulses for which the echo amplitude scales as $\Theta_1\Theta_2^2$ where $\Theta_{1(2)}$ refers to the area of the first (second) pulse. In many rare earth systems at low temperatures the dephasing times are $1-10\ \mu$ sec and this is the time scale of the pulse sequence required.

The actual experimental methods for observing echoes have taken several forms. Pulses can be generated by gating the amplitude of a cw laser [143] by frequency switching the laser [132] or the sample [134] into resonance with a portion of the inhomogeneous line as described under optical FID, or a combination of these [144]. The use of frequency switching allows heterodyne detection of the echo amplitude rather than the intensity, so that the signal decays as $\exp(-2\tau/T_2)$. Alternatively, one or more pulsed lasers can be used. The delay can be generated by optical delay lines which are particularly convenient on the picosecond timescale [145] and can also be used with nanosecond pulses but here the adjustable delays obtained by two separate lasers are more versatile [146]. In these cases the echo is detected directly with a suitably gated detector, or by angular resolution of the echo pulses using non-colinear excitation pulses. The intensity of the echo then decays as $\exp(-4\tau/T_2)$.

Because short pulses are used for excitation, and the elimination of inhomogeneous broadening does not depend on selection of a single homogeneously broadened packet, the photon-echo experiment is relatively immune from laser jitter problems. In fact, even with pulses from multi-mode pulsed lasers, echoes can be observed and used for accurate T_2 measurements. In the case of pulses gated from a cw laser, the effect of laser frequency jitter on the decay time is effectively removed if the Fourier transform width of the pulse is much greater than the jitter width. However, if heterodyne detection is used, the phase diffusion of the laser causes the phase of the heterodyne beat within the echo envelope to be randomized, making signal averaging techniques more difficult to apply, and for this reason amplitude gating of the laser and non-heterodyne detection are often preferred.

A significant difference between the use of pulses of length $\sim 1\ \mu sec$ gated from a cw laser, as opposed to nsec pulses from a Nd:YAG or nitrogen pumped dye laser, arises when hyperfine splittings or other sublevel structure is present. If, in such a case, the splittings are less than the Fourier width of the excitation pulses, a quantum-beat-like interference can occur between coherences excited on two transitions which share a common level [147]. The result is an echo decay which exhibits beats at frequencies determined by the ground and excited state hyperfine splittings and their sums and differences (Fig.31). The Fourier transform of the decay curve (Fig. 31) yields the hyperfine splittings [148].

A complication in photon echo measurements has been uncovered in recent years, particularly in the case of long dephasing times. This is the phenomenon of "instantaneous spectral diffusion" [149-152] which is an additional contribution to the dephasing which depends on the power of the exciting pulses. The origin of the dephasing is the presence of neighboring excited ions produced by the echo pulses. These create a different crystal field due to magnetic-dipole [150] or electric-dipole [151] interactions which shift the optical transition frequency. The effect depends most strongly on the power of the second pulse in the echo sequence, since this creates a different local field in the second (rephasing) half of the echo sequence compared to the first (dephasing) half as shown in Fig. 32. In practice this makes it necessary to extrapolate T_2 measurements to zero power, particularly in the case of long dephasing times. Bai and Kachru [153] found that in the case of $YAlO_3{:}Pr^{3+}$, the echo decay time also depends on the power of the first pulse and they proposed that this came from the presence of nonequilibrium phonons produced in the relaxation of the excited states.

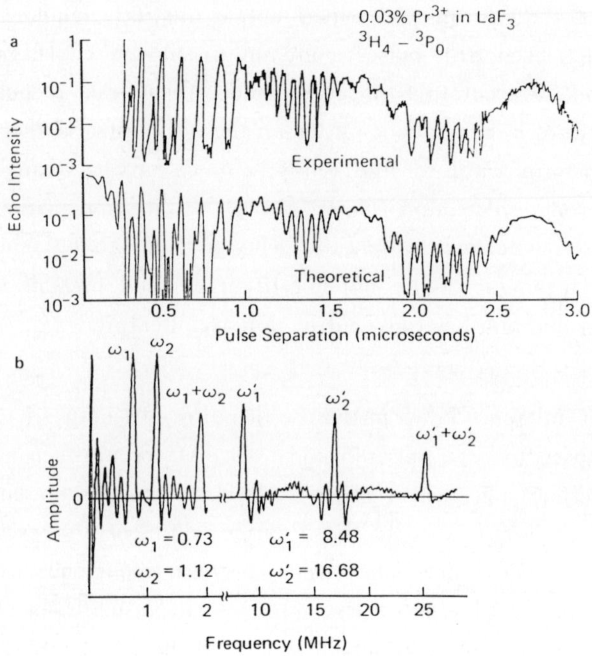

Figure 31 a) Photon echo modulation produced by short preparation pulses. b) The Fourier transform of a), showing peaks at the hyperfine splitting frequencies in ground and excited states.

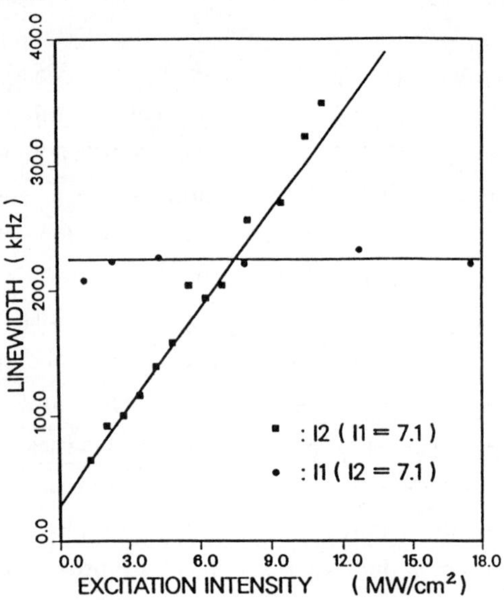

Figure 32 Excitation-intensity-dependent photon echo decays in $YLiF_4$:Tb^{3+} at 1.3K showing the effect of instantaneous spectral diffusion (ref. 150) I_1 and I_2 are the intensities of the first and second echo excitation pulses.

The above discussion has assumed a simple model in which dephasing can be accurately represented by a single exponential decay with time constant T_2. If the dephasing is due to interaction with a reservoir with a correlation time on the order of the echo decay time, this approximation may fail, and very nonexponential decays result. This situation occurs in the optical dephasing of rare earth ions due to the presence of the frozen core of nuclear spins surrounding the rare earth; slow spectral diffusion resulting in a nonexponential decay. These effects are well known in electron spin-echo measurements [154,155] and have recently been observed in photon echoes [156]. The example of $LaF_3{:}Tm^{3+}$ is shown in Fig 33. The non-exponential behavior is more pronounced in an external magnetic field which slows the mutual fluorine spin flips.

1. Stimulated Photon Echoes. The stimulated, or three pulse photon echo [157] can be thought of as a $(\pi/2, \pi)$ photon echo pulse sequence in which the π pulse is broken into two $\pi/2$ pulses by the insertion of a long delay, T in its center (Fig. 34). The resulting pulse sequence is two $\pi/2$ pulses separated by a time τ followed by a third $\pi/2$ pulse after an additional delay T. The first pulse creates a coherent superposition, and after an evolution time τ, the phases of these superposition states accumulate according to the offset from the laser frequency within the inhomogeneous line. The second pulse puts ions with in-phase coherence into the excited state, and those that have accumulated a π phase shift back into the ground-state, thus storing the accumulated phase information in the form of a population difference which varies periodically as a function of $\Delta\omega$ across the inhomogeneous line. This frequency grating is subject to decay due to population decay (T_1 for a two-level system) *and* due to spectral diffusion which tends to smear out the grating. This effect is analogous to the delayed FID experiment in which the hole is filled in by spectral diffusion. Again by analogy with the delayed FID, the amplitude of the grating can be read out after a delay T by applying a short pulse. The peaks in the population distribution will all have accumulated a multiple of 2π of phase after an additional delay time τ following the readout pulse. At that time a refocusing of the coherence takes place, and an echo is emitted. The amplitude of the echo gives information about dephasing during the two τ delays and spectral diffusion or population decay during the T delay. Stimulated echoes have been reported in $LaF_3{:}Pr^{3+}$ [158,159] and have been used to measure spectral diffusion in $YAlO_3{:}Pr^{3+}$ [160].

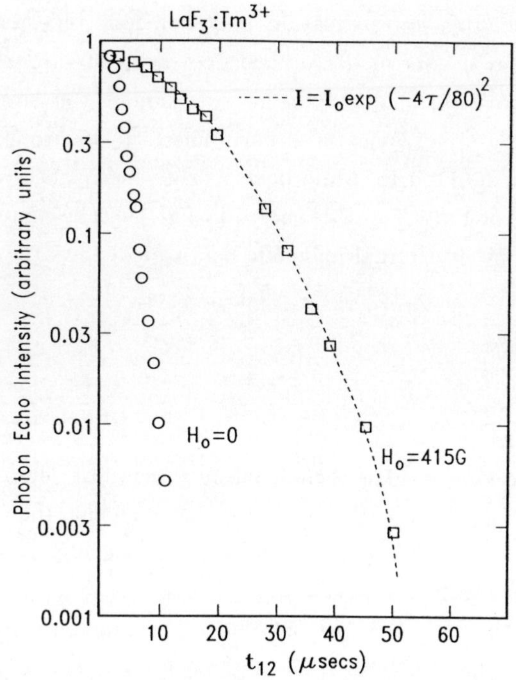

Figure 33 Non-exponential photon echo decay in $LaF_3:Tm^{3+}$. This is attributed to dephasing by a distribution of ^{19}F spin-flip rates.

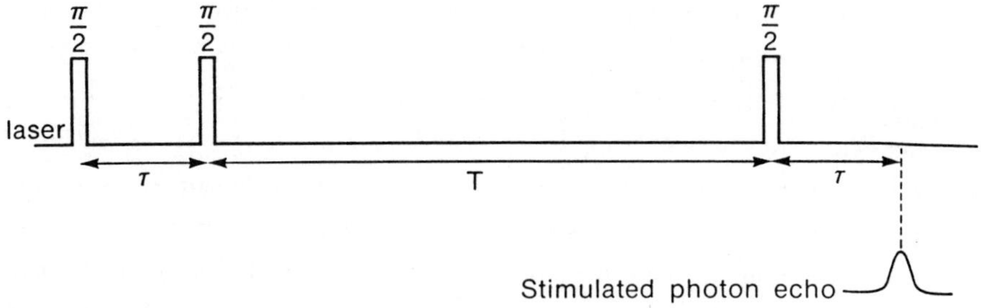

Figure 34 Pulse echo sequence for the measurement of a stimulated photon echo.

2. Accumulated Photon Echoes. If an excited state has a long lifetime or relaxes to a long lived intermediate state, the resulting population storage time, makes it possible for multiple repetitions of the first two pulses of a stimulated echo sequence to accumulate a much deeper population frequency grating than that from an individual pair of pulses [161,162]. If such pulse pairs with delay time τ and repetition period T are applied, each of the two pulses can act as a readout pulse to stimulate an echo from the grating which is built up. The amplitude of the echo still reflects the coherence lost during the two evolution periods of time τ, and thus can be used to measure T_2.

The main utility of this type of echo arises from the ability to build up large population differences via the accumulation effect and thus to observe relatively intense echo signals, even though only small area pulses might be available. This has been found particularly useful when picosecond pulses are used. If slow spectral diffusion is present on the timescale of the storage level relaxation time, however, it can cause the accumulated frequency grating to be washed out, and lead to a value of T_2 from the accumulated echo decay that is shorter than that which would be measured with a two pulse echo or optical FID. In this sense this experiment is equivalent to the determination of T_2 from a holeburning experiment which uses the same storage level as a population reservoir.

3. Photon Echo Nuclear Double Resonance (PENDOR). This technique involves the detection of a reduction in photon echo amplitude when an rf field is applied on resonance to a sublevel (e.g., hyperfine [163,164], or superhyperfine [143,165]) transition. There are two somewhat different experimental realizations of this idea.

In the first, [166,167] an rf pulse is applied during the time between the second and third pulses of a *three pulse* stimulated echo sequence. (Fig 34). Resonances are observed as reductions of the echo signal as the frequency of the rf is swept. There are two mechanisms for signal generation. First, in the case of resonances of host nuclei, i.e., superhyperfine resonances, the rf-induced spin flips of these nuclei tend to randomize the local field strengths. This will smear out the population frequency grating, just as spectral diffusion would do, and thus lead to a loss of echo amplitude. Second, the rf can remove population from the sublevels which form the echo, to ones which are not in resonance with the laser.

The second type of PENDOR experiment consists of the application of an rf pulse during the entire *two pulse* photon echo sequence [164]. The predominant mechanism for generation of the double resonance signal is the generation of sidebands on the sublevels at the rf Rabi frequency resulting in a modulation of the echo decay curve. As the rf frequency is swept, the echo amplitude exhibits resonances at the sublevel frequencies. Both ground and excited state hyperfine splittings can be obtained in this way, although in contrast to ODMR (see below) the two resonances cannot be easily distinguished.

The main advantage of PENDOR techniques over holeburning ODMR is that they are generally applicable to both excited and ground state resonances irrespective of branching considerations. Like ODMR, they measure sublevel spectra with a resolution determined by the inhomogeneous width of hyperfine transitions, not by optical T_2 or laser frequency jitter. However, they do require a reasonably long optical T_2 in order that echoes can be observed, and also that the rf transitions substantially affect local fields on the scale of the homogeneous linewidth on a timescale which is reasonable for the rf transition rates.

4. Stark Modulated Photon Echo. The population grating created by the two photon-echo preparation pulses provides an elegant way of precisely measuring small frequency shifting perturbations with sub-homogeneous linewidth resolution [168,169]. This is illustrated by the Stark effect on the $^7F_0 \rightarrow ^5D_0$ transition of Eu^{3+}:$YAlO_3$. Since the photon echo can be observed for pulse separations much larger than T_2, it is possible to generate frequency gratings which have a periodic spacing smaller than the homogeneous linewidth [170].

In the case of a centrosymmetric crystal there are two groups of ions with equal and opposite dipole moments. An applied electric field E_s shifts the ground and excited state energy levels through interaction with the electric dipole moments of these states and hence leads to a linear frequency shift, $\Omega = \delta\vec{\mu} \bullet \vec{E}_s/h$, for the two sets of ions in opposite directions, where $\delta\vec{\mu}$ is the difference between the ground and excited state dipole moments. If the electric field is applied for a time t′ after the first pulse, the two sets of ions accumulate equal and opposite phase shifts $\Phi = \pm\Omega t'$ during the time interval, t′. The result of the second pulse is that an excited state population grating is created in the frequency domain with a period τ^{-1} where τ is the time interval between the pulses. (Fig 35a). Since there are two sets of ions with opposite phase shifts, we formally get two population gratings, one with the phase $\Phi = +\Omega t'$ and one with the phase $\Phi = -\Omega t'$ with respect to the population grating that is caused without an applied field.

The electric field amplitude of the photon echo is proportional to the macroscopic polarization caused by rephasing of the ions at a time τ after the center of the second pulse and carries the optical phase of the population grating [169,171]. Thus the amplitude of the detected echo, $E(\tau)_{tot}$, is given by sum of the field amplitudes, $(1/2)E(\tau)e^{+i\Omega t'}$ and $(1/2)E(\tau)e^{-i\Omega t'}$ of the two groups of ions with opposite phase shifts:

$$E(\tau)_{tot} = 1/2[E(\tau)e^{+i\Omega t'} + E(\tau)e^{-i\Omega t'}] \tag{26}$$

The echo intensity is proportional to the square of the total field amplitude and is:

$$I_{echo} = (c\varepsilon_0/2)|E_{tot}(\tau)|^2 = I_0(1 + \cos(2\Omega t'))/2 \tag{27}$$

where I_0 is the photon echo intensity when no electric field is applied, and $c\varepsilon_0$ is the product of the light velocity and the vacuum dielectric constant. We see that the electric field induced modulation of the photon echo intensity results from the constructive and destructive interference of the echo components from the two sets of ions and depends on the magnitude of their relative phase shift. The minima involve a null and occur due to total destructive interference of the two echo components; the first one occurs when the field induced phase difference between the two sets of ions $2 \times \Omega t'$ is equal to π and subsequent minima are at odd multiples of π. Since the nulls can be located precisely, small Stark shifts can be determined with high accuracy. The minimum magnitude of a field induced frequency shift which can be measured in this way is determined by the periodic spacing of the population grating, (τ^{-1}) and hence the method is most sensitive for large pulse spacings.

Along with the excited state population grating there is a ground state population grating which can be probed by by applying an additional laser pulse. The effect of this pulse is that a new phase-coherent population grating is produced in the excited state which has the same electric-field dependence as the grating that was caused by the two initial echo-preparation pulses. In the course of the free induction decay at the time τ after the third excitation pulse the excited ions rephase and a stimulated echo appears with the same electric-field dependence as the two-pulse echo. If the excited state has

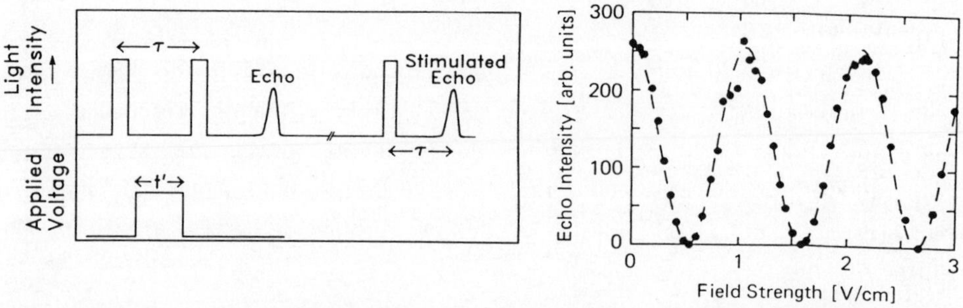

Figure 35 a) Pulse sequence for electric field induced echo modulation. b) Photon-echo modulation as a function of applied electric field for the $^7F_0 \leftarrow \rightarrow {}^5D_0$ transition of $YAlO_3{:}Eu^{3+}$ at 1.6K. (from ref. 168).

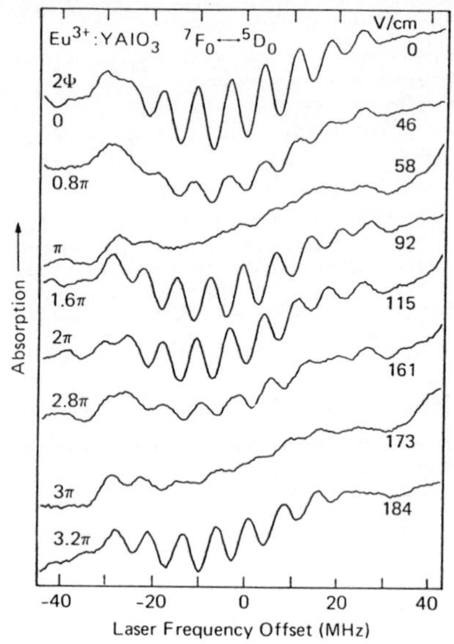

Figure 36 Effect of an electric field on the population grating created in a photon echo experiment on $YAlO_3{:}Eu^{3+}$ using 40ns pulses. The cancellation of the grating at electric field induced phase shifts of π and 3π can be seen.

a relaxation path into a long living off-resonance reservoir state, the ground state population grating can persist as a spectral hole pattern and a stimulated echo can be generated for times which are much longer than the excited state lifetime.

Measurements were made at 1.6K on the $^7F_0 \rightarrow {}^5D_0$ transition of $Eu^{3+}:YAlO_3$ at 17191.5 cm^{-1}, using dye laser pulses with widths of 40 nsec to 1μsec. gated by an acousto-optic modulator. The homogeneous dephasing time at 1.6 K is 58μs (i.e. the homogeneous linewidth is 5.4 kHz), and photon echoes were measured for pulse separations from 100 ns to 150μs. An electric field was applied to the sample between the two echo-preparation pulses while the time delay of the pulses was kept constant (Fig. 35a). As a function of the applied electric field the echo intensity showed modulation with a period that depended on the duration of the electric field pulse. Fig. 35b shows the results obtained with a preparation pulse of width of 400 nsec and a separation τ of 14.4μs. For an electric field pulse width of 14μs, the first echo minimum occurred at a voltage of 444 mV corresponding to a field strength in the mirror plane of 0.535 V/cm. From a least squares fit of eq. 27 to the experimental data, a Stark coefficient of 33.7 kHz/Vcm^{-1} was obtained. For a larger pulse separation of 140.4μs, the field at the first minimum dropped to only 54 mV/cm, since the ions can accumulate the phase shift over a longer period of time and hence less frequency shift is needed for a phase difference of π. At this pulse separation, the first minimum corresponds to a frequency shift of 1.9 kHz which is about one third of the homogeneous linewidth. The resolvable Stark shift is even much less than this because it corresponds to the separation between successive data points i.e. is about an order of magnitude smaller. The population grating written in the line also shows a modulated amplitude as a function of field strength as shown in Fig 36. The measurement of such a population grating is difficult because it requires very good phase and frequency stability in the laser.

The technique of Stark modulated echoes provides much higher resolution than could be obtained from the Stark splitting of a spectral hole. In part this is because of the laser frequency jitter but also the presence of the two frequency gratings and the null detection helps considerably.

IX. SUBLEVEL SPECTROSCOPY

IX.A Optically Detected Magnetic Resonance

The optical detection of magnetic resonance (ODMR) has a long and distinguished history. It has its origins in the concept of optical pumping of gas phase atoms. Changes in the intensity and polarization of emitted light were proposed as the basis for optically detected magnetic resonance by Bitter [172] and Brossel and Kastler [173] and reported by Brossel *et al.* [174] and Brossel and Bitter [175] in mercury vapor. The application to solids was made by Geschwind et al. [176] with experiments on ruby in which they used selective reabsorption of fluorescence rather than polarization changes in the emitted light as in the earlier gas-phase experiments. ODMR techniques were later applied to organic triplet states [177] and this proved to be an extremely useful structural probe of the organic solid state.

Rare-earth systems are ideally suited to ODMR studies because of the large number of fluorescing levels and the almost ubiquitous presence of hyperfine structure. In an optical pumping cycle there is normally branching between an excited state sublevel and two or more ground state sublevels. This redistributes population within the *ground state* sublevels *i.e.* burns a hole, and the application of rf or microwaves tends to restore the equilibrium population - there is an increase in absorption or light emission *i.e.* a 'hole-filling' resonance. It is also possible to aid the branching process by application of rf or microwaves at the resonance frequency of *excited state* sublevels. This increases holeburning and produces a decrease in optical absorption or light emission *i.e.* a 'hole-burning' resonance. In the case of axial symmetry the sublevel spin-projection is often a good quantum number and little or no branching occurs in the absence of the rf or microwave field. The sign of the absorption change thus shows whether the resonance observed is between excited state or ground state sublevels [178]. An example is shown in Fig 37. A number of measurements of this kind have been made to determine the hyperfine, superhyperfine, and nuclear Zeeman splittings [3]. When excited state resonances are measured, one normally thinks of the level directly excited by the laser, but ODMR resonances can also be observed in intermediate levels to which the system relaxes and which are perhaps inaccessible to direct optical excitation because of their resonance frequency [179]. In these cases the rf or microwave resonance in an intermediate level modifies the optical pumping cycle to induce a redistribution of population among the ground state levels. Optical pumping can be used to polarize nuclear or electron spin sublevels, preparatory to the application of a coherent transient pulse sequence. The sublevel system then evolves and for example, produces an echo. In order to detect this echo with incoherent probe light, the coherence must be converted into population changes, [180] that can be detected as changes in a transmitted light beam or the level of fluorescence. In the vector model picture this is done by rotating the Bloch vector from the x-y plane where the echo is formed to the z-axis which corresponds to changes in population. This technique was used by Shelby *et.al* [181] to measure free induction decay and spin echoes in the pseudoquadrupole components of the ground state of $LaF_3:Pr^{3+}$

IX.B. Sublevel Coherence

In sublevel coherence experiments one is dealing with a three (or more) level system. Light is used to detect sublevel coherence created by rf or microwave fields, to produce this coherence by a short optical pulse of sufficient bandwidth or in combination with the low frequency radiation to induce both optical and sublevel coherence.

An early experiment of the first kind was on ^{87}Rb vapor by Mathur, *et.al.* [182] in which they observed microwave induced Rabi oscillations on the 6.835 GHz hyperfine transition by monitoring transmission changes of light from a resonant lamp. Quantum beats in incoherent fluorescence [183] have also been reported many times.

In another group of experiments both optical and sublevel coherence is produced. Shoemaker and Brewer [184] showed what is now called a coherent Raman beat, by irradiating a pair of degenerate levels with a laser which induced both optical and (degenerate) sublevel coherence. The degeneracy was rapidly lifted with a Stark field and the resulting coherent Raman emission was heterodyned with the exciting laser yielding beats at the sublevel splitting frequency. These Raman beat experiments were the forerunner of a number of related coherent sublevel experiments. The basic theory is contained in the treatment by Brewer and Hahn [185]. Mlynek and Lange [186] treated the case of a 3-level system of so-called V-type and Λ-type *i.e.* with a sublevel

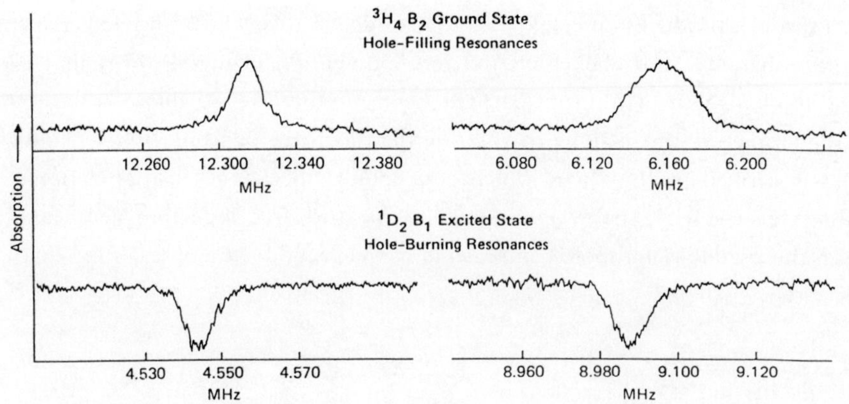

Figure 37 Optically detected nuclear resonance in $YPO_4{:}Pr^{3+}$ showing hole-burning and hole-filling resonances.

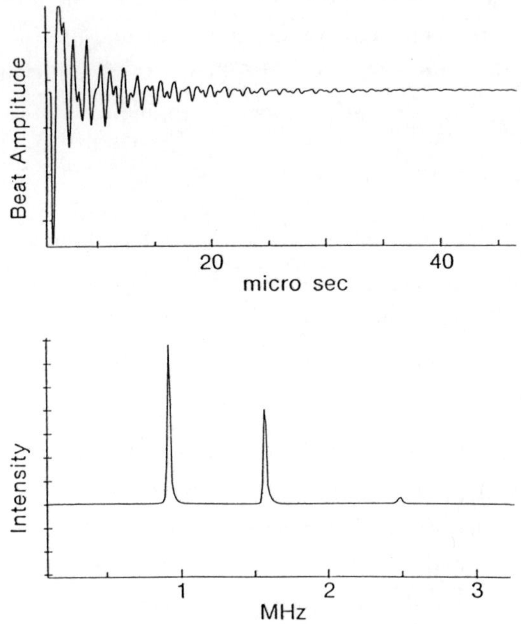

Figure 38 a) Quantum beat FID on the $^3H_4(1) \leftarrow \rightarrow {}^1D_2(1)$ transition of $YAlO_3{:}Pr^{3+}$ b) Fourier transform of the beat signal showing the excited state hyperfine splittings of 0.923 MHz and 1.565 MHz and their sum. show

splitting in the excited state or the ground state respectively. They used a short optical preparation pulse with sufficient bandwidth to span the sublevel splitting. By sweeping the resonance frequency, or by using a cw probe laser they observed beats corresponding to the sublevel splittings. Instead of a short pulse to excite sublevel coherence, a cw laser which is amplitude modulated at the splitting frequency [187] or an rf modulated Stark field [188] can be used. Another method uses a train of short pulses whose bandwidth covers the splitting and whose repetition rate, or a harmonic thereof, matches the sublevel splitting [189-191]. This is called 'periodic impact excitation'. The advantage of short pulse excitation is one of multiplexing *i.e.* a single pulse can yield multiple splitting frequencies. This is illustrated by a 'quantum beat FID' experiment in $YAlO_3:Pr^{3+}$ where sublevel coherence was produced by a pulse gated from a cw laser by an acousto-optic modulator [192]. Detection was made by resonant coherent Raman scattering with a cw probe laser which beats with the scattered light at the sublevel splitting frequencies. The resulting beat signal is complex but when Fourier transformed gives the individual resonance frequencies and widths (Fig. 38). Quantum beat echo experiments in which sublevel coherence is produced by a train of short laser pulses have been carried out by Mishina *et.al.* [193].

The advantage of all these techniques, some of them purely optical, is that the resolution obtained is that of the sublevel transition linewidths rather than either the optical linewidth or the laser linewidth. For example, Kohmoto *et.al.* [193] used quantum beat FID to measure electron-spin Zeeman splittings of Tm^{2+} in SrF_2 with a resolution of ~ 15 MHz using excitation into a short-lived optical level whose linewidth was many orders of magnitude greater than that of the sublevels.

An important double resonance technique which grew out of the coherent Raman beat experiments is Raman heterodyne detection of nmr [195,196], epr [197], or endor [198]. This is a 3-wave mixing phenomenon in which two fields couple three levels as shown in Fig. 39. As normally implemented an rf or microwave field near frequency ω_{12} connects levels (1) and (2) and an optical field at Ω_{23} connects (2) and (3). The transitions 1↔2, 2↔3, and 3↔1, should all be allowed. This creates coherence on the transition 1↔3, and generates a coherent optical Stokes' field at the sum frequency $\Omega_{13} = \Omega_{23} + \omega_{12}$. For another set of ions in the inhomogeneous optical linewidth Γ_{13}^{inh}, the anti-Stokes' difference frequency $\Omega_{23} - \omega_{12}$ is created. The Stokes' and anti-Stokes' fields beat with the laser to give a signal at ω_{12}. This can be de-modulated

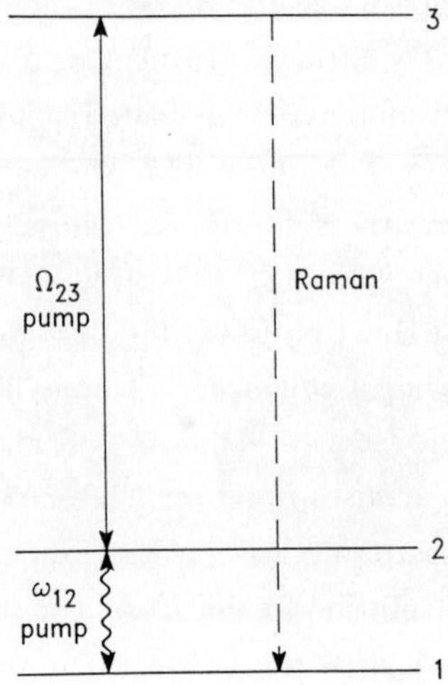

Figure 39 Energy level scheme for Raman heterodyne measurement of sublevel spectroscopy.

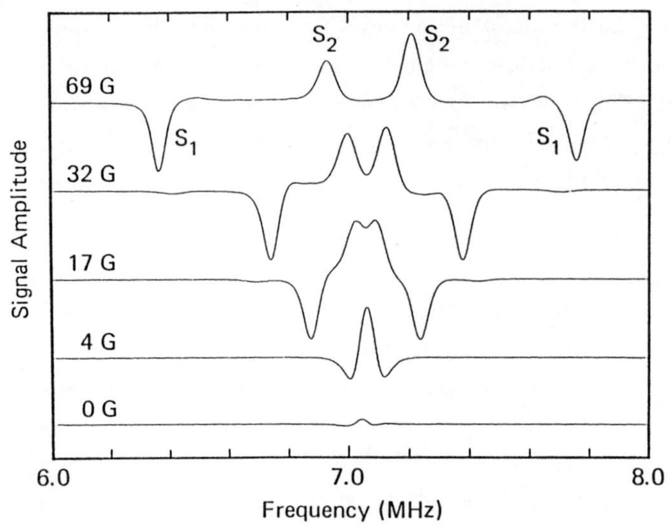

Figure 40 Raman heterodyne spectrum of the ground state hyperfine transition of $YAlO_3:Pr^{3+}$ showing symmetry determined interference effects which are removed by an external magnetic field (ref. 193). S_1 and S_2 are the signals from two magnetically inequivalent sites.

to give a cw signal which traces out the inhomogeneous profile of the 1↔2 resonance of half-width σ_{ω}. Alternatively a pulsed (rf) source at ω_{12} can be used to create a sublevel echo. Wong *et.al* [196] analyzed the theory of the Raman heterodyne process and it is instructive to look at their final expression for the absorptive heterodyne signal strength:

$$I_s = a_1 \chi_{12} \chi_{23} \chi_{31} (\rho_{22}^o - \rho_{11}^o) \exp\left(-\frac{\delta\Omega}{\Gamma_{1,3}^{inh}}\right)^2 \exp\left(-\frac{\delta\omega}{\Gamma_{1,2}^{inh}}\right)^2 \tag{28}$$

where a_1 is a constant, the χ's the Rabi frequencies of the respective transitions $(\rho_{22}^o - \rho_{11}^o)$ the population difference between the sublevels, δ_{ω}, δ_{Ω} are the frequency offsets from the peak of the inhomogeneous distributions of the transitions 1↔2, and 2↔3 and $\Gamma_{1,2}^{inh}$ is the inhomogeneous width of the sublevel transition 1-2. From eq. (28) we see that the signal is proportional to the laser intensity times the amplitude of the rf or microwave field, times the population difference between the sublevels. This population difference can be controlled by optical pumping with the laser.

This technique provides high sensitivity because of the heterodyne detection, high resolution provided by the rf or microwave sources and low noise because the detection frequency is shifted away to ω_{12} where laser noise is generally low. Another interesting and important result from eq. (28) is that the product $\chi_{12}\chi_{23}\chi_{31}$ contains a product of three matrix elements which appear linearly and hence can show destructive interference between the signals from different sites in a crystal depending on the overall crystal symmetry [199,200]. These symmetry determined cancellation effects have been observed, in $YAlO_3:Pr^{3+}$ and $LaF_3:Pr^{3+}$ and can be unbalanced by applying an external field, for example [201]. This is shown in Fig. 40.

X. CONCLUSION

We have attempted to show how the application of a variety of nonlinear spectroscopic techniques can be used to extract very detailed information about the nature and dynamics of solids even in the presence of unavoidable and often large inhomogeneous line broadening. These techniques rely heavily on the properties of modern tunable lasers such as their power, tunability and narrow optical bandwidths. The progress in this field has been rapid and will continue to be so as lasers continue to improve and the curiosity of the researcher remains insatiable.

REFERENCES

1. Y. R. Shen, *The Principles of Nonlinear Optics* (Wiley, New York, 1984).
2. M. D. Levenson and S. Kano, *Introduction to Nonlinear Laser Spectroscopy* (Academic Press, New York, 1988).
3. R. M. Macfarlane and R. M. Shelby in *Spectroscopy of Solids Containing Rare Earth Ions*, A. A. Kaplyanskii and R. M. Macfarlane, eds. (North Holland, Amsterdam, 1987), p. 51.
4. *CRC Handbook of Laser Science and Technology*, Vol. 1, M. J. Weber, ed. (CRC Press, Boca Raton, 1982); Supplement 1 (1991).
5. J. E. Geusic, H. M. Marcos and L. G. Van Uitert, *Appl. Phys. Lett.* **4**, 182 (1964).
6. G. H. Dieke and H. M. Crosswhite, *Appl. Opt.* **2**, 675 (1963).
7. B. G. Wybourne, *Spectroscopic Properties of Rare Earths* (Wiley Interscience, New York, 1965).
8. G. H. Dieke, *Spectra and Energy Levels of Rare-Earth Ions in Crystals* (Wiley Interscience, New York, 1968).
9. C. A. Morrison and R. P. Leavitt in *Handbook of the Physics and Chemistry of Rare Earths*, K. A. Gschneidner, Jr. and L. Eyring, eds., Vol. 5 (Amsterdam, 1982), p. 463.
10. W. T. Carnall, G. L. Goodman, K. Rajnak and R. S. Rana, *J. Chem. Phys.* **90**, 3443 (1989).
11. H. M. Crosswhite and H. Crosswhite, *J. Opt. Soc. Am.* **B1**, 246 (1984).
12. W. T. Carnall, H. Crosswhite and H. M. Crosswhite, Argonne National Lab Report No. 60439 (1977).
13. R. M. Sternheimer *Phys. Rev.* **146**, 140 (1966).
14. M. A. Teplov *Soviet Phys. JETP* **26**, 872 (1968).
15. B. Bleaney *Physica* **69**, 317 (1973).
16. J. M. Baker and B. Bleaney *Proc. Roy. Soc. (Lond.)* **A245**, 156 (1958).
17. F. Bloch, *Phys. Rev.* **70**, 460 (1946).
18. L. Allen and J. H. Eberley, *Optical Resonance and Two-Level Atoms* (Wiley, New York, 1975).
19. A. Schenzle and R. G. Brewer, *Phys. Rev.* **A14**, 1756 (1976).
20. R. P. Feynman, F. L. Vernon and R. W. Hellwarth, *J. Appl. Phys.* **28**, 49 (1957).
21. B. S. Mathur, H. Tang, R. Bulos and W. Happer, *Phys. Rev. Lett.* **21**, 1035 (1968).
22. R. G. Brewer and E. L. Hahn, *Phys. Rev.* **A8**, 464 (1973).
23. P. Hu, S. Geschwind and T. M. Jedju *Phys. Rev. Lett.* **37**, 1357 (1976).

24. J. Mlynek and W. Lange, *Opt. Commun.* **30**, 337 (1979).
25. E. W. Van Stryland and R. L. Shoemaker *Phys. Rev.* **A20**, 1376 (1979).
26. T. Endo, S. Nakanishi, T. Muramoto and T. Hashi, *Opt. Commun.* **43**, 359 (1982).
27. J. Mlynek, N. C. Wong, R. G. DeVoe, E. S. Kintzer and R. G. Brewer, *Phys. Rev. Lett.* **50**, 993 (1983); N. C. Wong, E. S. Kintzer J. Mlynek, R. G. DeVoe and R. G. Brewer, *Phys. Rev.* **B28**, 4993 (1983).
28. R. M. Shelby, A. C. Tropper, R. T. Harley and R. M. Macfarlane, *Opt. Lett.* **8**, 304 (1983).
29. M. Mitsunaga, E. S. Kintzer and R. G. Brewer, *Phys. Rev.* **B31**, 6947 (1985).
30. T. Mishina, M. Tanigawa, Y. Fukuda and T. Hashi, *Opt. Commun.* **62**, 166 (1987).
31. K. Holliday, X. -F. He, P. T. H. Fisk and N. B. Manson, *Opt. Lett.* **15**, 983 (1990).
32. A. G. Redfield, *Phys. Rev.* **98**, 1787 (1955); *IBM J. Res. Dev.* **1**, 19 (1957).
33. R. G. DeVoe and R. G. Brewer, *Phys. Rev. Lett.* **50**, 1269 (1983).
34. A. Szabo and Muramoto, *Phys. Rev.* **A38**, 5928 (1988).
35. A. G. Yodh, J. Golub, N. W. Carlson and T. W. Mossberg, *Phys. Rev. Lett.* **53**, 659 (1984).
36. A. Schenzle, M. Mitsunaga, R. G. DeVoe and R. G. Brewer, *Phys. Rev.* **A30**, 325 (1984).
37. A. I. Burshtein, A. A. Zharikov and V. S. Malinovskii, *Sov. Phys. JETP* **69**, 1164 (1989).
38. V. S. Malinovskii, *Opt. Spectrosc.* **70**, 399 (1991).
39. S. Huang, S. Quan, W. Qin and J. Yu, *J. Lumin.* **53**, 231 (1992).
40. G. E. Barasch and G. H. Dieke, *J. Chem. Phys.* **43**, 988 (1965); L. A. Riseberg and H. W. Moos, *Phys. Rev. Lett.* **19**, 1423 (1967); L. A. Riseberg and H. W. Moos, *Phys. Rev.* **174**, 429 (1968).
41. M. J.Weber, *Phys. Rev.* **157**, 262 (1967); M. J. Weber, *Phys. Rev.* **171**, 283 (1968).
42. P. M. Selzer, D. L. Huber, D. S. Hamilton, W. M. Yen and M. J. Weber, *Phys. Rev. Lett.* **36**, 813 (1976).
43. R. M. Macfarlane and R. M. Shelby, *J. Lumin.* **36**, 179 (1987).
44. D. E. McCumber and M. D. Sturge, *J. Appl. Phys.* **34**, 1682 (1963).
45. D. E. McCumber, *J. Math. Phys.* **5**, 221 (1964); *Phys. Rev.* **133**, A163 (1964).
46. B. DiBartolo, *Optical Interactions in Solids* (John Wiley, New York, 1968).
47. D. Hsu and J. L. Skinner, *J. Chem. Phys.* **81**, 1604 (1984); *ibid.* **81**, 5471 (1984); *J. Chem. Phys.* **83**, 2097 (1985); *ibid.* **87**, 54 (1987).
48. J. L. Skinner, *Ann. Rev. Phys. Chem.* **39**, 463 (1988).

49. R. M. Macfarlane in *Lasers, Spectroscopy and New Ideas*: A Tribute to Arthur L. Schawlow (Springer-Verlag, Berlin, 1981).
50. C. Wei, S. Huang and J. Yu, *J. Lumin.* **43**, 161 (1989); C. Wei, K. Holliday, A. J. Meixner, M. Croci and U. P. Wild, *J. Lumin.* **50**, 89 (1991).
51. A. M. Stoneham, *Rev. Mod. Phys.* **41**, 82 (1969).
52. B. B. Laird and J. L. Skinner, *J. Chem. Phys.* **90**, 3880 (1989); J. L. Skinner, B. B. Laird and L. Root, *J. Lumin.* **45**, 6 (1990).
53. R. M. Macfarlane, A. Cassanho and R. S. Meltzer, *Phys. Rev. Lett.* **69**, 542 (1992).
54. J. D. Axe Jr., *Phys. Rev.* **136**, A42 (1964); M. C. Downer and A. Bivas, *Phys. Rev.* **B28**, 3677 (1983).
55. N. Bloembergen, *Phys. Rev. Lett.* **2**, 84 (1959).
56. W. Lenth and R. M. Macfarlane, *J. Lumin.* **45**, 346 (1990).
57. A. J. Silversmith, W. Lenth and R. M. Macfarlane, *Appl. Phys. Lett.* **51**, 1977 (1987).
58. J. Y. Allain, M. Monerie and H. Poignant, *Electron. Lett.* **26**, 262 (1990).
59. G. R. Atkins, M. G. Sceats and S. B. Poole, *J. Non.-Cryst. Solids* **140**, 123 (1992).
60. F. Auzel, *C. R. Acad. Sci. (Paris)* **262**, 1016 (1966).
61. W. Lenth, A. J. Silversmith and R. M. Macfarlane, *Advances in Laser Science III*, A. C. Tam, J. L. Gole and W. C. Stwalley, eds., QIP Conf. Proc. **172**, 8 (1988).
62. R. A. McFarlane, *Appl. Phys. Lett.* **54**, 2301 (1989); R. A. McFarlane, M. Robinson and S. A. Pollack, *Proc. SPIE* **1223**, 294 (1990).
63. T. Hebert, W. P. Risk, R. M. Macfarlane and W. Lenth, *Advanced Solid State Lasers*, H. P. Jenssen and G. Dubé, eds. (Opt. Soc. Am., Washington DC, 1990), p. 379.
64. J. S. Chivian, W. E. Case and D. D. Eden, *Appl. Phys. Lett.* **35**, 124 (1979).
65. M. E. Koch and W. E. Case, *AIP Conf. Proc.* **146**, 142 (1986).
66. A. W. Kueny, W. E. Case and M. E. Koch, *J. Opt. Soc. Am.* **B6**, 639 (1989).
67. N. Pelletier-Allard and R. Pelletier, *Phys. Rev.* **B36**, 4425 (1987).
68. N. Pelletier-Allard and R. Pelletier, *J. Lumin.* **48/49**, 867 (1991).
69. M. F. Joubert, S. Guy and B. Jacquier, to be published.
70. N. J. Krasutsky, *J. Appl. Phys.* **54**, 1261 (1983).
71. R. M. Macfarlane, R. Wannemacher, T. Hebert and W. Lenth, Conference on Lasers and Electro-optics (CLEO-90) Technical Digest, Optical Society of America, p. 250; T. Hebert, R. Wannemacher, R. M. Macfarlane and W. Lenth, *Appl. Phys. Lett.* **60**, 2592 (1992).

72. H. Ni and S. C. Rand, *Opt. Lett.* **16**, 1424 (1991).

73. U. Oetliker, M. J. Riley, P. S. May and H. U. Güdel, *J. Lumin.* **53**, 553 (1992).

74. R. M. Macfarlane, A. J. Silversmith, F. Tong and W. Lenth, *Proceedings of the International Conference on Laser Science and Laser Materials*, Shanghai, July 25-27, 1988, edited by Z. J. Wang and Z. M. Zhang (World Scientific, Singapore, 1989), p. 24.

75. *Persistent Spectral Holeburning: Science and Applications* ed. W.E.Moerner, Springer Verlag (Heidelberg), 1988.

76. N. Bloembergen, E.M. Purcell, and R.V. Pound, *Phys.Rev.*, **71**, 679 (1948).

77. P.H. Lee, and M.L. Skolnick, *Appl.Phys.Lett.*, **10**, 303 (1967).

78. B.M. Kharlamov, R.I. Personov, and L.A. Bykovskaya, *Opt.Commun.*, **12**, 191 (1974).

79. A.A. Gorokhovskii, R. Kaarli, and L.A. Rebane, *JETP Lett.*, **20**, 216 (1974).

80. A. Szabo, *Phys.Rev.*, **B11**, 4512 (1975).

81. G.J. Small in *Spectroscopy and Excitation Dynamics of Condensed Molecular Systems* eds V.M. Agranovich and R.M. Hochstrasser, North Holland (Amsterdam) 1983 p. *Phys.Rev.*, **B11**, 4512 (1975).

82. S. Volker in *Relaxation Processes in Molecular Excited States* ed. J. Fünfschilling, Kluwer Academic Publishers, (1989) p.113

83. R. Jaaniso, and H. Bill, *Europhys.Lett.*, **16**, 569 (1991).

84. K. Holliday, C. Wei, M. Croci, and U. Wild *J. Lumin.*, **53**, 227 (1992).

85. K. Hirao, S. Todokori, and N. Soga, *(to be published)*

86. A. Szabo, *U.S.Patent 3,896,420 (1975)*; G. Castro, D. Haarer, R.M. Macfarlane, H.P. Trommsdorff *U.S.Patent 4,101,976 (1978)*; W.E. Moerner, W. Lenth, and G.C. Bjorklund in ref. 75.

87. T.W. Mossberg, *Opt.Lett.*, **7**, 77 (1982); T.W. Mossberg *U.S. Patent* 4,459,682 (1984).

88. C.E. Wieman, and L. Hollberg, *Rev.Sci.Instrum.*, **62**, 1 (1991).

89. A. Szabo and R. Kaarli, *Phys. Rev.*, **B44**, 12307 (1991); M. Mitsunaga, N. Uesugi and K. Sugiyama, *Opt.Lett.*, **18**, 1256 (1993).

90. C.M. Jefferson, and A.J. Meixner, *Chem.Phys.Lett.*, **189**, 60 (1992).

91. A. Szabo and M. Kroll, *Opt.Commun.*, **18**, 224 (1976); T. Muramoto, S. Nakanishi, and T. Hashi, *Opt.Commun.*, **21**, 139 (1977).

92. Y.P. Wang, D.P. Landau, R.S. Meltzer, and R.M. Macfarlane, *J.Opt.Soc.Am.*, **89**, 946 (1992).

93. R. Wannemacher, H.E. Smorenburg, Th. Schmidt, and S. Volker, *J.Lumin.*, **53**, 226 (1992).

94. G.C. Bjorklund, *Opt.Lett.*, **5**, 15 (1980).

95. G.C. Bjorklund, W. Lenth, and C. Ortiz, *SPIE*, **298**, 107 (1981).

96. R.M. Macfarlane in *Lasers, Spectroscopy and New Ideas: A Tribute to Arthur L. Schawlow*, eds. M.D. Levenson, and W.M. Yen, Springer Verlag (Berlin) 1981 p.

97. L.E. Erickson *Phys.Rev.*, **B16**, 4731 (1977).

98. R.M. Macfarlane, R.M. Shelby, and D.P. Burum, *Opt.Lett.*, **6**, 593 (1981).

99. R.M. Macfarlane and J.C. Vial, *Phys.Rev.*, **B36**, 3511 (1987).

100. M.B. Schulz and C.D. Jeffries, *Phys.Rev.*, **149**, 270 (1966); E. S. Sabisky and C. H. Anderson, *Phys.Rev.*, **B1**, 2028 (1970).

101. P. W. Anderson, B. I. Halperin and C. M. Varma, *Phil.Mag.*, **25**, 1 (1972); W. A. Phillips, *J. Low Temp. Phys.*, **7**, 351 (1972).

102. R.M. Macfarlane, and R.M. Shelby, *J.Lumin.*, **36**, 179 (1987).

103. R.M. Macfarlane, and R.M. Shelby, *Cryst.Latt.Def. and Amorph.Mat.*, **12**, 417 (1985).

104. S. Völker, and R.M. Macfarlane, *IBM J. Res. and Dev.*, **23**, 547 (1979).

105. K. Holliday, and N.B. Manson, *J. Phys. Condens.Matt.*, **1**, 1339 (1989).

106. C. von Borczyskowski, A. Oppenlander, H. P. Trommsdorff and J. -C. Vial, *Phys. Rev. Lett.* **65**, 3277 (1990).

107. R. M. Macfarlane, in *Optical Properties of Excited States in Solids*, ed. B. Di Bartolo, Plenum Press (New York) 1992. p.399.

108. R. M. Macfarlane, *Opt. Lett.* **18**, 829 (1993).

109. R. M. Macfarlane, R. J. Reeves and G. D. Jones, *Opt. Lett.* **12**, 660 (1987).

110. N. J. Cockroft, T. P. J. Han, R. J. Reeves, G. D. Jones and R. W. G. Syme, *Opt. Lett.* **12**, 36 (1987).

111. R. J. Reeves, G. D. Jones and R. W. G. Syme, *Phys. Rev.* **B40**, 6475 (1989).

112. R. J. Reeves and R. M. Macfarlane, *J. Opt. Soc. Am.* **B9**, 763 (1992); R. J. Reeves and R. M. Macfarlane, *Phys. Rev.* **B47**, 158 (1993).

113. A. Winnacker, R. M. Shelby and R. M. Macfarlane, *Opt. Lett.* **10**, 350 (1985).

114. R. M. Macfarlane, *J. Lumin.* **38**, 20 (1987).

115. W. E. Moerner, *Jap. J. Appl. Phys.* **28**, Suppl.28-3, 221 (1989).

116. S. Machida, K. Horie and T. Yamashita, *J. Appl. Phys.* **60**, 286 (1991).

117. Suzuki, T. Nishi, T. Shimada and H. Hiratsuka, *J. Appl. Phys.* **73**, 374 (1992).

118. Wang, L. Hu, H. He, J. Rong, J. Xie, J. Zhang, L. Zhao, Z. Lu, X. Zhang, D. Zhang, X. Min, Y, Ni and P. Ye, *J. Opt. Soc. Am. B* **9**, 800 (1992).

119. A. Winnacker, R. M. Shelby and R. M. Macfarlane, *J. Phys. Colloque* **C7**, 543 (1985).

120. R. M. Macfarlane and J. -C. Vial, unpublished

121. J. Zhang, S. Huang, W. Qin, D. Gao and J. Yu, *J. Lumin.* **53**, 275 (1992).

122. A. Oppenlander, J. -C. Vial, R. M. Macfarlane and J.-P Chaminade, *J. Lumin.* **42**, 331 (1989).

123. R. M. Macfarlane and R. M. Shelby, *Opt. Lett.* **9**, 533 (1984).

124. K. Hirao, S. Todoroki, K. Tanaka, N. Soga, T. Izumitani, A. Kurita and T. Kushida, *J. Non-Cryst. Sol.* **152**, 267 (1993).

125. R.M. Macfarlane, and R.M. Shelby, in *Laser Spectroscopy VI*, eds. H.P. Weber and W. Luthy, (Springer-Verlag, Berlin, 1983) p. 113.

126. Y.P. Wang, D.P. Landau, R.S. Meltzer, and R.M. Macfarlane, *J.Opt.Soc.Am.*, **B9**, 946 (1992).

127. C.A. Walsh, M. Berg, L.R. Narasemhan, K.A. Littau, and M.D. Fayer, *J.Chem.Phys.*, **86**, 77 (1987); Y.S. Bai and M.D. Fayer, *Phys.Rev.*, **B39**, 1066 (1989).

128. N. Bloembergen, *Physics*, **15**, 386 (1949).

129. W.B. Mims in *Electron Paramagnetic Resonance*, ed. S. Geschwind (Plenum, New York, 1972) p.263

130. R.M. Shelby, C.S. Yannoni, and R.M. Macfarlane, *Phys. Rev. Lett.* **41**, 1739 (1978).

131. J. L.Hall in Atomic Physics 3 eds., S. J. Smith and G. K. Walters (Plenum, New York), p. 615 (1973);

132. R. G. Brewer and A. Z. Genack, *Phys. Rev. Lett.* **36**, 959 (1976).

133. R. G. DeVoe, A. Szabo, S. C. Rand and R. G. Brewer, *Phys. Rev. Lett.* **42**, 1560 (1979).

134. R. G. Brewer and R. L. Shoemaker, *Phys. Rev.* **A6**, 2001 (1972).

135. D. M. Boye, R. Wannemacher, Y. P. Wang, W. Grill, J. E. Rives and R. S. Meltzer, *J. Lumin.* **45**, 431 (1990).

136. J. L. Hall and T. W. Hansch, *Opt. Lett.* **9**, 502 (1978); M. Zhu and J. L. Hall, *J. Opt. Soc. Am.* **B10**, 802 (1993).

137. T. Muramoto, S. Nakanishi and T. Hashi, *Opt. Commun.* **24**, 316 (1978).

138. R. M. Shelby and R. M. Macfarlane, *J. Lumin.* **31/32**, 839 (1984).

139. R. Wannemacher, R. S. Meltzer and R. M. Macfarlane, *J. Lumin.* **45**, 307 (1990).

140. E. L. Hahn, *Phys. Rev.* **80**, 580 (1950).

141. N. A. Kurnit, I. D. Abella and S. R. Hartmann, *Phys. Rev. Lett.* **13**, 567 (1964).

142. W. H. Hesselink and D. A. Wiersma, in "Spectroscopy and Excitation Dynamics of Condensed Molecular Systems", eds. V. M. Agranovitch and R. M. Hochstrasser, North Holland, Amsterdam, (1983) p. 249.

143. P. F. Liao, R. Leigh, P. Hu and S. R. Hartmann, *Phys. Lett. A* **41**, 285 (1972).

144. R. M. Macfarlane, R. M. Shelby and R. L. Shoemaker, *Phys. Rev. Lett.* **43**, 1726 (1979).

145. W. H. Hesselink and D. A. Wiersma, *Chem. Phys. Lett.* **54**, 227 (1978).

146. J. B. Morsink, T. J. Aartsma and D. A. Wiersma, *Chem. Phys. Lett.* **49**, 34 (1977).

147. D. Grischkowsky and S. R. Hartmann, *Phys. Rev. B* **2**, 60 (1970).

148. Y. C. Chen, K. P. Chiang and S. R. Hartmann, *Phys. Rev. B* **21**, 40 (1980).

149. D. R. Taylor and J. P. Hessler, *Phys. Lett.* **50A**, 205 (1974); D. R. Taylor, J. R. Marko and I. G. Bartlet, *Sol. State Commun.* **14**, 295 (1974).

150. G. K. Liu and R. L. Cone, *Phys. Rev.* **B41**, 6193 (1990).

151. J. Huang, J. M. Zhang, A. Lezama and T. W. Mossberg, *Phys. Rev.Lett.* **63**, 78 (1989).

152. S. Kroll, E. Y. Xu, M. K. Kim, M. Mitsunaga and R. Kachru, *Phys. Rev.* **B41**, 11568 (1990).

153. Y. S. Bai and R. Kachru, *Phys. Rev.* **B46**, 13735 (1992).

154. W. B. Mims, *Phys. Rev.* **168**, 370 (1968).

155. P. Hu and S. R. Hartmann, *Phys. Rev. B* **9**, 1 (1974).

156. J. Ganem, Y. P. Wang, D. Boye, R. S. Meltzer, W. M. Yen and R. M. Macfarlane, *Phys. Rev. Lett.* **66**, 695 (1991).

157. N. A. Kurnit and S. R. Hartmann, *Interaction of Radiation with Solids* (Plenum Press, New York) p.693. (1970).

158. Y. C. Chen, K. P. Chiang and S. R. Hartmann, *Opt. Commun.* **29**, 181 (1979).

159. J. B. Morsink and D. A. Wiersma, *Chem. Phys. Lett.* **65**, 105 (1979).

160. M. K. Kim and R. Kachru, *Phys. Rev.* **B44**, 9826 (1991).

161. W. H. Hesselink and D. A. Wiersma, *Phys. Rev. Lett.* **43**, 1991 (1979).

162. W. H. Hesselink and D. A. Wiersma, *J. Chem. Phys.* **75**, 4192 (1981).

163. K. Chiang, E. A. Whittaker and S. R. Hartmann, *Phys. Rev. B* **23**, 6142 (1981).

164. R. M. Shelby, R. M. Macfarlane and R. L. Shoemaker, *Phys. Rev. B* **25**, 6578 (1982).

165. D. P. Burum, R. M. Macfarlane and R. M. Shelby, *Phys. Lett. A* **90**, 483 (1982).

166. P. Hu, R. Leigh and S. R. Hartmann, *Phys. Lett. A* **40**, 164 (1972).
167. P. F. Liao, P. Hu, R. Leigh and S. R. Hartmann, *Phys. Rev. A* **9**, 332 (1982).
168. Y. P. Wang and R. S. Meltzer, *Phys. Rev.*, **B45**, 10119 (1992).
169. A. J. Meixner, C. M. Jefferson and R. M. Macfarlane, *Phys. Rev.*, **B46**, 5912 (1992).
170. F. Shimizu, K. Umezu and H. Takuma, *Phys. Rev. Lett.*, **47**, 825 (1981); S. Saikan, A. Imaoka and T. Kushida, *Phys. Rev.*, **B39**, 11207 (1989).
171. C. M. Jefferson and A. J. Meixner *Chem. Phys. Lett.*, **189**, 60 (1992).
172. F. Bitter, *Phys. Rev.*, **76**, 833 (1949).
173. J. Brossel and A. Kastler, *Compt. Rend.*, **229**, 1213 (1949).
174. J. Brossel, P. Sagalyn and F. Bitter, *Phys. Rev.*, **79**, 196,225 (1950).
175. J. Brossel and F. Bitter, *Phys. Rev.*, **86**, 308 (1952).
176. S. Geshwind, R. J. Collins and A. L. Schawlow, *Phys. Rev. Lett.*, **3**, 545 (1959).
177. A. L. Kwiram, *Chem. Phys. Lett.*, **1**, 272 (1967); M. Sharnoff, *J. Chem. Phys.*, **46**, 3263 (1967); J. Schmidt, I. A. M. Hesselmann, M. S. de Groot and J. H. van der Waals, *Chem. Phys. Lett.*, **1**, 434 (1967); J. Schmidt, and J. H. van der Waals, *Chem. Phys. Lett.*, **2**, 640 (1968).
178. R.M. Shelby, D.P. Burum and R.M. Macfarlane, *Opt.Commun.*, **52**, 283 (1984).
179. L.E. Erickson, *Phys.Rev.*, **B32**, 1 (1985).
180. C.B. Harris, D.S. Sinti, M.A. El-Sayed, and A.H. Maki, *Chem.Phys.Lett.*, **4**, 409 (1969).
181. R.M. Shelby, C.S. Yannoni, and R.M. Macfarlane, *Phys.Rev.Lett.*, **41**, 1739 (1978).
182. B.S. Mathur, H. Tang, R. Bulos, and W. Happer, *Phys.Rev.Lett.*, **21**, 1035 (1968).
183. J.N. Dodd, R.D. Kaul, and D.M. Warrington, *Proc.Phys.Soc.*, (London) **84**, 176 (1964).
184. R.L. Shoemaker, and R.G. Brewer, *Phys.Rev.Lett.*, **28**, 1430 (1972).
185. R.G. Brewer, and E.L. Hahn, *Phys.Rev.*, **A8**, 464 (1973).
186. J. Mlynek, and W. Lange, *Opt.Commun.*, **30**, 337 (1979).
187. J. Mlynek, K.H. Drake, G. Kersten, D. Fröhlich, and W. Lange, *Opt.Lett.*, **52**, 283 (1984).
188. T. Endo, S. Nakanishi, T. Muramoto, and T. Hashi, *Opt.Commun.*, **43**, 359 (1982).
189. W.E. Bell, and A.L. Bloom, *Phys.Rev.Lett.*, **6**, 280 (1961).
190. Y. Fukuda, Y. Takagi, and T. Hashi, *Phys.Lett.*, **48A**, 183 (1974).
191. J. Mlynek, W. Lange, H. Harde, and H. Burggraf, *Phys.Rev.*, **A24**, 1099 (1981).

192. R.M. Shelby, A.C. Tropper, R.T. Harley, and R.M. Macfarlane, *Opt.Lett.*, **8**, 304 (1983).

193. T. Mishina, M. Tanigawa, Y. Fukuda, and T. Hashi, *Opt.Commun.*, **62**, 166 (1987).

194. A. T. Kohmoto, Y. Fukuda, M. Tanigawa, T. Mishina, and T. Hashi, *Phys.Rev.*, **B28**, 2869 (1983).

195. J. Mlynek, N.C. Wong, R.G. DeVoe, E.S. Kintzer, and R.G. Brewer, *Phys.Rev.Lett.*, **50**, 993 (1983).

196. N.C. Wong, E.S. Kintzer, J. Mlynek, R.G. DeVoe, and R.G. Brewer, *Phys.Rev.*, **B28**, 4993 (1983).

197. K. Holliday, X.-F. He, P.T.H. Fisk, and N.B. Manson, *Opt.Lett.*, **15**, 983 (1990).

198. N.B. Manson, X.-F. He, and P.T.H. Fisk, *Opt.Lett.*, **15**, 103 (1990).

199. M. Mitsunaga, E.S. Kintzer, and R.G. Brewer, *Phys.Rev.Lett.*, **52**, 1484 (1984); D.R. Taylor, *Opt.Commun.*, **52**, 204 (1984); E.S. Kintzer, M. Mitsunaga, and R.G. Brewer, *Phys.Rev.*, **B31**, 6958 (1985).

200. Y. Takahashi, K. Ishikawa, Y. Kukuda, T. Yabuzaki, and T. Hashi, *Phys.Rev.*, **B43**, 7527 (1991).

201. M. Mitsunaga, E.S. Kintzer, and R.G. Brewer, *Phys.Rev.*, **B31**, 6947 (1985).

ADVANCES IN SOLID STATE LASERS

Allister I. Ferguson

Department of Physics and Applied Physics
University of Strathclyde
Glasgow G4 0NG
Scotland, UK

ABSTRACT

The advances in solid state lasers over the past few years have been very dramatic. This has been brought about by a combination of developments in new laser diodes, new laser gain media and nonlinear optical materials which provide frequency conversion and tunability, and new ideas concerning methods of mode-locking to produce extremely short light pulses. Some of these advances will be described and illustrated by work from my own group and that of many others. This is far from a comprehensive summary but more illustrative to show the major trends.

I. INTRODUCTION

The very rapid progress which has taken place in solid state laser technology over the past few year is due to the culmination of a number of technical innovations covering a wide range of disciplines. High power laser diodes are now one of the most efficient ways of converting electrical energy into optical energy. Continuous-wave (cw) diodes with average powers of up to 20W are now available either as 1cm bars or with fibre delivery. They may also be operated in a quasi-cw mode which is suitable for pumping Q-switched solid state lasers. Laser diodes make ideal pump sources for solid state laser gain media since they can be tuned to absorption bands and can be efficiently coupled into the mode volume of the laser gain medium. The resulting diode pumped laser has, in general, greatly improved spatial and spectral characteristics than the laser diode itself. Furthermore, the diode pumped laser may be broadly tunable and can normally be mode-locked or Q-switched. The trend towards diode pumping has led to the development of many new laser gain media and the revitalisation of older gain media which had previously been rejected because of the limited size, thermal effects and optical quality.

Many of the modes of operation exhibited by flashlamp pumped lasers are also shown by diode pumped lasers but often with superior performance. One of the most common modes of operation is single frequency operation. This is generally the mode of operation required for high resolution spectroscopy. The all-solid-state nature of the gain medium and the pump laser means that cw single frequency solid state lasers have exceptionally narrow linewidths. Linewidths in the region of a few kilohertz are quite routine whereas much narrower linewidths may readily be obtained with some effort.

The long upper state lifetime of many of the solid state gain media used in diode pumping makes them ideal candidates for storing energy which can then be extracted in a

Nonlinear Spectroscopy of Solids: Advances and Applications
Edited by B. Di Bartolo and B. Bowlby, Plenum Press, New York, 1994

single giant pulse by use of Q-switching techniques. These techniques can be readily adapted to provide single frequency Q-switched pulses.

Advances in mode-locking of solid state lasers over the very recent past have been quite spectacular. Media which were once thought capable of producing pulses with duration no shorter than a few picoseconds have now been shown to be capable of producing pulses of around 10fsec duration. These advanced mode-locking techniques rely on the use of optical nonlinearities to provide the pulse compression mechanism.

For many applications the wavelength of operation of the laser with desirable properties is different from that required by the experiment. This inevitably means that the wavelength has to be shifted in some way. The classic way of doing this is to use harmonic generation. This requires the use of high laser powers and nonlinear crystals with the appropriate properties. Fortunately, in recent years there have been exciting advances in both nonlinear materials and in manipulating the properties of the laser beam either in enhancement cavities or by mode-locking. This permits very efficient generation of the second and higher harmonics.

Harmonic generation, of course, only generates frequencies which are harmonics of the fundamental laser. Wide tunability can be provided by the use of optical parametric oscillation. Although this technique has been around since the earliest days of the laser it has only recently 'come of age' with advances in lasers and nonlinear materials.

An alternative to optical parametric oscillation is a range of materials which themselves emit over a wide range and hence produce tunable laser radiation. This class of laser is generally called a vibronic laser with Ti:sapphire being probably the most successful example. At present Ti:sapphire has to be pumped by ion lasers operating in the green region of the spectrum for high power. Although it has been pumped by frequency doubled diode pumped systems it gives only low power. A new range of vibronic materials based on Cr^{3+} doped into fluoride hosts has now become available. These materials can be directly pumped by diodes operating in the red region of the spectrum. These materials can be operated as single frequency sources or mode-locked to produce femtosecond pulses. The compact, all-solid-state design that such a system offers can be expected to have a widespread impact on many areas of science and technology.

In this paper I will describe, in more detail, some of the progress summarised above. The aim is to give a flavour for the exciting developments rather than to provide a comprehensive survey. The reader is referred to many of the excellent reviews which have been produced in recent years for more details [1]. My approach will be to provide some background information and to illustrate the principles with some examples from the work of my group and that of others.

In section 2 development of diode lasers as pump sources for solid state lasers is discussed. This is based on commercial sources since these are the only devices to which most of us will have access. In section 3 some simple diode pumped solid state lasers are described with Nd:YAG being the typical material. Single frequency operation and Q-switching of these devices will be discussed. Section 4 deals with a review of techniques for mode-locking solid state lasers. This includes brief discussion of active mode-locking, passive mode-locking and the new techniques based on the optical Kerr effect. Section 5 is concerned with nonlinear frequency conversion. It deals with harmonic generation in enhancement cavities both for single frequency lasers and mode-locked lasers. Some recent progress on the provision of wide tunability using optical parametric oscillation will be discussed. In section 6 tunable gain media will be considered. This includes single frequency operation and mode-locking using the Kerr nonlinearity. The prospects for diode pumping will be discussed. In section 7 we finally discuss an exciting application for an all-solid-state mode-locked laser for systems in the area of biological imaging using confocal fluorescence microscopy.

II. DIODE LASERS AS PUMP SOURCES

The GaAlAs laser diode system which operates in the region of 800nm is one of the most efficient devices for the conversion of electrical energy into optical energy. A typical electrical to optical efficiency is in the region of 30%. The beam quality from high power laser diodes is quite good but usually far from diffraction limited. The area of the device has to be made large using multi-elements, or just a large area, in order to avoid damage to the facets of the diode. The relatively narrow spectral emission means that the laser can be tuned into the strong absorption bands of the laser gain medium. This gives excellent coupling efficiency and ensures that only a small volume of active gain medium is needed. This makes a diode pumped laser system compact, efficient and reliable. Laser diodes are at present quite expensive. However, there is nothing inherently expensive in these devices and so the price will certainly reduce as demand increases. Progress in diode lasers as pump sources is reviewed by Streifer [2]. Progress since that time has been quite rapid and can best be gauged by reference to the major manufacturers' catalogues.

A typical high power laser diode suitable for pumping a solid state laser would be packaged onto a Peltier cooler for heat extraction and temperature tuning of the emission wavelength. At present single arrays of about 500μm by 1μm give cw powers of up to 4W at around 800nm. This output can be coupled using high numerical aperture lenses or can be delivered using multimode fibres.

Higher powers are available from arrays of arrays mounted on a 1cm bar. Each array is separated by about 1mm and made up of a large number of emitters with a total aperture of 500μm. Individual emitters are limited to giving an output power of about 50mW. A 20W array thus contains up to 400 emitters. The spread in wavelength is typically a few nanometers. There are several designs for coupling a 1cm bar to end pump a laser rod using cylindrical optics. The pump light can also be coupled using a multi-mode fibre.

Although the technology of high power laser diodes is most fully developed for GaAlAs at 800nm, rapid progress is being made in going to shorter wavelengths and also to longer wavelengths. With the availability of laser diodes over such a large spectral region it is interesting to ask why they should not themselves be used directly as light sources in applications. The major limitation for diodes as spectroscopic sources lies in their susceptibility to facet damage.

Diode lasers are themselves useful as spectroscopic sources provided they can be used at low power. A single element device is capable of a few 10s of mW but is only available at wavelengths for which there is a large commercial market. When engineered properly they can certainly be configured to produce narrow linewidths [3]. Because of their minute size and the electronic pumping, laser diodes suffer from very high frequency noise which is very difficult to eliminate. In certain low noise experiments this is unacceptable.

Although laser diodes have been mode-locked to produce short pulses they are not widely accepted as good ultrashort pulse sources since they have to be operated at an average power such that peak power, when mode-locked, will not cause damage to the laser facets. The storage time and damage problems in semiconductor lasers is such that Q-switching is impractical.

The GaAlAs laser system is well suited to pumping of Nd doped materials. Progress in the development of red diodes at around 670nm means that it is now practical to consider pumping Cr^{3+} doped materials with high power.

III. DIODE PUMPED SOLID STATE LASERS

Early work on diode pumping of solid state laser materials was inspired by the fortuitous overlap between the emission region of semiconductor (GaAs/GaAlAs) laser diodes operating at around 800nm and the absorption peaks of the widely used laser material neodymium doped yttrium aluminium garnet (Nd:YAG), shown in figure 1. Taking this material as an example we examine the physical processes of diode pumping which leads to the high efficiencies displayed by this type of all-solid-state laser.

III.A. Diode Pumping of Nd:YAG

Figure 2 shows the energy level diagram of Nd:YAG. Under illumination by light at 810nm, the wavelength of the strong absorption peak of Nd:YAG, the Nd^{3+} ions are excited into pump bands just above the upper laser level. Non-radiative relaxation then occurs to the $^4F_{3/2}$ level. Low threshold laser action can then occur by one of two high gain transitions; the $^4F_{3/2}$ -$^4I_{11/2}$ transition resulting in emission at around 1·06μm or the

$^4F_{3/2}$ - $^4I_{13/2}$ transition producing emission at around 1·32μm. Stark splitting of these energy manifolds by the crystal field produces a series of doubly degenerate levels. It is between these stark levels that the actual laser transitions occur. The $^4F_{3/2}$ - $^4I_{9/2}$ transition produces emission at 0·946μm, but usually with poor efficiency due to the thermally populated lower level.

Since the diode pump excites the neodymium to a pump level that lies only slightly above the upper laser level the amount of waste heat generated is several times lower than that produced by flashlamps. Typically in excess of 90% of the diode laser pump is absorbed by the Nd:YAG rod, resulting in extremely high conversion to the Nd:YAG laser output. Optical conversion from the diode pump at 810nm to the 1·064μm high gain line of Nd:YAG can be as high as 50%.

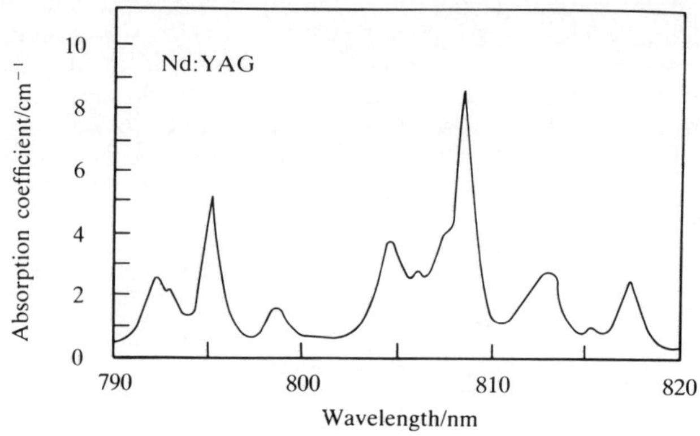

Figure 1. Absorption spectrum of Nd:YAG, showing a strong absorption peak around 810nm, suitable for excitation by GaAs/GaAlAs semiconductor diode lasers.

III.B. Advantages of the Diode-Pumped All-Solid-State Laser

Diode pumping of solid state laser materials offers major advantages over both flashlamp pumping schemes and direct use of laser diodes as sources of coherent radiation.

Comparison with flashlamp pumping. A typical arc lamp used to pump Nd:YAG will emit on several strong lines around the Nd:YAG absorption with a broadband blackbody radiation background. Around 5% of the energy supplied to the lamp will be emitted in the region of the Nd:YAG absorption. Owing to the high power densities of the arc lamp, water cooling is also normally required. Poor overlap between the volume of the Nd:YAG pumped by the flashlamp and the mode of the laser cavity also result in a drop in laser efficiency and causes excess heat withih the gain material which can induce thermal focusing and birefringence. The overall efficiency of a flashlamp pumped Nd:YAG laser is therefore approximately 0.5%. Other gain materials have been demonstrated to yield higher efficiencies by flashlamp pumping, but these still represent less than 10% conversion from electrical to optical power.

The current generation of diode laser used for pumping solid state materials yields efficiencies of around 30%, resulting in diode-pumped solid-state lasers with overall electrical to optical ('or wall plug') efficiency of around 15%. New technologies in semiconductor diode laser fabrication have demonstrated devices with 50% conversion from electrical to optical power. Such improvements in the basic diode pump will, in future, lead to even greater efficiencies being produced by diode-pumped all-solid-state lasers.

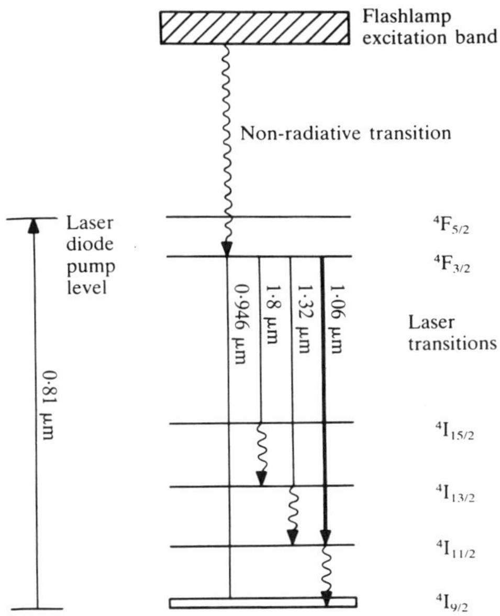

Figure 2 Energy level diagram for Nd:YAG, illustrating process of diode laser excitation and Nd:YAG laser transitions.

Perhaps the greatest advantage of diode pumping over conventional techniques is the reduction in noise which can be achieved. Cooling of flashlamp pumped systems and poor overlap between the pump and lasing volumes in the gain medium results in thermal fluctuations which cause frequency instabilities in the laser output. The reduced densities of waste heat in the efficient diode pumping process, combined with intrinsically more stable solid state diode pumps, resulting in improvements in performance of several orders of magnitude. Diode-pumped lasers have been demonstrated to yield 1Hz linewidths compared with around 100kHz for flashlamp pumped lasers.

III.C. Materials for All-Solid-State Lasers

For laser materials other than Nd:YAG the positions of the energy levels discussed in section 3A vary, producing different absorption and emission characteristics, though lasing is achieved by the same processes. While much of the early work on diode pumping of all-solid-state lasers concentrated on Nd:YAG, and this is still the most widely investigated holosteric system, the possibility of extending diode pumping to other gain materials has also been widely investigated. Diode laser pumping of neodymium-doped hosts other than YAG, trivalent rare earth dopants, novel tunable laser media, waveguide and fibre geometry lasers are all currently areas of great research activity.

Wide-ranging wavelength diversity of diode-pumped solid-state lasers can be achieved by employing laser materials based on other ion dopants [4]. Trivalent rare earth ions with significant absorption in the 750-850nm diode pumping region include Nd^{3+}, Pm^{3+}, Dy^{3+}, Er^{3+}, and Tm^{3+}. Recently there has been great interest in thulium, holmium co-doped yttrium lithium fluoride (Tm, Ho:YLF) and thulium, holmium co-doped yttrium aluminium garnet (Tm, Ho:YAG) lasers operating at around 2·1μm for eyesafe applications requiring atmospheric propagation including coherent laser radar [5].

III.D. Cavity Configurations for Diode-Pumping

The output from a laser diode is highly divergent and astigmatic. Several schemes have been investigated to match this radiation into the laser mode volume in the gain material to ensure efficient operation of the laser system. the most common schemes are longitudinal end pumping of the gain medium in either a rod or slab geometry.

Longitudinally pumped cavities. To date the most extensively used pumping geometry is longitudinal pumping. The output from the diode laser is collimated and beamshaped to achieve a circular profile before being focused down to form a pump spot on the laser rod. This technique allows good matching between the lasing spot size of the solid state laser cavity and the pump spot size in the gain medium. This spatial overlap between pump and lasing modes, known as mode-matching, is critical to the efficiency of the diode pumping process.

The earliest designs for end pumping geometries concentrated on monolithic cavity designs as shown in figure 3(a). This is the simplest form of miniature all-solid-state laser, which includes all the necessary mirror coatings and curvatures onto the ends of the gain medium. The laser rod therefore forms a stable cavity with the focus forming a lasing spot size of normally less than 100μm. This form of cavity produced very low thresholds of a few milliwatts and, at low powers, operated on a single frequency. The low output powers from these devices soon led to the development of diode-pumped lasers with cavity designs using external mirrors.

Two mirror cavity designs as shown in figure 3(b) and (c) consist of a laser rod mirror coated on one end and a discrete curved mirror which forms the lasing mode.

This is well matched to the pump spot for high-efficiency operation. Access to the cavity now allows the insertion of Q-switches, etalons or nonlinear crystals for pulsed, single frequency or frequency doubled operation.

The three mirror astigmatically-compensated cavity design shown in figure 3(d), similar to that widely used in dye lasers, has also become popular for diode-pumped lasers. One end of the laser rod is plane cut and mirror coated while the other end is cut at the Brewster's angle, for low loss within the laser cavity. The angle of the cavity fold can be adjusted to compensate for astigmatism produced at the Brewster surface. The laser mode size in the crystal is controlled by the choice of curved folding mirror and length of cavity arm, so that the pump spot can again be matched to it for high efficiency operation. An additional benefit of this design is a near collimated arm suitable for the insertion of mode-lockers or Q-switches into the cavity.

Longitudinal pumping geometries allow the development of all-solid-state lasers for a wide range of applications in low-power regimes including scientific sources for spectroscopy and the investigation of ultrafast phenomena, sources for optical communications and information processing, medical diagnostics and laser surgery, semiconductor processing, holography and high definition television.

Transversely pumped cavities. Transverse pumping geometries become attractive for power scaling of diode-pumped solid-state lasers into regimes beyond the 10W output level. Side pumping of laser rods and zig-zag laser slabs promise to yield average power

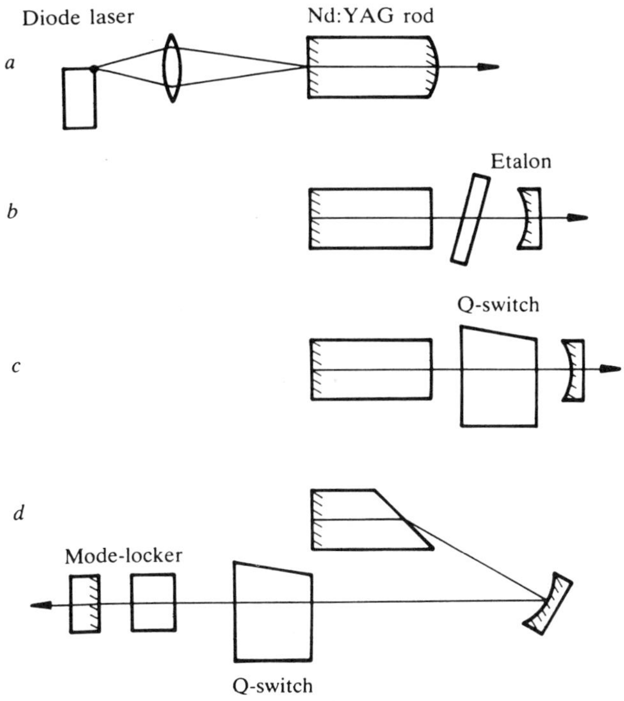

Figure 3 Evolution of the longitudinally pumped Nd:YAG laser. Original design (a) limited output power and lacked access to the cavity. By separating one of the mirrors it has been possible to introduce frequency selective etalons (b), Q-switches (c,d) and mode-lockers (d).

lasers from around 10W to several kilowatts. Side pumping of laser rods by close coupling of diode arrays is shown in figure 4. This technique allows pumping at much higher power levels than longitudinally pumped schemes, but suffers from lower efficiency due to poorer matching of the pump radiation into the laser mode. A more successful technique is the pumping of a zig-zag slab. By close coupling a diode laser with emitting areas matched to the bounce separation of laser mode in the rod, or by using optical fibres to transmit the pump light to these points, large densities of pump radiation

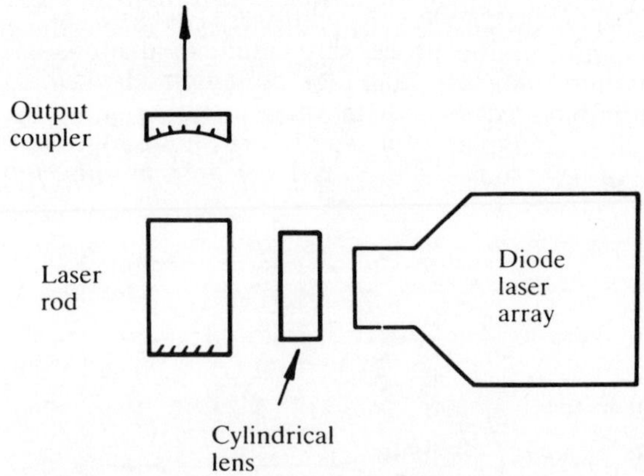

Figure 4 Transverse side pumping of a solid-state laser rod by a laser diode array

can be matched into the laser mode. Efficient conversion of the pump radiation to laser output reduces thermal effects, which are further minimised by pumping at many discrete points of the laser slab. Applications for high power diode-pumped solid-state lasers include the processing of industrial materials processing photolithography and atmospheric laser analysis.

For extremely high power solid-state lasers, diode-pumped systems based on transverse pumping of Brewster angled slabs or moving disk layers have been proposed [6]. Such systems may achieve outputs of many kilowatts for applications, after amplification, in nuclear fusion processes. These lasers will be based on future generations of extremely high power laser diodes, fabricated in two dimensional arrays of closely packed emitters.

III.E. Single Frequency Operation of All-Solid-State Lasers

Great effort has been employed to reduce the emission linewidth of laser output to achieve extremely narrow spectral purity since the first predictions of monochromatic output by lasers in the theory of Schawlow and Townes [7]. Diode pumping of solid-state lasers allows operation of devices very close to the fundamental linewidth given by the Schawlow-Townes equation:

$$\Delta\nu = \frac{h\nu}{2\pi\delta^2 P}, \tag{1}$$

where $\Delta\nu$ is the linewidth in Hertz, δ is the cavity decay time, P is the power output and $h\nu$ is the photon energy.

A laser cavity defines longitudinal modes with frequency spacing $\delta\nu = c/2l$. The mode at the centre of the gain profile will be the first to reach threshold. Homogeneous broadening would be expected for solid-state lasers whose line broadening mechanism of lattice vibrations has a characteristic time much shorter than the population relaxation time. For this broadening mechanism the laser mode would be expected to access gain at the line centre without burning a hole in the population inversion and produce extremely narrow linewidth output. Spatial hole burning occurs owing to the slow spatial diffusion of the population inversion, causing the field of a standing wave cavity to create a spatial variation of the gain in the active medium. As the pump power increases this causes the gain level of the first oscillating mode to be held at its threshold value at the antinodes of the field. Excess gain close to the nodes of the standing wave encourages other modes to oscillate. Spatial hole burning therefore causes the spectral linewidth to broaden. Higher pump powers encourage more modes to oscillate and further broaden the output. To ensure narrow single frequency operation of the all-solid-state laser the effects of spatial hole burning must be overcome.

The earliest single longitudinal mode all-solid-state lasers were diode-pumped monolithic Nd:YAG lasers. The laser cavity consists of a solid Nd:YAG rod of around 5mm in length, one end plane and the other curved with mirrors coated onto each end. These devices exhibited linewidths of less than 50kHz threshold of 2.2mW and slope efficiencies of 25% [8]. However, soon after the discovery of the single frequency monolithic Nd:YAG laser it was realised that single frequency operation could only be achieved at very lower power levels of less than 1mW before the effect of spatial hole burning caused multifrequency operation.

A similar and perhaps the simplest approach to single-frequency operation from a solid-state laser was to reduce the length of the gain medium and cavity length so that only one longitudinal mode lies within the gain spectrum. These microchip lasers employ Nd:YAG chips whose length are less than 1mm. Such lasers can produce single frequency operation with linewidth less than 5kHz at output levels up to 20mW [9]. Higher power single frequency operation of monolithic diode-pumped lasers has been demonstrated employing twisted mode cavity designs [10].

Although many of the materials for diode pumping including Nd:YAG are homogeneously broadened gain media there has been little success in achieving single frequency operation in any other standing wave configurations even when using intercavity etalons to reduce the effects of spatial hole burning. The use of unidirectional ring laser oscillators to completely eliminate spatial hole burning has proved the most successful approach to high average power single frequency diode-pumped all-solid-state lasers. Unidirectional operation of the ring is commonly achieved by using the Faraday effect. It is well known that the Faraday effect produces a non-reciprocal polarisation rotation. Faraday rotation can be combined with an optical rotator so that the optical rotations add in one direction while they will cancel in the opposite direction. The inclusion of such a device in a cavity with polarising elements such as the Brewster surface of a laser rod results in low loss in only one direction and hence unidirectional laser operation.

Unidirectional ring operation in a Nd:YAG laser cavity is easily accomplished by applying a magnetic field to the laser rod to employ the small Faraday effect in Nd:YAG. The small optical rotation required to compensate for the Faraday rotation in one direction can be introduced by operating the ring laser in an out-of-plane geometry. This concept has been incorporated into a monolithic structure by [11] called the monolithic nonplanar ring oscillator as shown in figure 5. The linewidth of an actively stabilised diode laser pumped nonplanar ring oscillator based on neodymium-doped gadolinium-gallium garnet (Nd:GGG) has been measured to be less than 3Hz [12]. Diode-pumped solid-state lasers can now therefore achieve linewidths only 20 times that of the fundamental shot noise limit, as given by the Schawlow-Townes equation.

Single frequency lasers based on this design can be used to injection seed high power Q-switched lasers. Diode-pumped single frequency lasers based on other cavity designs can also be Q-switched to provide high peak power single frequency output.

IV. MODE-LOCKING OF SOLID-STATE LASERS

One of the most exciting developments in laser physics over the past few years has been the development of mode-locking techniques for solid state lasers. There are three main classifications for mode-locking. Active mode-locking involves insertion of a modulator into the laser cavity. This modulator is driven to provide amplitude or phase modulation at the cavity mode spacing frequency. This locks the modes together and hence produces a train of pulses with a repetition rate equal to the modulation frequency.

In the passive scheme a saturable absorber is placed in the cavity. The high intensity noise spikes will experience less loss than the lower intensity light. These spikes will therefore be amplified at the expense of low light levels. Eventually all the light will be

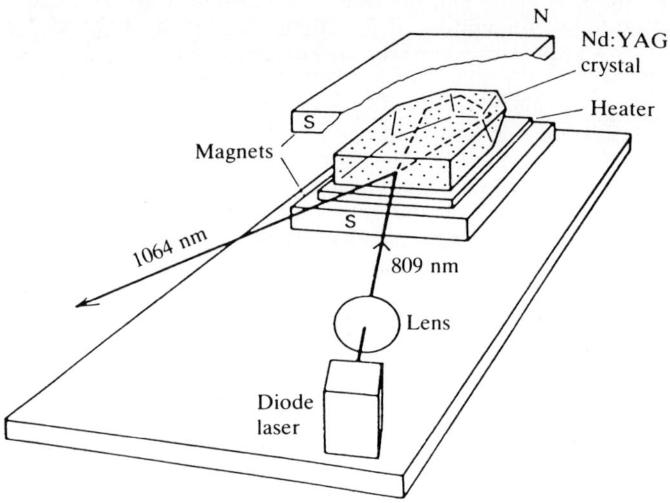

Figure 5 The single frequency non-planar ring oscillator [11]

compressed into just a single pulse per round trip. This will therefore provide a mode-locked pulse train.

The technique of nonlinear self mode-locking relies on the use of the very fast optical Kerr effect in which the intensity of light modifies the refractive index. Techniques based on the Kerr nonlinearity have provided a breakthrough in mode-locking in the sense that systems which were once thought to have reached the lower limit in pulse duration were shown to be capable of much shorter pulse durations when these techniques were employed. The key to the nonlinear mode-locking techniques is that the pulse compression mechanism gets stronger as the pulses become shorter in contract to active mode-locking discussed above. A tutorial review of mode-locking of solid-state lasers has been published [13, 14].

4.A. Active Mode-Locking

A schematic diagram of an actively mode-locked laser is shown in figure 6. The amplitude or phase modulator is placed in the cavity and driven to provide an amplitude or phase modulation at the cavity mode spacing frequency. This system has been modelled in detail by Keuizenga and Siegman [15]. In their model they propagate a Gaussian pulse around the cavity and look for a steady-state solution. The resulting pulse duration τ_p is given by

$$\tau_p \propto (g / \delta)^{1/4} (1 / f_m \Delta f)^{1/2} \tag{2}$$

where g is the saturated gain, δ is related to magnitude of the loss or phase modulation, f_m is the modulation frequency and Δf is the bandwidth of the gain medium. From this equation it can be seen that the pulse duration is relatively insensitive to the gain and

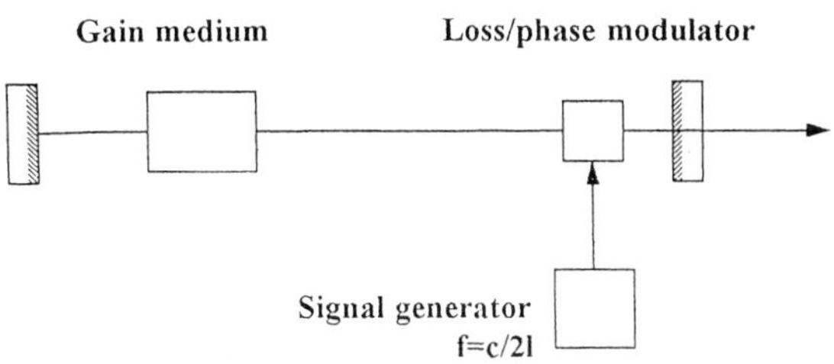

Figure 6. Schematic diagram of an actively mode-locked laser

modulation depth, these quantities appearing only to the quarter power. The gain bandwidth is a characteristic of the gain medium but appears only as the square root. The only other design parameter available is the modulation frequency. Higher modulation frequency gives shorter pulses but again varies only as the square root.

These rather weak dependencies on the various parameters arises from the fact that as the pulses become shorter they operate on a part of the modulation curve which becomes flatter. The pulse compression mechanism therefore gets ever weaker and eventually the pulse duration is limited to a level which may be much less than could be sustained by the available bandwidth of the gain medium.

A typical frequency modulation (FM) mode-locked system is shown in figure 7. In this case Nd:YLF has been used as the gain medium. Nd:YLF is a superior material to Nd:YAG for mode-locking because the bandwidth is a factor of 2 greater than Nd:YAG. It is therefore capable of generating much shorter pulses. The phase modulator is a simple tuned resonance device and can operate from a few megahertz up to about 1GHz. The Nd:YLF is Brewster angled at one end and the turning mirror is angled for astigmatic compensation. When the cavity length is matched to the modulation frequency the system

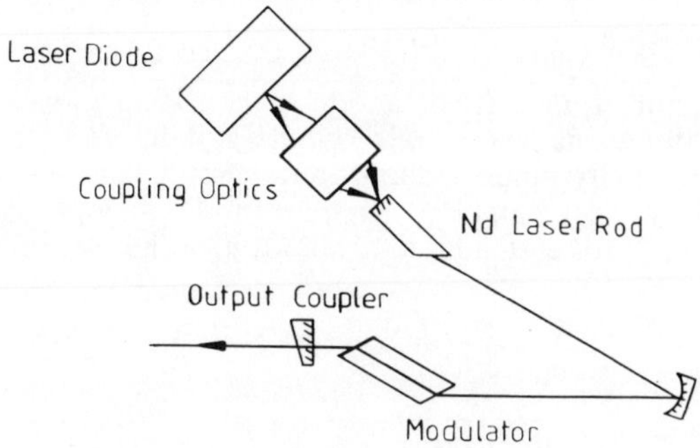

Figure 7. FM mode-locked Nd:YLF laser

mode-locks. The shortest pulses to have been generated with such a system is no shorter than about 10psec. This is much shorter than has been generated using flashlamp pumped systems but is still well in excess of the sub-picosecond duration which the available bandwidth makes possible.

The pump laser diodes are tuned into the Nd:YLF absorptions around 792nm. For this particular system two 3W laser diodes were coupled together using a polarization beam splitter. This provided an output power close to 2W at 1047nm [16].

IV.B. The Kerr Nonlinearity

The Kerr effect was discovered in 1875 by John Kerr who was a student of Lord Kelvin. He found that for isotropic media there was a change in the refractive index which was proportional to the square of an applied DC field. The optical Kerr effect is a modification to the refractive index which is proportional to the instantaneous intensity of an applied optical field.

If the field is spatially inhomogeneous, such as in a Gaussian beam, there will be a spatial variation in the refractive index. There will therefore be an induced lens which will be either positive or negative depending on the sign of the nonlinearity.

If, on the other hand, we have a pulse of light there will be a temporal variation in the intensity. This will give rise to a refractive index which will vary throughout the pulse. Remembering that frequency is the time derivative of the instantaneous phase we see that we obtain a time varying frequency. This is called self-phase modulation and gives rise to a frequency chirp.

IV.C. Additive Pulse Mode-Locking (APM)

The technique of additive pulse mode-locking (APM) or coupled cavity mode-locking (CCM) is illustrated in figure 8 [17]. In this scheme the output from the laser is coupled into a second cavity which contains a nonlinear element. After having undergone

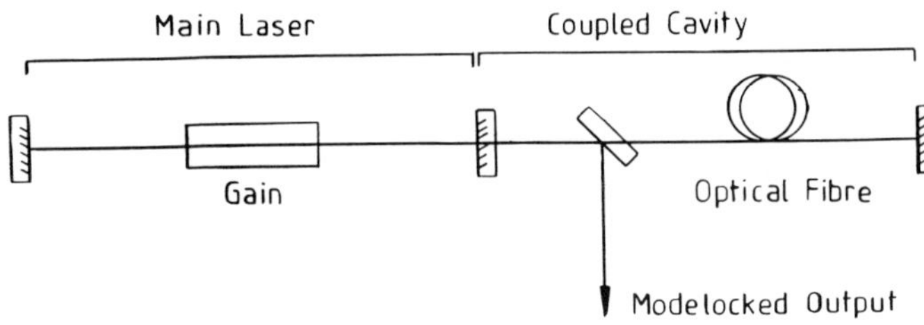

Figure 8. Schematic diagram of additive pulse mode-locked (APM) laser

propagation through this element the light is returned to the main cavity where it interferes with the light in that cavity. This method of mode-locking was pioneered by Mollenauer and Stolen [18] for the case of solitons in optical fibres at around 1500nm. It has only recently been discovered that the techniques did not in fact depend crucially on soliton propagation in a fibre and has consequently been generalised to apply to several types of laser. The details of this mode-locking is still the subject of research and debate. The basic idea is that light entering the nonlinear arm will undergo self-phase modulation due to the Kerr nonlinearity. This will produce a frequency chirp on the pulse. If this is returned to the main cavity such that constructive interference occurs at the centre of the pulse there will be destructive interference at the wings. This will have the effect of

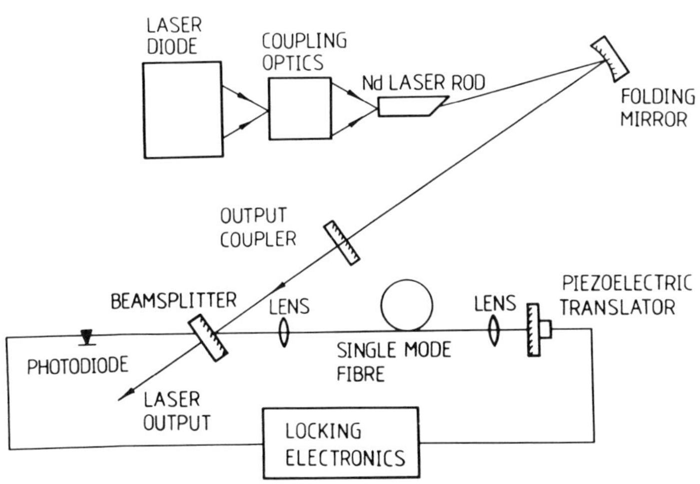

Figure 9. An APM diode pumped Nd:YLF laser

shortening the pulse which returns to the main cavity. In going round the main cavity it will be amplified and then re-enter the nonlinear cavity where it will produce even more self-phase modulation. This compression process will continue until some limit sets in. The usual limit is caused by dispersion in either the laser cavity or in the nonlinear cavity. By controlling the dispersion it is possible to utilise a very large fraction of the gain bandwidth. Pulses produced by this method have often been found to be the shortest ever to have been produced by the gain medium in question.

A schematic diagram of an APM diode pumped Nd:YLF laser is shown in figure 9. In this scheme the diode pumped laser is similar to the actively mode-locked laser system shown earlier with the exception that there is no modulator in the cavity [19]. The output from the laser is coupled into a single mode optical fibre of about 2m length. The output from the fibre is coupled back into the laser with a delay corresponding to the round trip time in the external cavity being twice that of the main cavity. The phase of the return signal is monitored on a photodiode. By use of servo electronics it is possible to return the light with the appropriate phase to ensure mode-locking. For a fibre length of a few metres on input average power of about 200mW is needed to obtain reliable mode-locking. When the system is pumped with two 3W diodes an average output power of 1.5W is available for use. The measured pulse duration from this system has been as low as 1.5psec. The peak power available from such a system is therefore around 10kW.

IV.D. Kerr Lens Mode-Locking (KLM)

The second of the novel mode-locking techniques discussed here is that of Kerr lens mode-locking (KLM) [20]. In this case the laser is designed such that there is a focal region where a medium which exhibits the Kerr nonlinearity can be placed. If the cavity is carefully designed the resulting self-focusing produced by the Kerr lens will form a region in the cavity where the spot size of the high intensity, self-focused radiation will be smaller than the CW spot. If an aperture is placed in this region it is possible to discriminate against the CW radiation and force the laser to switch over to the mode-locked mode of operation. Once the laser has switched to this mode the self-focusing will become even more dominant and the pulses will become even more compressed. The pulse duration will collapse until the entire available bandwidth has been utilised or, more usually, until the dispersion in the cavity limits the pulse duration.

In the KLM system the active medium may also be the Kerr medium but this does not have to be the case. Increased design flexibility is made possible by the provision of a second focus where a Kerr medium can be placed. An example of a KLM diode pumped Nd:YLF system is shown in figure 10 [21]. In this design a second focus is provided and a piece of high refractive index glass is placed at the focus. The size of the pump spot from the laser diodes and the requirement that the Nd:YLF mode be well matched to the pump mode forces the design to require an intracavity focus for the Kerr medium. An aperture is provided for suppression of the CW mode of operation but it was not always needed. The average output power from this system was about 1W for two 3W diode pumps. By careful alignment of the cavity and adjustment of the aperture it was possible to force Kerr lens mode-locking. The KLM system will not start spontaneously. It is generally necessary to tap a mirror to cause a perturbation which initiates the mode-locking. Once started the KLM laser is stable and requires little attention. Pulses obtained from this system are about 6psec [10]. It is probably that this pulse duration is limited by dispersion in the cavity. It should be possible to obtain pulses below 2psec with proper dispersion compensation.

V. NONLINEAR FREQUENCY CONVERSION

The range of applications for solid state lasers can be greatly extended by the use of

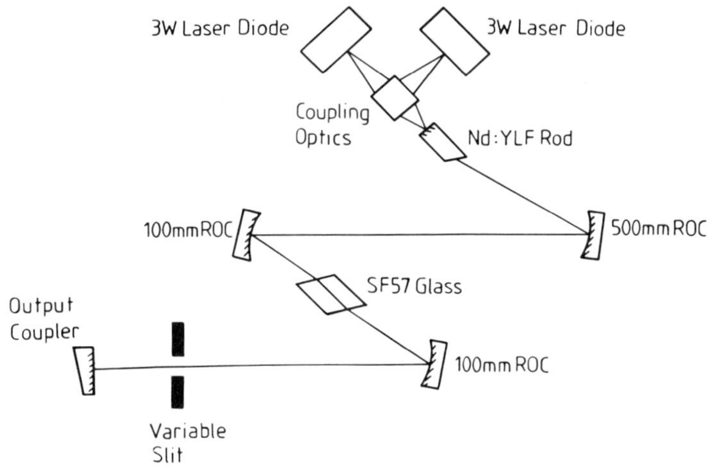

Figure 10. A Kerr lens mode-locked diode pumped Nd:YLF laser

nonlinear optical techniques to extend the spectral coverage of laser systems. Harmonic generation was one of the first nonlinear optical effects to have been discovered and continues to be one of the most important methods of frequency conversion. Harmonic generation has been given a new lease of life by the discovery of a new range of nonlinear optical materials and by the development of techniques for resonant cavity enhancement. The combination of these factors together with improvements in beam quality of modern lasers means that harmonic generation can now be very efficient.

A second nonlinear effect which has been revitalised by materials and laser developments has been optical parametric generation. This effect has been known almost since the invention of the laser itself but it is only now that optical parametric oscillators (OPOs) are beginning to be accepted. The improvements in the coherence of the pump lasers together with nonlinear materials marks OPOs one of the most rapidly developing areas in nonlinear optics. OPOs now enable virtually all of the UV, visible and near infrared regions of the spectrum to be accessed.

Much of the excitement in nonlinear optics has been caused by a new generation of nonlinear materials. These new materials are non-hygroscopic, have high damage threshold and good phase-matching characteristics. Materials which are now widely available include potassium titanyl phosphate (KTP), lithium triborate (LBO), potassium niobate (KNB) and beta barium borate (BBO).

V.A. Harmonic Generation

Harmonic generation is a nonlinear optical process in which the fundamental beam is doubled in frequency using the second order nonlinear susceptibility of non-centrosymmetric crystals. Until recently the efficiency of harmonic generation has been very low. The conversion efficiency is given by

$$\varepsilon = P_{2\omega} / P_{\omega} = \gamma P_{\omega} \tag{3}$$

where P_{ω} is the power at the fundamental and $P_{2\omega}$ is the second harmonic power [22]. The coefficient γ includes parameters such as the nonlinear coefficient, spot sizes and several other important parameters. In general, γ ranges from about $10^{-5}W^{-1}$ to $10^{-3}W^{-1}$. This means that for a fundamental laser of 1W the output at the second harmonic will range from 10μW up to 1mW giving efficiencies of 10^{-3}% up to 0.1^{-1}%. This is impractically low for many applications.

Improved conversion efficiency can be obtained by mode-locking the fundamental laser such that the average power is maintained and the energy comes in the form of a pulse train. Roughly speaking the harmonic efficiency increases by the duty cycle. If we mode-lock a laser of 10psec pulses at 10nsec repetition period the conversion efficiency could rise by as much as a factor of 10^3, all other things being equal. If the laser can be Q-switched the harmonic efficiency can also be greatly increased.

If on the other hand the laser must be operated in the CW mode the efficiency can be greatly enhanced by the use of resonant enhancement cavities. This idea goes back to 1966 when it was shown that it was possible to build up the radiation inside a cavity [23]. A schematic of typical system is shown in figure 11. Light from the laser is fed through a partially transmitting input mirror and propagates around the cavity. A nonlinear crystal is located at an intracavity focus. If the phase of the light passing round the cavity is appropriately adjusted it will constructively interfere with the light entering the cavity. The value of reflectivity of the input coupler can be chosen such as to balance the cavity losses and conversion to second harmonic. In this case all the light enters the cavity. Enhancements in the region of 100 have been reported but 20 to 30 is quite typical. Since the second harmonic is proportional to the square of the intracavity power harmonic generation can be very efficient using this method.

In order to maintain the cavity on resonance a servo locking scheme is required. A scheme which works very well and is widely adopted is the FM sideband locking scheme in which a phase modulator produces FM sidebands [24]. These sidebands are of equal amplitude in anti-phase. When they beat together on a square law detector the resulting

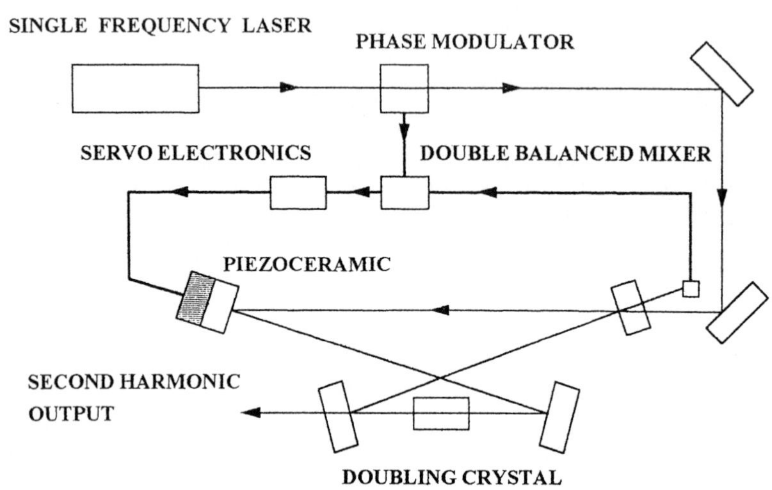

Figure 11. Resonant enhancement cavity for high efficiency harmonic generation

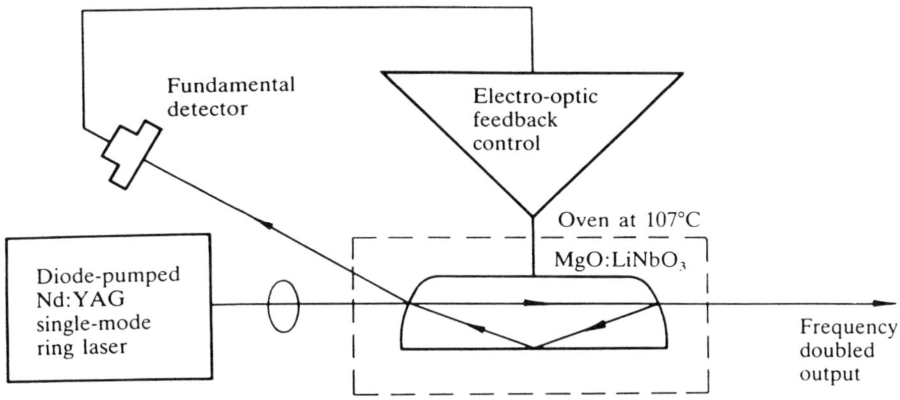

Figure 12. A monolithic resonant enhancement cavity [25]

heterodyne beats will be in anti-phase and therefore cancel. If, on the other hand, the phase modulated light is reflected from the cavity, which has frequency dependent reflectivity or phase shift, the beats will be unbalanced and can be readily detected using phase sensitive techniques. The resulting error signal is very attractive for locking purposes. The slope of the locking signal is determined by the finesse of the cavity whereas the capture range is set by the modulation frequency. This gives a very robust and tight lock.

If the intracavity losses can be kept low in a resonant enhancement scheme such that the single pass conversion efficiency through the crystal exceed the round trip losses, very high conversion efficiencies are possible. In CW mode the best conversion efficiencies have been obtained when the doubler is in the form of a monolithic cavity with mirrors coated directly onto the doubling crystal. Such a cavity is shown in figure 12. In this system a diode pumped Nd:YAG laser giving an output power of 52mW has been doubled

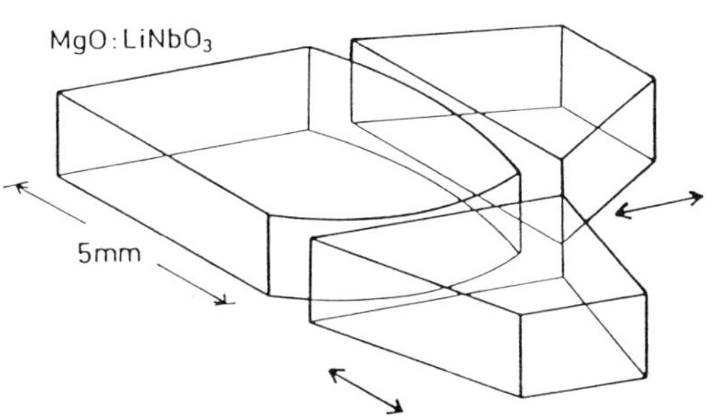

Figure 13. A monolithic resonator with total internal reflection (MOTIRR)

to 523nm with a conversion efficiency of 56% corresponding to 30mW generated [25].

A variation of the theme of resonant enhancement cavities is shown in figure 13. In this scheme there are no coatings at all applied to the doubling crystal. The cavity is formed by total internal reflection. The pump light is coupled in and the harmonic is coupled out using frustrated total internal reflection [26]. A conversion efficiency of up to 50% has been obtained with only 10mW of pump power. This remarkable scheme has also been used to make a triply resonant OPO with a threshold of less than 1mW [27].

The notion of efficient harmonic generation in a resonant enhancement cavity has been extended to the case of mode-locked pulse trains in my own laboratory [28]. In this case a pulse of light from a mode-locked laser enters the cavity which contains a doubling crystal at a focus. After propagating round the cavity the pulse meets a second pulse coming from the mode-locked laser provided that the cavity lengths are matched. If the phase of the pulse which has gone round the cavity is appropriately adjusted it will add in phase with the next pulse entering the cavity. Eventually the intracavity power will build up in a manner analogous to the single frequency case. If the frequency domain picture is preferred it can be viewed as matching the free spectral range of the enhancement cavity to

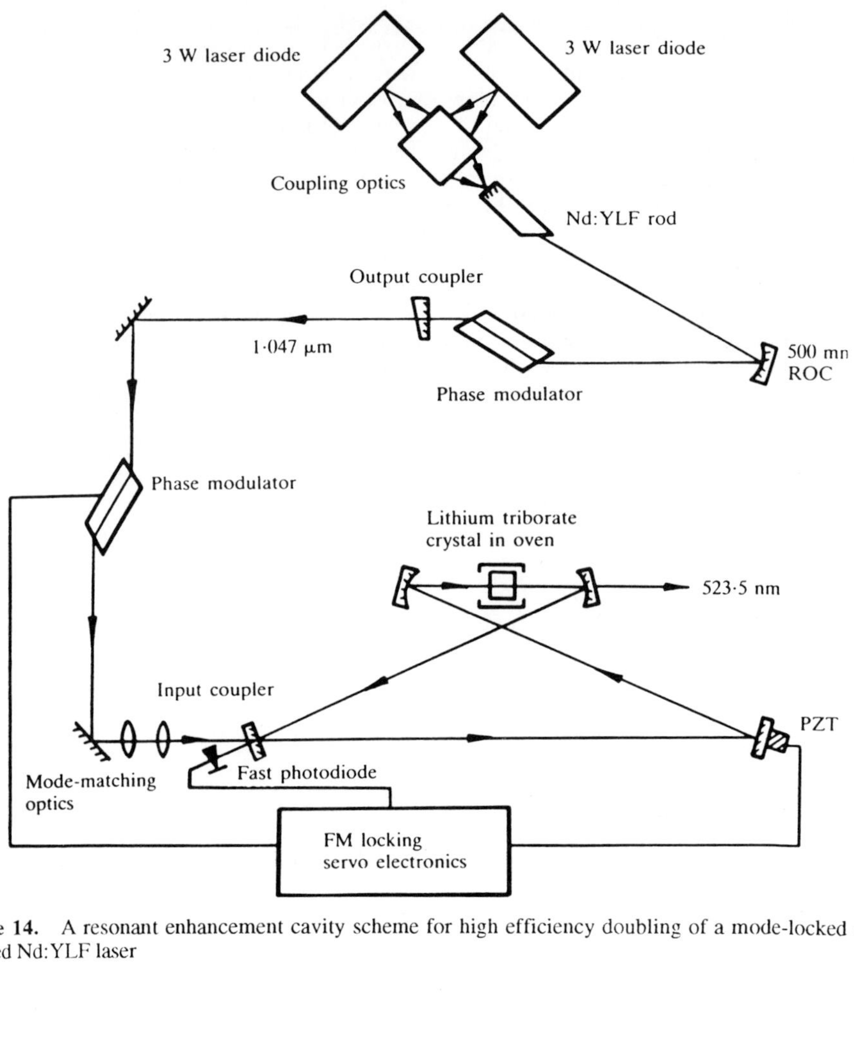

Figure 14. A resonant enhancement cavity scheme for high efficiency doubling of a mode-locked diode pumped Nd:YLF laser

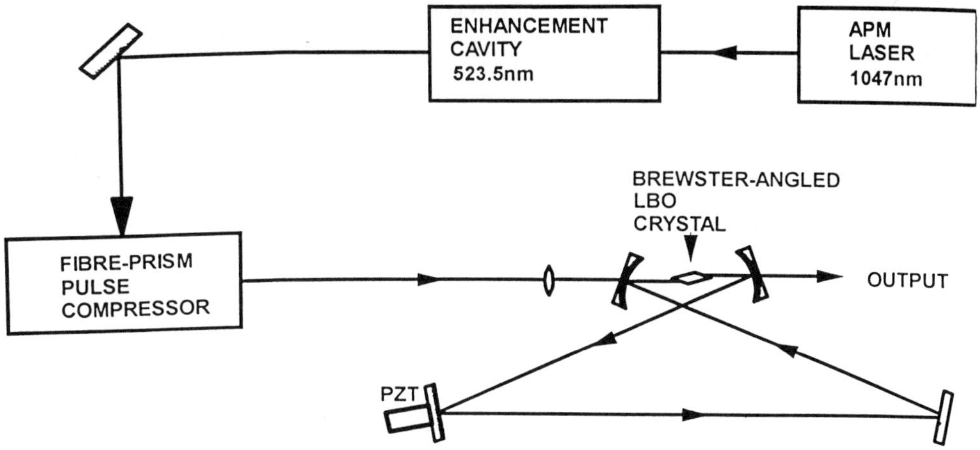

Figure 15. Synchronously pumped lithium triborate optical parametric oscillator (LBO OPO)

the mode spacing of the pulse train. In either case it is important that the phase be adjusted for enhancement. This is accomplished in an identical way to the single frequency case, again using FM sideband locking. Efficiencies approaching 100% have been achieved using this method.

A schematic diagram of a frequency doubled Nd:YLF laser using an enhancement cavity is shown in figure 14. In this case the laser is FM mode-locked but KLM or APM mode-locked systems work equally well. The doubling crystal in this case was LBO which non-critically phase matches for 1047nm at a temperature of about 140°C. With 1.8W incident on the cavity a usable output power of 850mW has been obtained at 523nm. This was obtained using only 6W of pump power and represents one of the most efficient methods of generating short pulses in the visible region. This system has been used to pump Ti:sapphire which gave an output of about 100mW [29].

Once second harmonic has been achieved with efficiency there is no difficulty in going to another stage of harmonic generation. Fourth harmonic generation of mode-locked Nd:YLF has been demonstrated using a non-optimised piece of BBO. In excess of 40mW as generated at 262nm in this experiment [30].

V.B. Optical Parametric Oscillation

The optical parametric oscillator has become one of the most effective methods of producing tunable coherent radiation. In an OPO the pump beam is used to generate a signal and idler beam in a three wave mixing process. The frequency of the signal and idler must sum to equal the frequency of the pump. This is the energy matching condition. The signal and idler frequencies are determined by the need to phase-match the interaction. This is the momentum matching condition. Tuning of the OPO is accomplished by altering the phase-matching by either orientation of the crystal or by changing the temperature of the crystalThe operation of an OPO is intimately connected with the coherence properties of the pump beam since this directly effects the signal and idler frequencies. It is therefore crucially important that the pump beam be coherent. Many OPO materials have a threshold for oscillation which is close to or even above the damage threshold. The threshold energy can be greatly reduced by use of synchronous pumping. The pump beam in this case is a mode-locked pulse train and hence is coherent. The OPO cavity length is matched to the repetition rate of the pump pulses. The signal and idler therefore appear as a train of short pulses.

Figure 15 shows a recently developed synchronously pumped OPO based on a diode pumped Nd:YLF laser [31]. The OPO crystal consists of an LBO crystal which is Brewster cut. The light propagates through the crystal such that $\theta = 90^{o}$, $\phi = 0^{o}$. When the cavity length is carefully adjusted a doubly resonant threshold of less than 3mW has been obtained. By changing the crystal temperature it is possible to tune the OPO away from degeneracy and to force singly resonant operation. Under these conditions the output power is very stable. The singly resonant threshold has been measured to be 70mW and the output power was 100mW when pumped at 250mW.

The temperature tuning range of LBO under non-critical phase matching conditions is quite remarkable. Figure 16 shows the experimental and calculated tuning curve for an LBO OPO pumped at 523nm. By changing the crystal temperature from 120°C to 200°C it is possible to tune from 650nm out to 2700nm. In certain regions it is possible for the device to operate on four different wavelengths. These curves were generated using just three mirror sets.

It is also possible to provide a widely tunable OPO source by pumping with the third harmonic of Nd:YAG. Using LBO as the OPO crystal a tuning range from 420nm to 2300nm has been obtained using a quasi-CW array pumped Q-switched Nd:YLF laser [32]. Starting with just 70mJ of diode energy, tunable output with about 1mJ of energy has been obtained.

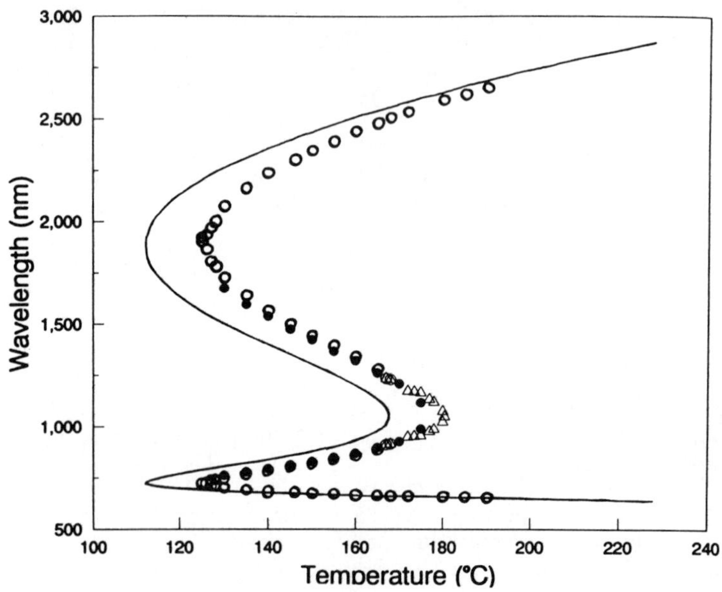

Figure 16. Tuning range for non-critical phase matching of an LBO OPO

VI. TUNABLE SOLID STATE LASERS

Although OPOs do provide wide tunability their major limitation is that they require very well controlled pump lasers and that they do not store energy. Tunable gain media do provide storage of energy for Q-switching and amplifier applications and their output wavelength is characteristic of the lasing ion rather than being intimately connected to the pump frequency.

A typical schematic diagram of a vibronic laser is very similar to that of a dye laser. The pump laser excites high in the excited state band. The system relaxes rapidly due to photon interactions. Laser action is produced by emission from the bottom of the upper level which has a lifetime of some microseconds down to the unpopulated ground state band. The energy difference is made up by phonons. The pump bands are therefore very broad and emission is also broad. This means that vibronic laser provide wide tunability and are capable of producing very short pulses.

The most widely accepted vibronic laser system is Ti:sapphire. The laser tunes from about 680nm out to 1100nm. Ti:sapphire is usually pumped with an argon ion laser operating in the green region. Although Ti:sapphire has been pumped by a doubled diode system [29] the power is low and the tuning very limited. A minimmum of 4W gives a reasonably wide tuning range in Ti:sapphire but the extremes can only be accessed by pump powers in the region of 20W.

VI.A. Single Frequency Operation

Solid state tunable laser gain media make ideal single frequency narrow linewidth lasers. A unidirectional ring with active stabilization in a well-designed Ti:Al_20_3 laser cavity will have in linewidth of around 100kHz. This can be reduced to about 1kHz fairly readily since all the frequency fluctuations are at relatively low frequency. Laser systems have been designed with beam ranges under electronic control of up to 40GHz.

VI.B. Mode-Locked Ti:Sapphire

The broad bandwidth and high power available from Ti:sapphire has meant that there has been considerable efforts to induce mode-locking. Early attempts based to active mode-locking were very disappointing. The technique KLM was discovered in Ti:sapphire [33] and works exceptionally well. A schematic diagram of a KLM Ti:sapphire laser is shown in figure 17. The gain medium is at a focus and provides the Kerr lens. There is provision for two intracavity prisms. Prisms arranged in this way provide anomalous dispersion. It is well known that the combination of anomalous dispersion and self-phase modulation are required for soliton-like effects. By careful choice of low dispersion prisms and by use of external cavity chirp removal using prisms, pulses of less than 20fsec have been reported [34]. Figure 18 shows a colinear autocorrelation of a 19fsec optical pulse together with its spectrum. The optical fringes in an autocorrelation can be used for calibration.

VI.C. Other Tunable Gain Media

Ti:sapphire has shown the way forward in tunable laser systems. This has led to intense research to fund tunable gain media which can be diode pumped. Prime candidates for this are the Cr^{3+} doped fluoride Cr:LiCAF, Cr:LiSAF and Cr:LiSCAF. In many ways the properties of these materials match Ti:sapphire except that they absorb at 670nm where red laser diodes of high power are becoming available. They also have longer upper state lifetimes which is good for Q-switching at regenerative amplification. They open up the possibility of widely tunable operation from a diode pumped laser system. They have already been mode-locked and produce pulses comparable to those of Ti:sapphire [35].

VII. TWO-PHOTON CONFOCAL MICROSCOPY

The high peak power available for a relatively low average power of a KLM Ti:sapphire laser make it an attractive source for two-photon excitation of fluorescent biological dyes. We conclude this overview of advances in solid-state lasers with the description of an application of mode-locked lasers to the field of confocal laser-scanning microscopy.

The use of fluorescent dyes for observing biological samples has become a widely used biomedical technique. The dyes can be attached to ligands, such as antibodies, that have high affinities to specific components of a cell or tissue. When optically excited, the dyes fluoresce, enabling the structure and organization of the stained components to be readily visualized. Many fluorochromes exist which change their spectral characteristics when they bind to specific ion species such as protons or calcium ions. By taking quantitative measurements of the fluorescence of these dyes when they have been introduced into live cells, it is possible to monitor the physiological parameters of a live cell such as pH and calcium concentrations [36].

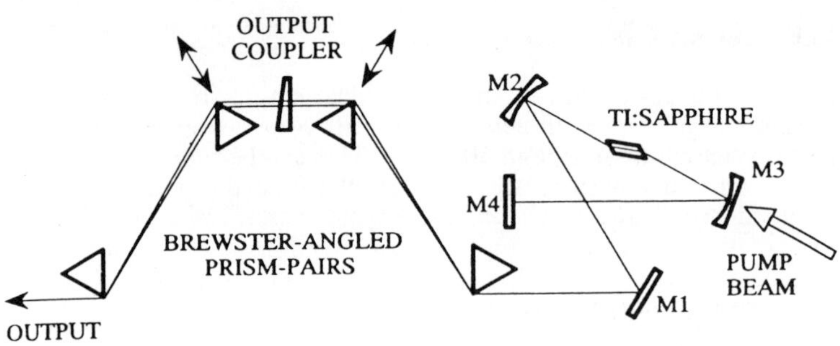

Figure 17. Kerr lens mode-locked Ti:Al_2O_3

The development of the technique of laser scanning confocal microscopy has considerably extended the scope of fluorescence microscopy [37]. The remarkable optical sectioning capabilities of these devies enables details to be revealed from the inner regions of cells or tissue with little or no interference from structures that may be above or below the plane of focus. At present CW argon-ion lasers are most often used for conventional single-photon excitation of such dyes in confocal microscopes. However, a major problem associated with CW illumination in fluorescence microscopy when applied to living cells or tissues is phototoxicity. This can often severely limit the time for which living tissue can be viewed before it is killed by the incident illumination. An alternative approach that may alleviate this problem is to illuminate stained samples using a short pulse mode-locked laser as presented by Denk *et al* [38]. If femtosecond pulses are focused down through a microscope system the resulting peak power at the focus can be so intense that two-photon absorption in the dye may be high enough to produce relaxation fluorescence at a detectable level. By ensuring that the illumination wavelength is around twice the wavelength of the excitation peak of the fluorochrome, single-photon absorption can be negligible.

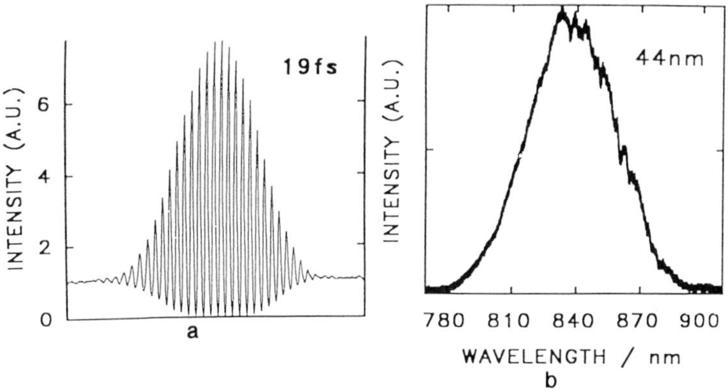

Figure 18. Autocorrelation of a KLM Ti:Al_2O_3 laser

VII.A. Confocal Microscopy

The impact of the development of confocal laser scanning microscopy (LSM) on the biological field has been dramatic. Over a standard microscope, LSM offers unprecedented rejection of out-of-focus interferene and improved resolution. The technique enables optical sections through samples to be obtained from which sharp '3D' images may be reconstructed. The improved resolution is a direct result of the confocal imaging arrangement of the illumination and imaging optics.

The idea of confocal microscopy was first discussed by [39], and the first usable laser scanning devices appeared around 1970 [40]. Currently, confocal microscopy is finding applications in a wide diversity of fields including the life sciences and the optoelectronic/semiconductor industry.

Confocal imaging can be achieved by illuminating a sample with laser light through a standard optical microscope system, and imaging the fluorescence from the sample back through the same optics. By the appropriate positioning of an aperture in the image plane it is possible to reject nearly all of the out-of-focus light coming back from the sample. The result is a vast improvement in the visibility of structures within thick specimens. To record a complete image of the sample the illuminating laser beam is raster-scanned across the sample by either mechanical or acousto-optic beam deflectors. The image signal is measured as a function of beam position using a photomultiplier tube and a full image can be built up in a digital frame store and displayed on a video screen.

A schematic of a confocal laser scanning microscope (LSM) similar to that used in these experiments is shown in figure 19. In these experiments the mode-locked laser beam is introduced into the microscope system by means of a dichroic mirror, that is highly reflecting at the laser wavelength but highly transmitting to the dye fluorescence. A second transmission filter is included to further minimize the entry of fundamental laser light into the detection system.

Standard illumination systems in most commercial LSMs include argon-ion and helium-neon lasers. However, in biological applications the choice of illumination wavelength is largely dictated by the availability of a suitable fluorescent dye for the application required. Most dyes currently available suffer from a major problem in that under high illumination intensities they can become cytotoxic. Dyes also tend to bleach under the long averaging times required to build up an adequate image from single-photon excitation. A further parameter that also has to be considered is that each dye has its own

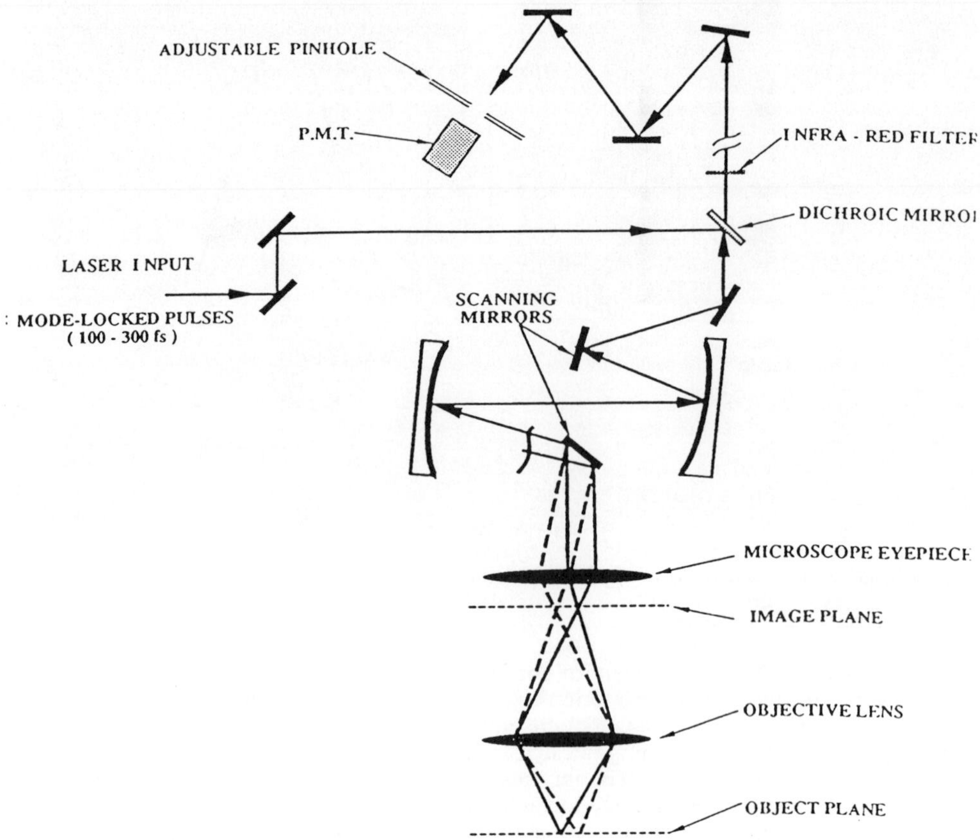

Figure 19. Schematic of a standard commercial Confocal Laser Scanning microscope (BIORAD MRC-500)

optimum absorption peak wavelength, and hence it would prove advantageous to tune the laser wavelength to optimize the dye fluorescence.

The efficiency of two-photon processes is peak power dependent. By delivering a modest average power in the form of ultrashort laser pulses, it is possible to excite dyes by two-photon rather than single-photon absorption. Hence, dyes with single-photon absorptions in the visible would require mode-locked near-IR excitation wavelengths for two-photon processes; a task ideally suited to Ti:sapphire lasers. Whilst dye toxicity problems are not entirely eliminated by using mode-locked illumination, the region over which damage can occur in the vertical plane of the sample can be very sharp, owing to the quadratic spot-size dependency of the two-photon absorption process. Hence, although photodamage and bleaching will still occur, these effects will be confined to a very thin xy plane, whereas with single-photon illumination they will extend throughout the illuminated region of the specimen along the z axis.

The idea of exploiting the small interaction region to observe samples safely can be exploited in reverse, by using two-photon processes to change the optical properties of photochromic materials. This scheme has been proposed as a basis for a high-density, three-dimensional optical memory.

Experimental results from a commercial LSM under such mode-locked illumination from a KLM $Ti:Al_2O_3$ laser producing 270fsec pulses conditions are shown in figure 20. The images shown were averaged and normalized over $\geq$ 100 scans, but visible fringes were recorded on each individual scan. On occasions where the laser dropped out of mode-locking, the fluorescence signal disappeared and immediately returned when the mode-locking was reinitiated by applying a mechanical perturbation to a laser cavity mirror. Figure 20 shows the cell images for a sample of blood from the clawed toad

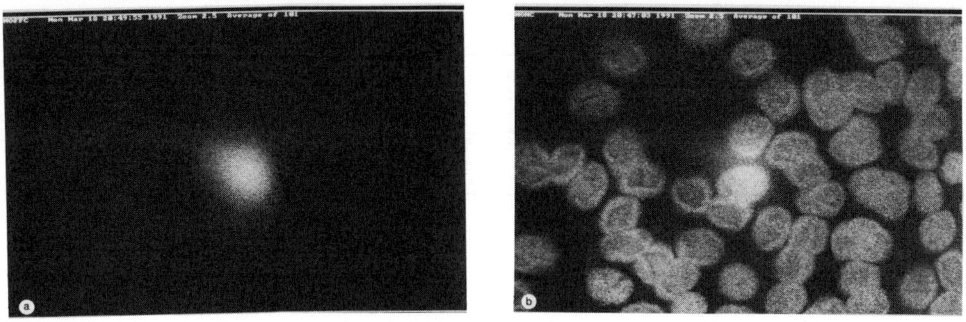

Figure 20. A laser scanning microscopy image of Xenopus erythrocytes under (a) cw illumination and (b) mode-locked illumination. The scale base represents 10μm.

Xenopus laevis stained with DAPI. To give an idea of the power required under confocal illumination to obtain these images, the measured throughput just before the objective lens was <5mW. The central bright spot seen on the images is an artefact that arose because some of the incident laser beam was reflected back in the microscope optics and was not completely blocked off in the barrier filter. This artefact is present in both CW and mode-locked conditions and is independent of the specimen.

VIII. CONCLUSION

The rapid developments in solid-state lasers and nonlinear optical devices over the past few years have been extremely exciting. The new materials and new devices give a renewed insight into the operation of coupled nonlinear optical systems.

The progress in device physics has now made a jump ahead of the applications. In the next few years we are likely to see a broadening of the range of applications to which this exciting new technology is applied.

REFERENCES

1. See for example
   T Y Fan and R L Byer, IEEE J. Quant. Electron. **QE-24**, 895 (1988).
   G P A Malcolm and A I Ferguson, Contemp. Phys. **32**, 305 (1991).
   D W Hughes and J R M Barr, J. Phys. D: Appl. Phys. **25**, 563 (1992).
2. W Streifer, D R Scifres, G L Harnagel, D F Welch, J Berger and M Sakamoto, IEEE J Quant. Electron **QE24,** 882 (1988).
3. C E Wieman and L Hollberg, Rev. Sci. Instr. **62**, 1 (1991).
4. W Koechner, Solid-State Laser Engineering, 3rd Ed. New York, Springer-Verlag (1992).
5. T Y Fan, G Huber, R L Byer and P Mitzscherlich, Opt. Lett. **11,** 204 (1987).
6. S Basu and R L Byer, Appl. Optics, **29**, 1765 (1990).
7. A L Schawlow and C H Townes, Phys. Rev. **112**, 1940 (1958).
8. B Zhou, T J Kane, G J Dixon and R L Byer, Opt. Lett. **10**, 62 (1985).
9. J J Zayhowski and A Mooradian, Opt. Lett. **14**, 24 (1989).
10. K Wallermouth, Opt. Lett. **15**, 903 (1990).
11. T J Kane and R L Byer, Opt. Lett. **10**, 65 (1985).
12. T Day, E K Gustafson and R L Byer, Opt. Lett., **15**, 221 (1990).
13. G P A Malcolm and A I Ferguson, Opt. Quant. Electron, **24**, 705 (1992)
14. See special issue of IEEE J. Quant. Electron. **QE-28**, No. 10 (1992).
15. D J Kuizenga and A E Siegman, IEEE J. quant. Electron. **QE-6**, 694 (1970).

16. G P A Malcolm, M Ebrahimzadeh and A I Ferguson, IEEE J. Quant. Electron **QE28**, 1172 (1992).
17. See for example H A Haus, J G Fugimoto and E P Ippen, IEEE J. Quant. Electron. **QE28**, 2086 (1992).
18. L F Mollenauer and R H Stolen, Opt. Lett. **9**, 13 (1984).
19. G P A Malcolm, P F Curley and A I Ferguson, Opt. Lett. **15**, 1303 (1990).
20. D E Spence, P N Kean and W Sibbett, Opt. Lett. **16**, 42 (1991) and D K Negus *et al* Digest of Meeting on Advanced Solid State Lasers (Optical Society of America, Washington DC 1991) PDP-4.
21. G P A Malcolm and A I Ferguson, Opt. Lett. **16**, 1967 (1991).
22. See for example, R L Byer, Parametric Oscillators and Nonlinear Materials in Nonlinear Optics, Academic Press, London (1977).
23. A Ashkin, G D Boyd and J M Dziedzic, IEEE J. Quant. Electron, **QE2**, 109 (1966).
24. R W P Drever *et al*, Appl. Phys. **B31**, 97 (1983).
25. W J Kozlovsky, C D Nabors and R L Byer, IEEE J. Quant. Electron. **QE24**, 913 (1988).
26. K Fiedler, S Schiller, R Paschotta, P Kürtz and J Mlynek, Conference on Lasers and Electro-Optics, 1993, Vol 11, OSA Technical Digest Series (Optical Society of America, Washington DC, 1993) CPD17.
27. S Schiller and R L Byer, Conference on Lasers and Electro-Optics, 1993, Vol 11, OSA Technical Digest Series (Optical Society of America, Washington DC, 1993) CPD14.
28. M A Persaud, J Tolchard and A I Ferguson, IEEE J. Quant. Electron, **QE-26,** 1253 (1990).
29. G P A Malcolm and A I Ferguson, Opt. Commun. **82,** 299 (1991).
30. G P A Malcolm, M A Persaud and A I Ferguson, Opt. Lett. **16**, 983 (1991).
31. A Robertson, G P A Malcolm, M Ebrahimzadeh and A I Ferguson, Conference on Lasers and Electro-Optics, 1992, Vol 10, OSA Technical Digest Series (Optical Society of America, Washington DC, 1992) CPD15.
32. Y Cui, D E Withers, C F Rae, C J Norrie, Y Tang, B D Sinclair, W Sibbett and M H Dunn, Opt. Lett. **18**, 122 (1993).
33. D E Spence, P N Kean and W Sibbett, Opt. Lett. **16**, 42 (1991).
34. F Krausz *et al*, IEEE J. Quant. Electron **QE28**, 2097 (1992).
35. J M Evans, P M W French and J R Taylor, Opt. Lett. **17**, 1605 (1992) and Opt. Lett. **17**, 877 (1992).
36. R Y Tsien, Methods of Cell Biol. 30, Part B, 127 (1989).
37. J G White, W B Amos and M Fordham, J. Cell Biol. **105**, 41 (1987).
38. W Denk, J H Strickler and W W Webb, Science: Reports **248**, 73 (1990).
39. M Minsky, Res. Scanning **10**, 128 (1988).
40. M D Egger, Rev. Trends Neurosci. **12**, (1989).

QUANTUM NOISE IN PARAMETRIC AMPLIFICATION

I. Abram and J.A. Levenson

FRANCE TELECOM / CNET / PAB
Laboratoire de Bagneux
92220 Bagneux, France

ABSTRACT

Optical parametric emission, whereby an intense "pump" beam incident on a non-linear crystal is down-converted into radiation at the sub-harmonic frequency ($\omega/2$), has proved to be an efficient means for amplifying light and producing photon statistics with non-classical distributions. In this course, we present the principles of optical parametric amplification, we review some recent experimental and theoretical investigations on the quantum fluctuations of parametric light and we discuss the potential applications that these studies suggest.

1. INTRODUCTION

When a classical pendulum is set into motion, there are several ways in which energy can be transferred to the pendulum to maintain and increase its motion. The simplest way is to drive its displacement periodically. When the frequency of the driving force is the same as that of the pendulum, **resonance** occurs and the amplitude of oscillation of the pendulum increases, as energy is transferred to the pendulum. The emission of light by lasers and coherent sources can be understood in terms of a similar mechanism. The radiation modes can each be considered as a harmonic oscillator in which the electric and magnetic fields oscillate sinusoidally. The laser oscillator, then, drives the electric field associated with each mode, thus injecting energy in the field.

A slightly more complicated way to drive a pendulum is to vary its length periodically with time, so that its frequency varies with the same periodicity. For the case in which the frequency has small sinusoidal variations around some central value ω_0, the equation of motion of such a pendulum has the form

$$\frac{d^2x}{dt^2} + \omega_0^2(1 + \alpha \cos \omega_p t)x = 0 \tag{1.1}$$

Nonlinear Spectroscopy of Solids: Advances and Applications
Edited by B. Di Bartolo and B. Bowlby, Plenum Press, New York, 1994

A simple analysis [1] shows that such a system, displays the phenomenon of **parametric resonance** which means that, when the pendulum frequency is a sub-harmonic of the "pumping" frequency (for example, when $\omega_0 = \omega_p/2$) , energy is transferred from the **pump** to the pendulum. By using nonlinear optical materials that permit us to modulate the frequency of the modes of the electromagnetic field at sub-femtosecond periods, it has been possible to realize, over the past 30 years, **optical parametric oscillators** that can be used as tunable light sources [2].

The practical use of parametric light sources has proved to be somewhat more difficult than that of the lasers, because of their extreme sensitivity to fluctuations. On the other hand, the study of their quantum mechanical properties has produced truly spectacular results.

In this course, we focus on one particular aspect of the quantum mechanical manifestations of parametric light, namely the quantum statistics that result when a coherent beam is amplified by optical parametric emission. Over the past 15 years it has been shown both theoretically and experimentally that the light emerging from a parametric emitter is "nonclassical" in the sense that its photon fluctuations can be lower than those that quantum mechanics imposes on coherent (classical) light. These observations suggest that by carefully engineering the quantum noise properties of parametric light it is possible to overcome many of the limitations that are inherent in laser light regarding optical measurements and optical information transmission and detection.

The outline of this course is as follows: In Section 2, we review the basic ideas of nonlinear optics regarding parametric generation and amplification and we survey the material properties and the different experimental configurations that are possible. In Section 3, we present some basic notions on quantum noise, its origin in the Heisenberg uncertainty principle, the observation of the shot noise of an ideal laser in direct and in homodyne detection and the graphical representation of shot noise. In Section 4, we examine the quantum noise properties of a parametric emitter operated as a linear amplifier, while in Section 5 we present the nonclassical noise properties of phase-dependent optical parametric amplification through recent experimental and theoretical investigations. Finally, in Section 6, we discuss the possibilities of "engineering" the quantum noise of light to overcome the limitations imposed by shot noise.

2. WHAT IS OPTICAL PARAMETRIC AMPLIFICATION

2.1 Definitions

An **optical parametric emitter** is a device that has two essential ingredents: a transparent, birefringent **electro-optic crystal** (such as KH_2PO_4, $KTiOPO_4$, or $LiIO_3$) and an incident laser **pump beam** that provides the energy of the emission process. The basic property of the electro-optic crystal is that its refractive index is slightly modified when an external voltage (i.e. an electric field) is applied to it. Thus, if the pump beam traversing the crystal is intense enough, its oscillating electric field produces a small sinusoidal modulation of the refractive index of the crystal at the frequency of oscillation of the pump field ω_p. As in the mechanical parametric oscillator, this rapid modulation causes a transfer of energy from the pump beam to a pair of lower frequency modes, commonly called **signal** and **idler**, thus producing an emission of light. Because of energy conservation, the signal and idler frequencies obey $\omega_s + \omega_i = \omega_p$. If the signal and idler beams are identical in frequency (i.e. $\omega_s = \omega_i = \omega_p/2$), in polarization and in direction, they are indistinguishable and the parametric emission process is said to be **degenerate**. When there is an external input to the signal and/or idler modes, then the energy transfer serves to increase the intensity of the input beam and the device is called an **optical parametric amplifier**.

In this Section we give a brief description of the physics that underlies parametric emission and we introduce the formalism that can be used to account for the quantum effects.

2.2 Review of Quantum Nonlinear Optics

When a beam of light is incident on a material medium, the oscillating electromagnetic field of the beam polarizes the electronic cloud of the atoms or molecules in the material. The oscillating macroscopic polarization thus induced in the medium can, in turn, emit light (or more generally interact with the electromagnetic field) according to Maxwell's equations. The Hamiltonian that describes this situation, can be written as,

$$H = H_{field} + H_{mat} + H_{int} \tag{2.1}$$

where

$$H_{field} = \frac{1}{2}\int_V \{E^2(r) + B^2(r)\}\, dr \tag{2.2}$$

is the Hamiltonian of the free field, with $E(r)$ and $B(r)$ being the electric and magnetic fields respectively at point r, in the absence of the material medium. Integration is performed over the volume V of the cavity in which the field is quantized. H_{mat} is the Hamiltonian of the electrons and nuclei that compose the material medium, while

$$H_{int} = \sum_j q_j r_j E(r_j) \tag{2.3}$$

is the Hamiltonian that describes the interaction between the electric field and the collection of charged particles that compose the medium, with q_j and r_j being the charge and the position respectively of the j-th particle.

To simplify the notation and the algebra, we shall consider a simple one-dimensional geometry for the electromagnetic field, involving only plane waves, such that the electric field is polarized along the x-axis, the magnetic field along the y-axis and propagation occurs along the z-axis. This geometry, which is representative of many experimental situations, permits us to ignore the vector nature of the electromagnetic field and treat all quantities as if they were scalars. Of course, whenever the physical situation requires other geometries, we shall consider explicitly the particular features of these geometries. An additional feature we shall invoke to simplify the formalism is the decomposition of the electromagnetic field into its normal modes. The modes are, essentially, the spatial Fourier components of the field distribution and are extended over all space. Superpositions of a large number of modes can describe short pulses of light that move in space as the relative phases of the modes evolve in time. Thus, even though the modal representation is rather cumbersome in the description of propagative phenomena, it greatly facilitates the discussion of quantum mechanical effects, as it brings forth quite naturally the energy levels of the individual modes, that is the photons. This is done by introducing for the mode of wavevector k two quantum operators $b_k^\dagger$ and b_k that respectively create and annihilate a single elementary excitation in the mode, i.e. a photon.

These modal creation and annihilation operators follow the Bose commutation relation, that is,

$$[b_{k'}, b_k^\dagger] = \delta_{k',k} \tag{2.4}$$

where $\delta_{k',k}$ is the Kronecker delta-symbol. This commutation relation implies that each mode may be considered as a harmonic oscillator whose eigenstates correspond to an infinity of equidistant energy levels, each state containing a definite number of photons. Strictly speaking, when describing traveling-wave experiments an infinity of modes must be considered. However, very often consideration of only a few modes may be sufficient to gain a basic insight on the quantum features of parametric emission and it is not necessary to use cumbersome propagative formulations of quantum optics. In this course, then we shall develop an analysis of the quantum phenomena of parametric emission based on a very small number of modes, but we shall assume that the conclusions apply also in propagative situations.

The Hamiltonian (2.1) acquires a particularly simple form when we consider only radiation whose frequency is far from the electronic resonances of the medium i.e. far from the eigenvalues of H_{mat}. Experimentally, in that case, the material is transparent to the incident radiation. No energy is transferred from the field to the medium for any significant amount of time, and the field induces only virtual transitions among the eigenstates of H_{mat}. As the characteristic time for observation is very long compared with the characteristic time for the virtual transitions, the material degrees of freedom can be removed from the Hamiltonian through the procedure of **adiabatic elimination**. An effective Hamiltonian can thus be obtained, such that it involves only the electromagnetic field. In this effective Hamiltonian, the deformation of the electronic cloud of the medium by the incident electromagnetic field and the interaction of this induced polarization back onto the electromagnetic field appear as a self-interaction of the field that can be expressed as a power series in the electric field. The effective Hamiltonian is thus of the form,

$$H = \frac{1}{2}\int_V \{E^2(r) + B^2(r)\}\, dr + \frac{1}{2}\chi^{(1)}\int_V E^2(r)\, dr + \frac{2}{3}\chi^{(2)}\int_V E^3(r)\, dr + \cdots \tag{2.5}$$

where the coefficients $\chi^{(n)}$, that embody the optical response of the medium, are called the **optical susceptibilities** and are, in principle, tensor quantities. In the simplified geometry we consider here, however, only one tensor component is important and, thus, the susceptibilities can be treated as scalars. Also, to simplify the notation in writing eq. (2.5), we have ignored the frequency dependance of the susceptibilities that arises from the resonances of the material medium. This approximation is reasonable in the transparency region of the medium (i.e. far from the material resonances) where the frequency variation of the χ's is very slow. However, in situations in which even this slow frequency dependance of $\chi^{(1)}$ is important, namely in the problem of phase-matching which we shall discuss shortly, we shall examine its consequences explicitly. Use of the classical coefficients $\chi^{(n)}$ to replace the quantum mechanical degrees of freedom of the material implies that the correlations among the virtual absorption and emission processes of the molecules in the medium are neglected. In the effective Hamiltonian (2.5), then, the field is described quantum mechanically, while the material medium is described classically and plays simply the role of a material support on which light interacts with light.

For a transparent medium, the series expansion of the Hamiltonian (2.5) in powers of the electric field converges rapidly, as the ratio of two successive orders of the susceptibility is of the order of the intra-atomic electric field binding the electron,

$$\frac{\chi^{(n)}}{\chi^{(n+1)}} \approx E_{atom} \sim 3 \times 10^8 \ V/cm \tag{2.6}$$

a quantity which is much larger than the applied optical field, as it would correspond to incident intensities of the order of 10^{15} W/cm^2. The Hamiltonian (2.5) can thus be thought as an expansion in powers of a small parameter which is the ratio of the applied optical field to the atomic field, ($|E/E_{atom}| \ll 1$). In this paper, we shall consider systems that display only a first- and a second- order susceptibility, and assume that the effects of the higher orders are negligible, an assumption that is verified for all known electro-optic crystals.

As can be seen in the Hamiltonian (2.5), the first-order (or, linear) susceptibility $\chi^{(1)}$ is the coefficient of E^2 and thus represents simply a re-scaling of the energy of the electric field. Physically, it gives rise to the refractive index of the medium and, thus, it can be incorporated in the definition of E by introducing the dielectric constant of the medium, $\epsilon = 1 + \chi^{(1)}$ or, equivalently, its refractive index $n = \sqrt{\epsilon}$. This can be done by writing the electric field in terms of the modal (photon) creation and annihilation operators as

$$E(z,t) = i \sum_k \left(\frac{\hbar\omega_k}{2\epsilon V}\right)^{1/2} \{b_k e^{ikz - i\omega_k t} - b_k^\dagger e^{-ikz + i\omega_k t}\} \tag{2.7}$$

and the magnetic field as

$$B(z,t) = i \sum_k \left(\frac{\hbar\omega_k}{2V}\right)^{1/2} \{b_k e^{ikz - i\omega_k t} - b_k^\dagger e^{-ikz + i\omega_k t}\} \tag{2.8}$$

with the frequency of the k-th mode given by

$$\omega_k = \frac{c}{\sqrt{\epsilon}} |k| \tag{2.9}$$

It can readily be verified that these definitions for the electric and magnetic field operators permit us to express the **linear medium** Hamiltonian (i.e. the sum of the first two terms in eq. 2.5) as

$$H_{LIN} = \sum_k \hbar\omega_k b_k^\dagger b_k \tag{2.10}$$

The second-order susceptibility, $\chi^{(2)}$, is the nonlinearity on which we shall focus in this course. Because of symmetry considerations, this coefficient vanishes for media that possess a center of inversion. This means that gases, liquids, or amorphous solids, which are statistically isotropic, display no second-order nonlinearity. Tus, the only materials that have a non-zero $\chi^{(2)}$ are crystals that belong to one of the 21 space groups with no center of inversion.

Introducing the modal decomposition of the electric field (2.7) into eq. (2.5), we can write also the "nonlinear" part of the effective Hamiltonian in the modal representation. We note that in direct space, the optical nonlinearity is local, that is it involves the product of three electric field operators (E^3) all at the same point in space r. This means that when we go to the modal representation (which is essentially a Fourier transformation) the nonlinear interaction will involve a triple convolution of the modal operators, thus coupling all modes to each other. However, because the field energy due to the nonlinearity is very small compared with that of the free field, an important approximation can be invoked at this point and only interactions that are stationary with respect to the free-field evolution need be retained. The nonlinear part of the Hamiltonian thus becomes

$$H_{NL} = -i\frac{2}{3}\chi^{(2)} \sum_{k_p,k_s,k_i} \left(\frac{\hbar\omega_p}{2\epsilon V}\frac{\hbar\omega_s}{2\epsilon V}\frac{\hbar\omega_i}{2\epsilon V}\right)^{1/2} \int_V \{b_p b_s^\dagger b_i^\dagger e^{i(k_p-k_s-k_i)z} - b_p^\dagger b_s b_i e^{-i(k_p-k_s-k_i)z}\}\, dr \tag{2.11}$$

where we have retained only the triplets of modes whose frequencies are resonant, that is, they satisfy

$$\omega_p = \omega_s + \omega_i \tag{2.12}$$

and thus correspond to transitions between states of the electromagnetic field with the same free-field energy.

At this point, it is important to consider the frequency-dependance of the refractive index (or, equivalently, of $\chi^{(1)}$). As the wavevector is related to the frequency by

$$k(\omega) = n(\omega)\frac{\omega}{c} \tag{2.13}$$

the dispersion of the refractive index (i.e. its slight dependance on the frequency) means that the triplets of modes that satisfy energy conservation (eq. 2.13) do not generally satisfy wavevector conservation (also called **phase-matching condition**)

$$k(\omega_p) = k(\omega_s) + k(\omega_i) \tag{2.14}$$

The problem that arises when phase-matching is not satisfied is that the exponentials in eq. (2.11) produce oscillations which get averaged out to a very small value (and vanish as $V \to \infty$) when spatial integration is performed. Physically, this means that, for macroscopic propagation lengths inside a nonlinear crystal (typically of the order of a millimeter) and for large phase mismatch, the interacting triplets of modes interfere destructively, so that the overall effect of the nonlinear interaction is very small. On the other hand, if the frequency variation of the refractive index of the crystal is chosen so that phase-matching is satisfied, then the interacting triplets interfere constructively as they propagate in the crystal and the efficiency of the nonlinear interaction increases with the length of propagation.

One trick that is often used to achieve phase-matching in a particular frequency band, is to choose an electro-optic crystal that is also birefringent and to orient its optic axis with respect to the propagation of light so that the refractive indices associated with the ordinary and extraordinary waves have the appropriate ratio. For the sake of concreteness, we shall discuss birefringent phase-matching for the case in which the two lower-frequency modes are degenerate in frequency and correspond to sub-harmonic generation (i.e. $\omega_s = \omega_i = \omega_p/2$). Two configurations are then possible:

(1) the pump mode is polarized as an extraordinary wave (i.e. polarized in the plane containing the optic axis and the propagation direction) while the signal and idler modes are both polarized as ordinary waves (i.e. orthogonal to the pump mode) and are thus truly degenerate. For those signal and pump frequencies for which, respectively, the ordinary and extraordinary refractive indices are equal, that is $n_o(\omega_p/2) = n_e(\omega_p)$, the energy conservation relation (2.12) gives immediately the wavevector conservation relation (2.14), since

$$n_e(\omega_p)\frac{\omega_p}{c} = n_o(\omega_p/2)\left(\frac{\omega_s}{c} + \frac{\omega_i}{c}\right) \tag{2.15}$$

This configuration is called **type I phase-matching.**

(2) the pump is extraordinary, and the signal and idler modes are ordinary and extraordinary respectively. In that case phase-matching is achieved for the frequencies for which the extraordinary index at the pump frequency equals the arithmetic

mean of the ordinary and extraordinary indices at the signal frequency, i.e. $n_e(\omega_p) = (n_o(\omega_p/2) + n_e(\omega_p/2))/2$, so that

$$n_e(\omega_p)\frac{\omega_p}{c} = \frac{1}{2}\left(n_o(\omega_p/2)\frac{\omega_s}{c} + n_e(\omega_p/2)\frac{\omega_i}{c}\right) \tag{2.16}$$

This configuration is called **type II phase-matching** and the signal and idler modes are distinguishable, since they can be separated from each other at the exit of the crystal by use of a polarizer.

A type II crystal, phase-matched for sub-harmonic generation ($\omega_s = \omega_i = \omega_p/2$) can operate either as a degnenrate or as a non-degenerate parametric emitter, depending on the way that the polarization of the input and/or output beams is analyzed. In particular, when considering separately the two orthogonal polarizations that are parallel to the crystal axes, so that the signal and idler modes can be measured separately, the parametric emitter is non-degenerate. On the other hand, when considering a mode whose polarization is oriented at 45° with respect to the crystal axes, the electric field of this mode is a linear combination of the signal and idler fields so that, for this mode, the parametric emitter can be considered to be degenerate. Thus, throughout this course, whenever we need examples of parametric emitters, we shall refer to a type II crystal, typically $KTiOPO_4$ (KTP), and consider simply the modes that are polarized along the crystal axes, for the non-degenerate configuration, or at 45° with respect to the crystal axes, for the degenerate configuration.

2.3 Origin of the quantum effects

Because phase-matching is usually achieved over a relatively narrow frequency bandwidth, we can consider that there is only a single triplet of modes for which the nonlinear interaction is efficient and, thus, the nonlinear Hamiltonian (2.11) can be simplified as,

$$H_{NL} = -i\frac{2}{3}\chi^{(2)}\left(\frac{\hbar\omega_p}{2\epsilon V}\frac{\hbar\omega_s}{2\epsilon V}\frac{\hbar\omega_i}{2\epsilon V}\right)^{1/2}(b_p b_s^\dagger b_i^\dagger - b_p^\dagger b_s b_i) \tag{2.17}$$

For type I phase-matching and when the signal and idler frequencies are identical we have $s = i$ since, in that configuration, the two modes are indistinguishable.

Inspection of this Hamiltonian gives us readily an insight into the origin of the quantum mechanical effects in parametric emission.

According to the first term in eq. (2.17), in which the annihilation of a pump photon is accompanied by the creation of one signal and one idler photons, each photon of the pump beam may undergo "fission" to give birth to a pair of twin photons sharing between them the energy of the parent photon. The simultaneous generation of the twin photons implies that the signal and idler modes must display strong quantum mechanical correlations. In particular, if the signal and idler modes are distinguishable (for example, if they have different polarizations) their fluctuations in photon numbers must be correlated, since for every photon in the signal beam, there must be a twin in the idler beam. On the other hand, if the two modes are degenerate, then twin emission populates only the even occupation number states of the degenerate mode. The situation of twin photon emission contrasts with that of a laser in which each elementary emission process involves only one photon. Thus, if the laser emission occurs in two modes simultaneously, each elementary emission process may contribute photons to one mode or the other but not to both, implying that there cannot be any correlations in the photon number fluctuations of the two modes populated by laser emission. If, on the other hand, the laser emits in a single mode, that mode will evolve into a superposition that involves both the even and the odd occupation number states. These simple observations indicate that the quantum statistics of parametric light must be "nonclassical", in the sense that it cannot be reproduced by the statistics of a coherent (classical) laser beam.

Another way of viewing the nonclassical quantum statistics of the light produced by parametric emission is to consider the structure of the full nonlinear Hamiltonian (2.17). The second term in eq. (2.17) indicates that the inverse of the parametric emission process can also take place: Two lower-frequency photons can "fuse" to produce a photon of twice the energy. This is the well-known process of frequency doubling (also called Second Harmonic Generation). The overall Hamiltonian, thus, indicates that when two beams one at frequency ω_p and one at frequency $\omega_p/2$ co-propagate in a phase-matched electro-optic crystal (and have the appropriate polarizations) both processes can occur. Because of the "fission" process, energy can be transferred from the pump to amplify the signal beam while the "fusion" process causes energy to flow from the signal to the pump beam, thus de-amplifying the signal. The net direction in which energy will flow depends on the relative phases of the two co-propagating beams. For one phase the signal beam will be amplified, while for its quadrature it will be de-amplified. If the input signal beam fluctuates, then the fluctuations will also be amplified (or de-amplified) in the same phase-dependent way. Thus, as we will discuss in Section 4, if the input beam is a coherent laser beam with quantum-limited intensity fluctuations (i.e. shot noise), then the output beam will display fluctuations which, in one quadrature, will be reduced below the shot-noise level. Such sub-shot-noise fluctuations cannot be attained by a classical laser beam.

The nonclassical photon statistics of parametric light can be observed by single-photon correlation techniques [3, 4]. However, unlike many other quantum effects that become unobservable at high light intensities, the sub-shot-noise fluctuations of parametric light persist even when the output of the parametric emitter is an intense beam as, for example, when the parametric crystal is used to amplify an incoming laser signal.

The sub-shot-noise quantum statistics of parametric light have elicited a lot of interest over the past few years, because the quantum fluctuations of a light beam constitute the ultimate limitation in the precision of an optical measurement. Thus, as the sources of "technical noise" are progressively overcome and quantum-limited detection becomes possible in technologically important fields, such as optical communications, the ideas for controlling and engineering the quantum fluctuations of a light beam become relevant as they may offer additional means for increasing the signal-to-noise ratio in optical measurements and for reducing the rate of errors in optical communications.

Another quantum mechanical feature that arises from the production of twin photons, is that a measurement of one of the characteristics of a signal photon, immediately determines the corresponding characteristic of its idler twin, even if the two photons have been separated spatially by a very large distance. This feature leads to Einstein-Podolsky-Rosen-type correlations that bring forth the non-local character of quantum states and can be used to test the conceptual foundations of quantum mechanics. Several papers have been published on these quantum mechanical pairwise correlations of parametric photons, detected individually. We shall not discuss these effects any further in this course; the reader is referred to Refs. [5] for further discussion of the single-photon quantum correlations in parametric emission.

3. WHAT IS QUANTUM NOISE

All light detectors that are available today work on the principle of the photoelectric effect. Each photon that is incident on the detector ionizes an atom (for the case of a phototube) or produces an electron-hole pair (for the case of a semiconductor photodiode). The stream of electrons that is thus produced in the photodetector is subsequently analyzed by electric means, such as voltage or current measurements. Because of this principle of operation, photodetectors measure essentially the number of photons that are present in the incident beam. This means that if the incident beam is in a "photon-number" eigenstate, the photodetector will determine the eigenvalue of that state and give a unique answer.

Unfortunately, no practical device exists yet that can emit a photon-number eigenstate. Our "cleanest" light sources, the lasers, when they are perfectly stabilized, they produce coherent radiation. A coherent beam of light is a state of the radiation field in which the values of the electric and magnetic fields vary sinusoidally in time and space at a relatively well-defined frequency and phase, just like a classical stable oscillating field. Photon-number eigenstates, on the other hand, do not behave at all like classical oscillating fields, since their expectation values for the electric and magnetic field operators are always zero, with no temporal evolution.

A coherently excited mode can be represented [6] as a "displaced" vacuum state,

$$|\alpha\rangle = e^{\alpha b^\dagger - \alpha^* b}|0\rangle \tag{3.1}$$

since the one-point expectation values of such a state have all the properties of the corresponding observables of a classical oscillating field. In particular, using the definition of the electric field operator in terms of photon creation and annihilation operators (2.7), together with the fact that a quantum mechanical coherent state is an eigenstate of the annilation operator,

$$b\,|\alpha\rangle = \alpha\,|\alpha\rangle \tag{3.2}$$

it can readily be shown that $|\alpha\rangle$ presents a non-zero expectation value for the electric field operator, and this value varies sinusoidally in time as the state evolves under the free-field Hamiltonian

$$\langle\alpha|E|\alpha\rangle = 2|\alpha|\left(\frac{\hbar\omega}{2\epsilon V}\right)^{1/2}\sin(\omega t - kz + \phi) \tag{3.3}$$

just like a classical stable oscillating field. Thus, the absolute value of α gives the electric field amplitude (relative to that of the vacuum field which is $(\hbar\omega/2\epsilon V)^{1/2}$), while its phase ϕ gives the phase of the electric field at $t = 0$ and $z = 0$. Equivalently, the eigenvalue α can be thought as being related to the average number of photons that are present in the coherently excited mode since

$$\langle\alpha|b^\dagger b|\alpha\rangle = |\alpha|^2 \tag{3.4}$$

The average number of photons in the mode is related to the light intensity through a simple multiplicative factor

$$I = |\alpha|^2\left(\frac{\hbar\omega}{2\epsilon V}\right) \tag{3.5}$$

As a coherent state is an eigenstate of the annihilation operator (i.e. it is almost an eigenstate of the electric field) it is described as a superposition of photon-number eigenstates.

$$|\alpha\rangle = e^{-|\alpha|^2/2} \sum_{n=0}^{\infty} \frac{\alpha^n}{\sqrt{n!}} |n\rangle \tag{3.6}$$

This means that when a coherent beam is incident on a photodetector that measures the number of photons, we will not obtain a unique answer: Each time the measurement is performed, the answer will be different, and the values obtained will exhibit a statistical spread. This uncertainty in the values obtained in the measurement is often referred to as **quantum noise**. Clearly, quantum noise is not due to any technical imperfection of the lasers or of the photodetector, but is simply an expression of the fact that the observables that are relevant in the emission and detection processes are quantum mechanically "incompatibile" that is, their operators do not commute.

The existence of quantum noise limits the precision with which an optical measurement can be performed. In this section we shall examine the manifestation of quantum noise in two important measurement configurations: (1) **Direct detection**, which is implemented to determine the intensity of an incoming optical signal, and (2) **homodyne detection**, which is used to measure the amplitude and phase of a coherent optical signal. These two measurement configurations are widely used in technological applications, such as optical communications, to receive signals in which information is coded by **Amplitude Modulation**, for the case of direct detection, or by **Phase Modulation**, for the case of homodyne detection.

3.1 Quantum Noise in Direct Detection

In **direct detection** the beam of light to be measured is allowed to impinge directly on the photocathode of a detector and the resulting stream of electrons is analyzed.

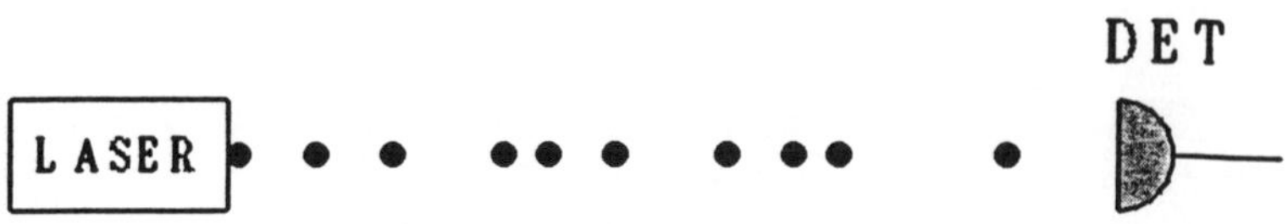

Figure 1. Direct detection: The laser produces a stream of photons which are received directly by the photodetector and are converted into an electric current.

Because of quantum noise, the photons of the incident beam do not arrive on the detector at regular time intervals. The statistics of photon arrivals is imprinted on the stream of photoelectrons and can be determined through two different experimental techniques.

3.1.1 Photon Counting. At low incident light intensities (typically a few million photons per second) digital counting electronics can be used to actually count the photoelectrons. The experiment consists, then, of setting a time-interval on a clock and counting the photoelectrons that are produced in the detector during the clock time. Each time the experiment is performed, the number of photons counted is different, but by repeating the experiment a large number of times, a histogram can be established giving the frequency of occurrance of each photon number. For a coherent laser beam, emerging from a perfectly stabilized laser, the histogram corresponds to a **Poisson distribution** that is peaked at $\overline{N}$, the number of photons that correspond to the mean intensity of the beam, and has a width $\Delta N = \sqrt{\overline{N}}$ or, equivalently, a variance $\Delta N^2 = \overline{N}$. The statistics of the arrivals of the photons of a coherent laser beam is similar to that of many classical processes that involve the flow of discrete and independent particles, such as the arrivals of gun-shot on a target or rain-drops on a window pane. Since the quantum noise of a coherent beam attests to the "granularity" of energy flow, it is often called **shot noise**.

The quantum mechanical representation of coherent states readily gives the photon statistics associated with shot noise. Using the representation of a coherent state in the photon-number basis (3.6), the probability of obtaining m photons in a measurement can be calculated as

$$P_m = |\langle m|\alpha\rangle|^2 = e^{-|\alpha|^2}\frac{|\alpha|^2}{m!} \tag{3.7}$$

This equation indicates that the number of photons in a coherent beam follows Poisson statistics, with the mean number of photons given by $\overline{N} = |\alpha|^2$ (see eq. 3.4).

3.1.2 Continuous-Intensity Measurements. When the incident light intensities is large, digital counting electronics can no longer follow the photon arrival rate. It is thus not possible to obtain the full statistical distribution of the incoming photons by directly counting the photoelectrons in real time. It is, nevertheless, possible to obtain some information on the quantum statistics of the incident beam by adopting frequency-domain techniques, such as the analysis of the power spectrum of the photocurrent. This can be carried out in the laboratory by introducing the output of the photodetector into an electronic instrument called **spectrum analyzer**, which gives the electrical power that is carried by each frequency component of the photocurrent. As the electrical power is proportional to the square of the current, and each frequency component of the stream of photoelectrons corresponds to a different photon arrival rate, the spectrum analyzer gives the variance in the fluctuations (that is, the square of the width of the distribution) of each photon arrival rate. For the case of a coherent laser beam the noise power measured at each frequency is proportional to the incident intensity and is the same for all frequencies. This underscores two important features of shot noise: First, that the noise variance is proportional to the mean number of photons, as is expected for Poisson statistics, and second, that the spectrum of shot noise is white, since its power is independent of the measurement frequency.

Following the engineering practice, in this course we shall treat the noise in terms of its variance (i.e. the power measured directly on the spectrum analyzer) rather than its standard deviation (i.e. the width of the statistical distribution), which is the square root of the variance. Thus, the quantum noise associated with the direct detection of a coherent beam can be calculated as

$$\Delta N^2 = \overline{N^2} - \overline{N}^2 = \tag{3.8}$$
$$= \langle\alpha|\, b^\dagger b b^\dagger\, b|\alpha\rangle - (\langle\alpha|\, b^\dagger b\, |\alpha\rangle)^2 \tag{3.9}$$

To calculate this variance, the first term must be put to normal order, that is all the creation operators have to be placed to the left of the annihilation operators. Normal order is useful, because it permits us to use the eigenvalue equation (3.2) for coherent states. Using the Bose commutation relation (2.4) every time the order of a creation and an annihilation operator is interchanged, we have

$$\Delta N^2 = \langle\alpha|\, b^\dagger b^\dagger b b\, |\alpha\rangle + \langle\alpha|\, b^\dagger b\, |\alpha\rangle - (\langle\alpha|\, b^\dagger b\, |\alpha\rangle)^2 = \tag{3.10}$$
$$= |\alpha|^4 + |\alpha|^2 - |\alpha|^4 = |\alpha|^2 = \overline{N} \tag{3.11}$$

Thus, the full quantum mechanical calculation shows also that, for shot noise, the variance of the fluctuations is equal to the mean, as expected for Poisson statistics.

In this context, the Signal-to-Noise Ratio (SNR) for a measurement will be defined also in terms of <u>noise power</u>, as

$$SNR = \frac{(\overline{N})^2}{\Delta N^2} \tag{3.12}$$

For the direct detection of a coherent signal, then, the SNR due to shot noise can be calculated as

$$SNR_{DIR} = \frac{|\alpha|^4}{|\alpha|^2} = |\alpha|^2 = \overline{N} \tag{3.13}$$

that is, it is proportional to the signal intensity.

3.2 Quantum Noise in Homodyne Detection

The manifestations of quantum noise when measuring the phase of an optical signal can be described by referring to a well-known optical interferometric setup, the Mach-Zehnder interferometer.

A **Mach-Zehnder interferometer** is a device whose operation is based on two beamsplitters (see Fig. 2).

An incoming laser beam is separated by the first beamsplitter (BS1) into two beams, which we shall call "signal" and "reference". These two beams are then recombined on the second beamsplitter (BS2), to produce two new beams (which we shall call collectively "output") whose electric field amplitudes correspond, respectively, to the sum and the difference of the electric field amplitudes of the signal and reference beams. The two output beams are sent each to a photodetector and the two resulting photocurrents are introduced into a differential amplifier whose electrical output is proportional to the difference of the two optical outputs of the Mach-Zehnder interferometer. If the signal and reference arms of the interferometer are set equal in length, the two beams arrive at the second beamsplitter **in-phase**: the interference is constructive for one output and destructive for the other and, in this situation, the differential current will be maximum. If a phase delay is introduced between the signal and reference arms, the electrical output will vary as the cosine of that phase delay. On the other hand, if the

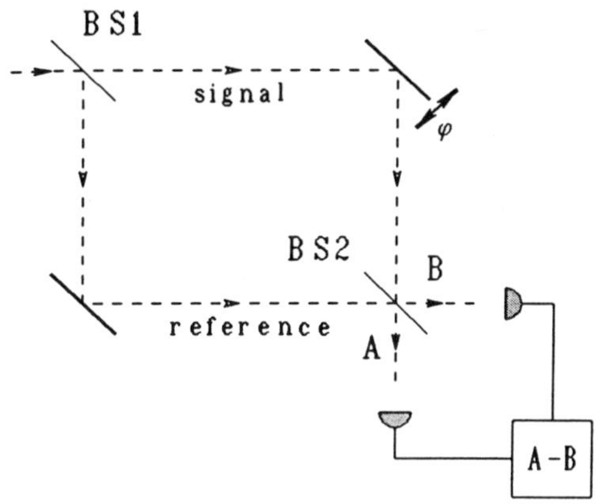

Figure 2. Mach-Zehnder interferometer: Its operation is described in the text below.

two arms differ by a pathlength of $\lambda/2$, the signal and reference beams arrive at the second beamsplitter **in-quadrature**: The two output beams have the same intensity and the differential amplifier will give a null electrical output. In this configuration, if a phase delay is introduced in the signal arm, the two output beams will no longer have the same intensity and the differential amplifier will give a non-zero electrical output, proportional to the sine of the signal phase delay.

In the engineering jargon, the setup consisting of the second beamsplitter the two photodetectors and the differential amplifier is called a **balanced homodyne detector** while the reference arm which provides the phase reference for the interferometric detection is called a **local oscillator**.

We can describe the operation of a balanced homodyne detector quantum mechanically by examining the transformations that are produced on the electromagnetic field operators by each component of the detector. At the input of the detector, we have two modes, the Signal (S) and the Local Oscillator (LO); the creation/annihilation operators associated with these two modes will be designated as $b_S^\dagger/b_S$ and $b_{LO}^\dagger/b_{LO}$, respectively. At the beamsplitter, the Signal and Local Oscillator modes are mixed, so that the annihilation operators for the two output modes are

$$b_1 = \frac{\sqrt{2}}{2}(b_S + b_{LO}) \tag{3.14}$$

$$b_2 = \frac{\sqrt{2}}{2}(b_S - b_{LO}) \tag{3.15}$$

with similar equations for the creation operators. The detection of each output beam by a photodiode is associated with the corresponding photon number operator. This feature permits us to relate the electrical output of each detector to the Signal and Local Oscillator operators as

$$b_1^\dagger b_1 = \frac{1}{2}(b_S^\dagger b_S + b_{LO}^\dagger b_{LO} + b_S^\dagger b_{LO} + b_{LO}^\dagger b_S) \tag{3.16}$$

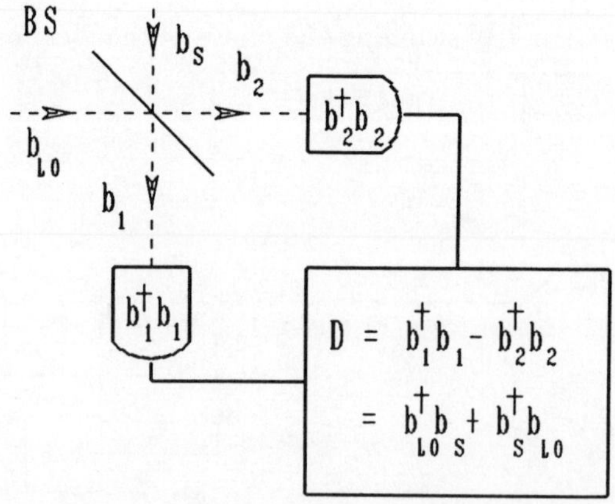

Figure 3. Balanced Homodyne Detector: Its operation is described in the text below.

$$b_2^\dagger b_2 = \frac{1}{2}(b_S^\dagger b_S + b_{LO}^\dagger b_{LO} - b_S^\dagger b_{LO} - b_{LO}^\dagger b_S) \tag{3.17}$$

Thus, the operator that corresponds to the differential detection of the photocurrent is,

$$D = b_1^\dagger b_1 - b_2^\dagger b_2 = b_S^\dagger b_{LO} + b_{LO}^\dagger b_S \tag{3.18}$$

If both the LO and S are coherent beams, the mean photocurrent produced by the balanced homodyne detector is

$$\overline{I_{ele}} = \langle \alpha_{LO}\alpha_S|D|\alpha_{LO}\alpha_S\rangle = 2|\alpha_{LO}\alpha_S|\cos(\phi_S - \phi_{LO}) \tag{3.19}$$

This equation indicates that the output of the homodyne detector is proportional to the amplitude of the signal beam (rather than being proportional to the intensity as in direct detection) and varies cyclically as the relative phase between the signal and the local oscillator is scanned. At the same time, this equation underscores the large sensitivity of homodyne detection in receiving very weak signals: by having a very intense local oscillator the differential photocurrent can be be brought up to any value that is necessary to overcome the "technical noise" of the subsequent electronics. Finally, this equation indicates that a null photocurrent is obtained when the signal and local oscillator phases are in quadrature ($\phi_S - \phi_{LO} = \pi/2$). Conversely, the homodyne measurements relative to the phase that gives a null current correspond to measuring the signal phase relative to the quadrature of the local oscillator.

The sensitivity of balanced homodyne detection is limited by the quantum noise of the signal and reference beams: each photon incident on the beamsplitter will be channelled towards one or the other photodetector (but not both simultaneously) so that the output of the differential amplifier will exhibit the fluctuations that correspond to the random photon arrivals in each of the two detectors. The statistics of these fluctuations can be analyzed through counting or spectral analysis techniques, as in the case of direct detection. By scanning the phase of the local oscillator, it is possible to measure the quantum noise power that is associated with each phase of the signal beam. For a coherent laser beam, the quantum noise is independent of the signal phase and corresponds to the Poisson statistics of shot noise. If the signal and local oscillator beams are represented by coherent states $|\alpha_S \alpha_{LO}\rangle$, the fluctuations of the differential photocurrent that are due to the quantum noise of the incoming beams can be written as,

$$\Delta I^2{}_{ele} = \langle \alpha_{LO}\alpha_S | D^2 | \alpha_{LO}\alpha_S \rangle - \overline{I_{ele}}^2 = \tag{3.20}$$

$$= \langle \alpha_{LO}\alpha_S | \, b_S^\dagger b_{LO} b_S^\dagger b_{LO} + b_{LO}^\dagger b_S b_{LO}^\dagger b_S + b_S^\dagger b_{LO} b_{LO}^\dagger b_S + b_{LO}^\dagger b_S b_S^\dagger b_{LO} \, | \alpha_{LO}\alpha_S \rangle - \overline{I_{ele}}^2 \tag{3.21}$$

Setting the operators to normal order by use of the Bose commutation relation (2.4) and writing only the terms that do not get cancelled by $\overline{I_{ele}}^2$, the variance of the photocurrent fluctuations can be calculated as

$$\Delta I^2{}_{ele} = \langle \alpha_{LO}\alpha_S | \, b_S^\dagger b_S + b_{LO}^\dagger b_{LO} \, | \alpha_{LO}\alpha_S \rangle = |\alpha_{LO}|^2 + |\alpha_S|^2 \tag{3.22}$$

Since the LO intensity is much larger than the signal intensity, eq. (3.22) can be approximated as,

$$\Delta I^2{}_{ele} \approx |\alpha_{LO}|^2 \tag{3.23}$$

That is, the noise variance in independent of the signal and local oscillator phase, in contrast to the value of the mean photocurrent.

3.3 Graphical Representation of Shot Noise

The electric field of a coherent signal measured in a balanced homodyne experiment, may be expressed as the sum of two quadratures oscillating in-phase and 90 degrees out of phase with respect to the local oscillator, and thus may be represented as a vector in the Cartesian plane defined by the two quadratures $\hat{e}_1$ and $\hat{e}_2$.

$$E = \hat{e}_1 |\alpha_S| \cos(\theta) + \hat{e}_2 |\alpha_S| \sin(\theta) \tag{3.24}$$

where $\theta = \phi_S - \phi_{LO}$ is the phase of the signal relative to the local oscillator, while $r = |\alpha_S|$, the length of the vector, corresponds to the signal field amplitude. The choice of units in eq. (3.24) corresponds to setting $|\alpha_{LO}| = \frac{1}{2}$ in eq. (3.19). The graphical representation of a coherent field according to eq. (3.24) (Fig. 4) is called a **phasor diagram**.

Alternatively, the phasor diagram can be obtained by re-writing the differential detection operator (3.18) as two normalized quadrature operators, one for in-phase detection of the signal,

$$D_1 = \frac{b_S + b_S^\dagger}{2} \tag{3.25}$$

and one for in-quadrature detection

$$D_2 = \frac{b_S - b_S^\dagger}{2i} \tag{3.26}$$

The two Cartesian coordinates of the phasor correspond, then, to the expectation values of these operators over the input coherent signal state $|\alpha_S\rangle_{in}$

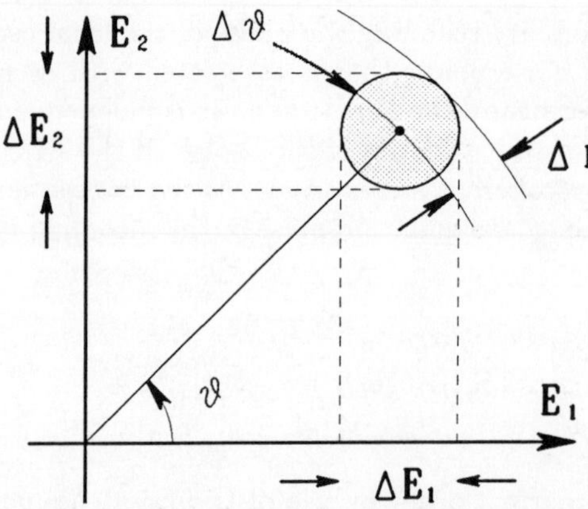

Figure 4. Phasor diagram for a coherent electromagnetic field.

When the signal is subject to noise fluctuations, its phasor fluctuates randomly in length and direction. This "fuzziness" in the definition of the phasor may be represented by a "spot" attached to the free end of a mean vector: at any instant, the phasor may take any value for which its free end is inside the "noise spot". The extent of the spot along the radial direction Δr corresponds to the magnitude of the amplitude noise, while the extent along the angular coordinate $\Delta\theta$ corresponds to the phase noise. The projection of the spot along the two Cartesian coordinates ΔE_1, ΔE_2 gives the noise in the two quadratures. In classical optics, the noise spot can be as small as desired, since the amplitude and phase of a classical signal can be measured with arbitrary accuracy. In quantum optics, however, the existence of quantum noise dictates that the noise spot must have a minimum area, set by the Heisenberg uncertainty principle, which be expressed in terms of the fluctuations in the two quadratures as

$$(\Delta E_1)(\Delta E_2) \geq \frac{1}{4} \tag{3.27}$$

For the case of a coherent signal, the fluctuations in the two quadratures can be obtained from the variance of the differential photocurrent (3.23) as

$$\Delta E_1 = \Delta E_2 = \sqrt{\Delta I_{ele}^2} = \frac{1}{2} \tag{3.28}$$

where we converted eq. (3.23) to the units of $\hat{e}_1$, $\hat{e}_2$ by setting $|\alpha_{LO}| = \frac{1}{2}$. Since the quantum noise is independent of the signal phase, the noise spot is circular. Clearly, the radius of the spot for a coherent state is the smallest allowed by quantum mechanics as it minimizes the Heisenberg inequality (3.27). For a light beam that contains excess noise (beyond shot noise) such as thermal fluctuations or "technical noise", the noise spot is larger than the disk of radius $\frac{1}{2}$.

In addition to carrying an immediate visual image of the quantum noise involved in homodyne detection, the phasor diagram of Fig. 4 contains also all the information associated with the noise of direct detection. Direct detection, that is, the determination of the photon number of a coherent beam corresponds, essentially, to the measurement of the square of the length of the phasor. Using this feature, it is straight-forward to verify that the graphical representation of shot noise in Fig. 4 (in which the length of the phasor is $\sqrt{\overline{N}}$ while the radius of the noise disk is independent of the photon number) does indeed correspond to Poisson statistics. The mean photon number and the width of the photon statistics can be obtained from the expression,

$$\overline{N} + \Delta N = |r + \Delta r|^2 \tag{3.29}$$

The mean photon number is the term that is of order $(\Delta r)^0$ (order zero)

$$\overline{N} = |r|^2 = |\alpha_S|^2 \tag{3.30}$$

while the width of the photon number distribution is of order $(\Delta r)^1$.

$$\Delta N = 2r\Delta r = 2|\alpha_S|.\frac{1}{2} = \sqrt{\overline{N}} \tag{3.31}$$

giving the expected mean and width for Poisson distribution of shot noise. This analysis indicates that a super-Poissonian photon distribution (that is, a beam with photon fluctuations in excess of shot noise) would be characterized by a noise spot whose radial extent $\Delta r > \frac{1}{2}$ is larger that one half. On the other hand, sub-Poissonian photon statistics, which can occur only for non-classical states, would correspond to $\Delta r < \frac{1}{2}$.

4. QUANTUM NOISE IN LINEAR AMPLIFICATION

A linear amplifier "magnifies" an input signal by a factor which is independent of the amplitude and the phase of the input signal but, at the same time, it degrades the SNR, as we shall show in this Section. In order to investigate the noise characteristics of linear amplification, we shall examine the case of a non-degenerate parametric amplifier in which the idler mode has no input. Furthermore, adopting the restrictive definition for amplification, whereby the output beam must have the same frequency and the same polarization as the input, we shall examine more particularly the configuration in which only the signal channel output is detected, while the idler output is rejected. A schematic of this configuration is given in Fig. 5, below.

4.1 A Quantum Description of Non-degenerate Parametric Amplification

We can now examine the quantum mechanical description of parametric emission for the case in which the signal and idler modes are non-degenerate. For the sake of concreteness, we shall examine the case of a type II parametric crystal in which the signal and idler modes are non-degenerate only in polarization. We shall assume that the pump is a coherent laser beam and is intense enough so that it can be considered to be un-depleted throughout the parametric emission process. Under this assumption, we can replace the pump operators in the Hamiltonian (2.17) by the corresponding

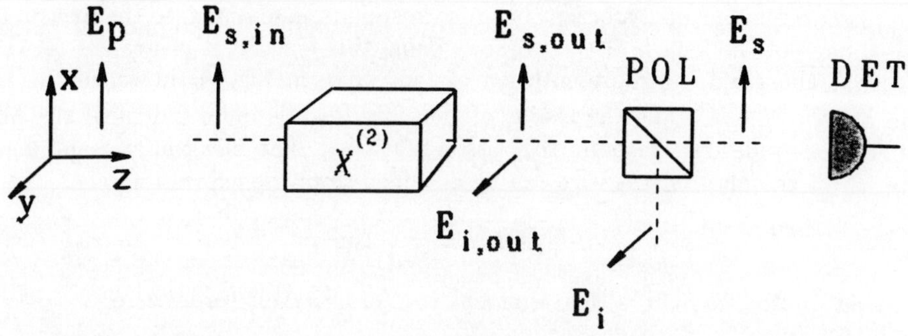

Figure 5. Nondegenerate parametric amplification in a type-II nonlinear crystal with only a signal input. E_p is the pump field. When the input signal $E_{s,in}$ is polarized parallel to the crystal axes, the amplification gain is independent of the signal phase. At the output of the amplifier, both signal and idler modes carry energy.

electric field amplitude, so that the Hamiltonian for degenerate parametric emission can be written as,

$$H_{NDEG} = i\hbar\gamma(b_s^\dagger b_i^\dagger - b_s b_i) \tag{4.1}$$

where we have introduced the constant γ

$$\gamma = \frac{2}{3}\chi^{(2)}\left(\frac{\omega_s}{2\epsilon V}\frac{\omega_i}{2\epsilon V}\right)^{1/2} E_p \tag{4.2}$$

to re-group all c-number constants, and we have chosen the phase of the pump beam as a reference. This choise implies that the phase of any input or output beam will be measured relative to this "internal phase reference" of the parametric crystal which is set by the pump beam. This feature is not used in this Section, but will be very important in the next Section.

A signal beam propagating in a nonlinear crystal in the presence of an intense coherent pump beam will evolve under the Hamiltonian (4.1). The equations of motion for the signal beam operators can be obtained through the Heisenberg equation as

$$\frac{db_s}{dt} = \frac{i}{\hbar}[H_{NDEG}, b_s] = \gamma b_i^\dagger \tag{4.3}$$

$$\frac{db_s^\dagger}{dt} = \frac{i}{\hbar}[H_{NDEG}, b_s^\dagger] = \gamma b_i \tag{4.4}$$

These equations have the well-known formal solution

$$b_s(t) = e^{(it/\hbar)H_{NDEG}}\, b_s(0)\, e^{-(it/\hbar)H_{NDEG}} \tag{4.5}$$

$$b_s^\dagger(t) = e^{(it/\hbar)H_{NDEG}}\, b_s^\dagger(0)\, e^{-(it/\hbar)H_{NDEG}} \tag{4.6}$$

where $e^{(it/\hbar)H}$ is the time-evolution operator. Expanding the exponential operators, eq. (4.5) may be re-written as a power series in $(it/\hbar)$ as

$$b_s(t) = b_s(0) + \left(\frac{it}{\hbar}\right)[H_{NDEG}, b_s(0)] + \frac{1}{2!}\left(\frac{it}{\hbar}\right)^2 [H_{NDEG}, [H_{NDEG}, b_s(0)]] + \cdots \quad (4.7)$$

which can be readily calculated, by evaluating the successive commutators, as

$$\begin{aligned} b_s(t) &= b_s(0) + (\gamma t)\, b_i^\dagger(0) + \frac{(\gamma t)^2}{2!}\, b_s(0) + \frac{(\gamma t)^3}{3!}\, b_i^\dagger(0) + \cdots = & (4.8) \\ &= \cosh(\gamma t)\, b_s(0) + \sinh(\gamma t)\, b_i^\dagger(0) & (4.9) \end{aligned}$$

This equation permits us to relate the signal and idler operators at the input of the parametric crystal to those obtained at the output, after propagation over a distance z, i.e. after an interaction time $t = z/v$ (where v is the group velocity of light at the signal frequency) as

$$\begin{aligned} b_{s,out} &= \cosh(\gamma z/v)\, b_{s,in} + \sinh(\gamma z/v)\, b_{i,in}^\dagger & (4.10) \\ b_{s,out}^\dagger &= \cosh(\gamma z/v)\, b_{s,in}^\dagger + \sinh(\gamma z/v)\, b_{i,in} & (4.11) \end{aligned}$$

Note that the parametric interaction mixes the signal and idler modes.

4.2 Amplification and Excess Noise: Direct Detection

We can now describe the process of non-degenerate parametric amplification, when the input signal is a coherent beam represented by a state $|\alpha_s\rangle_{in}$, corresponding to a mean number of input photons $N_{s,in} = |\alpha_{s,in}|^2$, while the idler input is in the vacuum state $|0_i\rangle_{in}$. The number of signal photons at the output of the amplifier can be written as,

$$N_{s,out} = {}_{in}\langle 0_i \alpha_s|\, b_{s,out}^\dagger b_{s,out}\, |\alpha_s 0_i\rangle_{in} \quad (4.12)$$

Using the relationships (4.10) and (4.11) between the input and output operators, eq. (4.12) can be written as

$$\begin{aligned} N_{s,out} &= {}_{in}\langle 0_i \alpha_s|\cosh^2(\gamma z/v) b_{s,in}^\dagger b_{s,in} + \sinh^2(\gamma z/v) b_{i,in} b_{i,in}^\dagger + \\ &\quad + \cosh(\gamma z/v)\sinh(\gamma z/v)\,(b_{s,in}^\dagger b_{i,in}^\dagger + b_{s,in} b_{i,in})|\alpha_s 0_i\rangle_{in} = & (4.13) \\ &= \sinh^2(\gamma z/v) + |\alpha_s|^2\cosh^2(\gamma z/v) & (4.14) \end{aligned}$$

Similarly, the number of idler photons produced in parametric emission can be obtained by interchanging the indices s and i in eqs. (4.12-4.14) as,

$$N_{i,out} = \sinh^2(\gamma z/v) + |\alpha_s|^2 \sinh^2(\gamma z/v) \quad (4.15)$$

The first term in eqs. (4.14) and (4.15) is independent of the input intensity and gives the number of photons that the parametric amplifier emits even in the absence of an input. This emission corresponds to a spontaneous initiation of the parametric emission process (called **parametric fluorescence**) that can only be accounted for within the framework of a quantum mechanical formalism. We note that the number of spontaneously generated idler photons is identical to the number of spontaneous signal photons, as would be expected from the twin-photon nature of parametric emission.

The correlation in the number of signal and idler photons in parametric fluorescence has been confirmed in many experiments [7, 8].

In situations in which the parametric crystal is used as an amplifier, the number of photons of the input signal mode is relatively large, ($N_{s,in} = |\alpha_{s,in}|^2 \gg 1$) so that this first term is small when compared with the next term which is proportional to $|\alpha_{s,in}|^2$. Thus, when considering the amplification of a light beam with finite input intensity, the spontaneous parametric emission term can be neglected. In that case the signal output of the parametric amplifier can be written as,

$$N_{s,out} = N_{s,in} G_{PIA} \tag{4.16}$$

where

$$G_{PIA} = \cosh^2(\gamma z/v) = 1 + \sinh^2(\gamma z/v) > 1 \tag{4.17}$$

is the gain of amplification. We note that when the idler mode is not excited, the parametric amplifier operates as a **linear amplifier** in which the intensity of the output is proportional to the intensity of the input signal, independently of its phase, much the same way as in a laser. We shall call this configuration Phase-Independent Amplification (PIA).

We now turn briefly to the idler output of the parametric amplifier. Eq. (4.15) indicates that even if the idler mode has no input in this configuration ($\alpha_i = 0$), the parametric emitter still produces a significant number of idler photons in its output. Indeed, in addition to the spontaneously generated idler photons, the presence of a finite signal input induces emission at the idler output, given by

$$N_{i,out} = N_{s,in} \sinh^2(\gamma z/v) \tag{4.18}$$

Comparison of this expression (4.18) with eqs. (4.16 - 4.17) indicates that the number of idler photons produced in the course of the amplification of the signal is precisely equal to the number of signal photons that have been added to the amplified beam by the parametric emitter through the twin-photon emission process.

The noise characteristics of the amplified beam, measured in direct detection, can be obtained by calculating the variance in the number of signal photons detected at the output of the parametric amplifier as

$$\Delta N^2_{s,out} = {}_{in}\langle \alpha_s 0_i |\ b^\dagger_{s,out} b_{s,out} b^\dagger_{s,out} b_{s,out}\ | 0_i \alpha_s \rangle_{in} - N^2_{s,out} \tag{4.19}$$

Using the input-output relations for a non-degenerate parametric amplifier (4.10) and (4.11) and writing only the terms in which the input signal and idler operators appear an even number of times to reduce the clutter of the equation (since the odd terms will have a zero expectation value when the idler mode is in the vacuum state) the photon number variance (4.19) can be written as

$$\begin{aligned} \Delta N^2_{s,out} = {}_{in}\langle \alpha_s 0_i | \quad & C^4 b^\dagger_{s,in} b_{s,in} b^\dagger_{s,in} b_{s,in} + S^4 b_{i,in} b^\dagger_{i,in} b_{i,in} b^\dagger_{i,in} \quad + \\ + \quad & C^2 S^2 (b^\dagger_{s,in} b^\dagger_{i,in} + b_{s,in} b_{i,in})(b^\dagger_{s,in} b^\dagger_{i,in} + b_{s,in} b_{i,in}) \quad + \\ + \quad & C^2 S^2 b^\dagger_{s,in} b_{s,in} b_{i,in} b^\dagger_{i,in} + C^2 S^2 b_{i,in} b^\dagger_{i,in} b^\dagger_{s,in} b_{s,in} \quad |0_i \alpha_s\rangle_{in} - \\ & - N^2_{s,out} \end{aligned} \tag{4.20}$$

where we have used the shorthand notation $C = \cosh(\gamma z/v)$ and $S = \sinh(\gamma z/v)$ to make the formula more readable. Putting the creation and annihilation operators to normal order by use of the Bose commutation relation (2.4) and performing all the cancellations, eq. (4.20) reduces to

$$\Delta N^2_{s,out} = {}_{in}\langle \alpha_s 0_i | C^4 b^\dagger_{s,in} b_{s,in} + S^4 b_{i,in} b^\dagger_{i,in} + +C^2 S^2 (b^\dagger_{s,in} b_{s,in} + b_{i,in} b^\dagger_{i,in}) | 0_i \alpha_s \rangle_{in} = \tag{4.21}$$

$$= (S^4 + C^2 S^2) + |\alpha_{s,in}|^2 (C^4 + C^2 S^2) \tag{4.22}$$

The first term in parentheses, which is independent of α_s corresponds to the noise associated with the spontaneous parametric fluorescence. However, since in the amplifying configuration the input signal contains a macroscopic number of photons ($|\alpha_{s,in}|^2 \gg 1$), this first term can be neglected as it is very small when compared with the next term, which is proportional to the number of input signal photons. The noise of the amplified signal is, therefore,

$$\Delta N^2_{s,out} = |\alpha_{s,in}|^2 \cosh^2(\gamma z/v)(\cosh(\gamma z/v) + \sinh(\gamma z/v)) = \tag{4.23}$$

$$= |\alpha_{s,in}|^2 G_{PIA}(2G_{PIA} - 1) \tag{4.24}$$

Clearly, because of the noise that is added by the amplifier to the signal, the SNR at the output of the amplifier

$$SNR_{out} = \frac{N^2_{s,in} G^2_{PIA}}{N_{s,in} G_{PIA}(2G_{PIA} - 1)} = N_{s,in} \frac{G_{PIA}}{2G_{PIA} - 1} \tag{4.25}$$

is degraded with respect to the SNR at the input which, for a coherent beam, is (eq. 3.13)

$$SNR_{in} = N_{s,in} \tag{4.26}$$

The degradation of the SNR in the course of amplification is usually expressed in terms of the **Noise Figure** (NF) of the amplifier, defined as

$$NF = \frac{SNR_{in}}{SNR_{out}} = \frac{\Delta N^2_{out}}{\Delta N^2_{in}} \frac{1}{G} \tag{4.27}$$

For a PIA, the NF can readily be calculated as

$$NF = \frac{2G_{PIA} - 1}{G_{PIA}} = 2 - \frac{1}{G_{PIA}} \tag{4.28}$$

This equation indicates that for large gains ($G_{PIA} \gg 1$) the NF approaches $NF = 2$ or, in the engineering parlance, 3 dB ($= 10\log_{10} 2$). This means essentially that amplification of light by a non-degenerate parametric amplifier degrades the Signal-to-(Quantum)-Noise Ratio of a coherent beam by a factor of 2. In other words, direct detection of the amplified beam will be noisier than the detection of the same beam without amplification.

The noise figure of a non-degenerate parametric amplifier according to eq. (4.28) has been verified experimentally [9] (see Fig. 6).

4.3 Excess Noise in Homodyne Detection

The mean value and the fluctuations of the amplifier output measured by a homodyne detector is given by the expectation value and the variance of the quadrature

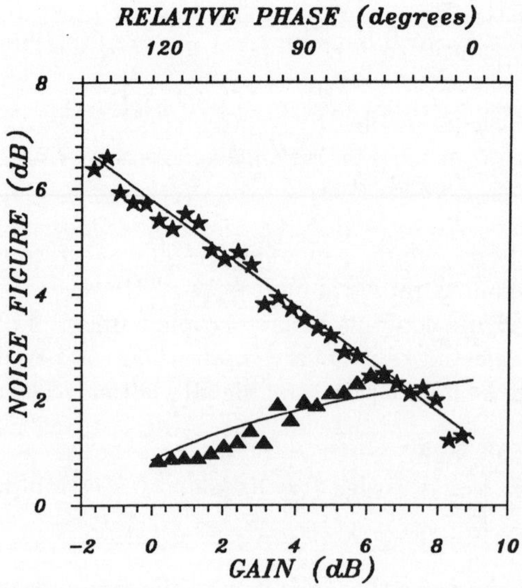

Figure 6. Noise Figure for a parametric amplifier operated in a phase-independent configuration (triangles) or in a phase-dependent configuration (stars). The noise figure of the phase-independent configuration follows the curve for classical amplifiers. In the amplified quadrature of the phase-dependent configuration (signal phase = 0 degrees or Gain = 9 dB) the noise figure drops below the classical value.

operators D_1, D_2 (eqs. 3.25 and 3.26) calculated over the output state of the amplifier. Alternatively, it is possible to calculate expectation values over the input state $|\alpha_s 0_i\rangle_{in}$ if the quadrature operators are expressed in terms of the operators for the input signal and idler modes of the amplifier. Using the input-output relations for non-degenerate parametric amplification, eqs. (4.10) and (4.11), we have

$$D_1 = \frac{b_{s,out} + b^\dagger_{s,out}}{2} = \cosh(\gamma z/v)\left(\frac{b_{s,in} + b^\dagger_{s,in}}{2}\right) + \sinh(\gamma z/v)\left(\frac{b_{i,in} + b^\dagger_{i,in}}{2}\right) = \tag{4.29}$$

$$= \sqrt{G_{PIA}}\left(\frac{b_{s,in} + b^\dagger_{s,in}}{2}\right) + \sqrt{G_{PIA}-1}\left(\frac{b_{i,in} + b^\dagger_{i,in}}{2}\right) \tag{4.30}$$

and

$$D_2 = \sqrt{G_{PIA}}\left(\frac{b_{s,in} - b^\dagger_{s,in}}{2i}\right) + \sqrt{G_{PIA}-1}\left(\frac{b_{i,in} - b^\dagger_{i,in}}{2i}\right) \tag{4.31}$$

where we have used the definition (4.17) for G_{PIA}, the power gain for the parametric amplifier in the Phase-Independent configuration. The expectation values for the two quadratures, then, are

$$E_1 = {}_{in}\langle\alpha_s 0_i|D_1|0_i\alpha_s\rangle_{in} = \sqrt{G_{PIA}}\,|\alpha_{s,in}|\cos\theta \tag{4.32}$$

and

$$E_2 = {}_{in}\langle\alpha_s 0_i|D_2|0_i\alpha_s\rangle_{in} = \sqrt{G_{PIA}}\,|\alpha_{s,in}|\sin\theta \tag{4.33}$$

Thus, amplification of the signal intensity by a gain of G_{PIA} produces an increase in the mean amplitude of the signal channel by a factor of $\sqrt{G_{PIA}}$ in both quadratures which, in turn, is reflected in the result of the homodyne measurement. This change can be represented by an increase in the length of the phasor by the factor $\sqrt{G_{PIA}}$. Similarly, the variance of the distribution of values that can be obtained when measuring the two quadratures can be calculated as,

$$\begin{aligned}
\Delta E_1^2 &= {}_{in}\langle\alpha_s 0_i|D_1^2|0_i\alpha_s\rangle_{in} - E_1^2 = && (4.34)\\
&= \frac{G_{PIA}}{4}\,{}_{in}\langle\alpha_s|\; b_{s,in}^\dagger b_{s,in}^\dagger + b_{s,in}b_{s,in} + b_{s,in}^\dagger b_{s,in} + b_{s,in}b_{s,in}^\dagger\;|\alpha_s\rangle_{in} - E_1^2 + \\
&+\frac{G_{PIA}-1}{4}\,{}_{in}\langle 0_i|\; b_{i,in}^\dagger b_{i,in}^\dagger + b_{i,in}b_{i,in} + b_{i,in}^\dagger b_{i,in} + b_{i,in}b_{i,in}^\dagger\;|0_i\rangle_{in} = && (4.35)\\
&= \frac{2G_{PIA}-1}{4} && (4.36)
\end{aligned}$$

and

$$\Delta E_2^2 = \frac{2G_{PIA}-1}{4} \tag{4.37}$$

Thus, after linear (phase-independent) amplification, the noise in each of the two quadratures of homodyne detection is

$$\Delta E_1 = \Delta E_2 = \frac{\sqrt{2G_{PIA}-1}}{2} > \frac{1}{2} \tag{4.38}$$

and is thus larger than in the non-amplified signal.

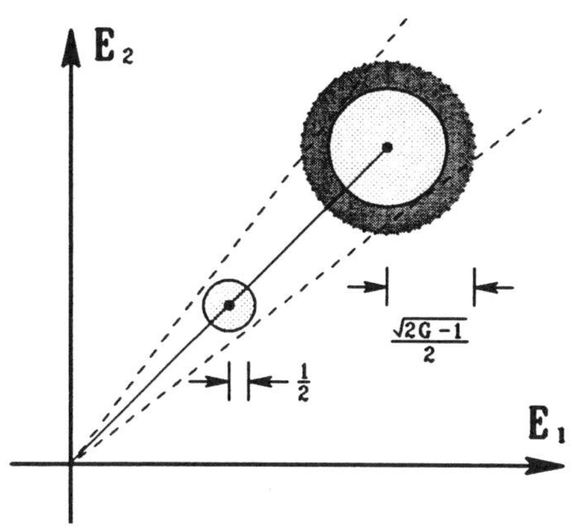

Figure 7. Phasor diagram for a coherent signal after linear amplification.

The graphical representation of this noise corresponds still to a circular disk, since the variances in the two quadratures are equal. However, the radius of the noise disk is now larger. This increase in the area of the noise disk indicates that, for a finite gain of amplification ($G_{PIA} > 1$), the Heisenberg inequality is no longer minimized

$$(\Delta E_1)(\Delta E_2) = \frac{2G_{PIA} - 1}{4} > \frac{1}{4} \tag{4.39}$$

In other words, the amplified beam contains excess noise, beyond the minimum required by the quantum mechanical uncertainty. At the same time, the increase of the extent of the noise disk in the radial direction (Δr) in the phasor diagram of Fig. 7, indicates that the photon number fluctuations of the amplified beam are larger than those of shot noise, as was calculated in eq. (4.24).

The origin of this excess noise can be traced to the input-output relations for non-degenerate parametric amplification (eqs. 4.10 and 4.11) which mix the signal and idler modes. This mixing, however, does not manifest itself in the mean value of the amplified signal, but only in its variance: If the idler mode has no input, emission into that mode does not contribute to the increase in the mean value of the amplitude in the signal mode, as can be seen in eqs. (4.32) and (4.33). However, the noise associated with emission into the idler mode (i.e. the second term in eq. 4.35) produces random phase fluctuations of the macroscopic polarization that drives the signal mode. Thus, the noise of the idler is mixed with the signal mode output giving rise to excess fluctuations in the amplified signal and to the noise figure of NF = 2 calculated in eq. (4.28).

It should be noted that the same considerations regarding the output noise hold also for traditional optical amplifiers, such as lasers. The similarity lies in the fact that, in addition to the amplifying (or "lasing") mode, lasers have also a very large number of "non-lasing" modes in which spontaneous emission is, nevertheless, possible. The possibility of spontaneous emission into these modes produces excess noise in the amplifying mode, much like parametric emission into the idler mode introduces fluctuations in the output of the signal mode. The existence of a bath of unobserved modes (such as the idler or the non-lasing modes) that introduce excess noise in the output signal, is a basic characteristic of classical linear amplifiers. The noise introduced by this bath is clearly a limitation in the performance of linear amplifiers.

For laser amplifiers, the presence of "technical" noise increases further the noise figure. In fact, some of the best optical amplifiers that are commonly in use today, the Erbium doped optical fibers used as in-line amplifiers in optical fiber communications, have a noise figure of 4 dB, slightly above the limit of 3 dB imposed by Quantum Mechanics.

5. QUANTUM NOISE IN PHASE-DEPENDENT AMPLIFICATION

The phase-dependance of the amplification process and the simultaneous and correlated emission of pairs of photons in parametric emission impart nonclassical quantum statistics to parametric light, as discussed in Section 2.3. In this Section we shall examine the quantum noise characteristics of degenerate parametric amplification which occurs when the signal frequency is the sub-harmonic of the pump ($\omega_s = \omega_i = \omega_p/2$). Experimentally, this can be achieved in two distinct situations
(1) in a type I phase-matched crystal, so that the signal and idler are truly indistin-

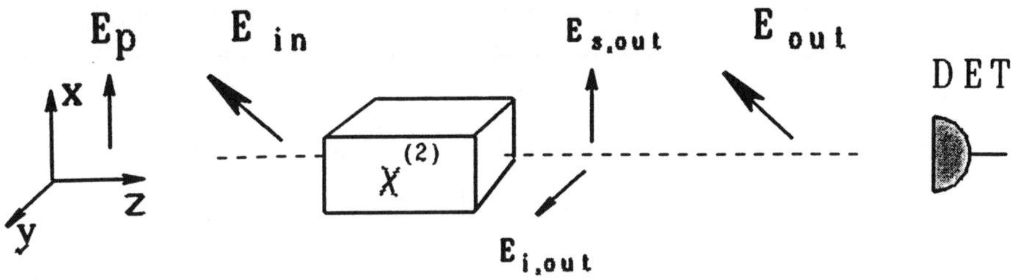

Figure 8. Degenerate parametric amplification in a type-II nonlinear crystal. E_p is the pump field. When the input field, E_{in}, is polarized at 45° to the crystal axes, the gain of the amplifier depends on the signal phase. The output field, E_{out}, is also polarized at 45° to the crystal axes.

guishable or
(2) in a type II crystal for the mode which is polarized at 45° with respect to the axes of the crystal and is thus a linear superposition of the (orthogonally polarized) signal and idler modes with equal weights.

5.1 A Quantum Description of Degenerate Parametric Amplification

The Hamiltonian for degenerate parametric emission can be obtained from the non-degenerate Hamiltonian (4.1) simply by dropping the s and i subscripts

$$H = i\hbar\gamma'(b^\dagger b^\dagger - bb) \tag{5.1}$$

For type II phase-matching in which the degenerate mode at 45° (see Fig. 8) is related to the non-degenerate signal and idler modes by

$$b = \frac{b_s + b_i}{\sqrt{2}} \tag{5.2}$$

or

$$b = \frac{b_s - b_i}{\sqrt{2}} \tag{5.3}$$

the constant γ' of the degenerate Hamiltonian is related to the γ of the Hamiltonian (4.1) by $\gamma' = \gamma/2$ so that for the type II configuration the Hamiltonian for the degenerate mode at 45° is

$$H_{DEG} = i\hbar\frac{\gamma}{2}(b^\dagger b^\dagger - bb) \tag{5.4}$$

The equations of motion for the operators of a signal beam co-propagating in a phase-matched nonlinear crystal with an intense classical pump can be written as,

$$\frac{db}{dt} = \frac{i}{\hbar}[H_{DEG}, b] = \gamma b^{\dagger} \tag{5.5}$$

$$\frac{db^{\dagger}}{dt} = \frac{i}{\hbar}[H_{DEG}, b^{\dagger}] = \gamma b \tag{5.6}$$

so that the input-output relations for the operators of the degenerate mode can be obtained as,

$$b_{out} = \cosh(\gamma z/v)\, b_{in} + \sinh(\gamma z/v)\, b_{in}^{\dagger} \tag{5.7}$$

$$b_{out}^{\dagger} = \cosh(\gamma z/v)\, b_{in}^{\dagger} + \sinh(\gamma z/v)\, b_{in} \tag{5.8}$$

5.2 Elimination of Excess Noise in Direct Detection: Noiseless Amplification

We can now describe the process of degenerate parametric amplification, when the input signal is a coherent beam represented by a state $|\alpha\rangle_{in}$, corresponding to a mean number of input photons $N_{in} = |\alpha|^2$. The photon number at the output of the amplifier can be written as,

$$N_{out} = {}_{in}\langle\alpha|b_{out}^{\dagger}b_{out}|\alpha\rangle_{in} \tag{5.9}$$

Using the relationships (5.7) and (5.8) between the input and output operators, eq. (5.9) can be written as

$$\begin{aligned} N_{out} &= {}_{in}\langle\alpha|\cosh^2(\gamma z/v)b_{in}^{\dagger}b_{in} + \sinh^2(\gamma z/v)b_{in}b_{in}^{\dagger} + \\ &\quad + \cosh(\gamma z/v)\sinh(\gamma z/v)(b_{in}^{\dagger}b_{in}^{\dagger} + b_{in}b_{in})|\alpha\rangle_{in} = \end{aligned} \tag{5.10}$$

$$\begin{aligned} &= {}_{in}\langle\alpha|\cosh(2\gamma z/v)b_{in}^{\dagger}b_{in} + \sinh^2(\gamma z/v) + \\ &\quad + \frac{1}{2}\sinh(2\gamma z/v)(b_{in}^{\dagger}b_{in}^{\dagger} + b_{in}b_{in})|\alpha\rangle_{in} = \end{aligned} \tag{5.11}$$

$$= \sinh^2(\gamma z/v) + |\alpha|^2\left\{\cosh(2\gamma z/v) + \sinh(2\gamma z/v)\cos(2\phi)\right\} \tag{5.12}$$

As in the case of non-degenerate parametric amplification, the first term in eq. (5.12) corresponds to the spontaneously initiated parametric fluorescence and can be neglected when the the number of photons of the input is relatively large, ($N_{in} = |\alpha|^2 \gg 1$) as is the case when the parametric crystal is used as an amplifier. Thus, the output of the parametric amplifier can be written as,

$$N_{out} = N_{in}G_{\phi} \tag{5.13}$$

where

$$G_{\phi} = \cosh(2\gamma z/v) + \sinh(2\gamma z/v)\cos(2\phi) \tag{5.14}$$

$$= e^{2\gamma z/v}\cos^2\phi + e^{-2\gamma z/v}\sin^2\phi \tag{5.15}$$

is the gain of the amplifier, which depends on ϕ the phase of the input beam relative to the "internal reference" of the amplifier. Eqs. (5.13 - 5.14) indicate that a degenerate parametric amplifier performs a **phase-dependent amplification** of the input signal.

In particular, when the input signal beam is in-phase with the reference ($\phi = 0$) the gain is maximum,

$$G_0 = e^{2\gamma z/v} > 1 \tag{5.16}$$

and is greater than one, denoting a net amplification of the signal. On the other hand, if the input signal beam is in quadrature with the "internal reference" ($\phi = \pi/2$), the gain is less than one,

$$G_{\pi/2} = e^{-2\gamma z/v} = \frac{1}{G_0} < 1 \tag{5.17}$$

meaning that the output is de-amplified relative to the input. This phase dependance of the gain (and the possiblility of de-amplification that it entails) is an important result that underlies the nonclassical quantum statistics of parametric light. It must be stressed that parametric de-amplification of a coherent light beam is quite different from the classical attenuation of the beam. In parametric de-amplification energy is removed from only one quadrature of the coherent signal (while, at the same time, energy is introduced in the other quadrature), while in ordinary attenuation energy is removed from both quadratures because the beam is subject to a loss mechanism.

The fluctuations in the photon number at the output of the parametric amplifier can be calculated in terms of the variance

$$\Delta N_{out}^2 = {}_{in}\langle\alpha|b_{out}^\dagger b_{out} b_{out}^\dagger b_{out}|\alpha\rangle_{in} - N_{out}^2 \tag{5.18}$$

Using the input-output relation (5.7 - 5.8) this variance can be expressed as,

$$\begin{aligned}\Delta N_{out}^2 = {}_{in}\langle\alpha| \quad & C_2^2 b_{in}^\dagger b_{in} b_{in}^\dagger b_{in} + \tfrac{1}{4}S_2^2(b_{in}^\dagger b_{in}^\dagger + b_{in}b_{in})(b_{in}^\dagger b_{in}^\dagger + b_{in}b_{in}) \quad + \\ + \quad & \tfrac{1}{2}C_2S_2 b_{in}^\dagger b_{in}(b_{in}^\dagger b_{in}^\dagger + b_{in}b_{in}) + \tfrac{1}{2}C_2S_2(b_{in}^\dagger b_{in}^\dagger + b_{in}b_{in})b_{in}^\dagger b_{in} \quad |\alpha\rangle_{in} \\ & -N_{out}^2\end{aligned} \tag{5.19}$$

where we have used the shorthand notation $C_2 = \cosh(4\gamma z/v)$ and $S_2 = \sinh(4\gamma z/v)$.

Setting the creation and annihilation operators to normal order, and eliminating the terms that are cancelled by N_{out}^2, and neglecting the noise due to spontaneous parametric fluorescence, we have

$$\begin{aligned}\Delta N_{out}^2 &= |\alpha|^2\{C_2^2 + S_2^2 + 2C_2S_2\cos(2\phi)\} = &(5.20)\\ &= N_{in}\{\cosh(4\gamma z/v) + \sinh(4\gamma z/v)\cos(2\phi)\} &(5.21)\\ &= N_{in}\{G_0^2\cos^2\phi + \frac{1}{G_0^2}\sin^2\phi\} &(5.22)\end{aligned}$$

Thus, the fluctuation of the number of photons in the amplified signal is also phase-dependent. In particular, if the input signal is in-phase with the internal reference of the parametric amplifier ($\phi = 0$), then, the photon number fluctuations measured by the photodetector are maximum and are given by

$$(\Delta N_{out}^2)_{\phi=0} = N_{in}G_0^2 = N_{out}G_0 \tag{5.23}$$

Clearly, as is indicated by the last equality in eq. (5.23), the photon number fluctuations at the output of the amplifier have a variance that is larger by a factor of G_0 from that of the shot-noise fluctuations of a coherent beam of the same mean intensity N_{out}. Nevertheless, the SNR at the output of the amplifier is the same as at the input,

$$SNR_0 = \frac{(N_{in}G_0)^2}{N_{in}G_0^2} = N_{in} \tag{5.24}$$

since both the signal and the input noise are amplified by the same multiplicative factor, the gain. For in-phase parametric amplification, therefore, the noise figure of the amplifier is unity

$$NF_0 = 1 \tag{5.25}$$

implying that in-phase amplification does not degrade the SNR of an incoming signal. On the other hand, if the input signal is in-quadrature with the internal reference of the amplifier ($\phi = \pi/2$) the variance of the photon number fluctuations is,

$$(\Delta N_{out}^2)_{\phi=\pi/2} = \frac{N_{in}}{G_0^2} = \frac{N_{out}}{G_0} \tag{5.26}$$

and is, therefore, smaller by a factor $1/G_0$ than the Poissonian variance of a coherent beam of intensity N_{out}. Such a sub-Poissonian distribution of photon numbers is a nonclassical effect that cannot be attained by any of the traditional sources of light. Such sources produce, at best, coherent light which is subject to shot noise fluctuations. We note that in this configuration too, the SNR of the input signal is maintained

$$SNR_{\pi/2} = \frac{(N_{in}/G_0)^2}{N_{in}/G_0^2} = N_{in} \tag{5.27}$$

since both the input signal and the input shot noise are de-amplified by the same factor $1/G_0$. Thus, for in-quadrature de-amplification the noise figure of the degenerate parametric amplifier is also unity

$$NF_{\pi/2} = 1 \tag{5.28}$$

indicating that parametric de-amplification does not degrade the SNR. For intermediate phases, however, the NF is generally greater than one. Thus, parametric amplification does degrade the SNR of the input signal, except when the input signal is in-phase or in-quadrature with the internal reference of the amplifier.

The quantum noise characteristics of phase-dependent amplification in direct detection have been investigated experimentally [9, 10] for the degenerate mode of a type-II phase-matched parametric crystal, at the output of which, however, a polarizer was used to separate the signal and idler modes. In this configuration, only the signal photons were detected, while the idler photons are rejected, as shown schematically in Fig. 9.

The 50 % loss associated with the rejection of the idler mode at the output introduces noise in the measurement of the signal output. In this course we shall only give the equations for the intensity and noise of the signal output in the experimental configuration of Fig. 9, without a detailed derivation. The reader is referred to Refs. [9, 10] for the details of these experiments and a thourough theoretical analysis of their results. The number of photons received by the detector placed after the polarizer that separates the two polarizations is,

$$N_{out} = \frac{N_{in}}{2} G_\phi \tag{5.29}$$

where G_ϕ is the phase-dependent gain defined in eq. (5.14), while the factor of 1/2 comes from the 50 % loss due to the rejection of the idler polarization. The variance in the number of photons of the signal output is,

$$\Delta N_{out}^2 = \frac{N_{in}}{2} G_\phi \frac{G_0 + 1/G_0}{2} \tag{5.30}$$

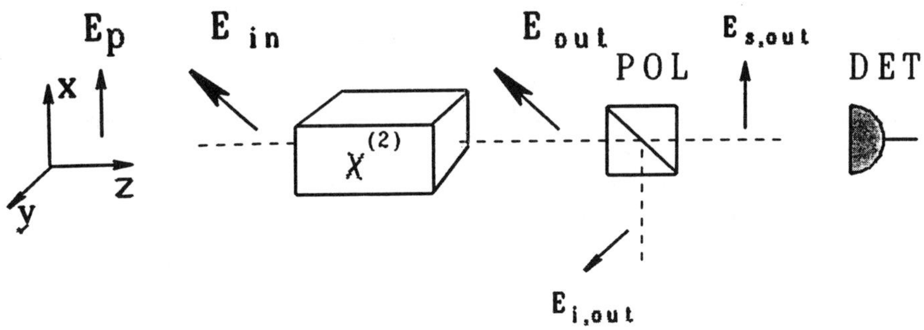

Figure 9. Experimental configuration for phase-dependent parametric amplification in a type-II nonlinear crystal with only signal detection. E_p is the pump field. As in Figure 8, the input field E_{in} is polarized at 45° so that it has a projection on both axes of the crystal. At the output, a polarizer separates the two polarization channels ("signal" and "idler") before detection.

which is clearly larger than the variance of a Poisson distribution. This noise variance gives a noise figure for the amplifier-plus-polarizer setup,

$$NF_\phi = \frac{G_0 + 1/G_0}{G_\phi} \tag{5.31}$$

This variation of the Noise Figure of a phase-dependent parametric amplifier has been observed experimetally [9] (see Fig. 6). It should be noted that, in the experimental conditions of Ref. [9], the noise figure of the parametric amplifier is clearly below that of a classical linear amplifier, for gains larger than 7 dB. For these gains, therefore, the phase-dependent parametric amplifier operates in the non-classical regime. It should be noted that, because of the 50 % loss, the noise figure in the amplifying quadrature ($\phi = 0$) is no longer unity (as in eq. 5.25) but is given by

$$NF_0 = 1 + \frac{1}{G_0^2} \tag{5.32}$$

Thus, if the amplifer is not pumped and its gain is unity the noise figure goes to 2, corresponding to the 50 % loss associated with the detection of one only channel. On the other hand, at large parametric gains the noise figure in the amplifying quadrature tends towards $NF_0 \rightarrow 1$ indicating nonclassical noiseless amplification. As for the case of a degenerate parametric amplifier, in-phase amplification does not degrade the SNR of an incoming signal.

5.3 Quantum Noise Reduction in Homodyne Detection: Squeezing

When a homodyne detector preceded by a degenerate parametric amplifier, the two quadrature operators of the detector (3.25) and (3.26), have to be transformed through the input-output relation of the parametric amplifier. Assuming, for simplicity, that the "internal phase reference" of the parametric amplifier is in-phase with the local oscillator that provides the phase reference for the homodyne detector, we can use the input-output relations (5.7) and (5.8). Thus, the transformed operators are,

$$D_1 = \frac{b_{out} + b_{out}^\dagger}{2} = e^{2\gamma z/v}\left(\frac{b_{in} + b_{in}^\dagger}{2}\right) = \tag{5.33}$$

$$= \sqrt{G_0}\left(\frac{b_{in} + b_{in}^\dagger}{2}\right) \tag{5.34}$$

and

$$D_2 = e^{-2\gamma z/v}\left(\frac{b_{in} - b_{in}^\dagger}{2i}\right) = \frac{1}{\sqrt{G_0}}\left(\frac{b_{in} - b_{in}^\dagger}{2i}\right) \tag{5.35}$$

where G_0 is the power gain for the amplification of the in-phase component of the input beam, defined in eq. (5.16).

When a coherent beam $|\alpha\rangle_{in}$ is amplified by the degenerate parametric amplifier and is then detected by the homodyne detector, the mean amplitude measured by the detector can be calculated by taking the expectation value of the transformed quadrature operators over the input state of the amplifier. Thus, in one quadrature

$$E_1 = {}_{in}\langle\alpha|D_1|\alpha\rangle_{in} = \sqrt{G_0}\,|\alpha|\cos\theta \tag{5.36}$$

the signal amplitude is amplified by a gain $\sqrt{G_0}$, while in the other quadrature

$$E_2 = {}_{in}\langle\alpha|D_2|\alpha\rangle_{in} = \frac{1}{\sqrt{G_0}}\,|\alpha|\sin\theta \tag{5.37}$$

the signal amplitude is de-amplified by a gain $1/\sqrt{G_0}$.

The range of possible outcomes in these measurements is given by the variance of the two quadratures, which can be calculated as,

$$\Delta E_1^2 = {}_{in}\langle\alpha|D_1^2|\alpha\rangle_{in} - E_1^2 = \tag{5.38}$$

$$= \frac{G_0}{4}\,{}_{in}\langle\alpha|\; b_{in}^\dagger b_{in}^\dagger + b_{in}b_{in} + b_{in}^\dagger b_{in} + b_{in}b_{in}^\dagger \;|\alpha\rangle_{in} - E_1^2 = \tag{5.39}$$

$$= \frac{G_0}{4} \tag{5.40}$$

and

$$\Delta E_2^2 = {}_{in}\langle\alpha|D_2^2|\alpha\rangle_{in} - E_2^2 = \tag{5.41}$$

$$= -\frac{1}{4G_0}\,{}_{in}\langle\alpha|\; b_{in}^\dagger b_{in}^\dagger + b_{in}b_{in} - b_{in}^\dagger b_{in} - b_{in}b_{in}^\dagger \;|\alpha\rangle_{in} - E_2^2 = \tag{5.42}$$

$$= \frac{1}{4G_0} \tag{5.43}$$

indicating that the quantum noise in quadrature 1 (which is in-phase both with the local oscillator and with the internal reference of the amplifier) is amplified, while in

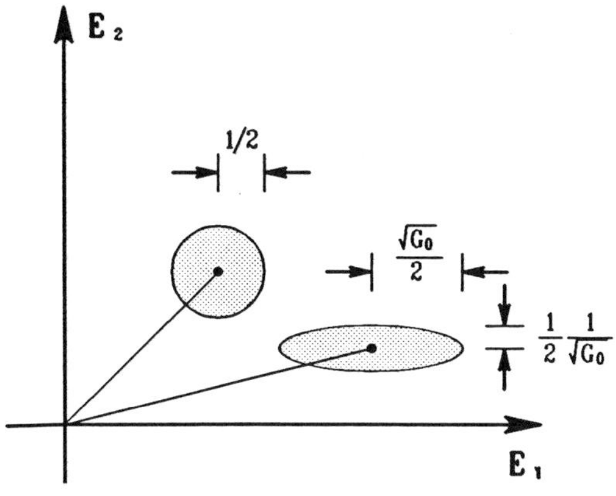

Figure 10. Phasor diagram for phase-dependent amplification.

the other quadrature it is de-amplified. The noise spot is thus deformed from a disk to an ellipse, but its area remains unchanged: after parametric amplification, the noise in the two quadratures still minimizes the Heisenberg inequality

$$(\Delta E_1)(\Delta E_2) = \frac{\sqrt{G_0}}{2}\frac{1}{2\sqrt{G_0}} = \frac{1}{4} \tag{5.44}$$

The fact that the amplified beam remains a minimum-uncertainty wavepacket indicates that the amplifier adds no noise to the signal, but simply redistributes the quantum noise between the two quadratures while keeping the noise power to the minimum value dictated by the Heisenberg uncertainty relation. Because of the particular shape of the noise spot of these quantum-limited states, they are called **squeezed states**.

In practical terms, the fact that the extension of the noise spot in quadrature 2 is de-amplifed to a value below 1/2, indicates that a homodyne measurement of that quadrature will produce a quantum noise level which is below the level of shot noise. This is clearly a nonclassical noise distribution since, even with a perfectly stabilized coherent lase beam, it is impossible to reduce the photon noise below the shot-noise level. In the other quadrature, on the other hand, the quantum noise will be much higher than shot noise.

The elimination of amplification noise in degenerate parametric amplification stems, essentially, form the fact that in degenerate parametric amplification ther is no bath of unobserved modes to introduce noise in the amplifed signal. All the photons that are produced by the parametric amplifier contribute to the increase of both the amplitude and the fluctuations of the amplified output. Furthermore, as degenerate parametric emission proceeds through the emission of pairs of twin photons in the single output mode, the correlations that exist between the twins are manifested as a reduction (or squeezing) in the quantum noise, below the shot-noise level.

The "squeezing" of quantum noise in one quadrature below the shot noise limit has been observed experimentally by performing a spectral analysis of the differential photocurrent of homodyne detection, as a function of the relative phase of the amplified signal and the local oscillator. The noise of the photocurrent as a function of the relative

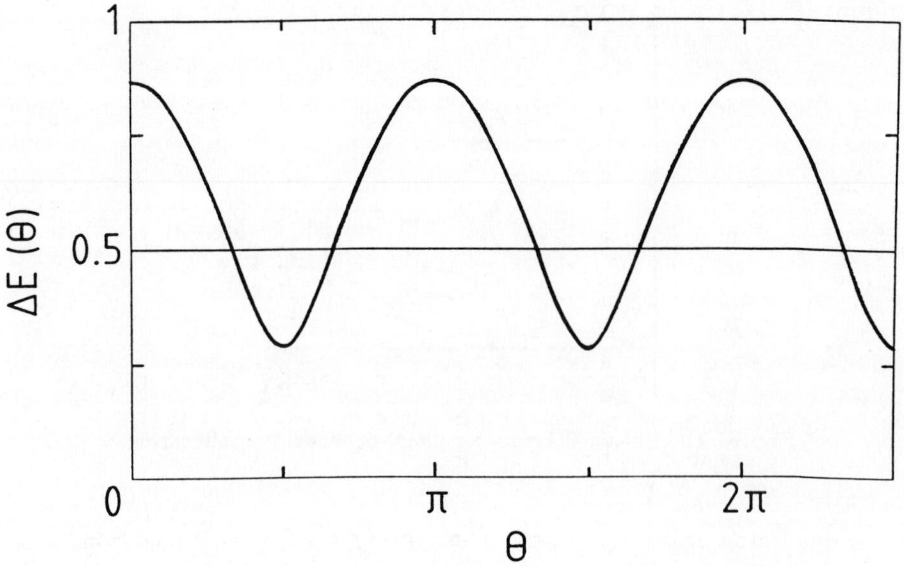

Figure 11. Phase dependence of quantum noise in parametric amplification of gain 5 dB. The shot noise level corresponds to the horizontal line $\Delta E = 0.5$. Squeezing occurs whenever the quantum noise level of the parametric light drops below the shot-noise level. In this figure, in quadrature 1 ($\theta = 0,\ \pi,\ 2\pi$) there is excess noise ($\Delta E = 0.88 > 0.5$) while in quadrature 2 ($\theta = \pi/2,\ 3\pi/2$) the noise is squeezed ($\Delta E = 0.28 < 0.5$).

phase is,

$$\Delta E^2(\theta) = \Delta E_1^2 \cos^2\theta + \Delta E_2^2 \sin^2\theta = \tag{5.45}$$

$$= \frac{1}{4}\{G_0 \cos^2\theta + \frac{1}{G_0}\sin^2\theta\} \tag{5.46}$$

Squeezing, then, is manifested as a decrease of the quantum noise level in one quadrature, below a pre-calibrated level given by shot noise, as in Fig. 11.

Squeezing was observed experimentally for the first time in 1985 [11]. The strongest quadrature squeezing measured to date is approximately 4 dB that is, almost 60 % below the level of shot noise [12]. However, in these experiments, the measurement was limited essentially by the efficiency of the experimental system, and when correcting for residual losses in the material components of the setup it was estimated that the actual degree of squeezing attained by the parametric amplifier was of the order of 10 dB, that is, a factor of 10 below the shot-noise level.

6. POTENTIAL USES FOR NONCLASSICAL PARAMETRIC AMPLIFICATION

The modification of the quantum statistics upon parametric amplification of a "macroscopic" optical signal has suggested the possibility of "engineering" the quantum noise of light to produce light with a nonclassical noise distribution such that it can overcome the limitations imposed by shot noise on optical measurements. This possibility is relevant in applications in which the improvement of optical and electronic components has permitted the elimination of "technical noise" and has led to operation in the shot-noise-limited regime.

It should be noted, however, that the very feature that gives parametric amplification its nonclassical characteristics, namely the phase-dependence of its gain, is precisely the one that gives rise to technical difficulties in the implementation of nonclassical amplification. Efficient parametric amplification requires a very high degree of control of the signal phase relative to the pump, as well as elimination of losses and of mode-mismatches. At present, a reasonable control of all these parameters can be achieved only in specially prepared experimental setups. Even in these situations, however, the reduced efficiencies of the available components severely limit the observation of the nonclassical effects of parametric amplification, as was mentioned in Section 5 for the case of the squeezing experiments [12]. And, of course, in most "real life" situations, outside the laboratory, even this limited control is very difficult to achieve. Thus, at present, the technical difficulties of implementing parametric amplification outweight the practical advantages that can be attained by reducing the quantum noise of a measurement below the shot noise limit. However, the constant improvement of optical hardware – lower loss optical fibers, higher quantum efficiency detectors and emitters, stronger optical nonlinearities – bring operational systems closer to the quantum noise limit and make them more compatible with the operation of parametric amplification. In this Section, we shall examine briefly how the implementation of phase-dependent amplification can improve the operation of some present-day systems, if the technical difficulties associated with the control of the signal phase can be mastered. We shall examine more particularly the implications of the two aspects of nonclassical parametric amplification that we examined in Section 5, namely noiseless amplification and squeezing.

6.1 Noiseless Amplification

The most straight-forward use of phase-dependent amplification would consist of replacing present-day linear optical amplifiers (that are used, for example, as in-line repeaters in optical communications or as pre-amplifiers before detection) and thus avoid the 3 dB penaly in the SNR degradation upon amplification.

In addition to this obvious improvement, this situation offers an additional advantage that arises from the fact that in a multi-component system, in which an amplifier is followed by other noisy elements, the noise figure of the overall system is determined essentially by the noise figure of the leading amplifier. In particular, if the leading amplifier has gain G_1 corresponding to a noise figure NF_1, while the optical components following the amplifier are characterized by a noise figure NF_2, the noise figure of the overall setup (see appendix) is

$$NF_{1+2} = NF_1 + \frac{1}{G_1}(NF_2 - 1) \tag{6.1}$$

For large gains of the leading amplifier ($G_1 \gg 1$), the noise contribution of the other components to the overall noise figure can become negligible, so that only the noise figure of the leading amplifier is important. Thus, if the leading amplifier is noiseless ($NF_1 = 1$), the overall setup will also be essentially noiseless.

The significance of the noise figure of the leading amplifier can be seen most clearly if the noise figure of the elements following the leading amplifier is $NF_2 < 2$. In that case, use of a linear phase independent amplifier will only degrade further the SNR of an incoming signal, since such an amplifier will bring the overall noise figure to $NF_{1+2} = 2$. On the other hand, if the leading amplifier is noiseless ($NF_1 = 1$) then, the following elements will not degrade the SNR and the signal will reach its destination with no extra noise.

An interesting use of noiseless amplification is in the realization of an optical tap, which is a device that can channel a small fraction of a signal-carrying beam towards a terminal or distribute an optical signal to two separate terminals. A paradigm for the coupler performing the signal distribution is a 50/50 beamsplitter. Since the beamsplitter attenuates a coherent beam of initial intensity N_{in} to $N_{in}/2$, while maintaining its Poissonian character, it degrades the SNR by a factor of 2. In other words, a 50/50 beamsplitter has a noise figure of $NF_2 = 2$. Clearly, if the beamsplitter is preceded by a noiseless amplifier of high gain, then the noise figure of the overall setup consisting of the amplifier plus the beamsplitter will be $NF_{1+2} = 1$, meaning that both terminals will receive the optical signal with no degradation in the signal-to-noise ratio.

6.2 Squeezed Light

The partition noise of an optical tap can also be reduced by introducing into the "unused port" of the beamsplitter a very weak beam of squeezed light, corresponding to a phasor in which the noise spot is an ellipse centered at the origin, so that the coherent amplitude of the squeezed beam is practically zero. The operation of the squeezed light assisted beamsplitter can be understood by examining the noise characteristics of the two output beams in which the information-carrying signal is mixed with the auxiliary squeezed beam. Clearly, the two output beams will have one quadrature in which photon noise is lower than shot noise. Thus, if detection is performed only in that quadrature, the SNR of the two output beams can be made as high as that of the incoming signal-carrying beam, indicating that squeezed light assisted optical tapping can be essentially noiseless [13].

The principle of squeezed light assisted beamsplitting can be used also to enhance the sensitivity of optical interferometers, as these instruments can be described as particular geometrical arraangements of beamsplitters. Thus, two independent experiments have shown that if squeezed light is introduced in the "unused port" of a Mach-Zehnder [14] or a polarization [15] interferometer, the precision in the measurement of the relative optical phase of the two arms of the interferomenter is enhanced beyond the limit imposed by shot noise. The shot noise limit for a phase measurement is

$$\Delta\phi = 1/\sqrt{N} \tag{6.2}$$

where N is the number of photons arriving on the detector during the measurement. In these experiments, improvements of the SNR by a factor of 2 have been obtained. Further noise reduction would require essentially technical improvements of the system efficiency (elimination of losses and better mode-matching) as the estimated squeezing produced by the parametric amplifier is of the order of a factor of ten.

The rapid technical improvements of optical communication systems bring the closer to the "quantum limit", where error rates and channel capacity are determined essentially by the shot noise level of the information-carrying signal. In this context, use of squeezed light to carry information seems to be an alternative that is not subject to the limitations imposed by the shot noise. However, squeezed light communications over long distances will suffer from a practical problem that arises from attenuation and mode matching losses that the signal undergoes during its propagation. The random deletion of photons from the propagating signal destroys squeezing, as it brings back Poissonian statistics. Squeezed light communication can conceivably be implemented in compact networks, in which fiber losses are very small so that the nonclassical statistics of the light signals can be maintained. Several theoretical papers analyzing squeezed light communications have appeared [16, 17], indicating that the cannel capacity of a squeezed-light system (i.e. the amount of information it can transmit) is increased only by a factor of 2 with respect to the capacity of a coherent system. At the same time, the number of photons necessary to define a bit (i.e. a 0 or a 1 in binary digital communications) with a 10^{-9} error rate is lower by only 6 photons when squeezed light is used, from the value of about 45 photons needed to define a bit for coherent light [18]. At present, therefore, this modest improvement in the sensitivity of detection does not seem to be worth the price of the increased complexity of squeezed light systems, especially since similar improvements in the error rates can be made by using more sophisticated coding techniques. However, as optical communications evolve towards higher capacities, higher bit rates and lower error rates, we may arrive at a point at which the enhancement produced by squeezed light may well be worth the effort.

APPENDIX: THE NOISE FIGURE OF A COMPOSITE SYSTEM

In order to derive the formula for the noise figure of a setup in which a leading amplifier is followed by another noisy optical component, we shall examine the special case in which a classical PIA of gain G_1 and noise figure

$$NF_1 = 2 - \frac{1}{G_1} \tag{A.1}$$

is followed by a second PIA of gain G_2 and noise figure NF_2, related to the gain in a way analogous to eq. (A.1). The gain of the overall setup is

$$G_{1+2} = G_1 G_2 \tag{A.2}$$

and the corresponding noise figure is,

$$NF_{1+2} = 2 - \frac{1}{G_{1+2}} = 2 - \frac{1}{G_1 G_2} = 2 - \frac{1}{G_1} + \frac{1}{G_1}\left(1 - \frac{1}{G_2}\right) \tag{A.3}$$

Thus, using eq. (A.1), the noise figure of the compositte system can be written in terms of the noise figures of its two components as,

$$NF_{1+2} = NF_1 + \frac{1}{G_1}(NF_2 - 1) \tag{A.4}$$

This formula is valid for all situations in which a leading amplifier (linear or phase-dependent) is followed by another optical component (amplifier or attenuator), even though in this appendix the derivation was carried out for the special case of two linear amplifiers.

REFERENCES

1. L.D. Landau and E.M. Lifshitz, *Course of Theoretical Physics* Vol. 1: Mechanics, Ch. 5 (Pergamon Press, Oxford, 1975).
2. R.L. Byer, "Parametric Oscillators and Nonlinear Materials" in *Nonlinear Optics*, ed. by P.G. Harper and B.S. Wherrett (Academic Press, New York, 1977).
3. J.G. Rarity and P.R. Tapster, J. Opt. Soc. Am. **6**, 1221 (1989).
4. Z.Y. Ou, X.Y. Zou, L.J. Wang, and L. Mandel, Phys. Rev. A **42**, 2957 (1990).
5. Z.Y. Ou and L. Mandel, Phys. Rev. Lett. **61**, 50 (1988);
   P. Grangier, M.J. Potasek, and B. Yurke, Phys. Rev. A **38**, 3132 (1988);
   J.G. Rarity and P.R. Tapster, Phys. Rev. Lett. **64**, 2495 (1990);
   J.G. Rarity and P.R. Tapster, Phys. Rev. A **41**, 5139 (1990);
   J.G. Rarity, P.R. Tapster, E. Jakeman, T. Larchuk, R.A. Campos, M.C. Teich, and B.E.A. Saleh, Phys. Rev. Lett. **65**, 1348 (1990).
6. R.J. Glauber, Phys. Rev. **130**, 2529 (1963).
7. D.C. Burnham and D.L. Weinberg, Phys. Rev. Lett. **25**, 84 (1970).
8. I. Abram, R.K. Raj, J.L. Oudar, and G. Dolique, Phys. Rev. Lett. **57**, 2516 (1986).
9. J.A. Levenson, I. Abram, P. Grangier, and T. Rivera, J. Opt. Soc. Am. (in press).
10. J.A. Levenson, I. Abram, T. Rivera, P. Fayolle, J.C. Garreau, and P. Grangier, Phys. Rev. Lett. **70**, 267 (1993).
11. R.E. Slusher, L.W. Hollberg, B. Yurke, J.C. Mertz, and J.F. Valley, Phys. Rev. Lett. **55**, 2409 (1985).
12. L.A. Wu, H.J. Kimble, J.L. Hall, and H. Wu, Phys. Rev. Lett. **57**, 2520 (1986).
13. J.H. Shapiro, Opt. Lett. **5**, 351 (1980).

14. M. Xiao, L.A. Au, and H.J. Kimble, Phys. Rev. Lett. **59,** 278 (1987).

15. P. Grangier, R.E. Slusher, B. Yurke, and A. LaPorta, Phys. Rev. Lett. **59,** 2153 (1987).

16. H.P. Yuen and J.H. Shapiro, IEEE Trans. Inform. Theory, **IT-24,** 657 (1978); also, **IT-26,** 78 (1980);
J.H. Shapiro, H.P. Yuen and J. Machado Mata, IEEE Trans. Inform. Theory, **IT-25,** 179 (1979).

17. Y. Yamamoto, and H.A. Haus, Rev. Mod. Phys. **58,** 1001 (1986).

18. R.E. Slusher and B. Yurke, J. Lightwave Technology **8,** 466 (1990).

TWO- AND THREE-PHOTON SPECTROSCOPY OF SOLIDS

Dietmar Fröhlich

Institut für Physik
Universität Dortmund
44221 Dortmund, Germany

ABSTRACT

Two- and three-photon spectroscopy is very useful for the study of electronic properties of solids. After a rather general discussion of the "additional degrees of freedom" of these nonlinear techniques as compared to linear spectroscopy, the theoretical background of two-photon absorption and the polarization dependence for the special case of cubic symmetry are discussed. The polariton concept is briefly outlined. It is shown that the Maxwell boundary conditions have to be taken into account in the analysis of special cases in k-space spectroscopy. Several examples of two- and three-photon spectra in external fields will be presented. The effect of uniaxial stress and the Zeeman and Stark effect are studied in CuCl. As a further example experiments on two-photon magnetooptics of Landau levels in GaAs-AlGaAs multiple quantum wells are presented. Detailed information on experimental setups using nanosecond and picosecond tunable lasers will be given.

I. INTRODUCTION

Nonlinear optical techniques are intensively used for the study of electronic properties of solids. The electronic structure can be deduced from resonances, which are measured with various spectroscopic methods. The study of dynamical properties has led to the development of ultra-short laser pulses in the femtosecond range. They allow to investigate ultra-fast dephasing processes. Dynamic processes can also be measured in the frequency domain, which then leads to a demand for lasers of high spectral resolution and frequency stability. Various aspects of these techniques in the time and frequency domain are discussed in the contribution by Macfarlane [1].

The mere spectroscopic aspects of nonlinear spectroscopy like two-photon absorption were first treated theoretically by Maria Göppert-Mayer [2] more than 60 years ago. More than 30 years later, the experiments of Hopfield, Worlock and Park [3] opened the field of two–photon spectroscopy. They studied two–photon transitions in alkali halides. For a review of the early two–photon measurements we refer to Mahr [4]. The first three–photon spectra of excitonic transitions were again measured in alkali halides. Beerwerth and Fröhlich [5] were able to resolve resonances on the upper polariton branch and the longitudinal exciton of the lowest 1S excitons in KI and CsI by three–photon excitation spectroscopy (TPS). The coherent method of three–photon difference–frequency generation (TP–DFG) was applied to study resonances on the lower polariton branch [6]. The various methods of "spectroscopy in k space" allow a detailed study of the polariton dispersion also in non–cubic crystals [7]. Since in many cases the two– and three–photon resonances are very narrow (down to $\Delta E \sim 0.1$ meV), measurements in external fields lead to clearly resolved splittings and/or shifts of lines, which are then used to deduce fundamental parameters like g values, deformation potentials, exchange energy values, effective masses. Examples of nonlinear spectra of semiconductors and insulators in external fields are presented in this contribution.

In chapter II the "additional degrees of freedom" are discussed in detail. In the following more theoretical chapter expressions for the two–photon absorption constant and two–photon polarization selection rules for linearly polarized light are derived for cubic crystals. In chapter IV the dispersion relations for transverse polaritons and the longitudinal exciton are deduced from a classical model and the wave equation for the electric field. For the detailed analysis of some nonlinear techniques as, e.g., three–photon sum–frequency generation (TP–SFG) it turns out that the Maxwell boundary conditions, known from classical linear optics, have to be taken into account. These boundary conditions even lead to a three–photon band in reflection, as will be discussed in chapter V. In chapter VI and VII various examples of two– and three–photon experiments in different substances and in external fields will be presented followed by a chapter on experimental setups.

II. ADDITIONAL DEGREES OF FREEDOM IN NONLINEAR SPECTROSCOPY

One rather obvious motivation to do nonlinear spectroscopy is the fact that in optical experiments, where more than one photon is necessary to induce an electronic transition, there are "additional degrees of freedom" as compared to linear spectroscopy. The specific use of these degrees of freedom leads to specific information about electronic properties of solids, which are not (easily) accessible to classical spectroscopic experiments (one–photon absorption).

As known from classical mechanics and quantum mechanics general symmetry considerations lead to conservation laws and selection rules:

1. From homogeneity in time one easily derives the energy conservation law. For a one–photon process between a ground state E_g and an excited state E_f we get

$$E_f - E_g = \pm\hbar\omega_1 \ , \tag{1}$$

where + and – refer to absorption or emission processes, respectively. For a two–photon process we get

$$E_f - E_g = \pm\hbar\omega_1 \pm \hbar\omega_2 \ , \tag{2}$$

where the additional degrees of freedom are obvious. (++) and (– –) processes refer to two–photon absorption and emission, respectively. Mixed signs (+–) and (–+) are known as Raman processes, where one can distinguish between Stokes– (photon of higher energy with + sign) and anti–Stokes–Raman effect (photon of higher energy with – sign). It is clear that for three–photon processes there are even more possibilities:

$$E_f - E_g = \pm\hbar\omega_1 \pm \hbar\omega_2 \pm \hbar\omega_3 \ . \tag{3}$$

2. From isotropy of space, which means that the system under consideration has full rotational symmetry, one derives the law of conservation of angular momentum. For electric dipole transitions this leads in atomic physics to the well known selection rules for one photon transitions

$$\Delta l = \pm 1, \ \Delta m = 0, \pm 1 \ , \tag{4}$$

where l refers to the angular quantum momentum number and m to its component on the quantization axis. For one–photon processes one gets the well known selection rules

$$\text{s}\leftrightarrow\text{p, p}\leftrightarrow\text{s and d, d}\leftrightarrow\text{p and f, etc. ,} \tag{5}$$

where s, p, d, f, refer to angular momentum quantum numbers for a one–electron system like hydrogen (l = 0, 1, 2, 3). By applying these rules twice one immediately derives dipole selection rules for two–photon processes

$$\begin{gathered} \Delta l = 0, \pm 2, \ \Delta m = 0, \pm 1, \pm 2 \ , \\ \text{s}\leftrightarrow\text{s and d, p}\leftrightarrow\text{p and f, etc.} \end{gathered} \tag{6}$$

By comparing Eq. (4) to (6) one immediately recognizes that one– and two–photon processes between the same levels exclude each other. Extending the selection rules to three–photon processes one gets

$$\begin{gathered} \Delta l = \pm 1, \pm 3, \ \Delta m = 0, \pm 1, \pm 2, \pm 3 \ , \\ \text{s}\leftrightarrow\text{p and f, p}\leftrightarrow\text{s, d and g, etc.} \end{gathered} \tag{7}$$

By comparing Eq. (4) to (7) one recognizes that two– and three–photon processes exclude each other and that one–photon transitions are simultaneously three–photon allowed but not vice versa. As an example for one– and two–photon transitions we present in Fig. 1 experimental results on the classical exciton system in Cu_2O. In Cu_2O P excitons are one–photon and S,D excitons are two–photon allowed. The weak 1S line in the one–photon spectrum is due to a quadrupole transition. For details we refer to Ref. [8].

In solids we do not have the full rotational symmetry, since only rotations by discrete angles (e.g. 90°, 60°) leave the crystal invariant. We expect that the states cannot (in general) be classified by an angular momentum quantum number. Nevertheless one can again introduce quantum numbers, which are derived from very general group–theoretical considerations. In a cubic crystal (O_h symmetry) we get quantum numbers

$$\Gamma_1^\pm(1), \ \Gamma_2^\pm(1), \ \Gamma_3^\pm(2), \ \Gamma_4^\pm(3), \ \Gamma_5^\pm(3) \ . \tag{8}$$

$\pm$ refer to the symmetry of states with respect to inversion (parity of state). The numbers in parentheses give the degeneracy of the state. Since the highest degeneracy (without spin) is 3, we expect that fivefold d–states will split in a cubic crystal field. It turns out that a d–state splits into a $\Gamma_3^+(2)$ and $\Gamma_5^+(3)$ state. Since crystals of O_h symmetry have a center of inversion, parity is a good quantum number. As we will discuss in detail later one– and two–photon transitions exclude each other in this case.

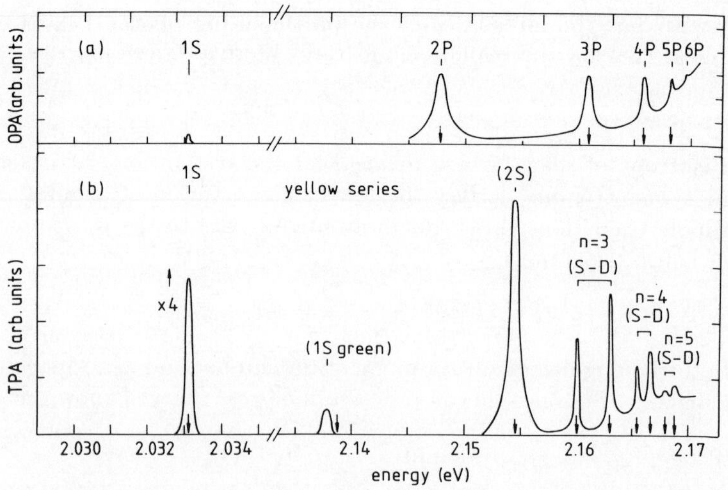

Figure 1. One– and two–photon spectra of Cu_2O. (a) One–photon absorption (OPA) at 4.2 K. (b) Two–photon absorption (TPA) at 4.5 K. The arrows indicate calculated exciton energies [8].

Again all one–photon transitions are three–photon allowed but not vice versa. In cubic crystals there is no polarization dependence for a one–photon dipole transition. For two– and three–photon transitions, however, there is a very pronounced polarization dependence, which allows to distinguish between different excited states.

3. From homogeneity of space one derives the law of conservation of momentum. In crystalline solids there is no homogeneity of space but (only) translational symmetry. Nevertheless, there is a conservation law for the quasi–momentum $\hbar\vec{k}$. As in the case of energy conservation we get for one–, two– and three–photon transitions

$$\begin{aligned} \hbar\vec{k}_f - \hbar\vec{k}_g &= \pm\hbar\vec{k}_1 \ , \\ \hbar\vec{k}_f - \hbar\vec{k}_g &= \pm\hbar\vec{k}_1 \pm \hbar\vec{k}_2 \ , \\ \hbar\vec{k}_f - \hbar\vec{k}_g &= \pm\hbar\vec{k}_1 \pm \hbar\vec{k}_2 \pm \hbar\vec{k}_3 \ . \end{aligned} \tag{9}$$

$\hbar\vec{k}_i$ (i = 1,2,3) refer to the momenta of the photons in the crystal, which we will later term "polaritons". Once the photon is in the crystal, its k vector is determined by the dispersion relation. For one–photon processes energy ($\hbar\omega$) and k vector in the crystal are firmly coupled. In two– and three–photon processes, however, one has the unique possibility to tune the total energy and the total k vector in certain limits independently of each other. One can thus do "spectroscopy in k space". This additional degree of freedom of nonlinear spectroscopy is caused by the fact that the photon energies add up as scalars, whereas their momenta add up as vectors.

4. Making use of the polarization dependence in two– and three–photon processes one can excite transverse or longitudinal states in the crystal with proper combinations of k vectors and orientations with respect to the crystalline axes. One can thus study elementary excitations in the solid, where the total polarization $\vec{P}$ is either perpendicular or parallel to the total k vector. The excited states are termed transverse polaritons or longitudinal excitons, respectively. Without perturbation one can excite in linear spectroscopy only states with a transverse polarization.

5. In the previous the intensities of the participating photon beams were of no influence on the nonlinear spectra. The intensities had to be large enough to enable the detection of the (in most cases) rather weak signals. The intensities had to be small enough to prevent any real or virtual population of states. Even the coupling of empty states by an intense laser field might lead to drastic changes of linear and nonlinear spectra as measured by a weak test field. With this kind of experiments the optical Stark effect in solids can be studied [9, 10]. In this contribution we are only concerned with nonlinear optical spectroscopy, where high intensity lasers are still a weak probe of the system. Renormalization effects due to high intensities are neglected.

III. THEORY OF TWO–PHOTON TRANSITIONS

III.A. Two–Photon Absorption Constant

After the rather general discussion in the last chapter of the additional degrees of freedom in nonlinear spectroscopy we will now write down explicitly an expression for the two–photon absorption constant $\alpha^{(2)}(\omega_1)$ for a tunable beam of photon energy $\hbar\omega_1$ which is induced by a high intensity laser beam of photon energy $\hbar\omega_2$. From this expression one can deduce two–photon selection rules which allow to discriminate between different excited states by proper choice of the polarization directions of the incoming laser beams (chapter III.B.).

As was already shown by Göppert–Mayer [2] the two–photon transition probability can be calculated from time–dependent second–order perturbation theory. The interaction Hamiltonian of the particle with a classical electro–magnetic field is derived as usual by replacing the kinetic energy term by $\frac{1}{2m}(\vec{p}-e\vec{A})^2$ where $\vec{A}=\vec{A}_1+\vec{A}_2$ is the total vector potential for both photons. For transverse fields one gets the terms $\frac{e}{m}\vec{A}\cdot\vec{p}$ and $\frac{e^2}{2m}\vec{A}^2$ which have to be considered in a perturbational approach. In the dipole approximation ($\vec{k}\cdot\vec{r}\ll 1$) matrix elements of the A^2–term vanish. In the quadrupole approximation, however, there might be two–photon transitions due to the A^2–term already in first–order perturbation theory. Fröhlich [11] has found out that for a typical two–photon transition in alkali halides the first–order contribution of the A^2 is smaller by about 10^4 than the second–order contribution of the $\vec{A}\cdot\vec{p}$–term. From the second–order transition probability one can then calculate a two–photon absorption constant [12]

$$\alpha^{(2)}(\omega_1)=C\sum_f\left|\sum_i\left\{\frac{\langle \mathrm{f}|\vec{e}_2\cdot\vec{p}|\mathrm{i}\rangle\langle \mathrm{i}|\vec{e}_1\cdot\vec{p}|\mathrm{g}\rangle}{E_i-E_g-\hbar\omega_1}+\frac{\langle \mathrm{f}|\vec{e}_1\cdot\vec{p}|\mathrm{i}\rangle\langle \mathrm{i}|\vec{e}_2\cdot\vec{p}|\mathrm{g}\rangle}{E_i-E_g-\hbar\omega_2}\right\}\right|^2\cdot \delta(E_f-E_g-(\hbar\omega_1+\hbar\omega_2))\,, \tag{10}$$

$$C=(8\pi^3\hbar e^2N_2)/(c\,m^4n_1n_2\omega_1\omega_2)\,.$$

As indicated by the δ function we consider a transition from a ground state g to discrete final states f, which might be degenerate. The sum over i has to take into account all eigenstates of the system, which lead to non–vanishing dipole matrix elements. $\vec{e}_1$ and $\vec{e}_2$ refer to the polarization vectors of both photon beams, n_1 and n_2 to their refractive indices. As expected the absorption constant for photons of energy $\hbar\omega_1$ is proportional to the number density N_2 of photons of energy $\hbar\omega_2$. For transitions to continuum states one has to consider the density of final states instead of the δ function. The resonance denominators in Eq. (10) are typical for second–order processes. One expects a resonance enhancement if the photon energy of one photon (e.g. $\hbar\omega_1$) is close to an intermediate level E_i. In this case one can to a good approximation replace the sum over intermediate levels just by one level in order to calculate $\alpha^{(2)}(\omega_1)$. This was done by Mahan [13] for excitons in alkali halides with use of the so-called two-band model. In this model the first photon induces a virtual transition (no energy conservation) to S excitons composed of holes in the uppermost valence band and electrons in the conduction band, whereas the second photon just changes the envelope quantum number from S to P. For the sum of the photon energies ($\hbar\omega_1 + \hbar\omega_2$) one expects energy conservation independently of the individual choice of photon energy. Energy conservation and the resonance enhancement are demonstrated in Fig. 2, where the two-photon spectrum of KI is measured with three different laser beams and a tunable source (flashlamp and monochromator).

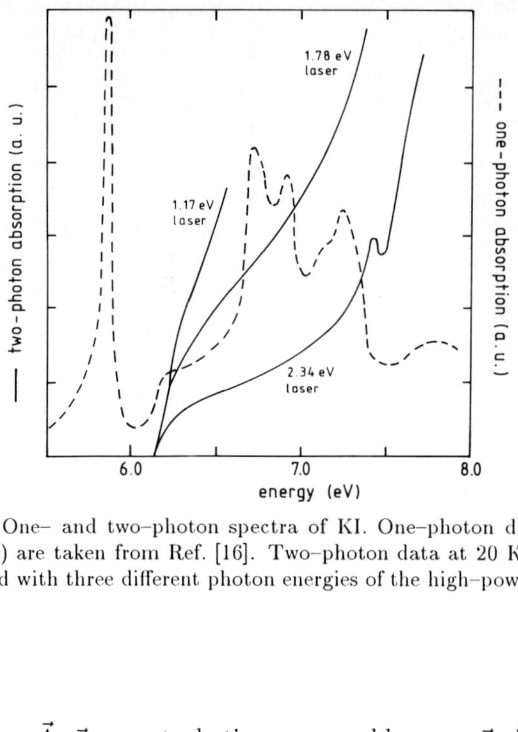

Figure 2. One– and two–photon spectra of KI. One–photon data at 10 K (dashed line) are taken from Ref. [16]. Two–photon data at 20 K (solid line) are measured with three different photon energies of the high–power laser [11].

Instead of using $-\vec{A}\cdot\vec{p}$ as perturbation, one could use $-e\vec{r}\cdot\vec{E}$ which corresponds to a transformation from the velocity gauge to the length gauge. The physical results should of course be independent of the choice of gauge. This point is extensively studied in the literature [14]. It is shown that the choice of gauge might be of great relevance if for computational reasons the sum over the complete set of intermediate states is transferred to a few states which are e.g. favored because of a small resonance denominator. Recently Pasquarello and Quattropani [15] have calculated two–photon transition rates to excitons in quantum wells by taking explicitly into account the sum over all the intermediate states, using a variational procedure. They also discuss the important point, how to get gauge–invariant results.

III.B. Polarization Selection Rules

As pointed out already in the last chapter selection rules for multi–photon transitions can be gained from symmetry considerations. As known from the classical Raman effect there is a characteristic polarization dependence even in cubic crystals, which allows the symmetry assignment of the phonons or electronic levels considered. For two–photon absorption one expects in general the same selection rules. Inoue and Toyozawa [17] were the first to derive the detailed polarization dependence for two–photon processes for all 32 point groups. Bader and Gold [18] extended the analysis to double–valued representations and Denisov and Makarov [19] took into account that levels which are simultaneously one– and two–photon allowed are split into transverse and longitudinal states, which in the case of excitons are called transverse polaritons and longitudinal excitons, respectively. Doni et al. [12] have derived expressions for the two–photon absorption coefficient in crystals using explicitly the Wigner–Eckart theorem for finite symmetry groups. They showed that the transition rate can be separated into a geometrical factor, which is solely determined by the polarization vectors of the photons, and a dynamical factor which contains the transition matrix elements and the photon frequencies. As discussed in detail by Pasquarello and Quattropani [20] in crystals of lower than cubic symmetry it is not always possible to express the transition rate as a simple product of dynamical and geometrical factors. Pasquarello and Quattropani have given a rather general symmetry analysis of the multiphoton transition rate. They study in detail the polarization dependence for the point groups O_h, T_d and C_{6v}. They include in their analysis of the transition rate the transformation properties under permutation of photon indices. For the special case of two–photon absorption in O_h one can easily demonstrate this point by considering Eq. (10). The two–photon absorption coefficient $\alpha^{(2)}(\omega_1)$ is proportional to

$$\left| \sum_i \left\{ \frac{\langle \mathrm{f}|\vec{e}_1 \cdot \vec{p}|\mathrm{i}\rangle\langle \mathrm{i}|\vec{e}_2 \cdot \vec{p}|\mathrm{g}\rangle}{E_i - E_g - \hbar\omega_2} + \frac{\langle \mathrm{f}|\vec{e}_2 \cdot \vec{p}|\mathrm{i}\rangle\langle \mathrm{i}|\vec{e}_1 \cdot \vec{p}|\mathrm{g}\rangle}{E_i - E_g - \hbar\omega_1} \right\} \right|^2 . \qquad (11)$$

By introducing [17, 18]

$$\begin{aligned} \Lambda(\omega_\beta) &= \sum_i \frac{|\mathrm{i}\rangle\langle \mathrm{i}|}{E_i - E_g - \hbar\omega_\beta} \qquad (\beta = 1, 2) , \\ \text{and} \quad \Lambda_+ &= \Lambda(\omega_2) + \Lambda(\omega_1) , \\ \Lambda_- &= \Lambda(\omega_2) - \Lambda(\omega_1) \end{aligned} \qquad (12)$$

we can rewrite Eq. (11) as

$$|\langle \mathrm{f}|\vec{e}_1 \cdot \{(\vec{p}\Lambda_+\vec{p})_S + (\vec{p}\Lambda_-\vec{p})_{AS}\} \cdot \vec{e}_2|\mathrm{g}\rangle|^2 , \qquad (13)$$

where S and AS denote the symmetric and antisymmetric part of the tensor operator. We introduce the tensor operator

$$\hat{T} = (\vec{p}\Lambda_+\vec{p})_S + (\vec{p}\Lambda_-\vec{p})_{AS} . \qquad (14)$$

The selection rules are determined by the symmetry properties of this operator. Assuming a ground state of Γ_1^+ symmetry one gets from the decomposition of the two–photon transition operator $\vec{e}_1\hat{T}\vec{e}_2$ into its irreducible components immediately the polarization selection rules:

$$
\begin{array}{lll}
 & \vec{e}_1 \hat{T} \vec{e}_2 = & \\
\Gamma_1^+ : & & \frac{1}{3}(e_{1x}e_{2x} + e_{1y}e_{2y} + e_{1z}e_{2z})(p_x\Lambda_+p_x + p_y\Lambda_+p_y + p_z\Lambda_+p_z) \\
\Gamma_{31}^+ : & + & \frac{1}{2}(e_{1x}e_{2x} - e_{1y}e_{2y})(p_x\Lambda_+p_x - p_y\Lambda_+p_y) \\
\Gamma_{32}^+ : & + & \frac{1}{6}(e_{1x}e_{2x} + e_{1y}e_{2y} - 2e_{1z}e_{2z})(p_x\Lambda_+p_x + p_y\Lambda_+p_y - 2p_z\Lambda_+p_z) \\
\Gamma_{41}^+ : & + & \frac{1}{2}(e_{1y}e_{2z} - e_{1z}e_{2y})(p_y\Lambda_-p_z - p_z\Lambda_-p_y) \\
\Gamma_{42}^+ : & + & \frac{1}{2}(e_{1z}e_{2x} - e_{1x}e_{2z})(p_z\Lambda_-p_x - p_x\Lambda_-p_z) \\
\Gamma_{43}^+ : & + & \frac{1}{2}(e_{1x}e_{2y} - e_{1y}e_{2x})(p_x\Lambda_-p_y - p_y\Lambda_-p_x) \\
\Gamma_{51}^+ : & + & \frac{1}{2}(e_{1y}e_{2z} + e_{1z}e_{2y})(p_y\Lambda_+p_z + p_z\Lambda_+p_y) \\
\Gamma_{52}^+ : & + & \frac{1}{2}(e_{1z}e_{2x} + e_{1x}e_{2z})(p_z\Lambda_+p_x + p_x\Lambda_+p_z) \\
\Gamma_{53}^+ : & + & \frac{1}{2}(e_{1x}e_{2y} + e_{1y}e_{2x})(p_x\Lambda_+p_y + p_y\Lambda_+p_x)\ .
\end{array}
\tag{15}
$$

Eq. (15) gives all the information on the polarization dependence. There is a geometrical factor containing only the components of the polarization vectors and a dynamical factor which contains the matrix elements. One also recognizes that a transition to Γ_4^+ vanishes for equal photon energies independently of the polarization of the photon beams.

For the angular dependence of the different two–photon allowed transitions one derives from Eq. (15):

$$
\begin{array}{ll}
\Gamma_1^+ \rightarrow \Gamma_1^+ : & (e_{1x}e_{2x} + e_{1y}e_{2y} + e_{1z}e_{2z})^2 = (\vec{e}_1 \cdot \vec{e}_2)^2\ , \\
\Gamma_1^+ \rightarrow \Gamma_3^+ : & (e_{1x}^2e_{2x}^2 + e_{1y}^2e_{2y}^2 + e_{1z}^2e_{2z}^2) - (e_{1y}e_{2y}e_{1z}e_{2z} + e_{1z}e_{2z}e_{1x}e_{2x} + e_{1x}e_{2x}e_{1y}e_{2y})\ , \\
\Gamma_1^+ \rightarrow \Gamma_4^+ : & (e_{1y}e_{2z} - e_{1z}e_{2y})^2 + (e_{1z}e_{2x} - e_{1x}e_{2z})^2 + (e_{1x}e_{2y} - e_{1y}e_{2x})^2 = (\vec{e}_1 \times \vec{e}_2)^2\ , \\
\Gamma_1^+ \rightarrow \Gamma_5^+ : & (e_{1y}e_{2z} + e_{1z}e_{2y})^2 + (e_{1z}e_{2x} + e_{1x}e_{2z})^2 + (e_{1x}e_{2y} + e_{1y}e_{2x})^2\ .
\end{array}
\tag{16}
$$

Transitions to Γ_1^+ and Γ_4^+ depend only on the mutual orientation of the polarization vectors, whereas the Γ_3^+ and Γ_5^+ transitions also depend on the orientation with respect to the crystalline axes.

The polarization dependence can directly be derived by use of the tables of Koster et al. [21]. For O_h symmetry the coupling coefficients (Clebsch–Gordan coefficients) for the coupling of two dipole operators (Γ_4^- symmetry) lead to the geometrical factors given in Eq. (15). The polarization selection rules can be extended to include circularly polarized light [22, 23]. Measurements with linearly and circularly polarized light allow a unique decomposition of any two–photon spectrum into the different symmetry contributions.

The polarization selection rules for three–photon transitions in O_h were recently derived by Pasquarello and Andreani [24]. They apply these rules to excitons in alkali halides.

IV. EXCITON POLARITONS

Up to now we have not taken into account that besides point symmetry in a crystalline solid there is also translational symmetry. As discussed in chapter II point symmetry leads to quantum numbers $\Gamma_i^{\pm}$ (i=1...8 in O_h including spin) whereas translational symmetry leads to the vector quantum number $\vec{k}$. Although the photon wave vector for light in the visible and ultraviolet spectral range is small compared to the size of the Brillonin zone ($|\vec{k}|_{\text{phot.}} \sim (1/1000)\ \pi/a$, a: lattice constant) it leads to a modification of Eq. (10) by another δ function ($\delta(\vec{k}_f - \vec{k}_1 - \vec{k}_2)$) which guarantees $\vec{k}$ conservation. $\vec{k}_f$ refers to the k vector of the final state and $\vec{k}_1$, $\vec{k}_2$ refer to the k vectors of the photon beams in the crystal. For two–photon transitions in centrosymmetric systems one expects no drastic effects in k–space spectroscopy, since the dispersion of the final states, which is in the case of excitons determined by their total mass, can usually be neglected.

The coupling of the electromagnetic field to a discrete state in a crystal leads to new eigenstates, which are termed polaritons. In the region of $(\omega, \vec{k})$–space where light in the crystal is in resonance with a discrete level (e.g. exciton or phonon) the dispersion of the coupled system shows the characteristic features of anti–crossing as depicted in Fig. 3. The concept of mixed modes of the electromagnetic field and crystal excitations was first introduced by Huang [25] for polar phonons in ionic crystals. Pekar [26] and Hopfield [27] investigated the exciton–photon coupling which leads to the concept of polaritons. For review articles we refer to Ref. [28, 29, 30].

The dispersion of the polaritons and the longitudinal exciton can be derived from a classical equation of motion for the polarization $\vec{P}$ and the Maxwell equations:

$$\ddot{\vec{P}} + (\Gamma/\hbar)\dot{\vec{P}} + \omega_T^2 \vec{P} = \epsilon_0 (f/\hbar^2)\vec{E}\ . \tag{17}$$

f accounts for the coupling strength of the oscillator of resonance frequency ω_T and damping Γ to the electric field. With a plane–wave ansatz

$$\vec{P}(\omega) = \epsilon_0 \chi(\omega) \vec{E}_0 e^{i(\vec{k}\cdot\vec{r} - \omega t)} \tag{18}$$

for $\vec{P}$ where $\chi(\omega)$ denotes the frequency dependent linear susceptibility one derives

$$\chi(\omega) = \frac{f}{\hbar^2\omega_T^2 - \hbar^2\omega^2 - i\hbar\omega\Gamma}\ , \tag{19}$$

which then leads to a one-oscillator equation for the dielectric function $\epsilon(\omega)$

$$\epsilon(\omega) = \epsilon_b + \frac{f}{\hbar^2\omega_T^2 - \hbar^2\omega^2 - i\hbar\omega\Gamma}\ . \tag{20}$$

ϵ_b is the so–called background dielectric constant, which takes into account the oscillators at higher energies. In the spectral region of rather close lying oscillators one has to use a multi–oscillator formula which is known as Sellmeyer formula:

$$\epsilon(\omega, k) = \epsilon_b\ + \sum_j \frac{f_j}{\hbar^2\omega_{T_j}^2 - \hbar^2\omega^2 - i\hbar\omega\Gamma_j}\ . \tag{21}$$

If one includes spatial dispersion, ω_T in Eq. (21) has to be replaced by $\omega_0 + \frac{\hbar^2 k^2}{2M}$, where $M = m_e + m_h$ is the total exciton mass, and $\epsilon(\omega, k)$ is then k–dependent.

From the Maxwell equations one derives the general wave equation

$$\vec{\nabla} \times \left(\vec{\nabla} \times \vec{E}\right) + \frac{\epsilon(\omega)}{c^2}\frac{\partial^2 \vec{E}}{\partial t^2} = 0\ , \tag{22}$$

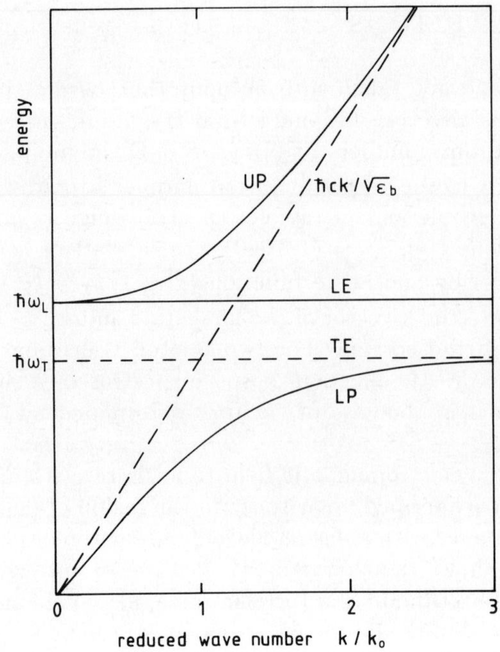

Figure 3. Schematic diagram of one–oscillator polariton. LP and UP mark the lower and upper polariton branches, respectively. LE and TE mark the resonances of the longitudinal and transverse exciton. $k_0 = \sqrt{\epsilon_b}\omega_T/c$ is the wave number at the crossing point of the TE and the light line ($\hbar ck/\sqrt{\epsilon_b}$, dashed line).

which leads for plane waves to

$$\vec{k} \times \left(\vec{k} \times \vec{E}_0\right) + \frac{\epsilon(\omega)\omega^2}{c^2}\vec{E}_0 = 0 , \tag{23}$$

$$\vec{k}\left(\vec{k} \cdot \vec{E}_0\right) - k^2\vec{E}_0 + \frac{\epsilon(\omega)\omega^2}{c^2}\vec{E}_0 = 0 . \tag{24}$$

In the simple case of cubic crystals, where ϵ does not depend on the polarizations of the photon field, one can distinguish solutions of Eq. (24) for transverse polaritons and the longitudinal exciton.

1. Transverse polariton $\left(\vec{k} \cdot \vec{E}_0 = 0\right)$:

$$\epsilon(\omega) = \frac{k^2c^2}{\omega^2} . \tag{25}$$

2. Longitudinal exciton $\left(\vec{k}(\vec{k} \cdot \vec{E}_0) = k^2\vec{E}_0\right)$:

$$\epsilon(\omega) = 0 . \tag{26}$$

The dispersion for two oscillators is depicted in Fig. 4. Neglecting damping one can rewrite Eq. (21) in product form as was shown by Kurosawa: [31]

$$\epsilon(\omega) = \epsilon_b \prod_i \frac{\omega_{L_i}^2 - \omega^2}{\omega_{T_i}^2 - \omega^2} . \tag{27}$$

The vanishing of the numerator ($\omega = \omega_{L_i}$) yields the resonances of the longitudinal excitons and the poles ($\omega = \omega_{T_i}$) the transverse resonances. k–space spectroscopy allows to excite resonances on the lower and upper polariton branch and even on the longitudinal exciton. Resonances on the upper polariton branch and the longitudinal exciton of CuCl were first measured by two–photon absorption (TPA) [32]. Measurements by two–photon sum–frequency generation also yield resonances on the upper polariton branch [33]. The first measurements on the lower polariton branch in CuCl were done by hyper–Raman scattering (HRS). For a review we refer to Hönerlage et al. [29]. HRS takes advantage of one–photon (close to lowest allowed exciton) and two–photon resonance enhancement (virtual transition to biexciton). The direction and energy of the outcoming photon is then a clear finger–print of the excitation left in the crystal (upper or lower polariton or longitudinal exciton). Since in HRS three photons are involved, we expect in a third–order perturbational approach two resonance denominators, where the incoming photons $\hbar\omega_1$ and $\hbar\omega_2$ are close to E_{ex} (lower exciton) and $\hbar(\omega_1 + \omega_2)$ is close to E_{Bi} (biexciton). Instead of the observation of the spontaneous emission in HRS one can use a second beam to induce the decay of the virtually excited biexcitons [34, 35], which then leads to a strong coherent signal. This technique is known as nondegenerate four–wave mixing. Even if there are no resonances for the intermediate processes, this technique works well, although the outgoing signals are much weaker. Fröhlich et al. [6] have termed this technique three–photon difference–frequency generation (TP–DFG). The different techniques are depicted in Fig. 4. As seen in Fig. 4 three–photon sum–frequency generation (TP–SFG) leads only

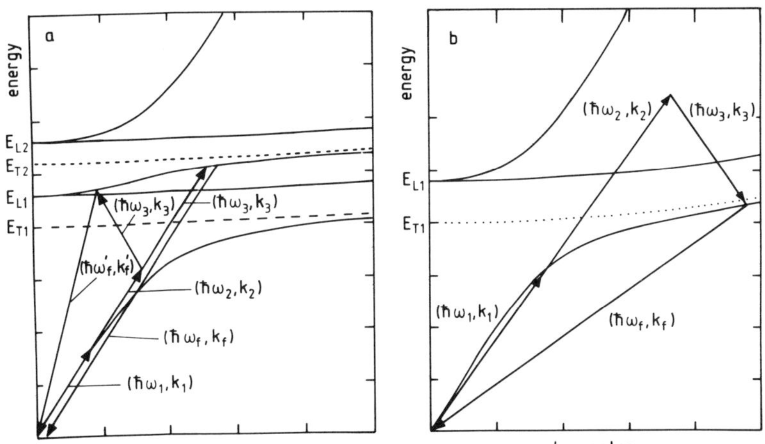

Figure 4. Schematic diagram for illustration of TP–SFG (a) and TP–DFG processes (b). In Fig. 4a the dispersion of two oscillators with spatial dispersion is depicted.

to resonances on the upper polariton branch, since each of the incoming photons originates from the lower polariton branch, which shows a bending to the right in the region of normal dispersion. The tuning curves, which are shown for the special case of equal photons in TPA in Fig. 5 never cross the lower branch. In TP–DFG, however, one can reach resonances on the lower branch. The resonances on the longitudinal exciton (LE) can be excited by two– and three–photon processes. They cannot be detected directly by a coherent process like SFG and DFG, since the induced longitudinal polarization does not couple to the photon field in a transition to the ground state. The LE in CuCl can be detected by TPA as was shown by Fröhlich et al. [32] and by HRS as discussed in Ref. [29]. The first three–photon spectra of resonances on the upper polariton branch

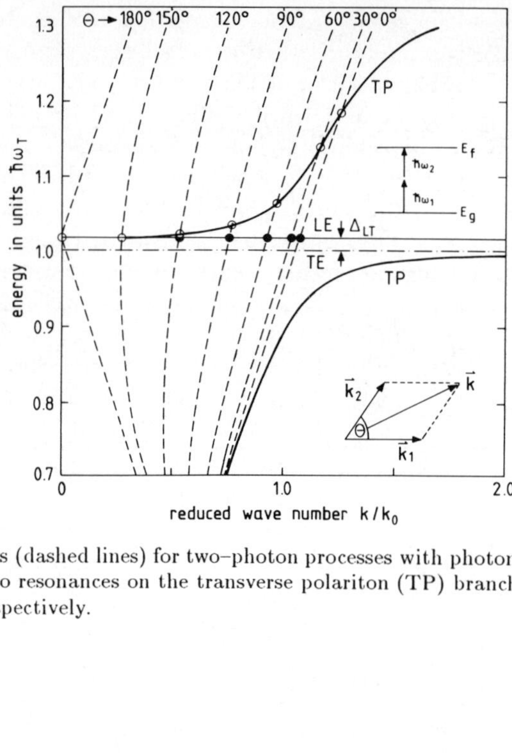

Figure 5. Tuning curves (dashed lines) for two–photon processes with photons of equal energy. Open and closed circles refer to resonances on the transverse polariton (TP) branches and the longitudinal exciton (LE) branch, respectively.

and the longitudinal exciton in alkali halides [5] were detected via the luminescence of the self–trapped exciton, which exhibits a Stokes shift of about 2 eV. This technique might be termed three–photon (excitation) spectroscopy (TPS). Contrary to the coherent processes of TP-SFG and TP-DFG the luminescence light is emitted in the total solid angle and any information on the polarization of the excited state is lost. There is another important difference between TPS and the coherent techniques (TP–SFG and TP–DFG). In TPS the emitted light is observed out of the whole crystal, since in the spectral region of the luminescence there is no absorption. In the case of coherent processes in a region of strong resonances, where the absorption length is in the range

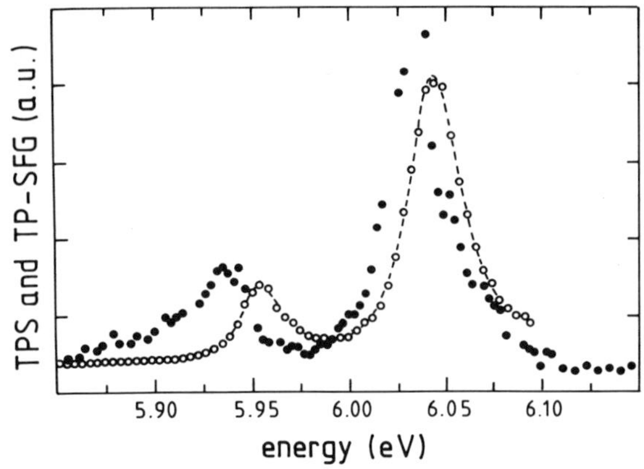

Figure 6. Three-photon excitation (TPS) (open circles) and three-photon sum-frequency spectrum (TP-SFG) (closed circles) of two resonances on the upper polariton branch in KI. The dashed line represents a symmetric Lorentzian fit to the TPS spectrum.

of 100 nm to 1 μm, the emitted light is expected only from a region close to the surface. If one considers the polariton as an exciton–photon particle one monitors in TPS the decay via the exciton part and in TP–SFG and TP–DFG the decay via its photon part. There is a difference in the shape and position of the same resonance if detected by TPS and TP–SFG as shown in Fig. 6. In the following chapter we will show that this is due to the Maxwell boundary conditions (MBC), which lead to drastic changes of reflection and transmission coefficients in the resonance region, thus affecting the TP–SFG signals.

V. NONLINEAR SPECTROSCOPY AND MAXWELL BOUNDARY CONDITIONS

The MBC for a nonlinear medium were first derived by Bloembergen and Pershan [36]. They have shown that the solution of the Maxwell equations with the proper boundary conditions leads to harmonic waves even in reflection. This effect was demonstrated by Ducuing and Bloembergen [37] in GaAs and more recently by Mizutani and Nagasawa [38] in CuCl. These experiments were not performed at normal incidence, which has significant experimental advantages but complicates the analysis considerably. Fröhlich et al. [39] have recently performed three–photon reflection measurements for perfect normal incidence. They were able to detect a three–photon reststrahl band in KI. For the special case of bulk crystals (multiple reflections neglected) Fresnel–type formulas for a nonlinear reflection factor R_{nl} for the front surface and a nonlinear transmission factor T_{nl} for the back surface are derived. As discussed in Ref. [40] the difference between TP–SFG and TPS spectra of the same resonances on the upper polariton branch (Fig. 6) can be explained by taking T_{nl} into account. In the following we will briefly derive the expressions for R_{nl} and T_{nl}. We start from the classical wave equation with a nonlinear source term

$$\nabla^2 \vec{E} - \frac{\epsilon(\omega)}{c^2}\frac{\partial^2 \vec{E}}{\partial t^2} = \frac{1}{c^2 \epsilon_0}\frac{\partial^2 \vec{P}_{nl}}{\partial t^2}\,, \tag{28}$$

where $\epsilon(\omega)$ is the complex dielectric function. Effects of spatial dispersion, which would lead to an additional wave vector dependence of ϵ, are neglected.
We consider

$$P_i^{nl}(\omega) = \epsilon_0 \chi^{(3)}_{ijkl} E_j E_k E_l\,. \tag{29}$$

For the special case of third harmonic generation (THG) we get:

$$P_{nl}(\omega) = P^0_{nl}(\omega) e^{i(3k_l z - 3\omega_l t)}\,, \tag{30}$$

where $\omega_l = 1/3\ \omega$ and $k_l = n(\omega_l)\omega_l/c$ refer to the incoming laser beam. $n(\omega_l)$ is a slowly varying function of ω_l and real, since the laser photon energy is in the spectral region of normal dispersion and since there is no linear absorption.

$$P^0_{nl} = \epsilon_0 \chi^{(3)}(\omega) E_0^3(\omega) \tag{31}$$

refers to the amplitude of the nonlinear polarization. In the interesting region of a strong resonance we consider for the complex nonlinear susceptibility only one resonance at the transverse exciton energy $\hbar\omega_T$ with a damping Γ.

$$\chi^{(3)}(\omega) \sim (\hbar^2\omega_T^2 - \hbar^2\omega^2 - i\hbar\omega\Gamma)^{-1}\,. \tag{32}$$

As discussed in detail by Bloembergen [41], for the solution of Eq. (28) one has to consider the general solution of the homogeneous wave equation $E_h(\omega)$ and a particular solution of the inhomogeneous equation $E_{ih}(\omega)$. The situation is sketched in Fig. 7. Incoming laser photons of wave vector $\vec{k}_l$ lead to a wave $E_{ih}(\omega)$ with wave vector $\vec{k}_{ih}(3\omega_l) = 3\vec{k}(\omega_l)$ inside the crystal. The homogeneous wave $E_h(\omega)$ with complex wave vector $\vec{\tilde{k}}(\omega)$ from the front is damped out. The solutions are thus given by:

$$\begin{aligned} E_h(\omega) &= E_h^0 e^{i(-\tilde{k}(\omega)z - \omega t)}\,, \\ E_{ih}(\omega) &= E_{ih}^0 e^{i(3k(\omega_l)z - 3\omega_l t)}\,, \\ E_{ih}^0 &= \frac{1}{\epsilon_0} P^0_{nl} \frac{1}{n^2(\omega_l) - \tilde{n}^2(\omega)}\,. \end{aligned} \tag{33}$$

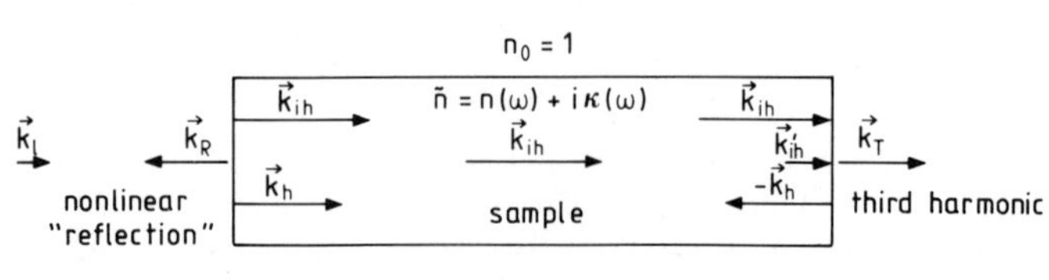

Figure 7. Schematic of experimental situation. Sketch of wave vectors inside and outside of sample. $\vec{k}_l$, incoming laser; $\vec{k}_T$ and $\vec{k}_R$ third harmonic transmitted and reflected waves, respectively; $\vec{k}_h$ and $\vec{k}_{ih}$ third harmonic homogeneous and inhomogeneous waves, respectively.

As stated before $\tilde{k} = \tilde{n}(\omega)\omega/c$ is complex with $\tilde{n}^2 = \epsilon(\omega)$ and $k = n(\omega_l)\omega_l/c$ is real. The transmitted intensity I_T is calculated from the Poynting vector outside the crystal

$$I_T = \frac{1}{2}\epsilon_0 c|E_T^0|^2 . \tag{34}$$

From the classical MBC one gets:

$$\begin{aligned} E_T^0 &= E_h^0 + E_{ih} , \\ E_T^0 &= -\tilde{n}E_h^0 + nE_{ih} . \end{aligned} \tag{35}$$

The MBC enforce a reflected wave (homogeneous solution) back into the crystal. From Eq. (35) we derive:

$$E_T^0 = \frac{\tilde{n}+n}{\tilde{n}+1}E_{ih}^0 . \tag{36}$$

We thus get from Eq. (34) and Eq. (36):

$$I_T = \left|\frac{\tilde{n}(\omega)+n(\omega_l)}{\tilde{n}(\omega)+1}\right|^2 \frac{1}{2}\epsilon_0 c\,|E_{ih}^0|^2 . \tag{37}$$

We define the nonlinear transmission coefficient

$$\begin{aligned} T_{nl} &= \left|\frac{\tilde{n}(\omega)+n(\omega_l)}{\tilde{n}(\omega)+1}\right|^2 \\ &= \frac{(n(\omega)+n(\omega_l))^2+\kappa^2(\omega)}{(n(\omega)+1)^2+\kappa^2(\omega)} , \end{aligned} \tag{38}$$

where $\tilde{n}(\omega) = n(\omega)+i\kappa(\omega)$. As indicated in Fig. 7 there is another wave E'_{ih} with wave vector $k'_{ih}(3\omega_l) = 2k(\omega_l) - k(\omega_l) = k(\omega_l)$, which is due to reflection of the laser beam at the back of the crystal. Because of the polariton dispersion the two inhomogeneous waves lead to resonances at different photon energies $\omega = 3\omega_l$. The resonance at $k(\omega_l)$ is weaker, because a reflection of the laser beam is necessary. At the front surface the MBC enforce a reflected amplitude E_R^0. We get

$$\begin{aligned} E_R^0 &= E_h^0 + E_{ih}^0 , \\ -E_R^0 &= \tilde{n}E_h^0 + nE_{ih}^0 . \end{aligned} \tag{39}$$

From Eq. (39) we deduce:

$$E_R^0 = \frac{\tilde{n}-n}{\tilde{n}+1}E_{ih}^0 . \tag{40}$$

This leads to a nonlinear reflection coefficient R_{nl}

$$\begin{aligned} I_R &= \frac{1}{2}\epsilon_0 c\,|E_R^0|^2 \\ &= R_{nl}\frac{1}{2}\epsilon_0 c\,|E_{ih}^0|^2 , \\ R_{nl} &= \left|\frac{\tilde{n}(\omega)-n(\omega_l)}{\tilde{n}(\omega)-1}\right|^2 = \frac{(n(\omega)-n(\omega_l))^2+\kappa^2(\omega)}{(n(\omega)+1)^2+\kappa^2(\omega)} . \end{aligned} \tag{41}$$

For the dielectric function we assume a two-oscillator Sellmeyer formula:

$$\epsilon = \tilde{n}^2 = \epsilon_b + \frac{f_1}{\hbar^2\omega_{T1}^2 - \hbar^2\omega^2 - i\hbar\omega\Gamma} + \frac{f_2}{\hbar^2\omega_{T2}^2 - \hbar^2\omega^2} . \tag{42}$$

For KI all parameters besides Γ are taken from Ref. [42] ($\epsilon_b = 1.64$, $\hbar\omega_{T_1} = 5.847$ eV, $\hbar\omega_{T_2} = 6.87$ eV, $f_1 = 5.78$ (eV)^2, $f_2 = 43.76$ (eV)^2). In Fig. 8 we present the experimental results. The only fit parameter is the damping $\Gamma = 40$ meV. The rather weak resonance near the longitudinal energy E_L is caused by the reflection of the laser beam at the back of the crystal ($k_{ih} = k(\omega_l)$). In the analysis this reflected wave is included. Since the reflection coefficient for the laser can easily be calculated from the refractive index $n(\omega_l)$, there is no additional free parameter. The agreement between the experimental results and the fit can certainly be improved by a refined theory of the dielectric function. From a careful analysis of the line shape of three–photon reflection and transmission data, one should then get information on the frequency dependence of the polariton damping.

It is straightforward to derive an expression for the nonlinear transmission and reflection for thin films [40], where the film thickness L is no longer large compared to the absorption length $l_\alpha = 1/\alpha$ (α: absorption coefficient). In alkali halides l_α is in the resonance region 100nm – 1μm. For thin films one has to apply the MBC to the front

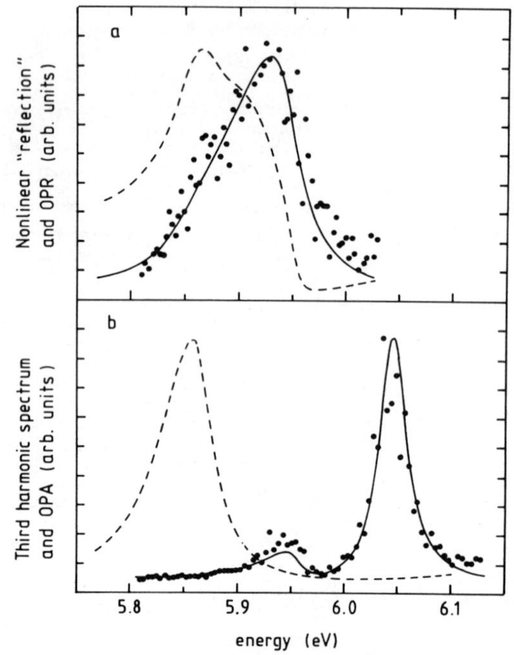

Figure 8. Three–photon and one–photon spectra of KI at 1.5 K. a) Circles and solid line represent experimental results and fit for the three–photon reflection spectrum, respectively. The one–photon reflection (OPR) spectrum (dashed line) is taken from Ref. [43]. b) Circles and solid line represent experimental results and fit for third harmonic spectrum, respectively. The one–photon absorption (OPA) spectrum (dashed line) is taken from Ref. [16].

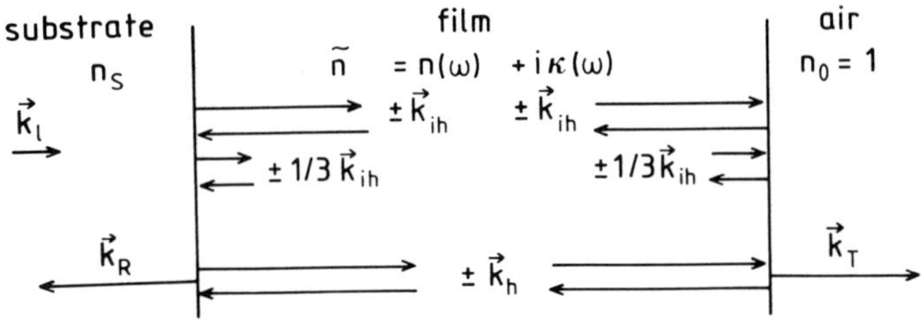

Figure 9. Schematic of experimental situation. Sketch of wave vectors inside and outside of a thin sample on a substrate. The situation refers to transmission geometry.

and back surface simultaneously. This situation is depicted in Fig. 9. In our experiments the film is evaporated on a transparent substrate with refractive index n_s(e.g. KI film on KCl). There are two types of experiments: i) The substrate points towards the laser, which is called "transmission geometry". ii) The film points towards the laser, which is called "reflection geometry". In Fig. 10 we present results for measurements of a KI film (L = 436 nm) on KCl for both geometries [40].

The rather large width of the reflection spectrum in the transmission geometry is probably caused by lattice distortions of the film at the interface to the substrate, which are expected to affect strongly the reflection spectra since the penetration depth in the reststrahl region is small.

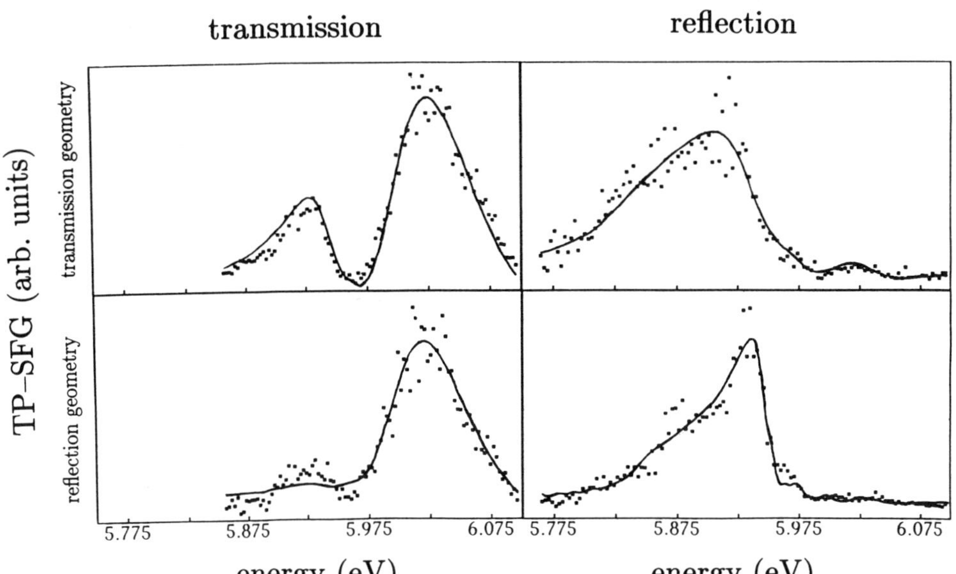

Figure 10. Three-photon transmission and reflection spectra of a film of KI (L=436nm) on a substrate of KCl at 1.5 K. The upper spectra refer to transmission, the lower spectra to reflection geometry. The solid lines result from a fit taking into account the MBC [40].

VI. EXAMPLES OF NONLINEAR SPECTROSCOPY

This chapter is devoted to two– and three–photon experiments without external perturbation. In the first part we consider transitions which are one–photon forbidden, polariton effects can thus be neglected. In the second part there is a strong dipole coupling between the ground state and the excited state which leads to a pronounced polariton structure.

VI.A. Band Gap and Valence Band Assignments from TPA

In this chapter we will briefly discuss two examples, where TPA leads to important information on fundamental band parameters. In alkali bromides TPA leads to a new assignment of the band gap and in CuCl a transition to a lower valence band (Γ_3 symmetry) can be detected, which is forbidden in linear spectroscopy.

Hopfield et al. [3] were the first to do two–photon spectroscopy. They got similar results in KI as shown in Fig. 2 for a ruby laser. In agreement with previous assignments of the band gap they interpreted the onset of the TPA as transitions to P excitons close to the band gap. In alkali bromides, however, the onset of TPA starts about 0.6 eV lower than the previously assigned band gap as shown in Fig. 11. The reason for this wrong assignment is evident: In one–photon absorption the band gap is hidden under the 1S exciton of the split–off valence band, since in alkali bromides the spin–orbit splitting (about 0.5 eV) is close to the exciton binding energy (about 0.6 eV). The main advantage of TPA in this case is due to the fact that transitions to the S excitons are parity forbidden [44].

In the second example it is just the other way around. In CuCl there is another valence band of Γ_3 symmetry expected (Γ_8 with spin, Fig. 12). One–photon transitions from this valance band to a conduction band Γ_1 (Γ_6 with spin) are forbidden. In TPA, however, this transition is allowed. In addition it exhibits the characteristic Γ_3–polarization dependence. The results are presented in Fig. 13. From these TPA data the first optical assignment of the Γ_3 valence band was possible [45]. An energy difference of 1.79 eV to the uppermost valence band is deduced from these measurements.

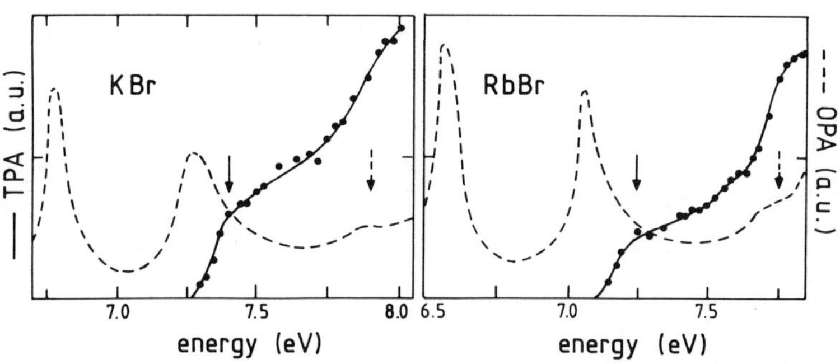

Figure 11. One–photon (dashed lines [16]) and two–photon spectra (solid lines [44]) of KBr and RbBr. Solid and dashed arrows indicate new and old band gap assignments, respectively.

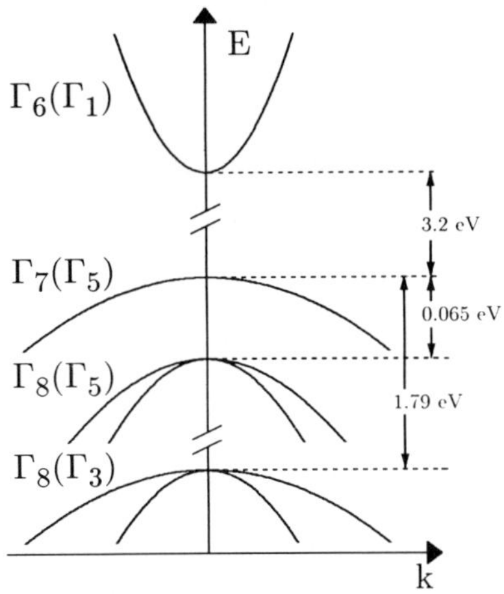

Figure 12. Schematic band structure of CuCl at Γ point. Γ values in parentheses refer to symmetry assignments without spin.

VI.B. K–Space Spectroscopy

As mentioned before two– and three–photon spectroscopy allow to do k–space spectroscopy. This was first demonstrated for the upper polariton branch and the longitudinal exciton in CuCl by Fröhlich et al. [32]. As shown schematically in Fig. 5 one can excite resonances from k=0 to a maximal k value, if one uses photons of equal energy. From the selection rules for a $\Gamma_1 \rightarrow \Gamma_5$ transition one can easily derive compact formulas for the polarization dependence of the transverse polariton (TP) and the longitudinal exciton (LE) resonances [48].

$$\begin{aligned} f_{TP} &= f_0(\omega_{TP}) \left(\vec{b} \times \vec{s}\right)^2 \\ f_{LE} &= f_0(\omega_{LE}) \left(\vec{b} \cdot \vec{s}\right)^2 \end{aligned} \tag{43}$$

with $\vec{b} = \vec{e}_1 \otimes \vec{e}_2$ defined as a symmetric cross product ($b_x = e_{1y}e_{2z} + e_{1z}e_{2y}$, $b_y = \cdots$, $b_z = \cdots$ s. Eq. (15)) and $\vec{s} = (\vec{k}_1 + \vec{k}_2)/|\vec{k}_1 + \vec{k}_2|$. The frequency dependence of the reduced transition rate $f_0(\omega)$ can be calculated as shown by Boggett and Loudon [49] and in Ref. [48]. In Fig. 14 we show experimental results which are quantitatively described by Eq. (43). k–space spectroscopy of polaritons by two–photon processes can only be done in systems which lack a center of inversion. As discussed before all one–photon transitions are also three-photon allowed (but not vice versa). One should thus be able to do k–space spectroscopy on polaritons and longitudinal excitons in crystals with an inversion center (e.g. O_h symmetry) by use of three–photon processes. This was first shown for alkali halides by Beerwerth and Fröhlich [5]. In the first experiments

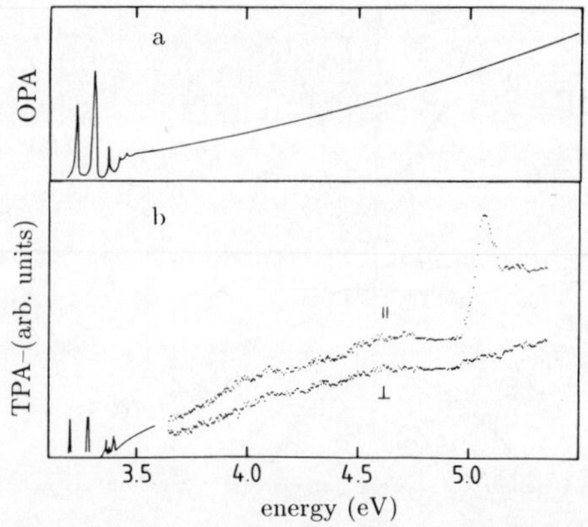

Figure 13. One–photon absorption (OPA) and two–photon absorption (TPA) of CuCl. a) One–photon data from Ref. [46] and [47]. b) Two–photon data at 4.3 K. || configuration, both polarization vectors $\vec{e}_1$ and $\vec{e}_2$ parallel to [100]; ⊥ configuration, $\vec{e}_1$ in [100] and $\vec{e}_2$ in [010] direction.

transitions to the upper polariton branch and the longitudinal exciton were detected via the subsequent luminescence of the self–trapped exciton. As discussed before resonances on the upper polariton branch can also be detected by TP–SFG (three–photon sum–frequency generation) whereas for resonances on the lower branch one has to do TP-DFG (three–photon difference–frequency generation) [6]. By use of a two–oscillator Kurosawa dispersion relation [42]

$$\frac{\hbar^2 c^2 k^2}{E^2} = \epsilon_b \frac{E_{L_1}^2 - E^2}{E_{T_1}^2 - E^2} \frac{E_{L_2}^2 - E^2}{E_{T_2}^2 - E^2} \tag{44}$$

one can fit the experimental results for KI as shown in Fig. 15. From this figure the longitudinal and transverse energy values $E_{T_1} = (5.847 \pm 0.001)$eV and $E_{L_1} = (5.946 \pm 0.001)$eV and thus the longitudinal–transverse splitting $\Delta_{LT} = E_{L_1} - E_{T_1} = (99 \pm 1)$meV are gained with high accuracy. $E_{L_2} = 8.81$eV and $E_{T_2} = 6.97$eV and $\epsilon_b = 1.64$ describe the influence of higher oscillators. In CsI there are three rather close lying polariton resonances. As shown by Beerwerth et al. [50] the dispersion can be fitted by a six–oscillator Kurosawa formula.

VII. NONLINEAR SPECTROSCOPY IN EXTERNAL FIELDS

In this chapter we will discuss examples of nonlinear spectroscopy in external fields. Besides results on CuCl we will present recent measurements of Landau transitions in GaAs–AlGaAs multiple quantum wells. In many cases the linewidth of resonances measured by two–photon or three–photon techniques is much smaller than the linewidth of corresponding resonances in one–photon absorption. This is partly due to the fact that nonlinear measurements are taken on bulk crystals. They are much less sensitive to surface effects and strain than thin samples, which are in many cases necessary for linear optical experiments. If one has to do reflectivity measurements the linewidth

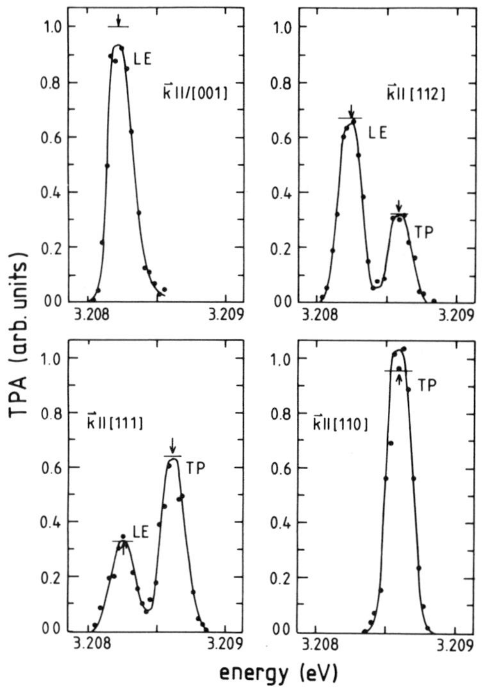

Figure 14. Two–photon absorption of longitudinal exciton (LE) and transverse polariton (TP) of Z_3 exciton in CuCl for different orientation of the total k vector in 180° configuration [48]; polarization of laser and flash–lamp along $[1\bar{1}0]$; arrows indicate theoretical values of Eq. (43) with $f_0(\omega_{LE}) = 1$ and $f_0(\omega_{TP}) = 0.96$.

problem is even more evident as seen in the case of CuCl in Fig. 16. The linewidth of the resonances for the nonlinear data ranges from 0.1 meV (B to G) to about 10 meV(H), whereas the linewidth of one–photon reflectivity is about 10 meV [51]. The increasing linewidth for the higher energy resonance is a signature of an increasing damping due to phonon emission processes. The study of these effects by nonlinear optical methods is certainly another interesting problem.

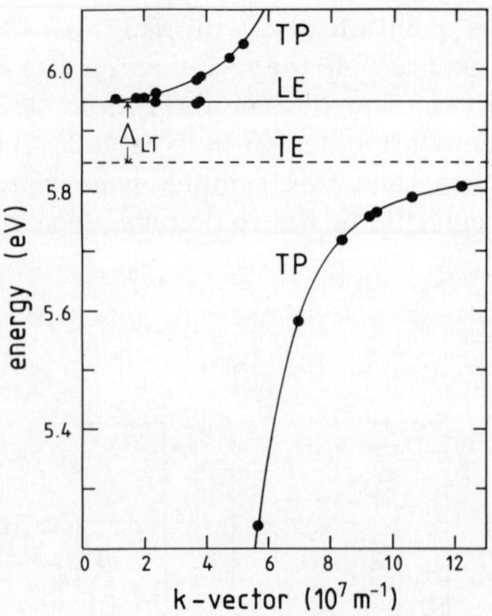

Figure 15. Resonances on the upper and lower polariton branches (TP) and the longitudinal exciton (LE) in KI at 1.5 K. Solid lines, fit to experimental results (full circles); dashed line, transverse exciton (TE); Δ_{LT}, longitudinal–transverse splitting.

VII.A. Nonlinear Spectroscopy of CuCl in External Fields

In the following we will show some recent results of two– and three–photon data of CuCl in a magnetic field, an electric field and under uniaxial stress. Before we present the experimental results, we will briefly describe the excitonic properties of CuCl.

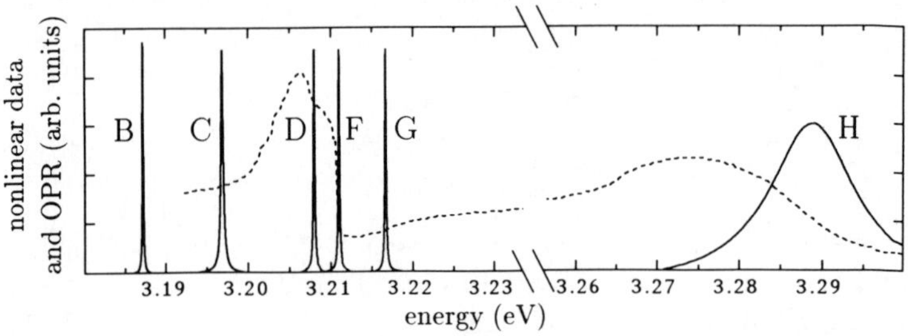

Figure 16. Spectra of resonances in CuCl at 5 K. Solid line, nonlinear resonances; dashed line, linear reflection spectrum from Ref. [51].

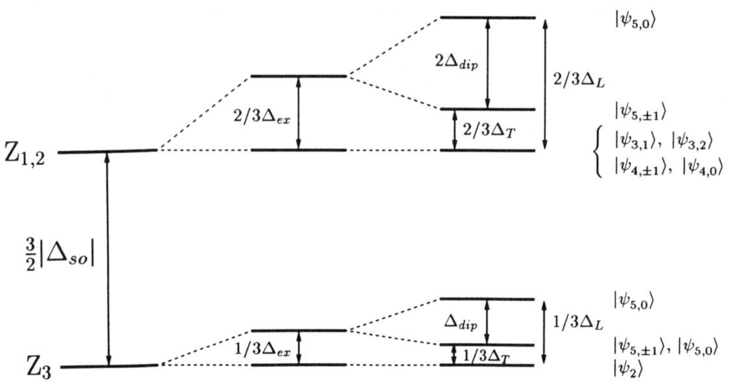

Figure 17. Schematic diagram of 1S excitons in CuCl with successive inclusion of interactions: (a) spin–orbit interaction, b) short–range, c) long–range part of electron–hole exchange interaction.

CuCl crystallizes in zincblende structure (T_d symmetry). Due to the fact that in T_d there is no center of inversion, there is a strong p–d hybridization of the upper valence bands. This leads to an inverted valence band structure, as shown in Fig. 12. Excitons due to the transition from the twofold Γ_7 valence band to the twofold Γ_6 conduction band are called Z_3 excitons, whereas the fourfold Γ_8 valence band leads to the $Z_{1,2}$ excitons. The lower Γ_8 valence band leads to the Γ_3 excitons which were discussed in chapter VI.A., but will not be further considered in this chapter. As discussed in more detail in Ref. [52] one can easily derive the symmetry of the S excitons in CuCl by use of the tables of Koster et al. [21]

$$
\begin{aligned}
Z_3 \text{ excitons:} \quad & \Gamma_7(2) \otimes \Gamma_6(2) \otimes \Gamma_1 &=& \quad \Gamma_2(1) \oplus \Gamma_5(3) \\
Z_{1,2} \text{ excitons:} \quad & \Gamma_8(4) \otimes \Gamma_6(2) \otimes \Gamma_1 &=& \quad \Gamma_3(2) \oplus \Gamma_4(3) \oplus \Gamma_5(3) \\
\Gamma_3 \text{ excitons:} \quad & \Gamma_8(4) \otimes \Gamma_6(2) \otimes \Gamma_1 &=& \quad \Gamma_3(2) \oplus \Gamma_4(3) \oplus \Gamma_5(3)
\end{aligned} \tag{45}
$$

For S excitons the symmetry of the envelope is Γ_1 whereas for P excitons one would have to multiply by Γ_5 which then leads to a large number of excited states. The corresponding exciton wave functions can be derived from Eq. (45) in the jj–coupling scheme. We first discuss Z_3 and $Z_{1,2}$ excitons. For this case excitons with Γ_5 symmetry are one–photon allowed. They are singlet–triplet mixed states (orthoexcitons) and lead to polaritons. The excitons of Γ_2, Γ_3 and Γ_4 symmetry are pure triplet states (paraexcitons). Optical transitions to these excitons without external field are therefore spin forbidden. In the case of Γ_3 excitons we get different results. The twofold Γ_3 excitons are singlet–triplet mixed states and thus orthoexcitons. They are two–photon allowed. The Γ_4, Γ_5 excitons are pure triplet states (paraexcitons) and again spin forbidden. In Fig. 17 we present the level scheme for the case that only diagonal terms of the Hamiltonian are taken into account as discussed by Andreani et al. [53]. As shown in Fig. 17 the detailed exciton structure is determined by the spin–orbit coupling (Δ_{so}), the short–range exchange interaction (Δ_{ex}), which determines the splitting between ortho– and paraexitons and the long–range exchange interaction (Δ_L), which leads to the

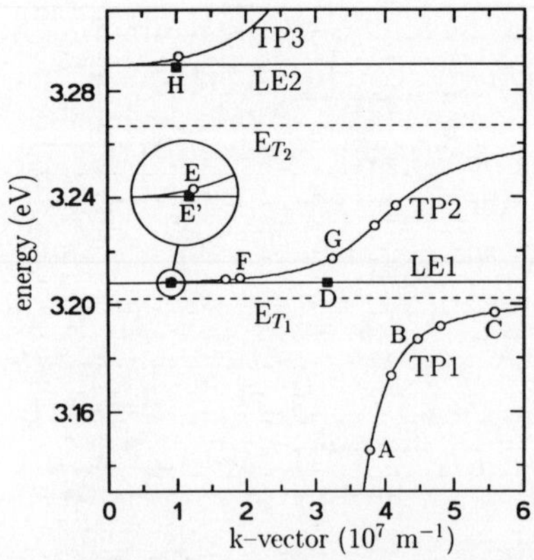

Figure 18. Polariton dispersion of CuCl with resonances on the polariton branches TP1–TP3 (open circles) and the longitudinal exciton branches LE1 and LE2 (full squares). The dashed lines represent the transverse energies E_{T_1} and E_{T_2}. A to H mark some resonances for purpose of reference.

splitting of the transverse ($M = \pm 1$) and longitudinal ($M = 0$) orthoexcitons. Taking the coupling to the electromagnetic field into account we get a polariton structure which can be described by a two–oscillator polariton dispersion [52].

$$\frac{\hbar^2 c^2 k^2}{E^2} = \epsilon_b + CE^2 + \frac{f_1 E_{pl}^2}{E_{T_1}^2 - E^2} + \frac{f_2 E_{pl}^2}{E_{T_2}^2 - E^2} . \tag{46}$$

The fit parameters used in Eq. (46) and shown in Fig. 17 are given in Table 1. In Fig. 18 we present the polariton dispersion as measured by two– and three–photon techniques. Some of the marked resonances are shown in Fig. 16.

Table 1. Fit parameters of two–oscillator polariton dispersion for CuCl [52].

ϵ_b	3.52	E_{pl} $[eV]$	5.901
E_{T_1} $[eV]$	3.2023	E_{T_2} $[eV]$	3.2668
f_1	$6.17 \cdot 10^{-3}$	f_2	$1.72 \cdot 10^{-2}$
E_{L_1} $[eV]$	3.2080	E_{L_2} $[eV]$	3.2898
E_{para} $[eV]$	3.2000	E_0 $[eV]$	3.2398
$\Delta_{so}^{(1)}$ $[meV]$	-40.4	$\Delta_{so}^{(2)}$ $[meV]$	-39.8
Δ_T $[meV]$	9.8	Δ_L $[meV]$	35.1
Δ_{ex} $[meV]$	18.2	Δ_{dip} $[meV]$	8.4
C $[eV]^{-2}$	0.076		

1. Zeeman Effect of Exciton–Polaritons in CuCl. In the following we will briefly present recent experimental results of nonlinear optical spectra of excitons in CuCl in an external magnetic field. The rather small linewidth (about 100 μeV) of resonances on the lower polariton branch TP1, the longitudinal exciton LE1 and the polariton branch TP2 allows to determine accurately their splitting and shift as function of an external magnetic field. Experimental results are shown in Fig. 19 for the polariton resonances A to G (see Fig. 18). The splitting of the polariton resonances shows a spectacular k-dependence. On the lower polariton branch one even observes a change of sign. For a k value of about $4.5 \cdot 10^7$ m^{-1} the splitting goes to zero. In Fig. 20 the field–induced repulsion of the longitudinal exciton (LE) and the paraexciton (PE) is shown. The experimental results can be quantitatively explained by considering the Zeeman interaction

$$H_{ze} = (l_{h,z} + 2(\sigma_{h,z} + \sigma_{e,z}))\,\mu_B B\ , \tag{47}$$

where l_z and σ_z are the z–components of the angular momentum operators with $l = 1$ and $\sigma = 1/2$ (e,h refer to conduction electron and valence hole, respectively). Many of the characteristic electronic properties of CuCl are caused by the hybridization of the halogen 3p and the copper 3d levels, which leads to the inverted valence band structure (Fig. 12). As discussed in detail in Ref.[52] there is a field–induced mixing between Z_3 and $Z_{1,2}$ excitons. From the diagonalization of H_{ze} (Eq. (47)) one gets the exciton energies E_{T_1} and E_{T_2} and the oscillator strengths f_1 and f_2, which depend on the magnetic field and the quantum number $M = \pm 1$. These values have then to be used to calculate the dispersion (Eq. (46)) for both polariton components ($M = \pm 1$). The rather surprising k–dependent splitting of the polaritons (Fig. 19) and the repulsion of the longitudinal exciton and the paraexciton (Fig. 20) are fully explained by this model.

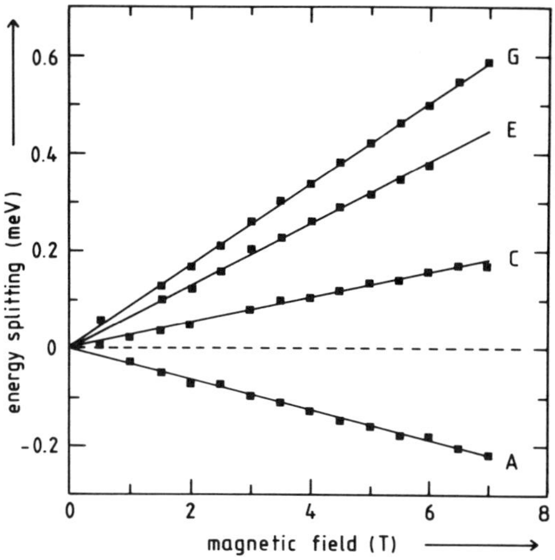

Figure 19. Magnetic field dependent splitting of resonances A to G on the transverse polariton branches TP1 and TP2 as indicated in Fig. 18. Note the sign reversal of the splitting on the lower branch TP1 between resonances A and C.

2. Stark Effect of Exciton–Polaritons in CuCl. In the following we will present the first nonlinear optical Stark effect measurements on CuCl [54]. Again the main advantage of two–photon measurements as compared to one–photon electro–reflectivity data is due to the small linewidth of the polariton and the longitudinal exciton resonances in two–photon experiments. In Fig. 21 we present our experimental results which are taken on a crystal which was oriented along [1 1 1], $[1-\sqrt{3}, 1+\sqrt{3}, -2]$ and $[1+\sqrt{3}, 1-\sqrt{3}, -2]$. The k vectors of the laser beams (dye laser and Nd-YAG laser) are parallel $[1-\sqrt{3}, 1+\sqrt{3}, -2]$ and the electric field is applied along [1 1 1]. As can be seen from the two–photon selection rules (Eq. (15) and Eq. (43)) this orientation allows to excite selectively transverse polaritons and the longitudinal exciton by proper choice of the laser polarizations $\vec{e}_1$ and $\vec{e}_2$. The vector $\vec{b}$ (Eq. (43)) accounts for the orientation of the exciton dipole moment. For $\vec{e}_1 \parallel [1\ 1\ 1]$

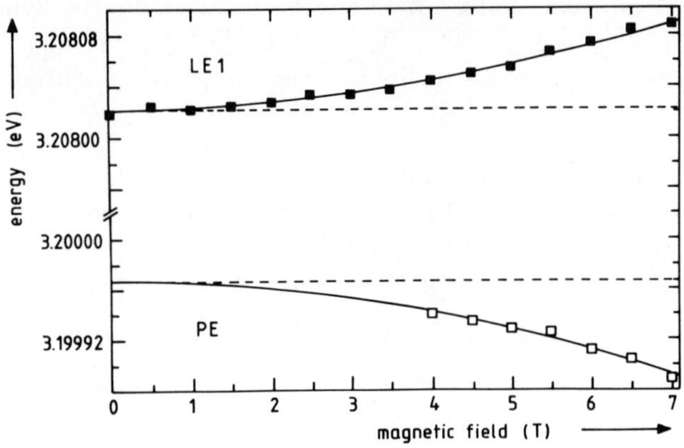

Figure 20. Repulsion of TPA resonances (open and full squares) of the Z_3 paraexciton (PE) and the longitudinal Z_3 exciton (LE1). Solid curves are theoretical fits, dashed curves mark unperturbed positions of LE1 and PE.

and $\vec{e}_2 \parallel [1+\sqrt{3}, 1-\sqrt{3}, -2]$ the vector $\vec{b}$ is parallel to $[1+\sqrt{3}, 1-\sqrt{3}, -2]$. For this choice of laser polarizations one thus excites only a transverse polariton, the dipole moment of which is perpendicular to the field direction ($\vec{b} \perp \vec{F}$). The other transverse polariton component with dipole moment parallel to the field direction ($\vec{b} \parallel \vec{F}$) is excited with $\vec{e}_1$ and $\vec{e}_2 \parallel [1\ 1\ 1]$. For $\vec{e}_1$ and $\vec{e}_2 \parallel [1+\sqrt{3}, 1-\sqrt{3}, -2]$ one can excite the longitudinal exciton ($\vec{b} \parallel [1-\sqrt{3}, 1+\sqrt{3}, -2] \parallel k$).

The resonances in Fig. 21 show a linear and a quadratic shift of the peak energy. The sign of the linear shift depends on the crystal orientation with respect to the field direction. One expects such an effect in zincblende structure, since for $\vec{F} \parallel [1\ 1\ 1]$ it makes a difference if the dipole moment is parallel or antiparallel to the field direction. The quadratic effect is caused by the envelope Stark effect which is due to the field induced repulsion by P excitons ($n \geq 2$).

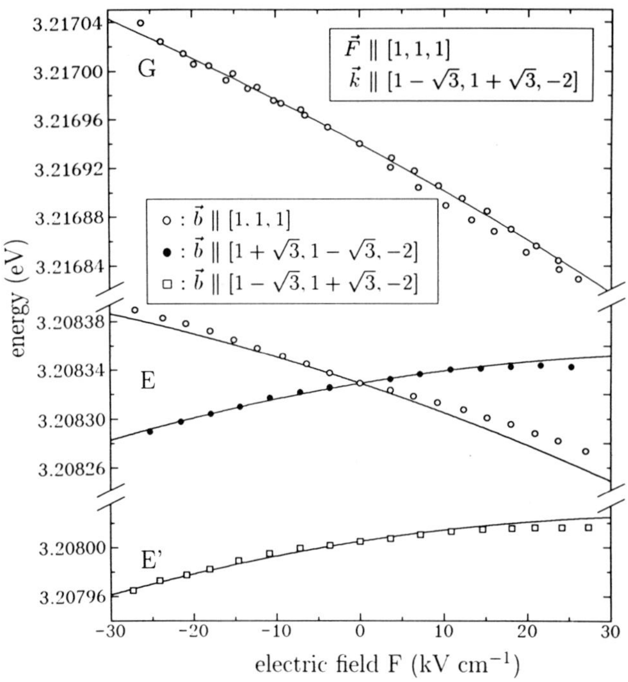

Figure 21. Splitting and shift of 1S resonances in an electric field for different orientation of $\vec{b}$. Solid lines represent theoretical energy values derived from the Hamiltonian (Eq. (48)) using d_F and a_0 given in the text and zero-field parameters from Table 1.

The solid lines in Fig. 21 are a fit to a theory which is based on the following Hamiltonian:

$$\begin{aligned} H &= H_{so} + H_{ex} + H_F \\ H_{so} &= \Delta_{so} \vec{l}_h \cdot \vec{\sigma}_h \\ H_{ex} &= \left[\frac{1}{4} - \vec{\sigma}_e \cdot \vec{\sigma}_h\right] \left[\Delta_{ex} - \Delta_{dip}\left(1 - 3\frac{(\vec{k}\cdot\vec{b})^2}{|\vec{k}|^2|\vec{b}|^2}\right)\right] \\ H_F &= d_F \left[(l_{h,x} l_{h,y} + l_{h,y} l_{h,x})\, F_z \; + \; c.p.\right] \; - e\, \vec{F} \cdot (\vec{r}_e - \vec{r}_h) \end{aligned} \tag{48}$$

H_{so} accounts for spin–orbit interaction, H_{ex} and H_F for exchange and electric field interaction, respectively. The values for Δ_{ex} and Δ_{dip} are given in Table 1. H_F consists of two terms. The first term leads to the linear Stark effect of the valence bands, where d_F is the dipole matrix element of the Γ_5 valence band states. The second term causes a mixing of S and P envelope states. The corresponding matrix element is proportional to the exciton Bohr radius a_0, which is a fit parameter. The eigenvalues and eigenvectors are then used to calculate the polariton dispersion using the two–oscillator Sellmeyer formula (Eq. (46) and Table 1). As a result of the fit we get $d_F = (-6.65 \pm 0.08) \cdot 10^{-9}$ e·cm and $a_0 = (5.8 \pm 0.3)$Å. One can also study the electric field effects on the 2S and 2P excitons, where one expects changes of the oscillator strength due to a mixing of S and P envelope functions.

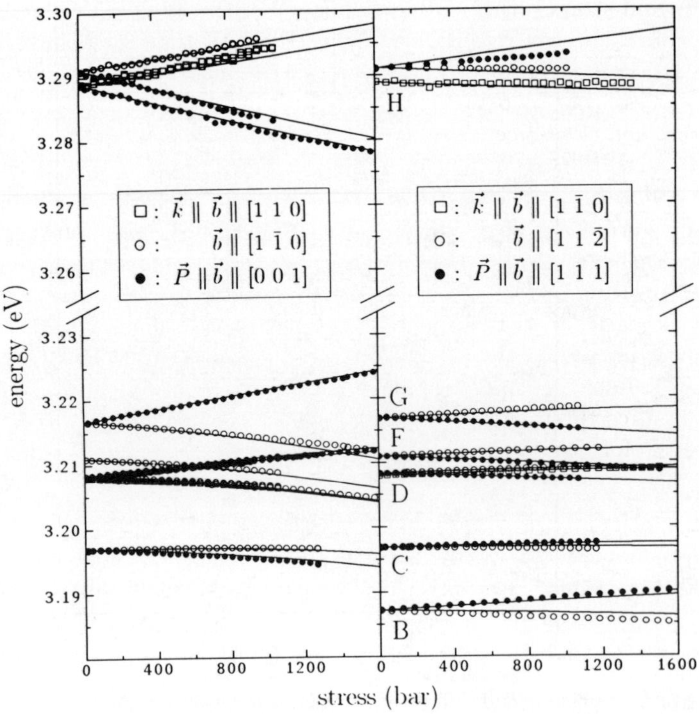

Figure 22. Stress-dependent shift and splitting of resonances in CuCl at 5 K for different orientation of $\vec{b}$, stress $\vec{P}$ parallel to [001] and [111].

It should finally be mentioned that AC electric fields with a frequency of 300 Hz were used, since polarization effects were observed in DC-fields, which lead to much smaller field-induced effects.

3. Exciton–Polaritons in CuCl under Uniaxial Stress. As a last example we will now present measurements under uniaxial stress. Again the small linewidth of the resonances in nonlinear spectroscopy leads to a clearly resolved stress dependence of different resonances as shown in Fig. 22. The solid lines are a fit to a theory which is based on a Hamiltonian similar to Eq. (48) where H_F is replaced by the terms $H_1 + H_2$ which describe the stress dependence of the exciton states.

$$
\begin{aligned}
H &= H_{so} + H_{ex} + H_1 + H_2 \, , \\
H_1 &= a\, Tr(\epsilon) \; - \; b\left[(l_{hx}^2 - l_h^2/3)\, \epsilon_{xx} + c.p.\right] \\
&\quad - d\, \left[(l_{h,x} l_{h,y} + l_{h,y} l_{h,x})\, \epsilon_{xy} \; + \; c.p.\right] \, , \\
H_2 &= \tilde{a}\, Tr(\epsilon)\, \vec{l}_h \vec{\sigma}_h \, .
\end{aligned}
\tag{49}
$$

H_1 is the well known Pikus–Bir Hamiltonian [55, 56]. a, b, d are the hydrostatic, tetragonal and trigonal deformation potentials. H_2 takes the stress-dependence of the spin-orbit splitting into account. The matrix elements are calculated in the basis of the twelve 1S excitons (jj-coupling scheme). The resulting 12 x 12 matrix which has nondiagonal exchange and stress terms is diagonalized. For the analysis of the experimental data the eigenvalues and eigenvectors are then used to calculate the polariton dispersion and the resonances taking energy and k–vector conservation for the particular two– or three–photon process into account. The hydrostatic deformation potential a and the parameter $\tilde{a}$ are taken from recent measurements by Reimann and Rübenacke [57], who studied the shift of the longitudinal Z_3 and $Z_{1,2}$ excitons under hydrostatic pressure up to 40 kbar. Therefore there is only one adjustable parameter for each stress direction. The tetragonal deformation potential b is determined from measurements with stress parallel [0 0 1] and the trigonal deformation potential d from measurements with stress parallel [1 1 1]. In Table 2 we present our results and results of other authors, which were gained by linear optical methods. The advantage of nonlinear as compared to linear optical experiments for the determination of deformation potentials is even more evident in alkali halides as was recently shown for KI by Fröhlich et al. [58]. In this case the stress–induced splitting of paraexcitons was studied, which led to an accurate determination of deformation potentials.

Table 2. Deformation potentials of CuCl. Numbers in parentheses give experimental errors.

	Fröhlich *et. al.* [59]	Koda *et. al.* [51]	Blacha *et. al.* [60]
a [eV]	-0.31(1)	≅0	-0.45(3)
b [eV]	1.34(1)	1.18	—
d [eV]	-1.65(3)	-1.81	-0.86(8)
ã [eV]	0.02(1)	—	0.03(2)

VII.B. Two–Photon Absorption to Landau Levels

Besides a magnetic field induced splitting of excitons, which we have discussed for CuCl, one expects also two–photon transitions between Landau levels. For a recent review of nonlinear magnetooptics in solids we refer to [61]. First measurements of two–photon photoconductivity on rather low–gap materials (PbTe and InSb) are reported by Button et al. [62]. Bassani and Girlanda [63, 64] discuss in detail different cases of two–photon selection rules. They derive explicit expressions for the two–photon absorption constant for $\Delta n = 0$ and $\Delta n = \pm 1$ transitions (Δn: difference of Landau quantum numbers of conduction and valence band). Of special interest are experiments on II-VI and III-V semiconductors, since they allow to determine fundamental band parameters as effective masses and g values [65, 66]. First measurements are reported by Fröhlich et al. [67]. As an example we will present recent two–photon magneto–absorption measurements in multiple quantum wells (MQW) by Engbring and Zimmermann [68]. For a recent review on optical properties of MQW we refer to [69]. In Fig. 23 we show a schematic sketch of two–photon selection rules for one– and two–photon transitions in a two–band model [68]. One can discuss the absorption mechanism in the Landau or magnetoexciton picture. Since we use for our two–photon experiments a CO_2 laser ($\hbar\omega \sim 0.1$ eV) as a high–power laser, the tunable dye laser ($\hbar\omega \sim 1.5$ eV) is close to the 1S exciton resonance. We therefore consider in the expression for the two–photon absorption constant $\alpha^{(2)}$ (Eq. (10)) only the 1S exciton as an intermediate state, which

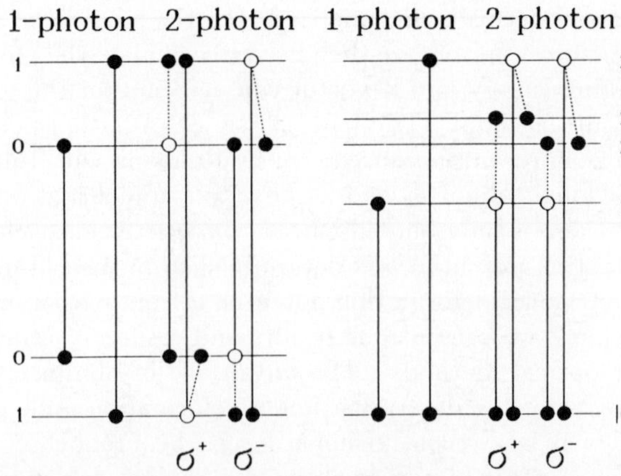

Figure 23. Selection rules for transitions between valence band and conduction band in the two-band model both for one-photon absorption and for two-photon absorption (schematically). Solid and dotted lines represent matrix elements due to the dye laser and the CO_2 laser, respectively. Open circles mark the virtual intermediate states $|i\rangle$ considered. The interband selection rule is independent of the dye laser polarization. The intraband transitions due to the CO_2 laser lead to a two-photon selection rule of $\Delta n = n_e - n_h = \pm 1$ in the Landau picture (left part) or to $nP^{\pm}$ in the magnetoexciton picture (right part) depending only on the CO_2 laser polarization (σ^+, σ^-).

corresponds to a $\Delta n = 0$ transition in the Landau picture. Because of the large resonance denominator the contribution of the CO_2 laser to the interband transition is much smaller. The CO_2 laser induces the intraband transition from the nS state to an $n'P^{\pm}$ state depending on its polarization (σ^+ or σ^-). In the Landau picture this corresponds to a cyclotron transition either in the valence or in the conduction band. It is shown in Fig. 23, that this leads to two-photon selection rules $\Delta n = \pm 1$, which depend only on the CO_2 laser polarization. Experimental results are shown in Fig. 24. A close inspection of the data shows that there is an additional rather small dependence on the polarization of the dye laser. As discussed in detail in Ref.[68] there are several reasons for this dependence. The main point is that the dye laser polarization defines the spin of the excited electron and hole. One thus expects a spin splitting for different dye polarizations. It is of special interest to compare the field dependence of one- and two-photon data from the same sample. The analysis of these data leads to a consistent set of material parameters as shown in Ref.[68].

VIII. EXPERIMENTAL METHODS FOR TWO–AND THREE–PHOTON SPECTROSCOPY

Two–photon processes were treated by Maria Göppert–Mayer already in 1931 [2] but it took 30 years to have them demonstrated by optical means. Kaiser and Garrett [70] observed a blue fluorescence in $CaF_2{:}Eu^{2+}$ after irradiation with red light of a ruby laser. The square–dependence of the fluorescence intensity on the laser intensity proved that they had seen two–photon transitions. Hopfield et al. [3] were the first to do two–photon spectroscopy. They used a ruby laser and the continuous spectrum of a Xenon lamp (and monochromator) to measure a two–photon spectrum of KI. The experimen-

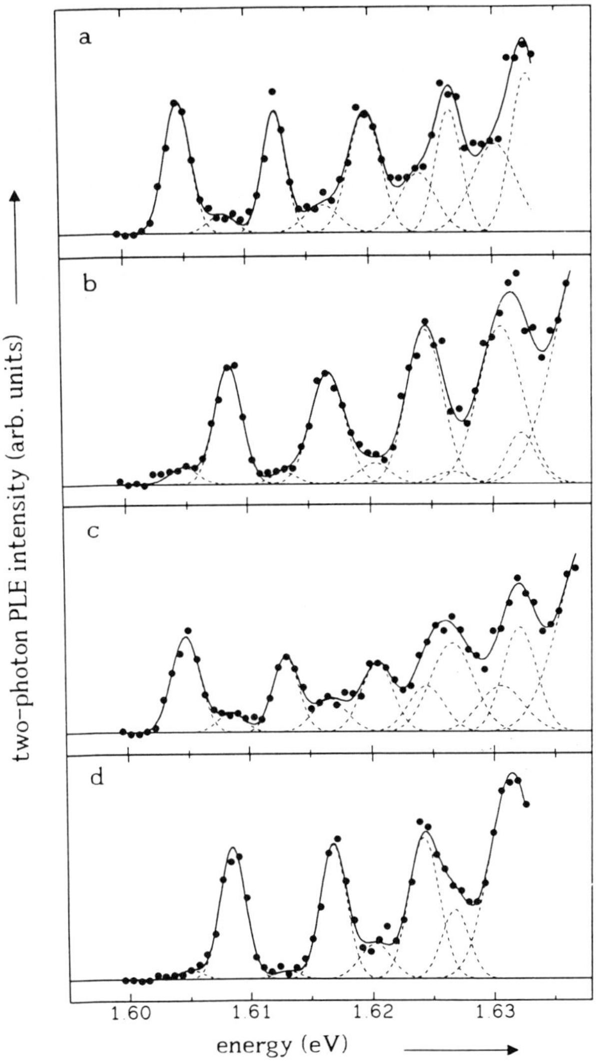

Figure 24. Two-photon excitation spectra of GaAs–$Al_{0.3}Ga_{0.7}As$ sample with well width 5.8 nm in a magnetic field of 5 T for different polarization combinations of dye and CO_2 laser (dye, CO_2 laser). a) (σ^-, σ^-); b) (σ^-, σ^+); c) (σ^+, σ^-); d) (σ^+, σ^+). Solid lines are the result of a fit by a superposition of Gaussians (dashed curves).

tal techniques were improved by Fröhlich and Mahr [71] by the use of a Q–switched laser and a pulsed flashlamp. The first automatic two–photon spectrophotometer was reported by Staginnus et al. [72]. For further literature on the early experiments we refer to the review article by Mahr [4]. In the experiments mentioned so far one measures a two–photon absorption constant $\alpha^{(2)}(\omega_1)$ as derived from second–order perturbation theory (Eq. (10)). $\alpha^{(2)}(\omega_1)$ measures the absorption for a tunable source of photon energy $\hbar\omega_1$ in the presence of a fixed frequency laser of photon energy $\hbar\omega_2$. $\alpha^{(2)}(\omega_1)$

can also be derived from the imaginary part of $\chi^{(3)}(\omega_1)$ [4]. $\alpha^{(2)}(\omega_1)$ is given by

$$\begin{aligned} \alpha^{(2)}(\omega_1) &= (1/d)\, ln\,(I_0/(I_0 - \Delta I)) \\ &\approx (1/d)\, \Delta I/I_0 \quad (\alpha^{(2)} d \ll 1)\ . \end{aligned} \tag{50}$$

I_0 is the intensity of the transmitted light for the weak tunable source without high-power laser (intensity I_L). ΔI is the change in transmitted intensity induced by the high-power laser. Since in most cases $\Delta I/I_0 \ll 1$, one can use the approximation given in Eq. (50). Since $\alpha^{(2)}$ is proportional to N_2 (number density of photons of energy $\hbar\omega_2$, Eq. (10)) one has to measure the intensity of the high-power laser to correct for intensity fluctuations. A typical setup for this technique is shown in Fig. 25. With this technique one measures a rather small signal ($\Delta I \sim 10^{-1} I_0$) on the large background (I_0). Any fast fluctuations of I_0 might cause problems, since it is difficult to correct for them by electronic means. Some details concerning this point are discussed in a review by Fröhlich and Sondergeld [73]. For three-photon absorption one could also define an absorption constant $\alpha^{(3)}$ which would then be proportional to the square of the intensity of a fixed frequency laser. $\alpha^{(3)}$ can be derived from third-order perturbation theory or from the imaginary part of $\chi^{(5)}(\omega_1)$. Unless there is a strong resonance enhancement the expected third-order effects are too small to be detected as a change of transmission (ΔI-effect). In this case one monitors the simultaneous absorption of two or three photons by the subsequent emission of one photon, which has in most cases less energy than the sum of the incoming photons due to relaxation processes. The main advantage of this technique, which might be termed "nonlinear excitation spectroscopy", is due to the fact that one does not have to measure a small signal (ΔI) on a

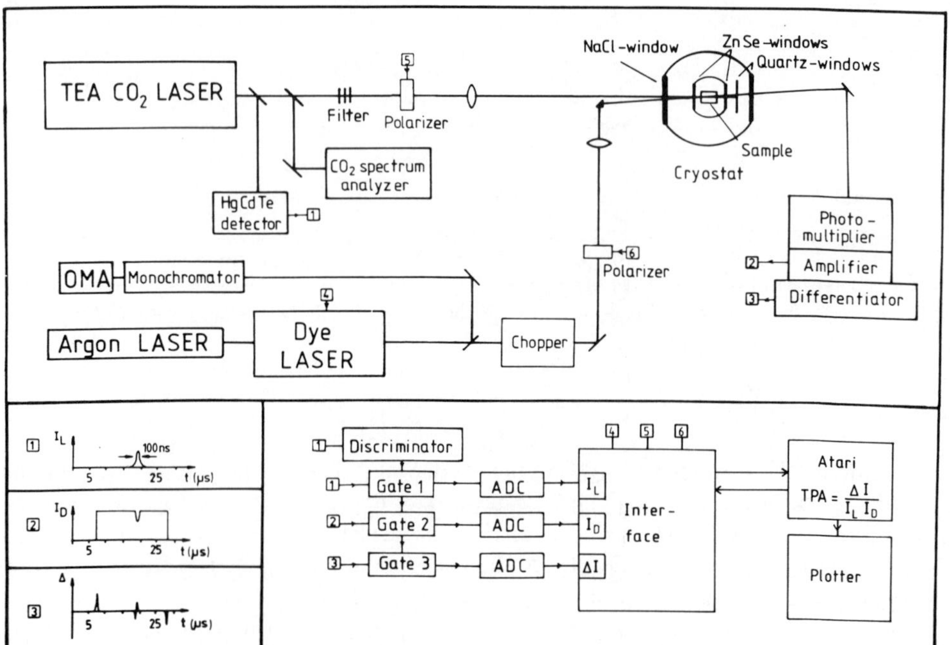

Figure 25. Schematic diagram of experimental setup for two-photon absorption measurements.

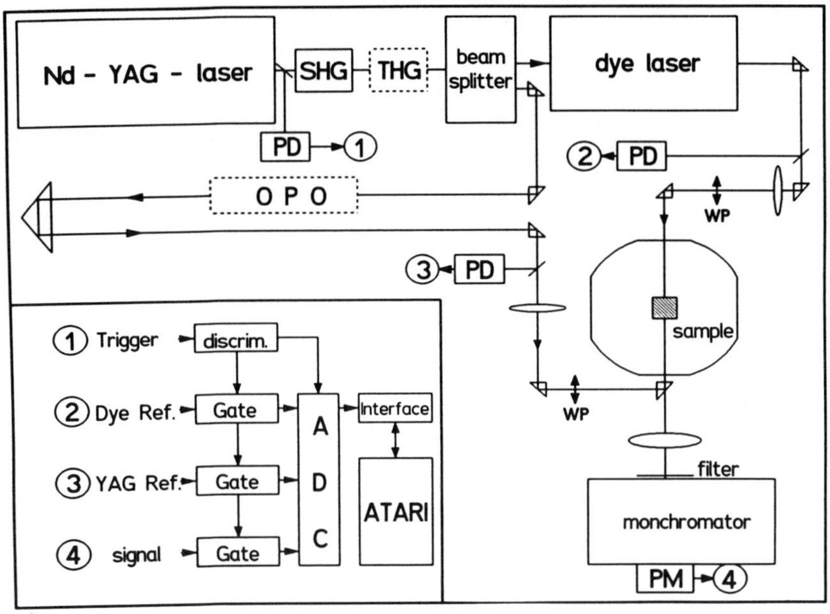

Figure 26. Schematic diagram of experimental setup for three-photon measurements. SHG, second-harmonic generator; THG, third-harmonic generator; OPO, optical parametric oscillator; PD, photodiodes for reference; WP, quarter or half-wave plate; PM, photomultiplier; ADC, analog-to-digital converter.

large background (I_0). This technique was applied to observe three–photon absorption in alkali halides [50]. A typical setup for this technique is shown in Fig. 26.

As discussed before (chapter IV) there are also coherent techniques like TP–SFG and TP–DFG. In this case the emitted beam is expected in a small solid angle. The direction of emission is determined by the phase–matching conditions. In order to discriminate against the intense laser beam one can use a setup of 4 Pellin–Broca prisms as shown in Fig. 27. Due to the dispersion of the quartz prisms the laser beam can be blocked after passing two prisms. This setup can also be used to measure the three–photon reflection for perfect normal incidence of the laser beam as discussed in chapter V.

For the measurements of two–photon absorption to Landau levels in multiple quantum wells a special setup with a picosecond tunable dye laser and a synchronized CO_2 laser was used (Fig. 28). The synchronization was achieved by plasma switching on a germanium plate. The 150 ns pulse of the high–power CO_2 laser (up to 1 J with amplifier) is incident at Brewster angle with perfect horizontal polarization on the germanium plate. A Nd–YAG laser pulse (pulsewidth 60 ps), which then is perfectly synchronized to the dye pulse, excites a plasma in the germanium plate, which leads to an instantaneous rise of the reflectivity for the 10.6 μm pulse of the CO_2 laser. The finite lifetime of the plasma results in a pulsewidth of about 250 ps for the 10.6 μm pulse. The shortening of the high–power CO_2 pulse was necessary to avoid heating and damage of the quantum well samples. The measurements are done with circularly

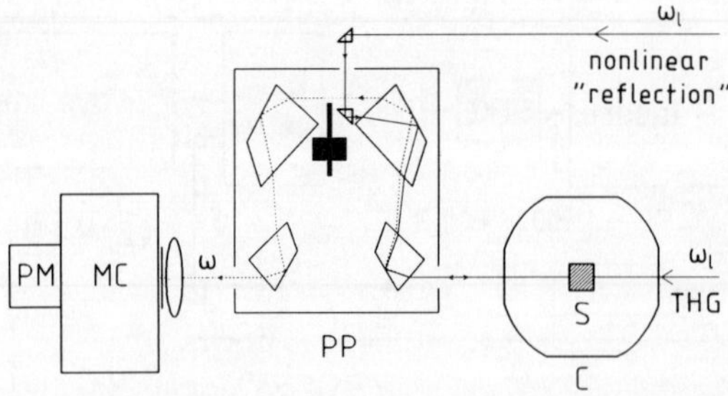

Figure 27. Schematic diagram of part of experimental setup for detection of coherent processes. S, sample; C, cryostat; PP, Pellin–Broca prisms; MC, monochromator; PM, photomultiplier; solid and dashed lines mark incoming laser beams and outgoing signal beams, respectively.

polarized light. The TPA signal is detected via the luminescence of the lowest exciton. A monochromator and suitable filters are used to suppress scattered light from the high–power dye laser.

In all cases of nonlinear spectroscopy where more than one photon beam is used, it is of great importance to guarantee a stable overlap of the different laser beams in the sample. The overlap of e.g. the beams from a CO_2 laser (10.6 μm) and a dye laser (800 nm) can be adjusted by detection of the sum–frequency at 743.9 nm from a polycrystalline ZnSe sample, which is transparent for both laser beams. Optical parts like mirrors and lenses have to be mounted with the laser sources on mechanically stable benches and tables, which allow a decoupling from vibrations of the building.

In the setups discussed in this chapter lasers of low repetition rate (10 Hz) but high intensity (up to 10^8 W cm^{-2}) were used. The signal–to–noise ratio can certainly be improved by the use of lasers of higher repetition rate.

IX. CONCLUDING REMARKS

This contribution is certainly not meant as a review on two– and three–photon spectroscopy of solids. The author has selected mainly experimental results from his group to demonstrate the far–reaching possibilities of these nonlinear techniques. They can certainly be applied to other materials besides crystalline semiconductors and insulators. Organic crystals, glasses and low dimensional systems down to quantum dots are also studied by nonlinear methods as outlined in other contributions of this course.

As mentioned already in Chapter II we have only discussed experiments where the laser intensity is high enough to allow the detection of nonlinear processes, but low enough not to have any influence on the electronic system. For high intensities drastic changes of linear and nonlinear spectra can be observed. Since these effects are caused by the strong electric field of the laser they are termed "optical Stark effect". As compared to the classical Stark effect, where a constant electric field is applied to the sample, there is again an "additional degree of freedom" in the optical Stark effect, since one can choose the optical frequency and thus the photon energy to be resonant or nearly resonant with electronic transitions. For further information on various aspects of the optical Stark effect especially in semiconductors we refer to Ref. [9, 10].

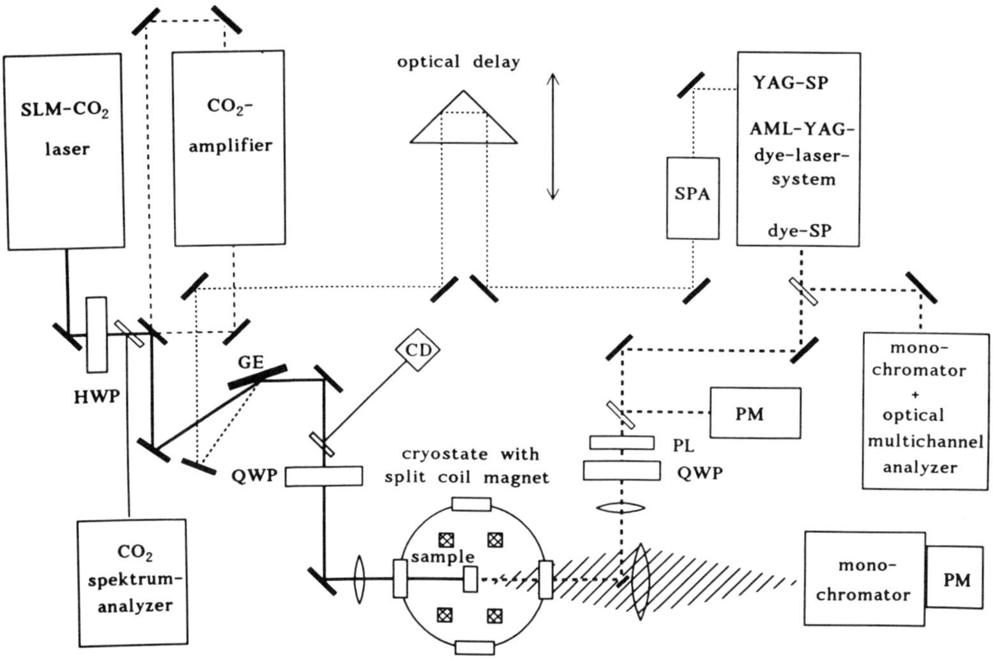

Figure 28. Schematic diagram of experimental setup for magnetooptical measurements of multiple quantum wells. AML–YAG, active mode–locked YAG laser; SPA, single pulse amplifier; dye–SP, dye single pulse; SLM–CO_2 laser, single longitudinal mode CO_2 laser; GE, germanium plate; PL, polarizer; QWP, quarter–wave plate; HWP, half–wave plate; PM, photomultiplier; CD, control diode for reference.

ACKNOWLEDGMENTS

The author is grateful to his co–workers with whom he has worked over years on various projects of nonlinear spectroscopy in solids. Without their dedicated commitment most of the experiments would not have been done. Technical help for preparation of the manuscript by G. Biese is appreciated. Many of the projects were supported by the Deutsche Forschungsgemeinschaft.

REFERENCES

[1] R. L. Macfarlane. High resolution nonlinear spectroscopy of rare earth ions in solids. *This volume, page ....* (1993).

[2] M. Göppert-Mayer. Über Elementarakte mit zwei Quantensprüngen. *Ann. Phys.* **9**, *273* (1931).

[3] J. J. Hopfield, J. M. Worlock, and Kwangjai Park. Two–quantum absorption spectrum of KI. *Phys. Rev. Lett.* **11**, *414* (1963).

[4] H. Mahr. Two–photon absorption spectroscopy. In H. Rabin and C. L. Tang, editors, "Quantum Electronics", volume 1 A, pages 285–361. Academic Press, New York (1975).

[5] F. Beerwerth and D. Fröhlich. Three–photon spectroscopy of exciton polaritons in alkali halides. *Phys. Rev. Lett.* **55**, *2603* (1985).

[6] D. Fröhlich, St. Kirchhoff, P. Köhler, and W. Nieswand. Three–photon difference–frequency spectroscopy of polaritons in alkali halides. *Phys. Rev.* **B40**, *1976* (1989).

[7] M. Fiebig, D. Fröhlich, and Ch. Pahlke-Lerch. Multiphoton polariton spectroscopy on ZnO. *phys. stat. sol. (b)* **177**, *187* (1993).

[8] Ch. Uihlein, D. Fröhlich, and R. Kenklies. Investigation of exciton fine structure in Cu_2O. *Phys. Rev.* **B23**, *2731* (1981).

[9] D. Fröhlich, Ch. Neumann, B. Uebbing, and R. Wille. Experimental investigation of three–level optical Stark effekt in semiconductors. *phys. stat. sol. (b)* **159**, *297* (1990).

[10] R. Zimmermann and D. Fröhlich. Optical Stark Effect in Quantum Wells, *in:* F. Henneberger, St. Schmitt–Rink, and E. O. Göbel, "Optics of Semiconductors and Nanostructures", pages 51–74. Akademie Verlag, Berlin (1993).

[11] D. Fröhlich. 2–Photonenspektroskopie in Festkörpern, *in:* "Festkörperprobleme X", pages 227–255. Vieweg Verlag, Braunschweig (1970).

[12] E. Doni, R. Girlanda, and G. Pastori Parravicini. A note on the polarization dependence of two–photon absorption coefficient in crystals. *phys. stat. sol (b)* **65**, *203* (1974).

[13] G. D. Mahan. Theory of two–photon spectroscopy in solids. *Phys. Rev.* **170**, *825* (1968).

[14] A. Quattropani and R. Girlanda. Two–photon absorption in atoms and solids. *Riv. Nuovo Cimento* **6**, *1* (1983).

[15] A. Pasquarello and A. Quattropani. Two–photon transitions to excitons in quantum wells. *Phys. Rev.* **B42**, *9073* (1990).

[16] K. Teegarden and G. Baldini. Optical absorption spectra of the alkali halides at 10 K. *Phys. Rev.* **155**, *896* (1967).

[17] M. Inoue and Y. Toyozawa. Two–photon absorption and energy band structure. *J. Phys. Soc. Japan* **20**, *363* (1965).

[18] T. R. Bader and A. Gold. Polarization dependence of two–photon absorption in solids. *Phys. Rev.* **171**, *997* (1968).

[19] M. M. Denisov and V. P. Makarov. Polarization dependence of the two–photon absorption coefficient in crystals. *J. Phys. C: Solid State Phys.* **5**, *2651* (1972).

[20] A. Pasquarello and A. Quattropani. Polarization dependence of multiphoton transitions. *Phys. Rev.* **B43**, *3837* (1991).

[21] G. F. Koster, J. O. Dimmock, R. G. Wheeler, and H. Statz. "Properties of the Thirty–Two Point Groups". M.I.T. Press Cambridge (Mass.) (1963).

[22] D. Fröhlich, B. Staginnus, and S. Thurm. Symmetry assignments of two–photon transitions in TlCl. *phys. stat. sol.* **40**, *287* (1970).

[23] E. Doni, R. Girlanda, and G. Pastori Parravicini. A note on the polarization dependence of two–photon absorption coefficient in crystals, extension to circularly polarized light. *phys. stat. sol (b)* **88**, *773* (1978).

[24] A. Pasquarello and L. C. Andreani. Interpretation of three–photon spectra in alkali halides. *Phys. Rev.* **B41**, *12230* (1990).

[25] K. Huang. On the interaction between the radiation field and ionic crystals. *Proc. Roy. Soc.* **A208**, *352* (1951).

[26] S. I. Pekar. The theory of electromagnetic waves in a crystal in which excitons are produced. *Sov. Phys. JETP* **6**, *785* (1958).

[27] J. J. Hopfield. Theory of the contribution of excitons to the complex dielectric constant of crystals. *Phys. Rev.* **112**, *1555* (1958).

[28] R. Loudon. Nonlinear optics with polaritons. Proceedings of the International School of Physics "Enrico Fermi", Course LXIV. In N. Bloembergen, editor, "Nonlinear Spectroscopy", pages 296–318. North Holland, Amsterdam, New York, Oxford (1977).

[29] B. Hönerlage, R. Lévy, J. B. Grun, C. Klingshirn, and K. Bohnert. The dispersion of excitons, polaritons and biexcitons in direct-gap seminconductors. *Physics Reports* **124**, *161* (1985).

[30] F. Bassani and L. C. Andreani. Exciton–polariton states in insulators and semiconductors. Proceedings of the International School of Physics "Enrico Fermi", Course XCVI. In U. W. Grassano and N. Terzi, editors, "Excited State Spectroscopy in Solids", pages 1–36. North-Holland, Amsterdam, New York, Oxford (1987).

[31] T. Kurosawa. Polarization waves in solids. *J. Phys. Soc. Japan* **16**, *1298* (1961).

[32] D. Fröhlich, E. Mohler, and P. Wiesner. Observation of exciton polariton dispersion in CuCl. *Phys. Rev. Lett.* **26**, *554* (1971).

[33] D. C. Haueisen and H. Mahr. Nonlinear electronic dispersion in CuCl. *Phys. Rev. Lett.* **26**, *838* (1971).

[34] B. Hönerlage, R. Lévy, and J. B. Grun. Four–wave mixing and hyper–Raman scattering in CuCl. *Opt. Commun.* **43**, *443* (1982).

[35] R. Lévy, F. Tomasi, J. Y. Bigot, and K. B. Grun. Nonlinear spectroscopy of CuCl using nondegenerate four–wave mixing. *J. Lumin.* **35**, *79* (1986).

[36] N. Bloembergen and P. S. Pershan. Light waves at the boundary of nonlinear media. *Phys. Rev.* **128**, *606* (1962).

[37] J. Ducuing and N. Bloembergen. Observation of reflected light harmonics at the boundary of piezoelectric crystals. *Phys. Rev. Lett.* **10**, *474* (1963).

[38] G. Mizutani and N. Nagasawa. A study on the optical nonlinearity due to the Z_3 and $Z_{1,2}$ excitons of CuCl by the reflected four–wave mixed light generation. *Opt. Commun.* **50**, *31* (1984).

[39] D. Fröhlich, P. Köhler, W. Nieswand, and Ch. Pahlke. Three–photon reststrahl bands near exciton resonances. *Phys. Rev. Lett.* **66**, *2136* (1991).

[40] D. Fröhlich, J. Kauschalek, W. Nieswand, and Ch. Pahlke. Three–photon spectroscopy on thin films. *Phys. Rev.* **B45**, *11016* (1992).

[41] N. Bloembergen. "Non–linear Optics". Benjamin, New York (1965).

[42] D. Fröhlich, St. Kirchhoff, P. Köhler, and W. Nieswand. Magnetic field splitting and dispersion of the lower 1S–exciton polariton in KI and RbI. *phys. stat. sol. (b)* **158**, *267* (1990).

[43] G. Baldini, A. Bosacchi, and B. Bosacchi. Exciton–phonon interaction in alkali halides. *Phys. Rev. Lett.* **23**, *846* (1969).

[44] D. Fröhlich and B. Staginnus. New assignment of the band gap in the alkali bromides by two–photon spectroscopy. *Phys. Rev. Lett.* **19**, *496* (1967).

[45] D. Fröhlich and H. Volkenandt. Determination of Γ_3 valence band in CuCl by two–photon absorption. *Solid State Commun.* **43**, *189* (1982).

[46] M. Cardona. Optical properties of the silver and cuprous halides. *Phys. Rev.* **129**, *69* (1963).

[47] S. Nikitine, J. Ringeissen, and J. L. Deiss. Étude du spectre excitonique de CuCl. Influence du champ electrique. *J. Phys. Radium* **23**, *890* (1962).

[48] D. Fröhlich, E. Mohler, and Ch. Uihlein. Polariton fusion in CuCl. *phys. stat. sol. (b)***55**, *175* (1973).

[49] D. Boggett and R. Loudon. Theory of exciton–polariton generation by two–photon absorption. *J. Phys.* **C 6**, *1763* (1973).

[50] F. Beerwerth, D. Fröhlich, P. Köhler, V. Leinweber, and A. Voß. Three–photon spectroscopy of excitons and polaritons in alkali halides. *Phys. Rev.* **B38**, *4250* (1988).

[51] T. Koda, T. Murahashi, T. Mitani, S. Sakoda, and Y. Onodera. Effects of uniaxial stress on excitons in CuCl. *Phys. Rev.* **B5**, *705* (1972).

[52] D. Fröhlich, P. Köhler, W. Nieswand, and E. Mohler. Dispersion of the exciton polariton branches in CuCl and their k–dependent splitting in a magnetic field. *phys. stat. sol. (b)* **167**, *147* (1991).

[53] L.C. Andreani, F. Bassani, and A. Quattropani. Longitudinal–transverse splitting in Wannier excitons and polariton states. *Nuovo Cimento* **10D**, *1473* (1988).

[54] G. Biese, D. Fröhlich, W. Nieswand, and E. Mohler. Two–photon spectroscopy of 1S excitons in an AC electric field. J. Lumin. To be published.

[55] G. E. Pikus and G. L. Bir. Effect of deformation on the hole energy spectrum of germanium and silicon. *Sov. Phys. Solid State* **1**, *1502* (1959).

[56] G. L. Bir and G. E. Pikus. Effect of deformation on the energy spectrum and electrical properties of semiconductors of the InSb type. *Sov. Phys. Solid State* **3**, *2221* (1962).

[57] K. Reimann and St. Rübenacke. Two–photon–excited luminescence in CuCl and CuBr under hydrostatic pressure. To be published.

[58] D. Fröhlich, W. Nieswand, and St. Rübenacke. Determination of deformation potentials by three–photon piezoabsorption of paraexcitons in KI. *Phys. Rev.* **B47**, *6736* (1993).

[59] D. Fröhlich, W. Nieswand, and St. Rübenacke. Nonlinear spectroscopy of 1S excitons in CuCl under uniaxial stress. J. Lumin. To be published.

[60] A. Blacha, S. Ves, and M. Cardona. Effects of uniaxial strain on the exciton spectra of CuCl, CuBr, and CuI. *Phys. Rev.* **B27**, *6346* (1983).

[61] D. Fröhlich. Nonlinear magnetooptics in solids. In E. Doni, R. Girlanda, G. Pastori Parravicini, and A. Quattropani, editors, "Progress on Electron Properties of Solids", pages 255–274. Kluwer Academic Publishers (1989).

[62] K. J. Button, B. Lax, M. H. Weiler, and M. Reine. Multiphoton magneto–optical resonance in PbTe and InSb. *Phys. Rev. Lett.* **17**, *1005* (1966).

[63] F. Bassani and R. Girlanda. Two–photon transitions in solids in a magnetic field. *Opt. Commun.* **1**, *359* (1970).

[64] R. Girlanda. Two–photon absorption in semiconductors in a magnetic field. *Nuov. Cim.* **6B**, *53* (1971).

[65] Ch. Neumann, A. Nöthe, and N. O. Lipari. Two–photon magnetoabsorption of ZnTe, CdTe, and GaAs. *Phys. Rev.* **B37**, *922* (1988).

[66] A. Nöthe and G. Spata. Influence of coulomb correlation on Landau states in CdTe. *Phys. Rev.* **B39**, *8301* (1989).

[67] D. Fröhlich, R. Wille, W. Schlapp, and G. Weimann. Two–photon magnetoabsorption in multiple quantum wells. *Phys. Rev. Lett* **61**, *1878* (1988).

[68] J. Engbring and R. Zimmermann. Parametrization of S– and P–magnetoexcitons in quantum wells. *phys. stat. sol. (b)* **172**, *733* (1992).

[69] R. Cingolani and K. Ploog. Frequency and density dependent radiative recombination processes in III–V semiconductor quantum wells and superlattices. *Adv. Phys.* **40**, *535* (1991).

[70] W. Kaiser and C. G. B. Garrett. Two–photon excitation in CaF_2:Eu^{2+}. *Phys. Rev. Lett.* **7**, *229* (1961).

[71] D. Fröhlich and H. Mahr. Two–photon spectroscopy in anthracene. *Phys. Rev. Lett.* **16**, *895* (1966).

[72] B. Staginnus, D. Fröhlich, and T. Caps. Automatic 2–photon spectrophotometer. *Rev. Scient. Instr.* **39**, *1129* (1968).

[73] D. Fröhlich and M. Sondergeld. Experimental techniques in two–photon spectroscopy. *J. Phys. E: Scient. Instr.* **10**, *761* (1977).

NONLINEAR SPECTROSCOPY OF THE ELECTRON-HOLE PLASMA IN SEMICONDUCTORS

C. Klingshirn

Fachbereich Physik der Universität[1]
Erwin Schrödinger Straße 1
D 67663 Kaiserslautern, Germany

ABSTRACT

The intrinsic optical properties of semiconductors and insulators are governed in the vicinity of the fundamental absorption edge by the exciton states. Under intermediate excitation densities, optical nonlinearities arise which are due to coherent and incoherent interaction processes in the excitonic system. At the highest excitation densities a transition to a new collective state of electrons and holes occurs which is known as electron-hole plasma. In this contribution the properties of this electron-hole plasma are explained and some selected experimental examples are given for its nonlinear spectroscopy in three-dimensional semiconductors and in quasi two- and one-dimensional structures, respectively.

I. INTRODUCTION, OR WHAT IS AN ELECTRON-HOLE PLASMA?

The linear optical properties of semiconductors are governed in the vicinity of the fundamental absorption edge by excitons. These excitons may be either free, bound to some defects, or localized by disorder e.g. in mixed crystals. The properties of excitons including structures of reduced dimensionality have been reviewed many times e.g. [1 to 7] including this and previous courses of the School of Atomic and Molecular Spectroscopy [8 - 10], so that we can be rather short on this topic here. In Fig. 1a we show schematically the band-structure (Fig. 1a) and the optical transitions close to the global band-extrema for a direct and indirect gap semiconductor. It is obvious that in an absorption or emission process two particles are always created or annihilated, namely an electron in the conduction band and a hole in the valence band. In this sense optical transitions in semiconductors are always two particle transitions. The negatively charged electron and the positively charged

---

[1]New address: Institute für angewandte Physik der Universität
Postfach 6980, D 76128 Karlsruhe

hole form a series of bound states below the gap. These states are known as excitons. The ground state of a semiconductor with a completely filled valence band and a completely empty conduction-band has total momentum **K** = 0 and is considered as the zero of energy in the exciton dispersion of Fig.1b. Excitons are analogous to the hydrogen or the positronium atom, with the difference that the excitonic Rydberg energy Ry^* lies in semiconductors between 200 meV $\geq Ry^* \geq$ 5 meV and the excitonic Bohr radius a_o ranges from 1 nm to 20 nm. Both facts are consequences of the small reduced mass μ of electron and hole ($1m_o > \mu \geq 0.05\ m_o$) and of the dielectric constant ϵ ($15 \geq \epsilon \geq 5$) of the semiconductor in which electron and hole move. This type of exciton is called Wannier exciton [11]. It is characterized by a Bahr radius a_o which is larger than the lattice constant

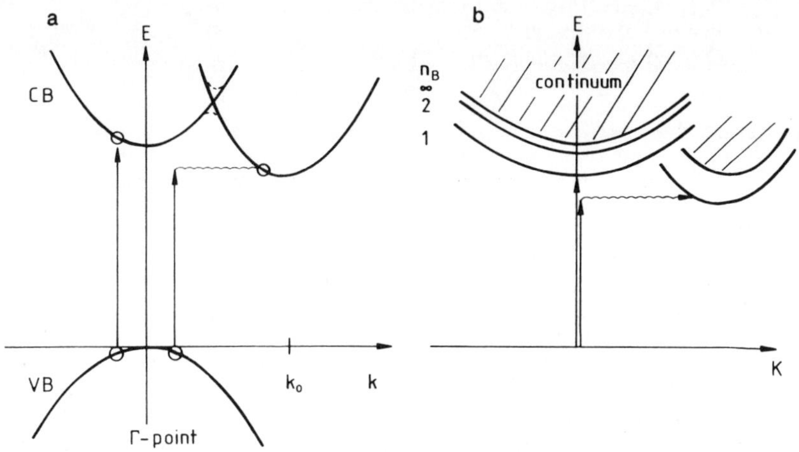

Figure 1. Schematic drawing of the bandstructure, i.e. of the one particle states for a direct and an indirect gap semiconductor (a) and the resulting dispersion for the excitons i.e. of the excited two particle states (b).

and allows therefore the application of the effective mass approximation in contrast to the so-called Frenkel excitons [12]. They occur preferentially in insualtors and are so small that the electron and hole are in one unit cell. We concentrate in the following exclusively on Wannier excitons.

In direct gap semiconductors with dipole allowed band-to-band transition singulet exciton states with S-type envelope function couple around **K** = 0 strongly to the electromagnetic radiation field resulting in characteristic resonances in the reflexion and absorption spectra (see Fig. 2a,b). They are properly described in the polariton picture [2-10]. In indirect semiconductors, the absorption coefficient is much smaller due to the participation of a momentum conserving phonon, and stays generally below $10^2 cm^{-2}$

compared to peak values up to $10^{4...6}cm^{-1}$ in the exciton resonances of direct gap materials. Consequently no excitonic structures appear in the reflection spectra of indirect gap transitions. In addition, the absorption sets in at values shifted by one or more phonon energies from the exciton resonance to higher or lower energies, depending if the phonon is created or (at higher lattice temperatures) absorbed. The excitonic absorption spectra of indirect gap semiconductors have threshold character since states with various $\hbar K$-values can be reached due to the phonon participation (see Fig.2c). Many examples of exciton spectra and further details of semiconductors can be found e.g. in [13] and references given therein.

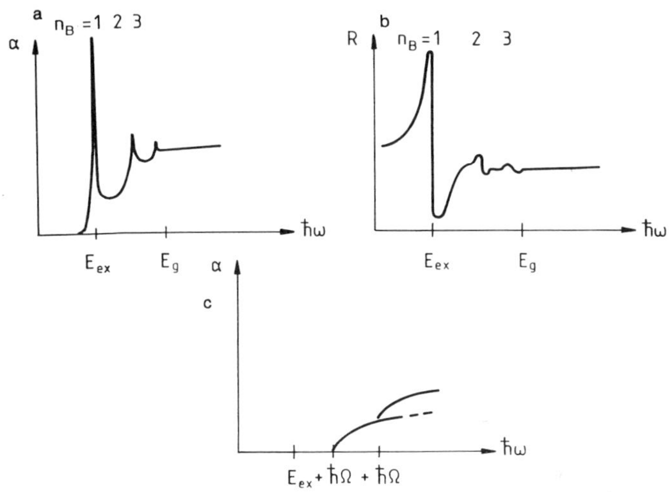

Figure 2. Schematic drawing of the absorption (a) and reflection spectra (b) of a direct gap semiconductor with dipole allowed band-to-band transition and of the absorption spectrum of an indirect gap semiconductor (c).

Concerning the nonlinear optical properties and spectroscopy of semiconductors, a rather general scenario has been found for all semiconductors. It is shown in Fig.3. In the low excitation limit i.e. in the regime of linear optics, the spectra are determined by non-interacting excitons. If the excitation intensity is increased, we come to the so-called intermediate density regime. In this regime, excitons are still good quasi particles but they start to interact with each other e.g. via scattering processes or the formation of excitonic molecules (biexcitons). The spectroscopy of coherent and incoherent optical nonlinearities of semiconductors in this regime has been treated in this school by various lectures [10] and also e.g. in [4-7,9]. We proceed therefore to the highest excitation levels. In this case, the density of electron hole pairs becomes at least in a fraction of the excited volume so high, that the average distance between electrons and holes is comparable to the excitonic Bohr radius. Under such conditions it is no longer possible to say, that a certain electron is bound to certain hole. Instead one has a new collective phase of electrons and hole in which the excitons cease to exist as individual quasi-particles. This new collective phase is called electron-hole plasma (EHP). The transition form a low density gas of excitons to the EHP is connected with significant changes of the optical properties. The description of these and of related phenomena is the topic of the next two sections. In order to keep the length of the article finite, we concentrate mainly on quasi-stationary excitation conditions

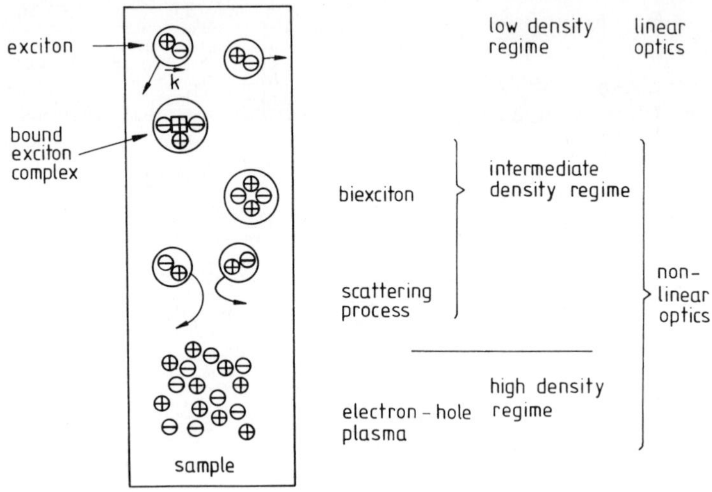

Figure 3. General scenario of the effects occuring in semiconductors with increasing density.

and give in the last section a short outlook to dynamic properties. Since the phase-relaxation time T_2 of carriers in an EHP is much shorter than their lifetime T_1, the optical nonlinearities presented in the following are essentially due to an incoherent population of excited states with density n_p. The description of optical nonlinearities by an expansion of the susceptibility $\chi(E)$ in a power series of the electric field, which is used in several contributions to this school, is adequate for coherent processes but fails here. We had rather to consider $\chi(n_p)$. Only for reasons of comparison, one may eventually express $\chi(n_p)$ as an effective $\chi_{eff}^{(3)}$. For the argument of finite length we can also present only some selected examples and cite only a small fraction of the relevant literature. We apologize for this shortcoming. Reviews for the theoretical description of the properties of the EHP are found e.g. [5,6,14-16] experimental data are compiled in [17,18].

II. PROPERTIES OF THE ELECTRON-HOLE PLASMA

In this section we present the many-particle and renormalization effects of an EHP, the phase-diagram, which can be expected under quasi-equilibrium conditions, and the optical properties.

II.A Many particle effects in the electronic system of semiconductors

There are two main many particles effects in the electron-hole plasma. One is the disappearance of the exciton resonances mentioned already in the introduction, the other is the renormalization of the width of the forbidden gap accompanied with a filling of the bands according to Fermi-Dirac statistics.

Let us adress the first question. How can exciton states disappear from the optical spectra? One contribution comes from screening. It is especially important in three dimensional semiconductors. We can understand this effect in a rather classical picture. Imagine the electron and the hole orbiting around their common center of mass under the influence of their Coulomb attraction. If the electron-hole pair density is increased, it will occur that other carriers penetrate with a certain probability in the electron-hole pair orbit and "cut" the field lines. This phenomenon reduces the attraction between electron and hole and consequently also the binding energy of the exciton. Mathematically this effect can be described by a transition from a Coulomb potential to a screened Coulomb potential

$$V(|r_e - r_h|) = (4\pi\epsilon\epsilon_o)^{-1}e^2/|r_e - r_h| \Rightarrow V'(|r_e - r_h|) = (4\pi\epsilon'\epsilon_o)^{-1}e^2\exp(-\kappa|r_e - r_h|)/|r_e - r_h| \quad (1)$$

The quantity $\kappa(n_p)$ is an inverse screening length and increases with increasing electron-hole pair density n_p. It can be shown that no more bound state excists in bulk semiconductors, if κ and a_o fulfill the inequality (2)

$$\kappa(n_p)a_o \geq 1.19 \quad (2)$$

and this means, that the exciton does no longer exist as a bound state. Without going into details, it should be mentioned, that the dielectric function ϵ appearing in (1) changes also with increasing n_p.

The other effect, which makes excitons disappeare with increasing n_p is phase-space filling. This effect is especially important in systems of reduced dimensionality since field lines propagating through the barrier material are difficult to screen. The phase-space filling argument goes basically as follows. The wavefunction of an exciton can be written schematically as in (3)

$$\phi_{\vec{K},n_B,l,m} = \Omega^{-\frac{1}{2}} e^{i\vec{K}\vec{R}} \varphi_e(r_e)\varphi_n(r_h)\Psi_{n_{B,l,m}}(r_e - r_h) \quad (3)$$

There is a normalization factor $\Omega^{1/2}$, a plane wave term for the total exciton momentum $K = k_e + k_h$ and the center of mass coordinate R, wavepakets of electron and hole formed from a superposition of Bloch-states and finally the Hydrogen-like relative motion of electron and hole, characterized by the usual quantum numbers. In order to form the wavepakets and to describe the relative motion of them with a finite radius a_o, Bloch-functions from a certain range of the Brillouin zone are necessary. This range increases with decreasing a_o. If we populate by the excitation these states with electrons and holes these wave-functions are blocked due to Pauli's principle and can no longer be used to build up the exciton wave function. As a consequence of this filling or blocking of phase space, the excitons disappeare. Both the screening and the phase-space blocking contribute to the disappearance of the excitons with increasing n_p with relative contributions depending on the materials and on the dimensionality of the problem as stated above [19].

Now we proceed to the other aspect namely the band-gap renormalization. We forget for the moment the excitons and consider a semiconductor with a band-gap E_g at low excitation. We populate the conduction and valence-bands with an equal density n_p of electrons and holes. If theses carriers are randomly distributed, the width of the gap would not change, since the repulsive energies between equal carriers would be exactly compensated by the attractive energies of opposite carriers. Indeed, the carriers are not randomly distributed in an EHP due to exchange- and correlation effects. The exchange effect (or Pauli's principle) forbids that equal carriers with parallel spin at sitting in the same unit cell or close together. In a random distribution this situation would occur more and more frequently with increasing n_p. Since the exchange excludes this situation it increases the average distance between equal carriers and thus reduces their average Coulomb repulsion. A reduction of a repulsive term is equivalent to a lowering of the energy i.e. here to a reduction of the gap with increasing n_p.

The other contribution to the band gap renormalization comes from the correlation effect. If we sit on one carrier and look around, we shall see with higher probability in the near vicinity a carrier of opposite charge than an equal one, i.e. the probability to find carriers is correlated. This effect is due to the Coulomb attraction between different carriers. It is a reminder of the binding of electron-hole pairs to excitons at low densities and contributes also to the decrease of E_g with increasing n_p. Some universial formula for the band-gap renormalization ΔE_g are given in literature [16,20,21].

$$\Delta E_g = E'_g(n_p) - E_g \qquad (4)$$

In these formulas the energies are given in units of the excitonic Rydberg Ry^* and the density is expressed in the dimensionless quantity r_s which is proportional to $n_p^{-1/3}$ and compares essentially the average distance between electron-hole pairs with the excitonic Bohr radius. In Fig.4 we give the graphic repesentation of the universial formula according to [20,21]. We see that $E_g'(n_p)$ is a strictly monotonously decreasing function with increasing n_p. It has been found that these formula describe the behaviour of the more covalent bound semiconductors like Si, Ge, GaAs or GaP quite well. The experimental data points given in Fig.4 for the more ionic bound II-VI compounds are lying on the average about one half Ry^* lower, for reasons that will be given later.

In order to calculate the thermodynamic properties of the EHP it is essential to note that there is apart from the band-gap renormalization a positive energy term, which comes from the kinetic energy of the carriers. For nondegenerate parabolic bands one finds for zero temperature a simple relation between the average kinetic energy E_{kin} and the quasi Fermi energy $E^{e,h}{}_F$ counted from the band extrema

$$E_{kin} = 3/5 E^{e,h}{}_F \tag{5}$$

II.B. Phase diagram of the electron-hole liquid

After having outlined in the preceeding subsection the band-gap renormalization in

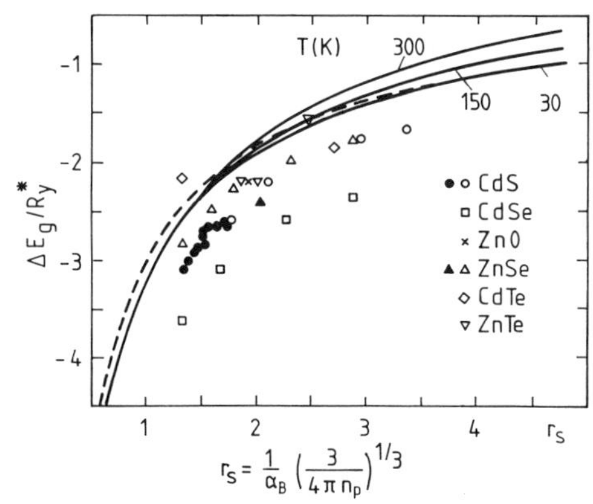

Figure 4. The dependence of the gap on the density of electron-hole pairs in normalized units according to [20,21] and data for II-VI semiconductors according to [22].

an EHP, we concentrate now on the chemical potential $\mu(n_p T_p)$ of the electron-hole pair system, on its dependence on n_p and on the consequence for the phase diagram, which follows from it. In Fig.5 we summarize what we have learned already. We show the bandstructure of an unexcited direct gap semiconductor, the renormalization of the gap and the state filling in the presence of an EHP. Since we are then far away from thermodynamic equilibrium, we have to introduce quasi Fermi levels for electrons and holes to describe their distributions in their respective bands. A thermal equilibrium between the carriers, which allows to define a plasma temperature T_p is usually reached in times shorter than a ps but T_p can be higher than the lattice temperature T_L since the times to establish thermal equilibrium between the carriers and the lattice can be considerably longer and exceed eventually the carrier lifetime T_1. Furthermore the carrier-phonon interaction depends on the material via the various carrier-phonon coupling processes, on the excess energy and on n_p. Examples for the cooling of carriers after pulsed excitation

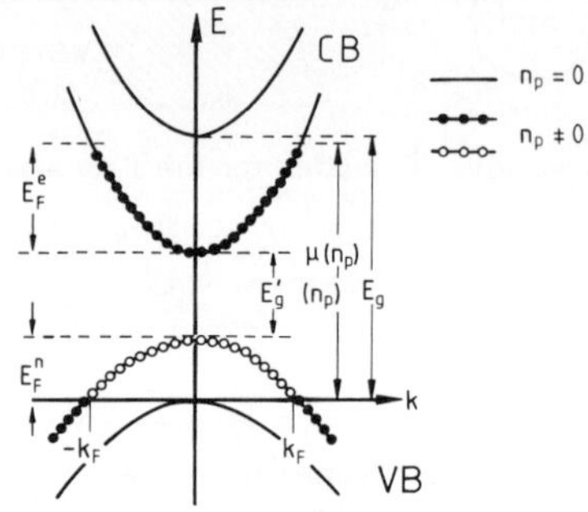

Figure 5. Schematic drawing of the band structure of an unexcited and of a highly excited semiconductor. Compare to Fig. 1a.

are given e.g. in [23-25]. The chemical potential of the electron-hole pair system is defined as the sum of the reduced gap and the quasi-Fermi energies $E^{e,h}_F$

$$\mu(n_p,T_p) = Eg'(n_p) + E^e_F(n_p,T_p) + E^h_F(n_p,T_p) \tag{6}$$

Depending on T_p,n_p and on the effective density of states, the quasi Fermi-levels can be situated in the bands, describing a degenerate population or in the forbidden gap for a non-degenerate one. In Fig.6a we show $E'_g(n_p)$ and $\mu(n_p)$ with T_p as a parameter. If we neglect for the moment again exciton effects we see that μ reaches at zero temperature E_g as soon as we create one electron-hole pair and goes above $E'_g(n_p)$ for higher densities. At higher temperatures $\mu(n_p,T_p)$ stays below $E_g'(n_p)$ up to a certain density. The situation $\mu(n_p,T_p) > E_g'(n_p)$ describes a degenerate plasma in which the population of at least one type of carriers is degenerate. The above condition gives simultaneously the limit for population inversion between conduction and valence band. The condition $\mu(n_p,T_p) > E_g'(n_p)$ is a necessary prerequisite for stimulated emission by direct band-to-band recombination in an EHP. It should be noted, that there are many other processes in semiconductors which exhibit optical gain, also excitonic ones in the intermediate density regime. For some recent reviews of this topic see e.g. [26-28] and references therein. If excitonic features are more realistically included in Fig.6a, $\mu(n_p, T=0)$ would start from the lowest exciton level rather than from E_g but the conclusions drawn in the following would remain essentially the same. So we do not got deeper into this aspect but refer the reader to [16,29,30].

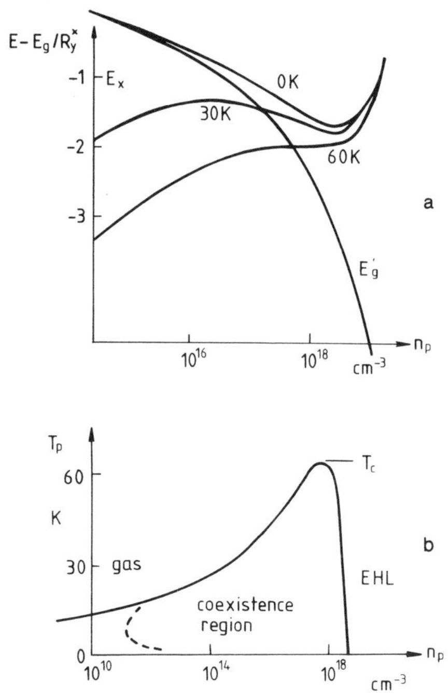

Figure 6. Schematic drawing of the renormalized gap $E_g'(n_p)$ and of the chemical potential $\mu(n_p, T_p)$ for vavrious plasma temperatures (a) and the resulting phase diagram for the EHP (b). Numerical values apply for CdS according to [16].

It is now important to note, that $\mu(n_p, T_p)$ shows in a certain temperature intervall $0 < T_p \leq T_c$ a region with a negative slope between a maximum and a minimum. The energy distance of the minimum for $T_p = 0K$ from the lowest free exciton energy is the binding energy φ of the EHP with respect to a gas of free excitons. It has been found both experimentally and theoretically that most semiconductors have a positive binding energy of the EHP i.e.

$$E_{exciton} - \mu(n_p^{min}, T_p = 0) = \varphi > 0 \qquad (7)$$

The positive and negative slopes of μ as a function of n_p are caused by the interplay of band-gap renormalization and band filling. Evidently band-filling dominates over renormalization for all temperatures at sufficiently high values of n_p.

For the thermodynamic description of the EHP in quasi-thermodynamic equilibrium (i.e. assuming that the carriers have infinite life-time once they are excited) one can consider the connection of μ with the thermodynamic potentials

$$\mu(n_p, Tp) = \partial F/\partial n_p \qquad (8)$$

where F is the free energy of the electron-hole pair system. It is in turn connected with the inner energy U and the entropy S of the electron-hole pair system by

$$F = U - TS \qquad (9)$$

We do not follow this formal approach [15,16], but a more intuitive one. It is not difficult to understand that a density region with $\partial\mu/\partial n_p|T_p$ is unstable. A consequence is that a phase-separation develops for $T_p < T_c$ (for $T_c > T_p$ the above conditions is no longer fulfilled, see Fig.6a) in a very analogous way as in a real or van der Waals gas. This means that the electron-hole pairs will separate into a gas-like low density phase of excitons and free carriers and into a liquid-like, high density EHP phase. The phase diagram which is shown in Fig.6b can be obtained using Maxwell's construction. The phase diagram in Fig.6b is identical to the one of a real gas, if one consideres that one gives there usually the volume occupied by the gas since the number of molecules is fixed, while we have for the EHP usually a given volume which is excited and we can vary the average number of electron-hole pairs in the system by varying the pump intensity.

The properties, which we can expect for an EHP in the liquid state (a so-called electron-hole liquid EHL) and which we can read from Fig.6 are the following:

- The plasma should be bound, i.e.$\varphi > 0$.
- At constant $T_p < T_c$ a phase-separation should set in, when we hit with increasing n_p the low-density border of the coexistence region.
- the density of the EHL should remain constant for increasing n_p at constant T_p in the coexistence region and only the volume fractions filled with gas and EHL should change
- the liquid density n^ℓ_p should decrease with increasing T_p from a value $n^\ell_p(T_p = 0)$ to $n^\ell_p(T = T_c)$
- above T_c there should be no more phase separation and n_p can vary at constant T_p with the pump power
- if the experimental conditions allow to fill the whole excited volume with $n_p > n^\ell_p$ the chemical potential μ should increase rapidly. This phenomenon corresponds to a huge Fermi pressure of the EHP $p = n_p^2\, \partial F/\partial n_p$ which results in turn in a rapid expansion of the EHP into the sample if not prevented by the boundary conditions.

A more detailed theoretical investigation of the phase diagram indicates some minor modification. At very low densities and temperatures, independent electrons and holes will hardly meet each other to form an exciton. Consequently the phase diagram can have a modification given by the dotted line. This phenomenon is also known as entropy ionisation [30,31]. The other modification occures around the critical point. Some calculations show that there may be two subsequent transitions with increasing density, namely a first one from an exciton gas to a gas-like plasma phase and then a transition to the liquid phase [14,18,32].

Furthermore, we should like to note that the transition to a EHL is a first order phase transition in a pumped or driven system [15] and we should repeat at the end of this subsection that the statements concerning the phase diagram of Fig.6b are the result of quasi-equilibrium assumptions. Indeed it will take a finite time until the carriers condense

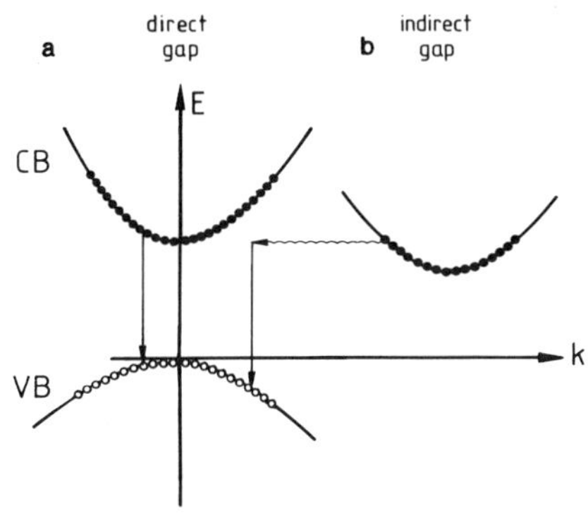

Figure 7. Schematic drawing of the renormalization processes in a direct (l.h.s.) and an indirect gap semiconductor (r.h.s.) containing and EHP.

spatially into liquid-like droplets (EHD) with a surrounding gas phase. As a consequence of this effect, it is a good choice to look for an EHL in materials with long carrier lifetime. These are the indirect gap materials like Si, Ge, GaP or indirect $Al_{1-y}Ga_yAs$. In direct gap materials like CdS, GaAs etc the life-time may be so short, that no clear phase-separation can develop.

II.C. Optical properties of the electron-hole plasma

The properties of an EHP can be investigated by many techniques comprising optical ones and others. In a school on nonlinear spectroscopy we concentrate on the optical methods and refer the reader for the other ones to [17,18] and references given therein. Indeed we have mentioned already one of the non-linear optical properties connected with an EHP. The excitonic features disappear from the optical spectra of those parts of the semiconductor, which are filled with an EHP. If the semiconductor is not completely filled by an EHP, excitonic and plasma features coexist.

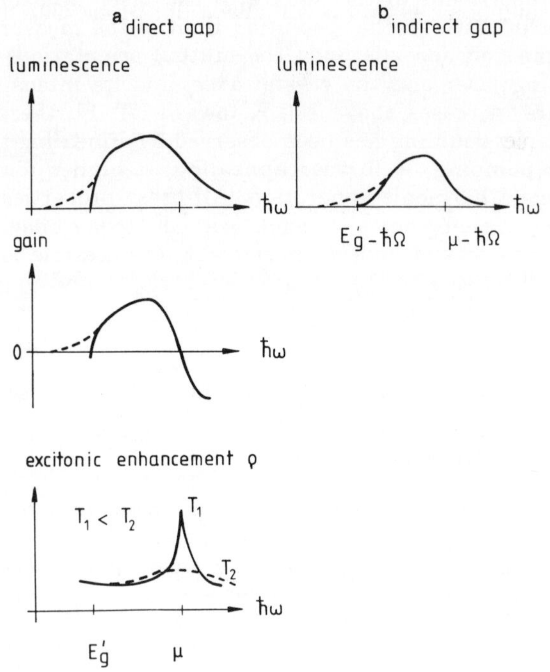

Figure 8. Schematic drawing of the luminescence and gain spectra of a direct gap semiconductor including some corrections (r.h.s.) and for an indirect gap semiconductor including some corrections (l.h.s.). For explanations see text.

Since the optical properties of the EHP are different for direct and indirect gap materials, we present them separately in Figs. 7 and 8. In Fig.7 we show the situation for a dengereate EHP in a direct and an indirect gap semiconductor, respectively. In a direct gap semiconductor one has recombination under **k**-conservation. So one expects as a first guess for the spectra of luminescence $I_{lum}(\hbar\omega)$ and of optical gain $g(\hbar\omega)$ a square-root increase above the reduced gap under the assumption of a k-independent transition matrix elememt and a decay at higher energies goverened by the Fermi-functions of electrons and holes i.e.

$$I_{lum}(\hbar\omega) \sim (\hbar\omega - E'_g(n_p))^{1/2} f_e \cdot f_h \tag{10a}$$

$$g(\hbar\omega) \sim (\hbar\omega - E'_g(n_p))^{1/2}(f_e + f_h - 1) \tag{10b}$$

with $\hbar\omega \geq E'_g$ and

$$f_{e,h} = (\exp(E - E^{e,h}_F)/ k_B T + 1)^{-1} \tag{11}$$

It should be mentioned that the transition from gain to absorption occurs always at $\hbar\omega = \mu(n_p,T_p)$ as long as there is no drift of the EHP with velocities comparable to the Fermi-velocity. For a discussion of this latter point, which will not be treated here see e.g. [33] and references therein. For the indirect gap material, a momentum conserving phonon is involved. As a consequence all transitions contribute to the emission at a certain energy $\hbar\omega$ which are separated by $\hbar\omega \pm \hbar\Omega$ in the cases of emission (+) or absorption (-) of a phonon with energy $\hbar\Omega$ independent of their wave vector. The summation over all possible transitions gives a power 3/2 increase of the luminescence above $E'_g(n_p)$ assuming again a constant transition matrix element. The high energy tail of the luminescence is again governed by the Fermi-functions

$$I_{lum}(\hbar\omega) \sim (\hbar\omega - E'_g(n_p) + \hbar\Omega)^{3/2} F(f_e,f_h) \tag{10c}$$

Though population inversion can be easily reached in the EHP of indirect gap semiconductors like Si or Ge, gain is usually not observed and we shall not discuss it here. The optical transition probability is so low in theses materials due to the participation of the phonon, that volumes of several cm length had to be completely fills by EHP (or EHL) to get considerable gain. This is not practical from an experimental point of view and is the main draw back in the application of these materials in light emitting devices [34]. An exception of the above rule should be mentioned namely indirect gap alloy semiconductors like $Ga_{1-x}Al_y$ As or $GaAs_xP_{1-x}$ [35,36] close to the indirect-direct crossover composition. They can show stimulated emission due to a coupling of the transition of the direct and the indirect conduction band minima via alloy disorder scattering even without the participation of phonons. We shall not go deeper into this aspect here, but refer the reader to [35,36,37] where in addition the topic of band gap renormalization is treated for the case that several partly non-equivalent minima in the conduction band structure are populated.

As mentioned already, the optical spectra of Fig.8 present only a zero order approximation. There are some corrections or modifications which have to be added and which are shortly outlined in the following.

The first one is the so-called Landsberg or final state damping [16,26,39]. After a recombination process there will be an empty state left in the sea of electrons in the conduction band and one occupied state in the sea of holes in the valence band. These states will rapidly be filled with other carriers by intraband scattering resulting in an energy dependent broadening Γ. This broadening is minimal around the quasi-Fermi levels and will increase to both sides. As a consequence of this damping, the gain spectra have to be folded with an energy dependent Lorentzian, which modifies especially the low energy edges in the way shown in Fig.8 by the dahed lines. The square root onset of the

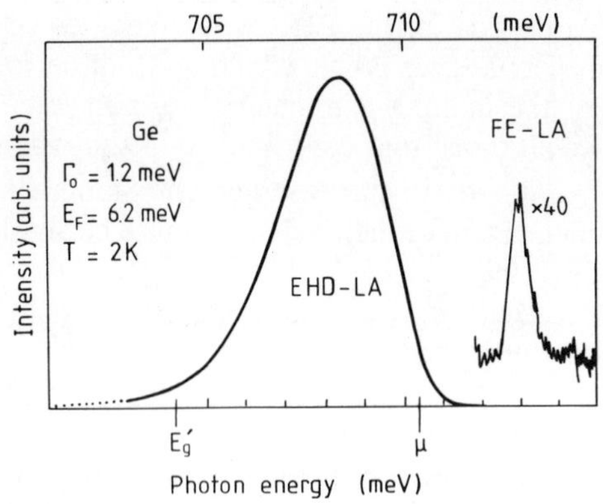

Figure 9. Luminescence spectra of Ge in the presence of an EHP, under simultaneous emission of LA phonons. According to [44].

spectra in direct gap materials is washed out and the spectra extend both in direct and in indirect gap semiconductors to values below $E'_g(n_p)$.

Another contribution to the broadening of the spectra influencing again preferentially their low energy wing is the so called Fermi sea shake up. This process has recently been reviewed and stressed in [39]. An electron hole pair recombines and excites either a single carrier to a higher (empty) state or a collective mode of the carriers i.e. a plasmon (or more precisely a plasmon phonon mixed state). Such inelastic scattering processes are also known in the intermediate and partly in the low density regime [26] where they are essential for optical gain. For the EHP they can to some extend be included in an effective damping Γ.

A third modification to the optical spectra is known as excitonic enhancement [16,26] or in quasi two-dimensional systems, including modulation doped structures also as Fermi-edge singularity or Mahan exciton or X-ray singularity see e.g. [40,41] and references therein. It means essentially the following. Electrons and holes sitting in a degenerate population far away from the quasi Fermi energies have no possibility to be scattered by small amounts of energy due to Pauli's principle. Carriers around $E_F^{e,h}$ in constrast have occupied and empty states available. As a consequence electrons- and holes can be scattered due to the mutual Coulomb interaction performing a fraction of an

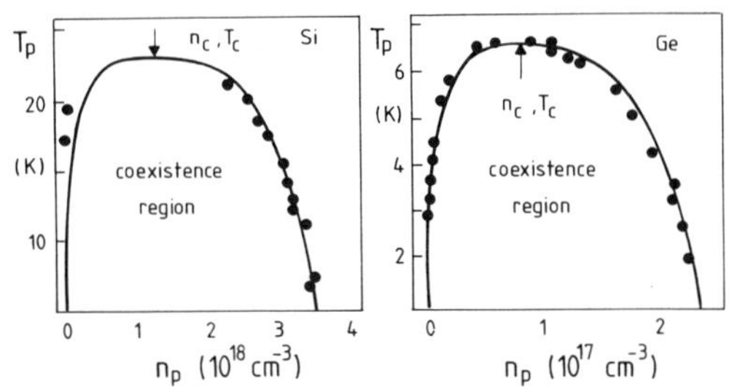

Figure 10. Phase diagrams of the EHL in Si and Ge according to [45] and [46], respectively.

exciton-like orbit. This fact enhances the optical transition probability around μ and is a reminder of the exciton occuring at low densities. This is the reason, why this phenomenon is called excitonic enhancement. The enhancement influences the spectra around μ but it broadens and decreases rapidly with increasing T_p and can be often neglected for T_p above several ten K. The excitonic enhancement is usually given as a function $\varrho(\hbar\omega)$ which has to be multiplied to the luminescence and/or gain spectra. An example is shown in Fig.8a.

To summarize this short outline of the optical spectra of the EHP, we can state, that a detailed analysis of the luminescence and gain spectra allows to deduce the relevant parameters of the plasma like its density n_p, its temperature T_p, the position of the reduced gap $E'_g(n_p)$ and of the chemical potential $\mu(n_p,T_p)$. We shall see examples in the next section. More details about the EHP, its properties and their theoretical description can be found e.g. in [5-7,14-19,42,43] and references given therein.

III EXPERIMENTAL RESULTS

Now we shall present some selected examples of the properties of the electron-hole plasma stressing the optical nonlinearities.

III.A. Bulk semiconductors

This subsection is devoted to direct-gap and indirect-gap bulk materials

III.A.1. Indirect gap semiconducors

In Fig.9 we show a low temperature photoluminescence spectrum of Ge, more precisedly the bands which occur under emission of a LA phonon. At low excitation densities only the emission of the free exicton (FE-LA) around 0.713 eV is seen, apart

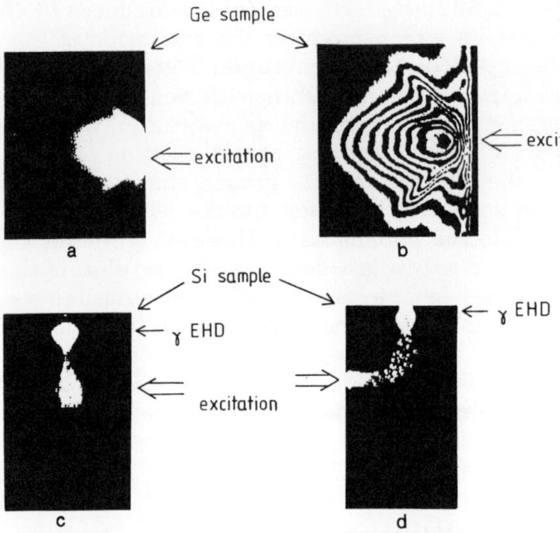

Figure 11. The motion of the EHD under the influence of a phonon-wind (a) and (b) or an inhomogeneous strain field in the sample (c) and (c). Redrawing according to [50] and [51], respectively.

eventually from some bound exciton lines. With higher excitation a new braod emission band appears, which is attributed to the recombination in an EHL. A lineshap analysis following the ideas developed above gives μ and E_g' and thus the quasi Fermi energies and n_p. The parameter Γ_0 used to describe the damping is assumed to be constant. The plasma temperature coincides here essentially with the lattice temperature T_L because of the long carrier life time, which allows to reach thermal equilibrium between the carriers and the phonons. The density of the EHP deduced from a series of spectra with incresing excitation is constant and only the relative contribution from exciton and EHP emission changes. Since the plasma is evidently bound, we get already strong hints, that it is really in a liquid-like state.

In a next step one tries to measure the phase diagram of the EHL. Results are shown in Fig.10 for Si and Ge. The high density side of the coexistence region can be deduced directly from the analysis of the EHP luminescence, the low side from the onset of the EHP emission with increasing generation rate, if the lifetime and the excited volume are known. The fact that the phase diagrams of Fig.10 and 6b look different is only caused by the linear and logarithic density scales. From Fig.10 one can read equilibrium densities of the EHL of Si and Ge at T = 0 of about $3.5 \cdot 10^{18}cm^{-3}$ and $2.3 \cdot 10^{17}cm^{-3}$, respectively. The critical temperature T_c for liquid formation are about 25 K and 6.5 K for these two materials.

It has been found, that the EHL occurs below T_c in the form of small droplets (EHD) containing of the order of 10^3 carrier pairs [14,17,18]. The size and the dynamics of these droplets are governed by nucleation, condensation of excitons and/or carriers into the droplets, recombination in the droplets and evaporation out of the droplets, the latter quantity being influenced by the size-dependent surface energy of the EHD. Comparable information concerning e.g. the phase diagram has been found also for other indirect gap materials like GaP, GaP:N [44,45,46,47] or more recently in indirect $Ga_{1-y}Al_yAs$ [48]. We shall not go into details of these materials, but select from the numerous beautiful experiments made with EHP in Si or Ge two, which are connected with the motion of the droplets.

The droplets can be attatched to some lattice defect, which acts e.g. as a nucleation center and then they are immobile [49]. In very pure samples, they are free and can move around e.g. in a Braunian motion like way when pushed by a random phonon population.

Under some experimental conditions (see Figs. 11a and b, which are redrawn from [50]) one can produce a spatially directed propagation of phonons. The sample is excited in a thin layer spot at the surface and EHD's are formed there. Since the carriers in the EHD recombine essentially nonradiative by phonon emission, the excitation spot acts as a source of a nonequilibrium phonon population which starts to expand into the sample. Since the group velocity v_g of the phonons is given by $\hbar^{-1}$ $\mathrm{grad}_{\hbar}$ $E(\boldsymbol{k})$ and since the surfaces of constant energy of phonons in $\boldsymbol{k}$-space are no spheres but structures of lower, namely O_h symmetry, one finds, that there are certain directions into which phonons propagate preferentially. If these phonons push now the EHD, they will propagate into these preferential directions, too. This phenomenon is seen for Ge in Fig.11a and b. In Fig.11a the sample is observed in the IR-light of the EHD emission and one can clearly observe well defined directions of preferential propagation. In Fig.11b contours of equal light intensity are shown, making the above effect even more clear.

In Fig.11c and d a motion of EHD's is shown, which is caused by an externally applied, inhomogeneous stress field. It is well known, that the width of the forbidden gap depends on strain described by the deformation potentials. It is even possible to lift partly the degeneracy of equivalent minima of the conduction bands in Si or Ge and to study the properties of the EHP under such conditions [45,51,52]. In the case of Fig.11c and d which gives the front and the side view of the sample, inhomogeneous stress has been applied in a way that a minimum of the band-gap is created inside the sample. The sample is again excited in a spot near the surface as indicated by an arrow and is observed in the IR emission light of excitons and carriers in the EHD. One can clearly see a stream of carriers towards the potential minimum where the electron-hole pairs condense into a macroscopic, so-called γ-drop of EHP. These γ-drops can have a diameter in the mm range and are completly filled with EHP. In some popular scientific journal these γ-drops consisting of electrons and holes have been adressed as "pure electricity". We leave it as a question of semantics, what pure electricity might be, but finish this subsection with the statement, that the existence of an electron-hole plasma in a liquid-like form is well established below T_c in indirect gap, bulk semiconductors with long carrier lifetime by partly very beautiful experiments.

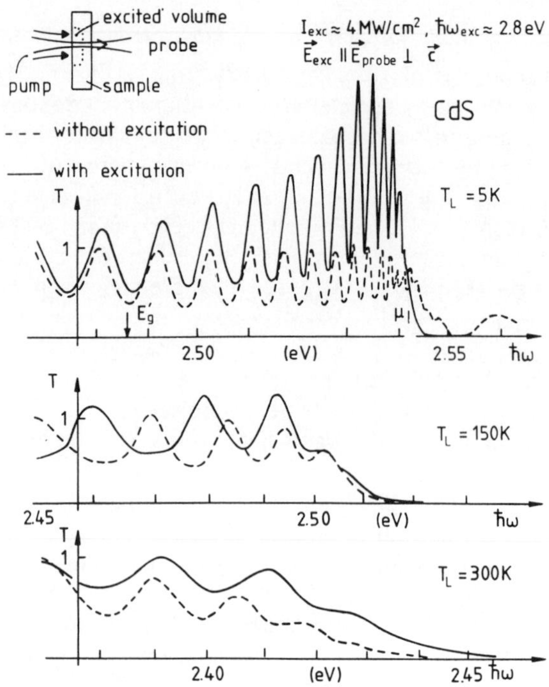

Figure 12. Transmission spectra of a thin CdS platelet without and with pump beam for three different lattice temperatures. According to [55].

III.A.2 Direct gap semiconductors

In direct gap, bulk semiconductors a degenerate EHP is usually connected with strong optical amplification reaching values up to 10^4 cm^{-1} i.e. $1\mu m^{-1}$. As a consequence, luminescence spectra are often distorted by stimulated emission and it needs some experimental skill and tiny excited areas to get the spontaneous luminescence spectra [53,54]. A way out of this problem is to measure the spectra of the optical gain $g(\omega)$ directly by pump- and - probe beam techniques. We show two examples in the following. In Fig.12a-c we give the transmission spectra of a CdS platelet-type sample, which is a few μm wide for three different lattice temperatures and once without excitation and once with simultaneous ns pulsed excitation in the band-to-band transition region. The transmission spectra are modualted in the transparent part by Fabry Perot modes. At T_L = 5 K one sees without excitation the transmission minimum caused by the $n_B = 1$ $A\Gamma_5$ exciton resonance around 2.552 eV and a transparent window above, before other higher exciton transitions set in. Under strong pumping an EHP is created. As a consequence, the excitonic feature disappeares from the transmission (Fig.12a) and reflection spectra [55,56]. Optical gain evidently appeares at T_5 = 5 K between approximately 2.54 eV and

2.48 eV depending on the excitation conditions. A slight blue shift of the Fabry-Perot modes indicates a decrease of the refractive index in the EHP [57]. With increasing temperature the latter phenomenon persists and even increases but the gain decreases, since the effective density of states N_{eff}

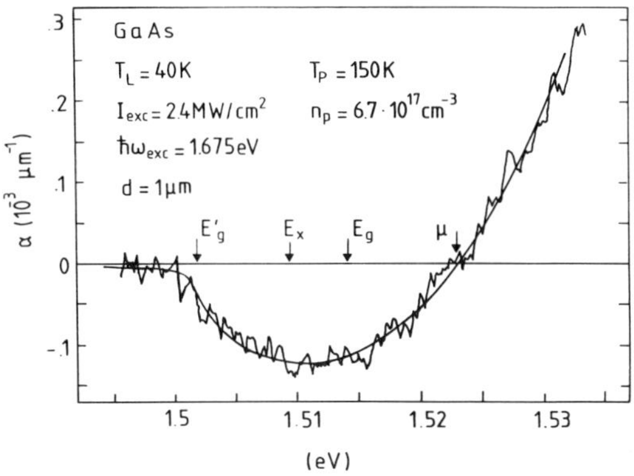

Figure 13. Transmission spectrum of a thin, highly excited GaAs layer. According to [60].

$$N_{eff} \sim (T_p \, m_{eff}^{e,h})^{d/2} \tag{12}$$

which gives the onset of a degenerate population in a band increases with the temperature to a power three halfs for a d = 3 dimensional semiconductor. At room temperature, one sees for CdS and the present excitation conditions only a bleaching of the absorption tail, which is described by the so-called Urbach or Urbach-Martienssen rule, see e.g. [58,59] and references therein. From spectra like in Fig. 12 it is possible to deduce the gain spectra $g(\hbar\omega)$ and to determine from them the plasma parameters as outlined in the preceeding section. Just to demonstrate that such gain spectra occur in many direct gap semiconductors, we give another example for GaAs in Fig. 13. The sample has been grown by MBE as a 1 μm thick layer sandwiched between $Al_{1-y}Ga_yAs$. The energies of the exciton and of the gap without excitation and the positions of the reduced gap and of the chemical potential under excitation are indicated. Though the overall appearance of the gain spectra at Fig. 12a and Fig. 13 is very similar, there are some differences which should be shortly painted out.

The plasma temperature T_p found for quasi-stationary ns excitation pulses in CdS or CdSe is typically only about 10 K or a few 10K above T_L. The difference T_p-T_L in Ga As is around 100 K under comparable excitation conditions, reflecting the stronger coupling of the carreirs to the lattice in the more ionic bound II-VI semiconductors as compared to the more covalent bound III-V materials. Additionally we see in Fig.12a a red-shift of the absorption edge and a blue-shift in Fig.13. Actually it depends on n_p, T_p and on the material parameters, if the band gap renormalization or the band gap filling dominate. It should be mentioned here, that the EHP forms also in most direct gap semiconductors a bound state as defined above [58].

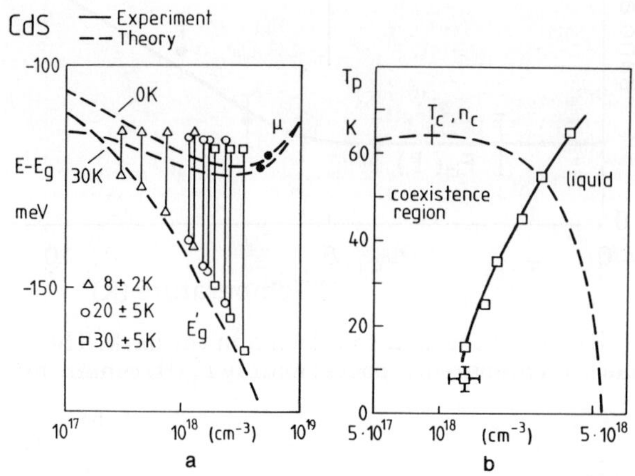

Figure 14. A comparison of experiment and theory for the EHP in CdS concerning the normalized gap and the chemical potential (a) and the high density boarder of the phase-diagram of the EHL (b). According to [16] and [56].

We adress now the question if a liquid-like plasma state occurs in direct gap materials. Since the calculated critical temperature is for CdS with 64 K [16] considerably higher than e.g. in GaAs, and since plasma temperatures below this value are easily accessible experimentally we concentrate here on CdS but mention that the results are also valid for other direct gap semiconductors. It has been found, that the width of the EHP gain spectra and thus n_p depend even at constant T_p strongly on the excitation intensity, see e.g. [33,56,58,62] and references therein. This is a first indication, that no liquid like state occurs. On the other hand, this effect allows to follow the band gap renormalization and the chemical patential continously over a wide range of densities. The agreement between experiment and a theory dedicated to CdS is very good as demonstrated in Fig. 14.

Comparing the universal formula for the band-gap renormalization in Fig.4 with data for various II-VI compounds, one finds an average shift of about 0.5 Ry* of the experimental points towards an increased binding. This effect is partly due to the influence of plasmon-phonon mixed states which enhance the binding of the EHP rather strongly in the more ionic bound II-VI compounds and which are not considered in the universal

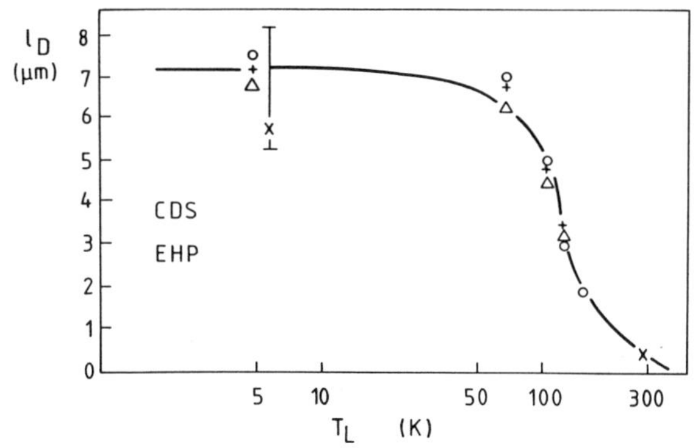

Figure 15. The diffusion lenth of the EHP in CdS as a function of the lattice temperature under strongly inhomogeneous excitation conditions. According to [55,56].

formula. Furthermore a certain discrepancy may come from the difference of the experimentally observed exciton binding energy and Bohr radii and the calculated quantitis [26].

In Fig. 14b we compare the experimentally observed temperature dependence of n_p under constant excitation conditions with the calculated high energy boarder of the coexistence region. The opposite trend of both curves shows clearly, that there is no liquid-like state of the EHP formed in CdS even for $T_p << T_c$. The reason for this discrepancy between the experimental results and the predictions of quasi-equilibrium thermodynamics is the short lifetime of the carriers with values around 200 ps [33]. If this short lifetime is taken into account it can be shown also theoretically e.g. for CdS and GaAs [56,63] that no spatial phase-separation into a liquid - and a gas phase can develop but only some density fluctuations with electron-hole pair clusters containing on the average less than 10^2 carrier pairs.

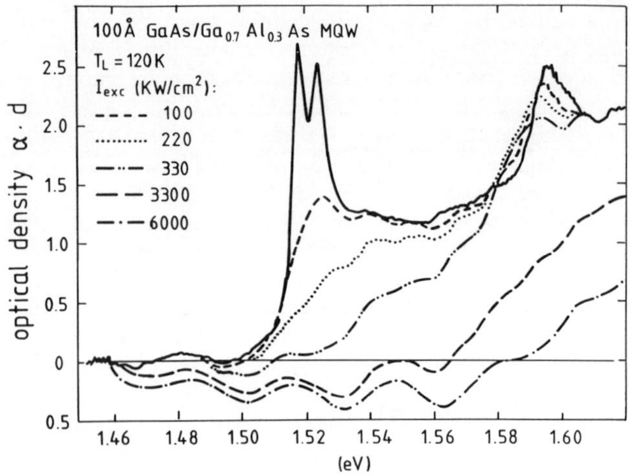

Figure 16. Transmission spectra of a GaAs/Al_{1-y} Ga_yAs MQW sample for various pump intensities. According to [66].

The short carrier lifetime reduces the spatial motion of the EHP to the nm range. Under strongly inhomogeneous excitation conditions which result in a considerably spatial gradient of μ or n_p, drift length of the EHP in CdS around 7 μm have been found with a trend to decrease once the EHP is no longer degenerate. See Fig.15. In CdSe the low temperature value is with 12 μm slightly higher [56] and it reaches about 60 μm in InSb where T_1 is about 700 ns [64]. These values have to be compared with mm range travelling distance of the EHP in indirect gap materials as outlined above. Some care has to be taken, to measure with spatially resolved methods really the expansion of the carriers of the EHP and not only the scattered luminescence light.

III.B. Systems of reduced dimensionality

In this subsection we start with semiconductor structures in which the motion of the carriers is quantized in one dimension, so-called quasi two-dimensional systems like single- and multiple quantum wells (SQW, MQW) and superlattices (SL). Then we proceed to structures providing quantization in two directions, so-called quasi one-dimensional quantum wires. We shall not treat quantum dots here, since this topic is covered by other contributions to this school [65]. Details can be found there and in the references given therein.

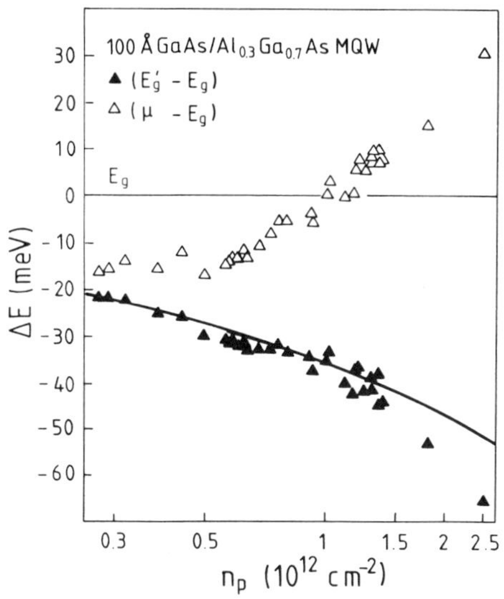

Figure 17. The density dependence of the reduced gap and of the chemical potential of the EHP in a $GaAs/Al_{1-y}Ga_y$ As MQW sample and the comparison of the first quantity with theory. According to [60,66].

III.B.1 Quantum wells

In Fig.16 we show a set of transmission spectra with increasing pump power for a typical $GaAs/Ga_{1-y}Al_yAs$ MQW structure. At low excitation the $n_z = 1$, $n_B = 1$ heavy-hole (hh) and light hole (lh) excitons are seen around 1.52 eV and the $n_z = 2$, $n_B = 1$ hh exciton at 1.60 eV. With increasing band-to-band excitation the $n_z = 1$ excitons disappear and optical gain is observed over a spectral range from 1.46 eV to 1.58 eV. The modulation in the gain spectra is due to Fabry-Perot modes in the sample structure. Otherwise the gain exhibits nicely the constant two-dimensional density of states. Deducing the reduced gap and the chemical potential from spectra like in Fig.16 results in Fig.17, where in addition very good agreement between experiment and theory [66,67] can be stated.

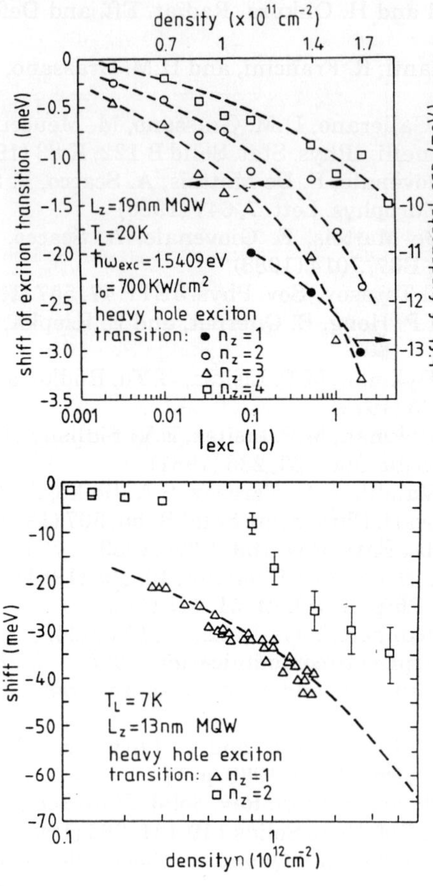

Figure 18. The shift of the various sub-band transitions as a function of the plasma density for two GaAs/$Al_{1-y}Ga_yAs$ MQW samples with different values of ℓ_z. According to [60,66].

So far the properties of the EHP seem to be rather similar in three- and quasi two-dimensional systems. A closer look, however, reveals some differences. One sees, that the $n_z = 2$ excitons are slightly shifted, but still present at excitation conditions, where the $n_z = 1$ states are already gone. See e.g. the dotted line in Fig.16. This is a clear indication, that phase-space filling is more important for the disappearance of the excitons in systems reduced dimensionality [19], since the Coulomb-screening acts on populated and

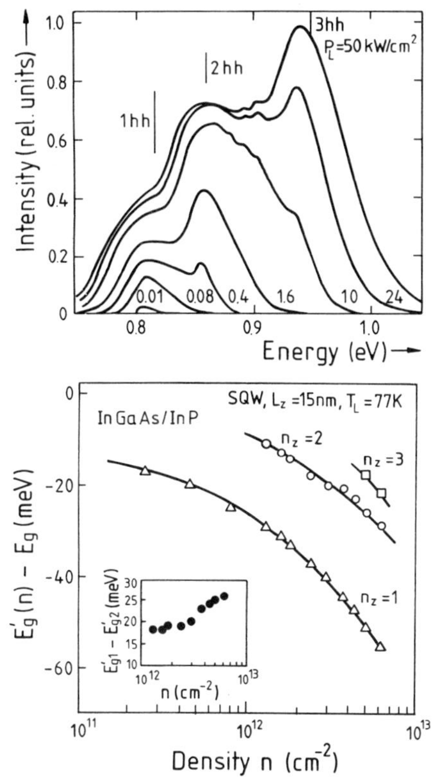

Figure 19. CW-luminescence spectra of a $In_{1-y}Ga_yAs/InP$ SQW mesa structure as a function of the excitation intensity (a) and the resulting density dependence of the subband structure. According to [54] and [69], respectively.

unpopulated states in the same way [68]. Furthermore, the sub-band structure does not show a rigid shift, see Fig.18. The renormalization the second subband accellerates, when it starts to be populated. This is the case for the sample for Fig.18b at densities around 10^{12} cm^{-2}. To demonstrate that the above described findings are of rather general nature we show in Fig.19 the results of a beautiful luminescence experiment on a SQW of the $In_{1-y}Ga_yAs/InP$ system. In order to reduce lateral carrier diffusion and the distortion of the

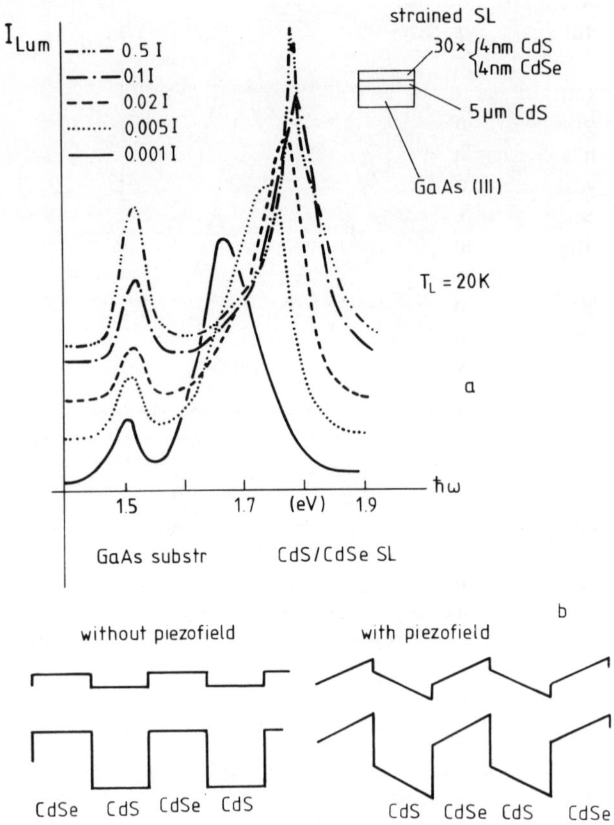

Figure 20. Luminescence spectra from a hexagonal CdS/CdSe strained layer superlattice with increasing excitation (a) and a schematic drawing of the band structure without and with piezofield (b). According to [70].

luminescence spectra by stimulted emission, small mesa-structures have been etched. Fig.19a shows in a cw experiment with increasing pump power the gradual shift of the reduced gap and the filling of higher sub-levels. From Fig.19b it is again possible to deduce a considerably reduced shift of the higher sublevels compared to the fundamental one.

It should be mentioned, that all III-V laser diodes exploit the stimulated recombination in a degenerate EHP for operation. The main difference is, that the carriers are created via injection by a forward biased, highly doped p-n junction in contrast to optical pumping used here. For a short review of the gain-processes in the II-VI semiconductor structures see e.g. [27,58] and references therein.

As a last example for quasi two-dimensional systems, we show in Fig.20 the emission of a hexagonal, strained-layer CdS/CdSe SL grown by hot wall epitaxy on a GaAs substrate with a 5 μm CdS buffer layer. The emission of this structure coincides spectrally at the lowest excitation intensities almost with the one from the GaAs substrate and shifts with increasing pump power to the blue by an amount of up to 400 meV. At the highest pump intensities stimulated emission occurs. The shift of the emission results from a screening of the piezo-field, which is characteristic for strained-layer SL, by the excited carriers. See also [69,70,71]. In contrast to zinc-blende materials, this effect is by almost one order of magnitude larger in the hexagonal structures. More details on this effect, on its dynamics, on the laser transition and a calculation of the band-structure are found in [70].

III.B.2 Quantum wires

In the present trend to investigate the properties of semiconductors of reduced dimensinality, the obviously next step after quasi two-dimensional systems are quasi one-dimensional so-called quantum wires. Two possibilities to produce them are shortly outlined in Fig.21. More details are found e.g. in [43,72] and a third method will be adressed below.

The most straight forward possibility is a lateral structuring of MQW either by intermixing with ion implantation [73] or by etching as shown in Fig.21a. The present state of technology allows to produce in y-direction lateral structures of the order of 50 to 100 nm only, while the dimension in z-direction is usually about 10 nm. Due to the l^{-2} dependence of the quantization energy in this type of square-well potentials, the energy steps due to quantization in z-direction are much larger than the ones in y-direction as indicated schematically in Fig.21b. Another way to produce wires is to etch a V - shaped grove into the substrate made from the barrier material with the help of masking techniques and orientation dependent selective etching, Fig.21c. In two further steps a thin layer of well material is deposited, followed by another layer of barrier material. It has been found that the thickness of the well is largest in the deepest areas of the V-grove and decreases on both sides. As a consequence one gets again a rectangular potential in z-direction (Fig.21d) with rather large quantization energies and a more smooth potential in y-direction caused by the y-dependence of the well thickness. This potential can be considered approximately as a harmonic one, resulting in an equal spacing at least for the lowest sublevels.

The EHP has been investigated in both types of structures [43,74,75]. We present here in Fig.22 and 23 results for quantum wires of the type described by Fig.21a. In Fig.22a the luminescence under quasistationary excitation with ns pulses is shown for increasing pump power. The gradual filling of the states with $n_z = 1$ and $n_y = 1,2,3...$

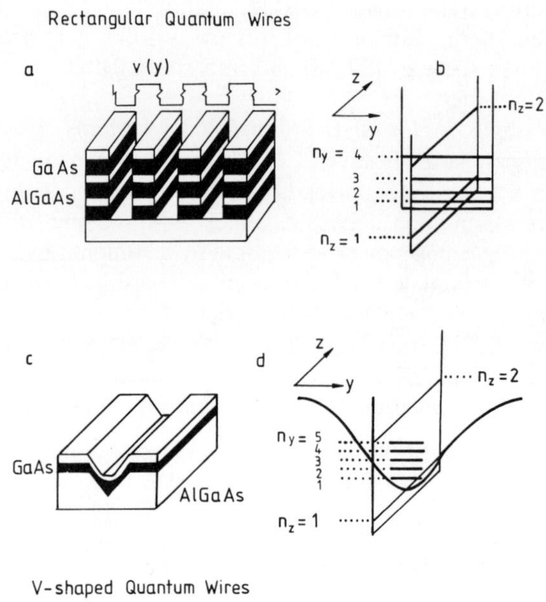

Figure 21. Schematic drawing of quantum wires (a,c) and of their energy levels (b,d). According to [43b].

followed by the n_z =2 transition can be nicely observed. In Fig.22b the temporal evaluation of the luminescence after strong ps excitation is shown. Shortly after the pulse (0 and 50 ps) one finds a similar population as in Fig.22a for the highest pump intensity. The carriers recombine from all populated states, but the higher states feed in addition the lower laying ones. As a consequence the population in the higher states decays faster than in the lower ones. An analysis of spectra like in Fig.22 gave also first results for the band-gap renormalization in these quantum wires. Good agreement is obtained for a value $l_y \approx$ 50 nm as seen in Fig.23.

In the last example of this subsection a rather complex structure is presented, a so-called quantum well wire superlattice (QWWSL). On a highly indexed surface of GaAs it is possible to grow a GaAs/AlAs SL which shows a rod-like structure in one dimension as sketched in an idealized way in the inset of Fig.24. Narrow GaAs/AlAs SL are known to be indirect both in real space and in k-space [42]. The quantization energy for the electrons in GaAs at the Γ points gets with decreasing l_z so large, that it shifts the electron levels over the indirect conduction band minima in AlAs. Consequently the electrons are transferred into the indirect minimum in the AlAs, while the holes are still in the Γ-point

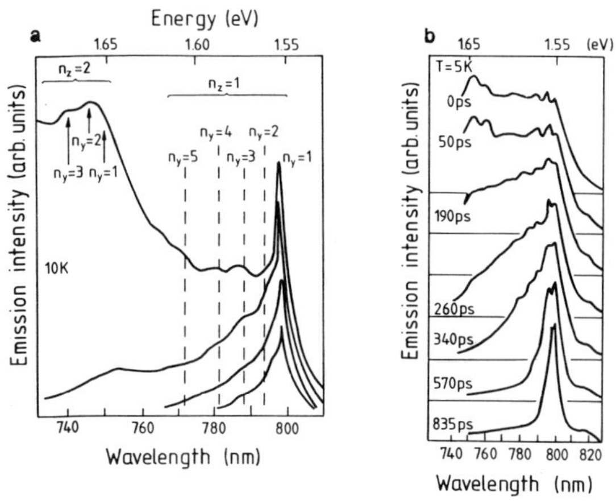

Figure 22. Luminescence form a GaAs/Al_{1-y}GaAs quantum wire sample under quasi stationary excitation for increasing pump intensities (a) and for different delay times after ps excitation (b). According to [74] and [75], respectively.

maximum of the valences band of GaAs. As a consequence of this band structure one gets again a rather long carrier life-time and the chance that a liquid-like plasma state may develop. Indeed such a condensation into droplets has not been found until now in simple narrow GaAs/AlAs SL but in the structure shown in Fig.24. The reason is eventually the presence of nucleation centers caused by fluctuations in the wire width.

Fig.25 gives emission spectra for constant temperature and increasing excitation and vice verse. The spatial separation of electrons and holes results in a coupling to the

LO-phonons and to LO-phonon satellites, which is unusually strong for III-Vcompounds. With incresing pump power a plasma emission line appeares, which fulfills below a certain temperature all criteria, which one usually imposed on an EHL. Some data points from the high energy side of the coexistence region are shown in Fig.24. T_c seems to lay slightly

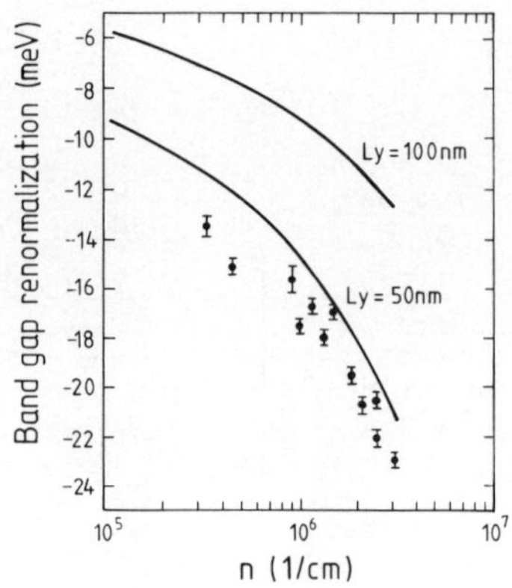

Figure 23. The measured and the calculated band gap renormalization for the EHP in a quantum wire sample. According to [43b,76].

below 20 K. Above T_c the plasma density shows the opposite trend. It is not jet clear, why the EHP has in this case (and also for some indirect gap bulk materials) a rather constant density for $T_p > T_c$. Further work for the clarification of this question is in progress [78].

As mentioned already above, the topic of quantum dots will not be covered in this contribution. Apart from dots precipitated in organic or inorganic matrices there are also dots defined by an electrical potential produced via gate electrodes and others made from MQW by lateral structuring in two dimensions. Information concerning these latter two kinds can be found in [65,72,79] and references therin.

IV CONCLUSIONS AND OUTLOOK

To summarize we can state, that the properties of the EHP in bulk materials are quite well understood, including the transition to a liquid like phase below T_c. As is evident from the literature which we cited here, the main part of work in this field has been done in the seventies and in the first half of the eighties.

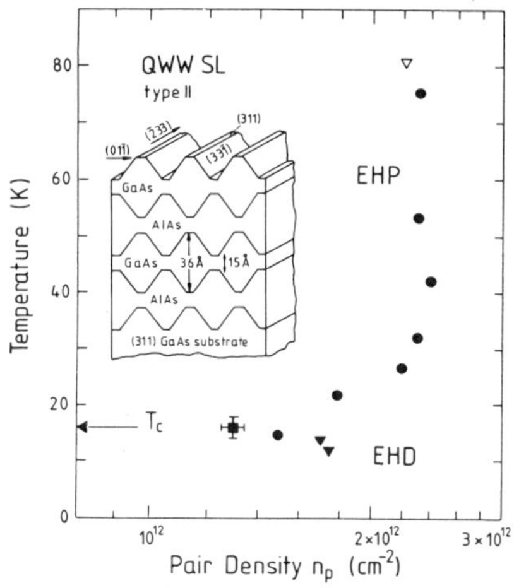

Figure 24. The phase diagram of the EHL in a GaAs/AlAs quantum well wire superlattice. Inset: structure of the sample. According to [37,77].

Presently the interest focusses on semiconductor structures of reduced dimensionally like quantum wells, ~ wires and ~ dots. As outlook we want to mention two aspects. One is, that the scientific activities tend not only into the direction to understand the optical properties and the electronic eigenstates of high density systems, but also the transport properties including conductivity measurement along the planes and wires and perpendicular revealing new exciting aspects through resonant tunneling effects or in a magnetic field through the quantum Hall effect [79]. Other work aims towards (ultra-) fast optical and electrooptical switches and modulators. See e.g. [80] and references therein.

The other outlook, which we want to give, concerns time resolved spectroscopy, which we did not treat explicitely in this contribution.

A large fraction of the older results may be summarized in the following statements. In direct gap semiconductors, the effective life-time of the plasma T_1 lies in the hundred to a few hundred ps regime where the shorter times refer to a degenerate EHP [23-25,33]. For indirect gap materials T_1 ranges from several ten ns eg. in highly excited GaP [47] almost up to the ms regime for a EHL in ultrapure Ge kept away from the surface [14,17,18]. With increasing density non-radiative Auger-like recombination processes become increasingly important affecting also the performance of III-V laser diodes.

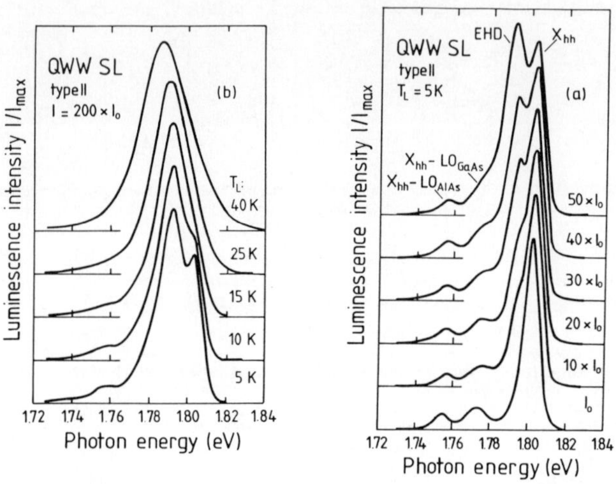

Figure 25. Luminescence spectra of the sample of Fig.24 at constant lattice temperature and increasing excitation (a) and vice versa (b). According to [37,77].

For the future we can expect in this field investigations of the coherent properties of an EHP. These experiments require some experimental skill and effort, especially if carried out directly in the time domain since T_2 times well below 1p have to be anticipated. Eventually it is more favourable to switch to experiments in the frequency domain like non-degenerate four wave mixing, which allows e.g. to measure T_2 down to 30fs using ns pulses as shown e.g. in [83] or like spectral hole burning experiments [82]. First results e.g. for the EHP in CdSe have been reported already. From the theoretical point, the first elementary interaction processes are investigated, which occur immediately after a dense population has been created by a short pulse. Before the coherence is lost and thermalisation starts, theory predicts e.g. some coherent oscillations between the carrier and the LO phonon system, which diappeare however very rapidly, among others because of the strong anharmonicity of the lattice vibrations [84]. So it does not need much prophecy to predict exciting new results in the EHP research even for the near future.

Acknowledgements

The author wants to thank many colleagues for fruitful and stimulating discussions about plasmas. Without being able to be complete I should like to mention by name Prof.Dr.H.Haug (Frankfurt). Prof.Dr.R.Zimmermann (Berlin), Prof.Dr.E.O.Göbel (Marburg), Prof.Dr. A.Forchel (Würzburg), Prof.Dr.S.Schmitt-Rink (†) (Marburg), Prof.Dr.S.W.Koch and Prof.Dr.N.Peyghambarian (Tucson) and Prof.Dr.D.S.Chemla (Berkely). The results from our group presented here have been worked out by several of my former or present corworkers. Again without trying to reach completeness I should like to mention by name Dr.K.Kempf, Dr.H.-E. Swoboda, Dr.F.A.Majunder, Dr.Ch.Weber and Dr.H.Kalt. The work has been supported by the Deutsche Forschungsgemeinschaft.

References

[1] R.S.Knox, Theory of Excitons, Solid State Physics, Suppl.5; F.Seitz and D.Turnbull eds, Academic Press, New York (1963)

[2] Excitons, K.Cho ed, Topics in Current Physics 14, Springer, Berlin (1979).

[3] Excitons, E.I.Rashba and M.D.Sturge eds., Modern Problems in Condensed Matter Sciences 2, North Holland, Amsterdam (1982).

[4] B.Hönerlage, R.Lévy, J.B.Grun, C.Klingshirn and K.Bohnert, Physics Reports 124,161 (1985).

[5] H.Haug and S.W.Koch, Quantum Theory of Optical and Electronic Properties of Semiconductors, World Scientific, Singapore (1990).

[6] N.Peyghambarian, S.W.Koch and A.Mysyrowicz, Introduction to Semiconductor Optics, Prentice Hall, New Jersey (1993).

[7] C.Klingshirn, Semiconductor Optics, in preparation.

[8] R.S.Knox, in Collective Excitations in Solids, B.Di Bartolo ed, NATO ASI Series B 88, 183 (1983);
P.J.Dean, ibid. p 247; R.Grasser and A.Scharmann ibid. p 317.

[9] C.Klingshirn in Optical Properties of Excited States in Solids, B.Di Bartolo ed. NATO ASI Series B 301 119 (1993).

[10] See the contributions to this volume by D.Broido, D.Fröhlich and J.M.Hvam.

[11] G.H.Wannier, Phys.Rev. 52 191 (1937).

[12] J.Frenkel, Phys.Rev. 37,1276 (1931).

[13] Landolt-Börnstein, New Series, Group III, Vol 17, O.Madelung, M.Schulz and H.Weiss, eds, Springer, Berlin (1982).

[14] T.M.Rice, Solid State Physics 32, 1 (1988).

[15] S.W.Koch, Dynamics of First-Order Phase Transitions in Equilibrium and Nonequilibrium Systems, Lecture Notes in Physics, 207, Springer Berlin (1984).

[16] R.Zimmermann, Many Particle Theory of Highly Excited Semiconductors. Teubner Texte zur Physik 18, Teubner, Leipzig (1988).

[17] J.C.Hensel, T.G.Phillips and G.A.Thomas, Solid State Physics 32,88 (1988).

[18] Electron-Hole Droplets in Semiconductors, C.D.Jeffries and L.V.Keldysh eds. Modern Problems in Condensed Matter Sciences, Vol.6, North Holland, Amsterdam (1985).

[19] D.S.Chemla, I.Bar-Joseph, J.M.Kuo, T.Y.Chang, C.Klingshirn, G.Livescu and DAB Miller; IEEE J.QE. 24 1664 (1988).

[20] P.Vashista and R.K.Kalia, Phys.Rev. B25, 6492 (1982).

[21] R.Zimmermann, phys.stat.sol.b 146,371 (1988).

[22] C.Klingshirn, Festkörperprobleme/Advances in Solid State Physics 30,335 (1990) and J.Cryst.Growth 117,753 (1992).

[23] E.O.Göbel and O.Hildebrand, phys.stat.sol.b 88,645 (1978).

[24] C.V.Shank, R.L.Fork, R.F.Leheny and J.Shah, Phys.Rev.Lett. 42,112 (1979) v.d.Linde and R.Lambrich, ibid.p 1090.

[25] D.S.Chemla, S.Schmitt-Rink and DAB Miller, in Optical Nonlinearities and Instabilities in Semiconductors
H.Haug ed.,Academic Press, New York (1988).

[26] C.Klingshirn and H.Haug, Phys.Reports 70 315 (1981).

[27] C.Klingshirn, in " Spectroscopy of Solid State Laser Type Materials" B.Di Bartolo ed. Ettore Majorana International Science Series, Physical Sciences 30 485, Plenum Press, New York (1987).

[28] C.Klingshirn,
Advanced Materials for Optics and Electronics 2 (1993) in press

[29] G.Manzke, V.May and K.Henneberger
phys.stat.sol. b125 693 (1984) and
H.Stolz, R.Zimmermann and G.Röpke
phys.stat.sol.b 105 585 (1981).

[30] W.Ebeling, W.D.Kraeft, D.Kremp and K.Kilimann, phys.stat.sol.b 78 241 (1976).

[31] M.Combescot, Sol.State Commun., 30, 81 (1979)
M.Combescot, and C.Benoit à la Guillaume, ibid. 46, 579 (1983).

[32] G.A.Thomas, J.B.Mock and M.Capizzi, Phys.Rev.B18 4250 (1978).

[33] F.A.Majumder, H.-E. Swoboda, K.Kempf and C.Klingshirn, Phys.Rev. B32 2407 (1985).

[34] M.M.Rieger, P.Vogl, and J.A.Majowski
Europhysics Conference Abstract 17A, SC XV Regensburg, (1993) to be published in Physica Scripta.

[35] H.Kalt, A.L.Smirl and T.F.Bogges, phys.stat.sol.b 150 895 (1988)
M.Rinker, H.Kalt and K.Kohler, Appl.Phys.Lett 57 284 (1990)
H.Kalt and M.Rinker, Phys.Rev.B45 1139 (1992).

[36] M.Fieseler, R.Schwabe and J.Staehli, phys.stat.sol.b 159 411 (1990)
S.Gürtler, R.Schwabe, J.L.Staehli und K.Unger
ibid. 173 441 (1992)
P.P.Paskov, Europhys.Lett. 20 143 (1992).

[37] H.Kalt, Festkörperprobleme/Advances in Solid State Physics 32.p

[38] Proc.21th Intern.Conf.Phys.Semiconductors, Ping Jiang and Hou-Zhi Zheng eds, World Scientific, Hongkong (1993).

[39] M.S.Skolnick, K.J.Nash, P.E.Simmonds, D.J.Mowbray, T.A.Fisher, M.K.Saber, D.M.Whittaker, S.J.Bass and R.S.Smith in Ref.[38] p 41.

[40] H.Kalt, K.Leo, R.Cingolani and K.Ploog, Phys.Rev.B40 12017 (1989).

[41] M.Combescot and Ch.Tanguy in Ref.[38] p 141.

[42] E.O.Göbel and K.Ploog, Progr.Quant.El.14 289 (1990).

[43] a) C.Cingolani and K.Ploog, Adv.Phys.40 535 (1990)
b) R.Cingolani, Europhysics Abstracts 17ASC XIV Regensburg (1993) to be published in Physica Scripta.

[44] R.W.Martin, Sol.State Commun.19 373 (1976) and ibid 22 523 (1977).

[45] A.Forchel, B.Laurich, G.Moersch, W.Schmid and T.L.Reinecke, Phys.Rev.Lett. 46 678 (1981).

[46] G.A.Thomas, T.M.Rice and J.C.Hensel, Phys.Rev.Lett. 33, 219 (1974)
T.L.Reinecke and S.C.Ying ibid. 35 311 (1975).

[47] D.Hulin, M.Combescot, N.Bontemps and A.Mysyrowicz

Physics Lett. A 61 349 (1977)
H.Maaret, J.Barrau, M.Brousseau, J.Collet, J.Mazzaschi, Sol.State Commun., 25, 601 (1978).
G.Beni and T.M.Rice, Sol.State Commun. 23 871 (1977)
R.Schwabe, F.Thuselt, H.Weinert,R.Bindemann and K.Unger, phys.stat.sol. b 89 561(1978)
D.Bimberg, M.S.Skolnick and L.M. Sander; Phys.Rev. B14 2231 (1979).
U.Woggon, F.Henneberger and H.Weinert, phys.stat.sol.b 143 611 (1987), and References therein.

[48] H.Kalt, K.Reimann, W.W.Rühle, M.Rinker, and E.Bauser,Phys.Rev. B 42 7058 (1990).

[49] R.M.Westervelt, J.C.Culbertson and B.S. Black, Phys.Rev.Lett. 42 267 (1979).

[50] M.Greenstein and J.P.Wolfe, Solid State Commun. 33, 309 (1980).

[51] R.S.Markiewicz, J.P.Wolfe and C.D.Jeffries, Phys.Rev.B15 1988 (1977).

[52] P.L.Gourley and J.P.Wolfe, Phys.Rev. B24 5970 (1981).

[53] V.G.Lyssenko, V.I.Revenko, T.G.Tratas and V.B.Timofeev, Sov.Phys.JETP 41 173 (1975).

[54] E.Lach, V.D.Kulakovskii, A.Forchel, T.L.Reinecke, J.Straka, D.Grützmacher und G.Weinmann, phys.stat.sol.b 159 125 (1990).

[55] H.-E.Swoboda, F.A.Majumder, V.G.Lyssenko, C.Klingshirn and L.Banyai, Z.Physik B 70 341 (1988).

[56] K.Bohnert, M.Anselment, G.Kobbe, C.Klingshirn, H.Haug, S.W.Koch, S.Schmitt-Rink and F.F.Abraham, Z.Physik B 42 1 (1981).

[57] A.Kreissl, K.Bohnert, V.G.Lyssenko, and C.Klingshirn, phys.stat.sol. b114 537 (1982).

[58] C.Klingshirn, J.Crystal Growth 117 753 (1992)
and Advanced Materials of Optics and Electronics 2 (1993) in press.

[59] M.V.Kurik, phys.stat.sol.a 8,9 (1971)
J.G.Liebler, S.Schmitt-Rink and H.Haug, J.Luminesc. 34 1 (1985)

[60] C.Klingshirn, Ch.Weber, D.S.Chemla, DAB Miller, J.E.Cunningham, C.Ell and H.Haug, NATO ASI series B 194 353 (1989).

[61] G.Beni and T.M.Rice, Phys.Rev.B18 768 (1978).

[62] O.Hildebrand, E.O.Goebel, K.M.Romanek, H.Weber, and G.Mahler Phys.Rev.B 17 4775 (1978).

[63] H.Haug, and F.F. Abraham, Phys.Rev.B 23 2960 (1981).

[64] D.J.Haan, H.A.MacKenzie, H.A.Al Attar and W.J.Firth, Optics Lett. 10 187 (1985).

[65] See the contributions to this book by D.Broido, Ch.Flytzanis, R.Reisfeld, U.Woggon and references given therein.

[66] K.H.Schlaad, Ch.Weber, D.S.Chemla, J.Cunnigham, C.V.Hoof, G.Borghs, G.Weinmann, W.Schlapp, H.Nickel and C.Klingshirn, phys.stat.sol.b 159 173 (1990) and Phys.Rev.B 43 4268 (1991).

[67] S.Schmitt-Rink, C.Ell and H.Haug, Phys.Rev.B 33 1183 (1986).

[68] H.Schweizer, A.Forchel, A.Hangleiter, S.Schmitt-Rink, J.P.Löwenau and H.Haug, Phys.Rev.Lett. 51 698 (1983).

[69] V.D.Kulakovskii, E.Lach, A.Forchel and D.Grützmacher, Phys.Rev. B40 8087 (1989).

[70] W.Langbein, H.Kalt, M.Grün, M.Hetterich and C.Klingshirn, to be published in J.Cryst.Growth.

[71] J.Cibert, Europhysics Conf.Abstr. 17A, SC XCVII (1993), to be published in Physica Scripta and D.Broido, this book.

[72] F.Daiminger, L.V.Butov, Ch.Gréus, J.Straka and A.Forchel in Ref.[38] p1293 K.H.Ploog, R.Nötzel and O.Brandt in Ref.[38] p 1297.

[73] A.Forchel, G.Tränkle, U.Cebulla, H.Leier and B.E.Maile, NATOASI series B194 361 (1989).

[74] R.Rinaldi, R.Cingolani, M.Ferrara, H.Lage, D.Heitmann and K.Ploog Phys.Rev. B47 (1993) in press.

[75] R.Cingalani, H.Lage, L.Tapfer, H.Kalt, D.Heitmann and K.Ploog, Phys.Rev.Lett. 67 891 (1991).

[76] Ben Yu-Kang and S.Das Sarma, Phys.Rev. B45 1750 (1992).

[77] H.Kalt, R.Nötzel, K.Ploog and H.Giessen, phys.stat.sol. b 173 389 (1992).

[78] A.Wörner, short seminar at this school.

[79] Physics of Nanostructures
J.H.Davics and A.R.Long eds. Proc.39th Scottish Universities Summer School in Physics, St.Andrews (1991). SUSSP Publications, Edinburgh (1992).

[80] Optical Computing

B.S.Wherret and F.A.P.Tooley eds.
Proc 34th Scottish Universities Summer School in Physics, Edinburgh (1988), SUSSP Publications, Edinburgh (1989).

[81] A.Uhrig, A.Wörner, W.Langbein, U.Woggon, S.Gaponenko, M.Saleh, O.Wind, H.Kalt and C.Klingshirn in Ref.[38] p 1379.

[82] See e.g. the contributions by R.M.Macfarlane or by U.Woggon to this book and references given therein.

[83] K.Meissner, B.Flügel, H.Gießen, B.P. McGinnis, A.Paul, R.Binder, S.W.Koch, N.Peyghambarian, M.Grün, and C.Klingshirn; Phys.Rev.B. in press

[84] K.El Sayed, T.Wicht, H.Haug and L.Banyai, Z.Physik B86 345 (1992)
R.Zimmermann, phys.stat.sol. b173 129 (1992)
H.Haug ibid.p 139
L.Banyai, D.B.Tran Thoai, C.Remling and H.Haug, ibid p 149
S.Glutsch and P.Zimmermann, Phys.Rev. B45 5857 (1992).

THE TWO-PHOTON LASER

Daniel J. Gauthier and Hope M. Concannon

Department of Physics
Duke University
P.O. Box 90305
Durham, NC 27708
USA

ABSTRACT

We explain in simple terms the behavior of two-photon lasers and describe recent results that have lead to the realization of the first continuous-wave two-photon optical laser. We stress the differences between one- and two-photon lasers to develop an appreciation of their dynamics and the difficulties associated with achieving two-photon lasing. It will be shown that the fundamental light-matter interactions give them a unique threshold behavior and that competing nonlinear optical processes make it difficult to achieve lasing. Finally, we describe a new class of driven-atom gain media that are ideally suited for constructing two-photon lasers.

I. INTRODUCTION

Soon after the advent of the laser in the early 1960's, researchers discovered other types of quantum oscillators that have distinct and potentially useful characteristics. One such oscillator is known as the two-photon laser [1], so named because it is based on the two-photon stimulated emission process. The two-photon laser appeared to be an experimentalists dream because it was predicted [2] that it could produce two tunable frequencies, operate at high power and store large energies (recall that there were only fixed-frequency, modest power lasers in the early sixties). In addition, the two-photon laser challenged our understanding of the interaction of light with matter because it is a highly nonlinear, far from equilibrium system that cannot be analyzed using standard perturbation techniques. Unfortunately, the practical benefits of the laser have not been realized nor have the theories been tested because it has been difficult to achieve two-photon lasing [3] due to the lack of suitable gain media. In this chapter, we will explain in simple terms the behavior of two-photon lasers, discuss why it is difficult to

Nonlinear Spectroscopy of Solids: Advances and Applications
Edited by B. Di Bartolo and B. Bowlby, Plenum Press, New York, 1994

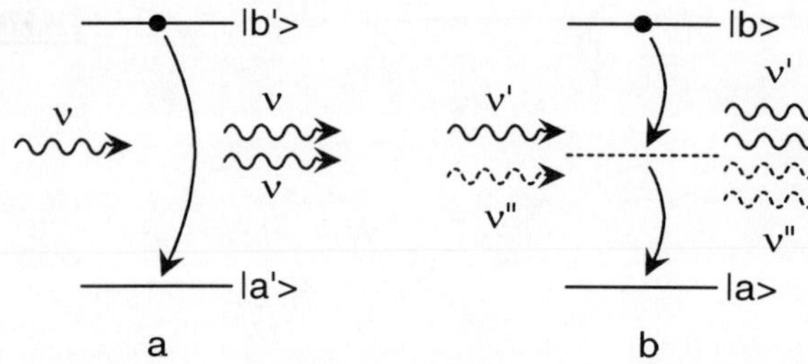

Figure 1 (a) The one-photon stimulated emission process. The states have opposite parity. (b) The two-photon stimulated emission process. The states have the same parity.

achieve two-photon lasing and describe recent work that led to the demonstration of the first continuous-wave two-photon optical laser.

An obvious question that might arise when reading the title of this chapter is: What is a two-photon laser? The answer to this question is quite simple if we compare and contrast the origin of gain in one- and two-photon lasers. In normal one-photon lasers, the gain is due to the one-photon stimulated emission process shown schematically in Fig. 1a. In this process, an incident photon can stimulate an atom from the upper state $| b' >$ to the lower state $| a' >$ and two photons are scattered by the atom. The scattered photons have the same frequency, phase and direction of the incident photon which give the laser its unique coherence properties. The stimulated emission rate is proportional to the incident photon flux and proceeds most efficiently when the frequency of the incident photon ω is equal to the transition frequency $\omega_{b'a'}$ and when the states have opposite parity (connected by an allowed electric dipole transition matrix element).

In contrast, the gain in two-photon lasers is due to the two-photon stimulated emission process shown schematically in Fig. 1b. In this process, two incident photons can stimulate an atom from the upper state $| b >$ to the lower state $| a >$ and four photons are scattered by the atom. Note that the frequencies ω' and ω'' of the incident photons can take on any value so long as $\omega' + \omega'' \simeq \omega_{ba}$, where ω_{ba} is the two-photon transition frequency and that the process proceeds most efficiently when the states have the same parity (not connected by an allowed electric dipole transition matrix element). The scattered photons have the same frequency, phase and direction of the incident photons which give the laser coherence properties different than those of normal one-photon lasers [4],[5]. Throughout the rest of the chapter, we assume for simplicity that there is only a single beam of light incident on the atom and hence that $\omega' = \omega'' \equiv \omega \simeq \omega_{ba}/2$. In this case, the two-photon stimulated emission rate is proportional to the square of the incident photon flux.

One consequence of the scaling of the two-photon stimulated emission rate with the incident photon flux is that the gain is intensity dependent. Consider a beam of light (intensity I_{IN}) incident on a collection of atoms that posseses a two-photon inversion. It will be amplified by an amount that is approximately given by

$$I_{OUT} \sim I_{IN} \ \exp(G^{(2)} I_{IN} L) \tag{1}$$

when the frequency of the incident beam is equal to one-half the two-photon transition frequency, where L is the length of the medium and $G^{(2)}$ is the two-photon gain coefficient. In deriving Eq. 1, we have assumed that the transition is not saturated and that the gain is very small. Note that the input intensity appears in the exponential and hence the gain vanishes when the input intensity is small. This is in contrast to a one-photon gain medium where the gain is independent of the intensity (until saturation sets in).

The smallness of $G^{(2)}$ presents the main stumbling block to achieving two-photon lasing. Seemingly, a natural solution would involve using high intensities as can be seen from Eq. 1. Unfortunately, several other competing nonlinear optical processes occur simultaneously in the gain medium, such as four-wave mixing, harmonic generation, anti-Stokes Raman scattering, self-focusing, and photo ionization, and become more prevalent at higher intensities. In addition, normal one-photon lasing or superfluorescence from the upper state $| b >$ to some other lower-energy intermediate state may deplete the inversion between the states $| b >$ and $| a >$. Typically, the competing processes dominate the two-photon stimulated emission process and prevent the occurrence of two-photon lasing.

Is there a two-photon gain medium that avoids these problems? To answer this question, it is necessary to develop a simple model that sets bounds on the operating parameters of the laser. This will quantify some of their differences from normal one-photon lasers and will focus our attention on the characteristics that must be optimized.

II. SIMPLE MODEL OF A TWO-PHOTON LASER

We have developed a model for two-photon lasers based on a set of self-consistent rate equations that predicts many of their crucial attributes without being overly complex. Although other models are required to make quantitative comparisons with experimental results, we feel that our model is useful for developing an intuitive understanding of two-photon lasers. Below, we present our model and use it to make predictions of the behavior of two-photon optical lasers. Later, we discuss effects ignored in our model and describe a new type of gain medium that was used to realize the first continuous-wave, two-photon optical laser.

A key test for any model of two-photon lasers is whether it predicts the novel threshold behavior of the laser. Briefly, the threshold condition for all lasers is that the round-trip gain must equal the round-trip loss. For one-photon lasers this criterion yields the well known result that lasing will commence when a uniquely defined minimum inversion (proportional to the gain) is attained via sufficient pumping. The situation is more complicated for the two-photon laser because the gain increases with increasing inversion density ΔN and with increasing cavity photon number q (until the atoms are saturated), so the threshold condition must be specified by two parameters. We define a threshold inversion density ΔN_{th} as the inversion needed to satisfy the threshold condition with a cavity photon number q_{sat} just sufficient to saturate the two-photon gain. When $\Delta N > \Delta N_{th}$, there is a corresponding cavity photon number (which is less than q_{sat}) that must be present in the cavity before the laser will turn on. Hence, if the laser is initially off it cannot turn on unless some perturbation, such as an externally injected field, brings it above the necessary value [1],[6],[7],[9],[10].

Our rate-equation model of the two-photon laser follows from the standard model of one-photon lasers with the exception that the one-photon stimulated emission rate

$$W^{(1)} = B^{(1)}q \; , \tag{2}$$

is replaced by the two-photon stimulated emission rate [11]

$$W^{(2)} = B^{(2)}q^2 , \tag{3}$$

where $B^{(1)}$ ($B^{(2)}$) is the one- (two-) photon rate coefficient. The gain coefficient introduced in the previous section is given in terms of the rate coefficient through the relation

$$G^{(2)} = \frac{V_c V_a B^{(2)} \Delta N}{c^2 \hbar \omega} , \tag{4}$$

where V_c (V_a) is the volume of the cavity (volume of the medium contained in the cavity). For simplicity, we assume that the two-photon laser operates in the degenerate mode, that the laser oscillates in a single, plane-wave mode, and that the cavity (population) decay rate γ_c (γ) is much smaller than the atomic coherence dephasing rate. Under these oversimplifying conditions, the behavior of the laser is described by the mean photon number in the cavity q and the mean population inversion density ΔN between the atomic levels that participate in the stimulated emission process. The first-order, coupled nonlinear differential equations governing the evolution of these quantities are given by

$$\frac{dq}{dt} = V_a B^{(2)} q^2 \Delta N - \gamma_c (q - q_{inj}(t)) , \tag{5}$$

and

$$\frac{d\Delta N}{dt} = -2B^{(2)} q^2 \Delta N - \gamma(\Delta N - \Delta N_o) , \tag{6}$$

where ΔN_o is the inversion density in the absence of the field due to the pump process, $\gamma \Delta N_o$ is the pump rate, and $q_{inj}(t)$ is the photon number injected into the cavity by an external source. We see from Eq. 5 that the photon number increases due to the two-photon stimulated emission process and by injection from the external source, and decreases due to linear loss through the cavity mirrors. We have ignored the possibility of two-photon spontaneous emission processes at the laser frequency because the emission rates are extremely small in the optical regime [12]. This approximation is not valid for two-photon masers where the stimulated and spontaneous rates are comparable [13]. From Eq. 6, we see that the inversion decreases in response to the stimulated emission process and due to other radiative (at frequencies distinct from the laser frequency) and non radiative decay mechanisms, and increases due to the pump process.

The steady-state behavior of the two-photon laser can be obtained readily from our model. From Eq. 6, we find that

$$\Delta N_{ss} = \frac{\Delta N_o}{1 + q_{ss}^2 / q_{sat}^2} , \tag{7}$$

where

$$q_{sat} = \sqrt{\gamma / 2B^{(2)}} \tag{8}$$

is the two-photon saturation photon number. The saturation intensity is given in terms of the saturation photon number through the relation $I_{sat} = c\hbar\omega q_{sat}/V_c$. Equation 7 is reminiscent of the steady-state inversion for a one-photon laser [14] except that the denominator is not linear in q_{ss}. In contrast, the steady-state solution for the photon number is very different from that of one-photon lasers. We find three solutions given by

$$q_{ss}^o = 0 , \tag{9}$$

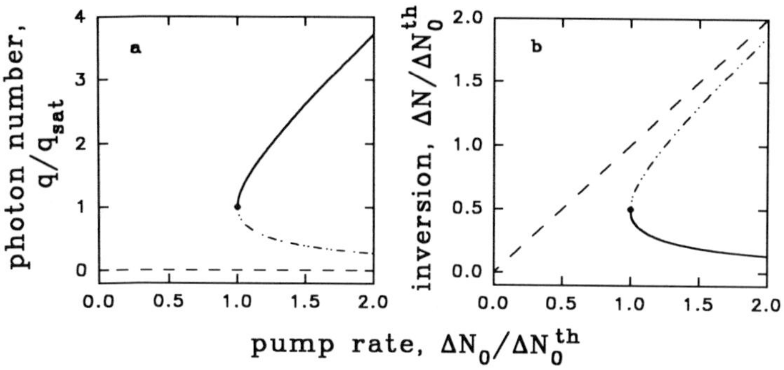

Figure 2 Steady state behavior of (a) the cavity photon number normalized to the saturation photon number and (b) the atomic inversion normalized to the threshold atomic inversion. The dashed, solid, and dot-dash lines are the $(q_{ss}^o, \Delta_{ss}^o), (q_{ss}^+, \Delta_{ss}^+), and (q_{ss}^-, \Delta_{ss}^-)$ solutions, respectively.

and

$$q_{ss}^{\pm} = \frac{\gamma V_a}{4\gamma_c}\left[\Delta N_o \pm \sqrt{\Delta N_o^2 - 16 q_{sat}^2 \gamma_c^2/\gamma^2 V_a^2}\right] , \tag{10}$$

for the case when $q_{inj}(t) = 0$. The physically meaningful (real) steady-state solutions represented by these equations are shown in Fig. 2 where they are plotted as a function of the pump rate. It is seen from Eq. 10 that

$$\Delta N_o^{th} = \frac{4 q_{sat}\gamma_c}{\gamma V_a} \tag{11}$$

is the minimum value of ΔN_o which admits a nonzero photon number. This yields $\Delta N_{ss} = 1/2\, \Delta N_o^{th}$ and $q_{ss}^{\pm} = q_{sat}$ at threshold, in agreement with the heuristic discussion of the threshold behavior presented earlier.

The discontinuous threshold behavior shown in Fig. 2 is indicative of a first-order phase transition, which is very different from the smooth turn-on behavior of normal one-photon lasers. Note from Fig. 2a that the two-photon gain is saturated at threshold, again in sharp contrast to the typical one-photon laser which operates very far below saturation. Figure 2b also shows that the inversion is never constant, unlike the behavior of one-photon lasers where the inversion clamps above threshold [14]. Rate equation models that assume the inversion remains constant have been investigated previously [10].

Based on past experience with one-photon lasers, we expect that the steady-state solutions may be unstable because the laser operates in the saturated regime. We have performed a linear stability analysis of the steady-state solutions and find that: 1) the zero-photon solution $(q_{ss}^o, \Delta N_{ss}^o)$ is always stable (two-photon spontaneous emission, neglected here, can destabilize this solution); 2) the $(q_{ss}^-, \Delta N_{ss}^-)$ solution, where the photon number decreases with increasing pump rate, is always unstable; 3) the $(q_{ss}^+, \Delta N_{ss}^+)$ solution is always stable for a 'good' cavity ($\gamma/\gamma_c > 1$); and 4) the $(q_{ss}^+, \Delta N_{ss}^+)$ solution is unstable for a 'bad' cavity ($\gamma/\gamma_c < 1$) for pumping just above threshold but stabilizes for higher pump rates [6], [7]. Ning and Haken [15] have found that the laser is always unstable for a bad cavity when coherent effects are taken into account.

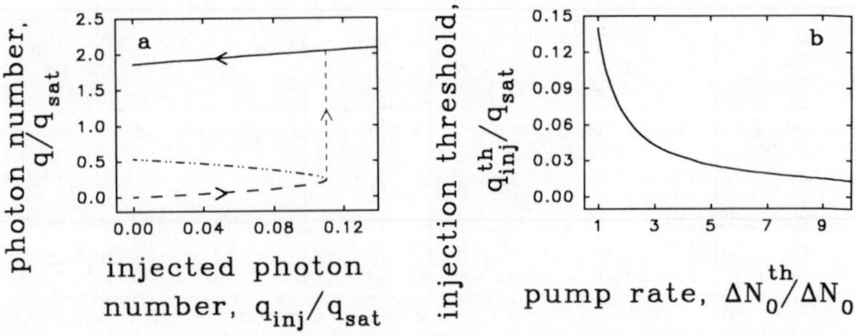

Figure 3 The injection threshold. (a) Steady-state photon number with a cw field injected into the cavity for a pump rate 20% above threshold. Threshold is defined as the point where the lowest solution ends. The arrows indicate the progression of the laser from off to on. (b) Variation of the injection threshold with the pump rate.

To address how an injected field can turn on the laser, we have determined the steady-state solutions of Eqs. 5 and 6 and their stability properties when a continuous-wave beam of photons is injected into the cavity ($q_{inj} \neq 0$). The stability of the solutions can be quite complex, especially in the bad-cavity limit. However, the analysis is straightforward if we consider the important case when the laser is initially 'off', $\Delta N_o > \Delta N_o^{th}$, and q_{inj} is increased slowly. This behavior is illustrated in Fig. 3a where we plot the three solutions for the steady-state photon number when $q_{inj} \neq 0$ and $\Delta N_o = 1.2\ \Delta N_o^{th}$. The low-power solution (dashed curve) increases as q_{inj} increases and is stable until it reaches a critical value $q_{inj}^{th} \simeq 0.11\ q_{sat}$. This point defines the minimum injected photon number necessary to initiate lasing. Increasing q_{inj} beyond q_{inj}^{th} forces the laser to switch from the low-power solution (now unstable) to the high-power solution (stable); the laser turns on. The behavior of the laser as q_{inj} decreases depends on the quality of the cavity. For a good cavity, the laser will continue to operate at high power as q_{inj} decreases. For a bad cavity, the laser will only continue to operate at high power if the solution is stable when $q_{inj} = 0$. We have studied the stability behavior for other values of the pump rate and find that the injection threshold q_{inj}^{th} decreases for higher values of the pump rate as shown in Fig. 3b. In practice, a pulse (peak photon number q_{inj}^{o}) is injected into the laser rather than a cw beam. When the laser can (cannot) adiabatically follow the temporal variation of the pulse, $q_{inj}^{o} = q_{inj}^{th}$ ($q_{inj}^{o} \geq q_{inj}^{th}$).

The transient behavior of the laser described above is explored by numerically integrating Eqs. 5 and 6 . Figure 4 shows how the laser responds to injected trigger pulses for a good cavity ($\gamma/\gamma_c = 2$) when the pump rate is greater than the threshold pump rate ($\Delta N_o = 1.2\ \Delta N_o^{th}$) and when there are no photons in the cavity initially. For a weak trigger pulse (peak amplitude $q_{inj}^{o} = 0.1\ q_{sat}$, Fig. 4a) the laser is not driven above threshold, while for a slightly stronger pulse (peak amplitude $q_{inj}^{o} = 0.12\ q_{sat}$, Fig. 4b) the laser is driven above threshold and it attains a constant amplitude after the injected pulse switches off. This is in good agreement with the injection threshold $q_{inj}^{th} \simeq 0.11\ q_{sat}$ calculated above.

The most important conclusion that we can draw from our model is that the intensity circulating in the laser resonator is always greater than or equal to the two-photon

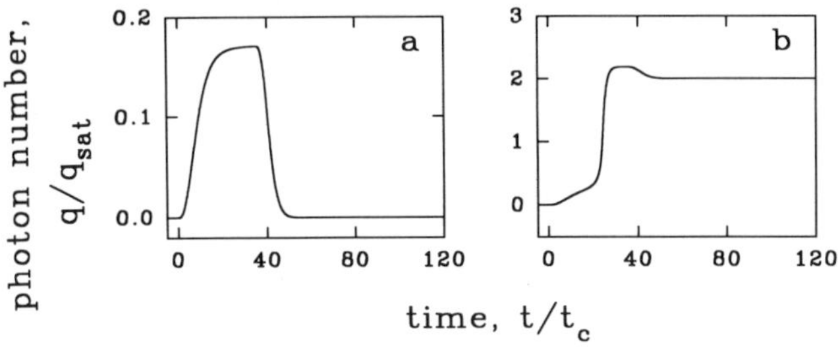

Figure 4 Transient evolution of the photon number for pumping 20% above threshold. (a) $q_{inj} = 0.1$ The trigger is not strong enough to reach threshold. (b) $q_{inj} = 0.12$ The laser is driven to the high-power state.

saturation intensity. Hence, the gain medium must have a low two-photon saturation intensity to avoid the high intensities that tend to magnify competing nonlinear optical processes. Fortunately, low saturation intensities go hand-in-hand with large two-photon gains as can be seen from Eqs. 4 and 8.

III. TWO-PHOTON GAIN IN A THREE-LEVEL ATOMIC SYSTEM

To estimate the size of the two-photon gain and saturation intensity, we will investigate the microscopic interaction of a field with a three-level atomic system using the density-matrix formalism developed in earlier chapters. Consider a field $\vec{E}(t) = [\vec{E}_o \exp(-i\omega t) + c.c.]$ interacting with the inverted $\mid b > \rightarrow \mid a >$ two-photon transition as shown in Fig. 5. Since these states have the same parity, they are not connected by a normal one-photon transition and hence the electric-dipole matrix element is zero. A real intermediate level $\mid i >$ with opposite parity is located near the virtual intermediate level of the two-photon transition (detuning Δ_{ia}) to enhance the two-photon transition rate. The electric-dipole matrix element of the $\mid b > \rightarrow \mid i >$ ($\mid i > \rightarrow \mid a >$)

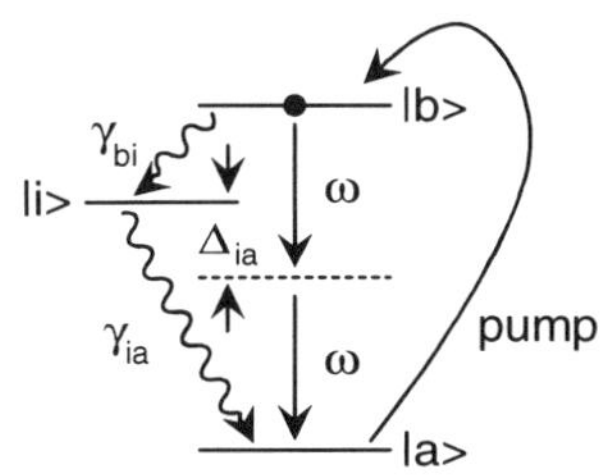

Figure 5 Two-photon gain in a three-level atomic system. The intermediate state $|i>$ enhances the two-photon transition rate.

transition is denoted by $\vec{\mu}_{bi}$ ($\vec{\mu}_{ia}$) and the corresponding Rabi frequency is denoted by $\Omega_{bi} = 2\vec{\mu}_{bi} \cdot \vec{E}_o/\hbar$ ($\Omega_{ia} = 2\vec{\mu}_{ia} \cdot \vec{E}_o/\hbar$). Population in the upper level can decay spontaneously to the intermediate level (rate γ_{bi}) which subsequently decays to the lower level (rate γ_{ia}). For simplicity, we assume that $\gamma_{ia} \gg \gamma_{bi}$ so that essentially no population builds up in the intermediate level.

The only nonzero elements of the density matrix are: the populations ρ_{aa} and ρ_{bb}; the one-photon coherences ρ_{bi} and ρ_{ia}; and the two-photon coherence ρ_{ba}. The equations-of-motion for the density matrix are obtained from Heisenberg's equation and are simplified using the one- and two-photon rotating-wave approximations. In addition, the equations are simplified by adiabatic elimination of the one-photon coherences (performed formally by setting $\dot{\rho}_{bi} = \dot{\rho}_{ia} = 0$) which is valid so long as $\Delta_{ia} \gg \gamma_{bi}$, Ω_{bi}, Ω_{ia}[17]. After some algebra, it is found that the equations-of-motion for the population inversion density $\Delta N = \eta(\rho_{bb} - \rho_{aa})$ and the slowly-varying two-photon coherence density $\sigma_{ba} = \eta\ \rho_{ba} \exp(i2\omega t)$ are given by

$$\frac{d\Delta N}{dt} = -i[(\Omega^{(2)})^* \sigma_{ba} - \sigma_{ab}\Omega^{(2)}] - \gamma(\Delta N - \Delta N_o)\ , \tag{12}$$

and

$$\frac{d\sigma_{ba}}{dt} = -i\frac{\Delta N\ \Omega^{(2)}}{2} - [i(\Delta_2 - \Delta_s) + \Gamma_{ba}]\sigma_{ba}\ , \tag{13}$$

where η is the atomic number density, $\gamma \simeq \gamma_{bi}$ is the inversion decay rate, $\Delta_2 = 2\omega - \omega_{ba}$ is the two-photon detuning, Γ_{ba} is the two-photon coherence dephasing rate, $\Omega^{(2)} = \Omega_{bi}\Omega_{ia}/\Delta_{ia}$ is the two-photon Rabi frequency, and $\Delta_s = (|\ \Omega_{ia}\ |^2 - |\ \Omega_{ia}\ |^2)/\Delta_{ia}$ is the AC Stark shift of the two-photon transition. Note that Eqs. 12 and 13 are nearly identical to the corresponding equations for a driven two-level atom. It can be seen from Eq. 13 that the two-photon transition frequency is intensity-dependent due to the AC Stark shift. These shifts can play an important role in the dynamics of two-photon lasers.

The results obtained from the density matrix formalism are connected to the rate equations used in the previous section under conditions when there is large dephasing of the two-photon coherence ($\Gamma_{ba} \gg \Delta_2$, Δ_s, γ, γ_c). In this case, the coherence can be adiabatically eliminated from Eq. 12 using Eq. 13 with $\dot{\sigma}_{ba} = 0$. By comparison with Eq. 6, it is found that the two-photon rate coefficient is given by

$$B^{(2)} = \frac{32\pi^2\ |\ \vec{\mu}_{bi} \cdot \hat{\epsilon}\ |^2 |\ \vec{\mu}_{ia} \cdot \hat{\epsilon}\ |^2\ \omega^2}{V_c^2 \hbar^2 \Delta_{ia}^2 \Gamma_{ba}}\ , \tag{14}$$

where $\hat{\epsilon}$ is the polarization unit-vector of the field. From Eq. 14, it is straightforward to determine the saturation photon number using Eq. 8 and the threshold inversion using Eq. 11. Note that the rate coefficient can be enhanced significantly when the real intermediate level $|\ i >$ is close to the virtual level of the two-photon process, that is, when Δ_{ia} is small. It was an understanding of this scaling that led to the development of the first microwave two-photon oscillator [13],[18] and the first two-photon optical laser [23].

What are numerical values for these parameters? Evaluating Eqs. 14, 8, and 11 using values that are typical for a two-photon optical gain medium (see Table 1) and a resonator with a loss-per-pass of 1%, we find that $B^{(2)} = 2.6 \times 10^{-12}$ s^{-1}, $q_{sat} = 4.4 \times 10^9$ photons (corresponding to $I_{sat} = 6.6 \times 10^4$ W-cm^{-2}), and $\Delta N_o^{th} = 1.8 \times 10^{13}$ atoms-cm^{-3}. We see that the intensity circulating in the resonator is modest in comparison to some present-day lasers, giving hope that competing nonlinear optical effects can be avoided (such as self-focusing and photo ionization). In addition, we see that the

Table 1 Parameters used to calculate the two-photon rate coefficient.

dipole moment, $\mid \vec{\mu}_{bi} \cdot \hat{\epsilon} \mid = \mid \vec{\mu}_{ia} \cdot \hat{\epsilon} \mid$	$ea_o = 2.4 \times 10^{-18}\text{cm}^{5/2}\text{g}^{1/2}\text{s}^{-1}$
population decay rate, γ	$10^8 s^{-1}$
coherence dephasing rate, Γ_{ba}	$1 \times 10^9\text{s}^{-1}$
intermediate level detuning, Δ_{ia}	$1 \times 10^{13}\text{s}^{-1}$
laser frequency, ω	$3 \times 10^{15}\text{s}^{-1}$
cavity length, L	5 cm
cavity volume, V_c	6×10^{-4} cm^3
cavity decay rate, γ_c (corresponds to 1% loss)	$6 \times 10^7\text{s}^{-1}$

threshold inversion density is quite low, so it should be straightforward to achieve two-photon lasing.

Unfortunately, nature is not so kind! We must consider the possibility that normal one-photon lasing can occur at a spectrally distinct frequency on the $\mid b > \rightarrow \mid i >$ transition. For sufficiently low one-photon gains, one-photon lasing can be avoided using a high-finesse cavity that selectively enhances the two-photon laser frequency and not the one-photon laser frequency. For high one-photon gain, superfluorescence will occur which will effectively destroy the inversion. The threshold for superfluorescence occurs when the single-pass one-photon gain $G^{(1)}L \sim 30$ which can be avoided only by reducing the number density (which also reduces the two-photon gain!). We estimate the single-pass one-photon gain on the $\mid b > \rightarrow \mid i >$ transition using

$$G^{(1)}L = \frac{4\pi\Delta N_o^{th} \mid \vec{\mu}_{bi} \cdot \hat{\epsilon} \mid^2 \omega L}{\hbar c \Gamma_{ba}} \tag{15}$$

and find that $G^{(1)}L = 640$. Hence the number density must be reduced by a factor of ~ 20 to just avoid superfluorescence and by a factor of ~ 100 to avoid significant depletion of the inversion. For a 100$\times$ decrease in the number density, the loss-per-pass in the resonator must be less than $\sim 10^{-4}$ to achieve two-photon lasing. Such low-loss resonators can be constructed using recently developed super-polished mirrors. Note, however, that the gain medium must be in contact with the mirrors that form the resonator because any other optics in the cavity will give too high of a loss.

IV. DRESSED-STATE TWO-PHOTON GAIN MEDIUM

One obvious way to increase the two-photon gain on the $\mid b > \rightarrow \mid a >$ transition without increasing the one-photon gain on the $\mid b > \rightarrow \mid i >$ transition is to decrease the detuning Δ_{ia} from the intermediate transition $\mid i >$. Two-photon lasing can be achieved using a resonator with 1% loss by decreasing Δ_{ia} by a factor of 100. In this case, $B^{(2)} = 2.6 \times 10^{-8}$ s^{-1}, $q_{sat} = 4.4 \times 10^7$ photons (corresponding to $I_{sat} = 6.6 \times 10^2$ W-cm^{-2}), and $\Delta N_o^{th} = 1.8 \times 10^{11}$ atoms-cm^{-3}.

The detuning from the intermediate level is not a parameter that can be adjusted easily, however. The atomic energy level structure is fixed and we know of no system in the optical regime that has such a small detuning. The breakthrough in the development of two-photon optical lasers occurred when Prof. Mossberg's group at the University of Oregon realized that it is possible to 'engineer' a near-ideal two-photon

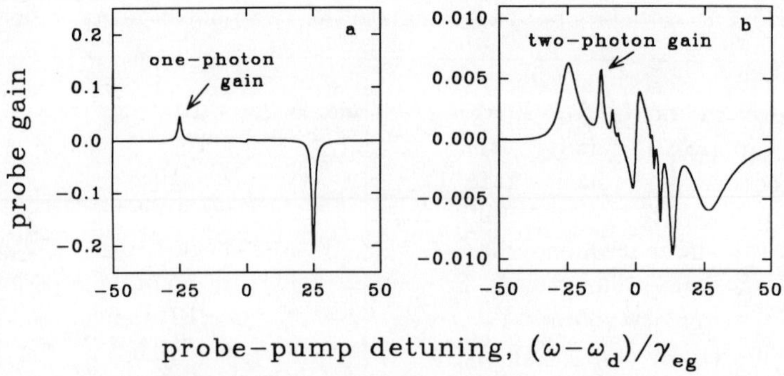

Figure 6 The response of a driven two-level atom with $\Omega'_d/\gamma_{eg} = 25$ and $\Delta_d = -0.4\Omega_d$ (a) The gain experienced by weak probe beam. (b) The gain experienced by a strong probe beam.

gain medium that possesses a small and adjustable detuning [19]. It consists of a two-level atom (one-photon transition) driven by an intense, near-resonant laser field. In the rest of this section, we will briefly describe the origin of two-photon gain in the driven-atom system.

Consider a two-level atom with a ground (excited) level denoted by $\mid g >$ ($\mid e >$) driven by an intense laser field $\vec{E}_d(t) = [\vec{E}_d \exp(-i\omega_d t) + c.c.]$. We are to think of the laser-driven atom as a composite system (a 'dressed' atom) rather than separate entities. The atomic states have opposite parity and are connected by an allowed electric-dipole moment $\vec{\mu}_{eg}$. The population decay rate (coherence dephasing rate) is denoted by γ_{eg} (Γ_{eg}) and we assume for simplicity that the system experiences only natural broadening so that $\Gamma_{eg} = \gamma_{eg}/2$. The interaction strength of the driven-atom system is characterized by the generalized Rabi frequency $\Omega'_d = \sqrt{\Omega_d^2 + \Delta_d^2}$, where $\Omega_d = 2\vec{\mu}_{eg} \cdot \vec{E}_d/\hbar$ is the resonant Rabi frequency and $\Delta_d = \omega_d - \omega_{eg}$ is the detuning of the driving field from the atomic resonance frequency ω_{eg}. The Rabi frequency can be of the order of 6×10^9 s^{-1} for a strong optical transition and a 1 W laser beam collimated to a diameter of ~1 mm.

It has been know for quite some time that the absorption spectrum of a weak probe beam interacting with a collection of dressed atoms has a spectrally narrow gain feature for nonzero atom-driving field detunings [20]. The gain feature shown in Fig. 6a occurs at a probe beam frequency $\omega = \omega_d - \Omega'_d$ for $\Delta_d < 0$, it's width is equal to the line width of the undriven transition Γ_{eg} and the maximum gain is equal in magnitude to $\sim 0.05\alpha_{eg}$, where α_{eg} is the line-center absorption coefficient of the undriven system. Most importantly, the strength of the gain is independent of the phase of the probe field; hence, the gain is not due to some type of parametric wave-mixing process. Rather, it is essentially identical to the familiar one-photon gain medium (several researchers have successfully constructed one-photon lasers using dressed atoms [21]).

The absorption spectrum of a dressed atom becomes complex when the probe beam intensity increases as shown in Fig. 6b. Several new, intensity-dependent gain features appear that occur at subharmonics of the generalized Rabi frequency, that is, for probe beam frequencies $\omega = \omega_d - \Omega'_d/n$ ($n = 1, 2, 3, \ldots$). The feature that occurs at $\omega =$

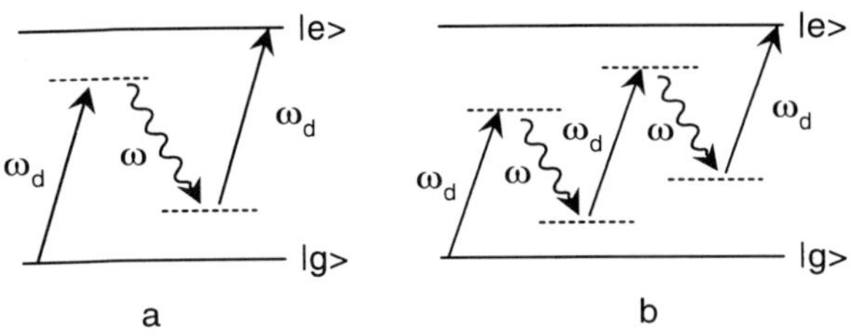

Figure 7 Perturbation-theory interpretation of the gain in a strongly driven atom. Hyper-Raman processes give rise to (a) one-photon and (b) two-photon gain in the driven-atom system.

$\omega_d - \Omega'_d/2$ arises from two-photon stimulated emission and was used to realize the first two-photon optical laser. The fact that this is a two-photon gain feature was not fully appreciated until recently - the early studies called it a 'subharmonic resonance'.

A crude understanding of the origin of the gain features can be obtained from a perturbation-theory analysis of the interaction of the probe beam and the driven atom. Of course, the analysis cannot fully explain the interaction because the pump beam is far too intense to be treated perturbatively. It is found that the one- and two-photon gain features arise from a type of hyper-Raman scattering process as shown in Fig. 7. In both processes, probe beam photons are created at the expense of pump beam photons when the fields induce a transition from the ground level $| g >$ to the excited level $| e >$. Hence, the 'upper' ('lower') laser level is actually the ground (excited) level of the atom for the driven-atom gain medium. The 'inversion' between the upper and lower laser levels is maintained by spontaneous emission from $| e > \rightarrow | g >$. Note that there is never an inversion between the excited and ground levels of the atom.

A quantitative (and intuitive) understanding of the origin of the gain features can be obtained using the dressed-state interpretation of the driving-field - atom interaction. The analysis properly accounts for the interaction of the intense pump field with the atom and treats as perturbations the probe field and spontaneous emission. We note that the study of driven two-level atoms transcends our present application of two-photon gain; it is a textbook example of the application of quantum electrodynamics and so we will only sketch the development of the theory. Our notation follows closely the recent and well-written text by Profs. Cohen-Tannoudji, Dupont-Roc, and Grynberg [22].

The dressed-state analysis differs from the standard semi-classical analysis in that the field mode of the driving field is treated quantum mechanically rather than classically. The reduced Hamiltonian for the driving-field - atom system can be broken up into 'bare' atom, 'bare' field, and interaction parts and is denoted by

$$\hat{H}_R = \hat{H}_{atom} + \hat{H}_{driving\ field} + \hat{V}_{interaction} \ . \tag{16}$$

The two-level atom Hamiltonian is given by

$$\hat{H}_{atom} = \hbar\omega_{eg} \mid e >< e \mid \tag{17}$$

which has eigenstates $\{ \mid g >, \mid e >\}$ and eigenenergies $\{0, \hbar\omega_{eg}\}$. The driving-field

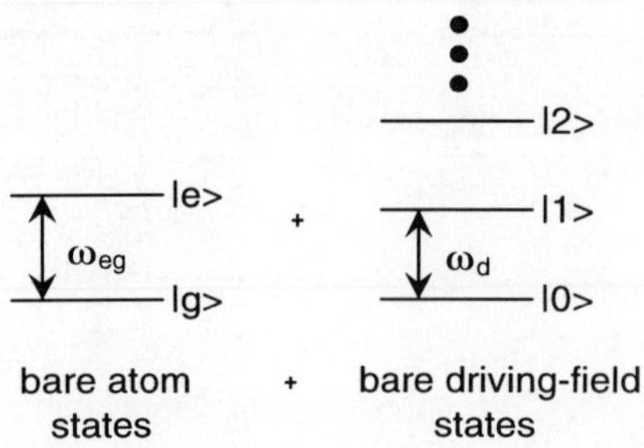

Figure 8 Energy eigenstates of the (a) bare atom and (b) the bare driving field.

Hamiltonian is given by

$$\hat{H}_{driving\ field} = \hbar\omega_d(\hat{a}_d^\dagger \hat{a}_d + \frac{1}{2}) \tag{18}$$

which has eigenstates $\{| n >\}$ and eigenenergies $\{\hbar\omega_d(n + 1/2)\}$, where $\hat{a}_d$ $(\hat{a}_d^\dagger)$ is the photon annihilation (creation) operator and n is the photon occupation number (the number of photons in the mode). The energy eigenstates of the uncoupled atom and field are shown schematically in Fig. 8a and 8b. It is seen that the driving-field eigenstates form an infinite ladder of equally spaced levels which is the underlying reason that the driven-atom system is a good two-photon gain medium.

The eigenstates of $\hat{H}_{atom} + \hat{H}_{driving\ field}$ are just the product states formed from the 'bare' atom and field states $\{| g, n >, \ | e, n >\}$, which are shown schematically in Fig. 9a. It is seen that the states $| g, n+1 >$ and $| e, n >$ are close in energy for a near-resonant driving field and are degenerate for resonant driving. The interaction part of the Hamiltonian couples the nearly degenerate levels and removes any degeneracies in the system. It is given by

$$\hat{V}_{interaction} = -\hat{\vec{\mu}} \cdot \hat{\vec{E}}_d = -\hbar g_d(| e >< g | + | g >< e |)(\hat{a}_d + \hat{a}_d^\dagger) \ , \tag{19}$$

where

$$\hat{\vec{\mu}} = \vec{\mu}_{eg}(| e >< g | + | g >< e |) \tag{20}$$

is the dipole-moment operator,

$$\hat{\vec{E}}_d = \tilde{\epsilon}_d \sqrt{\frac{8\pi\hbar\omega_d}{V_d}}(\hat{a}_d + \hat{a}_d^\dagger) \tag{21}$$

is the electric field operator, and

$$g_d = \vec{\mu}_{eg} \cdot \tilde{\epsilon}_d \sqrt{\frac{8\pi\omega_d}{\hbar V_d}} \tag{22}$$

is the atom-field coupling constant. The polarization unit vector of the driving field is denoted by $\tilde{\epsilon}_d$ and the volume of the driving-field mode is denoted by V_d. We have assumed, for simplicity, that $\vec{\mu}_{eg}$ is real. The nearly-degenerate product states are the only levels coupled by $\hat{V}_{interaction}$ where the interaction energies are given by

$$< e, n \mid \hat{V}_{interaction} \mid g, n+1 >= \hbar g_d\sqrt{n+1} \ . \tag{23}$$

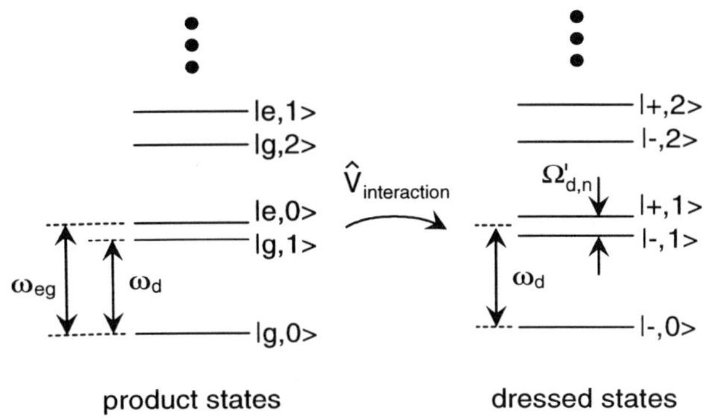

Figure 9 Energy eigenstates of (a) the uncoupled atom-field system (the product states), and (b) the coupled system (the dressed states).

While it is possible to solve for the dynamics of the interaction using the product-state basis, it is somewhat inconvenient because there are off-diagonal matrix elements of H_R.

A more natural approach is to use the dressed-state basis, that is, the basis formed by the energy eigenstates $\{|-,n>, |+,n>\}$ of H_R. They are a linear combination of the product states and are given by

$$\begin{aligned} |-,n> &= \cos\theta_n\,|g,n> - \sin\theta_n\,|e,n-1> \\ |+,n> &= \sin\theta_n\,|g,n> + \cos\theta_n\,|e,n-1> \end{aligned} , \tag{24}$$

where the rotation angle is defined through $\tan 2\theta_n = -g_d\sqrt{n}/\Delta_d$. The dressed states are shown schematically in Fig. 9b. We see that there is an infinite ladder of levels where each rung of the ladder is a doublet of levels. The doublet splitting is a function of the photon occupation number and is given by

$$\Omega'_{d,n} = \sqrt{n\, g_d^2 + \Delta_d^2}\ . \tag{25}$$

We now explore ways in which the infinite-dimensional problem can be simplified before considering the interaction of the dressed atom with the other modes of the electromagnetic field (including the probe field). The dimensionality of the problem can be reduced dramatically by including in the calculation the dressed levels that are selectively occupied by the driving field rather than the complete set of states. For a driving field produced by an intense laser (properly described quantum mechanically by a coherent state), only a narrow distribution of dressed levels high up the ladder are occupied. We say that the levels are 'locally periodic' since the relative populations and the splitting of a dressed-state doublet are essentially the same independent of the position on the ladder (that is, the value of n). In this case $g_d\sqrt{n} \sim g_d\sqrt{\bar{n}} \equiv \Omega_d$, $\tan 2\theta_n = -\Omega_d/\Delta_d$, and $\Omega'_{d,n} = \sqrt{\Omega_d^2 + \Delta_d^2} \equiv \Omega'_d$, where $\bar{n}$ is the mean photon number and Ω_d is the classical Rabi frequency of the field (a constant).

The interaction between the dressed atom and the other modes of the field can be treated straightforwardly when there is local periodicity because we only need to

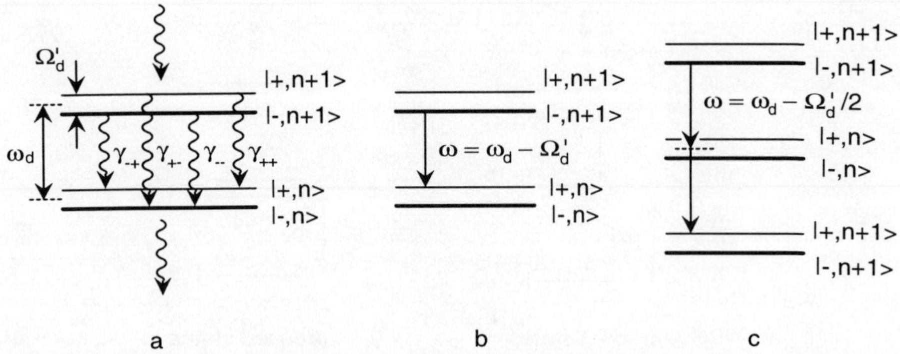

Figure 10 (a) The relative populations of the dressed-atom doublets become unbalanced due to spontaneous emission. One- (b) and (c) two-photon gain in the dressed-atom system.

consider transitions induced between one rung on the ladder and its nearby neighbors by the perturbation. The full dressed-atom - field interaction is just the sum of the local interactions. As usual, the strength of the interaction is related to the electric-dipole matrix elements between the dressed levels. It is found that all matrix elements are zero except for those that connect one rung of the ladder with its nearest neighbor. They are defined through the relation

$$\vec{\mu}_{ij}^{\pm} =< i, n \pm 1 \mid \hat{\vec{\mu}} \mid j, n > \quad i, j = +, -. \tag{26}$$

where $\vec{\mu}_{ij}^{+} = (\vec{\mu}_{ij}^{-})^{*}$. For example, it is found that $\vec{\mu}_{+-}^{+} = \vec{\mu}_{eg} \cos^2 \theta$. Note that the matrix elements are a function of the driving-field Rabi frequency and detuning through the rotation angle θ.

Coupling the dressed atom to all the unoccupied field modes leads to spontaneous decay of the levels. Considering the selection rules (Eq. 26), we see that each level on a rung of the ladder decays to both levels one step down the ladder at a rate

$$\gamma_{ij} = \gamma_{eg} \frac{\mid \vec{\mu}_{ij}^{+} \mid^2}{\mid \vec{\mu}_{eg} \mid^2} \quad i, j = +, - . \tag{27}$$

For example, it is found that $\gamma_{+-} = \gamma_{eg} \cos^4 \theta$. After several spontaneous emission events, the relative populations of a dressed-state doublet reach equilibrium in response to spontaneous decays into and out of the doublet. The relative population distribution is independent of n due to the local periodicity and is given by

$$\frac{\rho_{++}}{\rho_{--}} = \tan^4 \theta . \tag{28}$$

When the driving field is on resonance ($\Delta_d = 0$), both levels are equally populated since $\theta = \pi/4$. For off-resonance driving the populations become unbalanced. For example, $\rho_{--} > \rho_{++}$ when $\Delta_d < 0$ as shown in Fig. 10a.

A probe beam (frequency ω) interacting with the dressed atom will experience gain or absorption when the populations of the dressed levels are unbalanced just as a bare atom will absorb or amplify a beam of light. It can be seen from Fig. 10b that a probe beam interacting with the $| +, n > \rightarrow | -, n+1 >$ transition will be amplified due to one-photon stimulated emission when $\omega = \omega_d - \Omega'_d$ for $\Delta_d < 0$. This is precisely the spectral location of the gain feature shown in Fig. 6a. Analogously, one-photon absorption will occur on the $| -, n > \rightarrow | +, n+1 >$ transition.

An intense probe beam can induce multiphoton transitions among the dressed levels. In particular, a probe beam interacting with the $| -, n+1 > \rightarrow | +, n-1 >$ will be amplified due to two-photon stimulated emission when $\omega = \omega_d - \Omega'_d/2$ as shown in Fig. 10c. Most importantly, the $| +, n >$ and $| -, n >$ intermediate levels are nearly degenerate with the virtual level of the two-photon transition thereby enhancing significantly the two-photon rate coefficient. In addition, the two-photon gain is spectrally distinct from the regions of large one-photon gain; hence, a high-finesse resonator can selectively enhance the two-photon gain. It is precisely for these reasons that dressed atoms make a near ideal two-photon gain medium. Note that the driven-atom system can support n-photon gain due to n-photon stimulated emission from one rung on the ladder to n rungs below.

The two-photon rate coefficient can be determined using the density-matrix formalism and by treating the probe as a classical field (similar to the analysis presented above in Sec. III). Obviously, the local dynamics of the dressed levels is richer than the dynamics of the simple three-level system considered in Sec. III because there are two intermediate levels. One complication that arises from the two intermediate states is that there are two quantum mechanical pathways from the excited level $| -, n+1 >$ to the lower level $| +, n-1 >$. The two pathways are taken into account by adding the contributions from each pathway to the induced polarization. A somewhat more subtle effect is the presence of coherences between the intermediate levels. Under conditions when $\Omega'_d \gg \Gamma_{eg}$, these coherences have a negligible contribution and can be ignored (this is known as the secular approximation). From the detailed calculations, it is found that the rate coefficient is a complex function of the driving-field detuning and is maximized for a given Ω_d when $\Delta_d \sim -0.4\Omega_d$. At the optimum detuning

$$B^{(2)} \sim \frac{3\pi^2 \mid \vec{\mu}_{eg} \cdot \hat{\epsilon} \mid^4 \omega^2}{8V_c^2 \hbar^2 (\Omega'_d/2)^2 \Gamma_{eg}} . \tag{29}$$

We see that the dressed-state and three-level atom rate coefficients are similar with $\Delta_{ia} \rightarrow \Omega'_d/2$ by comparing Eqs. 29 and 14. Thus, the rate coefficient for a dressed atom can be considerably larger than a typical three-level atom because Ω'_d can be small. Note that there is an optimum value of Ω'_d for a particular experimental situation since the one- and two-photon gain features will start to overlap as Ω'_d becomes small. Using $\Omega'_d = 6 \times 10^9$ s^{-1} and the parameters listed in Table 1, we find that $B^{(2)} \sim 3.4 \times 10^{-7}$ s^{-1}, $q_{sat} = 1.2 \times 10^7$ photons (corresponding to $I_{sat} = 1.8 \times 10^2$ W-cm^{-2}), and $\Delta N_o^{th} = 4.8 \times 10^{10}$ atoms-cm^{-3}. Most importantly, we estimate that the one-photon dressed-atom gain that occurs at $\omega = \omega_d - \Omega'_d$ is only $G^{(1)}L \sim 0.1$. Therefore, it should be possible to realize a dressed-atom two-photon laser using a high-finesse resonator that selectively enhances the two-photon transition.

V. THE DRESSED-STATE TWO-PHOTON LASER

The experiments that were conducted to test the theoretical discussion outlined in the previous section used a unique experimental apparatus to mimic as closely as possible the theoretical assumptions [23],[24]. The 553.5-nm $^1S_o \rightarrow ^1P_1$ transition of

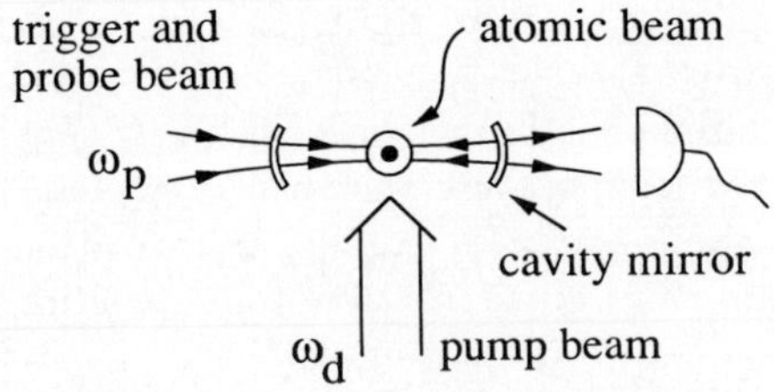

Figure 11 Schematic of the two-photon laser.

barium (natural linewidth $\gamma_{eg}/2\pi = 19$ MHz) approximated the two-level atom and an actively stabilized, tunable ring dye laser acted as the driving field. Typically, it was possible to achieve atomic number densities of 10^{10} atoms-cm^{-3} and a driving-field Rabi frequency of 2.6×10^9s^{-1}. Of paramount concern was the possibility that phase-matched multi-wave mixing processes could contribute a significant number of photons to the resonator and thus obscure two-photon lasing. In addition, broadening mechanisms that might reduce the two-photon gain, such as Doppler and collisional broadening, needed to be removed. Suppression of these effects was accomplished using a collimated, small-diameter ($\sim$ 2 mm) atomic beam of atoms that passed through the center of a high-finesse resonator where the atomic-beam, the cavity and the driving-field axes were mutually orthogonal, as shown in Fig. 11. The laser resonator and the atomic beam source were housed in a high-vacuum chamber to maintain the low-loss characteristics of the resonator (0.15% per pass). Note that the threshold parameters for the two-photon laser studied in the experiment are somewhat different from the ones used in the previous discussion because: the atomic constants are slightly different; there is only natural broadening; the resonator has lower loss; and the atoms only fill a small fraction of the resonator. Accounting for these effects, we find that $B^{(2)} \sim 3\times10^{-2}$ s^{-1}, $q_{sat} = 4.5\times10^4$ photons (corresponding to $I_{sat} = 0.8$ W-cm^{-2}), and $\Delta N_o^{th} = 1\times10^9$ atoms-cm^{-3} for $\Omega_d' = 2.7\times10^9$ s^{-1}.

Several preliminary experiments were performed to characterize the dressed-atom laser system. In the first experiment, the optical power emitted out one end of the cavity was measured as the cavity resonance frequency ω_c was tuned through the dressed-atom resonances. It was found that normal one-photon dressed-atom lasing occurred when $\omega_c \approx \omega_d - \Omega_d'$ and that no lasing occurred when $\omega_c \approx \omega_d - \Omega_d'/2$ (the predicted frequency of maximum two-photon gain). This measurement confirmed that the one-photon gain feature was spectrally removed from the region where high two-photon gain was expected and hence it would not obscure two-photon lasing. The next experiment measured directly the gain of the system by observing the transmission of an intense laser beam (approximately matched to the fundamental cavity mode and resonant with it) through the cavity. A pronounced, intensity-dependent gain feature was observed when $\omega_c \approx \omega_d - \Omega_d'/2$ which coincides precisely with the predicted location of the two-photon gain feature.

Based on the discussion of the previous sections, two-photon lasing should not occur in this system until both threshold conditions are satisfied. To determine whether there was a sufficient inversion density in the resonator to support two-photon lasing, ω_c was

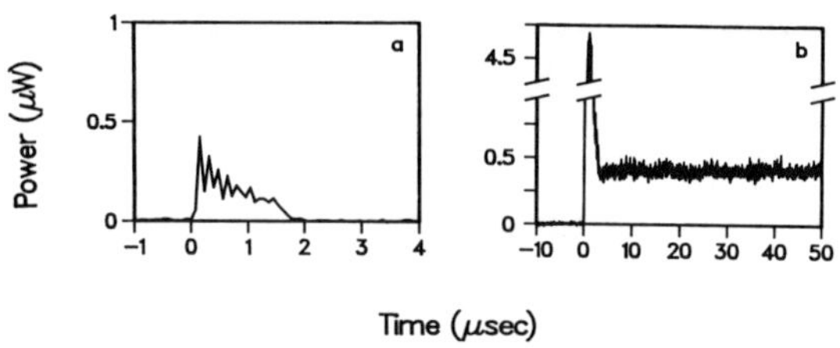

Figure 12 Time-resolved power emitted from the cavity before and after a 1.4-μsec trigger pulse is injected into the cavity. (a) The injected pulse is insufficient to satisfy the threshold condition. (b) The two-photon laser turns on.

adjusted to the peak of the two-photon gain and trigger pulses (duration ~ 1.4 μsec) resonant with the cavity were injected into it. The time-resolved cavity output power just prior to and following the injected pulse was repeatedly recorded for various trigger-pulse powers. As shown in Fig. 12a, the trigger-pulse power was insufficient to initiate two-photon lasing and the cavity output power decayed to zero after the injected pulse is turned off. For higher injected powers, the cavity output power remains high after the trigger pulse is turned off as shown in Fig. 12b. The trigger-induced transition to a state of nonzero cavity output power is entirely consistent with the expected threshold behavior of a two-photon laser as discussed in Sec. II (see Fig. 4). This demonstrates conclusively that two-photon lasing was achieved.

VI. DISCUSSION AND FUTURE DIRECTIONS

We are currently setting up the apparatus to study two-photon lasers at Duke where our main research directions will be to thoroughly characterize two-photon lasers and to develop new types of two-photon gain media. There are still many open questions about the behavior of two-photon lasers that we will be addressing, such as the laser

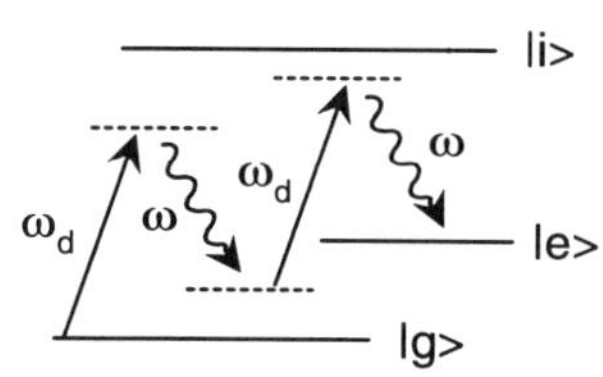

Figure 13 A new mechanism for two-photon gain: two-photon stimulated Raman scattering.

linewidth, coherence properties, instabilities in the output power, photon fluctuation noise, photon correlations and detailed studies of the threshold behavior, for example. We hope that these studies will give us a deeper understanding of the highly nonlinear interaction between light and matter and a better understanding of lasers in general.

We are searching for new two-photon gain media for two different reasons. The first reason is to have the ability to distinguish between the properties of two-photon lasers in general and those that are specific to a particular gain medium. The second reason is to make it easier to achieve two-photon lasing. The apparatus would be simplified significantly if a different source of atoms could be used, such as a vapor cell or atoms doped in a solid-state material. Toward this end, we have discovered a new two-photon gain medium that is based on a process that we call two-photon stimulated Raman scattering. The gain arises from scattering of laser photons from a three-level atom as shown in Fig. 13. We have found that the two-photon gain from this system is less susceptible to inhomogeneous broadening due to spatial variation in the pump beam intensity in comparison to the dressed-atom gain medium and that it should be possible to achieve two-photon lasing using a vapor cell. We also believe that it should be possible to develop a two-photon gain medium that is based on an analogous two-photon Brillouin process. This may pave the way for the realization of an optical fiber based two-photon laser.

Acknowlegments

We gratefully acknowledge collaborations on this work with Prof. T.W. Mossberg, Prof. M. Lewenstein, Prof. Y. Zhu, Dr. Q. Wu, S. Morin and fruitful discussions with R.W. Boyd and M.G. Raymer. This work was supported by the U.S. Army Research Office under Contract No. DAAL03-92-G-0286 and the National Science Foundation under Grant No. PHY-9357234.

References

[1] P. P. Sorokin and N. Braslau, IBM J. Res. Dev. **8**, 177 (1964); A. M. Prokhorov, Science **10**, 828 (1965).

[2] E. B. Gordon and Y. L. Moskvin, Zh. Eksp. Teor. Fiz. **70**, 1729 (1976); L. M. Narducci, W. W. Eidson, P. Furcinitti, and D. C. Eteson, Phys. Rev. A **16**, 1665 (1977); T. Hoshimiya, A. Yamagishi, N. Tanno, and H. Inaba, Japan. J. Appl. Phys. **17**, 2177 (1978); A. R. Bulsara and W. C. Schieve, Phys. Rev. A **19**, 2046 (1979); N. Nayak and B. K. Mohanty, Phys. Rev. A **19**, 1204 (1979); S. Ovadia and M. Sargent III, Opt. Comm. **49**, 447 (1983); S. Zhu and X. Li, Phys. Rev. A **36**, 3889 (1987);S. Songen, Chin. J. Lasers **14**, 1 (1987); P. Garcia Fernandez and L. Sainz de los Terreros, Opt. Comm. **61**, 151 (1987); W. Z. Cheng and H. Haken, Z. Phys. B **71**, 253 (1988); S. Swain, J. Mod. Opt. **35**, 103 (1988); I. Ashraf and M. S. Zubairy, Opt. Comm. **71**, 92 (1989); A. W. Boone and S. Swain, Opt. Comm. **73**, 47 (1989); A. W. Boone and S. Swain, Quantum Opt. 1, **27** (1989); N. Lu, F. Zhao, and J. Bergou, Phys. Rev. A **39**, 5189 (1989); G. Yang, Z. Huang and G. Hu, Commun. Thero. Phys. **14**, 397 (1990); N. Lu, Phys. Rev. A **42**, 6756 (1990); Y. Wang and Y. Guan, Commun Theor. Phys. **17**, 393 (1992).

[3] M. M. T. Loy, Phys. Rev. Lett. **41**, 473 (1978); H. Schlemmer, D. Frolich, and H. Welling, Opt. Commun. **32**, 141 (1980); B. Nikolaus, D. Z. Zhang, and P. E. Toschek, Phys. Rev. Lett. **47**, 171 (1981).

[4] H. P. Yuen, Appl. Phys. Lett. **26**, 505 (1975); H. P. Yuen, Phys. Rev. A **13**, 2226 (1976); O. Hirota, Elec. and Comm. in Japan, **60-A**, 10 (1977); O. Hirota, I. Shioya, S. Ikehara, and Y. Suematsu, Trans. of IECE of Japan, **E61**, 273 (1978); D. J. Rowe, Can. J. Phys. **56**, 442 (1978); L. A. Lugiato and G. Strini, Opt. Commun. **41**, 374 (1982); M. D. Reid and D. F. Walls, Phys. Rev. A **28**, 332 (1983); C. M. Savage and D. F. Walls, Phys. Rev. A **33**, 3282 (1986); M. Majeed and M. S. Zubairy, Phys. Rev. A **44**, 4688 (1991); G. Hu and C. Sha, Phys. Lett. A **159**, 47 (1991).

[5] M. S. Zubairy, Phys. Lett. **80A**, 225 (1980); M. P. Sharma and L. M. Brescansin, Phys. Rev. A **23**, 1893 (1981); M. S. Zubairy, Phys. Lett. **87A**, 162 (1982); A. Bandilla and H. Voigt, Opt. Comm. **43**, 277 (1982); N. Lu, Opt. Comm. 77, 203 (1990).

[6] M. Reid, K. J. McNeil, and D. F. Walls, Phys. Rev. A, **24**, 2029 (1981).

[7] A. W. Boone and S. Swain, Phys. Rev. A **41**, 343 (1990).

[8] P. Meystre and M. Sargent, *Elements of Quantum Optics*, (Springer-Verlag, New York, 1991).

[9] Z. C. Wang and H. Haken, Z. Phys. B - Condensed Matter **56**, 83 (1984).

[10] M. Schubert and G. Wiederhold, Exp. Tech. Phy. **27**, 225 (1979).

[11] R. W. Boyd, *Nonlinear Optics*, p. 16 (Academic, Boston, 1992); M. Schubert and B. Wilhelmi, *Nonlinear Optics and Quantum Electronics*, p. 219 (John Wiley & Sons, New York, 1986).

[12] See Holm and Sargent, Phys. Rev. A **33**, 1073 (1986) for a discussion of spontaneous emission effects.

[13] L. Davidovich, J. M. Raimond, M. Brune, and S. Haroche, Phys. Rev. A **36**, 3771 (1987); IEEE J. Quantum Electron. **24**, 1323 (1988).

[14] O. Svelto, *Principles of Lasers*, p. 59 (Plenum Press, New York, 1989).

[15] C. Z. Ning and H. Haken, Z. Phys. B. - Condensed Matter **77**, 157 (1989); C. Z. Ning, Ph.D. Thesis, University of Stuttgart, 1991, unpublished.

[16] H. R. Heatley, C. N. Ironside, and W. J. Firth, Opt. Lett. **8**, 628-630 (1993).

[17] L. Allen and C. R. Stroud Jr., Phys. Rep. **91**, 1 (1982).

[18] M. Brune, J. M. Raimond, and S. Haroche, Phys. Rev. A **35**, 154 (1987); M. Brune, J. M. Raimond, P. Goy, L. Davidovich, and S. Haroche, Phys. Rev. Lett. **59**, 1899 (1987).

[19] Y. Zhu, Q. Wu, S. Morin, and T. W. Mossberg, Phys. Rev. Lett. **65**, 1200 (1990).

[20] C. Cohen-Tannoudji and S. Reynaud, J. Phys. B. **10**, 345 (1977);

[21] G. Grynberg, E. LeBihan, and M. Pinard, J. Phys. (Paris) **47**, 1321 (1986); D. Grandclement, G. Grynberg, and M. Pinard, Phys. Rev. Lett. **59**, 40 (1987); G. Khitrova, J. F. Valley, and H. M. Gibbs, Phys. Rev. Lett. **60**, 1126 (1988); A. Lezama, Y. Zhu, M. Kanskar, and T. W. Mossberg, Phys. Rev. A **41**, 1576 (1990).

[22] C. Cohen-Tannoudji, J. Dupont-Roc, and G. Grynberg, *Atom-Photon Interactions* (John Wiley & Sons, Inc., New York, 1992).

[23] D. J. Gauthier, Q. Wu, S. E. Morin, and T. W. Mossberg, Phys. Rev. Lett. **68**, 464-467 (1992).

[24] M. Lewenstein, Y. Zhu, and T. W. Mossberg, Phys. Rev. Lett. **64**, 3131 (1990); J. Zakrzewski, M. Lewenstein, and T. W. Mossberg, Phys. Rev. A **44**, 7717 (1991); J. Zakrzewski and M. Lewenstein, Phys. Rev. A **45**, 2057 (1992).

DEFINITION OF COHERENCE

B. Bowlby

Department of Physics
Boston College
Chestnut Hill, MA 02167
USA

ABSTRACT

The idea of coherence dates to Young's Experiment where is was shown that two beams of light could interact and produce an interference pattern on a distant screen. While this phenomena supported the wave nature of light, another more interesting fact was uncovered. It was found that in order to produce an interference pattern, the two beams had to come from the same source. Also, light from some sources would not produce interference at all. The term coherent was applied to describe light that would produce interference in an experiment like Young's. As our understanding of electro-magnetic radiation grew, the idea of coherence developed beyond this original definition. This manuscript will provide a brief overview of the concept of coherence, from a classical description of Young's experiment, to a quantum mechanical definition of coherence.

I. INTRODUCTION

This manuscript began as a special seminar which I presented at Boston College in the spring of 1993. The subject has been covered quite extensively in various sources, and I have made use of a number of them [1-4] in this manuscript. I will present a brief overview of the subject, without going into the rigorous mathematical detail. The reader will be directed to the appropriate texts when details are omitted. Also note that we will be assuming monochromatic light throughout this treatment. The use of an 'analytic signal', such as by Born [2] and others, has been omitted for clarity and brevity.

II. FIRST ORDER COHERENCE

We begin our study with the simple, but illustrative example of the coherence at a fixed point when there are two sources present (see fig. 1). This problem is easily treated in any fundamental physics text, but its inclusion here allows us to define terms that will become important later on. The quantity that concerns us is the electric field at the point P due to the sources S_1 and S_2. At the point sources, the field is given by

Nonlinear Spectroscopy of Solids: Advances and Applications
Edited by B. Di Bartolo and B. Bowlby, Plenum Press, New York, 1994

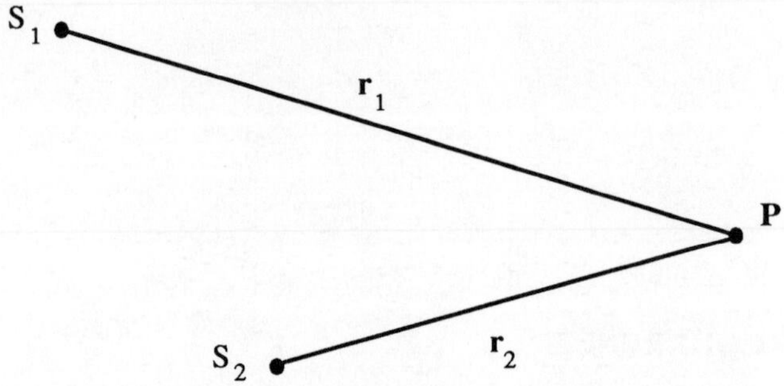

Figure 1. Coherence of light from two point sources S_1 and S_2

$$\mathbf{E}_{S_1} = \mathbf{E}_1 e^{-i\omega_1 t + i\phi_1(t)t}$$
$$\mathbf{E}_{S_2} = \mathbf{E}_2 e^{-i\omega_2 t + i\phi_2(t)t} \tag{1}$$

so that at a distant point P we have

$$\mathbf{E}_1(t) = u_1(\mathbf{r}_1)\mathbf{E}_{S_1}(t_1) \qquad t_1 = t - \frac{r_1}{c}$$
$$\mathbf{E}_2(t) = u_2(\mathbf{r}_2)\mathbf{E}_{S_2}(t_2) \qquad t_2 = t - \frac{r_2}{c}. \tag{2}$$

The functions $u_1(\mathbf{r}_1)$ and $u_2(\mathbf{r}_2)$ are in general proportional to 1/r and may include other effects as well. The electric field at P is the superposition of the two so

$$\mathbf{E}_\mathrm{P}(t) = \mathbf{E}_1 + \mathbf{E}_2 = u_1(\mathbf{r}_1)\mathbf{E}_{S_1}(t_1) + u_2(\mathbf{r}_2)\mathbf{E}_{S_2}(t_2). \tag{3}$$

The intensity at point P is given by

$$\begin{aligned} I_\mathrm{P}(t) &= \frac{1}{2}\varepsilon_0 c |\mathbf{E}_\mathrm{P}(t)|^2 \\ &= \frac{1}{2}\varepsilon_0 c |u_1(\mathbf{r}_1)|^2 |\mathbf{E}_{S_1}(t_1)|^2 + \frac{1}{2}\varepsilon_0 c |u_2(\mathbf{r}_2)|^2 |\mathbf{E}_{S_2}(t_2)|^2 \\ &\quad + \varepsilon_0 c u_1^*(\mathbf{r}_1) u_2(\mathbf{r}_2) \mathrm{Re}(\mathbf{E}_{S_1}^*(t_1)\mathbf{E}_{S_2}(t_2)) \\ &= I_1 + I_2 + \varepsilon_0 c u_1^*(\mathbf{r}_1) u_2(\mathbf{r}_2) \mathrm{Re}(\mathbf{E}_{S_1}^*(t_1)\mathbf{E}_{S_2}(t_1 + t_{21})) \end{aligned} \tag{4}$$

where I_1 and I_2 are the intensities at P due to S_1 and S_2 respectively when no other source is present and $t_{21} = t_2 - t_1$. It is the last term that leads to the interference effects. In order to observe the radiation at P, some kind of detection system is required. Whether the detector is a screen observed by the human eye, or some electronic system, a finite resolution time is required, and the signal detected will be the time average of the intensity during this resolution time. Denoting the time averages with bars we have

$$\bar{I} = \bar{I}_1 + \bar{I}_2 + \overline{\varepsilon_0 c u_1^* u_2 \mathrm{Re}(\mathbf{E}_1^*(t_1)\mathbf{E}_2(t_1 + t_{21}))}. \tag{5}$$

We find that unless $\omega_1 = \omega_2$ and $\phi_1(t) - \phi_2(t)$ is constant, the last term averages to zero, and no interference occurs. This is the reason that the two beams must come from the same source. While two separate sources may have the same frequency ω, there is in general no correlation between the phases of the two sources, and the interference term averages to zero. To continue we define

$$\Gamma_{12}(T) = \frac{1}{T}\int_0^T \mathbf{E}_1^*(t_1)\mathbf{E}(t_1+t_{21})dt_1 = \overline{\mathbf{E}_1^*(t_1)\mathbf{E}_2(t_1+t_{21})} \tag{6}$$

and

$$\gamma_{12}(T) = \frac{\Gamma_{12}(T)}{\sqrt{I_1}\sqrt{I_2}} \tag{7}$$

where $\Gamma_{12}(T)$ is called the mutual coherence or the 1st order correlation function and $\gamma_{12}(T)$ is called the 1st order degree of coherence. From the Swartz Inequality it can easily be shown that $0 \le |\gamma_{12}(T)| \le 1$. When $\gamma_{12}(T) = 1$ we say that the light is first order coherent. If $\gamma_{12}(T) = 0$ then the light is incoherent, while intermediate values of $\gamma_{12}(T)$ give what is called partial coherence. We then have

$$I = I_1 + I_2 + 2\sqrt{I_1}\sqrt{I_2}\,\mathrm{Re}(\gamma_{12}(T)) \tag{8}$$

where the bars have been dropped for ease of notation.

Turning our attention to the Young experiment we assume that I_1 and I_2 are constant throughout the region of interest and we define

$$\text{contrast} = \frac{I_{max} - I_{min}}{I_{max}}. \tag{9}$$

It is easy to see that if $\gamma_{12}(T) = 0$ then $I_{max} = I_{min}$ and the contrast is zero, while if $I_{min} = 0$ then the contrast is equal to one. In the general case we have

$$\text{contrast} = \frac{4\sqrt{I_1}\sqrt{I_2}\,|\gamma_{12}(T)|}{I_1 + I_2 + 2\sqrt{I_1}\sqrt{I_2}\,|\gamma_{12}(T)|}. \tag{10}$$

The exact form of $\gamma_{12}(T)$ depends on the lineshape of the source and is beyond the scope of this manuscript (see ref. 1). However, for the case where $I_1 = I_2$ and $\gamma_{12}(T) = 1$ (as in a typical Young type experiment) then the contrast = 1. Usually, however, $|\gamma_{12}(T)| < 1$ and the contrast is less than 1.

III. COHERENCE BETWEEN TWO POINTS ILLUMINATED BY AN EXTENDED SOURCE

We now move to a generalization of the above treatment. We will consider the coherence between two spatial points when the source is not a point source, but instead an extended body (see fig. 2). In this case we consider the source to be made up many independent emitting regions and sum the contributions of the various regions. At points P_1 and P_2 we have

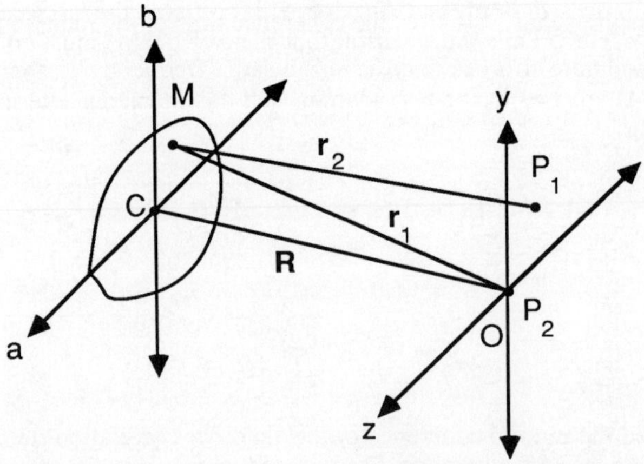

Figure 2. Diagram for coherence produced by an extended source. For convenience, P_2 is chosen to coincide with the yz origin.

$$\mathbf{E}_{1m}(t) = \frac{\mathbf{E}_m(t-\frac{r_1}{c})}{r_1} e^{i\omega_0(t-\frac{r_1}{c})}$$
$$\mathbf{E}_{2m}(t) = \frac{\mathbf{E}_m(t-\frac{r_2}{c})}{r_2} e^{i\omega_0(t-\frac{r_2}{c})}. \tag{11}$$

Summing over the various regions then gives

$$\mathbf{E}_1(t) = \sum_m \frac{\mathbf{E}_m(t-\frac{r_1}{c})}{r_1} e^{i\omega_0(t-\frac{r_1}{c})} = \sum_m \tilde{\mathbf{E}}_{1m}(t)$$
$$\mathbf{E}_2(t) = \sum_m \frac{\mathbf{E}_m(t-\frac{r_2}{c})}{r_2} e^{i\omega_0(t-\frac{r_2}{c})} = \sum_m \tilde{\mathbf{E}}_{2m}(t) \tag{12}$$

so that

$$\Gamma_{12}(0) = \overline{\mathbf{E}_1^*(t)\mathbf{E}_2(t)} = \overline{\sum_i \mathbf{E}_{1i}^*(t) \sum_j \mathbf{E}_{2j}(t)} = \overline{\sum_i \mathbf{E}_{1i}^*(t)\mathbf{E}_{2i}(t)}. \tag{13}$$

The last step is due to the fact that the emitting regions act independently. Expanding we have

$$\Gamma_{12}(0) = \sum_m \overline{\mathbf{E}_m^*(t-\frac{r_1}{c})\mathbf{E}_m(t-\frac{r_2}{c})} \frac{e^{i\omega_0(\frac{r_2-r_1}{c})}}{r_1 r_2}$$
$$= \sum_m \overline{\mathbf{E}_m^*(t)\mathbf{E}_m(t)} \frac{e^{i\omega_0(\frac{r_2-r_1}{c})}}{r_1 r_2} \tag{14}$$

where the last step assumes that $\frac{r_2-r_1}{c}$ << the coherence time of the radiation.

We now go to the limit of a very large number of emitting regions (atoms), so that the sum in (14) may be replaced by an integral. Since $\overline{\mathbf{E}_m^*(t)\mathbf{E}_m(t)}$ is proportional to the energy emitted by one region we can define

$$\overline{\mathbf{E}_m^*(t)\mathbf{E}_m(t)} = J(S)dS \tag{15}$$

where $J(S)$ is the energy emitted per unit area. We then have

$$\Gamma_{12}(0) = \int_S J(S)\frac{e^{ik_0(r_2-r_1)}}{r_1 r_2}dS \qquad k_0 = \frac{\omega_0}{c} \tag{16}$$

which gives

$$\begin{gathered}\gamma_{12}(0) = (J_1 J_2)^{-1/2}\int_S J(S)\frac{e^{ik_0(r_2-r_1)}}{r_1 r_2}dS \\ J_1 = \int_S \frac{J(S)}{r_1^2}dS \qquad J_2 = \int_S \frac{J(S)}{r_2^2}dS\,.\end{gathered} \tag{17}$$

If we assume that the source is a long way from the screen we can make use of several geometrical simplifications. The above reduces to (see ref. 2)

$$\gamma_{12}(0) = \frac{e^{-ik_0\left(\frac{y^2+z^2}{2R}\right)}\iint_S J(b,a)e^{ik_0\left(\frac{by+az}{R}\right)}dbda}{\iint_S J(b,a)dbda}\,. \tag{18}$$

Thus, the first order degree of coherence is proportional to the normalized Fourier transform of the energy spectrum. In general this integral can not be solved in closed form. However, for the special case of a circular source with radius ρ, the first order degree of coherence is given by

$$\gamma_{12}(0) = \left(\frac{2J_1(Z)}{Z}\right)e^{-ik_0\left(\frac{y^2+z^2}{2R}\right)} \tag{19}$$

Where $J_1(Z)$ is a Bessel function of the first kind and

$$Z = k_0\frac{\rho}{R}\left(y^2+z^2\right)^{\frac{1}{2}} = k_0\frac{\rho}{R}\overline{P_1P_2}\,. \tag{20}$$

This has a zero value when Z=3.83. This is the principle behind Michelson's method for the measurement of a star's diameter. The distance between two mirrors ($\overline{P_1P_2} = 2a$) is varied until the interference pattern they create disappears (see fig. 3). Then

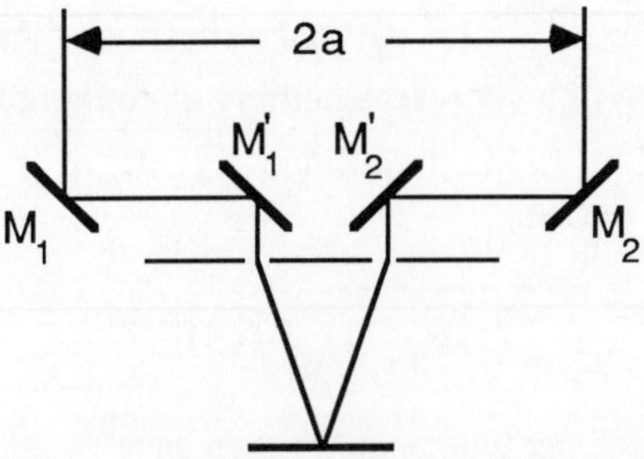

Figure 3. Michaelson's method for measuring a star's diameter. The distance 2a between mirrors M_1 and M_2 is varied until the interference pattern on the screen disappears. The distance between fringes on the interference pattern is determined by the distance between the two mirrors M_1' and M_2'.

$Z = 3.83 = k_0 \frac{\rho}{R}\overline{P_1P_2}$ so that $\alpha = \frac{\rho}{R} = \frac{1.22\lambda}{a}$. For a star with an apparent arc of 2α =.02 sec., the separation between the mirrors needs to be 7 meters. Because of the large extent of the mirrors, as well as atmospheric motions, Michelson's method for measuring star diameters was not widely used.

IV. THE INTENSITY INTERFEROMETER

A different way of measuring stellar diameters that did not suffer from the above shortcomings was developed by Hanbury Brown and Twiss. In their set-up it is the intensities at two points that are correlated, not the electric fields. The outputs of two photomultipliers are multiplied together and the result integrated and device is referred to as an intensity interferometer (see fig. 4). In the following discussion, the factor $\frac{1}{2}\varepsilon_0 c$ will bedropped from all intensity calculations. Since it is the intensities that are correlated, the quantity of interest in the intensity interferometer is $\overline{I_1(t+t_0)I_2(t)} = \overline{\mathbf{E}_1^*(t+t_0)\mathbf{E}_1(t+t_0)\mathbf{E}_2^*(t)\mathbf{E}_2(t)}$. We can write

$$\begin{aligned} \mathbf{E}_1(t) &= \mathbf{E}_1^{(r)}(t) + i\mathbf{E}_1^{(i)}(t) \\ \mathbf{E}_2(t) &= \mathbf{E}_2^{(r)}(t) + i\mathbf{E}_2^{(i)}(t) \end{aligned} \tag{21}$$

so that

$$\begin{aligned} \overline{I_1(t+t_0)I_2(t)} = {} & \overline{\left[\mathbf{E}_1^{(r)}(t+t_0)\right]^2\left[\mathbf{E}_2^{(r)}(t)\right]^2} + \overline{\left[\mathbf{E}_1^{(i)}(t+t_0)\right]^2\left[\mathbf{E}_2^{(r)}(t)\right]^2} \\ & + \overline{\left[\mathbf{E}_1^{(r)}(t+t_0)\right]^2\left[\mathbf{E}_2^{(i)}(t)\right]^2} + \overline{\left[\mathbf{E}_1^{(i)}(t+t_0)\right]^2\left[\mathbf{E}_2^{(i)}(t)\right]^2}. \end{aligned} \tag{22}$$

This relation reduces to (see reference 3)

$$\overline{I_1(t+t_0)I_2(t)} = \overline{I_1}\,\overline{I_2} + \left|\Gamma_{12}(t_0)\right|^2. \tag{23}$$

However, the quantity actually measured is due to the fluctuations

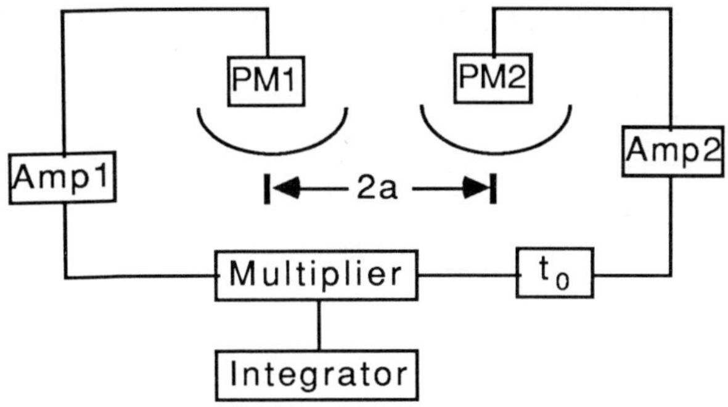

Figure 4. The intensity interferometer of Hanbury Brown and Twiss. Two concave mirrors focus light onto the photomultipliers PM1 and PM2. These signals are amplified by AMP1 and AMP2 and a time delay t0 is added to the signal from 2. The signals are then multiplied and integrated and the result recorded.

$$\Delta I_1(t) = I_1(t) - \overline{I_1} \qquad \text{and} \qquad \Delta I_2(t) = I_2(t) - \overline{I_2} \tag{24}$$

with $\overline{\Delta I_1(t)} = \overline{\Delta I_2(t)} = 0$. We have

$$\overline{I_1(t+t_0)I_2(t)} = \left[\overline{I_1} + \Delta I_1(t+t_0)\right]\left[\overline{I_2} + \Delta I_2(t)\right] = \overline{I_1}\,\overline{I_2} + \overline{\Delta I_1(t+t_0)\Delta I_2(t)} \tag{25}$$

so that

$$\overline{\Delta I_1(t+t_0)\Delta I_2(t)} = \left|\Gamma_{12}(t_0)\right|^2 = \overline{I_1}\,\overline{I_2}\left|\gamma_{12}(t_0)\right|^2 . \tag{26}$$

Similar to before, we define

$$\gamma^{(2)}(\mathbf{r}_1t_1\mathbf{r}_2t_2;\mathbf{r}_3t_3\mathbf{r}_4t_4) = \gamma^{(2)}_{1234} = \frac{\mathbf{E}^*(\mathbf{r}_1t_1)\mathbf{E}^*(\mathbf{r}_2t_2)\mathbf{E}(\mathbf{r}_3t_3)\mathbf{E}(\mathbf{r}_4t_4)}{\left\{\left|\mathbf{E}(\mathbf{r}_1t_1)\right|^2\left|\mathbf{E}(\mathbf{r}_2t_2)\right|^2\left|\mathbf{E}(\mathbf{r}_3t_3)\right|^2\left|\mathbf{E}(\mathbf{r}_4t_4)\right|^2\right\}^{1/2}} \tag{27}$$

as the normalized second order correlation function, or the second order degree of coherence. We then have

$$\gamma^{(2)}_{1221} = \gamma^{(2)}_{12} = \frac{\mathbf{E}_1^*\mathbf{E}_2^*\mathbf{E}_2\mathbf{E}_1}{I_1I_2} . \tag{28}$$

If the light source in the experiment is chaotic it can be shown that $\left(\gamma^{(1)}_{12}\right)^2 = \gamma^{(2)}_{12} - 1$ (see ref. 1). Substituting this in then gives $\overline{\Delta I_1(t+t_0)\Delta I_2(t)} = \overline{I_1}\,\overline{I_2}\left[\gamma^{(2)}_{12} - 1\right]$. The intensity interferometer of Hanbury Brown and Twiss was the first device to measure second order coherence or, more precisely, the lack of second order coherence. If the light was second order coherent, then $\gamma^{(2)}_{12} = 1$ and there would be no fluctuations in the intensity.

Higher orders of coherence can also be defined. We can generalize the form of (27) to

$$\gamma^{(n)}(\mathbf{r}_1t_1\ldots\mathbf{r}_nt_n;\mathbf{r}_{n+1}t_{n+1}\cdots\mathbf{r}_{2n}t_{2n})=\frac{\mathbf{E}^*(\mathbf{r}_1t_1)\ldots\mathbf{E}^*(\mathbf{r}_nt_n)\mathbf{E}(\mathbf{r}_{n+1}t_{n+1})\ldots\mathbf{E}(\mathbf{r}_{2n}t_{2n})}{\left\{|\mathbf{E}(\mathbf{r}_1t_1)|^2\ldots|\mathbf{E}(\mathbf{r}_nt_n)|^2|\mathbf{E}(\mathbf{r}_{n+1}t_{n+1})|^2\ldots|\mathbf{E}(\mathbf{r}_{2n}t_{2n})|^2\right\}^{1/2}}. \tag{29}$$

However, in practice there is very little use for these higher orders of coherence.

V. QUANTUM DEFINITION

Instead of photomultipliers and integrators, the intensity interferometer can also be constructed with photon counters and a coincidence counter. For this set-up it is insightful to employ the quantum mechanical definition of the electric field. This will necessarily alter the above definitions of coherence, and a new treatment will have to be developed. The following is a summary of the work of Glauber on the subject (see ref. 4).

For the quantized field, we consider $\mathbf{E}$ to be an operator given by

$$\begin{aligned} E &= i\left(\frac{h\omega}{V}\right)^{1/2}\left(ae^{i\left(\mathbf{k}\cdot\mathbf{r}-\omega t\right)}-a^+e^{-i\left(\mathbf{k}\cdot\mathbf{r}-\omega t\right)}\right) \\ &= E^+(\mathbf{r},t)+E^-(\mathbf{r},t) \end{aligned} \tag{30}$$

where E^+ and E^- are the annihilation and creation operators respectively. The matrix element for a transition in the photon counter from an initial state $|i\rangle$ to a final state $|f\rangle$ is $\langle f|E^+(\mathbf{r}t)|i\rangle$. The probability per unit time for this transition, according to Fermi's Golden Rule, is proportional to

$$\begin{aligned} \sum_f\left|\langle f|E^+(\mathbf{r}t)|i\rangle\right|^2 &= \sum_f\langle f|E^+(\mathbf{r}t)|i\rangle^*\langle f|E^+(\mathbf{r}t)|i\rangle \\ &= \sum_f\langle i|E^-(\mathbf{r}t)|f\rangle\langle f|E^+(\mathbf{r}t)|i\rangle=\langle i|E^-(\mathbf{r}t)E^+(\mathbf{r}t)|i\rangle. \end{aligned} \tag{31}$$

In order to measure coincidences we require the matrix element $\langle f|E^+(\mathbf{r}'t')E^+(\mathbf{r}t)|i\rangle$ so that this transition rate is proportional to

$$\sum_f\left|\langle f|E^+(\mathbf{r}'t')E^+(\mathbf{r}t)|i\rangle\right|^2=\langle i|E^-(\mathbf{r}t)E^-(\mathbf{r}'t')E^+(\mathbf{r}'t')E^+(\mathbf{r}t)|i\rangle. \tag{32}$$

In real life, however, the field is not fully specified. Because of this it is necessary to employ the density matrix formalism. Using this formalism the rates are proportional to

$$\begin{aligned} &Tr\left\{\rho E^{-}(\mathbf{r}t)E^{+}(\mathbf{r}t)\right\} \qquad \text{and} \\ &Tr\left\{\rho E^{-}(\mathbf{r}t)\ldots E^{-}(\mathbf{r}'t')E^{+}(\mathbf{r}'t')\ldots E^{+}(\mathbf{r}t)\right\} \end{aligned} \tag{33}$$

where ρ is the density matrix. We can then define

$$G^{(1)}(\mathbf{r}t,\mathbf{r}'t') = Tr\left\{\rho E^{-}(\mathbf{r}t)E^{+}(\mathbf{r}'t')\right\}$$

$$\begin{aligned} G^{(2)}(\mathbf{r}_1t_1,\mathbf{r}_2t_2,\mathbf{r}_3t_3,\mathbf{r}_4t_4) &= G^{(2)}(x_1x_2x_3x_4) = G^{(2)}{}_{1234} \\ &= Tr\left\{\rho E^{-}(x_1)\ldots E^{-}(x_n)E^{+}(x_{n+1})\ldots E^{+}(x_{2n})\right\} \\ G^{(n)}(x_1\ldots x_{2n}) &= Tr\left\{\rho E^{-}(x_1)\ldots E^{-}(x_n)E^{+}(x_{n+1})\ldots E^{+}(x_{2n})\right\}. \end{aligned} \tag{34}$$

Note that if the field is bounded, i.e. the number of photons is less than M, then $G^{(n)} = 0$ for n>M. This means that no real field can be coherent to all orders.

Similar to before, we now introduce the normalized correlation functions

$$g^{(1)}(\mathbf{r}t,\mathbf{r}'t') = \frac{G^{(1)}(\mathbf{r}t,\mathbf{r}'t')}{\left\{G^{(1)}(\mathbf{r}t,\mathbf{r}t)G^{(1)}(\mathbf{r}'t',\mathbf{r}'t')\right\}^{1/2}}$$

$$g^{(n)}(x_1\ldots x_{2n}) = \frac{G^{(n)}(x_1\ldots x_{2n})}{\prod_{j=1}^{2n}\left\{G^{(1)}(x_jx_j)\right\}^{1/2}} \tag{35}$$

where once again $0 \le |g^{(1)}(\mathbf{r}t,\mathbf{r}'t')| \le 1$. We can now give the quantum mechanical definition of coherence. A field is nth degree coherent if $|g^{(n)}(x_1\ldots x_{2n})| = 1$ for all n. For first order coherence we must have $|g^{(1)}(\mathbf{r}t,\mathbf{r}'t')| = 1$. Relating this to the Hanbury Brown and Twiss experiment we find that if the field is nth order coherent then $g^{(j)}(x_1\ldots x_jx_j\ldots x_1) = 1$ for j<n. From the definition of $g^{(j)}$ we have

$$G^{(j)}(x_1\ldots x_jx_j\ldots x_1) = \prod_{i=1}^{j} G^{(1)}(x_ix_i) \tag{36}$$

or, in other words, the correlation function factorizes which means that the counters act statistically independent. Light that is second order coherent would give a negative result, i.e. there would be no coincidences. This is precisely the result obtained earlier. This also allows us to give a stronger definition of coherence. If

$$\left|G^{(n)}(x_1\ldots x_{2n})\right| = \prod_{j=1}^{2n}\left\{G^{(1)}(x_jx_j)\right\}^{1/2} \tag{37}$$

i.e. if the nth correlation function factorizes, then the light is nth order coherent. Again, however, there is very little use for these higher orders of coherence in practice.

CONCLUSION

As our understanding of electro-magnetic radition and our technology has grown, our definition of coherence has grown as well. The simple definition of coherence as light that will produce interference effects in an experiment similar to Young's has grown into one where many orders of coherence have been defined. Although in practice there is little need for these higher orders of coherence, perhaps as our understanding continues to grow, they will become more useful as well.

Acknowledgments

I wish to thank Professor Di Bartolo for his encouragement in this paper, as well as fellow Boston College students ChunLai Yang, Hui Wang, and Gang Lei for their critical comments on the seminar.

REFERENCES

1. R. Loudon, The Quantum Theory of Light

2. M. Born and E. Wolf, Principles of Optics 4th Edition, Pergamon Press, London, England (1970)

3. M. Francon, Optical Interferometry, Academic Press, New York, New York (1966)

4. R. Glauber, Phys. Rev. **130**, 2529 (1963)

OPTICAL NONLINEARITIES OF COLOR CENTERS IN ALKALI HALIDES

G. Baldacchini

ENEA, Dipartimento Innovazione,
Settore Elettroottica e Laser,
C.R.E. Frascati
C.P. 65 - 00044 Frascati, Rome (Italy)

ABSTRACT

Two different classes of nonlinear optical phenomena are present in color centers in alkali halides.

One is the direct consequence of applying very intense electromagnetic fields, which results for instance in two and also three photon absorption with selection rules different from those of the much more usual one photon absorption. The other concerns the color centers with axial symmetry, whose dipole can be oriented in various directions under appropriate optical excitation. In this case the intensity of the exciting light plays a role only in the kinetics. Indeed the absorption of one photon not only excites centers like F_2, F_2^+, F_A, and F_H, but also reorients in the excited state their optical axes, producing at the same time a variation of the absorption itself.

In this context the newly discovered bistability of the $F_H(OH^-)$ center in KBr will be described in detail with the latest developments.

I. INTRODUCTION

In classical optics the polarization **P** of a dielectric medium traversed by an electromagnetic wave is proportional to the electric field **E** of the e.m. wave, and traditionally this concept is written as follows:

$$\mathbf{P} = \varepsilon_0 \chi^{(1)} \mathbf{E} \quad (1)$$

where ε_0 is the permittivity of vacuum and $\chi^{(1)}$ is the electric susceptibility of the medium. The polarization **P**, which is the induced dipole moment per unit volume of the medium, creates a second e.m. wave which sums up to the original one giving for the displacement field **D** the expression

$$\mathbf{D} = \mathbf{P} + \varepsilon_0 \mathbf{E} = \varepsilon_0 (\chi^{(1)} + 1) \mathbf{E} = \varepsilon \mathbf{E} \quad (2)$$

The susceptibility $\chi^{(1)}$ contains the electrodynamic characteristics of the medium, and,

Nonlinear Spectroscopy of Solids: Advances and Applications
Edited by B. Di Bartolo and B. Bowlby, Plenum Press, New York, 1994

through Eqs (1) and (2) describes all the familiar phenomena occurring in linear optics, i.e. absorption, index of refraction, birefringence, and dispersion.

However Eq. (1) is valid only in the limit of weak electric fields. In the case this is no longer true, in order to describe the observed effects it is necessary to expand the polarization in a power-series of the form

$$\mathbf{P}=\varepsilon_0\chi^{(1)}\mathbf{E}+\varepsilon_0\chi^{(2)}\mathbf{E}^2+\varepsilon_0\chi^{(3)}\mathbf{E}^3+..... \tag{3}$$

where the higher order susceptibilities $\chi^{(2)}$ and $\chi^{(3)}$ describes practically all the nonlinear phenomena known in modern optics. In particular $\chi^{(2)}$ is responsible of second harmonic generation and sum and difference frequency mixing, while $\chi^{(3)}$ attends to third harmonic generation, four-wave sum and difference frequency mixing, two photon absorption, and stimulated Raman scattering.

Although nonlinear optical effects have been predicted and observed long ago, for instance Raman scattering in 1927 and two photon absorption in 1931, only with the discovery of lasers sources in 1960 a broad and detailed investigation has been performed, and with such astonishing results that the field of optics has changed since then as never before [1,2].

Nonlinear optical behaviors have been measured in gases, liquid, and solids, but crystals represent the best known class of material, because they are used frequently in optical devices, and theoretical calculations are performed on them more easily. Indeed the crystalline symmetry is a good guideline for nonlinear optical properties; for example crystals with a center of symmetry have $\chi^{(2)}=0$, while they retain the third order nonlinear terms.

As far as crystals are concerned it is necessary to consider a further aspect, which has been neglected up to now in order to follow the traditional description of nonlinear optics. Usually nonlinear effects are produced by quadratic and cubic terms in Eq. 3 where the electric field is required to be very intense. However if the susceptibility $\chi^{(1)}$ is not a constant for a given system, and furthermore it depends upon the electric field of the e.m. wave, then also Eq. 2 can develop second and third order terms in the electric field, which this time is not required to be very intense. This new class of nonlinear phenomena happens in crystals containing axial centers, which can be reoriented by the electric field. By doing so the axial centers influence the propagation of the e.m. wave as in a feedback process.

Among axial centers color centers in alkali halides are a well known group of point defects with various spatial symmetries. They produce in the crystal host nonlinear optical effects originated both by the large intensity of the electric field and by the reorientational properties of some notable centers. In the first case nonlinear phenomena in color centers have been mainly studied to clarify the nature of the electronic transitions and states [3], while in the second case, besides the interest in basic research [4], there have been several attempts to use the dipole reorientation in various applications like holographic information storage [5], and polarization optical bistability [6].

In this paper both previous nonlinear effects in color centers will be described with an emphasis on those of the second type. In order to accomplish such task it is firstly necessary to describe briefly the main properties of color centers, which is made in § II. The classical nonlinear phenomena, i.e. the ones stimulated by intense e.m. fields, are described in § III, while the orientational ones are treated in § IV. Section V is devoted to the orientational bistability of the newly discovered $F_H(OH^-)$ centers, and future developments of the reorientational properties are discussed in § VI.

II. COLOR CENTERS IN ALKALI HALIDES

Color centers in alkali halide crystals have been discovered in the past century, and since then they have been studied in a systematic way up to our days [7]. The reason for

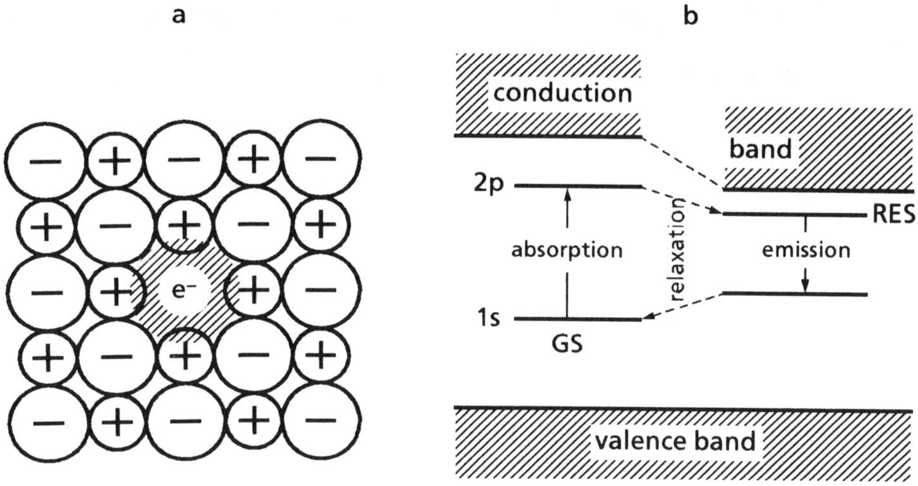

Figure 1. F Center in alkali halides: a) lattice model; b) main electronic states and optical cycle.

such unabated interest by the solid state community derives mainly from their intrinsic simplicity in comparison with other point defects in solids. So they have been used as prototype-models for more complex impurities or imperfections in insulating crystals. The study of electrical conductivity, diffusion, and optical properties of color centers in alkali halides has greatly improved the knowledge of these phenomena and other more complex ones in point defects in solids.

A simple description of color centers and their optical properties is presented in the following, referring to the abundant literature on this argument for a more detailed study [**8-12**].

The simplest color center, the F center, is an electron trapped in an anion vacancy. This electron moves around the positive charge of the anion vacancy, **Fig. 1a**, and so it forms hydrogen-like bound states 1s, 2p etc...., in analogy with the atomic case. The absorption from the ground state, 1s, to the first excited state, 2p, corresponds to the main absorption band of the F center, Fig. 1b. However, the analogy with the hydrogen atom ends here. Indeed, the new distribution of the electronic charge is not in electrostatic equilibrium with the local crystal fields, so that the first positive neighboring ions move outward in order to compensate for the charge variation. This process is usually called relaxation, because the lattice relaxes to its new equilibrium configuration, and at its completion the electron reaches the relaxed excited state (RES). From this peculiar state, the electron decays with the emission of light, the typical F-center luminescence, to the unrelaxed ground state and, eventually, after a further relaxation process, it reaches the ground state (GS) again. Usually the relaxation transitions, which bring the electron to new energy states, last $\sim 10^{-12}$ s, while the RES lifetime is of the order of 10^{-6} s. As a consequence in a continuous optical pumping experiment the GS and RES are the only states appreciably populated.

This simple descriptive model should not deceive the reader about the complexities of the F center, which have been amply discussed recently in two previous Erice Schools [13,14]. Indeed, although extended experimental and theoretical efforts have been made in the past and are still continuing today, there is not at moment a model able to explain at the same time all the parameters involved in the optical cycle. Nevertheless the F center possesses the simplest structure among color centers and, being one of the most known centers, is often used as a model case to understand other more complex centers.

The structures and the main optical properties of simple color centers, which are obtained by associating vacancies and foreign ions, are shown in **Fig. 2** along with

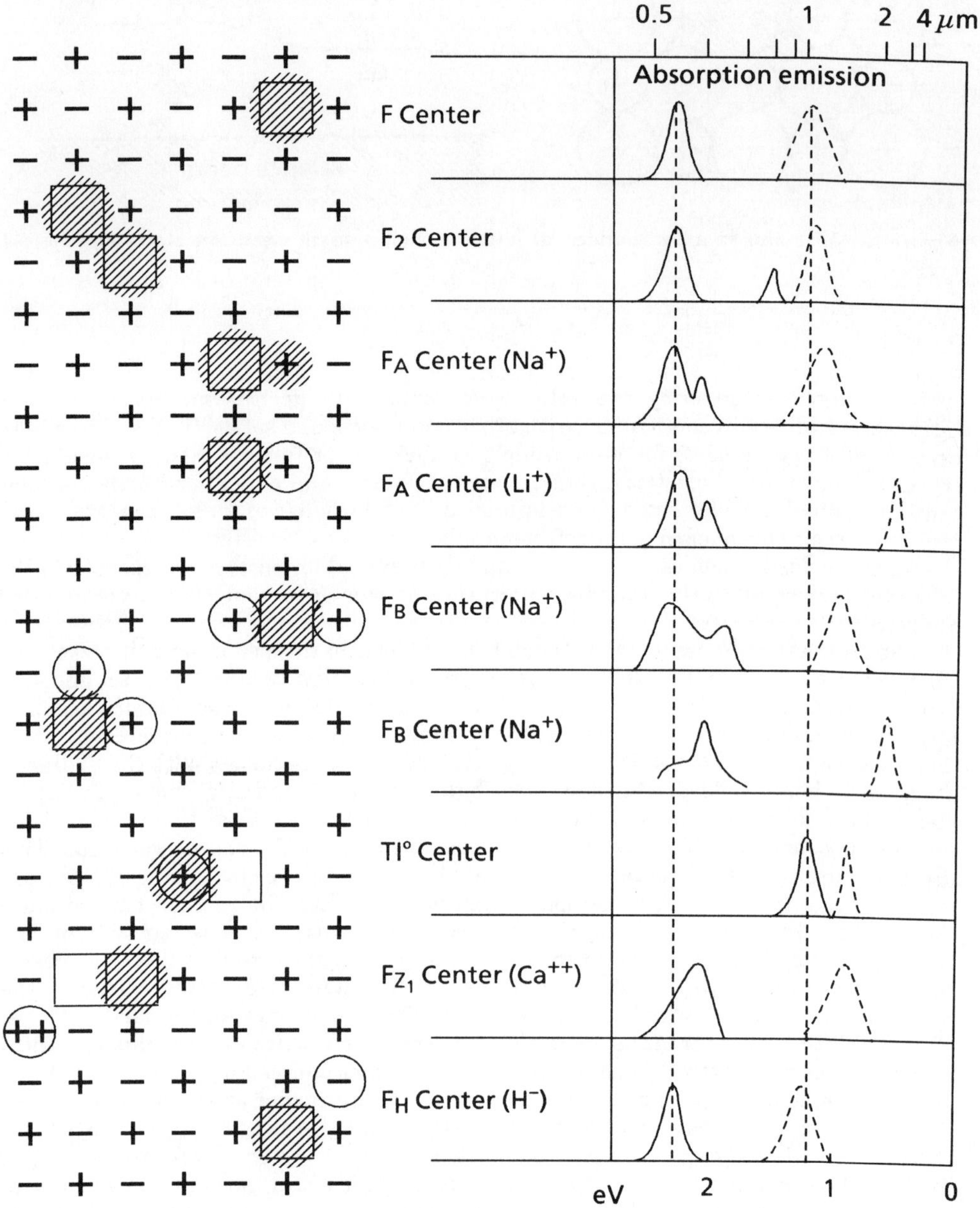

Figure 2. Structural models, absorption and emission behaviors of the F center and of simple aggregated centers at low temperature in KCl. The square is the halogen vacancy, the circle the foreign ion and the shaded area the approximate electron distribution.

those of the F center for a direct comparison. The process of aggregation or/and association with the impurity ions lowers the symmetry of the centers with respect to the complete spherical symmetry of the F center. This event reflects directly on the absorption and emission bands, as it is shown in Fig. 2, and also opens the way to nonlinear optical properties. For instance it is very clear that F_A centers have an axial symmetry, the direction of which, in timely conditions of temperature and optical pumping, can be changed from one to another of the cubic lattice axes [**15**]. The same is more or less true for all centers displayed in Fig. 2, which however, with the exception of F_2 centers, have been studied much less in this respect. For example it is hard to imagine how F_B centers can reorient themselves, and F_H centers apparently show some changes of the optical properties with respect to those of the F center, which become dramatic when the substitutional negative impurity is a molecular ion. Indeed, in this last case the emission properties change completely [**16**] and new orientational behaviors appear [**17**]. We will come back on this point in detail in § V.

III. CLASSICAL NONLINEAR OPTICS

The first color center investigated was the F center in KCl, where the absorption energy, $E_A = 2.31$ eV, is almost matched by twice the energy of the Nd-YAG laser, $\lambda_{laser} = 1.06\ \mu m \equiv 1.17$ eV. As a consequence a two-photon absorption, due to the $\chi^{(3)}$ term in Eq. 3, should be possible in this system. Moreover, due to the different selection rules, new excited states of the same parity as the ground state should be reached. So one of the first experiment was directed to solve the old problem of the nature of the RES, which has been the main topic of a previous Erice School [14].

It is well established that the lowest excited state of the F center is a 2p-like one, and theoretical calculations indicated and experimental results confirmed that during the process of relaxation the 2p and 2s states cross each other, with the final result that the 2s* state is the lowest one in the RES configuration, see **Fig. 3**. Because the two-photon absorption excites the 2s state, and not the 2p state as in the normal one-photon absorption, there was hope to observe a different emission, having in this way new informations about the nature of the RES. The experimental results [**19**] gave a two-peaked structure for the luminescence, which was ascribed to the two states 2p* and 2s*. Moreover also two lifetimes, different from the well known one of the F center, were measured for the two peaks. However in subsequent experiments [**20**] on the same system there was no indication that the two-photon excited fluorescence decays differently with respect to the one-photon excitation, and also, somewhat limited by spectral accuracy, there was a spectral coincidence of the two-photon and one-photon excited luminescences.

In conclusion the previous two-photon absorption experiments and subsequent ones did not add any new information on the nature of the RES, which is still debated today [14]. However they did show the complexities of this new spectroscopic technique, which is based mainly on the very high power of the laser sources used. Indeed not only small amount of impurities can be excited efficiently, but also other nonlinear phenomena can be generated adding confusion to an already complex picture.

However nonlinear absorptions of color centers, due to $\chi^{(2)}$ and $\chi^{(3)}$ terms in Eq. 3, have been experimentally observed by monitoring the emission radiation after excitation with powerful pulsed lasers [3].

As an example **Fig. 4** shows the spectral structure of the luminescence of a KCl crystal containing simple aggregated F centers upon excitation with a Nd:YAG laser at $\lambda = 1.06\ \mu m$ [**21**]. This frequency up-conversion phenomenon was originally attributed to a cooperative excitation of F_3 centers. Subsequent experiments have shown both a cubic dependence of the emission intensity upon the power of the exciting laser and a decrease of the emission bands upon further aggregation, which produces F_3 centers at the expense of the F_2 centers [**22**]. As a consequence the new red emission has been attributed to three-photon absorption, $3h\ \nu_{laser} = 3.51$ eV $\equiv 354$ nm, by higher excited

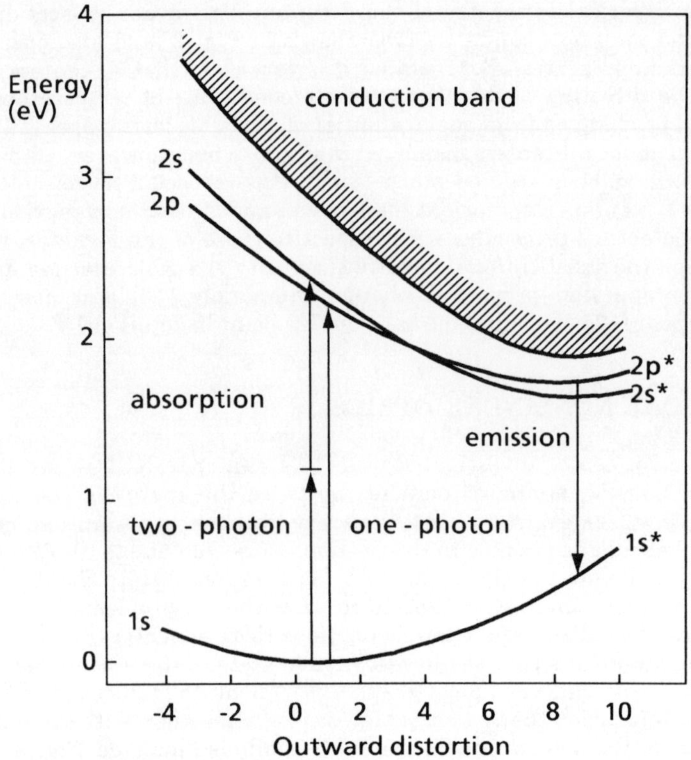

Figure 3. Energy diagram of the F center in KCl as a function of the position of the neighbouring positive ions, which shows a possible crossing of 2s and 2p levels [18]. One-phonon and two-phonon transitions are also indicated in the absorption band.

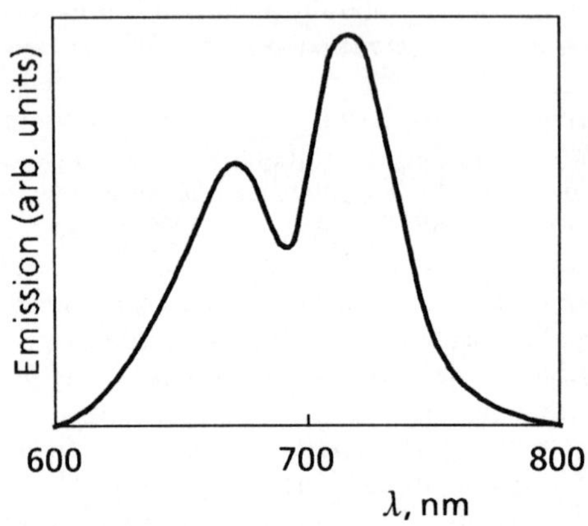

Figure 4. Emission spectrum of an additively colored KCl crystal, containing F, F_2, and F_3 centers, at 80 K and pumped at 1.06 μm [21].

states of F_2 centers. However this story, which is a good example of the intrinsic difficulties of optical experiments involving very high laser intensities and their interpretation, did not finish here. Indeed more recently new experiments were made supposedly on the same system, and a quadratic increase with the laser power, involving a two-photon absorption, was found [23]. On the basis of these new evidences this up-conversion has been attributed to a loose aggregation of a pair of F centers!

Second harmonic generation involving the term $\chi^{(2)}$ has been observed in a heavily coloured NaCl crystal [24]. When pumped with a Nd:YAG laser, $\lambda_{laser}=1064$ nm, the crystal emits a green light, $\lambda_{emi}=532$ nm, along the linepath of the pumping laser. The emission light is so intense to be visible to naked eyes as a bright lamp. The intensity of the emission is proportional to the square of the laser intensity, and its spectral lineshape is monochromatic in the resolution limit of the measuring apparatus. The second armonic emission depends upon the mutual orientations of the polarization plane of the exciting laser and the crystal axes, and its intensity decreases as the sample temperature increases above 100 K toward RT. Furthermore the maximum efficiency in frequency doubling has been observed by annealing the sample at 480°C prior of the laser pumping. A further annealing at higher temperatures destroys completely the second harmonic generation. All these properties led the authors to ascribe this nonlinear phenomenon to small colloids, today known as quantum dots, finely dispersed in the crystal. Indeed in this way it is possible to explain the lack of both a measurable absorption in the colloid spectral region and an inversion symmetry, that allows the second order nonlinear scattering.

IV. ORIENTATIONAL NONLINEAR OPTICS

Figure 5 shows the absorption of the F_A(II) centers in KCl:Li^+ at $T \simeq 2$ K for light linearly polarized along the axes x and y as a function of the time of pumping with an He-Ne laser linearly polarized at 45°, Fig. 5a, and subsequently along the axis x, Fig. 5b, and few mW of power [25]. Two dynamic phenomena are happening with two different time scales. The crystal is bleaching very quickly in the first pumping scheme, while its y-polarized absorption in the second pumping scheme is increasing with a time constant two order of magnitude longer. In order to understand what is going on it is necessary to go back to the properties of the F_A centers [15].

This peculiar point defect is formed by an F center and an alkali ion impurity in a <100> lattice position, see Fig. 2, and it can be of type I or type II depending on the impurity, on the temperature, as it has been discovered recently [26], and on other factors not yet completely known. Anyway F_A(II) centers reorient under optical pumping at all temperatures, while F_A(I) centers are practically frozen at low temperatures. The reorientation refers to the F-center impurity axis which can be aligned along one of the three crystallographic directions. A change of direction happens when a neighbor halogen ion jumps into the vacancy of the excited F center, leaving behind a new vacancy, a process which is equivalent to a 90°-angle rotation of the center axis.

The F_A center has a lower symmetry with respect to the F center and because of that the absorption band is split in two components. The one at higher energy, F_{A2}, is due to two degenerate transitions perpendicular to the center axis, while the other one at lower energy, F_{A1}, is parallel to the center axis or dipole moment, as it is often called.

The He-Ne laser is pumping mainly the F_{A1} band, and the rapid bleaching along the x direction, Fig 5a, is due to the excitation of the dipoles aligned along x and y, and their rotation along the z direction, where supposedely they cannot absorb light any more. Instead quite surprising is the slow increase of the absorption along y well after the completion of the first bleaching. However accurate studies have shown that this increase is due to the violation of the selection rules for absorption of F_A light. More exactly a light linearly polarized along y can also excite dipoles along z because of a

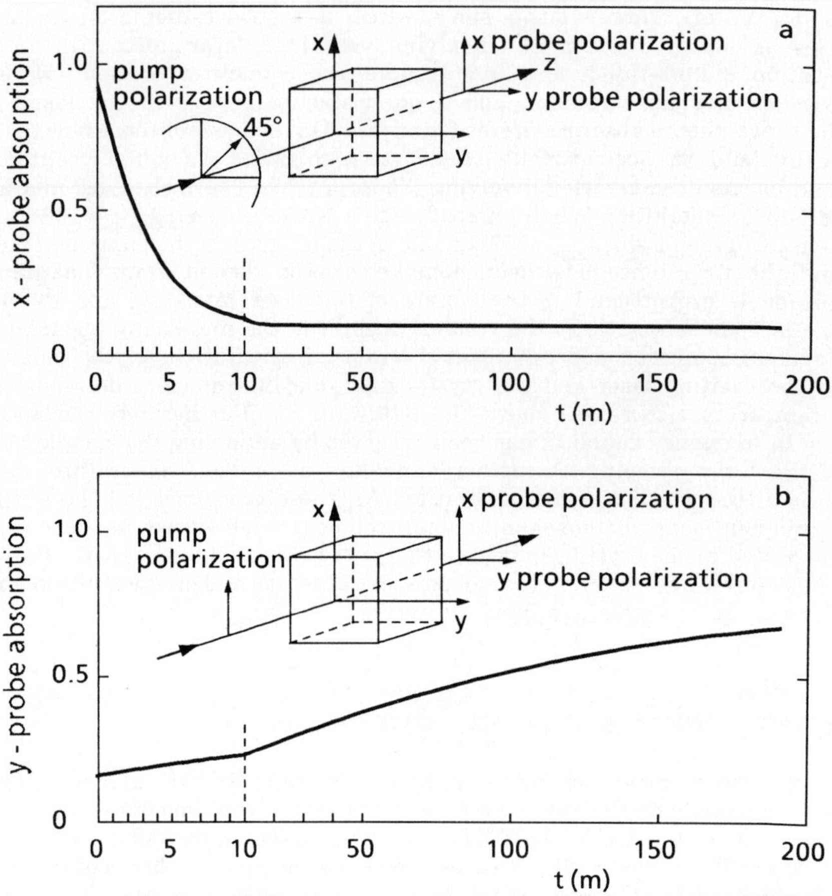

Figure 5. Absorption of F_A(II) centers in KCl:Li^+ at 2 K under two linearly polarized pumping schemes with a He-Ne laser [25]. Note the change in the time scale after 10 m.

partial overlapping of F_{A1} and F_{A2} bands, and of the existence of a small tilt angle of the dipole moment with respect to the crystallographic axes [27]. As a consequence the dipoles aligned along z, the direction of propagation of the pumping light, are also excited, and slowly are reoriented equally along x and y axes, but those in the x direction are again strongly excited and reoriented by the pumping beam. Eventually the dipoles along the z axis are slowly reoriented along the y axis as it is shown in Fig. 5b.

In conclusion the passage of a polarized light beam through a crystal containing F_A centers produces a sizeable variation of the optical properties of the crystal, and undergoes itself changes of intensity and also of polarization state, as it happens in nonlinear optical systems. In practice the pumping beam induces effects of dichroism and birefringence, which depends strongly on the wavelength and polarization state of the exciting light. These nonlinear effects can be taken properly into account by using complex theoretical formulations [**28,29**], to which the interested reader is referred, but here we will follow a much simpler descriptive approach.

Figure 6 shows the stady state rotation of the polarization plane of a He-He laser beam which passes through a crystal of KCl:Li^+ containing F_A(II) centers as a function

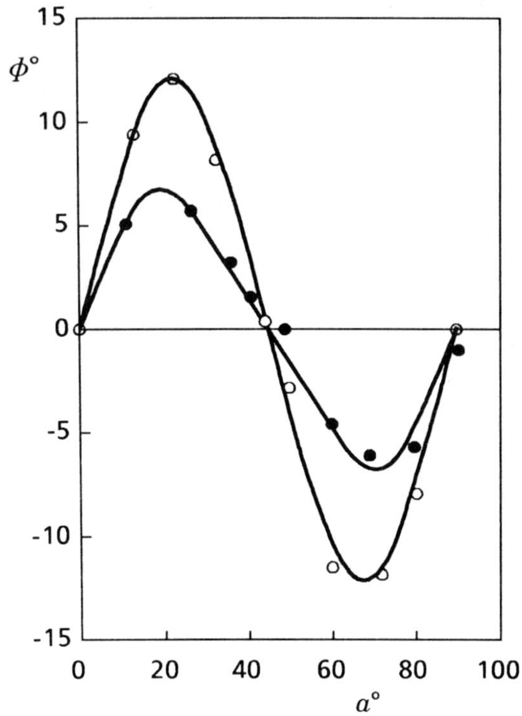

Figure 6. Rotation of polarization of an He-Ne beam by a KCl crystal with two different concentrations of F_A(Li) centers at T= 77 K. Open circles refer to a crystal about two time optical dense with respect to the dots [30].

of the angle α between the [100] axis and the E vector of the incident radiation [30]. Here ϕ is the angle between the **E** vectors of the input and output exciting light. The rotation of the polarization plane is well described by the following equation:

$$\frac{d\psi}{dz} = -\frac{3}{2\alpha_0} \cdot \frac{\sin 4\psi}{2+(\chi^{-1}+2)\sin 2\psi} \tag{4}$$

where $\psi=\alpha+\phi$, α_0 is the absorption contant in the absence of F_A(II)-center orientation, and $\chi=\alpha_2(\lambda)/\alpha_1(\lambda)$ with $\alpha_1(\lambda)$ and $\alpha_2(\lambda)$ the absorption cross sections for light polarized parallel and perpendicular to the center axis (in other words $\alpha_1(\lambda)$ refers to the absorption of the F_{A1} band and $\alpha_2(\lambda)$ to the F_{A2} band) at the given pumping wavelength. The solid curves in Fig. 6 have been calculated with Eq. (4), and the agreement with the experimental points is quite satisfactory. It is worthwhile to note that the rotation of the plane of polarization is zero for α=0°, 45°, 90°, i.e. for **E** parallel to [100] and [110] crystallographic directions, and that the angle ϕ is independent from the light intensity, which influences only the kinetics of the phenomenon [31].

The maximum rotation of the polarization plane in Fig. 6 for a given sample possesses a strong dependence from the absorption lineshape, as it is shown in **Fig. 7** where ϕ_{max} is reported as a function of wavelength [32]. The experimental points

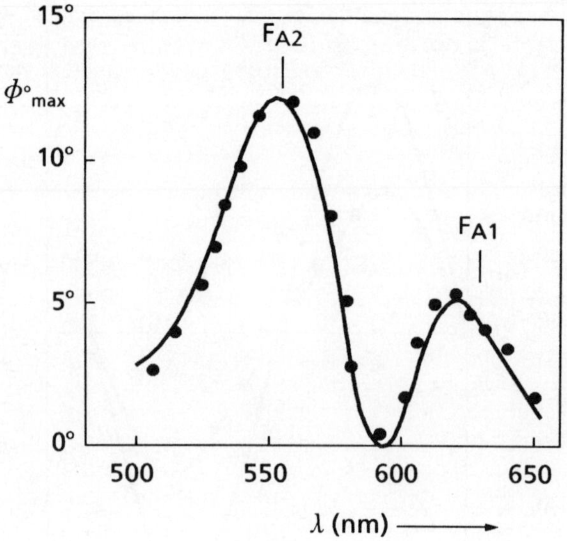

Figure 7. Maximum rotation of the polarization plane of the exciting light in KCl with F_A(Li) centers as a function of wavelength at T=77 K. The F_{A1} and F_{A2} center-band have been marked [32].

follows approximately the absorption profile of the F_{A2} and F_{A1} bands, and they are accounted for by the theoretical fitting, continuous curve, of the equation

$$\frac{\mathrm{tg}\, 2\psi(d)}{\mathrm{tg}\, 2\psi(0)} \cdot \left[\frac{\cos 2\,\psi(0)}{\cos 2\,\psi(d)}\right]^{\beta} = \exp\,(-\,d/\ell)\,, \tag{5}$$

where d is the thickness of the crystal, and β and ℓ are parameters related to the susceptibility of the absorbing centers. In the case of a small absorption Eq. 5 reduces to the simple form

$$\phi_{max} = \Big|\psi(d) - \psi(0)\Big|_{max} = \frac{d}{4\ell}\sqrt{\beta - 1}\,, \tag{6}$$

which shows explicitely that the rotation of the polarization plane is proportional to the absorption of the crystal, i.e. to the concentration of the F_A(II) centers. Just to have a much clear idea about the efficiency of such rotation effect, about 10° of maximum rotation are obtained in a crystal with 1 O.D. of absorption at low temperatures.

V. ORIENTATIONAL BISTABILITY OF F_H CENTERS

F_H centers in alkali halides, formed by association of F centers with negative foreign ions, have been studied since long time ago and, with the exception of CsBr doped with Cl^- impurities [33], the experimental results on absorption and emission have not shown sizeable changes with respect to the same properties of F centers [34], contrary to what happens instead for F_A centers, where foreign positive ions are associated with F centers. However quite different is the case when the substitutional ions are diatonic molecules like OH^- and CN^-, which produce interesting and novel physical properties [16,35]. We will limit our attention here to the case of F-center/OH^--molecule-pairs in KBr which displays unusual nonlinear optical behaviors.

Figure 8 shows the well known absorption and emission bands of pure F centers in KBr at various temperatures [**36**]. Doping the crystal with OH^- ions produces a decrease of the luminescence efficiency, expecially at temperatures above 50 K, see **Fig. 9**, which depends on the concentration of the diluite OH^- ions. However an almost complete quenching of the emission occurs when the OH^- ions are coupled to the F centers at all temperatures [37]. $F_H(OH^-)$ centers are produced by illuminating the F band at T=-30 °C, a process which produces the excitation of the F center and its migration through the crystal until an encounter with one molecular ion stops the wandering. In KBr, analogously to all crystals with NaCl structure, the OH^- molecule occupies a <200> neighboring position, and its association with F centers suppresses the electronic emission at 1.4 μm and the photoionization, and an extremely rapid radiationless return of the electron into its ground state occurs [16,**38**]. This optical cycle obviously produces an energy transfer into the OH^- molecular stretching mode which relaxes later on into the phonon bath. However while for the majority of the crystals with NaCl structure only a broadening and a red shift of the absorption band occurs at all temperatures, in the case of KBr an unusual strong temperature dependence of the absorption peak energy has been observed down to ~12 K, and at LHeT an optical and thermal bistable absorption band is displayed.

The optical absorption of a KBr:OH^- sample containing F centers, aggregated at -30°C and immediately cooled in the dark at liquid helium temperature (1.9 K) is

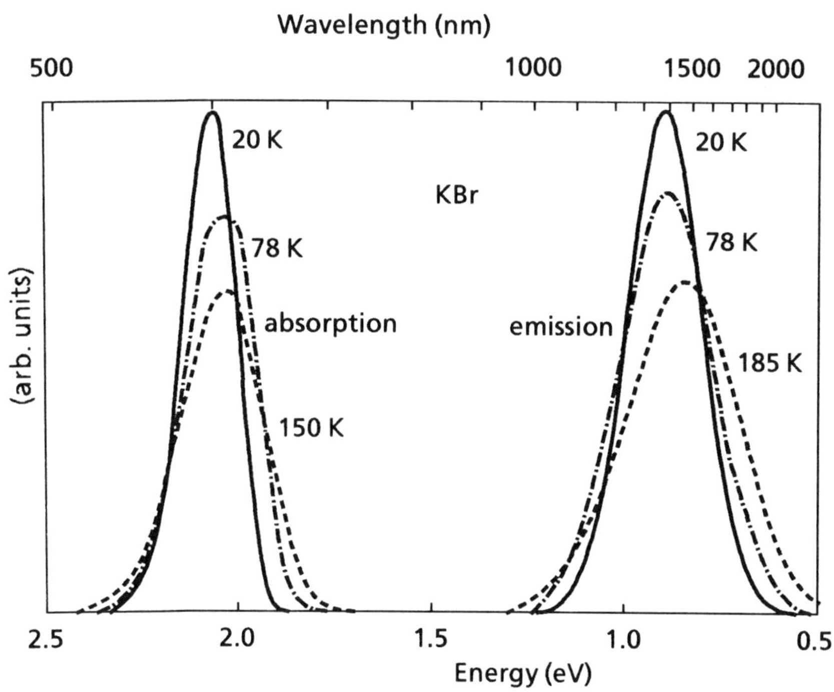

Figure 8. F-Center absorption and emission bands in KBr, as a function of temperature [36].

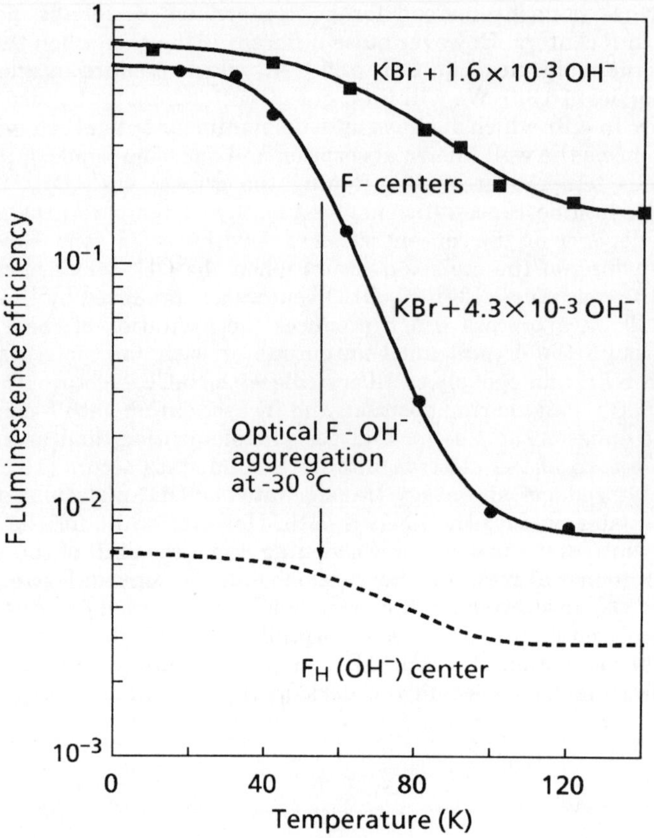

Figure 9. Efficiency of the F luminescence in a crystal of KBr containing OH^- ions before (continuous curves) and after optical F-OH^- aggregation, leading to the formation of $F_H(OH^-)$ centers [37]

reported in **Fig. 10** (curve I). The band peaks at ~588 nm on the blue side and with almost the same halfwidth (HW=0.17 eV) of the F band (shown with dotted line and peak at 599 nm). Weak optical irradiation with monochromatic ligth at 580 nm irrespectively of its polarization status changes the band into the red-shifted (peak at ~614 nm), broadened (HW=0.26 eV) band shown in the same figure (curve II). Similar irradiation at 650 nm reconverts the spectrum II back into the spectrum I. This bidirectional conversion can be repeated any number of times with no loss of area under the absorption. We are in presence of an optically bistable state. Due to the close position of the peaks of the bands I and II (ΔE=0.09 eV) and of their overlap, it was not possible to measure pure spectra of the F_H(I) and F_H(II) centers. However they have evaluated in **Fig. 11** by fitting the experimental curves to the sum of two gaussian bands, 1≡pure curve I and 2≡pure curve II, whose parameters have been found to be, F_H(I): peak=585 nm, HW=0.16 eV; F_H(II): peak=618 nm, HW=0.24 eV. After the conversion with irradiation at the wavelengths quoted above, we obtain a residual absorption of the band that we are destroying of approximately 20%.

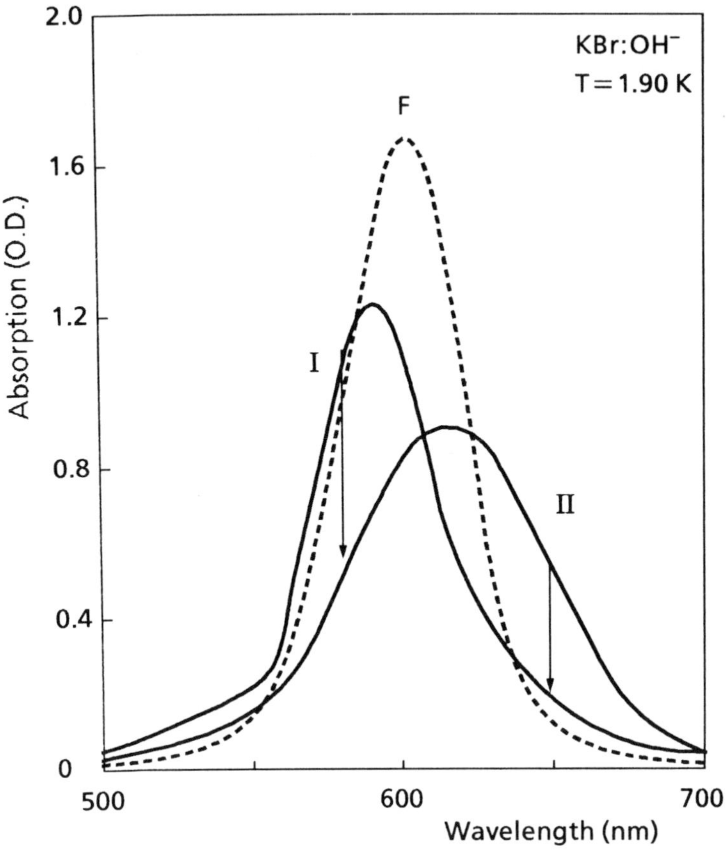

Figure 10. Optical absorption of the F center (dashed curve) and of the two configurations I and II of $F_H(OH^-)$ centers in KBr + 1.5% KOH at T = 1.9 K.

The $F_H(OH^-)$ center in KBr has a very peculiar temperature dependence. Warming up the crystal produces complex changes from one configuration into the other, as one can measure from the area under the two bands. If we prepare the crystal mostly in configuration I and we warm in the dark the sample (see **Fig. 12**), the configuration is stable up to about 30 K where the band F_H(I) decreases drastically, while the F_H(II) band increases and takes the place of the disappeared band. On the contrary, if we prepare the crystal at low temperature in the configuration II (see **Fig. 13**, note the expanded temperature scale), already at 5 K this configuration ceases to be stable and converts in the dark into the configuration I. As before this latter is stable up to 30 K and returns to the configuration II at higher temperatures. The absorption band, contrary to what occurs at LHeT, does not show any shape variation under optical pumping above 10 K, i.e. the optical bistability is limited as far as we know to the region of low temperatures.

A tentative interpretation of the temperature dependence of the F_H center absorption is illustrated in **Fig. 14**, by using for the ground state a simple two box potential well with unequal energies E(I) < E(II) for the two configurations. In the

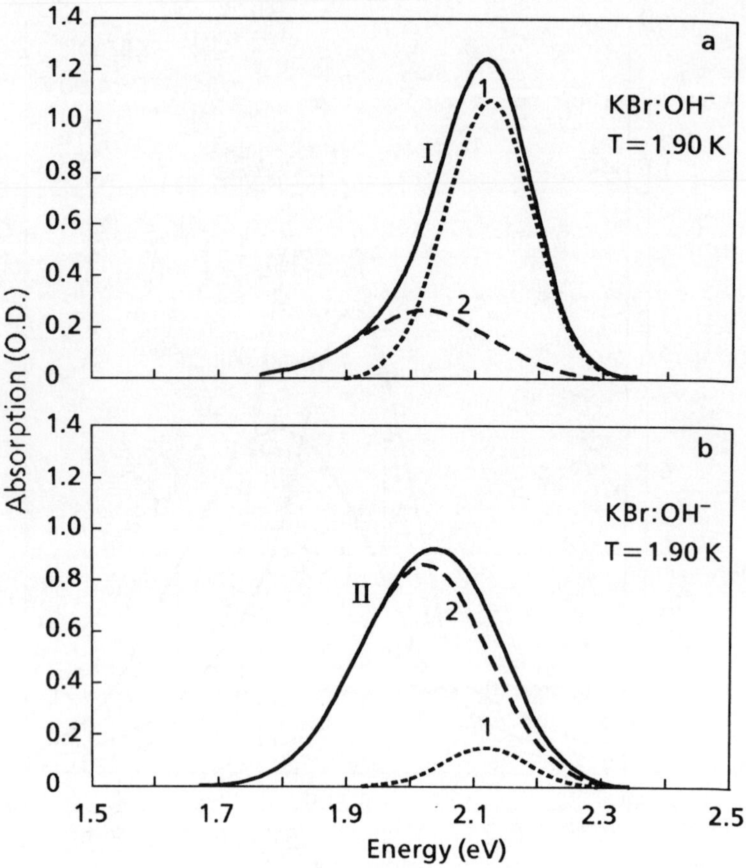

Figure 11. Optical absorption of the two bands I(a) and II(b) of $F_H(OH^-)$ centers in KBr + 1.5% KOH at T = 1.9 K, decomposed in two gaussian bands (dashed curves).

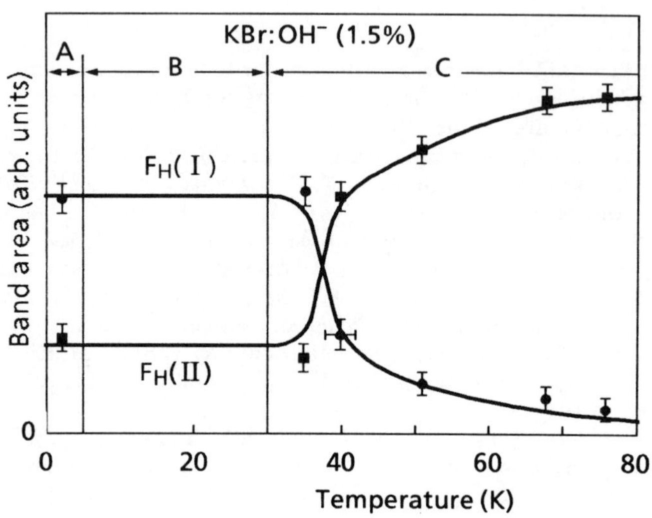

Figure 12. $F_H(I) \rightarrow F_H(II)$ thermal conversion in the dark. The crystal containing F_H centers has been prepared to possess mainly $F_H(I)$ centers by optical pumping at 1.9 K.

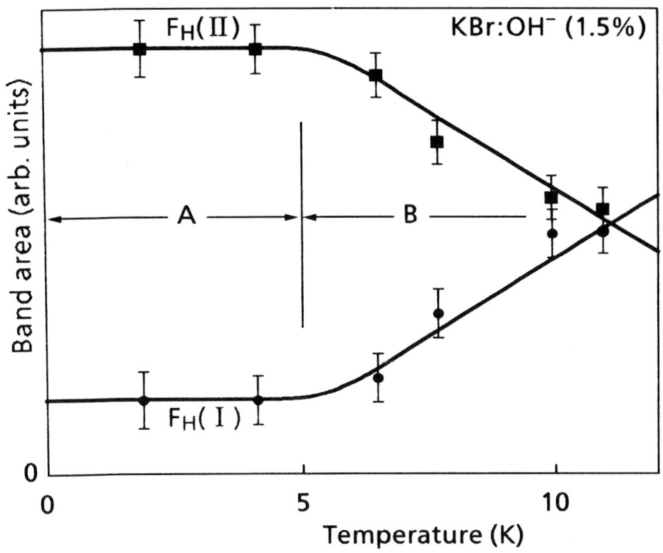

Figure 13. F_H(II)→F_H(I) thermal conversion in the dark. The crystal containing F_H centers has been prepared to posses mainly F_H(II) centers by optical pumping at 1.9 K.

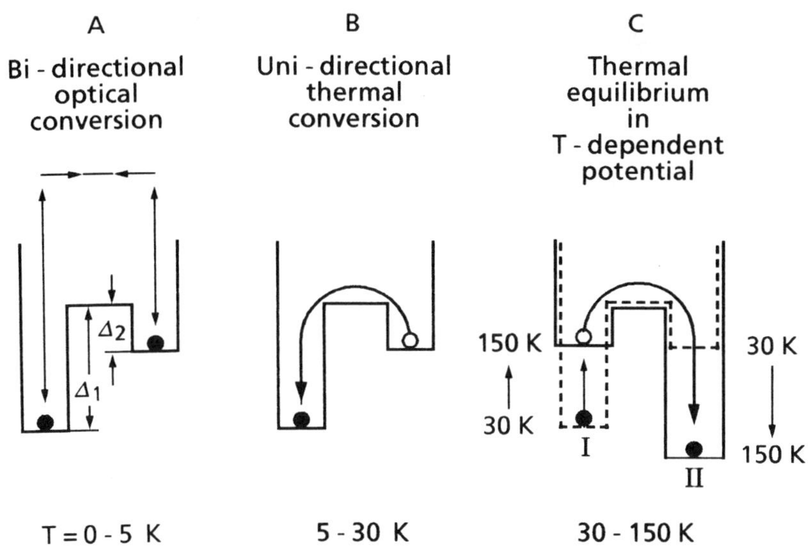

Figure 14. A box potential model with unequal energies for the two configurations F_H(I) and F_H(II).

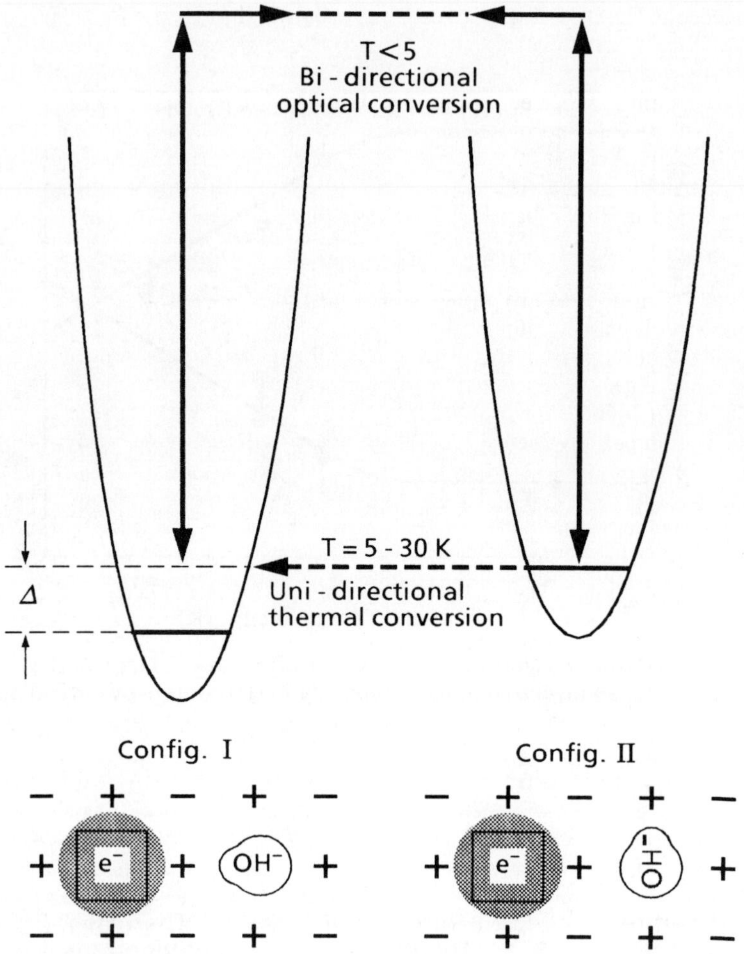

Figure 15. Tentative models for $F_H(OH^-)$ defect configurations I and II. A simplified energy potential diagram for reorientation of the OH^- between its two directions indicates schematically the optical (T=0-5 K) and thermal (T=5-30 K) conversion I⇄II.

temperature region *A* the energy barrier from well II to well I is $\Delta_2 > KT$, so that no thermal exchange but only bidirectional optical conversion, II⇄I, can take place. In the region *B* $\Delta_2 < KT$, and unidirectional thermal, II→I, conversion into the deeper well occurs. This state of matter is stable up to ~30 K where most probably temperature effects start to produce changes in the potential wells. It is assumed that the situation E(I) <E(II) in *A* and *B* regions changes gradually under temperature increase into the opposite one, such that for T≥100 K E(I)>E(II), and for even higher temperatures essentially only the configuration II is occupied, as it has been observed experimentally. From the temperature dependence of the F_H(II)→F_H(I) of Fig. 13 and of the F_H(I)→F_H(II) conversion of Fig. 12 one can derive an approximate value of the two barrier heigths, which are $\Delta_2 \simeq 2.8$ meV and $\Delta_1 \simeq 10$ meV.

However recent ENDOR experiments have shown that in the $F_H(OH^-)$ complex the F center and OH^- molecule occupy <200> neighboring positions [39]. As a consequence the six energy equivalent [100] orientations of the free OH^- ion split in presence of the F center in three levels: two single states of the OH^- molecule parallel to the pair axis with its dipole moment pointing either towards or away from the F center, and a fourfold degenerate state of the OH^- orientation in the four [100] directions perpendicular to the pair axis. It has been suggested that configurations I and II corresponds to one of these parallel and perpendicular orientations, respectively, as indicated in the lower part of Fig. 15 [17]. This model is very similar to the previous one of Fig. 14 in the temperature ranges *A* and *B*, with the difference that now the unidirectional thermal conversion occurs by phonon assisted tunneling through the well barriers, which can be ~0.1 eV high. In this context it is also possible to explain the experimental behavior in the temperature range *C*, since the configuration II is fourfold degenerate and so its population becomes predominant over the population of the nondegenerate level I for $KT>\Delta$. By using the temperature dependence of the $F_H(OH^-)$ absorption peak, which in the temperature range *C* moves monotonically from F_H(I) to F_H(II) configuration [17], an approximate value of the degeneracy $g=4.2$ and of the well energy difference $\Delta=5$ meV is obtained. This last value is not strongly dissimilar to that obtained by using the area of the absorption band, and the degeneracy agree fairly well with the expected one from the present microscopic model.

However optical detected ENDOR measurements of the GS of the $F_H(OH^-)$ center have shown recently that, besides a strong displacement, ~15%, of the K^+ ion sitting between the F center and the OH^- ion, the orientation of the OH^- dipole is parallel for configuration II (red center) and perpendicular for configuration I (blue center) [40]. These results are at odd with the previous assignements, and the thermal behavior of this peculiar center has been described by taking into consideration the entropy of the two configurations. Indeed from a close analysis it has been found that at high temperatures the entropy of the configuration II (red center) exceeds that of the configuration I (blue center), making in this way the red center the stable one in the temperature range *C*.

VI. CONCLUSIONS

Usually when we hear of nonlinear optical properties we are more prone to think about applicative rather than basic implementations. In general this is not the case with color centers in alkaly halides. Indeed optical nonlinear phenomena have been mainly used to unravel the complex nature of the centers themselves rather than used as tools in science and technology. However there have been notable exceptions and the actual situation show some signs of reversing this attitude.

The dishroism of F_A centers have been used to store erasable, high-resolution and high-contrast holograms [5].

The orientational bleaching of F_A centers can be used to produce optical bistable devices by using Fabry-Perot systems [41,6].

Reversible hole-burnings in inhomogeneously broadened zero-phonon lines of various color centers have been systematically studied for their applications as data elements in a frequency-domain optical information storages system [42,43].

LiF crystal with F_2^- centers have been operated at room temperature as Q-switches for the generation of ultrashort pulses [44].

In general nonlinear phototropic and saturable absorption properties of color centers make them promising for practical use as passive Q-switches, nonlinear filters, and optical isolators [45].

As far as the basic research is concerned, expecially that connected to the reorientational axial centers, there are several questions still open.

The alignement of F_A centers in RbCl:Li^+ and KF:Na^+ is anomalous and there is a thermal assisted conversion between the two configurations, I→II [26,46]. The case of KF:Na^+ seems of particular interest, because the center F_A(I) at LHeT can be oriented in a reversible way by polarized light, contrary to the actual knowledges.

Still unknown are the orientational behaviors of F_B centers, which together with F_A and other kind of color centers are used as active laser media in the near infrared [47].

F_H centers with halogen impurities are known only in a few cases, as shown in § V, and more experimental and theoretical studies are required [34].

F_H centers with molecular ions have shown interesting reorientational nonlinear optical phenomena. A complete model in KBr:OH^- is still missing, and it is not yet clear whether the thermal bistability is originated by a simple thermal excitation or by a phonon assisted tunneling process. Moreover also the system KI:OH^- shows similar bistabilities in a more hidden way [17]. Evidently more detailed experiments are needed in order to fully clarify the static and dynamic interrelations between the two or more orientational microscopic configurations, which may coexist.

In conclusion the whole field of nonlinear optical properties of color centers appears to be very promising both for basic and applicative research.

ACKNOWLEDGMENTS

I am grateful to Prof. M. Casalboni for showing and discussing with me unpublished measurements from his laboratory. Many thanks are also due to Prof. U.M. Grassano for a number of conversations and a critical reading of the manuscript.

REFERENCES

1. P.N. Butcher and D. Cotter, The Elements of Nonlinear Optics, Cambridge University Press, Cambridge, 1990
2. Non Linear Optics in Signal Processing, R.W. Eason and A. Miller eds, Chapman & Hall, London, 1993
3. U.M. Grassano, J. Phys.(Paris) suppl. 41, C6-359 (1980), and in Optical Properties of Excited States in Solids, B. Di Bartolo ed., Plenum Press, New York, 1992, p.643
4. A. Scacco in Optical Properties of Excited States in Solids, B. Di Bartolo ed., Plenum Press, New York, 1992, p.625
5. H. Blume, T. Baden, and F. Luty, Opt. Commun. 12, 147 (1974)
6. A.M. Brodin, S.A. Boiko, M. Ya. Valakh, and G.G. Tarasov, Phys. Stat. Solid B **150**, 557 (1988)
7. J. Teichmann and K. Szymborski, in Out of the Crystal Maze, L. Hoddeson, E. Braun, J. Teichmann, and S. Weart, eds, Oxford University Press, New York, 1992, Chap. 4
8. J.H. Shulman and W.D. Compton, Color Centers in Solids, Pergamon Press, Oxford, 1963
9. Physics of Color Centers, W.B. Fowler, ed., Academic Press, New York, 1968
10. A.M. Stoneham, Theory of Defects in Solids, Clarendon Press, Oxford, 1975.
11. Y. Farge and M.P. Fontana, Electronic and Vibrational Properties of Point Defects in Ionic Crystals, North-Holland Publ. Company, Amsterdam, 1979
12. W. Hayes and A.M. Stoneham, Defects and Defect Processes in Nonmetallic Solids, Wiley Interscience Publ., New York, 1985
13. G. Baldacchini in Advances in Nonradiative Processes in Solids, B. Di Bartolo ed., Plenum Press, New York, 1991, p. 219
14. G. Baldacchini in Optical Properties of Excited States in Solids, B. Di Bartolo ed., Plenum Press, New York, 1992, p. 255
15. F. Luty in Physics of Color Centers, W.B. Fowler ed., Academic Press, New York, 1968, Chap. 3
16. L. Gomes and F. Luty, Phys. Rev. **B30**, 7194 (1984)

17. G. Baldacchini, S. Botti, U.M. Grassano, L. Gomes, and F. Luty, Europhys. Lett. **9**, 735 (1989)
18. R.F. Wood and U. Opik, Phys. Rev. **179**, 783 (1969)
19. F. De Martini, G. Giuliani, and P. Mataloni,Phys. Rev. Lett., **35**, 1464 (1975)
20. K. Vogler, Phys. Stat. Solid B **107**, 195 (1981)
21. I.A. Parfianovich, V.M. Mestik, L.M. Sobolev and É.É. Penzina, Sov. Tech. Phys. Lett. **5**, 616 (1979)
22. M. Casalboni, R. Francini, U.M. Grassano, and R. Pizzoferrato, Phys. Stat. Solid B **117**, 493 (1983)
23. H. Hanzawa, Y. Mori and H. Ohkura, Radiat. Eff. and Defects in Solids **119-121**, 791 (1991)
24. M. Casalboni, F. Crisanti, R. Francini, and U.M. Grassano, Opt. Commun. **30**, 440 (1979)
25. G. Baldacchini, G.P. Gallerano, U.M. Grassano, M. Meucci, A. Scacco, F. Somma, A. Tiurine, and M. Tonelli, Phys. Stat. Solid B **122**, K 83 (1984)
26. G. Baldacchini, E. Giovenale, F. De Matteis, A. Scacco, F. Somma, M. Casalboni, and U.M. Grassano, Europhys. Lett. **7**, 647 (1988)
27. G. Baldacchini, F. De Matteis, E. Giovenale, A. Scacco, F. Somma, and U.M. Grassano, Phys. Rev. **B37**, 7014 (1988)
28. M.I. Dykman and G.G.Tarasov, Sov. Phys. JETP **47**, 557 (1978)
29. M. May, S. Debrus, J.P. Hong, B. Quernet, and E. Rzepka, Phys. Rev. **B38**, 3469 (1988)
30. M. Ya. Valakh, M.I. Dykman, M.P. Lisitsa, G.Yu. Rudko, and G.G. Tarasov, Solid State Commun. **30**, 133 (1979)
31. M. Ya. Valakh, M.I. Dykman, M.P. Lisitsa, E.V. Pidlisnyi, G.Yu. Rud'ko, and G.G. Tarasov, Sov. Phys. Solid State **23**, 236 (1981)
32. M.P. Lisitsa, M. Ya Valakh, G.G. Tarasov, S.A. Boiko, A.M. Brodin, A.M. Shpak, R. Voska, and I. Feldwari, Phys. State Solid **B150**, 907 (1981)
33. L. Pelser and G. Jacobs, Phys. Rev. **188**, 1324 (1969)
34. Y. Kondo and F. Luty, Solid State Commun. **40**, 325 (1981)
35. Y. Yang and F. Luty, Phys. Rev. Lett. **51**, 419 (1983)
36. W. Gebhardt and H. Kuhnert, Phys. Letts. **11**, 15 (1964)
37. L. Gomes and F. Luty, private communication(1985)
38. M. Casalboni, P. Prosposito, and U.M. Grassano, Sol. State Commun. **87**, 305 (1993)
39. M. Jordan, H. Söthe, J.M. Spaeth, and F. Luty, Proc. Int. Conf. on Defects in Insulating Crystals, Parma, Italy, 1988, p. 11
40. H. Söthe, J.M. Spaeth, and F. Luty, Rev. Solid. State Science **4**, 369 (1990); Radiation Effects and Defects in Solids **119-121**, 269 (1991)
41. G. Liftin, Abstracts XIIth Int. Quantum Electronic Conf., Munich, Germany, 1982, p. 134
42. R.M. Macfarlane and R.M. Shelby, Phys. Rev. Lett. **42**, 788 (1979)
43. W.E. Moerner, F.M. Schellenberg, G.C. Bjorklund, P. Kaipa, and F. Luty, Phys. Rev. **B32**, 1270 (1985)
44. V.P. Chebotayev, S.I. Marennikov, and V.A. Smirnov, Appl. Phys. **B31**, 193 (1983)
45. T.T. Basiev, S.B. Mirov, and V.V. Osiko, IEEE J. Quantum. Electron. **24**, 1052 (1988)
46. G. Baldacchini et al., Experiments in $KF{:}Na^+$ are still under way
47. W. Gellermann, J. Phys. Chem. Solids **52**, 249 (1991)

ELECTRONIC AND OPTICAL PROPERTIES OF LOWER-DIMENSIONAL SEMICONDUCTOR SYSTEMS

D. A. Broido

Department of Physics
Boston College
Chestnut Hill, MA 02167 USA

ABSTRACT

The tremendous advances in electronics technology have been made possible through the study and characterization of the optical and electronic properties of novel semiconductor materials, such as quantum wells, heterojunctions and superlattices. In this paper, we review some of the basic properties of these systems.

I. SEMICONDUCTOR QUANTUM WELLS AND SUPERLATTICES

A semiconductor is characterized (at zero temperature) by a filled valence band, and an empty conduction band separated by an energy gap. In III-V semiconductors such as GaAs, the gap is direct and the topmost valence bands possess a four-fold degeneracy at the center of the Brillouin zone, where the light and heavy hole bands coincide (Fig. 1). Some of the interesting consequences of this degeneracy are focussed on below.

Semiconductor quantum wells and superlattices consist of alternating thin layers of two different kinds of semiconductor material. Figure 2a illustrates the most common quantum well system, composed of GaAs and AlGaAs. The increase in Al concentration increases the energy gap. The growth can be along different crystallographic axes, typically a [100] or [111] axis. Figure 2b illustrates the band edges as a function of the distance through the material. The smaller energy gap material (GaAs) is sandwiched between the larger gap material (AlGaAs) essentially producing a square well potential energy profile. For sufficiently thin layers of the smaller gap material such that the layer thickness is smaller than the deBroglie wavelength of the electrons or holes, carriers become localized in the narrower gap material, and quantum confinement occurs. If the larger gap material is also correspondingly thin, a superlattice band structure results. A ladder of discrete energy levels occurs above the GaAs conduction band that depend on the electron effective mass, the layer thickness, and the height of the potential well. Due to the differences in the masses of the heavy and light holes in the degenerate valence band, two distinct ladders of energy levels occur. While the effective mass in the conduction band is essentially isotropic, this is not the case in the valence band. The heavy hole mass along the [100] direction is 0.34 m_0 while along the [111] direction it is 0.75 m_0, more than twice that along [100].[1]

The band structure of the quantum well or superlattice can be further modified by biaxial strain that occurs at the interface between two lattice mismatched materials. For [100] growth, the energy gap is increased (decreased) in the layer under biaxial compression (tension) (Fig. 3a). In addition, the degeneracy in the valence bands is lifted

Nonlinear Spectroscopy of Solids: Advances and Applications
Edited by B. Di Bartolo and B. Bowlby, Plenum Press, New York, 1994

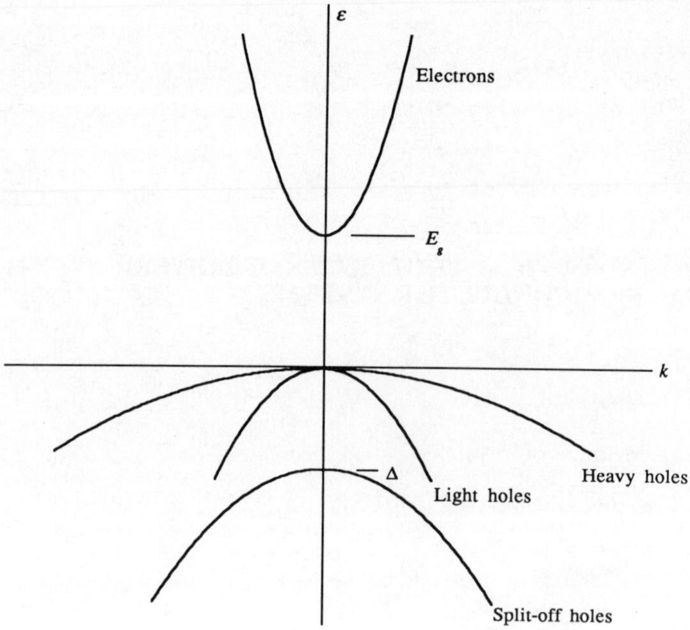

Fig. 1. Band structure of GaAs near the center of the Brillouin zone.

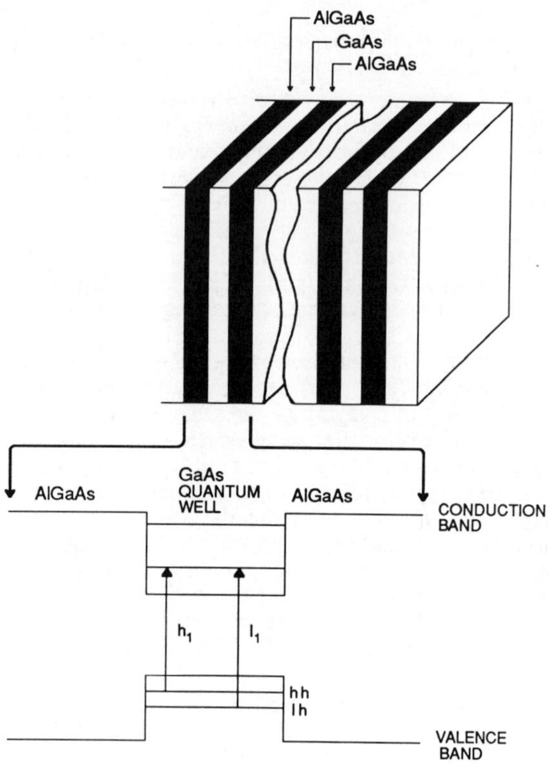

Fig 2. GaAs-AlGaAs quantum well. Inset illustrates the quantum confinement in the conduction and valence bands that results from the larger energy gap of the AlGaAs.

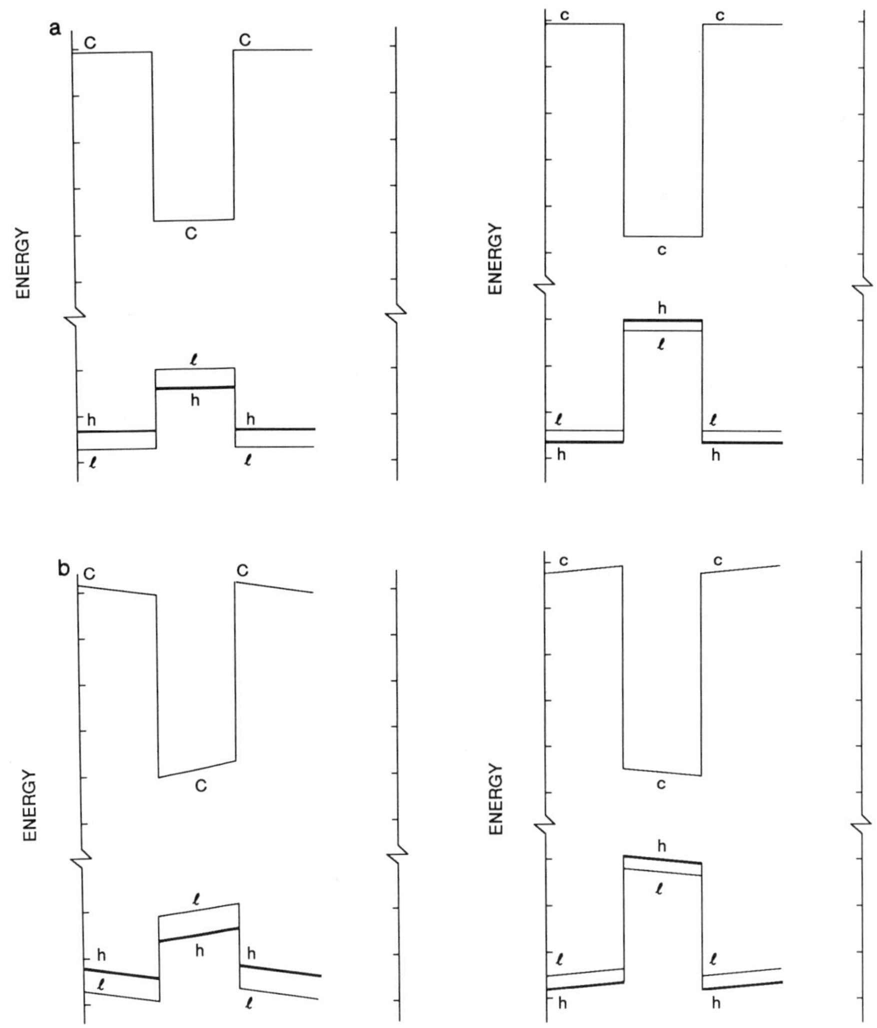

Fig. 3. a) Conduction and valence band edges for strained layer superlattice grown along the [001] direction. In the left (right) figure, the well material is under tensile (compressive) strain while the barrier material is under compressive (tensile) strain.
b) Conduction and valence band edges for strained layer superlattice grown along the [111] direction. For this growth direction, the strain induces an electric field, which tilts the band edges as shown. In the left (right) figure, the well material is under tensile (compressive) strain while the barrier material is under compressive (tensile) strain.

with the heavy hole band shifted up (down) relative to the light hole band. For growth along other crystallographic directions the piezoelectric effect occurs in which the strain induces oppositely directed electric polarizations in each layer. For [111] oriented systems the polarization and resulting electric fields are along the growth direction leading to linear potentials superimosed on the superlattice band edge profiles (Fig. 3b).[2] These internal electric fields can be screened by optical generation of electrons and holes.[2] The oppositely charged carriers move to opposite sides of the quantum well producing an opposing electric field that screens the original internal field. Such optically induced screening of piezoelectric fields has been observed and studied in detail through resonant Raman scattering experiments performed by Shanabrook et al.[3] on [111] grown GaSb/AlSb multi-quantum well systems. These experiments measured the intersubband transition energy, E_{21} as a function of optically induced carrier density. The downward shift in E_{21} with increasing carrier density provided evidence of the screening of the internal field which reduces the confinement energy. The rate at which E_{21} shifts downward depends on the efficiency of the screening carrier. Electrons and light holes possess relatively light masses and so have larger confinement energies; their wavefunctions are less affected by the presence of an electric field and therefore they are less efficient at screening the internal piezoelectric field. On the other hand, the considerably larger mass of the heavy holes causes the corresponding wavefunctions to be strongly localized in the presence of an electric field thus providing much more efficient screening. The interplay between the strain, piezoelectric field and the efficiency of the carrier screening of a given carrier type are crucial in determining the intersubband energy. Calculations[4] show that while at low carrier density the heavy holes are the dominant carrier type, as the density increases and the field is increasingly screened the light hole population increases substantially and causes the screening to become less efficient. It is essential to include this effect in order to obtain agreement with the experimentally measured intersubband energy.

II. EXCITONS IN QUANTUM WELLS

The onset of absorption in semiconductors is characterized by excitonic peaks. An exciton is formed when an electron is excited from the valence band to the conduction band and binds to the hole created in the valence band. The attractive Coulomb interaction between the electron and hole leads to an absorption spectrum consisting essentially of a discrete Hydrogenic levels below the the energy gap, E_g, and a continuum of levels above E_g. In quantum well systems, this absorption is modified due to the formation of electron and hole subbands. We consider first the case of simple nondegenerate parabolic bands. Even in the absence of excitonic effects selection rules exist. The transition strength depends of the square of electron-hole overlap matrix element. Due to the definite parity of the electron and hole wave functions, the dominant matrix elements are those for which the electron and hole subband indices are identical, and transitions between electron and hole subbands having opposite parity are forbidden. In the simplest model that includes excitonic effects the exciton is taken to derive from a single electron and single hole subband state. The exciton wavefunction is then made up of the product of electron and hole subband states and an exciton envelope function:

$$\Psi_{ij}^{ex} = F_{ij}(\rho)\psi_i(z_e)\psi_j(z_h) \tag{1}$$

$$\rho = \rho_e - \rho_h$$

Note that the in-plane motion is decoupled from the z-motion. The transition strength is determined by:

$$f_{ij} \sim |F_{ij}(\rho=0)|^2 \, |\langle\psi_i(z)\psi_j(z)\rangle|^2 \quad (2)$$

As a result of the cylindrical symmetry, the Exciton envelope function has the form of a two-dimensional hydrogenic wavefunction:

$$F_{ij}(\rho) \sim \rho^{|m|} e^{im\phi} \qquad m = 0,1,... \quad (3)$$

where the quantum number, m, characterizes the angular momentum symmetry i.e. m=0 corresponds to an exciton state with s-symmetry, m = ±1 to a state with p-symmetry, etc. It is evident from equations (2) and (3) that only those states with s-symmetry are finite at ρ=0. Two selectrion rules apply for excitons to be optically active:

1. Exciton envelope functions must have s-symmetry.
2. The electron and hole wavefunctions must have the same parity.

Taking into account the above selection rules, excitons such as 11H(s), 11L(s), and 13H(s) are allowed while 12H(s) or 11H(p,d, etc) are forbidden. The absorption spectrum in this simple model thus consists of a superposition of the absorption deriving from each subband pair. The absorption from each subband pair consists of a series of excitonic peaks with (s-symmetry) lying below a continuum of states for each pair of subbands.

There are two problems with the above simple model: 1) This model neglects the valence band mixing arising from the four-fold degeneracy at the valence band edge. 2) It neglects the intersubband scattering that occurs due to the Coulomb interaction between the electron and hole. In other words, the Coulomb interaction causes the electron and hole subband indices not to be good quantum numbers; the exciton does not derive from a single electron and hole subband pair. As a result of the intersubband scattering and the band mixing, only the lowest excitons are true bound states. All other excitons lie in the continuum and so are mixed exciton-continuum resonance states.

The valence subband and exciton mixing and can be tuned through the application of uniaxial stress. This is true because the stress shifts the heavy and light hole subbands at different rates. For stress applied along the growth axis the heavy hole subbands shift more rapidly than do the light hole subbands. For stress applied in the plane of the quantum well the opposite is true. Since the strength of the suband mixing depends in part on the energy separation between the heavy and light hole subbands the application of uniaxial stress can adjust this in a controlled manner. Evidence of mixing due to in-plane uniaxial stress was provided by photoluminescence excitation spectroscopy measurements performed by Koteles et al.[5] As the uniaxial stress is increased light hole excitons are passed through heavy hole excitons. As the lowest light hole (11L) passes in the vicinity of the onset of the 12H continuum a second peak is observed that grows in strength and then dissappears again as the light hole exction moves farther up in energy. This peak cannot be explained through the simple exciton model that ignores valence subband mixing and intersubband scattering, and a theory that includes these effects must be developed.

The Hamiltonian describing quantum well excitons can be written as:

$$H = H_e + H_h + V \quad (4)$$

where H_e describes the electrons in the quantum well, H_h describes the holes in the quantum well and includes the band mixing and uniaxial stress, and $V = e^2/\varepsilon|\mathbf{r}_e - \mathbf{r}_h|$ is the Coulomb interaction between the electron and hole. The exciton wave function is expanded in terms of electron and hole subband states:

$$\Psi_{ex} = \sum_{\lambda_e \lambda_h} \Phi^{\lambda_e}_{\lambda_h} \psi_{\lambda_e}(z_e) \psi_{\lambda_h}(z_h) \quad (5)$$

$$\lambda_e = n, \mathbf{k}_e, \sigma_e \quad , \quad \lambda_h = m, \mathbf{k}_h, \sigma_h$$

Here, n and m are the electron and hole subband indices, $\mathbf{k}_e$ and $\mathbf{k}_h$ are the in-plane wavevectors and σ_e and σ_h are the spin indices. This exansion leads to a set of coupled integral equations for the expansion coefficients:[6]

$$(\varepsilon_n(k) + \varepsilon_m(k) - \varepsilon)\Phi^l_{nm}(k) + \sum_{n'm'} \int_0^\infty V^{nn'}_{mm'}(k,k')\, \Phi^l_{n'm'}(k')dk' = 0 \quad (6)$$

Here, k is the magnitude of the relative electron-hole wavevector, and l is a quantum number associated with the conserved angular momentum of the exciton.[6]

In order to decribe the excitons observed in the experiment, it is sufficient to retain the first electron subband and the first four hole subbands (1H, 2H, 3H, and 1L). The resulting four coupled integral equations are solved using the modified quadrature method. The details of this procedure are described in Ref. 6. Figures 4 and 5 illustrate the excellent agreement obtained between theory and experiment both in terms of the exciton energies .vs. uniaxial stress and in terms of the form of the absorption spectrum in the region of strong exciton mixing.

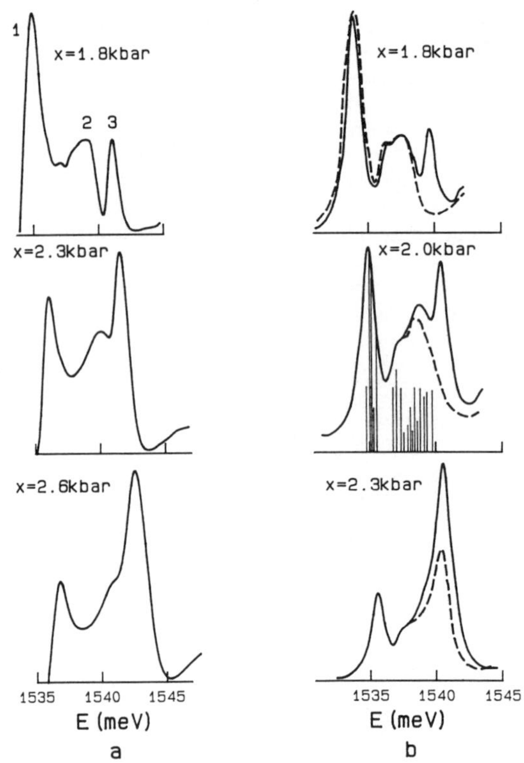

Fig 4. A portion of the experimental (a) and calculated (b) absorption spectra (arbitrary units) in the region of strong exciton mixing. The values of stress corresponding to the experiment and the theory are given in each figure. Further explanation is provided in the text.

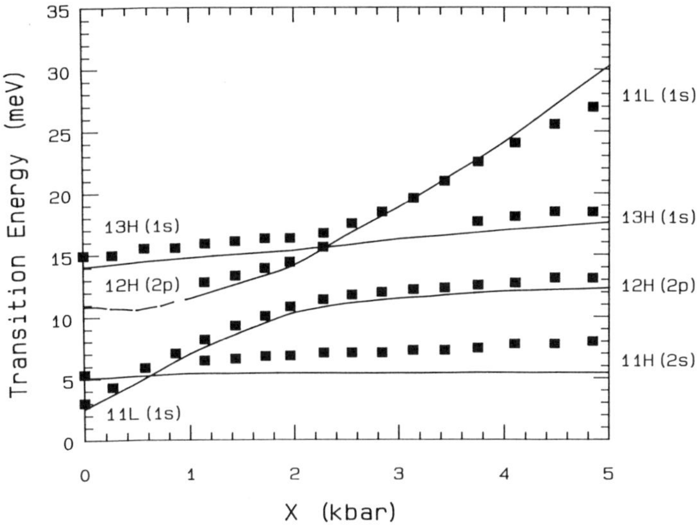

Fig 5. Comparison of several of the four lowest-lying excitonic features observed in Ref. 5 with the theory. A strong anti-crossing of the 11L(1s) and 12H(2p) excitons is evident.

III. THE FAR INFRARED SPECTRUM OF QUANTUM DOT STRUCTURES

Progress in semiconductor nanofabrication technology has led to the development of quantum dot structures, in which electrons are confined in all three spatial dimensions.[7-10] The extreme flexibility available in the fabrication of such structures is reflected in the capability of designing dots of different sizes and shapes, and in controlling the strength of the confinement independently from number of electrons confined.[7] Quantum dots can thus be viewed as artificially created atoms whose properties can be precisely controlled.

A common experimental method to probe the properties of electrons confined in quantum dots is far infrared (FIR) transmission spectroscopy measurements.[7-10] Since the confinement of electrons in quantum dots is analogous to that in atoms, it was expected that the FIR response of electrons in quantum dots would exhibit a rich structure. Furthermore, because of the complicated electron-electron interactions that take place, it was anticipated that the response should also depend on the number of electrons in the dot. However, recent measurements[7-10] have shown that the FIR response exhibits only a single strong peak. In addition, the far-infrared absorption frequency corresponding to this peak is essentially independent of the number of electrons in the structure.

It has been proposed[11,12] that the reason for these puzzling experimental results lies in the electronic confining potential of the nanostructure. If the potential that confines the electrons is assumed to be parabolic, then it has been shown that the electromagnetic radiation couples only to the center of mass of the entire many electron system. As a result, all the electrons in each dot respond as a single quasi-electron oscillating in a parabolic potential. The single absorption peak observed in experiment, is the single characteristic frequency of this quasi-electron in its assumed parabolic confining potential. Thus the frequency response of the electrons no longer depends on the electron-electron interaction.

Theoretical explanations of the single peak response in quantum dots have been based on the assumption that the confining potential for electrons is parabolic. If the confining potential for electrons is assumed to be parabolic then the full, many-electron Hamiltonian can be separated exactly into two distinct components depending on the center of mass and relative coordinates, respectively. This fact can be seen clearly by considering the Hamiltonian for a quantum dot with N electrons with a 2D parabolic confinement potential, which is given by

$$H=\sum_{i=1}^{N}\left[\frac{\mathbf{p}_i^2}{2m}+\tfrac{1}{2}m\omega_0^2\mathbf{r}_i^2\right]+\sum_{i>j}u(r_{ij}), \tag{7}$$

$$\mathbf{r}_{ij}=\mathbf{r}_i-\mathbf{r}_j,\quad r_{ij}=\left|\mathbf{r}_{ij}\right|,\qquad \mathbf{p}_i^2=\mathbf{p}_{xi}^2+\mathbf{p}_{yi}^2,\text{ and } \mathbf{r}_i^2=\mathbf{x}_i^2+\mathbf{y}_i^2.$$

Here $u(r_{ij})$ represents the coulomb potential that results among the many electrons. We note that the identity for the quadratic form of the momentum,

$$\sum_i \mathbf{p}_i^2=\frac{1}{N}\mathbf{P}^2+\frac{1}{N}\sum_{i>j}\mathbf{p}_{ij}^2,\quad \mathbf{p}_{ij}=\mathbf{p}_i-\mathbf{p}_j,\quad \mathbf{P}=\sum_i \mathbf{p}_i \tag{8}$$

completely separates the kinetic-energy term into center of mass and relative momenta (**P** and $\mathbf{p}_{ij}$), and the identity for the quadratic form of the position vector

$$\sum_i \mathbf{r}_i^2=N\mathbf{R}^2+\frac{1}{N}\sum_{i>j}\mathbf{r}_{ij}^2,\quad \mathbf{R}=\frac{1}{N}\sum_i \mathbf{r}_i \tag{9}$$

separates the parabolic confining potential into center of mass and relative coordinates (**R** and $\mathbf{r}_{ij}$). The electron-electron interaction term, $u(r_{ij})$, obviously only depends on the relative coordinates. Thus the separation is complete and we have explicitly

$$H=H_{cm}+H_{rel} \tag{10}$$

where,

$$H_{cm}=\frac{\mathbf{P}^2}{2Nm}+\tfrac{1}{2}Nm\omega_0^2\mathbf{R}^2, \tag{11a}$$

$$H_{rel}=\frac{1}{N}\sum_{i>j}\left[\frac{\mathbf{p}_{ij}^2}{2m}+\tfrac{1}{2}m\omega_0^2\mathbf{r}_{ij}^2+Nu(r_{ij})\right]. \tag{11b}$$

As a result of this complete separation of the Hamiltonian, we can express the wave function of the system of electrons (neglecting spin) into the product of two functions, one which depends only on the center of mass coordinates, and the other which depends only on the relative coordinates:

$$\Psi_{n,\{m\}}(\mathbf{R},\{\mathbf{r}_{ij}\})=\psi_n(\mathbf{R})\chi_{\{m\}}(\{\mathbf{r}_{ij}\}). \tag{12}$$

where n is the quantum number associated with the center of mass coordinate, and {m} are the quantum numbers associated with the relative coordinates of the many electrons. Since the long wavelength electromagnetic radiation will only couple with the center of mass of the system, only the center of mass will respond to the incident radiation. The complicated interaction between electrons, which depends on relative coordinates only, will not affect the response to the incident radiation. As a result, all the electrons in each dot respond as a single quasi-electron oscillating in a parabolic potential. Therefore, due to the unique separation associated with parabolic confining potentials, the frequency of the absorption peak corresponds to the frequency at which the center of mass of the electron system will oscillate in the confining potential, and this frequency is just the frequency associated with the bare confining potential in the absence of electron-electron interactions.

REFERENCES

1. B. V. Shanabrook, O. J. Glembocki, D. A. Broido, and W. I. Wang, Phys. Rev. B **39**, 3411, (1989).
2. D. L. Smith and C. Mailhot, Rev. Mod. Phys. **62**, 173 (1990), and references therein.
3. B. V. Shanabrook, D. Gammon, R. Beresford, W. I. Wang, R. P. Leavitt and D. A. Broido, in Proc. 20th Int.'l Conf. on the Physics of Semiconductors, Thesseloniki, pg. 901 (1990).
4. D. A. Broido, B. V. Shanabrook and D. Gammon, Final Program and Abstracts of Fall, 1990 Meeting of the Materials Research Society, Boston, pg. 58 (1990).
5. E. S. Koteles, C. Jagannath, J. Lee, Y. J. Chen, B. S. Elman, and J. Y. Chi, in Proceedings of the 18th International Conference on the Physics of Semiconductors, Stockholm, 1986, edited by O. Engstrom (World Scientific, Singapore, 1987).
6. D. A. Broido And S.-R. E. Yang, in Physical Review B **42**, pg. 11051 (1990).
7. Ch. Sikorski and U. Merkt, Phys. Rev. Lett. **62**, 2164 (1989).
8. J. Alsmeier, E. Batke, and J. P. Kotthaus, Phys. Rev. B **41**, 1699 (1990).
9. A. Lorke, J. P. Kotthaus and K. Ploog, Phy. Rev. Lett. **64**, 2559 (1990).
10. T. Demel, D. Heitmann, P. Grambow and K. Ploog, Phys. Rev. Lett. **64**, 788 (1990).
11. P. A. Maksym and T. Chakraborty, Phys. Rev. Lett. **65**, 108 (1990).
12. P. Bakshi, D. A. Broido and K. Kempa, Phys. Rev. B**42**, 7416 (1990).

RECOMBINATION DYNAMICS AND NONLINEAR SPECTROSCOPY OF QUANTUM DOTS

U. Woggon

Fachbereich Physik, Universität Kaiserslautern
E.-Schrödinger-Str., D-67663 Kaiserslautern
Germany

ABSTRACT

In a semiconductor material the excitation with light of energy higher than the bandgap results in the formation of a nonequilibrium concentration of electron-hole pairs. Returning back to the ground state, these excited carriers lose their phase coherence with the exciting light and their excess energy by relaxation and recombination processes, respectively. The analysis of these processes is an interesting subject of research. A great variety of experimental methods have been developed to determine the lifetime and the dephasing time. In this contribution some examples for the application of nonlinear optical methods, like time-resolved nonlinear absorption, spectral hole-burning, nondegenerate four-wave mixing, are given for the characterization of three-dimensionally confined semiconductor quantum dots. In particular, the usefulness of these methods both for the basic research as well as for the optimization of growth technologies of quantum dots will be shown. We concentrate on the materials CdS and CdSe embedded in glass. In comparison to the nonlinear optical methods, the results obtained by time-resolved luminescence will be discussed.

I. INTRODUCTION

Quantum dots (QD's) are three-dimensionally confined systems realized e.g. in II–VI semiconductors mostly by the inclusion of small spherical semiconductor nanocrystals in a dielectric matrix. The confining potential at the interface gives rise to a series of discrete, atomic-like energy levels for the electron-hole-pair states of the semiconductor. In recent years substantial progress has been achieved in the theoretical description of the quantum dot energy states and wave functions [1 – 5]. Attention has been paid to the more complex bandstructure of the semiconductor materials involving the nonparabolicity of the conduction band and the Luttinger parameters for the description of the valence band. At present, one of the most intensively discussed problems is the

Nonlinear Spectroscopy of Solids: Advances and Applications
Edited by B. Di Bartolo and B. Bowlby, Plenum Press, New York, 1994

influence of the interface configuration on the wave functions and charge distributions of the confined electrons and holes. Carrier localization and fast trapping processes seem to be essential both for the understanding of the basic electronic properties and the further application.

Nonlinear spectroscopy is a widely used experimental method to investigate the electronical and optical properties of quantum-confined semiconductors. For a first principle discussion of the experiments, the concept of the two-level system has been often used as the most simple model. The determination of the saturation intensity, of the nonlinear change of the refractive index, or of the recovery time of the nonlinear optical response is interesting for the comparison of different materials with respect to their possible application in optical switching devices. The investigation of the homogeneous linewidth and of the dephasing processes of the quantum-confined structures by nonlinear spectroscopy is an important problem for the basic research and for the application in ultrafast coherent nonlinear optics.

The analysis of the experimental results obtained by nonlinear spectroscopy requires the knowledge of the basic principles of the light–matter interaction for strong field strengths of the exciting electromagnetic wave. Therefore the paper is organized as follows: At first, we introduce the basic principles of the semiclassical description of the interaction light – matter for the case of strong electromagnetic fields. Because quantum dots possess discrete electronic states well separated in energy, the results expected within the two-level model are discussed and it will be shown, where the description has to go beyond this model. Then we present some selected experiments of nonlinear spectroscopy based on standard ns- and ps-laser systems performed at II – VI quantum dots and discuss, how we can analyse e.g. the technology process of growth of the semiconductor quantum dots. We summarize some experimental results obtained by nonlinear spectroscopy concerning the recently discussed problems of interface polarization and of mechanisms of homogeneous linebroadening. Finally, we report on time-resolved luminescence and compare the results for the electron-hole pair recombination dynamics derived from the different methods.

II. FUNDAMENTALS

The description of the nonlinear optical properties of matter combines in a semiclassical treatment the classical treatment of light with the quantum mechanical treatment of the medium (see [6 – 10]). The result of this analysis are coupled differential equations for the polarization of the electromagnetic field and for the population difference of the medium. Looking for characteristic parameters close to the experiments, the longitudinal and transversal relaxation times T_1 and T_2 or the saturation density N_S can be introduced and will be explained in the following.

In Maxwell's theory the propagation of an electromagnetic wave in z-direction inside a macroscopic medium (in absence of currents) results in

$$\left(\frac{1}{c^2}\frac{\partial^2}{\partial t^2} - \frac{\partial^2}{\partial z^2}\right)\mathbf{E}(z,t) = -\mu_0\frac{\partial^2}{\partial t^2}\mathbf{P}(z,t). \tag{1}$$

The classical macroscopic polarization is given by

$$\mathbf{D} = \epsilon_0\mathbf{E} + \mathbf{P}. \tag{2}$$

At low field strengths the linear interaction of the electromagnetic field with the medium produces a polarization $\mathbf{P}$ which is reflected by those phenomena like dispersion, refraction etc. However, the effect of the polarization of the medium back to the electromagnetic wave is hardly to describe within the classical concept. Nonlinear optical effects were introduced by a power expansion of the polarization based on the anharmonic oscillator model

$$\mathbf{P} = \epsilon_0 \chi \mathbf{E} + \epsilon_0 \chi^{(2)} \mathbf{E}^2 + \epsilon_0 \chi^{(3)} \mathbf{E}^3 + \dots . \tag{3}$$

More realistic and suited for analysing e.g. saturation effects and other incoherent or coherent phenomena under the action of a strong electromagnetic field, is the introduction of discrete levels (obtained by quantum mechanics) and their population. The two-level system is an atomic like ansatz going beyond the classical theory and considers discrete energy states. It should be applicable in three-dimensionally confined semiconductor systems exhibiting discrete energy states. The two-level model consists of two possible energy states E_1 and E_2 and the corresponding wavefunctions ψ_1 and ψ_2 interacting with light of the energy $\hbar\Omega = E_2 - E_1$. The solution of the time-dependent Schrödinger equation

$$i\hbar \frac{\partial \psi(\mathbf{r}, t)}{\partial t} = \hat{H} \psi(\mathbf{r}, t) \tag{4}$$

is then obtained by time-dependent perturbation theory with

$$\hat{H} = \hat{H}_0 + \hat{H}_W. \tag{5}$$

$\hat{H}_0$ is the time-independent part of the Hamiltonian giving the solution for the two-level system without the interaction with light. In dipole approximation the interaction operator $\hat{H}_W$ is equivalently to

$$\hat{H}_W = -d\, \mathbf{E}(z, t). \tag{6}$$

with the dipole moment d. For the following considerations, the light wave is assumed to be linearly polarized. The ansatz for the wave functions is

$$\psi(t) = c_1(t)\psi_1^0 + c_2(t)\psi_2^0 \tag{7}$$

where the $|c_i(t)|^2$ are the probabilities of finding the system in the state i. ψ_i^0 are the solutions of $\hat{H}_0$. The equations for the average polarization $< \mathbf{P} >$ and population difference $< \Delta n >$, the macroscopic measurable quantities, are given by the general expression for the quantum mechanical expectation value of the operator $\hat{A}$

$$< \hat{A} > = \mathrm{Tr}(\rho \hat{\mathrm{A}}) \tag{8}$$

where ρ is the density matrix. For the polarization, we obtain

$$< \mathbf{P} > = N < \hat{d} > = N\, \mathrm{Tr}(\rho \hat{\mathrm{d}}), \tag{9}$$

with N the dipole density and $< \hat{d} >$ is the quantum mechanical average of the operator of the dipole moment. For the population difference holds

$$< \Delta n > = n_2 - n_1 = \mathrm{Tr}(\rho \Delta \mathrm{n}). \tag{10}$$

n_2 is the population of state 2 and n_1 of state 1, respectively. The density operator is defined as the projector onto the attached state vectors $|\psi>$

$$\rho = |\psi >< \psi| \tag{11}$$

and in matrix notation therefore equivalently to

$$\rho = \begin{pmatrix} \rho_{11} & \rho_{12} \\ \rho_{21} & \rho_{22} \end{pmatrix} = \begin{pmatrix} |c_1|^2 & c_1^* c_2 \\ c_2^* c_1 & |c_2|^2 \end{pmatrix}. \tag{12}$$

The further treatment for the discussion of the interaction of light with matter is then the solution of the equation of motion for the density matrix. For the matrix elements of ρ the Schrödinger representation yields

$$\begin{aligned} \tfrac{d}{dt}\rho_{kk} + \tfrac{1}{T_k}(\rho_{kk} - \rho_{kk}^e) &= \tfrac{1}{i\hbar}[\hat{H}_W, \rho]_{kk} \\ \tfrac{d}{dt}\rho_{kl} + i\Omega\rho_{kl} + \tfrac{1}{T_{kl}}\rho_{kl} &= \tfrac{1}{i\hbar}[\hat{H}_W, \rho]_{kl} \end{aligned} \tag{13}$$

where ρ_{kk} and ρ_{kl} are the diagonal and nondiagonal elements, respectively. The energy $\hbar\Omega = E_2 - E_1$ corresponds to the transition energy from 1 to 2. Before obtaining the equation (13), one has to consider the influence of an external dissipative reservoir on the two-level atoms. This can be done by introducing an additional term into the equation of motion of ρ containing the relaxation processes (see e.g. [10]). $T_k := T_1$ is the energy relaxation time for the transition between 1 and 2 and comprises the transition probability. The diagonal elements of the density matrix ρ_{kk} converges to the thermodynamic equilibrium value for $t \longrightarrow \infty$. $T_{kl} := T_2$ is the phase relaxation time, representing the damping of the nondiagonal elements. The nondiagonal elements ρ_{kl} are going to zero for $t \longrightarrow \infty$, not necessarily connected with a change in energy. To obtain the macroscopic equations for $\mathbf{P}$ and Δn we can directly solve the equation of motion of the density matrix and determine the average values by $< \hat{A} >= \text{Tr} < \rho\hat{A} >$. Alternately, we can derive by further transformations from the equation of motion of ρ a system of differential equations for $\mathbf{P}$ and Δn. Going this way, we can write the equation of motion for $< \Delta n >$ and $< \hat{d} >$ with

$$\hat{H}_0 = \begin{pmatrix} E_1 & 0 \\ 0 & E_2 \end{pmatrix} \tag{14}$$

$$\hat{H}_W = -\begin{pmatrix} 0 & d_{12}\mathbf{E}(z,t) \\ d_{21}\mathbf{E}(z,t) & 0 \end{pmatrix} \tag{15}$$

giving

$$\frac{\partial}{\partial t} < \Delta n >= \frac{1}{i\hbar} < [\Delta n, \hat{H}] > - \frac{1}{T_1}(\Delta n - \Delta n_e). \tag{16}$$

$$\frac{\partial}{\partial t} < \hat{d} >= \frac{1}{i\hbar} < [\hat{d}, \hat{H}] > - \frac{1}{T_2} < \hat{d} > \tag{17}$$

Δn_e is the value of the population difference in the thermodynamic equilibrium. Inserting $\hat{H} = \hat{H}_0 + \hat{H}_W$, calculating the expressions for the commutators, and using the relation $\mathbf{P} = N < \hat{d} >$, we find the system of the optical Bloch equations given by Eq. (18) – (20) for the population difference and the polarization. They describe the coupling between the electromagnetic field and the matter, where the latter is represented by an atomic two-level system.

$$\frac{\partial}{\partial t}\Delta n + \frac{1}{T_1}(\Delta n - \Delta n_e) = \frac{2\,\mathbf{E}(z,t)}{\hbar\Omega\,\epsilon_0}\left[\frac{\partial}{\partial t}\mathbf{P}(z,t) + \frac{\mathbf{P}(z,t)}{T_2}\right] \tag{18}$$

$$\left(\frac{\partial^2}{\partial t^2} + \frac{2}{T_2}\frac{\partial}{\partial t} + \frac{1}{T_2^2} + \Omega^2\right)\mathbf{P}(z,t) = -\frac{2\Omega\epsilon_0}{\hbar}|d_{12}|^2\,\mathbf{E}(z,t)\,\Delta n \tag{19}$$

$$\left(\frac{1}{c^2}\frac{\partial}{\partial t^2} - \frac{\partial^2}{\partial z^2}\right)\mathbf{E}(z,t) = -\mu_0\frac{\partial^2}{\partial t^2}\mathbf{P}(z,t) \tag{20}$$

Eq. (20) is again the well known Maxwell wave equation (1). This system of differential equations introduces an equation of motion of the population and contains the action and reaction of field and matter.

By means of laser excitation of high intensity and/or short pulses, it is possible with the aid of Eq. (18) – (20) to investigate the time- and intensity-dependent changes of $\mathbf{P}$ and Δn as well as the reaction to the incident electromagnetic wave. For $t \longrightarrow \infty$ the thermodynamic equilibrium is achieved with $\Delta n \longrightarrow \Delta n_e$, $\mathbf{P} \longrightarrow 0$.

Eq. (18) for the population difference Δn consists of the recombination and generation term. This equation is the starting point for so called kinetic models using rate equation systems as shown later. $\mathbf{E} \cdot \partial\mathbf{P}/\partial t$ is the energy, given to the medium by the incident field per second and cm^3. If the electromagnetic field is resonant to the electronic state ($\Omega \approx \omega$), the division by $\hbar\Omega$ gives the number of photons, where the factor 2 considers the change of population difference by absorbing one photon. The term $\mathbf{P}/T_2$ considers the part of polarization which is lost by the relaxation and therefore does not result in a change in the population. It can be neglected for most of the problems.

The equation of polarization (19) is reminiscent of an equation of forced oscillations, whereas the coupling constant between $\mathbf{E}$ and $\mathbf{P}$, the population difference $\Delta n(z,t)$, is time-dependent.

The Eq. (20) is useful for approximations by means of the slowly varying envelope approximation (SVEA).

If we describe e.g. the electromagnetic field by

$$\mathbf{E}(z,t) = \frac{1}{2}\left[\mathbf{E}_0(z,t)e^{i(kz-\omega t)} + c.c.\right] \tag{21}$$

and the polarization by

$$\mathbf{P}(z,t) = \frac{1}{2}\left[\mathbf{P}_0(z,t)e^{i(kz-\omega t)} + c.c.\right] \tag{22}$$

and, going into the equation of polarization assuming $\Omega \gg 1/T_2^2$, we get

$$\left(-\omega^2 - \frac{2}{T_2}i\omega + \Omega^2\right)\frac{\mathbf{P}_0(z,t)}{2} = \epsilon_0\,\frac{-2\Omega}{\hbar}\,\Delta n\,\frac{\mathbf{E}_0(z,t)}{2}\,|d_{12}|^2. \tag{23}$$

From (23) follows

$$\mathbf{P}_0(z,t) = \epsilon_0\frac{2\,(E_1 - E_2)}{\hbar^2}\,\frac{|d_{12}|^2}{(\Omega^2 - \omega^2 - i\,\frac{2}{T_2}\omega)}\,\Delta n\,\mathbf{E}_0(z,t). \tag{24}$$

From these equations the Lorentzian line shape should be obtained in the limit of linear optics . If we define $\Gamma := 2/T_2$ and take

$$\frac{\Delta n}{\Delta n_e} = 1, \tag{25}$$

representing the limit of linear optical behaviour, we obtain

$$\mathbf{P}_0(z,t) = \epsilon_0 \underbrace{\frac{2\,(E_1 - E_2)}{\hbar^2} \frac{|d_{12}|^2 \Delta n_e}{\Omega^2 - \omega^2 - i\omega\Gamma}}_{\chi(\omega)} \mathbf{E}_0(z,t) \tag{26}$$

with

$$|d_{12}|^2 = e^2 |< \psi_1 | \mathbf{r} | \psi_2 > |^2 \tag{27}$$

and

$$\chi(\omega) = \epsilon(\omega) - 1. \tag{28}$$

Thus we finally have the well known formula for the dielectric function corresponding to

$$\epsilon(\omega) = 1 + \underbrace{\frac{\Delta n_e e^2}{m}}_{\omega_p^2} \underbrace{\frac{m\,(2\Omega)}{\hbar} | < \psi_1 | \mathbf{r} | \psi_2 > |^2}_{f_{12}} \frac{1}{\Omega^2 - \omega^2 - i\omega\Gamma} \tag{29}$$

giving

$$\epsilon(\omega) = 1 + \omega_p^2 \frac{f_{12}}{\Omega^2 - \omega^2 - i\omega\Gamma} \tag{30}$$

The result is, that the phase relaxation time T_2 is directly related to the homogeneous linewidth Γ for a Lorentzian absorption spectrum. T_1 is the decay time of the population difference into its equilibrium value.

Before discussing experiments for the determination of these parameters and their application to quantum dots, we consider in particular the effects of saturation and the saturation density (see also [10]). In the stationary case and SVEA, for the population difference we get

$$\frac{\partial}{\partial t} \Delta n + \frac{1}{T_1} (\Delta n - \Delta n_e) = \frac{1}{\hbar} \,\mathrm{Im}(\mathbf{P}_0^* \mathbf{E}_0). \tag{31}$$

With the absorption coefficient $\alpha = \omega/c\ \mathrm{Im}\chi(\omega)$, the absorption cross section $\sigma = -\alpha/\Delta n_e$, and the photon density $N(z,t)$, Eq. (31) can be written as

$$\frac{\partial}{\partial t} \Delta n + \frac{1}{T_1} (\Delta n - \Delta n_e) = -2\sigma \Delta n\, N(z,t). \tag{32}$$

For $\partial/\partial t\, \Delta n = 0$ we get

$$\Delta n = \frac{\Delta n_e}{1 + N\, 2\sigma T_1} = \frac{\Delta n_e}{1 + N/N_S} \tag{33}$$

With increasing light intensity the transition from 1 to 2 saturates for $N = N_S$ corresponding to $\Delta n = \Delta n_e/2$. For small intensities ($N \ll N_S$), the absorption is linear and $\Delta n \approx \Delta n_e$. At high intensity ($N \gg N_S$), the difference of the population of both states goes to zero and no further absorption process takes place. The saturation density $N_S = 2\,\sigma\, T_1$ is only determined by the properties of the two-level system and thus a further characteristic parameter for the nonlinear optical behaviour.

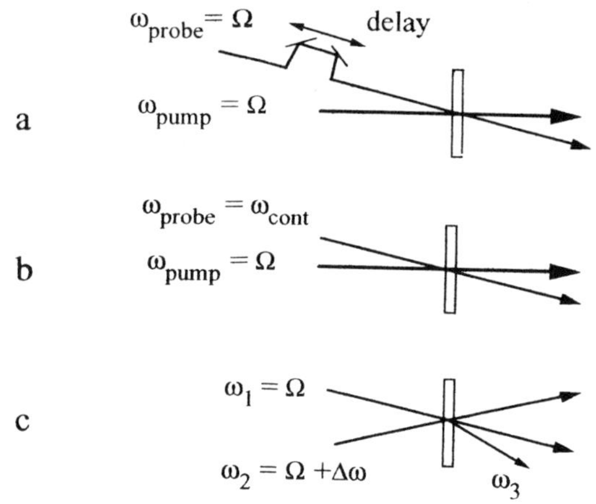

Fig. 1. *Methods of nonlinear spectroscopy. Pump- and probe configuration with the probe beam a) delayed in time, b) being a spectrally broad continuum, and c) Nondegenerate four wave mixing*

Summarizing, we have explained the quantities T_1, T_2, Γ, and the saturation density N_S as characterizing the interaction of light with matter in the most simple model of the two-level system. Their investigation is a main subject of nonlinear spectroscopy.

III. NONLINEAR SPECTROSCOPY OF QUANTUM DOTS – REPRESENTATIVE EXPERIMENTS

The experiments of nonlinear spectroscopy are directed to the investigation of the nonlinear change in the dielectric function $\epsilon(\omega, I)$ of the semiconductor represented by the change in the absorption coefficient or in the refraction index. Three examples of corresponding experimental methods are shown in Fig. 1. In Fig. 1 a) and b) in the pump-probe configuration the absorption of the sample can be measured in dependence on the intensity of the exciting pump beam by a probe beam of low intensity and e.g. a) delayed in time at a fixed energy or b) as a spectrally broad continuum measuring the whole spectrum. A special arrangement of a) is the single-beam experiment where the nonlinearity is tested by the intensity dependence of the transmission of the exciting light beam itself without the second probe beam. By the method of light induced gratings (Fig. 1c) the nonlinear change in the refraction index is determined by measuring the efficiency of the first order diffracted beam. In Fig. 1 c) a selfdiffraction experiment is considered using two intensive pump beams of different energies, the first resonant to the investigated electronic state whereas the second is tunable through the resonance (Nondegenerate Four Wave Mixing (NDFWM)).

Consider now three-dimensionally confined systems, where spherical quantum dots are embedded in a matrix. Assuming a low filling factor p

$$p = \frac{V_{QD}^{total}}{V_{glass}} \leq 0.01, \tag{34}$$

effects of differences in the dielectric constants of the semiconductor and of the host medium can be neglected in a first approximation. The single quantum dot absorption has been supposed to have a Lorentzian lineshape with the size dependent oscillator strength $f_{QD}(R)$ and energy states $E_{QD}(R)$

$$\alpha_{QD}(\omega, R) = C\ f_{QD}(R)\ \frac{\hbar\Gamma/2}{(E_{QD}(R) - \hbar\omega)^2 + (\frac{\hbar\Gamma}{2})^2}. \tag{35}$$

If you know the radius R of the dots, the size distribution function $S(R)$ and the size dependence of the wave functions and energy states the linear absorption spectrum of quantum dots can then be calculated by

$$\alpha(\omega, R) = p\ \frac{1}{\bar{V}_{QD}} \int dR \frac{4\pi}{3} R^3\ S(R)\ \alpha_{QD}(\omega, R). \tag{36}$$

The nonlinear change in absorption for a single quantum dot, increasing the intensity from $I = 0$ to I, is defined as

$$\Delta\alpha_{QD}(\omega, I) = \frac{1}{l} \int_0^l \alpha_{QD}(\omega, I)\ dx\ -\ \alpha_{QD}(\omega, 0) \tag{37}$$

with integration over the sample thickness l.

Considering in a first step the case of resonant excitation at a frequency $\omega_{pump} = \Omega = \omega_{probe}$. For dominant homogeneous broadening ($S(R) \approx \delta(R - R_0)$) the saturation of the quantum dot absorption is given by

$$\alpha(\Omega, I) = \alpha_0(\Omega, R_0)\ \frac{1}{1 + \frac{I}{I_S}} \tag{38}$$

with

$$I_S = \hbar\omega \frac{N_S}{\alpha_0\, T_1}, \tag{39}$$

whereas, in the opposite limit of dominant inhomogeneous broadening and $S(R)$ symmetrically around Ω with the maximum at R_0

$$\alpha(\Omega, I) = \alpha_0(\Omega, R_0)\ \frac{1}{\sqrt{1 + \frac{I}{I_S}}}. \tag{40}$$

The expected changes in transmission are illustrated in Fig. 2. In the two-level model the switching from the opaque state at low intensities to the transparent state of transmission T = 1 proceeds in a relatively narrow range of intensities of about one order of magnitude. The saturation intensity I_S determines the onset of noticeable bleaching. From its analysis the lifetime T_1 and the absorption cross section σ can be obtained. When adding e.g. relaxation and recombination processes involving a third or even more levels, the main result is a more gradual rise of the saturation curve with increasing intensity (Fig. 2, curve 3). Also for an inhomogeneously broadened two-level system a slower change in transmission can be observed due to the excitation into the tails of the neighbouring resonances.

The relation of homogeneous to inhomogeneous broadening is also essential for discussion of experiments measuring the spectrally resolved nonlinear absorption (configuration b) of Fig. 1). The pump frequency is resonant to a single quantum dot

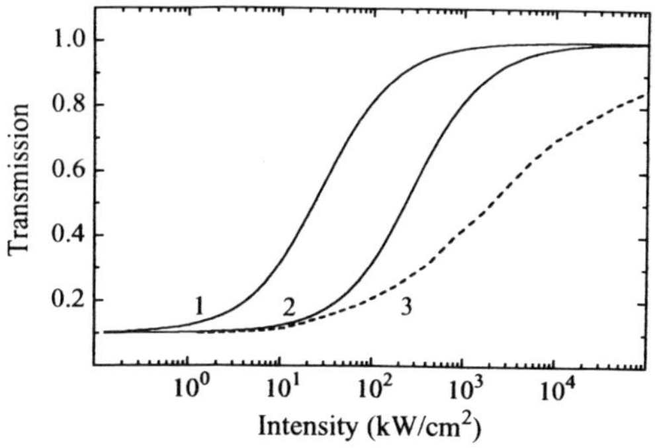

Fig. 2. *Transmission change according to the model of the homogeneously broadened two-level system with 1)* $I_S = 10\ kW/cm^2$, *2)* $I_S = 100\ kW/cm^2$, *3) more than two levels*

level $\omega_{pump} = \Omega_i$ whereas the probe beam ω_{probe} is a broad dye laser continuum. A homogeneously broadened Lorentzian line saturates uniformly over all frequencies. In a strongly inhomogeneous broadened system, represented in quantum dots e.g. by a wide distribution in sizes or in the shape of the confining potential for different dots, the change in α for the two-level system and Lorentzian line shape is

$$\Delta\alpha(\omega, R, \omega_{exc}, I_{exc}) =$$

$$-\frac{p}{\bar{V}_{QD}} \int_0^\infty dR \, \frac{4\pi}{3} R^3 \, S(R) \, \alpha_{QD}(\omega, R) \, \frac{\frac{I_{exc}}{I_S(\omega_{exc})}}{1 + \frac{I_{exc}}{I_S(\omega_{exc})}} \tag{41}$$

$$I_S(\omega_{exc}) = I_S(\Omega) \, \frac{\frac{\hbar^2}{4}\Gamma^2}{(\hbar\Omega - \hbar\omega_{exc})^2 + \frac{\hbar^2}{4}\Gamma^2} \tag{42}$$

$\Delta\alpha$ is a function of the excitation frequency. The maximum of $\Delta\alpha$ shifts with the pump energy. The width of the "burnt hole" is connected with the homogeneous linewidth Γ of the Lorentzian and can be exactly determined from the change of $\Delta\alpha$ if $S(R)$ and $E_{QD}(R)$ are known and the saturation follows the two-level behaviour. At not too high excitation intensities the homogeneous linewidth can be estimated measuring the spectral width of the homogeneous bleaching. The typical change in the absorption spectrum in the two limits of dominant homogeneous and inhomogeneous broadening is shown in Fig. 3, 4. Thus the investigation of the spectral change in $\Delta\alpha$ by a broad probe-beam continuum, e.g. in dependence on intensity, temperature, or on the parameter of the growth process, gives information about Γ and by this about the time T_2 and the dephasing mechanisms.

A further method to investigate the dephasing time is the Nondegenerate Four Wave Mixing (NDFWM). Two laser beams of different frequencies $\omega_1 \neq \omega_2$ and different wave vectors $\vec{k}_1 \neq \vec{k}_2$ produce a nonstationary intensity distribution with a temporal modulation of the maximum intensity with a frequency $\omega_3 = 2\omega_1 - \omega_2$. For comparison, the case $\omega_1 = \omega_2$, where the interference pattern is stationary, is nothing as the degenerate case or the well known method of light induced gratings. In DFWM the spatial periodic stationary intensity distribution produces inside the sample a periodic modulation in the excited carrier density and by this effect also in the refraction

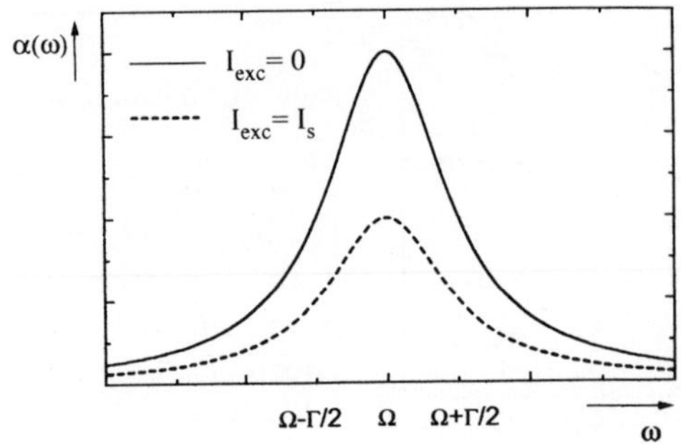

Fig. 3. *Saturation of a Lorentzian in the case of dominant homogeneous broadening,* $\Gamma_{hom} \approx \Gamma_{inhom}$

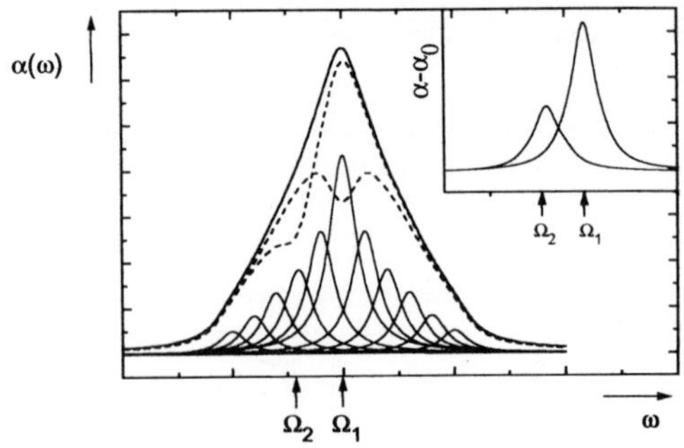

Fig. 4. *Saturation for the case of dominant inhomogeneous broadening,* $\Gamma_{hom} \ll \Gamma_{inhom}$

index and absorption coefficient. If light is now passing the sample, a diffraction pattern is found with a diffraction efficiency proportional to the contrast of the density modulation giving insight into e.g. the mechanisms of the nonlinearity or the diffusion constants or the lifetime of the excited carriers. If the frequency of one of the pump beam is detuned by a small difference $\Delta\omega$ (NDFWM) compared to the other one, then the wave-like modulated light intensity results in a "moving grating" structure. To obtain now maximum diffraction efficiency, the phase of the excited carriers forced by the incoming light field has to follow the modulation frequency of the moving grating. The strength of the decrease of the diffraction efficiency with increasing detuning is therefore a measure for the dephasing time T_2. Nondegenerate Four Wave Mixing is a useful tool for investigating isolated quantum dots in glasses because the diffusion processes are of minor importance, the expected T_2 times lie in the range below 1 ps and are also small compared to the lifetime T_1.

The theoretical description of the NDFWM for resonant excitation of an inhomogeneously broadened optical transition has been developed in [11, 12]. Starting directly from the equation of motion for the density matrix (Eq. 13)), ρ has been calculated in perturbation theory by an expansion in a power series of different orders of perturbation. For the output light intensity I at $\omega_3 = 2\omega_1 - \omega_2$ holds $I \sim |\chi^{(3)}|^2$

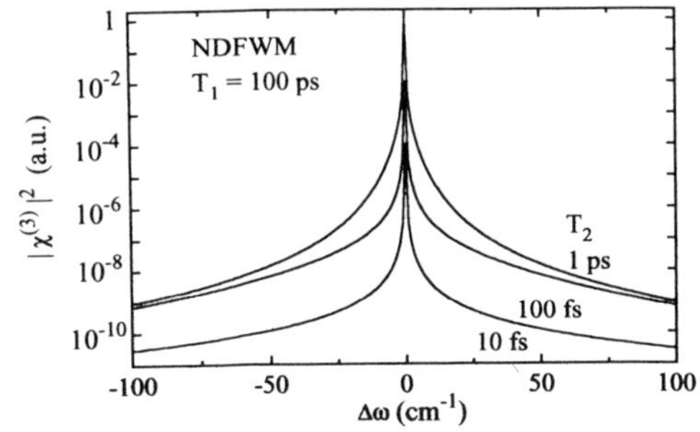

Fig. 5. *Intensity of the first diffracted order at the frequency ω_3 as a function of detuning $\Delta\omega = \omega_1 - \omega_2$ for different dephasing times T_2 = 1 ps, 100 fs and 10 fs and a constant T_1 = 100 ps*

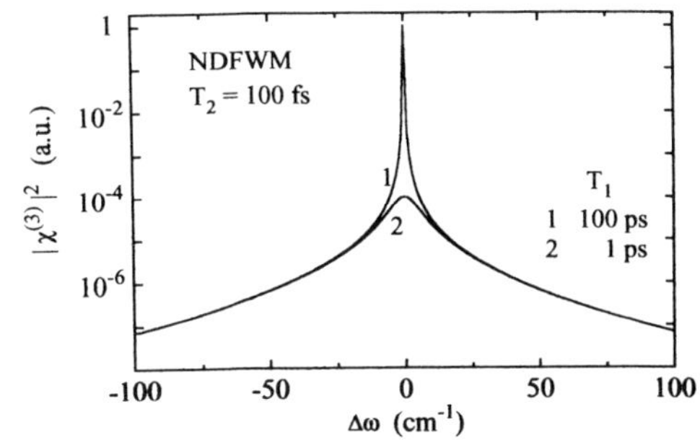

Fig. 6. *Intensity of the first diffracted order at the frequency ω_3 as a function of detuning $\Delta\omega = \omega_1 - \omega_2$ for the dephasing time T_2 = 100 fs and the lifetime $T_1 \gg T_2$ and $T_1 \approx T_2$*

with

$$|\chi^{(3)}|^2 = \frac{|K_3|^2}{(1+\Delta\omega^2 T_1^2)(1+\Delta\omega^2 T_2^2)} \tag{43}$$

where K_3 is

$$K_3 = -2\pi\, |d_{21}|^4\, n\rho_0\, S(\Omega)\, T_1\, T_2\, \hbar^{-3} \tag{44}$$

with d_{21} the dipol matrix element, n the density of two-level atoms, and $S(\Omega)$ the size distribution function normalized in energy. This result of theoretical calculation is illustrated in Fig. 5 and 6. The change around $\Delta\omega = 0$ is caused by the value of the lifetime T_1. A sharp peak is expected if T_1 is larger than 1 ps and large compared to T_2. The tails in the detuning curves were determined by the dephasing time T_2. The ratio of the signal intensity at zero detuning to the signal intensity for large values of $\Delta\omega$ is determined by the ratio in the times T_1 and T_2. For fast dephasing times T_2 the absolute value of the diffracted beam is generally decreased and the detection of signals needs high sensitivity. If the T_1 time is short and in the same order of magnitude as the T_2 time, then the contributions arising from the population grating and the polarization grating can no longer be clearly distinguished. The analysis of T_2 requires a careful fitting procedure to obtain unambiguous values of the times.

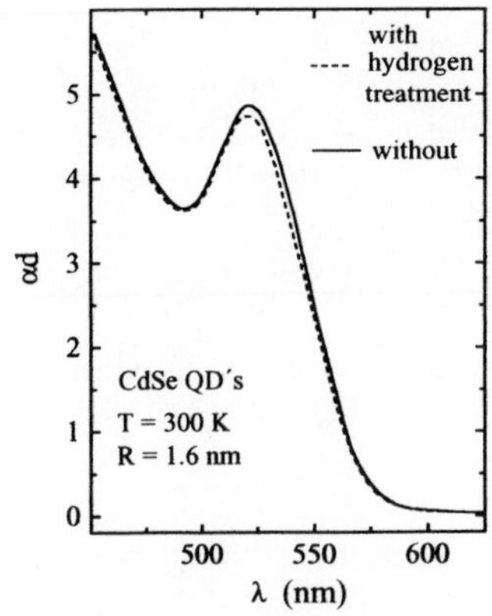

Fig. 7. *Linear absorption spectra of the nonpassivated (full line) and of the hydrogenated sample (dotted line) measured at room temperature*

IV. INTENSITY-DEPENDENT AND TIME-DEPENDENT NONLINEAR ABSORPTION OF CdSe–QUANTUM DOTS

When exciting quantum dots resonantly at $\omega_{pump} = \Omega$ and testing the nonlinear absorption by a second beam also at $\omega_{probe} = \Omega$, e.g. the saturation intensity I_S can be determined and compared with the bulk or other materials. Likewise a size dependence of the optical nonlinearity can be investigated or the involving of additional recombination, trapping, and transfer processes.

Since the beginning of the research on confinement effects in quantum dots the peculiarities introduced by the large surface to volume ratio of the QD have been the subject of intensive discussion. Evidently deviations from the simple two-level behaviour should appear involving more recombination channels or transfer processes. The existence of surface localized states has been reported in [13] investigating the time-behaviour and polarization of the luminescence. Theoretical studies of the surface polarization instabilities for Coulomb interacting electron-hole pairs predict for II – VI quantum dots in glass a transition from a volume to a surface trap state for a finite potential barrier of $V_B \approx 30$ Ryd [4]. It has been proposed e.g. by [14, 15, 16] that the dot – matrix interface has to take into consideration for the explanation of the temporal behaviour of the quantum dot recombination and of the photodarkening under prolonged laser exposure.

In the following, an example for the influence of the interface on the nonlinear optical properties of small CdSe quantum dots will be demonstrated. At resonant excitation an energy transfer of excited carriers into the glass matrix is supposed, which is advanced by the dot – matrix interface states. As a consequence, the observable bleaching should be smaller as expected in the ideal two-level system. To confirm this hypothesis the absorption change has been investigated in CdSe quantum dots after controlled surface passivation by hydrogenation [17]. Fig. 7 shows the linear absorption spectrum of a CdSe-sample of $R = 1.6$ nm cut into two pieces and one

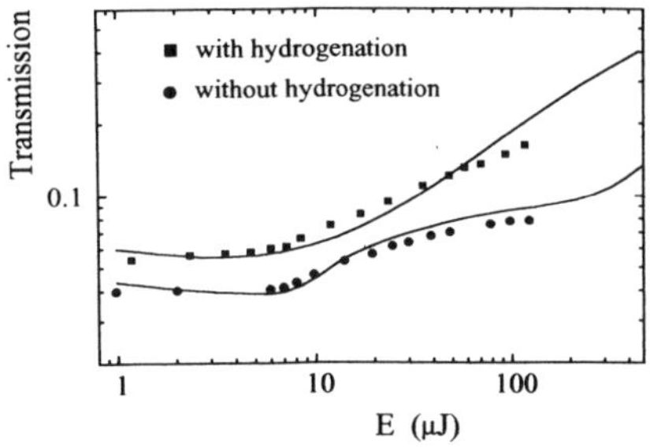

Fig. 8. *Measured probe-beam transmission versus pump energy of the non-passivated (full circles) and hydrogenated sample (full squares) at T = 300 K, for the calculated curves (solid lines) see text*

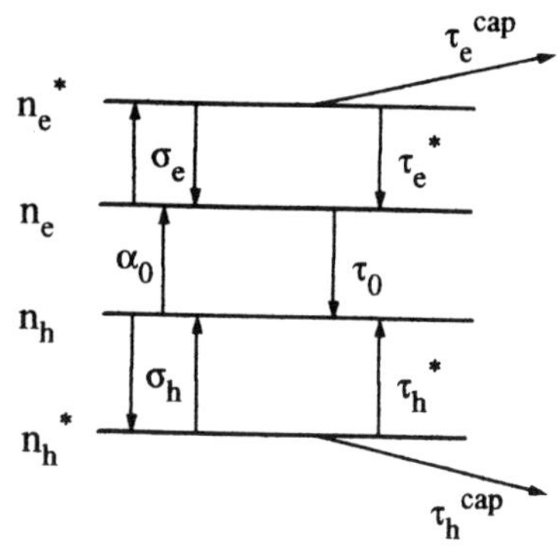

Fig. 9. *Kinetic model*

of them exposed additionally to a hydrogen atmosphere at 300 °C for 4 hours. The method of hydrogenation of defects is known from the passivation of Si-SiO_2 interfaces and a saturation of dangling bonds and nonbridging oxygen-bonds is expected. Only small differences occur in the linear absorption spectra (Fig. 7), which are attributed to a recharging of surface states by hydrogenation.

To study the absorption saturation and its time dependence, two-stage amplified 17 ps single pulses from a frequency doubled passive mode-locked YAG laser ($\lambda_{exc} = 539$ nm) have been used. In Fig. 8 the probe beam transmission versus pump intensity is shown for both samples measured at room temperature and at fixed frequency for pump and probe and a delay of the probe beam of 30 ps. All measurements were carried out after an initial exposure of each sample to 2000 laser shots. For the hydrogenated sample the observed switching factor between the transmission levels at low and high excitation intensity is significantly larger than that of the nonpassivated sample. To calculate the intensity dependence of the absorption change shown in Fig. 8, a three-level model (Fig. 9) has been used involving discrete quantum dot energy states and resonant excitation of electron-hole pairs. The processes occurring are (i) the absorption bleaching of the transition between the lowest electron and hole levels n_e and n_h, respectively, the population-dependent transition of electrons or holes to higher states, n_e^* and n_h^*, (induced absorption with the cross sections $\sigma_{e,h}$) and (iii) the carrier capture and transfer into interface or glass traps described by the capture time $\tau_{e,h}^{cap}$. Starting from a system of coupled differential equations according to Fig. 9, the transmission has been calculated in dependence on the pump intensity with the capture time as the parameter. For the calculations the cross section for the absorption of electrons has been assumed to be different from that of the holes. The results are shown in Fig. 10 with $\sigma_e = 10\,\sigma_h$ and $\tau_0 = 200$ ps. If the capture time $\tau_{e,h}^{cap}$ in the order of the lifetime of the lowest electron-hole pair transition (curve 1 in Fig. 10 with $\tau_0 =$ 200 ps and $\tau_{e,h}^{cap} = 100$ ps), then the intensity dependent change in transmission is to some extent two-level like. A transmission $T = 1$ can be reached at reasonable laser excitation intensities. Making the transfer process more efficient, a complete saturation is hardly achieved also at the highest intensities. For capture times below $\tau_{e,h}^{cap} \approx 2$ ps,

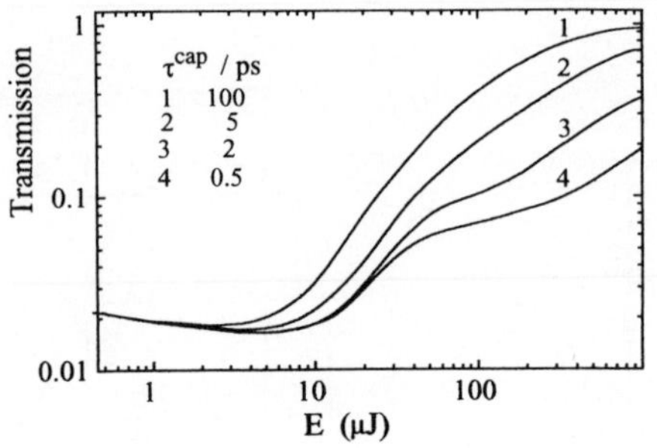

Fig. 10. *Calculated transmission versus pump energy for resonant excitation for different carrier capture times (σ_h = 8 x 10^{-17} cm ^{2}, σ_e = 8 x 10^{-16} cm^2, α_0 = 170 cm^{-1}, τ_0 = 200 ps)*

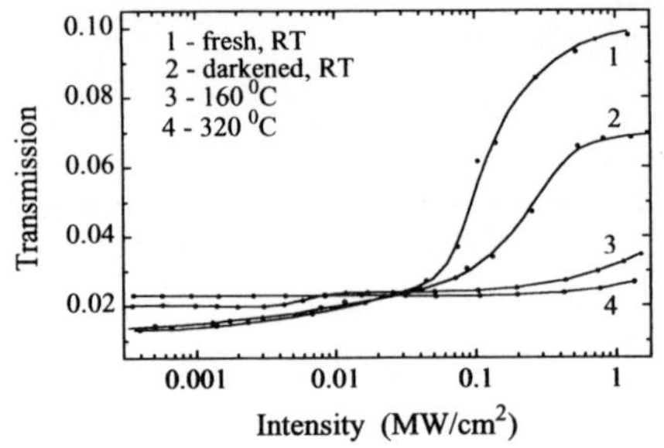

Fig. 11. *Probe beam transmission versus pump intensity of a fresh and a darkened glass and for heating at higher temperatures*

a new characteristic behaviour has been found. A plateau with a constant bleaching value over nearly one order of magnitude of the excitation intensity is obtained caused by the influence of the induced absorption process. The plateau in the saturation behaviour is a consequence that both kinds of carriers are captured and transferred. The corresponding range of intensity is determined by the carrier type having the larger σ and the smaller τ^{cap}.

In Fig. 8 two calculated curves are presented, explaining the experimentally observed results for the surface passivation. A qualitative agreement is achieved using for the hydrogenated sample $\tau_{e,h}^{cap}$ = 5 ps, σ_e = 5 x 10^{-16} cm^2, σ_h = 8 x 10^{-17} cm^2 and for the nonpassivated $\tau_{e,h}^{cap}$ = 0.8 ps, σ_e = 8 x 10^{-16} cm^2, σ_h = 8 x 10^{-17} cm^2. The plateau appears in the range of highest intensities used in the experiment. Fig. 11 shows the result when modulating the interface properties by a heat treatment procedure. For higher temperatures the saturation is prevented due to the high concentration of charged centers and the activation of migration and hopping processes both increasing the capture efficiency. The increase in temperature also influences the absorption cross section to the excited electron and hole states, taking into account the modification of the barrier potential. In the case of curve 3 and 4 of Fig. 11, the initial resonant ground state absorption is only an intermediate step before transferring the excited carriers out of the quantum dot. In this case absorption saturation can only be observed, if at least one of the carriers, the electron or the hole, populates the lowest level and is thus blocking the further absorption.

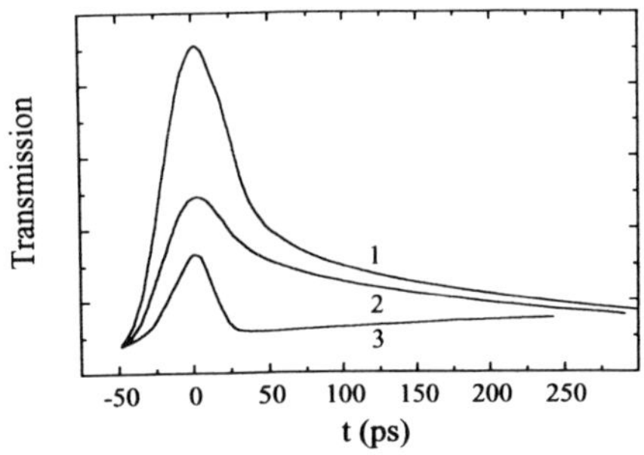

Fig. 12. *Time behaviour of quantum dot transmission after ps excitation calculated for capture times* $\tau_{e,h}^{cap} = 5$ *(curve 1), and 1 ps (curve 2), cross section* $\sigma_{e,h} = 8\ x\ 10^{-17}\ cm^2$. *Curve 3 was calculated for* $\tau_{e,h}^{cap} = 1$ *ps and* $\sigma = 4\ x\ 10^{-16}\ cm^2$.

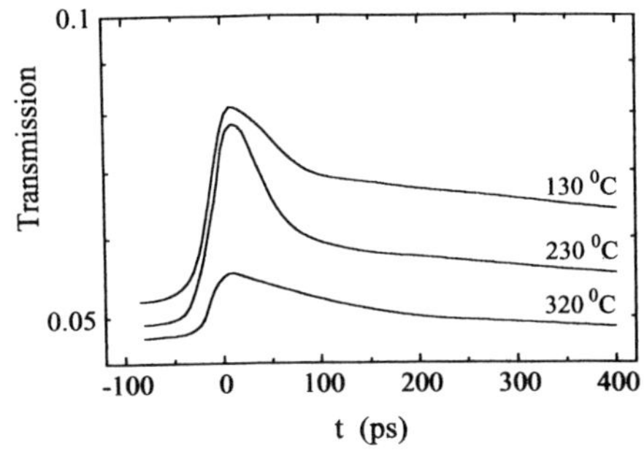

Fig. 13. *Temperature dependence of the time behaviour of the absorption recovery representing the mutual change in the cross section* σ *for electrons and holes*

For a discussion of the mutual influence of σ and τ^{cap}, the time behaviour of the transmission is calculated and shown in Fig. 12. Curves 1 and 2 are obtained assuming capture times of $\tau_{e,h}^{cap} = 5$ and 1 ps, respectively, and the same cross section for electrons and holes $\sigma_{e,h} = 8$ x 10^{-17} cm^2. The maximum bleaching value is smaller for $\tau^{cap} = 1$ ps, corresponding to a more efficient carrier capture. The decay of the fast components is similar and is determined rather by the short τ^{cap} than the lifetime τ_0. A further shortening of the fast component can be achieved only by an increase of the cross section σ (increasing induced absorption) as seen in curve 3 of Fig. 12. There the same capture time $\tau^{cap} = 1$ ps has been used, but for the parameter σ the value of 4 x 10^{-16} cm^2. No modification of the lifetime τ_0 are necessary to make the system faster. Such a behaviour cannot be described in the simple two level system. It takes into consideration that the quantum mechanical transition probability is not zero for intraband transitions in strongly quantum-confined systems. In Fig. 13 the experimental verification of the calculations from Fig. 12 are seen measuring the time-resolved absorption recovery. The probe beam testing the change in α at $\omega_{probe} = \Omega$ has been delayed in time over a range of 30 ps to 400 ps [17]. The curves obtained for heating the sample to different temperatures are in qualitative agreement with the calculations which predict a dependence of the decay time on the mutual change of the cross sections for electrons and holes and thus a mutual blocking of the ground state absorption.

Summarizing, a strong evidence for a fast energy and carrier transfer on a ps- and subps-time scale has been found. In the model discussed, modifications of the dot – matrix interface region are reflected by parametrizing σ and τ^{cap}, which corresponds to changes of the transition probability by influencing the potential barrier as well as by the formation and recharging of traps in the interface region.

V. DIFFERENTIAL ABSORPTION SPECTROSCOPY OF CdS- AND CdSe-QUANTUM DOTS

By means of differential absorption spectroscopy (DAS), i.e. the measurement of the spectrally resolved change in the absorption coefficient $\Delta\alpha$, the mechanisms of dephasing processes can be revealed. By energy selective excitation the energy positions of the involved electron-hole pair transitions can be identified and their homogeneous linewidth Γ determined. Already in two-dimensional systems like quantum wells (QW) or superlattices (SL), one is dealing with the investigation of different dephasing processes. The most important mechanisms considered are the scattering of the excited electron-hole pairs with phonons (LO-phonons or confined acoustic phonons), with interface potential fluctuations or localized states or, at higher excitation densities, the scattering by carrier–carrier interaction. In QD's, at present, the relative contribution of these mechanisms to the dephasing process is an open question. The recent research work is trying to identify the different dephasing processes from the experimental results of picosecond- or femtosecond-DAS [18, 19, 20], from the Stokes shift between absorption and luminescence [21], or from NDFWM applied to quantum dots [22] (see below).

An increasing electron-hole pair – LO-phonon coupling is expected if a polarization occurs caused by the differences in the radial distribution of the wavefunctions of electrons and holes.

Several reasons have been proposed giving rise to differences in the radial part of the wavefunctions and by this to an increase in electron–phonon interaction. Inclusion of Coulomb effects and the valence band coupling results in significant space charges inside the quantum dot [2, 3]. Besides this effect a consideration of an additional charge inside the quantum dot in the calculations yields pronounced Huang-Rhys factors for the electron-hole pair – LO-phonon coupling [5]. If introducing the dielectric surface polarization a drastic change of the wavefunctions of the pair states and the possibility of selftrapping of one of the carrier near the surface have been found [4]. Contributions from acoustic phonons and scattering at surface modes were also discussed [23, 24].

A further important competing dephasing mechanism is the scattering at interface potential fluctuations or at localized states. In particular for small QD's it has strong influence on the homogeneous linewidth because of the large ratio of number of atoms at the interface to those in the volume. While the interaction with phonons is mainly reflected by a temperature dependence, the scattering connected with the interface depends on the quality of the abrupt boundaries and therefore on the growth process of the quantum dots.

The analysis of Γ by DAS, e.g. in dependence on excitation intensity, growth, or temperature, gives more insight into the relative contributions of the different scattering processes to the phase relaxation time T_2.

In the following, we investigate the dephasing processes found at low excitation intensities excluding at first carrier-carrier scattering (see section VI.). From the temperature dependent change of the homogeneous linewidth we draw conclusions about the electron-phonon coupling and, analysing the growth- and annealing dependence

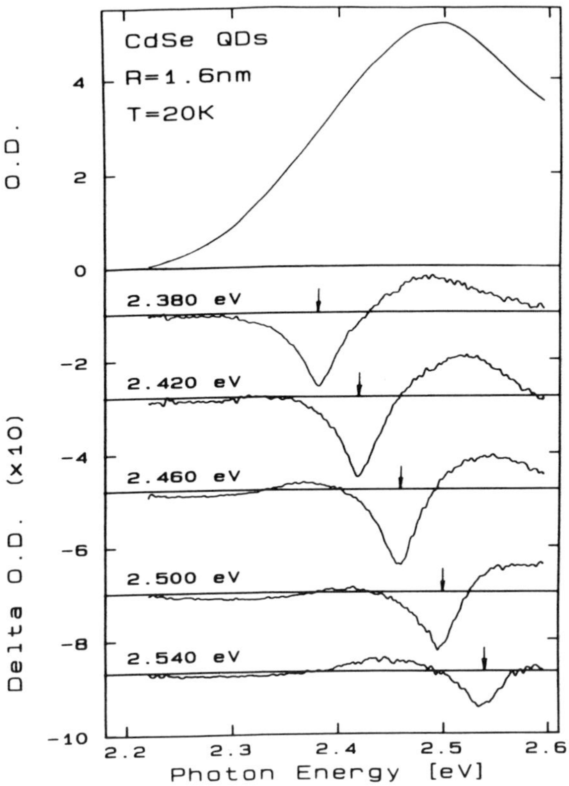

Fig. 14. *Linear and differential absorption spectra of CdSe-quantum dots of R = 1.6 nm measured at T = 20 K and an excitation intensity of 100 kW/cm^2. The pump energy is indicated by vertical arrows*

of the linewidth, we show the importance of interface related polarizations and space charges. As the result, we obtain that Γ can no longer be considered as a fixed value but more as parameter comprising different contributions and showing the behaviour according to the dominant component.

The nonlinear absorption measurements were performed using a standard pump-and-probe setup consisting of two dye lasers pumped by one nitrogen laser. The pulse duration and the repetition rate were 7 ns and 20 Hz, respectively. The narrow-band dye laser ($\leq$ 1 meV) has been focused to a spot of 0.1 mm diameter resulting in a maximum intensity of 10 MW/cm^2. For the probing a second weak continuum was used. Special care was taken to get a high sensitivity with respect to small changes of $\Delta\alpha$ below 0.01. Fig. 14 shows the change in absorption $\Delta\alpha$ in dependence on the pump energy of CdSe-quantum dots of R = 1.6 nm [25]. For these small sizes the energetic distance between the lowest electron-hole pair states is larger than 200 meV and therefore only the first electron-hole-pair transition is seen in the bleaching spectrum. For a laser excitation intensity of 100 kW/cm^2 spectrally narrow nonlinear resonances in the differential absorption have been found shifting with excitation energy within the spectral width of the linear absorption-peak. The spectral shape of the differential absorption agrees very well with that calculated in [1] including four-particle states. The narrow bleaching corresponds to the saturation of the one-electron-hole pair transition of the resonantly excited quantum dots, while the induced absorption on the low and high energy sides with respect to the absorption bleaching can be attributed to the creation of two-pair states from the one-pair states. The influence of power broadening on the lineshape has been checked by increasing the excitation intensity. A very high excitation of 3 MW/cm^2 is necessary to achieve a remarkable broadening.

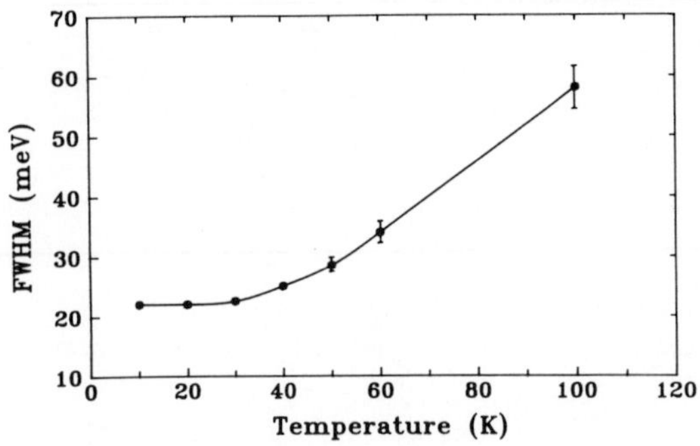

Fig. 15. *Temperature dependence of the full width at half of the maximum (2 Γ) of a stable narrow-line differential absorption spectra for a 2.3 nm* $CdS_{1-x}Se_x$ *sample and an excitation intensity of 300 kW/cm*^{2}

However the width of the structure remains also at these high intensities only a small fraction of the one of the broad-band linear spectrum as typically expected for strong inhomogeneously broadened systems. For CdSe-quantum dots of R = 1.6 to 2 nm we have determined the halfwidth of the nonlinear optical resonance to be 10 meV for a laser excitation of 30 kW/cm^2 and T = 20 K. Similar narrow-line, nonlinear absorption spectra are also known from small quantum dots embedded in organic matrices [13, 19].

In Fig. 15 we investigate the dependence of the width of the DAS on temperature to get more information about the contribution of scattering by acoustic or polar optical phonons to the homogeneous linebroadening. Between 10 and 40 K only a small increase in the linewidth with temperature is observed. The value T = 50 K can be considered as a threshold temperature. Below this temperature the scattering involving low frequency or confined acoustic phonons and should be the dominant mechanism besides the remaining constant contribution of the scattering at interface fluctuations. Above 50 K occurs a considerable increase of the linebroadening with temperature, now involving also scattering processes with phonons of larger energy, i.e. LO-phonons. The temperature dependence of the width of the spectral bleaching arising from the saturation of the lowest confined states verifies the electron–hole-pair coupling to phonons. The absolute value of the homogeneous linewidth Γ is about three times that of the bulk for temperatures around 50 K.

Up to now effects of the growth conditions on the linewidth of DAS have been reported mainly in connection with the inhomogeneous broadening given by the size distributions. A strong dependence of the homogeneous bleaching on the laser illumination history has been found when growing the quantum dots in the glass matrix by diffusion controlled normal growth avoiding the coalescence stage. Fig. 16 shows the sensitivity of the nonlinear absorption to strong laser illumination [26]. Spectrum 1 is recorded before strong laser exposure and shows a large and spectrally broad differential-absorption signal. A wide bleaching with two overlapping broad structures is seen. Spectrum 2 is recorded after laser exposure at 10 MW/cm^2 for some minutes. The differential absorption signal decreases, and now spectrally narrow resonances are obtained. We get from this spectrum 2 a well-resolved energetic distance of the two lowest electron-hole pair transitions $1S_{3/2} \longrightarrow 1s_e$ and $2S_{3/2} \longrightarrow 1s_e$ of 160 meV for R =2 nm. Spectrum 3 is recorded 20 minutes after spectrum 2 without any further changes. A gradual relaxation back to the feature of spectrum 1 has been found. Obviously it is possible to switch between two different states (or two different kinds of quantum dots) with different saturation intensities and homogeneous linewidth.

We propose as a possible explanation of this behaviour the following: The quan-

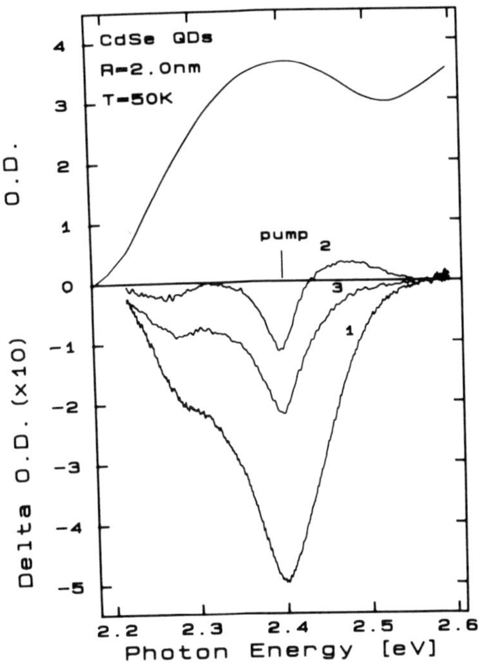

Fig. 16. *Differential absorption spectra of CdSe-quantum dots of R = 2.0 nm at an excitation of 100 kW/cm² before (spectrum 1) and after (spectrum 2) strong laser exposure. Spectrum 3 is recorded 20 minutes after spectrum 2 without any further changes at the same spot and shows a gradual relaxation back to the feature of spectrum 1.*

tum dots exhibiting the broad structures in the DAS, possess an interface configuration, which gives rise to the localization of one carrier near the interface with a high probability of trapping, whereas the other one is more or less delocalized inside the sphere of the quantum dot. The strong laser illumination introduces evidently changes in the charge states in the interface region, which are most pronounced for the small quantum dots grown near the nucleation stage. This growth stage can be realized by a low-temperature heat treatment over long times. In the stage of normal growth near nucleation a more or less continuous instable surface region exists determined by the ongoing diffusion process with a gradient in the ion concentration in the environment of the crystallite. The nonequilibrium state has been proved in TEM-measurements, where a further growth and construction of the surface has been observed during electron beam irradiation [27]. The further modification of the interfaces by strong laser illumination results in the formation of other confining potentials in the growth process near nucleation compared to the coalescence growth stage. This leads consequently to other wavefunctions for electron and holes with smaller charge separation in their radial parts. The new interface configuration introduced by laser exposure is not able to separate the excited electron and hole to the same extent as before and as a consequence, the homogeneous linewidth changes.

A clear correlation of the spectral change in the absorption coefficient α to the growth process is also demonstrated by Fig. 17 and 18 for the case of larger quantum dots [28]. The CdS quantum dots were grown using two different growth regimes. At first, shown in Fig. 17, the samples were developed in a diffusion controlled process with a preceeding nucleation step to promote the formation of nucleation centers. For this growth procedure a minimization of the average deviation in size has been expected

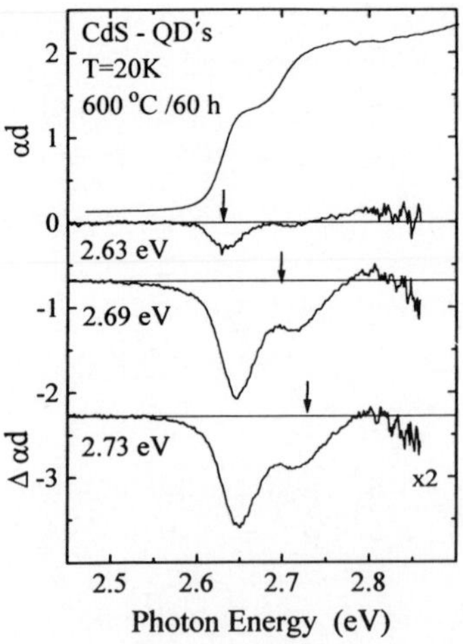

Fig. 17. *Linear and differential absorption spectra of CdS-quantum dots of R = 7.5 nm and a narrow size distribution grown at low temperature over long times with a preceeding nucleation step*

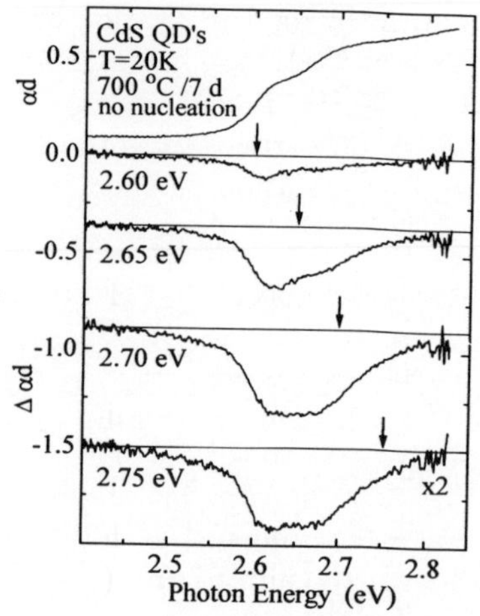

Fig. 18. *Linear and differential absorption spectra of CdS-quantum dots of similar sizes but larger distribution in sizes grown at high temperatures without nucleation*

giving small size distribution functions. As the result CdS quantum dots of $R = 7.5$ nm ± 1.5 nm have been grown. The linear absorption spectrum is characterized by two sharp steps attributed to the transitions from the two valence bands splitted by the spin–orbit interaction. The differential absorption spectrum shows a dominant homogeneous broadening with a half width at half of the maximum of approximately 20 meV.

The second growth process has been carried out with the same matrix material but without the nucleation step and at a high growth temperature of 700 °C. The sizes obtained are not remarkably larger, but in the linear absorption it is clearly seen that now the size distribution around the average size of $R \approx 9$ nm is considerably larger. If the homogeneous linewidth is not influenced by the growth, the spectral shape in the change of $\Delta\alpha$ of samples from different growth procedures should be similar at same excitation densities. In contrast to the results shown in Fig. 17 a smaller absolute value of the change in $\Delta\alpha$ has been found and a larger homogeneous bleaching. The structures can no longer be resolved in the DAS. Although the same low excitation intensity has been applied in the two experiments the width of the DAS of the two lowest hole states is evidently different and again attributed to growth dependent interface polarizations. The CdS quantum dots grown over 7 days at 700 °C were clearly developed in the stage of coalescent growth with a depletion region of semiconductor material at the interface. The CdS QD's obtained after a low temperature heat treatment exhibit a symmetric size distribution with an average distribution in sizes of ≈ 20 %. This comparably small and symmetric size distribution is a characteristic of the diffusion controlled growth stage with a gradient in the ion concentration in the interface region. This finding is in good agreement with the results discussed above for CdSe quantum dots.

Summarizing, the relation between homogeneous linewidth and the width of the homogeneous bleaching in the differential absorption spectra were discussed. Different from the two-level model, we have found the existence of near neighbouring hole states by resolving substructures in the DAS as well as a strong influence of the dot – matrix interface configuration on the dephasing time.

VI. NONDEGENERATE FOUR WAVE MIXING AT CdSSe-QUANTUM DOTS

After discussing of nonlinear optical methods investigating the nonlinear absorption we turn now to the study of the dephasing process via the nonlinear changes in the refraction index. Applying the method of nondegenerate four wave-mixing described in Sec. III, the influence of high carrier densities on the dephasing time T_2 has been studied with the aim to find (i) the carrier density, where the interparticle interaction starts to be the dominant process and (ii) the lowest limit of the time T_2 with increasing excitation intensities. It is expected to be in the range of some femtoseconds.

The experiments were performed using two excimer laser pumped dye lasers producing the light induced grating (see [22]). For the measurements of the diffraction efficiency, the beam of the laser 1 at ω_1 was usually fixed in frequency and the output light intensity at ω_3 was measured during the stepwise scanning of laser 2 with ω_2. For the fixed beam a dye laser with a beam expander and an etalon inserted in the oscillator cavity has been used giving a spectral band width of 0.02 cm^{-1}. The scanning beam has been supplied by a grazing incidence dye laser with the band width of 0.04 cm^{-1}. The spontaneous emission of the dye lasers was suppressed to $\sim$ 5 orders of magnitudes. The setup provides a maximum intensity of 50 MW/cm^2. The diffraction efficiency in dependence on the detuning $\Delta\omega$ has been measured at different intensities with attention to the wings of the curve towards larger values of the detuning $\Delta\omega$.

To minimize contributions of surface scattering, large $CdS_{0.7}Se_{0.3}$ - quantum dots were chosen with $R \approx 7$ nm corresponding approximately to two times the Bohr radius (OG 550). The excitation wavelength of the laser beam 1 was fixed at the maximum of the spectral change in the refraction coefficient at approximately 30 % transmission in the absorption tail.

Fig. 19 shows the first diffraction order versus detuning of the two laser frequencies with the pump intensity I_p as the parameter. The peak around $\Delta\omega = 0$ represents the lifetime T_1 of the underlying population grating. Values of T_1 larger than 20 ps can not be completely resolved within the experimental setup which is limited by the finite linewidth of the exciting laser beams and the spectral resolution provided by the diode array behind the grating spectrometer.

At excitation intensities around 10 MW/cm^2 the fitting procedure of the detuning curves according to [11] results in a T_2 time of about 200 fs. This is in a satisfying measure in agreement with the linewidth derived from DAS-measurements at comparable intensities. Above 10 MW/cm^2 the T_2 time decreases. For the maximum intensity reached in the experiment, the lowest limit has been determined to be 20 fs due to the efficient carrier – carrier scattering. Comparing the detuning curves at excitation in the first excited state of the QD and in the ground state transition, no remarkable difference has been found, suggesting the same mechanism of dephasing for both transitions, and supporting the assumption of the dominance of carrier – carrier interaction

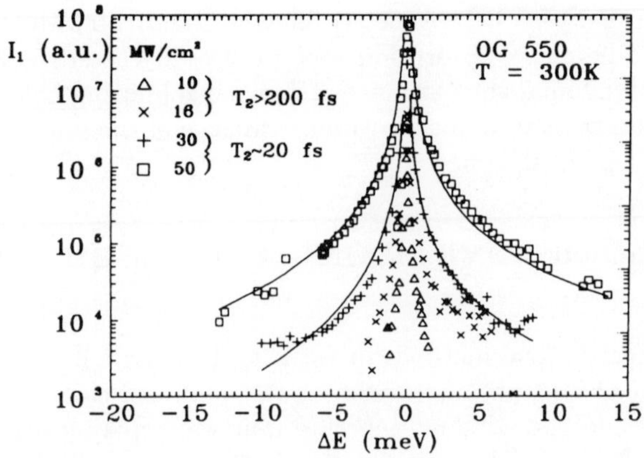

Fig. 19. *Intensity of the first diffraction order versus detuning of the two laser beams with the pump intensity as the parameter*

above 10 MW/cm^2. Especially the interlevel relaxation time seems to be very slowly.

At present the NDFWM experiments were confined to higher excitation intensities because of the higher signal of the first diffracted order and therefore more simple detection. However, it would be desirable to have an extension of the experiments towards lower intensities to compare the result with the linewidth from the differential absorption spectrum at low excitation intensities. An interesting application of NDF-WM as nonlinear optical method has been proposed in [29] calculating the modification of NDFWM-signal when the excited electron-hole pairs are trapped at defects.

VII. TIME-RESOLVED LUMINESCENCE

The application of nonlinear spectroscopy to determine T_1 and thus to characterize the recombination process has been shown using the two-beam configuration with the probe beam delayed in time. Another method already well established over the years is the analysis of the luminescence decay time by rate equations. The advantage of luminescence experiments is the detection over a wide spectral range and accordingly the identification of the different recombination processes by correlation to the trap energies. The limitations for the determination of T_1 from the luminescence decay curves are the width and energy of the exciting laser pulses, the time resolution of the detection system, and the knowledge about the levels involved in the energy relaxation process.

The measured photoluminescence is caused by the radiative decay of the excited electron-hole pair state generally after relaxation to the lowest energy states. The whole decay time comprises all components of recombination including radiative and nonradiative recombination, carrier capture and escape etc..

In the context of the discussion of the NDFWM-experiments in Sec. VI we have seen, that the fitting procedure to obtain T_2 is more unambiguous, if the lifetime T_1 is derived from an independent measurement. This is an additional reason to look at the information obtained for the CdSe quantum dots from time-resolved luminescence experiments.

Considering the data from literature (see e.g. [13, 30 - 34]) an uniform picture can be found concerning the fastest and slowest processes in the picosecond and mi-

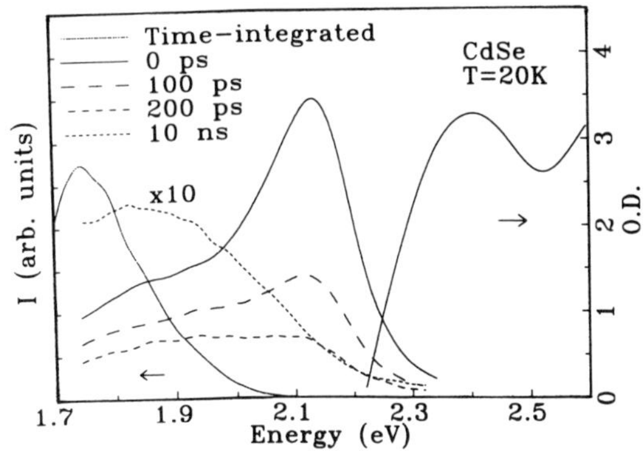

Fig. 20. *Absorption, time-integrated luminescence, and time-resolved luminescence spectra at different times after excitation of CdSe quantum dots with R = 2 nm*

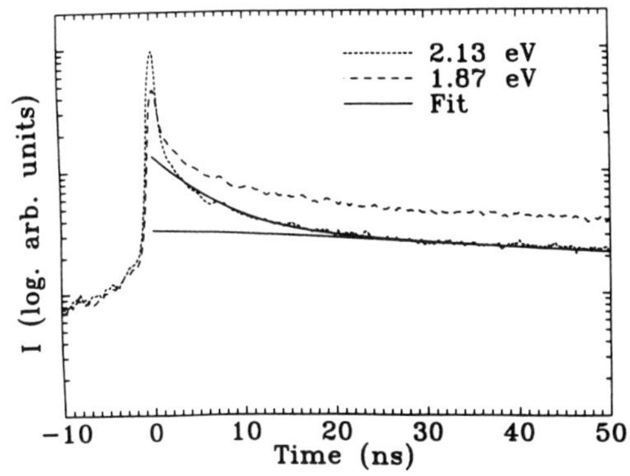

Fig. 21. *Luminescence decay on a 60 ns time-scale at the spectral positions of 2.13 eV and 1.87 eV. The two fit curves represent a one-exponential decay and a two-exponential decay behaviour, respectively.*

crosecond range, respectively. Recombination times of some tens of picoseconds up to 100 ps have been attributed to the direct electron-hole pair recombination influenced in different ways by the surrounding matrix material, which determines the trapping probabilities for both electrons and holes. The microsecond component reported e.g. in [13, 30, 35] is attributed to recombination after capture of the hole at deep vacancy trap states.

In particular for quantum dots embedded in a glass matrix the luminescence decay comprises also components in the nanosecond range shown in the following. Fig. 20 shows the luminescence spectra of CdSe quantum dots of $R = 2$ nm recorded at different times after ps-excitation [35]. The measurements of the luminescence decay have been performed using an excimer laser-pumped quenched transient dye laser system providing very intensive pulses of maximum fluence of 60 mJ/cm^2 with a duration of 70 ps and a repetition rate of 10 Hz. The experiment is specially designed for temporal resolution of the decay processes on a nanosecond scale. The laser photon energy was tuned to 2.504 eV thus exciting resonantly in the upper hole state.

From the measurements of the absorption recovery a value of T_1 around 30 ps has been determined for this sizes of quantum dots. In agreement with the result of time-resolved nonlinear absorption a very fast luminescence component has been found near the absorption edge with the maximum at 2.13 eV. It decays on a picosecond time-scale not resolvable within the exciting pulse-width of 70 ps. Added to this component is a broad emission band with a maximum at 1.87 eV. The spectral shift of the maximum of luminescence to lower energies with time confirms the relaxation to trap states. The time integrated spectrum of Fig. 20 is shifted further in the red and shows the μs-decay time.

The emission band at 1.87 eV dominates the luminescence in the nanosecond regime. This part of the luminescence is attributed to the recombination from the trap centers in the semiconductor–matrix interface region. It confirms the assumption that carriers may transfer rapidly into these states either directly from the resonantly excited pair state or via the ground state after relaxation. The temporal evolution of the nanosecond component can not be explained by a monoexponential decay. At least two components are involved into the recombination process as obtained from a two-exponential fit giving $\tau_1 = 5$ ns and $\tau_2 = 120$ ns, respectively (Fig. 21). The occurrence of more than one time is understandable considering the dot – matrix boundary as a local distribution of potential fluctuations across the surface giving a great variety in trapping centers as well as hopping processes between them.

In comparison to differential absorption measurements the luminescence analysis gives the possibility to investigate the electron-hole pair recombination over wide spectral and temporal ranges.

VIII. SUMMARY

By means of nonlinear spectroscopy we have analysed the energy and phase relaxation times of II – VI quantum dots. The results show a systematic correlation between the growth technology and the optic and electronic properties of the quantum dots. The nonlinear optical behaviour is influenced by the properties of the glass matrix determining the height and shape of the potential and creating new trapping centers in the matrix – dot interface region. The saturation of the quantum dot absorption can be controlled by surface chemistry. Controlling the fast energy transfer from the excited dot levels into the glass matrix by changing the charge states in the interface region, e.g. due to hydrogenation or the heat treatment, a longer lifetime of carriers, and hence, a larger occupation and a stronger bleaching can be achieved. In the theoretical model the transfer processes were modulated by changing the capture time τ_c. A blocking of the absorption is observed if the difference between the cross sections of the excited state absorption for electrons and holes is different. A shortening in the absorption recovery time can then be achieved by the increase of the efficiency of the excited state absorption by depopulating the ground state not by direct recombination but by induced absorption.

The interface configuration is held responsible for the observation of excitation dependent changes of the homogeneous linebroadening measured in the differential absorption spectra. Not only the longitudinal but also the transversal relaxation time depend on the charge localization in the interface region. There exists a great variety of samples with different surface construction determined by the matrix and the growth procedure and correspondingly different lineshapes in the nonlinear absorption.

The observed sharp nonlinear resonances give the possibility to investigate the homogeneous broadening and the size dependence of the energy of electron-hole-pair transitions. The temperature dependence of the hole width in differential absorption suggest that phonons are involved in the dephasing processes. However, scattering at surface states, caused by potential fluctuations could be as effective as the temperature dependent scattering.

The dephasing at higher excitation densities is dominated by carrier – carrier interaction yielding in a fast T_2 time of about 20 fs.

The time-resolved measurements of the luminescence are a valuable completion and the results consistent with those from nonlinear spectroscopy.

Acknowlegment

The author is grateful to C. Klingshirn, I. Rückmann, Al. L. Efros, S. V. Gaponenko, and H. Kalt for fruitful discussions and suggestions, J. Kolenda, J. Kornack, A. Uhrig, M. Saleh, O. Wind, and W. Langbein for carrying out carefully some of the presented experiments and M. Müller for the high quality CdSe samples and the support for growing CdS samples. The work has been supported by the Deutsche Forschungsgemeinschaft and by the Materialforschungsschwerpunkt der Universität Kaiserslautern.

References

[1] Y.Z. Hu, M. Lindberg, and S.W. Koch, Phys. Rev. **B 42**, 1713 (1990).

[2] S.W. Koch, Y.Z. Hu, B. Fluegel, and N. Peyghambarian, J. Cryst. Growth **117**, 592 (1992).

[3] A.I. Ekimov, F. Hache, M.C. Schanne-Klein, D. Ricard, C. Flytzanis, I. A. Kudryavtsev, T.V. Yazeva, A.V. Rodina, Al. L. Efros, J. Opt. Soc. Am. **B 10**, 106 (1993).

[4] L. Banyai, P. Gilliot, Y. Z. Hu, and S.W. Koch, Phys. Rev. **B 45**, 14136 (1992).

[5] S. Nomura and T. Kobayashi, Phys. Rev. **B 45**, 1305 (1992).

[6] N. Bloembergen, Nonlinear Optics, 3rd Ed. Benjamin, New York 1977.

[7] Y. Shen, The P rinciples of Nonlinear Optics, John Wiley, New York 1984.

[8] A. Yariv, Quantum Electronics, 3rd. Ed., John Wiley, New York 1989.

[9] P. Meystre and M. Sargent III, Elements of Quantum Optics, Springer Verlag Berlin,Heidelberg, New York 1990.

[10] M. Schubert, B. Wilhelmi, Einführung in die Nichtlineare Optik/T.II, BSB Teubner Verlagsgesellschaft, Leipzig 1978.

[11] T. Yajima and H. Souma, Phys. Rev. A **17**, 309 (1978).

[12] T. Yajima, H. Souma, and Y. Ishida, Phys. Rev. A **17**, 324 (1978).

[13] M.G. Bawendi, P.J. Carroll, W.L. Wilson, and L.E. Brus, J. Chem. Phys. **96**, 946 (1992).

[14] F. Hache, M.C. Klein, D. Ricard, and C. Flytzanis, JOSA **B 8**, 1802 (1991).

[15] B. van Wonterghem, S.M. Saltiel, T.E. Dutton, and P.M. Rentzepis, J. Appl. Phys. **66**, 4935 (1989).

[16] V. Esch, K. Kang, B. Fluegel, Y.Z. Hu, G. Khitrova, L.C. Liu, and L.C. Risbud, J. Nonlin. Opt. Phys. **1**, 25 (1992).

[17] U. Woggon, I. Rückmann, J. Kornack, M. Müller, J. Cesnulevicius, M. Petrauskas, and J. Kolenda, J. Cryst. Growth **117**, 608 (1992).

[18] P. Roussignol, D. Ricard, C. Flytzanis and N. Neuroth, Phys. Rev. Lett. **62**, 312 (1989).

[19] A.P. Alivisatos, A.L. Harris, N.J. Levinos, M.L. Steigerwald, and L.E. Brus, J. Chem. Phys. **89**, 4001 (1988).

[20] N. Peyghambarian, B. Fluegel, D. Hulin, A. Migus, M. Joffre, A. Antonetti, S.W. Koch, and M. Lindberg, IEEE J. Quant. Electr. **QE-25** ,2516 (1989).

[21] A. Uhrig, L. Banyai, Y.Z. Hu, S.W. Koch, C. Klingshirn, N. Neuroth, Z. Phys. B **81**, 385 (1990).

[22] A. Uhrig, A. Wörner, W. Langbein, U. Woggon, S. Gaponenko, M. Saleh, O. Wind, H. Kalt, C. Klingshirn, Proc. ICPS Beijing (1992); A. Uhrig, Thesis Universität Kaiserslautern 1993.

[23] M.C. Klein, F. Hache, D. Ricard, and C. Flytzanis, Phys. Rev. **B 42**, 11123 (1990).

[24] K. Misawa, H. Yao, T. Hayashi, and T. Kobayashi, J. Chem. Phys. **94**, 4131 (1991).

[25] U. Woggon, S. Gaponenko, A. Uhrig, W. Langbein, C. Klingshirn, Adv. Mat. Opt. Electr., in press

[26] U. Woggon, S. Gaponenko, W. Langbein, A. Uhrig, C. Klinghshirn, Phys. Rev. **B 47**, 3684 (1993).

[27] T. Yanagawa, H. Nakano, and Y. Sasaki, Appl. Phys. Lett. **59**, 1690 (1991)

[28] M. Saleh, U. Woggon, A. Uhrig, W. Langbein, C. Klingshirn, EPS Spring Meeting, Regensburg 1993; M. Saleh, Thesis Universität Kaiserslautern 1993.

[29] E. Hanamura, Phys. Rev. **B 46**, 4718 (1992).

[30] A. Schülzgen, J. Puls, V. Jungnickel, F. Henneberger, Proc. ICPS Beijing (1992).

[31] M. O'Neil, J. Marohn, G. McLendon, J. Phys. Chem. **94**, 4356 (1990).

[32] Y. Wang, A. Suna, J. McHugh, E. Hilinski, P. Lucas, R.D. Johnson, J. Chem. Phys. **92**, 6927 (1990).

[33] A. Eychmüller, A. Hasserlbarth, L. Katsikas, and H. Weller, Ber. Bunsenges. Phys. Chem. **95**, 79 (1991).

[34] M. Mitsunaga, H. Shinojima, and K. Kubodera, J. Opt. Soc. Am. **5**, 1448 (1988).

[35] O. Wind, H. Kalt, U. Woggon, C. Klingshirn, Adv. Mat. Opt. Electr., in press

SOLITONS IN OPTICAL FIBERS AND THEIR USE IN ULTRA LONG DISTANCE, HIGH BIT RATE TRANSMISSION

L. F. Mollenauer and J. P. Gordon

AT&T Bell Laboratories
Holmdel, NJ 07733

I. OPTICAL AMPLIFIERS

Long distance transmission using optical amplifiers can provide greater bandwidth at lower cost than that using conventional electronic regeneration. In the first place, the cost of regenerators tends to grow rapidly with increasing bit rate. More fundamentally, to take full advantage of the potential capacity of fibers, one would like to use wavelength division multiplexing (WDM). A separate set of regenerators is required for each channel, however, again multiplying the cost. Finally, electronic regenerators, in employing a great many active components that must work in perfect coordination, pose a serious potential for failure.

In contrast, optical amplifiers- especially the favored erbium-doped fiber amplifiers- are so broad-band that they place no practical limits on the bit rate of an individual channel (see Fig. 1), and they can easily handle several channels simultaneously, without crosstalk. (Their response, on the time scale of individual pulses, is strictly linear).

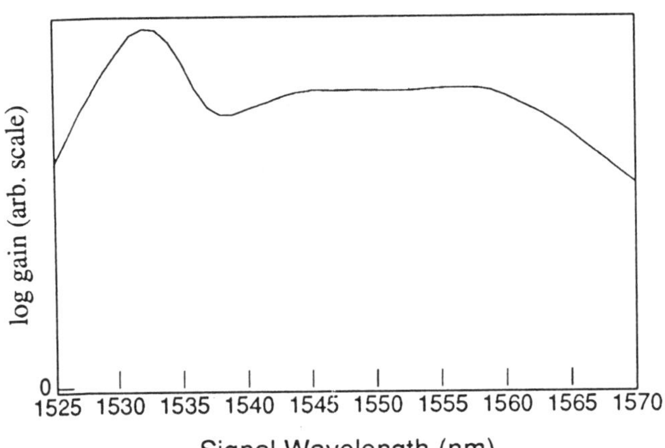

Figure 1 Gain vs. wavelength for a typical erbium fiber amplifier.

Nonlinear Spectroscopy of Solids: Advances and Applications
Edited by B. Di Bartolo and B. Bowlby, Plenum Press, New York, 1994

Several other properties make the erbium amplifiers nearly ideal: 1) When operated properly, their noise figure can be close to the quantum limit. (2) Their long-term saturation characteristics provide (at a slight compromise in the noise figure) for a natural and highly effective automatic gain control; that is, the time-average signal levels tend to be fixed by the rate at which the amplifiers are pumped. (3) The fiber amplifiers are highly compatible with the transmission line itself, so the problems of insertion loss, etalon effects, etc, associated with semiconductor optical amplifiers, are virtually non-existent. (4) Finally, the actual hardware - essentially just a few meters of fiber and a wavelength-dividing coupler (see Fig. 2)- is simple, rugged, and inexpensive.

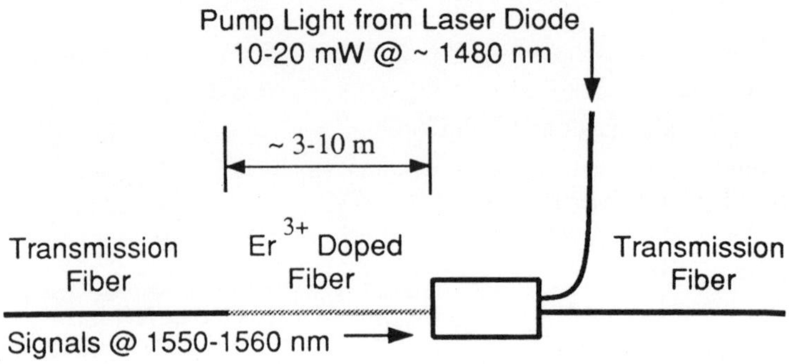

Figure 2 Erbium fiber amplifier. The wavelength-dependent coupler (small rectangle) efficiently couples pump light into the amplifier while allowing the longer-wavelength signals to continue along the transmission line.

The only potentially expensive item is the laser diode required for pumping each amplifier. But even here, because of the modest powers required (typically ~10 to 15 mW or less), and because of the technically accessible wavelength (~1480 nm), there is reason to expect an eventual reduction to modest cost. In sum, the erbium fiber amplifiers have uniquely provided a nearly ideal solution to the problem of overcoming loss. Their invention,[1,2] perhaps more than any other factor, is responsible for the present, intensive development of all-optical transmission lines.

I.A. The Nonlinear Schrodinger Equation

For ultra long distance transmission, however, overcoming loss is only half the story. It is equally important to choose a transmission mode resistant to the various dispersive effects of the fiber. The choice is best discussed in terms of the fundamental propagation equation, shown below:

$$-i\frac{\partial u}{\partial z} = \frac{\partial^2 u}{\partial t^2} + |u|^2 u - i(\alpha/2)u \tag{1}$$

In the present context, this equation, known as the nonlinear Schrödinger equation, just represents Maxwell's equations as specialized to a single mode fiber. Here u represents the pulse envelope function, z the distance of propagation along the fiber, t retarded time, and α the energy loss coefficient, all as measured in special "soliton" units, to avoid the appearance of scaling constants in the equation itself. The first term on the right (that involving the time derivative) represents the effects of the fiber's chromatic dispersion; acting by itself, this term always tends to broaden pulses in *time*. The second, or nonlinear term, is based on the fact that the fiber's index also depends on the light intensity; acting by itself, this second term, through the process of *self phase modulation*, always serves to broaden the pulse's *frequency spectrum*. The third and final term on the right represents the effects of energy loss (or gain). The amplifiers can act to make this term effectively zero, at least in a path-average sense.

I.B. Solitons

When the loss term has been reduced to zero, Eq. (1) has a special solution, known as the *soliton*:

$$u(z,t) = \text{sech}(t)\, e^{iz/2} \tag{2}$$

Note that, save for the phase term, which has no influence on the shape of the pulse, the soliton is independent of z, and hence is nondispersive in the time domain. Fourier transformation of Eq. (2) reveals that the soliton is also nondispersive in the frequency domain. The reason for the lack of dispersion in either domain is that for the soliton, the dispersive and nonlinear terms of Eq. (1) cancel each other out.

Now when it is stated thus, that the soliton involves the cancellation of terms in the propagation equation, one may get the impression that the soliton is delicate or unstable. Nothing could be further from the truth. The fact is that the soliton is the only stable solution to the propagation equation. Thus it can be shown, for example, that any minimum-bandwidth pulse of reasonable shape and of about the right peak intensity (a greater than 2 to 1 intensity ratio is allowable here), will evolve into an exact soliton as it propagates along the fiber. More generally, it has been possible to show, both theoretically and through real world experiment, that the soliton is in fact robust against all known fiber and other system defects of reasonable magnitude.

There is another view of the soliton which may make its robustness more intuitively obvious. That is, the peak of index change created by the soliton's intensity represents a potential well which travels right along with the soliton. Thus, the soliton can be viewed as a self-trapped pulse. The idea is represented in Fig. 3.

Figure 3 Runners on a mattress create a moving valley that pulls along slower runners and retards the faster ones. Similarly, in an optical fiber, high-intensity light creates a "valley" of higher dielectric constant material, which accelerates the slower, low-frequency components, and retards the faster, high-frequency ones. The resultant self-trapped light pulse, or soliton, can travel indefinitely long distances without changing its shape. (Cartoon idea by S. G. Evangelides.)

I.C. Soliton Units

Although we shall defer a derivation of the NLS equation and of the associated special "soliton" units[3] until the Appendix, the soliton units themselves are so important to all that follows that we cite formulas for them here. The unit time, t_c, is simply a measure of the soliton pulse width. From Eq. (2), one can easily show that it is related to the soliton's full width at half-maximum intensity, τ, as

$$t_c = \frac{\tau}{2\cosh^{-1}\sqrt{2}} = \frac{\tau}{1.762...} \tag{3}$$

The unit distance, z_c, is a characteristic length for effects of the dispersive term, and is given by

$$z_c = \frac{1}{(1.762...)^2} \frac{2\pi c}{\lambda^2} \frac{\tau^2}{D} \tag{4}$$

where c and λ are the light velocity and wavelength in vacuum, respectively, and where D is the dispersion constant of the reciprocal group velocity, expressed as a wavelength derivative. (Thus, D is often expressed as picoseconds of change in transit time, per nanometer change in wavelength, per kilometer of fiber length.) When D is expressed in those units, τ in picoseconds, and for $\lambda = 1556$ nm (corresponding to the longer wavelength erbium amplifier gain peak), Eq. (4) becomes

$$z_c \approx 0.251\,\tau^2/D \tag{4a}$$

where z_c is in kilometers. Note that for the pulse widths ($\tau \sim$ 15-50 ps) and dispersion parameters ($D \sim$ 0.3–1 ps/nm-km) most desirable for long distance soliton transmission, z_c is hundreds of kilometers. Finally, the unit of power, P_c, is just the soliton peak power, and is given by the formula

$$P_c = \frac{A_{eff}}{2\pi n_2} \frac{\lambda}{z_c} \tag{5}$$

where A_{eff} is the effective area of the fiber core, and where n_2, the nonlinear coefficient, has the polarization-averaged value of 2.7×10^{-8} μm^2/W for silica glass. Thus, for $A_{eff} \sim 50$ μm^2, and $\lambda = 1556$ nm, one has $P_c \approx 0.458/z_c$. Note that for z_c of hundreds of kilometers, the soliton power is just a few milliwatts.

I.D. Lumped Amplifiers and Solitons

From the discussion so far, one would probably tend to think of the soliton as involving continual balance between the dispersive and nonlinear terms of Eq. (1). Thus, it would be only natural to assume that *distributed* amplification, allowing for nearly uniform cancellation of the loss term, would be necessary for soliton transmission. Indeed, the Raman effect, which turns silica fiber into its own distributed amplifier, enabled the first experimental studies[4] of such transmission. More recently, however, we have come to realize that perfectly acceptable soliton transmission is possible through a chain of *lumped* amplifiers and dispersion shifted fiber segments. We have already shown (see Eq. (4a)), that for the pulse widths and fiber dispersion values of interest for long distance transmission, the characteristic dispersion length, z_c, is typically of order hundreds of kilometers. In that case, little or nothing happens to the pulse shape over the much shorter distance between amplifiers (the amplification period), no matter what the detailed behavior of the nonlinear term. [For $z \ll 1$ (where z is expressed in units of z_c), the fractional changes in pulse temporal and spectral widths, for uncompensated effects of the dispersive and nonlinear terms, respectively, are of order z^2. In this limit, then, there is no change to *first-order* in z, i.e., in "dz."] Thus, all that really matters is that the *path-average* pulse power, and hence the path-average value of the nonlinear term, over each amplification period, be such as to cancel the dispersive term. (Note that by the same principle, the dispersive term is not required to be constant either, i.e., only the path-average dispersion really matters.) Hence, there is no significant penalty for the use of lumped amplifiers spaced many tens of kilometers apart. This simple idea has been proven both analytically and through numerical simulation[5] (see Figs. 4 and 5), and well verified in real world experiment.[6,7]

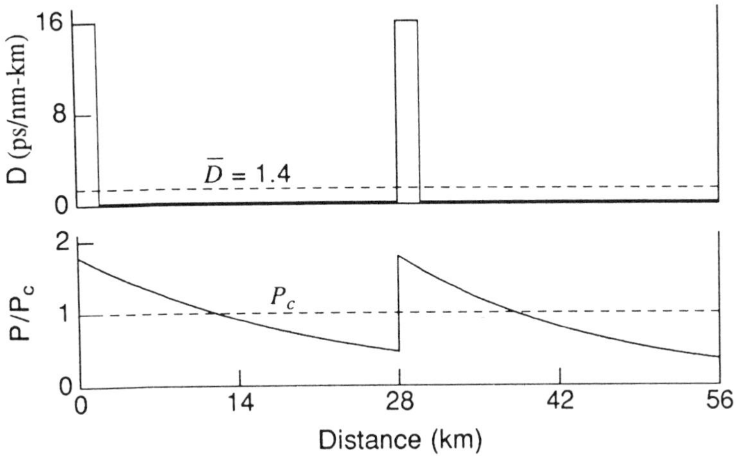

Figure 4 Sample of transmission line used for numerical test of the path-average soliton concept. As in our real-world experiments, the desired $\bar{D}$ is obtained by combining short lengths of high D fiber with dispersion-shifted fiber (for which $D \sim 0$), so there are large variations in D, periodic with the amplifier spacing, as well as in the pulse power.

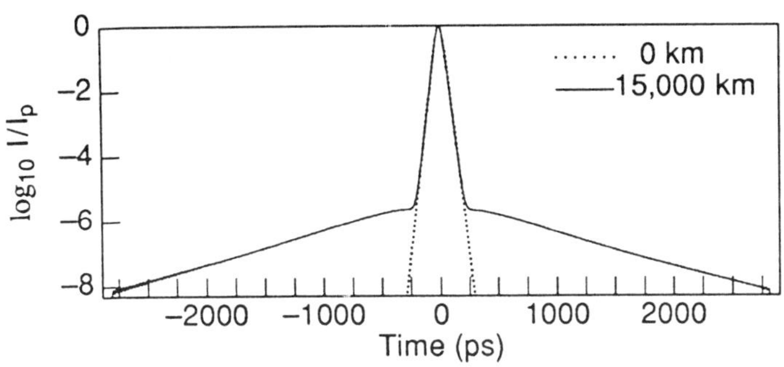

Figure 5 Solitons, for which $z_c \sim 440$ km, at input and after traversing 15,000 km of the transmission line of Fig. 4. Note that the defects (dispersive tails on the pulse) are extremely small. The defects have also been computed analytically.[8]

I.E. Other Potential Transmission Modes

One possible alternative to solitons is to transmit ordinary "square" pulses in the standard non-return-to-zero format at λ_0, the wavelength of zero dispersion in dispersion shifted fiber. This mode is often simply referred to, albeit somewhat incompletely, as "NRZ". Although NRZ is to first order nondispersive in the time domain, it is highly dispersive in the frequency domain, the pulse spectrum typically broadening to much greater than 100 GHz in the course of a transoceanic transmission. The large spectral broadening results, of course, because of the unbalanced nonlinear term in Eq. (1), and it absolutely prohibits the use of WDM with NRZ. Nevertheless, "error-free" transmission (bit error rate $\leq 10^{-9}$) has been demonstrated with NRZ over 14,000 km in a single channel at 5 Gbits/s.[9]

Another possibility would be to use phase or frequency shift keying in an ostensibly constant amplitude mode. It can be shown,[10] however, that such transmission succumbs to the combined effects of fiber nonlinearity and amplifier noise, long before any form of amplitude shift keying (in particular, either solitons or NRZ). Experiments[11] to about 2000 km tend to verify this prediction.

As we shall detail in this paper, solitons suffer from none of the above problems, and both WDM and polarization division multiplexing are possible with them. Perhaps the greatest advantage of all, however, comes from their unique ability to transit, without dispersion, a line containing many narrow-band frequency filters.

I.F. Transmission limits set by Spontaneous Emission Noise

The limits to distance and single-channel bit rate in soliton transmission tend to be set, above all other factors, by the various effects of noise.[12] In the prototypical system, consisting of amplifiers of gain G interspersed with fiber segments of loss $1/G$ (see Fig. 6), the path-average signal is maintained at the same high level in all segments. Thus, detector noise is overwhelmed, and the important noise is the accumulated or "amplified" spontaneous emission (ASE) created by the amplifiers. [In some cases, so-called "dispersive-wave radiation," or the energy left over when a pulse sheds its non-soliton part, can also make a significant contribution to the noise. Here, however, we shall concentrate on the inevitable ASE noise.] Although we shall later show how the proper use of narrow band filters can greatly reduce the limitations imposed by the ASE and other noise fields in the line, it is nevertheless instructive to first consider a broadband system, i.e., one in which the response of the transmission line is essentially flat over the entire bandwidth of the soliton.

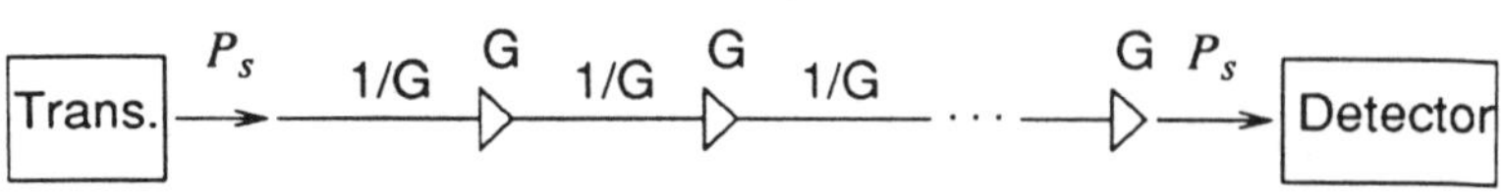

Figure 6 Prototypical all-optical transmission line, containing amplifiers of power gain G interleaved with fiber spans of loss $1/G$.

In the system of Fig. 6, each amplifier contributes a noise power per unit bandwidth of

$$P(\nu) = (G-1)\, n_{sp}\, h\nu \tag{6}$$

where G is the power gain of the amplifier, n_{sp} is the excess spontaneous emission factor (close to unity if the amplifier populations are highly inverted), and $h\nu$ is the photon energy. Note that power per unit bandwidth is just energy, so $P(\nu)$ is also known as the "equipartition energy." The physical meaning of $P(\nu)$ can perhaps be best appreciated by taking the unit time as the bit period T. (The unit bandwidth is then $1/T$.) Then the fact that $P(\nu)$ is just the expected value of the noise energy in each bit period, and in the corresponding bandwidth, becomes more immediately recognizable.

Since there is unity gain from the output of each amplifier in Fig. 6 to the receiver, $P(\nu)$ at the receiver is just the value given by Eq. (6) multiplied by N, the total number of amplifiers in the chain. To facilitate calculation of the signal to noise ratio, however, it is convenient to calculate the path-average noise in the final segment of the chain. (Recall that the soliton requires a fixed path-average value of the signal power.) The path-average noise at system output is:

$$P(\nu) = \alpha Z\, h\nu\, n_{sp} F(G) \tag{7}$$

where α is the fiber loss coefficient, Z the system length, and where the function

$$F(G) = \frac{1}{G}\left[\frac{(G-1)}{\ln G}\right]^2 \tag{8}$$

represents an important noise penalty for using high gain amplifiers (see Fig. 7). (Note that in Eq. (7), N has been written as $\alpha Z/\ln G$.) System designers, accustomed to the economics of regenerated systems, would like to place amplifiers no closer together than about 100 km. For transoceanic systems, however, the ~10 dB penalty for such large spacing is too great, and the spacing must be reduced to about one third that value (again, see Fig. 7).

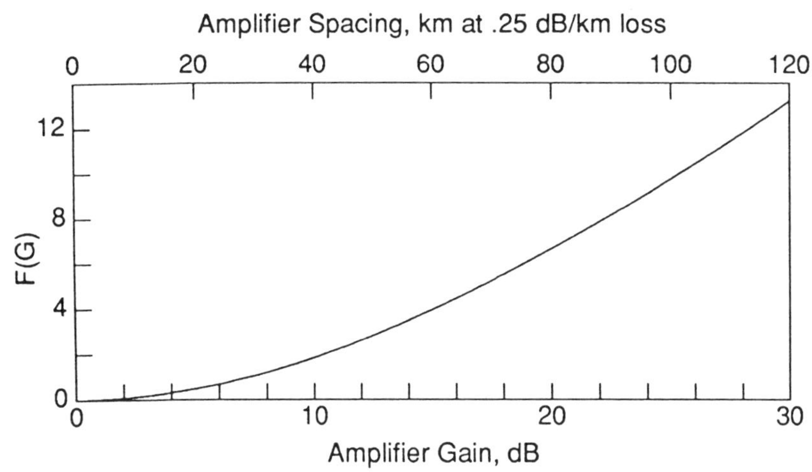

Figure 7 Penalty function, $F(G)$, as a function of amplifier gain.

For solitons, one important consequence of the ASE, known as the Gordon-Haus effect,[13] is as follows: Through the nonlinear term, the ASE modulates the optical frequencies (and hence the velocities) of the solitons at random. The resultant Gaussian distribution in pulse arrival times has the variance:

$$\sigma^2 = 4138\, n_{sp}\, F(G)\, \frac{\alpha}{A_{eff}}\, \frac{D}{\tau}\, Z^3 \tag{9}$$

where σ is in ps, n_{sp} and $F(G)$ are as defined above, the fiber loss factor α is in km^{-1}, the effective fiber core area A_{eff} is in μm^2, the group delay dispersion D is in ps/nm/km, τ is the soliton full width at half maximum intensity in ps, and Z is the total system length in thousands of km.

To get a feeling for the size of the effect, consider the example $Z = 9000$ km (trans-Pacific distance), $\tau = 20$ ps, $D = 0.5$ ps/nm/km, $A_{eff} = 50\ \mu m^2$, $\alpha = 0.048\ km^{-1}$, $n_{sp} = 1.4$, and $F = 1.19$ (~30 km amplifier spacing). Then the standard deviation, σ, is 11 ps. Note that the allowable spread of $\pm 6\sigma$ begins to fill the bit period as the bit rate approaches about 7.5 Gbits/s. In this way, in a broad-band system, the Gordon-Haus effect serves to set the upper limit on single channel bit rate.

II. EXPERIMENTS

To date, experiments on soliton transmission have been carried out in recirculating loops such as that shown in Fig. 8. Output of a mode-locked laser (MLL) is first amplified to the appropriate level for creating solitons in the loop. [The mode-locked laser is typically either a gain-modulated, external cavity semiconductor laser, or a modulator-driven, erbium fiber ring laser of a special design.[14] An acousto-optic (AO) modulator gates out a pulse train just long enough to fill the loop once. With the loop at unity gain, the train is allowed to recirculate as many times as necessary to cover the required distance. Samples of the pulse train, exiting the loop via the 1 dB fiber-fusion coupler, are detected and analyzed in one or more of several possible ways. One way which we have used consistently from the earliest experiments is to infer an average temporal behavior from the microwave spectrum of the detected pulse stream.

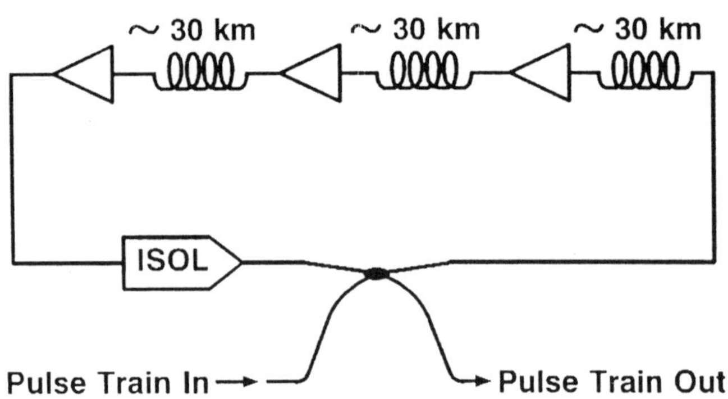

Figure 8 Recirculating loop used for soliton transmission experiments.

Figure 9 shows the measured effective pulse width, τ_{eff}, determined by such microwave spectral measurements,[6,7] as a function of total path. Note that the only sensible dispersion of the solitons is from Gordon-Haus timing jitter, so that the individual pulses are essentially dispersion free. The slight oscillation of the experimental points about the Gordon-Haus curve reflects the gradual conversion of the initially chirped and otherwise imperfect input pulses from the semiconductor diode laser into exact solitons. Thus it should come as no surprise that the pulse shapes look more nearly text-book perfect at about 5000 or 6000 km than they do at lesser distances.

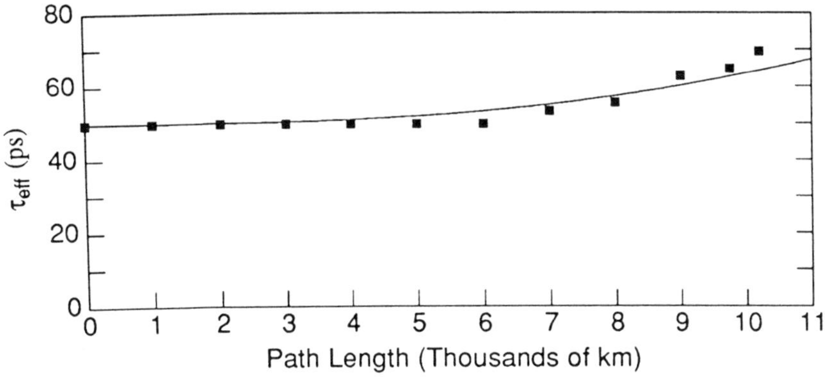

Figure 9 Measured effective pulse width as a function of total path.

It is instructive to consider a few typical numbers from the above experiment. For the cited pulse width and for the mean group delay dispersion parameter $D = 1.38$ ps/nm/km, the soliton peak power is ~0.6 mW on a path average basis, and is about 1.25 mW at the amplifier outputs. Potential saturation of the erbium amplifiers is determined only by the time-average power, which, for a typical bit pattern containing approximately equal numbers of ones and zeros, is a mere 0.1 mW. The amplifiers can easily handle many times that power level, so there is plenty of room for multiple channels in WDM. Finally, the measured signal to noise ratio, defined as the ratio of the soliton pulse energy to the equipartition energy, is greater than 250 to 1 at 9000 km.

In the experiment of Ref. 6, performed with the same loop and under the same conditions, we have verified the predicted interaction (attraction and repulsion) between closely spaced pairs of solitons. It is especially important that we see (as predicted) no significant interaction over a 10,000 km path between pulses separated by greater than five pulse widths.

III. BIT ERROR RATES - THEORY AND MEASUREMENT

Two types of error are possible in soliton transmission: first, energy or amplitude errors, when the ASE sufficiently reduces a "one" to make it look like a "zero" or vice versa, and second, timing errors, when a pulse arrives displaced entirely out of its normal bit period. Figure 10 shows the expected[12] error rate for soliton transmission over 9000 km, under the other conditions indicated, as a function of the parameter τ/D.

Note that the soliton energy, inversely proportional to τ/D, increases to the left of the graph. Note further that the energy errors, as one would expect, decrease exponentially with increasing soliton energy, and that they are independent of bit rate. Timing errors, on the other hand, *increase* with increasing soliton energy, as the D/τ term in Eq. (9) (Gordon Haus jitter) dictates, and of course, are dependent on bit rate.

Finally, note that for the smaller amplifier spacing (28 km), there is a reasonably large window of values of τ/D for which one could have a bit error rate (BER) less than 10^{-12}, while for the 80 km spacing, the best one could hope for would be an unacceptably large BER of 10^{-8}, and this for just one particular value of τ/D. The strong dependence of BER on amplifier spacing results, of course, from the penalty function $F(G)$ shown in Fig. 7.

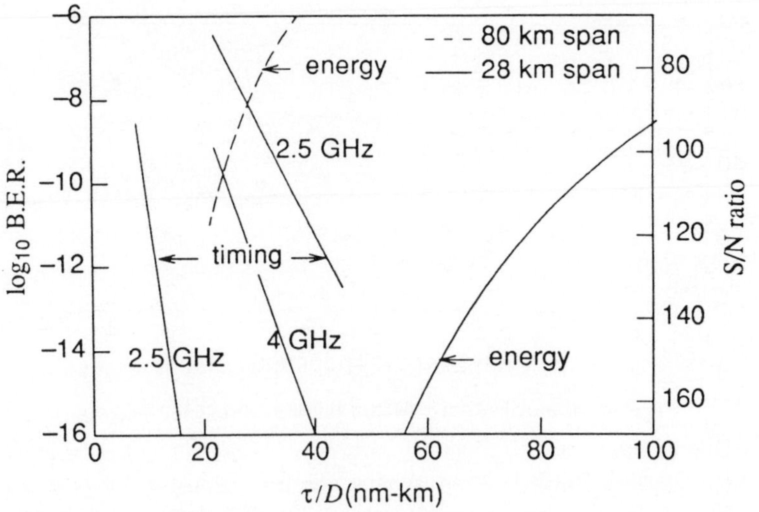

Figure 10 Predicted bit error rates as a function of τ/D for a 9000 km long, broad-band transmission line. (Adapted from Ref. 12.)

The ultimate proof of system viability rests with experimental measurement of bit error rates in the transmission of actual information. We have devised a special technique for making such measurements in the face of the intermittent output of a recirculating loop.[15] Our technique involves comparison of a received, 2^{12} bit long, pseudo-random word with its logical complement as received one word earlier, and as "remembered" by a fiber delay line. For details, see Fig. 11 and its caption. Figure 12 shows the results of one of the earliest measurements[15] made with that technique, where "error free" transmission (BER $\leq 10^{-10}$) was obtained at 2.5 Gbit/s over more than 11,000 km.

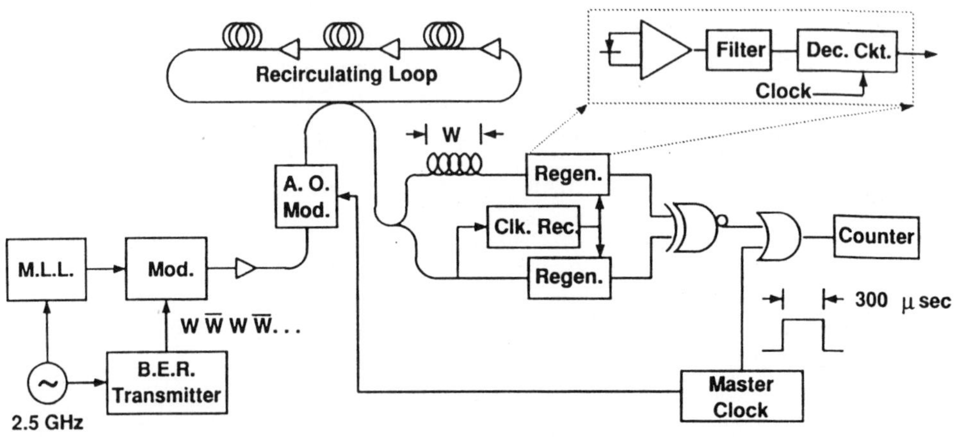

Figure 11 Apparatus used for BER measurements. On the transmitter side, the repeated pattern of a 2^{12} bit word, followed by its logical complement, is imposed on the output of a mode-locked laser by a high-speed electro-optic modulator. On the receiver side, the loop output is split, and a fiber coil delays one half by exactly one word. The two nominally complementary sets of data are then detected and regenerated, and compared by an ultrafast exclusive NOR gate. By counting the number of times the EX-NOR gate output goes "high" during the final round trip, one has good measure of the bit error rate at that distance.

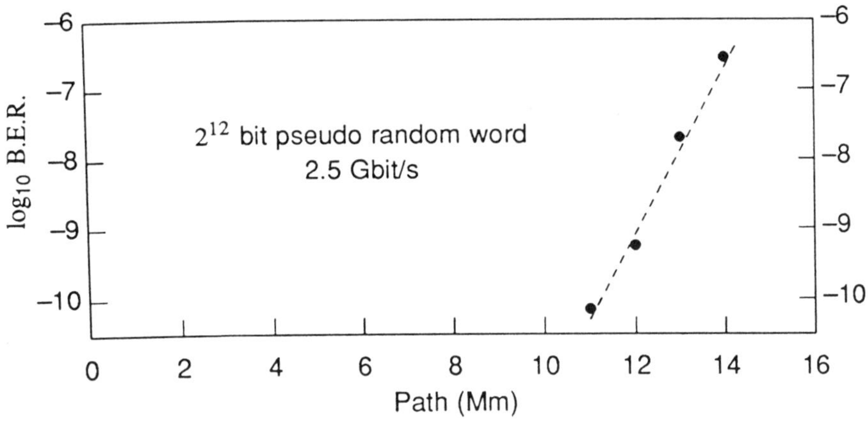

Figure 12 Measured bit error rates as a function of total path for a broad-band transmission line. (Note: 1 Mm = 1000 km.)

IV. FREQUENCY GUIDING FILTERS

Thus far we have discussed the characteristics and the performance of essentially broad-band transmission lines. In mid-1991, however, two groups independently suggested the idea that the Gordon-Haus jitter and other noise effects could be significantly suppressed in soliton transmission systems, simply through a narrowing of the amplifier gain–bandwidth.[16,17] In practice this means the use of narrow-band filters, typically one per amplifier. Figure 13 shows appropriate filter response curves in comparison with the spectrum of a 20 ps wide soliton. The fundamental idea is that any soliton whose central

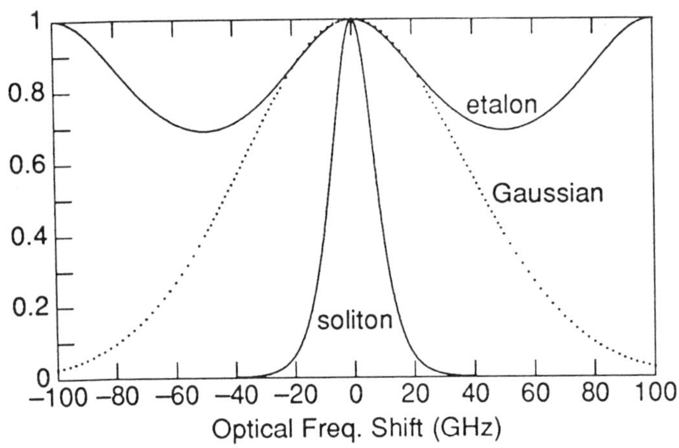

Figure 13 Intensity response curves of practical etalon guiding filter, Gaussian with same peak curvature, and spectrum of 20 ps soliton. The etalon mirrors have $R = 9\%$ and are spaced 1.5 mm apart.

frequency has strayed from the filter peak (as a result of the Gordon-Haus effect), will tend to be returned to the peak by virtue of the differential loss the filter induces across its spectrum. The resultant damping of the frequency jitter leads in turn to a corresponding damping of the jitter in pulse arrival times. The filters also cause a reduction in amplitude jitter. Consider, for example, a pulse with greater than normal amplitude; the pulse will be narrower in time, (see Eqs. 4 and 5), have greater bandwidth, hence experience greater loss from passage through the filter than the normal pulse, and, of course, the opposite will occur for a pulse of less than standard amplitude.

Since the major benefit comes from the filter response in the neighborhood of its peak, the etalon filter whose shallow response is also shown in Fig 13 provides almost as much benefit as the Gaussian filter. But the etalon, with its multiple peaks, has the great fundamental advantage that it is compatible with extensive WDM. The etalons also have the practical advantage that they are simple, cheap, and can be easily made in a rugged and highly stable form.

It should be understood that linear pulses can not traverse a long chain of such filters: after a sufficient distance, their spectra will be greatly narrowed, and the pulses correspondingly spread out in time. The solitons survive because they can regenerate, more or less continuously, the lost frequency components from the nonlinear term of the NLS equation. On the other hand, the amplifiers must supply a certain excess gain to compensate for the net loss even the solitons suffer from passage through the filters. As a result, noise components at or near the filter peak grow exponentially with distance. In order to keep the noise growth under control, the filters can be made only so strong, so the maximum possible benefit from them tends to be somewhat limited. For example, Fig. 14 shows the standard deviation of timing jitter, as a function of distance, for systems with the optimum strength filters (those experimentally observed[18] to produce the best BER performance), and for those with no filters. Note that at the trans-Pacific distance of 9,000 km, the filters reduce the jitter by a factor somewhat less than two times.

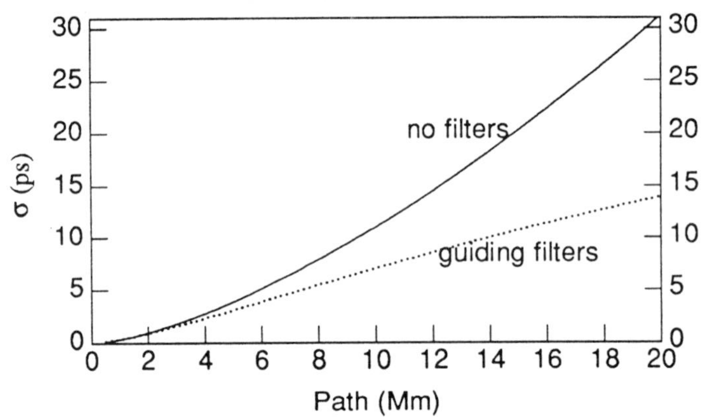

Figure 14 Standard deviation, σ, of pure Gordon Haus jitter as a function of total path, for a broad-band transmission line ("no filters"), and for one having optimum strength, fixed-frequency guiding filters (uncoated, 1.5 mm thick solid quartz etalons, one per 78 km, i.e., only one for each three amplifiers). The other pertinent transmission line and soliton parameters are: $D = 0.7$ ps/nm-km, $n_{sp} = 1.4$, and $\tau = 40$ ps.

Figure 15 shows the BER performance obtained with the same filters.[18] The BER data shown there for the 1010... pattern makes excellent fit to the combined, expected effects of pure Gordon-Haus jitter and energy errors. The random patterns, however, suffer additional timing jitter from a weak acoustic interaction among the solitons. To the extent that energy errors are pattern-independent, the difference can be attributed to an acoustic effect about 58% as strong as has been predicted theoretically.[19]

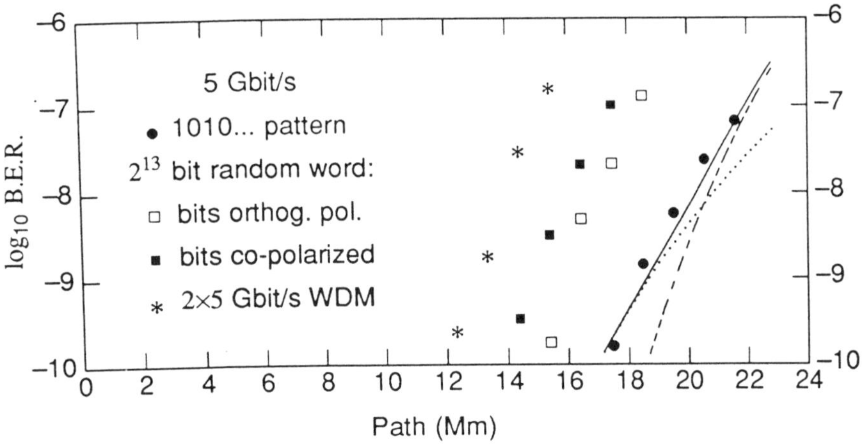

Figure 15 BER data obtained with the optimum fixed-frequency filters of Fig. 14. The curves of theoretical BER are: dashed, effect of pure Gordon Haus jitter; dotted, energy errors only; solid, their sum. With respect to the pseudorandom word data (squares), making adjacent bits orthogonally polarized reduces a (weak) interaction between them stemming from direct overlap. (From Ref. 18.)

IV.A. Sliding-Frequency Guiding Filters

Soon after doing the experiments of Fig. 15, we invented a simple way[20] to overcome the noise growth, and hence the limited performance, of a system of fixed-frequency filters. The trick is to "slide", i.e. translate, the peak frequency of the filters with distance along the transmission line (see Fig. 16). As long as the sliding is gradual enough, the solitons will follow, in accord with the same "guiding" principle that dampens the jitter. On the other hand, the noise, being essentially linear stuff, can follow only the horizontal path in Fig. 16. Thus, one creates a transmission line that is opaque to noise for all but a small, final fraction of its length, yet remains transparent to solitons. In consequence, the filters can be made many times stronger, and the jitter reduced by a correspondingly large factor, with the final result that the maximum bit rate can be increased at least several fold over that possible without sliding.

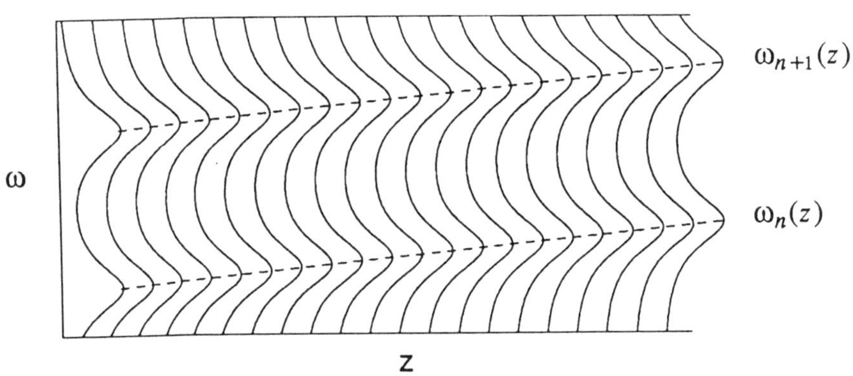

Figure 16 Transmission of sliding-frequency guiding filters versus z.

At this point it is worth while to take a look at the mathematical theory of a transmission line employing filters, whether sliding or not. When expanded in a Taylor series, the logarithm of the filter response function F takes the general form:

$$\ln F(\omega-\omega_f) = i\zeta_1(\omega-\omega_f) - \zeta_2(\omega-\omega_f)^2 - i\zeta_3(\omega-\omega_f)^3 + \cdots \tag{10}$$

where ω_f is the filter peak frequency, and where the constants ζ are all real and positive. While the first order term can be ignored (linear phase shift serves only to translate the pulses in time), and earlier analysis[16,17] has involved only the second order term, for the strong filters of interest here, the higher order terms have significant effects.

Nevertheless, it is useful to consider analytic solutions to the simplified propagation equation involving only the second order term of Eq. (10). Invoking the principle that if $u(t)$ and $\tilde{u}(\omega)$ are Fourier transforms of one another, then so are $\partial u/\partial t$ and $-i\omega\tilde{u}(\omega)$, we can write the modified NLS equation as:

$$\frac{\partial u}{\partial z} = i\left[\frac{1}{2}\frac{\partial^2 u}{\partial t^2} + u^* u^2\right] + \frac{1}{2}[\alpha - \eta(i\frac{\partial}{\partial t} - \omega_f)^2]u \tag{11}$$

where α is the gain required to overcome the loss imposed on the solitons by the filters, and where $\eta = 2\zeta_2$. (Both continuously distributed quantities α and η are easily converted into lumped, periodic equivalents.) Without filter sliding ($\frac{d\omega_f}{dz} \equiv \omega_f' = 0$), and where, for convenience, we set $\omega_f = 0$, the exact stationary solution is [19]

$$u = \sqrt{P}\,\mathrm{sech}(t)\exp(i\phi), \text{ where } \phi = Kz - \nu\ln\cosh(t) \tag{12}$$

and where the parameters (α, η, P, ν, K) must satisfy

$$\nu = \frac{3}{2\eta}\left[\left(1+\frac{8\eta^2}{9}\right)^{1/2} - 1\right] = \frac{2}{3}\eta - \frac{4}{27}\eta^3 + \cdots \tag{13a}$$

$$\alpha = (\eta/3)(1+\nu^2) \tag{13b}$$

$$P = (1+\eta^2)(1-\nu^2/2) \tag{13c}$$

$$K = (1/2)(1-\nu^2) + (\nu^2/3)(2-\nu^2) \tag{13d}$$

Note from Eq. (12) that the pulse's frequency is chirped, i.e., $\frac{\partial\phi}{\partial t} = -\nu\tanh(t)$. Because of this chirp, the rms bandwidth is increased by the factor $(1+\nu^2)^{1/2}$.

Numerical simulation involving real filters shows that the principal features (the chirp, extra bandwidth, and increased peak power) of the above solution are approximately preserved. The major differences lie in slight asymmetries induced in $u(t)$ and in its spectrum by the third-order filter term, and in the fact that, through a complex chain of events, those asymmetries cause the soliton mean frequency to come to rest $\delta\omega$ above the filter peak. Numerical simulation has also show that sliding (within certain limits given below) does not significantly alter this solution.

Sliding does, however, have the potential to alter the damping of amplitude and frequency fluctuations. Following is a brief summary of analysis given in Ref. 20: First, the soliton mean frequency Ω lags behind the filter frequencies by

$$\Delta\Omega \equiv \Omega - \omega_f = \delta\omega - \frac{3}{2\eta}\omega_f'. \tag{14}$$

Corresponding to the two normal modes of the coupled frequency and amplitude fluctuations, there are in general two damping constants

$$\gamma_1 = \frac{2}{3}\eta(1+\sqrt{6}\,\Delta\Omega) \quad \text{and} \quad \gamma_2 = \frac{2}{3}\eta(1-\sqrt{6}\,\Delta\Omega), \tag{15}$$

This implies a monotonic decrease of damping for both frequency and amplitude fluctuations with increasing $|\Delta\Omega|$, and, through Eq. (14), the existence of maximum allowable sliding rates for stability. Note, however, that because of the fact that the offset $\delta\omega$ is positive, a greater $|\omega_f'|$ can be tolerated with up-sliding than with down-sliding; thus, up-sliding is definitely preferred.

With respect to the variances in soliton energy and arrival time, we are primarily interested in the behavior for $\gamma z >> 1$, where the energy fluctuations have come to equilibrium with the noise. Then the energy variance can be written as

$$\frac{\langle \delta E_s^2 \rangle}{E_s^2} \approx \frac{N}{\gamma_E E_s} \tag{16}$$

where N is the spontaneous emission noise spectral density generated per unit length of the transmission line, E_s is the soliton pulse energy, and where the effective damping factor, γ_E, is expected to decrease monotonically with increasing $|\Delta\Omega|$. Similarly, for $\gamma z \gg 1$, the reduction factor[17] for variance in timing jitter can be written as

$$f(\gamma_t, z) \approx \frac{3}{(\gamma_t z)^2} \tag{17}$$

where γ_t is also expected to decrease monotonically with increasing $|\Delta\Omega|$. Although γ_E and γ_t are in general different, nevertheless, for Gaussian filters, both are expected to be approximately equal to $2\eta/3$ in the neighborhood of $\Delta\Omega=0$. For the more practical etalon filters like that shown in Fig. 13, the damping constants are about 81% of those of Gaussian filters having the same peak curvature.

To illustrate the many practical benefits of sliding, we now cite a specific example which has been thoroughly tested both in numerical simulation[20] and in a very closely related real world experiment.[21] Specifically, the soliton characteristics were, for the simulation (experiment), $\tau = 20$ (19) ps and $D = 0.4$ (0.45) ps/nm-km, at $\lambda = 1556$ nm, so $z_c = 250$ (200) km. The sliding rates were 5.6 (6-7) GHz/Mm, and in both cases the filters were exactly the etalons of Fig. 13, spaced apart by 28 (26) km (one at the output of each amplifier), so the filter strength parameters were $\eta = 0.6$ (0.52). Finally, the excess gain factors were 0.82 (0.89)/Mm. For the three-span recirculating loop used in the experiment, the three etalons were mechanically ganged together and driven by a common piezoelectric element. The piezo was driven by a linear ramp voltage, synchronized with the repeated transmissions.

The noise spectral density, $N(\omega, z)$, is computed as the sum of the noise contributions generated in each segment of the line, as modified by the excess gain factor and by the accumulated loss of the frequency-sliding filters. Figure 17 shows $N(\omega, z)$ as computed numerically for our example, and Fig. 18 shows the corresponding experimental result. Note that in both cases, after ~5 Mm, the peak spectral density saturates at a low value, whereas without sliding it would have risen by $e^{8.2} \sim 3600$ times in 10 Mm! (In a real experiment without sliding, the noise stops rising once it completely saturates the amplifiers. Its rise is still devastating, however, because in so saturating the amplifiers, it completely suppresses the signals.)

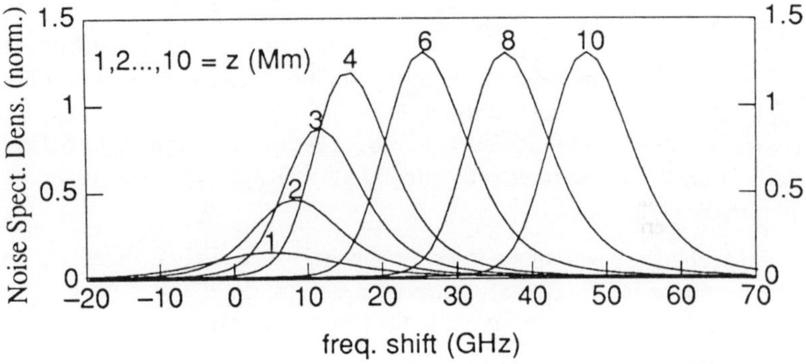

Figure 17 Computed noise spectral density, $N(\omega,z)$, at various distances along a transmission line with sliding filters, normalized to that density obtaining at 10 Mm with no filtering, for solitons of $\tau = 20$ psec, with $D = 0.4$ psec/nm/km ($z_c = 250$ km). Filtering, excess gain, and sliding rates (all per z_c), are $\eta = 0.6$, $\alpha = 0.205$, and $\omega_f' = 0.1$ (= 5.6 GHz/Mm), respectively. (From Ref. 20.)

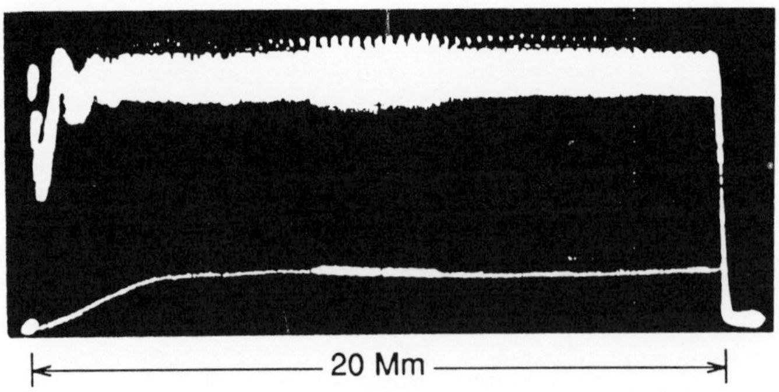

Figure 18 Experimentally observed evolution of noise (faint lower trace) and combined signal and noise (heavy upper trace) with distance, when the sliding-frequency guiding filters are used; note the close correspondence to the predictions of Fig. 17. Here the noise trace represents the total in about six channels, while the signal is a low 2.5 Gbits in only one channel; thus, the true S/N ratio of a 10 or 20 Gbit/s data stream to its own proper noise is much greater than seen directly here.

Figure 19 shows computed normalized standard deviations of the soliton energy ("ones") and of the noise energy in empty bit periods ("zeros") as a function of distance. These curves are obtained from Eq. (16) and the assumption that $\gamma_E = 0.32$ $(= 0.81 \times 2\eta/3)$ and the data of Fig. 18. Note that with the filtering, both standard deviations soon become clamped to small, indefinitely maintained values, corresponding to immeasurably small bit-error rates.

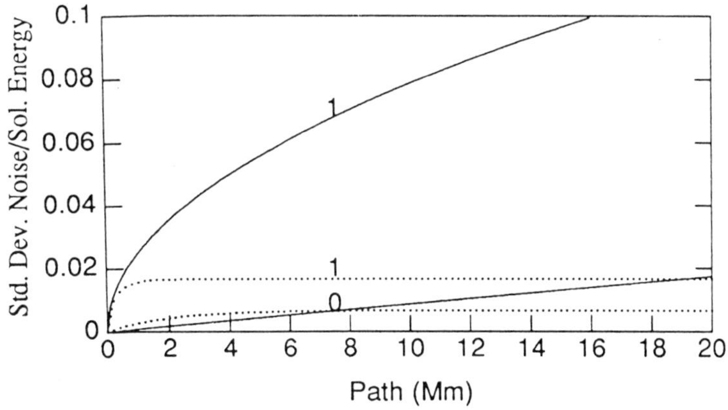

Figure 19 Standard deviations of the soliton energy ("ones") and of the noise in empty bit periods ("zeros"), versus distance, both normalized to the soliton energy itself, with the sliding filters dotted curves), and with no filtering at all, save for a single filter at z, passing only 8 noise modes (solid curves). Assumed fiber loss rate, fiber effective core area, amplifier spacing, and excess spontaneous emission factor are 0.21 dB/km, $50\,\mu m^2$, 28 km and 1.4, respectively. (From Ref. 20.)

Figure 20 shows the expected standard deviation in timing jitter, as a function of distance, based on Eqs. (9) and (17) and the assumption $\gamma_t = 0.32$. Note the dramatic improvement now obtained with filtering: at 10 Mm, the predicted standard deviations in timing jitter, for 0 Gbit/s (where the acoustic effect = 0) and 20 Gbit/s, are 1.50 and 1.86 ps, respectively. [It may also be helpful to note that the frequency fluctuations exhibit essentially the same saturation behavior as the dotted curves of Fig. 19. The variance of the timing jitter, however, is obtained by integrating those frequency fluctuations with respect to z. Since the timing variance thus soon becomes directly proportional to z, the dotted curves of Fig. 20 correspondingly rise with the square root of z.]

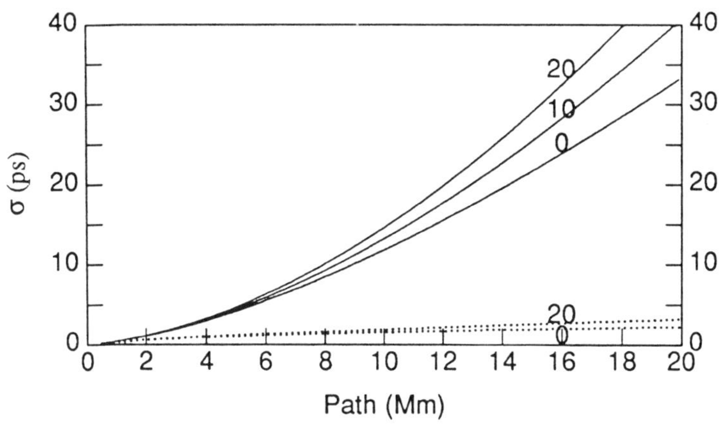

Figure 20 Standard deviation in soliton arrival times (σ), versus distance, with sliding filters (dotted curves), and with no filtering at all (solid curves). Jitter from the acoustic effect, which depends on bit rate (Gbit/s indicated by numbers next to each curve), is included. All assumed parameters are the same as in Figs. 17 and 19. (From Ref. 20.)

Figure 21 shows some closely related experimental results, viz., eye diagrams of the soliton pulses as received at ~10 Mm, made by triggering a sampling 'scope with appropriately gated intervals of the locally recovered clock. The small timing jitter can perhaps best be appreciated from the non-integrated eye (above), while the small jitter in amplitude can best be appreciated from the form of the eye (below) obtained following integration of the pulses by a low-pass microwave filter. [For the latter, a regular bit-pattern was used, in order to avoid certain small pattern-dependent effects of the integration.] When corrected for electronic jitter, the remaining jitter seen in these eye diagrams is entirely consistent with the predictions of Figs. 19 and 20.

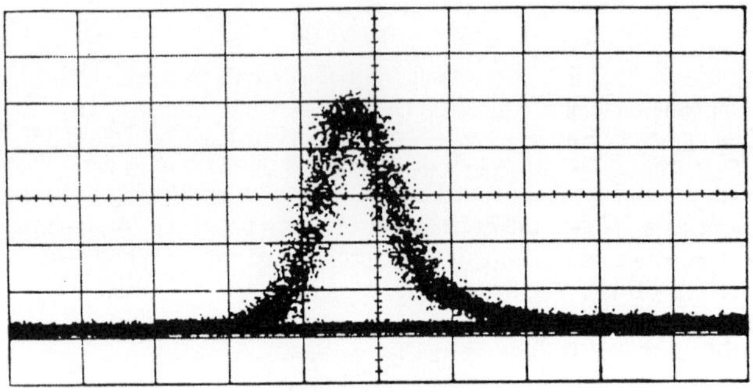

20 ps/division

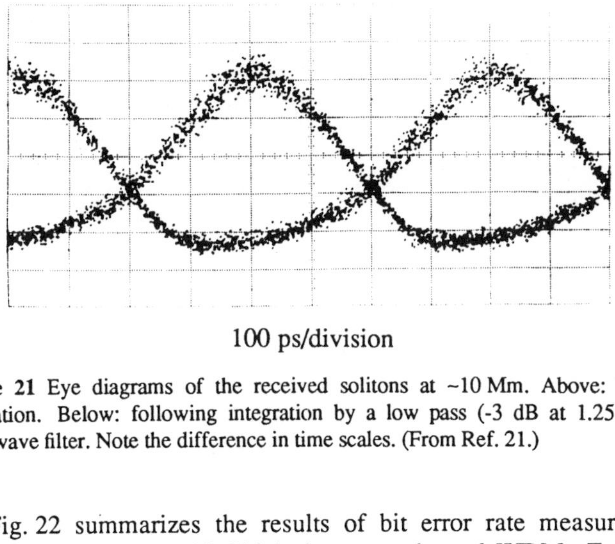

100 ps/division

Figure 21 Eye diagrams of the received solitons at ~10 Mm. Above: before integration. Below: following integration by a low pass (-3 dB at 1.25 GHz) microwave filter. Note the difference in time scales. (From Ref. 21.)

Finally, Fig. 22 summarizes the results of bit error rate measurements made at 10 Gbit/s, single channel, and at 20 Gbit/s, in a two-channel WDM. For these measurements, 2^{14} bit pseudo-random words were always used for the measured channel, while in the WDM experiments, both regular as well as random patterns were used in the potentially interfering channel. The safety margin on amplitude errors is large at all distances, so the errors here are thought to be entirely from timing. The error-free distances obtained here, while respectably large, are nevertheless considerably smaller than from the soliton timing jitter alone. Therefore "technical" jitter, as perhaps from the devices used in time-division demultiplexing the incoming signals from the incoming 10 Gbit/s rate to 2.5 Gbit/s, probably creates a significant (but ultimately avoidable) reduction in time-phase margin. Thus, it is not unreasonable to expect that eventually, the single channel bit rate can be extended, perhaps to 20 Gbit/s or beyond.

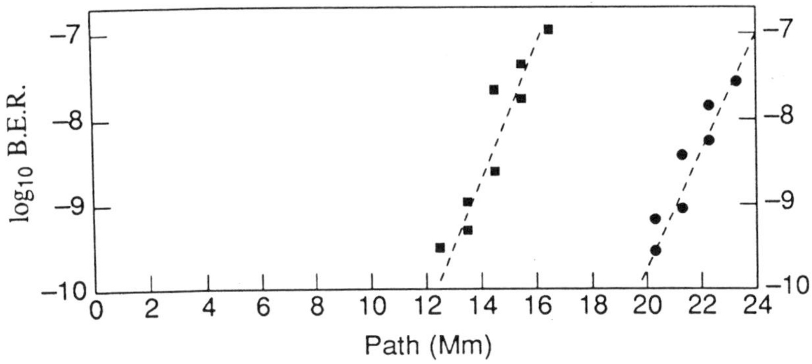

Figure 22 Measured B.E.R. versus path, with the sliding-frequency guiding filters, at 10 Gbit/s, single channel, and at 20 Gbit/s in a two channel WDM. The measured channel always contained a 2^{14} bit pseudorandom word. (From Ref. 21).

By now it should be abundantly clear that the sliding-frequency filters live up to their promise of greatly reducing jitter and thereby allowing for correspondingly increased bit rates. But the sliding filters bring other important benefits as well. Several of these stem from the fact that the sliding-frequency filters remove other noise fields just as efficiently as they do spontaneous emission. For example: (1) They make the system performance much less sensitive to defects in the input pulses. In a system without sliding filters, the dispersive wave radiation shed by imperfect pulses always remains in the same frequency band with the solitons, and hence can add significantly to the noise. Such noise tends to be generated entirely in the first fraction of a Megameter, however, so that the strong, sliding-frequency filters can completely remove it (long) before the receiving end. Thus, requirements on the pulse source can be significantly relaxed. (2) A similar "cleaning" effect applies to another potentially important source of noise, viz., the dispersive wave radiation shed by the solitons as they react to fiber birefringence, or "polarization dispersion." (In a system without sliding-frequency filters, this noise effect tends to accumulate directly with the distance, and can be significant even when the effective birefringence is well below the threshold at which it can begin to disperse individual pulses.[22]) (3) In a broad-band system, noise from well outside the signal bandwidth as well as that within it can easily grow to the point that it tends to overwhelm the amplifiers and quench the signal gain. In soliton transmission with sliding-frequency filters, however, the total noise soon settles into a narrow band and saturates at no more than a few percent of the (time-average) signal power. (4) The strong sliding-frequency filters allow solitons to be packed closer together. Neighboring solitons tend to interact through overlap of their tails, whereby they mutually induce frequency (velocity) shifts on each other.[23] By resisting the frequency shifts, the filters tend to stem the interaction. (The interaction, which is weak and always attractive for orthogonally polarized pulse pairs, is made much stronger by the potential interference of co-polarized pulses, and is attractive when the interference is constructive.) The net result is that the minimum allowable separation of co-polarized pulses can be reduced from about five pulse widths to about three. (5) As will be detailed in the following section, the strong damping constants permitted by filter sliding greatly reduce the various penalties of WDM, such as from soliton-soliton overlap at the system input[24].

Finally, it should be noted that the sliding-frequency guiding filters, in conjunction with the amplifiers and the (nonlinear) fiber itself, constitute a kind of distributed, polarization-independent, WDM-compatible regenerator, whose components (except, perhaps, for the amplifiers) are all *passive*. By contrast, filtering in the time domain[25] requires active devices, with all the same drawbacks of complexity, reduced reliability, high cost, and incompatibility with WDM that accompany electronic regeneration.

V. WAVELENGTH DIVISION MULTIPLEXING WITH SOLITONS

Wavelength division multiplexing has the potential to multiply the capacity of a soliton transmission system by a significant factor[26]. Indeed, we have already reported how, at several stages of the development, it has been possible to double the prevailing single-channel capacity with a two channel WDM (Figs. 15 and 22). As will be shown in this section, however, the multiplication factor is potentially considerably greater than just two.

Since soliton transmission is always at a finite dispersion ($D > 0$), solitons of a shorter-wavelength channel gradually overtake and pass through ("collide" with) those of a longer-wavelength channel. Of importance here is the collision length, L_{coll}, the distance two solitons must travel down the fiber in the course of passing through each other. If we define L_{coll} as beginning and ending when the solitons' half-power points coincide, then transparently $L_{coll} = (2\tau)/(D\Delta\lambda)$, where $\Delta\lambda$ is the channel wavelength separation. [For example, in the WDM of Fig. 22, where $\tau/D \sim 42$ nm-km, and where the channel separation was 0.8 nm, L_{coll} was just a little greater than 100 km.] The potential for WDM results from the fact that solitons- even the slightly imperfect path-average solitons- can emerge from such collisions virtually unscathed. That is, all their properties, such as shape, energy, frequency spectrum, etc, are left unaltered by the collision. Of course, the colliding solitons, as nonlinear pulses, do interact- they accelerate (attract) each other- but in a lossless and otherwise unperturbed fiber, the mutual acceleration is perfectly symmetric, so that the resultant temporary velocity shift returns exactly to zero following the collision (see Fig. 23). Thus, the only result of the collision is a mutual displacement in time, given by the z integral of the (reciprocal) velocity shift.

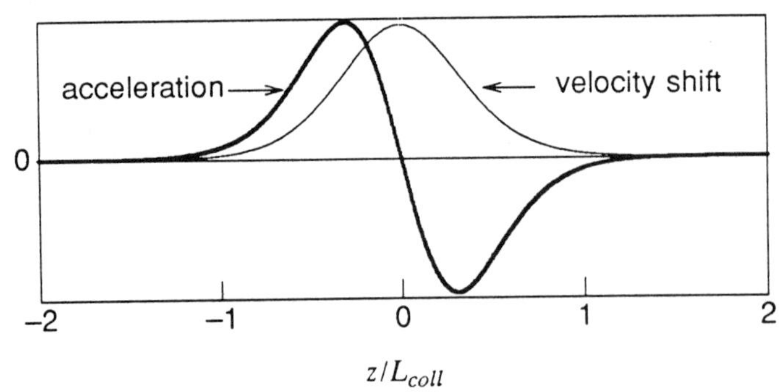

Figure 23 Collision-induced acceleration and velocity change, as a function of distance, to the faster of two colliding solitons, in a lossless fiber. (Effects on the slower soliton are of opposite sign.) (From Ref. 26.)

In the usual theoretical treatment, the colliding solitons are given dimensionless frequencies (radians per t_c) of $\pm\Omega$, which frequencies also play the role of reciprocal velocities (see Appendix). One can then easily show that the collison-induced time shift, in units of t_c, is $\delta t = \pm\Omega$, which can be translated into practical units as

$$\delta t = \pm 0.183\,(L_{coll}/z_c)^2\,\tau \tag{18}$$

where the coefficient 0.183 is the approximate value of $(1.762..)^{-3}$. Thus, for example, in the WDM experiment of Fig. 22 (where $\tau = 19$ ps, and $L_{coll}/z_c \sim 0.5$), each collision results in a displacement of $\delta t \sim 0.9$ ps. Although this may seem small, the accumulated effect of many such collisions could cause a problem in a broad-band system. As will be shown shortly, however, sliding-frequency filters strongly quench these displacements.

A more serious potential problem arises in the system using lumped amplifiers, where, for example, the collision may be centered about an amplifier. If L_{coll} is less than the amplifier spacing, the first half of the collision, taking place in a region of low signal intensity (small nonlinear effect), will yield lower acceleration than the second half, taking place in a region of high signal intensity (large nonlinear effect). Thus, the symmetry of the collision will be destroyed, and the collision can result in a large net velocity shift. That shift, multiplied by the remaining dispersion to the end of the system, can result in an unacceptably large shift in the pulse's arrival time, and hence in error.

On the other hand, when the collision length is two or more times greater than the amplifier spacing, the symmetry of the collision is sufficiently restored that the net velocity shift becomes negligibly small[26]. Figure 24 shows how this works for an example where the ratio of L_{coll} to the amplifier spacing is just 2.5.

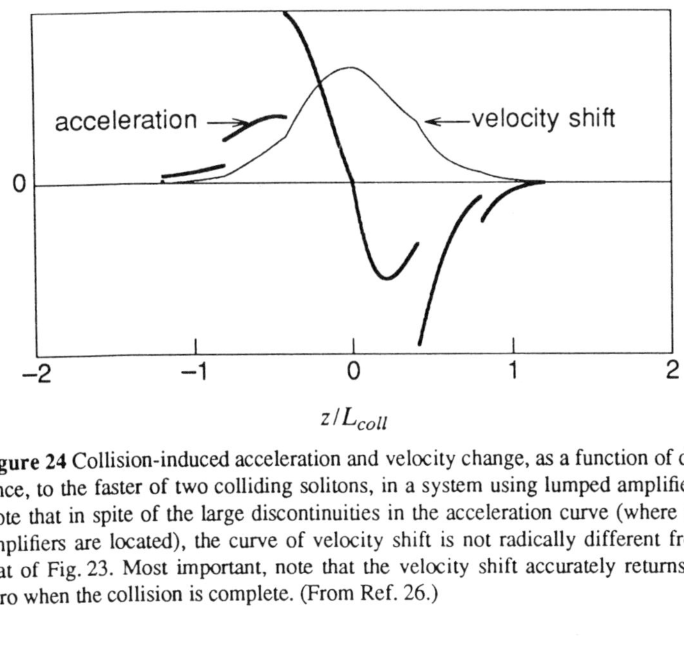

Figure 24 Collision-induced acceleration and velocity change, as a function of distance, to the faster of two colliding solitons, in a system using lumped amplifiers. Note that in spite of the large discontinuities in the acceleration curve (where the amplifiers are located), the curve of velocity shift is not radically different from that of Fig. 23. Most important, note that the velocity shift accurately returns to zero when the collision is complete. (From Ref. 26.)

Since the collision length is inversely proportional to the channel spacing, this condition implies that there is a maximum allowable channel spacing.

$$\Delta\lambda_{max} \approx \frac{\tau}{DL_{amp}} \tag{19}$$

where L_{amp} is the amplifier spacing. For example, the parameters of the experiment of Fig. 22 and Eq. (19) yield $\Delta\lambda_{max} \sim 1.6$ nm. Since, in that example, the adjacent channel spacing could be as small as ~0.4 nm, in principle one could have as many as five WDM channels. The problem of precisely equalizing the amplifier gains for 5 simultaneous channels has yet to be worked out, however.

In the previous section, we have already hinted that filters, and especially the strong, sliding-frequency filters, could have major benefits for WDM. Obviously, the filters will tend to quench collision-induced velocity shifts. For example, in a broad-band system, one must be careful to avoid the half-collisions that can occur when pulses overlap at the system input: the resultant large velocity shifts, when multiplied by the length (dispersion) of the entire line, yield unacceptably large time shifts, and a correspondingly large increase in bit error rate[24]. Filters, on the other hand, reduce the effective path for integrating the velocity shifts to $1/\gamma$, which, in the case of the strong sliding-frequency filters, can be as short as a few hundred kilometers. Thus the time shifts from pulse overlap at input are reduced at least twenty-fold or more from what they would be in a broad band system of trans-oceanic length. In the WDM experiment of Fig. 22, for example, pulse overlap at system input was observed to produce only a minor increase in bit error rate.[21]

Not so immediately apparent, however, is the fact[27] that filters can also eliminate the time displacements of Eq. (18). Nevertheless, the argument can be made quite simply, as follows: Without filtering, colliding solitons accelerate each other by $a_0(z)$, whose first integral is $v_0(z)$, and whose second integral is $t_0(z)$, as exemplified by the curves in Figs. 23 and 24. (The δt of Eq. (18) is just $t_0(\infty)$.) The only requirement is that the velocity shift return to zero when the collision is over, i.e., that

$$v_0(\infty) = \int_{-\infty}^{\infty} a_0(z)dz = 0 \tag{20}$$

The damping provided by the filters is just an acceleration proportional to -v. When the driving term a_0 is included, the complete equation of motion is:

$$\frac{dv}{dz} = a_0(z) - \gamma v(z) \tag{21}$$

Eq. (21) can be rewritten as:

$$v(z) = \frac{1}{\gamma}[a_0(z) - \frac{dv}{dz}] \tag{21a}$$

To get t, we simply integrate:

$$t(z) = \int_{-\infty}^{z} v(x)dx = \frac{1}{\gamma}\int_{-\infty}^{z} [a_0(x) - \frac{dv}{dx}]dx \tag{22}$$

We are primarily interested in $t(\infty)$:

$$t(\infty) = \frac{1}{\gamma}\int_{-\infty}^{\infty} a_0(x) - \frac{1}{\gamma}\int_{-\infty}^{\infty} \frac{dv(x)}{dx}dx \tag{22a}$$

The first term on the right = 0 by virtue of Eq. (20). The second term is just $v(\infty)$, which for a filtered system with no excitation beyond a certain point, must = 0. Therefore, $t(\infty) = 0$, Q.E.D.

It is easy to include the case of pulse overlap at system input in the above. In that case the lower limit of the integrals is not $-\infty$, but has some appropriate finite value (= 0 for exact pulse overlap at the system input). From the thus modified Eq. (22a), $t(\infty)$ takes the form of a non-zero velocity times the effective distance $1/\gamma$.

V.A. Polarization Division Multiplexing

It has been shown[28] that solitons maintain a very high degree of polarization over the very long paths of interest here. This property is unique to solitons. Of course, since practical fibers are somewhat birefringent, the state of polarization of the soliton stream evolves rather rapidly with distance. Nevertheless, at any given point along the path (and in particular, at the receiving end), there is only small variation from pulse to pulse in the state of polarization, determined only by the (spontaneous emission) noise. This fact should allow for polarization division multiplexing (PDM) with solitons, and by that means, the achievement of a further increase in single-channel bit rates.

To achieve PDM in practice, one would split the output of a mode-locked laser into two equal parts, impose separate data on each pulse stream with the use of electro-optic modulators, and then recombine such that pulses from the one stream are orthogonal to those from the other, and such that the two sets of pulses are interleaved in time. At the receiving end, an electronically controlled set[29] of wave plates would automatically convert the emerging, arbitrary states of polarization (which would also tend to vary slowly with time as temperature and other influences modify the fiber's birefringence) into two standard linear ones. These could then be separated (demultiplexed) by a polarizing beam splitter. For further technical details, see Ref. 28.

It has also been shown, through numerical simulation, that there is another great advantage to the PDM scheme just described. That is, the interaction between orthogonally polarized neighboring pulses is greatly reduced from that obtaining when their polarization states are the same. In particular, the minimum allowable pulse separation is reduced from the ~5 pulse widths cited earlier in this paper to about half that value. (It will be recalled that the strong, sliding-frequency guiding filters tend to accomplish about the same reduction, but it helps to have both effects at work.)

VI. CONCLUSION

The potential capacity and performance with solitons in trans-oceanic, all-optical transmission are truly exceptional: robust, error free transmission has already been demonstrated at a net rate of 20 Gbit/s per fiber, and it is reasonable to expect a several times increase in that rate in the near future. This exceptional capacity, in an essentially passive system, derives from the unique ability of solitons to transit a long chain of narrow-band frequency filters, where rate-limiting noise effects are severely suppressed. It also stems from the fact that both the solitons themselves and the simple filtering technique are compatible with extensive wavelength division multiplexing. At a deeper level, however, these wonderful properties stem from the fact that the soliton is the only stable solution to the propagation equation. As such, it is the only pulse allowing for transmission approaching the most fundamental limits.

Acknowledgment

We would like to thank S. G. Evangelides for the many invaluable numerical simulations that he performed to guide this work. Of equal importance are the extensive efforts of assistant M. J. Neubelt and post-doc E. Lichtman in preparing and running the many experiments reported on here. Neubelt is also to be thanked for his tireless efforts as administrator for the group's computer system. G. T. Harvey has almost single-handedly developed a fiber ring laser soliton source which has been of vital importance. Others whose assistance has been greatly appreciated are: M. Haner, J. R. Simpson, and B. M. Nyman.

APPENDIX: THE NLS EQUATION AND BASIC PROPERTIES OF OPTICAL SOLITONS

Apologia

This section constitutes a tutorial on the derivation of the nonlinear Schrödinger equation and related issues. It is intended for those who would like to delve a bit more deeply into the mathematical theory of solitons than was done in the text. It is adapted, primarily through minor changes in notation and with the addition of the immediately following section, from the Appendix to Ref. 3 of the text.

Dispersion relations and related velocities

Central to any physical problem involving waves is the dispersion relation, $\omega(k)$, where ω and k are the angular frequency and wavenumber (rate of change of phase with distance), respectively. From that relation, one can obtain the phase and group velocities, as follows. The phase of any frequency component is

$$\phi(z,t) = \omega t - kz$$

For an observer at fixed z, $d\phi/dt = \omega$. For an observer in a frame moving at velocity v, however,

$$\frac{d\phi}{dt} = \omega - kv \tag{A1}$$

For stationary ϕ, the terms in Eq. (A1) are zero. The corresponding solution, $v_p = \omega/k$, is the phase velocity at frequency ω.

Now consider a pulse. Its peak occurs where its Fourier components add most constructively. [The phases are all the same at the peak of a minimum-width pulse, but dispersion tends to gradually destroy that perfect alignment.] Thus, in a frame moving with the pulse, i.e., in one moving at the group velocity, v_g, the time rates of change of phase of the various frequency components are more alike than in any other frame. This condition is obtained when the derivative of Eq. (A1) with respect to k is set equal to zero:

$$\frac{d}{dk}\left(\frac{d\phi}{dt}\right) = \frac{d\omega}{dk} - v_g = 0 \tag{A2}$$

Thus the well-known result $v_g = d\omega/dk$ is obtained very simply. [Note that Eq. (A2) does not imply that $d\phi/dt$ itself is necessarily zero. In general, there is a certain phase slip with respect to the pulse peak; this phase slip is zero only in the special case where the phase and group velocities are identical.]

There is a certain symmetry, or interchangeability, between time and distance. [In the final analysis, there is no real distinction between observing a pulse develop in time at constant z, and taking a snapshot of it, i.e., observing it as a function of z at a fixed t.] In keeping with that fact, one can choose to write the dispersion relation as $k(\omega)$, and the phase of any frequency component by the sign convention opposite to that used above:

$$\phi(z,t) = kz - \omega t$$

For observation made at fixed time, $d\phi/dz = k$. For an observer in a time frame moving at reciprocal velocity v^{-1}, however,

$$\frac{d\phi}{dz} = k - \omega v^{-1} \tag{A1a}$$

Note that setting the derivative here equal to zero makes the reciprocal phase velocity $v_p^{-1} = k/\omega$. The reciprocal group velocity, v_g^{-1}, can be obtained analogously to the way v_g was obtained above. That is, in a time-frame moving with the pulse, i.e., with reciprocal group velocity v_g^{-1}, the derivative of Eq. (A1a) with respect to ω must be equal to zero:

$$\frac{d}{d\omega}\left(\frac{d\phi}{dz}\right) = \frac{dk}{d\omega} - v_g^{-1} = 0 \tag{A2a}$$

Thus one obtains the result $v_g^{-1} = dk/d\omega$, which we shall use shortly.

Derivation of the NLS equation

Consider a single transverse mode transmission line, such as an optical fiber, and denote distance down the line by z. Let an electromagnetic wave in the line be represented by a scalar function $U(z,t)$ proportional to the complex field amplitude, such that the power P in the line is given by

$$P = P_c \mid U \mid^2$$

The proportionality constant P_c can be considered as a power unit. For frequencies near some central frequency ω_0, the nonlinear Schrödinger equation results from conditions such that a dispersion relation of the form

$$k = k_0 + k'(\omega - \omega_0) + \frac{1}{2}k''(\omega - \omega_0)^2 + k_2 P \tag{A3}$$

adequately describes the propagation of monochromatic waves $U = u_0 exp(ikz - i\omega t)$ in the line. Equation (A4) has the form of a Taylor series expansion of $k(\omega,P)$ in the neighborhood of $(\omega_0,0)$. For solitons in fibers, the last two terms of Eq. (A3) are of comparable importance, k_2 is positive, k'' is negative, and succeeding higher order terms (e.g., $k'''(\omega - \omega_0)^3$, $k'_2(\omega - \omega_0)P$, etc.) can be neglected or adequately treated as perturbations. For the moment, we assume that the line has no loss or gain; i.e., that the constants k_0, k'', and k_2 are all real.

The expression for the reciprocal group velocity, namely

$$v_g^{-1} = \frac{\partial k}{\partial \omega} = k' + k''(\omega - \omega_0) \tag{A4}$$

identifies k' as the reciprocal group velocity at frequency ω_0, and k'' as its frequency dispersion constant. The dispersion parameter D used widely to describe fibers is the wavelength derivative of v_g^{-1}, and so is related to k'' by

$$D = -\frac{2\pi c}{\lambda^2} k''$$

The term $k_2 P$ in Eq. (A3) represents the primary nonlinear effect, self-phase modulation, resulting from the intensity dependence of the refractive index of the fiber. Note that the dependence of the group velocity on power is among the higher order terms not included in Eq. (A3).

If we now remove the central frequency and wave vector from U by defining

$$u(z,t) = U\, e^{i(\omega_0 t - k_0 z)} \tag{A5}$$

so that when U is written out explicitly, $u(z,t)$ becomes

$$u(z,t) = u_0\, e^{i[(k-k_0)z - (\omega-\omega_0)t]} \tag{A5a}$$

then the wave equation for u which is necessary and sufficient to reproduce Eq. (A3) exactly is

$$-i\frac{\partial u}{\partial z} = ik'\frac{\partial u}{\partial t} - \frac{1}{2}k''\frac{\partial^2 u}{\partial t^2} + k_2 P_c \,|\, u \,|^2 u \tag{A6}$$

This can be shown by inserting Eq. (A5a) in Eq. (A6).

The standard form of the nonlinear Schrödinger equation is generated from Eq. (A6) by transforming to a retarded time frame (new time proportional to $t - k'z$, so that a pulse at ω_0 is stationary in that frame) to eliminate the first t derivative, and choosing unit values of time and distance such that $k'' = -1$ and $k_2 = 1$ when measured in those units, and the power unit already mentioned. The appropriate new variables are

$$t' = (t - k'z)/t_c$$

$$z' = z/z_c \tag{A7}$$

where the unit values t_c, z_c, and P_c satisfy the relations

$$-t_c^2/z_c = k''$$

$$(z_c P_c)^{-1} = k_2 \tag{A8}$$

The resulting equation is the nonlinear Schrödinger equation

$$-i\frac{\partial u}{\partial z'} = \frac{1}{2}\frac{\partial^2 u}{\partial t'^2} + |\, u \,|^2 u \tag{A9}$$

Eq. (A9) is often referred to as a dimensionless form of the nonlinear Schrödinger equation. It is more helpful to think of it as having dimensions, with z', for example being a distance simply measured in units of z_c rather than in meters or kilometers or any other standard unit. Thus, $z'=2$ means a distance of 2 times z_c.

Small values of gain or loss can be accommodated by adding a term $-i(\alpha/2)u$ to the right side of Eq. (A9), where α is the coefficient of energy gain per z_c (negative values of α represent loss). If we now drop the primes on z and t (we shall do this consistently from now on), the result is Eq. (1) of the text.

There is an important arbitrariness left in the definitions of the three unit values z_c, t_c, and P_c, since there are only two relations (Eq. (A8)) that they must satisfy. One unit value may be chosen freely, and thus different real world fields can be represented by the same solution of Eq. (A9), and vice versa. In particular, if one solution of Eq. (A9) is $u(t,z)$, then different scalings of the same real world field give other solutions of the form $Ru(Rt,R^2z)$, where R is the ratio of the values of t_c. This scaling transformation of the solutions of Eq. (A9) can be verified by direct substitution.

Pulse motion in the retarded time frame

Another important transformation of the solutions of Eq. (A9) is that produced by a carrier frequency shift. Because the inverse group velocity dispersion constant has the value -1 in the soliton unit system, a frequency shift produces an inverse velocity shift of equal magnitude. Thus, for the same solution $u(t,z)$ as above, one finds yet other solutions, frequency shifted by Ω, of the form

$$u(t+\Omega z,z)e^{-i(\Omega t+\Omega^2 z/2)} \tag{A10}$$

These transformations apply to *any* solutions of Eq. (A9).

Solitons

We now turn our attention to solitons. The elementary single soliton solution of Eq. (A9) is just Eq. (2) of the text, which we repeat here for convenience:

$$u = sech(t)\,e^{iz/2} \tag{A11}$$

In the real world it has a peak power of P_c, since the maximum of $|u|^2$ is unity. As already indicated in the text, it has full width τ at half power of ~$1.76t_c$, since $sech^2(1/76../2) = 1/2$. It has a wavenumber shift of $(2z_c)^{-1}$ because of the nonlinearity, an effect that is important in understanding its interactions with perturbing non-soliton field components.

Any single real world soliton can be expressed in the form of Eq. (A11) by the appropriate choice of scale and frequency. Alternatively, application of the scale and frequency transformations to Eq. (A11) yields the more general form

$$u = A\,sech(A(t+\Omega z))\,e^{i(\phi-\Omega t)} \tag{A12}$$

where

$$\phi = (A^2-\Omega^2)z/2$$

The form given by Eq. 12 is necessary to the consideration of perturbation or multisoliton problems, such as soliton-soliton collisions in WDM, for example.

Relation to material properties of the fiber

We now discuss the relationships of the parameter values of Eq. (A3) with the material properties of the fiber. For a plane wave electromagnetic field traversing a uniform transparent medium, the wavenumber k is related to the refractive index n of the medium by $k = n\omega/c$. The index of quartz glass, of which the fibers are made, can be expressed as

$$n = n_0 + n_2 I \tag{A13}$$

where $n_0 = 1.45$ at $\lambda = 1.56\ \mu m$, where n_2 has the polarization-averaged value, already given in the text, of $2.7\times10^{-8}\mu m^2/W$), and where I is the intensity in watts/μm^2.

A single mode fiber consists of a core surrounded by a cladding of slightly lower index. We can express the wavenumber of the fiber mode in the form $k = \omega n_{eff}/c$, where n_{eff} is the effective index of the fiber. The linear part of n_{eff} varies, depending on frequency, in a range between the indices of the cladding and the core, so the low-field dispersion relation (k_0, k', k'') can be tailored somewhat by varying the fiber geometry; such modification accounts for the dispersion shifted fibers often referred to in the text. The nonlinear index is substantially independent of frequency and has the same value for both core and cladding. Its perturbing effect on the effective index of the fiber can be evaluated adequately as an average of $n_2 I$ over the fiber cross section, weighted by the intensity distribution I. If we thus write $n_{eff} = n_{0eff} + n_2 P/A_{eff}$, where $P = \int I dA$ is the power, we find the effective area given by

$$A_{eff} = \frac{P^2}{\int I^2 dA}$$

For typical fiber geometries the effective area is within a factor of two of the geometric core area. Thus we have the result

$$k_2 = \frac{\omega_0 n_2}{c A_{eff}} \tag{A14}$$

Inserting Eq. (A14) into Eq. (A8) yields Eq. (5) of the text.

Finally, we note that for pulse lengths much shorter than those considered in this work, for example pulses shorter than one picosecond, one must include consideration of higher order terms in the dispersion relation, and can no longer ignore the frequency dependence of k_2, or the effects of intrapulse Raman scattering.

REFERENCES

[1] R. J. Mears, L. Reekie, I. M. Jauncey, and D. N. Payne, "Low-noise erbium-doped fiber amplifier operating at 1.54 μm," *Electron. Lett.* **23,** 1026 (1987)

[2] E. Desurvire, J. R. Simpson, and P. C. Becker, "High-gain erbium-doped traveling-wave amplifier," *Optics Lett.* **12,** 888 (1987).

[3] L. F. Mollenauer, J. P. Gordon, and M. N. Islam, "Soliton propagation in long fibers with periodically compensated loss," *IEEE J. Quantum Electron.* **QE-22**, 157 (1986)

[4] L. F. Mollenauer and K. Smith, "Demonstration of soliton transmission over more than 4000 km in fiber with loss periodically compensated by Raman gain," *Opt. Lett.* **13**, 675 (1988)

[5] L. F. Mollenauer, S. G. Evangelides, and H. A. Haus, "Long distance soliton propagation using lumped amplifiers and dispersion shifted fiber," *J. Lightwave Technol.* **9**, 194 (1991)
Also see:
A. Hasegawa and Y. Kodama, "Guiding-center soliton in optical fibers," *Opt. Lett.* **15**, 1443 (1990) (Note: the term "guiding" here has no connection whatsoever with the "guiding-filters" of Refs. 20 and 21.)
and
K. J. Blow and N. J. Doran, "Average Soliton Dynamics and the Operation of Soliton Systems with Lumped Amplifiers," *Photonics Tech. Lett.*, **3**, 369 (1991)

[6] L. F. Mollenauer, M. J. Neubelt, S. G. Evangelides, J. P. Gordon, J. R. Simpson, and L. G. Cohen, "Experimental study of soliton transmission over more than 10,000 km in dispersion shifted fiber," *Opt. Lett.* **15**, 1203 (1990)

[7] L. F. Mollenauer, B. M. Nyman, M. J. Neubelt, G. Raybon, and S. G. Evangelides, "Demonstration of soliton transmission at 2.4 Gbit/s over 12,000 km," *Electron. Lett.* **27**, 178 (1991)

[8] J. P. Gordon, "Dispersive perturbations of solitons of the nonlinear Schrödinger equation," *J. Opt. Soc. Am. B* **9** (1992)

[9] N. S. Bergano, J. Aspell, C. R. Richardson, P. R. Trischitta, B. M. Nyman, and F. W. Kerfoot, "Bit error measurements of 14,000 km, 5 Gbit/s fiber-amplifier transmission system using circulating loop," *Electron. Lett.* **27**, 1889 (1991)

[10] J. P. Gordon and L. F. Mollenauer, "Phase noise in photonic communications systems using linear amplifiers," *Opt. Lett.* **15**, 1351 (1990)

[11] S. Saito, T. Imai, T. Sugie, N. Ohakawa, Y. Ichihashi, and T. Ito, "An over 2,200 km coherent transmission experiment at 2.5 GBits/s using erbium-doped-fiber amplifiers," post deadline paper PD2-1 in *Postdeadline Papers* of the Optical Fiber Communication Conference, San Francisco, California, 22-26 Jan., 1990; Optical Society of America, Washington, DC, publishers.

[12] J. P. Gordon and L. F. Mollenauer, "Effects of fiber nonlinearities and amplifier spacing on ultra long distance transmission," *J. Lightwave Technol.* **9**, 170 (1991)

[13] J. P. Gordon and H. A. Haus, "Random walk of coherently amplified solitons in optical fiber," *Opt. Lett.* **11**, 665 (1986)

[14] G. T. Harvey and L. F. Mollenauer, "Harmonically mode-locked fiber ring laser with an internal Fabry-Perot stabilizer for soliton transmission," *Opt. Lett.* **18**, 107 (1993)

[15] L. F. Mollenauer, M. J. Neubelt, M. Haner, E. Lichtman, S. G. Evangelides, and B. M. Nyman, "Demonstration of error-free soliton transmission at 2.5 Gbit/s over more than 14,000 km," *Electron. Lett.* **27**, 2055 (1991)

[16] A. Mecozzi, J. D. Moores, H. A. Haus, and Y. Lai, "Soliton transmission control," *Opt. Lett.* **16**, 1841 (1991)

[17] Y. Kodama and A. Hasegawa, "Generation of Asymptotically Stable Optical Solitons and Suppression of the Gordon-Haus Effect," *Opt. Lett.* **17**, 31 (1992)

[18] L. F. Mollenauer, E. Lichtman, G. T. Harvey, M. J. Neubelt, and B. M. Nyman, "Demonstration of error-free soliton transmission over more than 15,000 km at 5 Gbit/s, single-channel, and over 11,000 km at 10 Gbit/s in a two-channel WDM," *Electron. Lett.* **28**, 792 (1992)

[19] E. M. Dianov, A. V. Luchnikov, A. N. Pilipetskii, and A. M. Prokorov, "Long-range interaction of solitons in ultra-long communication systems," *Soviet Lightwave Communications* **1**, 235 (1991)

[20] L. F. Mollenauer, J. P. Gordon, and S. G. Evangelides, "The sliding-frequency guiding filter: an improved form of soliton jitter control," *Opt. Lett.* **17,** 1575 (1992) [Note: Because of an improper combination of sign conventions, the numerical simulations reported on in this *Letter* incorrectly yielded the offset $\delta\omega < 0$, so that we mistakenly recommended down-sliding. The mistake is corrected here.]

[21] L. F. Mollenauer, E. Lichtman, M. J. Neubelt, and G. T. Harvey, "Demonstration, using sliding-frequency guiding filters, of error-free soliton transmission over more than 20,000 km at 10 Gbit/s, single-channel, and over more than 13,000 km at 20 Gbit/s in a two-channel WDM," *Electron. Lett.* **29**, 910 (1993)

[22] L. F. Mollenauer, K. Smith, and J. P. Gordon, "Resistance of solitons to the effects of polarization dispersion in optical fibers," *Opt. Lett.* **14**, 1219 (1989)

[23] J. P. Gordon, "Interaction forces among solitons in optical fibers," *Opt. Lett.* **8,** 596 (1983)

[24] N. A. Olsson, P. K. Andrekson, J. R. Simpson, T. Tanbun-ek, R. A. Logan, and K. W. Wecht, "Bit-error-rate investigation of two-channel soliton propagation over more than 10,000 km, *Electronics Lett.* **27**, 695 (1991)

[25] M. Nakazawa, E. Yamada, H. Kubotra, K. Suzuki, "10 Gbit/s soliton transmission over one million kilometers," *Electron. Lett.* **27**, 1270 (1991)

[26] L. F. Mollenauer, S. G. Evangelides, and J. P. Gordon, "Wavelength Division Multiplexing with Solitons in Ultra Long Distance Transmission Using Lumped Amplifiers," *J. Lightwave Technol.* **9**, 362 (1991)

[27] A. Mecozzi and H. A. Haus, "Effect of filters on soliton interactions in wavelength-division-multiplexing systems," *Opt. Lett.* **17**, 988 (1992)

[28] S. G. Evangelides, L. F. Mollenauer, J. P. Gordon, and N. S. Bergano, "Polarization Division Multiplexing with Solitons," *J. Lightwave Technol.* **10**, 28 (1992)

[29] F. Heismann and M. S. Whalen, "Broadband reset-free automatic polarization controller," *Electron. Lett.* **27**, 377 (1991)

ENERGY TRANSFER MECHANISMS AND EXCITED STATE DYNAMICS OF Yb^{3+}, Tm^{3+} AND Ho^{3+} DOPED $Gd_3Ga_5O_{12}$ SINGLE CRYSTALS

A. Brenier, L. C. Courrol, C. Pédrini, C. Madej and G. Boulon

Laboratorie de Physico-Chemie des Matériaux Luminescents
Unité de Recherche Associée au CNRS n° 442
Université Claude Bernard Lyon 1
43, Boulevard du 11 Novembre 1918
69622 Villeurbanne Cedex, France

ABSTRACT

The sensitization of the Ho^{3+} 2μm laser emission by the Yb^{3+}-Tm^{3+}-Ho^{3+} ions in $Gd_3Ga_5O_{12}$ host is presented. We have analyzed both down and up-conversion processes and we have measured the up-conversion quantum yield. Various models are used to modelize the excited state dynamics. Especially, conversion of near infrared laser light into green emission has been obtained by a two step process: absorption from the fundamental states of the ions followed by a mechanism which we have called *looping mechanism*. The latter is composed of an absorption from the $^5I_7$(Ho) excited state and by a positive feedback process via Tm ions. A two ion (Tm-Ho) model is given to describe the dynamics and the gain and losses of the loop are evaluated.

I. INTRODUCTION

Sensitization of the Ho 2μm laser emission is usually obtained from the $^5I_7 \rightarrow {}^5I_8$ transition of Ho^{3+} ions in different hosts with an adequate sensitization. The usual sensitizers are Cr^{3+}, Er^{3+} and Tm^{3+} ions. The two former ones allow efficient flash lamp pumping while the latter is well adapted for laser diode pumping at around 800 nm. As can be seen in

Nonlinear Spectroscopy of Solids: Advances and Applications
Edited by B. Di Bartolo and B. Bowlby, Plenum Press, New York, 1994

Fig. 1, Ho-Tm sensitization by means of Yb^{3+} ions is also attractive because of the possibility of laser diode pumping between 920 and 975 nm where powerful laser diodes are now available and because Yb^{3+} ions have only one excited state ($^2F_{5/2}$) with no possibility of excited absorption. The different down-conversion channels are labelled 1,2,3 and 4 Fig.1. But the counterpart of the presence of sensitizers is the introduction of new energy losses. One of them is the Ho→Tm back transfer, labelled 5 in Fig. 1, which can be an important source of losses when the Tm concentration increases and which was presented in a previous work [1]. Another one source of losses is the up-conversion processes occuring from the Yb and Ho or Tm in levels towards the Tm and Ho visible levels as shown in Fig. 2.

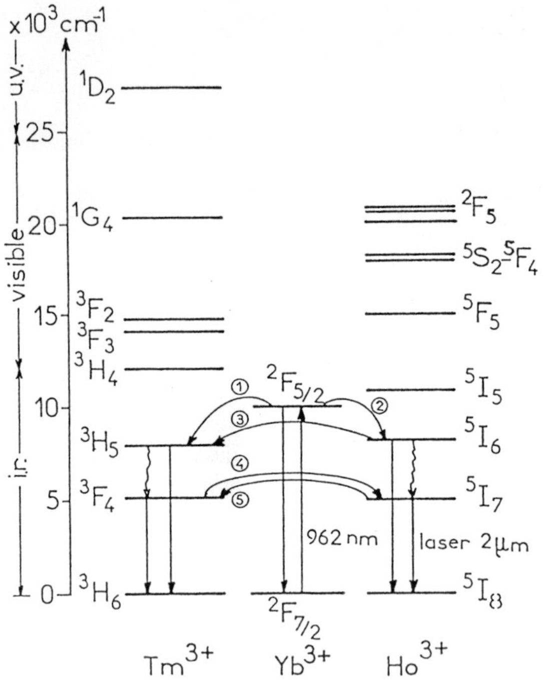

Fig. 1 Down-conversion processes in GGG:Yb:Tm:Ho sample under Yb ion pumping

We have shown that efficient Tm→Ho energy transfers occurs in $Gd_3Ga_5O_{12}$ (GGG) [2] and we have recently found that Yb→Tm and Yb→Ho energy transfers are leading to an enhancement of the Ho 2µm fluorescence. Due to the presence of 3 ions which two to them (Tm and Ho) have several involved electronic levels, the excited state dynamics under short pulse excitation into Yb ions is rather complicated one and both up and down conversion processes are observed.

To analyze the fluorescence dynamics, we are not allowed to use the rate equations because the Grant's procedure [3] can be applied only in the case of a very fast energy diffusion among donors and the decays were not found exponential. We therefore used standard models like Inokuti-Hirayama and Yokota-Tanimoto models which allow to describe the decay rates of the donors, but also other models like the Chandrasekhar's procedure [4] and the transfer function method [1] which permit to describe also the dynamics of the acceptors.

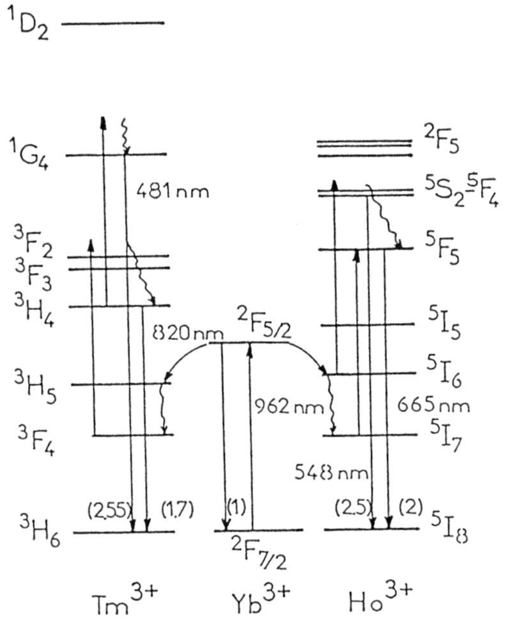

Fig. 2 Up-conversion processes in GGG:Yb:Tm:Ho sample under Yb ion pumping

We had measured with an integrating cavity the up-conversion energy losses quantum yield after infrared pumping into the Yb ion and we had shown that in the triply doped crystal it is weaker than in the Yb-Ho doubly doped one [5]. The reason of this could be the existence of efficient down-conversion mechanisms that bring back the up-converted energy in the $^5S_2$-$^5F_4$ (Ho) levels towards the 2μm emitting level, the $^5I_7$(Ho) manifold. If it is true, it means that a positive feed-back increases the population of the $^5I_7$ manifold and the purpose of this work was to point out this process. The excitation source chosen is a continuous laser one near 755 nm (obtained from a CW Ti:Sapphire laser) and corresponding to the $^5I_7 \rightarrow {}^5S_2$-$^5F_4$ (Ho) excited state absorption in such a way that a so-called looping mechanism occurs. The result is the conversion of the infrared excitation into a green emission from the $^5S_2$-$^5F_4$ (Ho) levels as shown in Fig.3.

All results given at 77 K in this paper are related with the triply Yb(5%)-Tm(5%)-Ho(0.5%) doped single crystalline sample which has been grown in our laboratory by using the Czochralski method. We also want to mention that only Tm and Ho ions are involved and we assume that at 77 K the Yb ions do not play a role.

II. DOWN-CONVERSION PROCESSES

- $^2F_{5/2}(Yb) \rightarrow {}^3H_5(Tm)$ and $^2F_{5/2}(Yb) \rightarrow {}^5I_6(Ho)$ energy transfers sketched by channels 1 and 2 in Fig. 1 : the time evolution of the Yb fluorescence is correctly described by the Yokota-Tanimoto expression with critical radius Ro equals to 10.9 Å A and 11.4 Å and

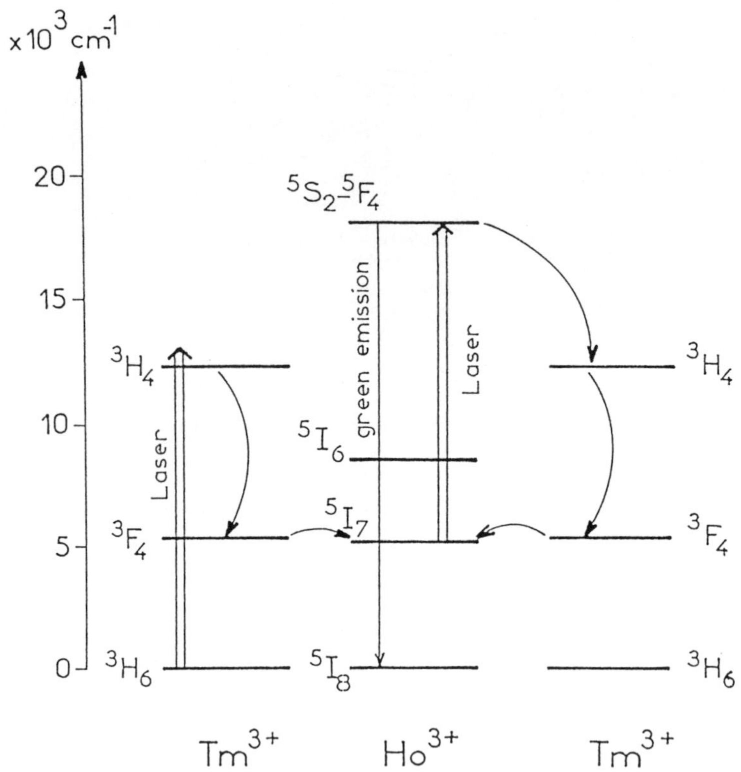

Fig. 3 Scheme of energy levels involved in the model of looping mechanism. The arrows represent the channels of energy mainly by cross-relaxation processes. The first step after non-resonant absorption in Tm ions feeds the $^5I_7$(Ho) level. The second step corresponds to resonant excited-state absorption in Ho ions of the laser pump and the so-called loop can be seen between $^5S_2$-$^5F_4$, $^3H_4$, $^3F_4$, $^5I_7$ and again $^5S_2$-$^5F_4$ levels of both Tm and Ho ions.

transfer efficiencies close to unity for GGG:5%Yb, 5%Tm and GGG:5%Yb, 3%Ho samples respectively.

- $^5I_6$(Ho)→$^3H_5$(Tm) energy transfers sketched by channel in Fig. 1 have been detected corresponding to a 90% quantum yield.

- $^3F_4$(Tm) $^5I_7$(Ho) energy transfer have been studied in details in a previous paper [1]. It has been shown that the two emitting levels are in thermal equilibrium. From the fits of the experimental temporal evolutions of the $^5I_7$ and $^3F_4$ fluorescences, we are able to determine the critical radium Ro of forward and backward energy transfers : they are of the same order of magnitude as the radii obtained by Dexter's classical way from the overlap of absorption and emission spectra, only in the case for which we take into account of diffusion process. As an example, for the back-transfer Ho→Tm, we have found in GGG:1%Tm, 0.5%Ho Ro=8.4 Å, 7.2 Å and 11.6 Å from spectroscopic data, with diffusion and without diffusion respectively.

III. UP-CONVERSION PROCESSES

The excitation of the Yb ions is also followed by anti-Stokes emissions coming from both upper Tm levels at 820nm ($^3F_4$) and 481nm($^1G_4$) and upper Ho levels at 548nm ($^5S_2$-$^5F_4$) and 665nm($^5F_5$). As it can be seen in Fig. 2, two or three-steps nearly resonant up-conversion processes have been observed and, in addition, the power dependence of the intensities of fluorescence at the top of their time evolution have been mentioned near the ground states. We expected that n should be integer but non-integer values, specially 2.5 for Ho^{3+} green emission and 2.55 for Tm^{3+} blue emission probably reveal a mixing of two and three-photon processes.

The up-conversion quantum yield from Yb ions towards a given Ho or Tm level was obtained using an integrating sphere. The results are gathered in Table 1.

TABLE 1 - Up-conversion quantum yields after infrared excitation into Yb ions.

CRYSTAL	LEVEL	QUANTUM YIELD %
5% Yb 0.5% Ho	$^5S_2$-$^5F_4$	2.7
5% Yb 3% Ho	$^5S_2$-$^5F_4$	4.4
5% Yb 5% Tm	$^3H_4$	6.0
5% Yb 5% Tm	$^1G_4$	0.03
5% Yb 5% Tm 0.5% Ho	$^3H_4$	5.1

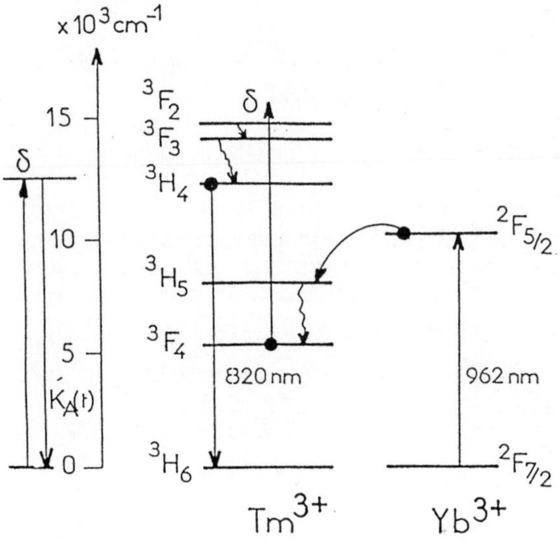

Fig. 4 a Yb-Tm up-conversion transfer in GGG:5%Yb:5%Tm sample:example of the $^3H_4 \rightarrow ^3H_6$Tm transition.

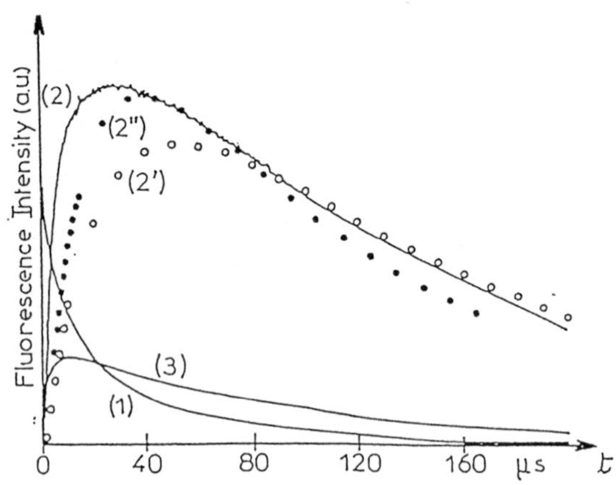

Fig. 4 b Time evolution of the $^3H_4 \rightarrow ^3H_6$ emission :

Curve (1) Experimental data of the $^3H_4 \rightarrow ^3H_6$ fluorescence after a direct d excitation into $^3H_4$ level. $K_A(t)$ is the response of the acceptor level to the excitation.
Curve (2) and circles (2') and (2") : after excitation into the Yb ions
Curve (2) : experimental data of the $^3H_4 \rightarrow ^3H_6$ fluorescence
Circles (2') : from a model which does not take into account the fluctuations of positions of donors and acceptors.
Circles (2") : from a model taaking account partially the fluctuations of positions of donors and acceptors.
Curve (3) : time evolution of the source e(t) used to obtain circles (2').

We turn now to the dynamics of up-conversion energy transfers in Yb-Ho and Tb-Tm bidoped samples. A tentative of modelisation of the initial rise-time evolution of all anti-Stokes fluorescences have been done. We only give in this paper the treatment of the $^3H_4 \rightarrow ^3H_6$ red transition as shown in Fig. 4. The population $N_A(t)$ of the $^3H_4$ acceptor level is described in the framework of the fluorescence transfer function method [1] in terms both of $K_A(t)$, the response of the acceptor level to an excitation, described by an Inokuti-Hirayama expression and of e(t) the source of excitation by the following relation :

$$N_A(t) = \int_0^t e(t-t')K_A(t')dt' \qquad (1)$$

If e(t) is taken as the product of $^2F_{5/2}$(Yb) and $^3F_4$(Tm) populations, the model is not able to reproduce the experimental data in Fig. 4. We think that it is due to the fact that our model supposes the same yield of up-conversion transfer for all Yb-Tm ion pairs whatever the interatomic distance. In fact, because of the fluctuations of ion distribution in the regions of the crystal where the ion pairs are closer, the coefficient of transfer are higher and a more important contribution to the up-conversion process is expected. Then we have used the Chandrasekhar's procedure to describe the source of excitation which can now be represented as :

$$e(t) = \iint a(\phi,\phi').N_{D_\phi}(t).N_{D'_{\phi'}}(t)d\phi d\phi' \qquad (2)$$

where N_{D_ϕ} and $N_{D'_{\phi'}}$ are the donor and acceptor populations created by the Yb ions belonging to one class of donor ions having the same total transfer rate f on each acceptor ion. The up-conversion probability $a(\phi,\phi')$ is an unknown function whose value was 1 in the previous model. Taking a larger probability $a(\phi,\phi') = \phi$ in (2) inserted in (1), we obtain the curves 2" in Fig.4 and the experimental rise times are now much better.

IV. EXCITED STATE ABSORPTION

The laser excitation spectrum of the Ho green emission (near 540 nm, transition $^5S_2$-$^5F_4 \rightarrow ^5I_8$) is shown in Fig. 5 (lower curve) at 77 K, because in this range of wavelengths, there is no peak of absorption from the Yb-Tm-Ho fundamental levels, there is no difficulty to attribute the absorption of the laser beam by the $^5I_7 \rightarrow ^5S_2$-$^5F_4$ (Ho) excited state transition, the $^5I_7$ level being fed first by a structureless vibronic absorption from the Tm ground state to the $^3H_4$(Tm), followed by a $^3H_4$ (Tm)+$^3H_6$(Tm)$\rightarrow$$^3F_4$(Tm)+$^3F_4$(Tm) cross-relaxation mechanism, followed by a$^3F_4$(Tm)+$^5I_8$(Ho)$\rightarrow$$^3H_6$(Tm)+$^5I_7$(Ho) transfer. We see that the excited state absorption cross-section peak occurs at 753.74 nm and so this wavelength was chosen for the pumping source.

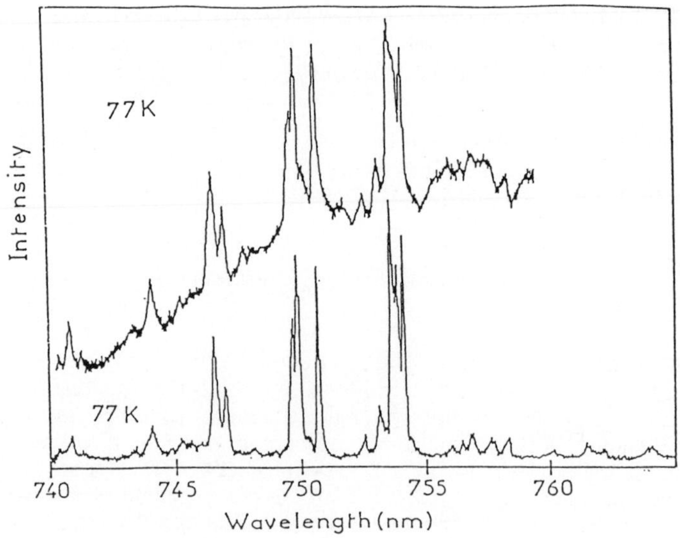

Fig. 5 Correlation between CW Ti:Sapphire laser excitation spectra by monitoring either 2 mm emission (upper curve) or green emission (lower curve) of Ho ions.

The intensity I(t) of the transmitted beam by the sample shows in Fig. 6 the time evolution of the excited state absorption. At time t=0, when the laser beam is turned on, there is only absorption from the fundamental state of the Tm ions because the $^5I_7$(Ho) level is empty. Then the $^5I_7$(Ho) level fills and the transmitted intensity decreases. At long time the transmitted intensity $I(t)$ becomes constant and its value depends of the incident power.

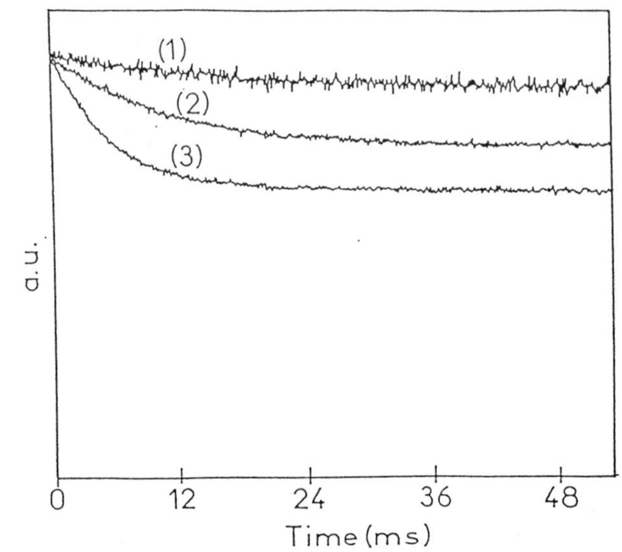

Fig. 6 Time evolution of the intensity I(t) of the transmitted laser beam at different incident power : (1) 0.113 W, (2) 0.580 W, (3) 1.17 W. The pump wavelength is 753.74 nm.

V. POSITIVE FEED-BACK AND LOOPING MECHANISM

We want to examine now what happens to the $^5I_7$(Ho) population when the laser beam is resonantly absorbed by the $^5I_7 \rightarrow {}^5S_2$-$^5F_4$(Ho) transition. The result is given by a laser excitation spectrum of the 2μm Ho emission because this last one originates from the $^5I_7$ level and its intensity is proportional to the $^5I_7$ population. Such a spectrum is given in Fig. 5 (upper curve). We can see that its peaks correspond exactly to the ones of the green emission laser excitation spectrum (Fig. 5 lower curve). More, we can see that the first spectrum appears in positive values. It means that when an excited state absorption occurs from the $^5I_7$ level it does not decrease its population but to the contrary, it increases it. This fact proves that a positive feed-back process occurs, the origin of it being a $^5S_2$-$^5F_4$(Ho)+$^3H_6$(Tm)$\rightarrow$ $^5I_7$(Ho)+$^3H_4$(Tm) transfer, followed by a $^3H_4$(Tm)+$^3H_6$(Tm)$\rightarrow$$^3F_4$(Tm)+$^3F_4$(Tm) cross-relaxation mechanism, followed by a $^3F_4$(Tm)+$^5I_8$(Ho)$\rightarrow$$^3H_6$(Tm)+$^5I_7$(Ho)transfer as briefly indicated in Fig. 3.

Let us start with one ion in the $^5I_7$(Ho) level. If by excited state absorption the ion is sent into the $^5S_2$-$^5F_4$(Ho) levels, it will be turned back on the $^5I_7$(Ho) level by the feed-back process with a certain efficiency and so a loop has been done. But the number of ions found at the end of the loop is generally different of one and can be called the gain of the loop. We understand that if the gain is larger than the losses, the dynamics of the system will be qualitatively different than in the case where it is smaller : in the first case, we have a photon avalanche. This phenomena has been studied more than ten years in monodoped crystals [6-7].

Gain and losses of the loop have been studied with the almost equivalent one-ion/three levels system represented in Fig. 7 and extensively used [7]. We have found that the threshold condition for photon avalanche was :

$$R_2(2H_{CR}+h-1) > 1/t_2+R_1+R_1R_2t_3(1-H_{CR}) \qquad (3)$$

where R_1 and R_2 are the absorption rates from the fundamental and excited states respectively, H_{CR} and h are the quantum yields of the cross-relaxation and 3→2 deexcitation mechanism respectively, t_2 and t_3 are the lifetimes of the spontanous deexcitations of the 2 and 3 levels respectively.

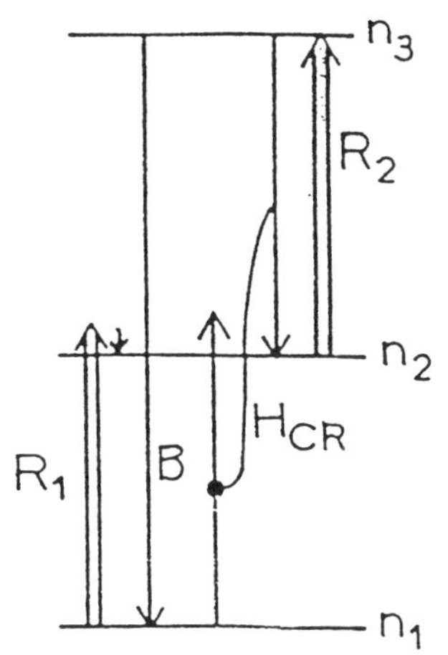

Fig. 7 Scheme of the one-ion/three-levels equivalent system.

Expression (3) has a very simple physical interpretation. Its left side is the net gain of the loop : when an ion is removed (-1) from the n_2 level by laser beam (this occurs with a yield R_2), it is returned back by the cross-relaxation process with a quantum yield $2H_{CR}$ and by the $3 \rightarrow 2$ deexcitation with a quantum yield h. The right side of (3) describes the losses from the n_2 level : $1/t_2$ for the spontaneous deexcitation, R_1 and $R_1R_2t_3(1-H_{CR})$ are losses of n_2 feeding by the laser pump due to accumulation of population into the n_2 and n_3 levels respectively. Numerically we have found that our system was working below the threshold because of the high value of R_1, the absorption from the fundamental state. This is due to the high value of the Tm concentration which is ten times the Ho one.

VI. CONCLUSION

We have summarized that efficient down-conversion energy transfers drive the pump energy into Yb ions towards both Tm and Ho ions. We have also studied up-conversion processes and measured the quantum yield of the energy losses. In addition, we have shown taht a good description of up-conversion dynamics must take into account fo the fluctuations of distribution of the donor and acceptor positions.

We have obtained green fluorescence from the $^5S_2$-$^5F_4$(Ho) level under pumping with a CW Ti:Sapphire laser at liquid nitrogen temperature in Ho-Tm-Yb doped $Gd_3Ga_5O_{12}$ due to a looping mechanism involving the two Ho-Tm species of ions. It is composed by an excited state absorption in the Ho ions followed by a positive feed-back process involving the Tm ions. We have calculated the gain and the losses of the looping mechanism from a theoretical expression leading, for the first time in the case of any value of the pump rate R_1 in the ground states of the ions, to the theoretical determination of the threshold for photon avalanche. We show that the threshold is an increasing function of the absorption from the ground state : weaker the last one is, easier it will be to obtain up-conversion via photon avalanche.

ACKNOWLEDGMENTS

We are grateful to Misses M. Blanchard, H. Defour and E. Jouffray, and to Mrs A. Lagriffoul, J. Y. Rivoire, B. Varrel and R. Perrin for efficient technical assistance. This work is supported in part by the Defense Ministry DGA/DRET and by Région Rhône-Alpes.

REFERENCES

1. A. Brenier, G. Boulon, C. Madej, C. Pédrini, L. Lou, J. of Luminescence **54** (1993) 271
2. A. Brenier, C. Madej, C. Pédrini, G. Boulon, J. Phys.: Condens. Matter **3,** (1991) 203 and 7887
3. J.C. Grant, Phys. Rev. **B4**, (1971) 648
4. S. Chandrasekhar, Rev. Mod. Phys. **15** (1943) 1
5. A. Brenier, L.C. Courrol, C. Pédrini, C. Madej, G. Boulon, Phys. Review B **49**, to be published, 1er janvier 1994.
6. J. S. Chivian, W. E. Case, D. D. Eden, Appl. Phys. Lett. **35** (2) (1979) 124.
7. U. Oetliker, M. J. Riley, P. S. May, H. U. Gudel, J. Lumin. **53** (1992) 553.

NANOPARTICLES IN AMORPHOUS SOLIDS AND THEIR NONLINEAR PROPERTIES

Renata Reisfeld*
Inorganic Chemistry Department,The Hebrew University of Jerusalem
91904 Jerusalem, Israel

ABSTRACT

Nonlinear behaviour of glasses and especially thin glass films doped by nanoparticles make these materials an interesting class for optoelectronic devices especially when connected with their waveguide properties. From theoretical point of view these materials provide models for three-dimensional quantum confinement and allow identification of the energy of the excited states. These energy states of the quantum dots are positioned between the discrete energy levels of the atoms or bands of molecules and the broad band of the condensed phase.

We shall describe the preparation of glasses doped by semiconductor metal and dye particles and discuss the origin of their nonlinear properties. Preparation of glass films doped by CdS, CuCl and CuBr quantum dots will be also presented and their properties analyzed. The relation between the energies of the electronic states and size of the particles will be discussed and their nonlinear properties reported.

I. INTRODUCTION

Nanoparticles of metals and semiconductors that have atomic arrangements, in agregates of the size ranging between molecular clusters and "infinite" solid-state arrays of atoms have distinctive properties determined by the extent of confinement of highly delocalized valence electrons. At this interface, the total number of atoms and the geometrical disposition of each atom can be used to significantly modify the electronic and photonic response of the medium. In addition to the novel inherent physical properties of the quantum-confined moieties, their "packaging" into nanocomposite bulk materials can be used to define the confinement surface states and environment, inter cluster interactions, the quantum-confinement geometry, and the effective charge-carrier density of the bulk.

There is strong spatial delocalization of valence electrons, and therefore a small crystallite must grow fairly large to achieve the limiting bulk electronic structure. The intermediate-sized clusters can have unique properties, characteristic of neither the molecular nor solid state limits. Examples include metal clusters (as predicted by free electron shell theory) and semiconductor crystallites which exhibit the bulk unit cell but have only partial band structure development. These latter crystallites are sometimes called quantum dots or nanoclusters, and are the subject of this chapter. Nanoclusters have optical spectra that can be tuned in wavelength simply by varying the crystallite size. They

Enrique Berman Professor of Solar Energy.

Nonlinear Spectroscopy of Solids: Advances and Applications
Edited by B. Di Bartolo and B. Bowlby, Plenum Press, New York, 1994

represent, in three-dimensions, an analogy to the quantum well semiconductor heterostructures that show a one-dimensional quantum size effect.

In highly polarizable materials such as intrinsic semiconductors, the availability of vacant states can result in extended anharmonic potential surfaces and correspondingly large nonlinear responses. Changes associated with quantum confinement, resulting in restriction of the electronic wave function to smaller and smaller regions of space, will have profound effects on the nonlinear displacement of charge with optical electromagnetic fields. The ability to generate nanostructure quantum-confined materials atom by atom in the context of recent scientific developments in nonlinear phenomena [1] is both an intriguing and potentially rewarding challenge.

For example, the future application could be found in electrooptic and optical switching. Because photons travel three order of magnitude faster than electrons in media, they have obvious advantages for signal transmission [2]. The combination of electronics and photonics that has been used in optical signal processing and particularly optical switches can be dramatically improved if the slowest member of the chain, that is, the electric signal used to induce refractive index changes necessary for optical switching, is eliminated. Recent theoretical and experimental results indicate a dramatic change in the linear and nonlinear optical (NLO) response as a function of decreasing the size of semiconductor atomic clusters [3, 4] to the point of quantum confinement. The use of the NLO response of these near "molecular" clusters provides a means to optically switch with photons alone at correspondingly faster response times.

An excellent example is the recent announcement by IBM Yorktown Heights of the development of experimental computer chips for transmitting and receiving data over fiber optic lines at speeds of one billion bits per second. Each chip has more than 8000 transistors with characteristic features as small as 1μm. One "receiver" chip holds 50 times more optical and electronic components than ever previously assembled on a chip for optoelectronic data transmission. Quantum-well confinement technology is used to create the light pulses for data transmission [5].

Electrons in recently developed devices can be confined to panes, lines or mathematical points quantum dots, corresponding to two, one and zero dimensions.

In addition to the exciting basic science quantum dots promise properties that could be harnessed for a range of electronic and optical applications. Arrays of densely packed dots could form a substrate for computers of unprecedented power. Dots could also constitute materials capable of absorbing and emitting light at predetermined wavelengths their designers specify, or could even serve as the basis of semiconductor lasers more efficient and precisely tuned than any now in existence.

Electrons confined in a plane have no freedom of motion in the third dimension. Those confined in quantum wire are free in only one dimension, and those confined in a quantum dot are not free in any dimension. For common semiconductors, the length scale for a free conduction electron is about 100 Angstrom (= 10 nanometer).

II. QUANTUM STRUCTURES AND ATOM ASSEMBLIES

The basic ideas of quantum confinement or quantum wells were introduced in the early 1970s. In these nanostructures, the dimensions of the wave function of the electron-hole pair (exciton) in the lowest excited state of the nanocluster are comparable to the physical size of the particle. This quantum confinement of the exciton means that the continuum band of energies becomes more molecular in character , with narrow ranges of energy and line structure in the optical spectra. For the chemist, less delocalization means less energy stabilization; a reflection of this is that the absorption band for direct transitions of nanosized semiconductor clusters is shifted to higher energies than in the extended bulk parent materials. The "photo charging" of the semiconductor particle, that is, photo inducing high electron concentrations into a narrow conduction band with a small effective density of states, also changes the band gap and the resulting absorption edge. In addition to NLO applications, the resulting phenomena can be used to carry out electron-transfer processes that are otherwise inaccessible to the bulk semiconductor.

Most importantly for NLO that strong nonlinear responses are observed near exciton resonances, a feature not available in comparable materials with no quantum confinement [6, 7]. In quantum structures the number of states that are available in the relatively small energy band can be very limited, so that adding or removing only a few carriers can substantially change the absorption edge and thus the optical coefficients (such as the index of refraction). Conversely, perturbation of the quantum cluster by light can modify the optical coefficients and the absorption edge. Small changes in the dielectric properties of the confining structure or the surface states of that cluster also can introduce large changes in optical properties. Quantum confinement can therefore be used to change the position and nature of the resonance absorption transitions, as well as to make the NLO response particularly susceptible to modification by optical and low-frequency electric fields. An additional and equally important property of quantum-confined materials is that ultra fast response times become accessible because of the increased importance of surface states.

The bulk NLO response depends on the number of polarizable electrons per unit volume so that a high-density collection of quantum nanoclusters is required, both for achieving a basic understanding of the quantum-structure phenomena and for device development. In designing such devices, one, by necessity, must consider not only the atomic construction of the individual quantum wires, boxes, sheets, and dots, but also the assembly of these into a macroscopic unit. Glasses and polymers have the advantage of being readily made into thin films for device application, but the disadvantage of having particle size distributions, which in present commercial samples typically vary by more than 10%. The consequence is that the exciton resonances are broadly distributed in energy so that wavelength-selective NLO phenomena are difficult to achieve [2].

The first hints that zero dimensional quantum confinement was possible came in the early 1980s, when A.I. Ekimov [8] and his colleagues at the Ioffe Physical-Technical Institute in St. Petersburg noticed unusual optical spectra from samples of glass containing the semiconductors cadmium sulfide or cadmium selenide. The samples had been subjected to high temperature; Ekimov suggested tentatively that the heating had caused nanocrystallites of the semiconductor to precipitate in the glass and that quantum confinement of electrons in these crystallites caused the unusual optical behavior.

How small must a nanocrystallite be for quantum confinement to be visible? In a vacuum the effects of confinement would begin to appear when the electron was trapped in a volume about 10 Angstrom across. That size implies an electron wavelength of 20 Angstrom and therefore an energy of about one fortieth of an electron volt.

The semiconductor physics comes to the aid of the nanotechnologist in the basic understanding of spectral behaviour. The wavelength of an electron depends on its energy and its mass. For a given wavelength, the smaller the mass, the larger the energy and the easier it is to observe the energy shift that confinement causes. The electrostatic potentials of the atoms in the crystalline lattice superimpose to provide a medium in which electron waves propagate with less inertia than they do in free space. The "effective mass" of the electron is thus less than its actual mass. In gallium arsenide the effective mass is about 7 percent of what it would be in a vacuum, and in silicon it is 14 percent. Quantum confinement (QC) in semiconductors occurs in volumes roughly 100 angstroms across.

In the QC regime hybrid molecular-solid state: the bulk unit cell is present, yet the spectra show three dimensional confinement and are in principle discrete. If the QC radius a_o, R $<a_o$ (the bulk exciton Bohr radius R), such as in R = 10 Angstrom in II-VI crystallites, the electron and hole are principally individually confined with minor electron-hole correlation and no hydrogenic exciton formation. This is the strong confinement limit. As an example let us take CdSe nanocrystallites of the radius of 16 Angstrom, the Bohr radius of the exciton in CdSe is 30 Angstrom [9].

In the weak confinement limit, when R is 2-4 times a_0, a hydrogenic exciton forms whose center of mass motion is quantized and exhibits discrete states within the crystallite [10]. CuCl crystallites provide the only well characterized examples in this regime: the a_0 here is 7 Angstrom. QCs show in principle only discrete transitions that are size tunable. Their optical properties are essentially molecular. Much of the QC nonlinear optical interest comes from the fact that the absorption coefficient at one wavelength might be much larger than in superlattices, and so the QC nonlinear effects per unit mass could be much larger.

The quantum-structure size that is desirable for optimum NLO response depends on the exciton diameter, the effective separation of hole and electron in the first excited state of the semiconductor. For large band gap semiconductors such as zinc sulfide (ZnS, E_g =3.6 eV) or cadmium sulfide (CdS, E_g = 2.58 eV), quantum confinement will not take place until the particle size is $\leq$ 15 Angstrom or $\approx$ 35 Angstrom, respectively, although for the small band gap indium antimonide (InSb, E_g = 0.23 eV) the exciton will be delocalized over $\approx 10^6$ atoms with a diameter of $\approx$ 1400 Angstrom. These numbers are "particle-in-the-box" estimates for quantum boxes or wells and different considerations apply to quantum wires or sheets.

The optical absorption threshold for nanocrystallites of this size shifts to higher energies - away from the red end of the spectrum - as the crystallite becomes smaller. This effect appears most elegantly in cadmium selenide clusters; the progression from deep red to orange to yellow as the diameter of the cluster declines can be clearly seen by the naked eye.

As seen above, the near band gap electronic absorption is predicted to be discrete and very sensitive to physical size.

Quantum confinement effects in semiconductor micro crystallites have been discussed by Efros and Efros [11] and the first quantitative effort using an idealized model for evaluation of the electronic levels of the quantum dots was performed by Brus [12]. He used wave-mechanical treatment to calculate the lowest eigenstate of an exciton confined in a spherical box, having either infinitely high walls or finite potential barrier. The calculations gave the correct trend of the blue shift of the lowest exciton level with the decrease of the size of the semiconductor particle. An improvement of the model was proposed by Fojtik et al. [13]. In their model the electron having the reduced mass $\mu = (1/m_e+1/m_h)^{-1}$ (where m_e and m_h are the effective masses of electron and hole, 0.19 and 0.8 for bulk CdS, respectively) moves around a massless hole placed at the center of the particle. The allowed energy levels are found by solving a semiclassical phase integral, as in the models of Brus [12].

Generally, the blue shift of the excition absorption peaks is discussed in two cases: one is valid for the excitons having an effective Bohr radius R much larger than the crystal size, a_o, $a_o/R<2$, and the other is for the exciton having an effective Bohr radius much smaller than the crystal size, $a_o/R>4$. These two regimes are named the 'electron-hole confinement' and the 'exciton confinement', for which energies are approximately given by, respectively,

$$\mathbf{E} = \mathbf{E_g} + (\hbar^2 \pi^2) / (2\mu \mathbf{a_o}^2), \qquad (1)$$

$$\mathbf{E} = \mathbf{E_g} + (\hbar^2 \pi^2) / (2\mathbf{M} \mathbf{a_o}^2) \qquad (2)$$

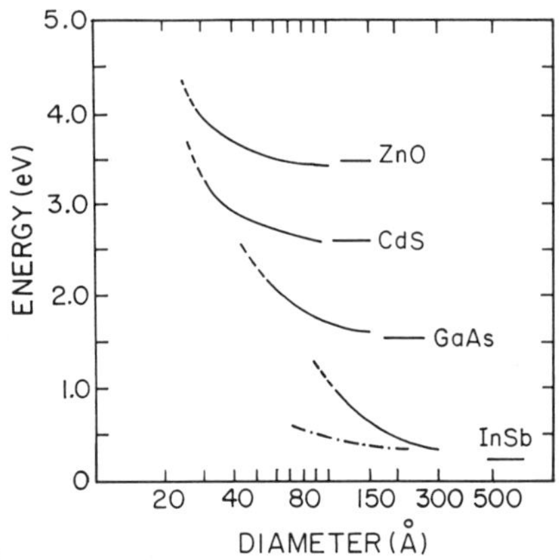

Fig. 1 Energy of the lowest excited electronic state as a function of crystallite diameter as calculated via wave function (13) from [12]. Short horizontal solid lines are the bulk band gap energies of the indicated materials. Dot-dashed line for InSb incorporates surface carrier charge density as described in the text.

where μ and M is the reduced mass ($\mu = m_e * m_h$) and the translational mass ($M = m_e + m_h$), respectively, of the exciton [13].

In general, the lower excited state energies can be calculated within the effective mass approximation using a different mass for the conduction and valence bands. In addition, the Coulomb interaction can be included to give an approximate expression for the lowest state energy [14]

$$E^* = E_g + \frac{\hbar^2 \pi^2}{2a_o^2}\left[\frac{1}{m_e} + \frac{1}{m_h}\right] - \frac{1.8e^2}{\varepsilon a_0} \qquad (3)$$

The expected behavior for different materials as calculated by Brus from formula (13) given in ref. [12] is shown in Figure 1 [14]. Small band gap III-V materials display their bulk energy gaps only at large diameter because of their very small electron effective mass. II-VI materials typically have larger masses and smaller dielectric coefficients. They display their bulk gaps at smaller diameters, and show one diameter at which significantly non zero effective mass and Coulomb terms accidentally cancel.

In both materials, transitions at much higher energies (typically in the far ultraviolet) involve localized discrete states near the center of the Brillouin zone [15]. These states, with energy insensitive to size, are present in the smallest crystallite. In a sense, they are σ to σ^* transitions within one unit cell.

Another approach is proposed by Schmidt and Weller [16]. Here the exciton is allowed to move inside the sphere defined by the radius of the quantum dot, i.e. both electron and hole are allowed to move, whereby the potential term is defined by the Coulomb interaction between the two particles V(r) is proportional to $e/\varepsilon|r_1-r_2|$, where r_1 and r_2 are the coordinates of the electron and the hole respectively.We start from the general Hamiltonian [17],

$$H = -(\hbar^2/2m_e)\nabla_1{}^2-(\hbar^2/2m_h)\nabla_2{}^2-e/\varepsilon|r_1-r_2| \qquad (4)$$

m_e and m_h are the reduced masses of the electron and hole and r_1 and r_2 are their corresponding coordinates where the first two terms are the kinetic energies of the electron and the hole respectively and the third term represents the Coulomb interaction between the two particles, screened by factor ε, the high frequency dielectric constant of bulk CdS.

The eigenvalues of the system represent the excited energy levels of the system counted from the lower edge of the conduction band. In the infinitely extended material the exciton exhibits hydrogen-like energy levels, proportional to $-(1/2n^2)$ with n=1, 2, 3, 4. As long as the crystal of CdS is much larger than the excitonic radius, these levels are bounded slightly below the conduction band edge.

When the electron and the hole are constrained in a small volume of radius a_o comparable to the radius of the exciton (about 19.3 Angstrom for CdS) all degeneracies are removed and the excitonic levels become blue-shifted in the respect to the conduction band edge.

Such a situation can be treated by the method used for two-electron atom. The Hamiltonian can be scaled by $r \rightarrow r/a_o$ to get the form.

$$H = H^0 + R(-1/r_{12}) \qquad (5)$$

where a_o, the radius of the quantum dot is treated as a perturbation term. The eigenvalues are then approximated by expansion around $R=a_o$;

$$E = 1/R^2(E^{(0)} + RE^{(1)} + (R^2)E^{(2)}...) \qquad (6)$$

where $E^{(0)}$ is the sum of kinetic energies of the electron and the hole and $E^{(1)}$ is the corresponding expectation value of $-1/r_{12}$.

Eigenfunctions are then a combination of hydrogen like functions, 1s,2s,3s for angular momentum L=0, and 2p,3p... for angular momentum L=1. For R smaller than the excitonic radius, the lowest state eigenfunction is just a product of the lowest-state functions of the electron and the hole. For larger values of R, however, the function of state should be expanded in terms of the compete orthogonal set of functions.

$$\Psi_a = \Sigma c_{a,b}\, \Phi_a(r_1)\Phi_b(r_2) \quad (7)$$

Predicted energy levels are calculated from Table 1 and Equation 6 in the first-order approximation and drawn in Fig. 2. These are compared with the energies calculated from

Table 1. Values of parameters for calculation of eigenvalues [17]

L	Configuration	$E^{(0)}$	$E^{(1)}$
0	1s1s	4.935	1.786
0	1s2s	7.776	1.780
0	2p2p	10.093	1.889
0	1s3s	12.511	1.777
0	2p3p	13.883	1.851
1	1s2p	5.926	1.620
1	2p1s	9.104	1.620
1	1s3p	9.715	1.724

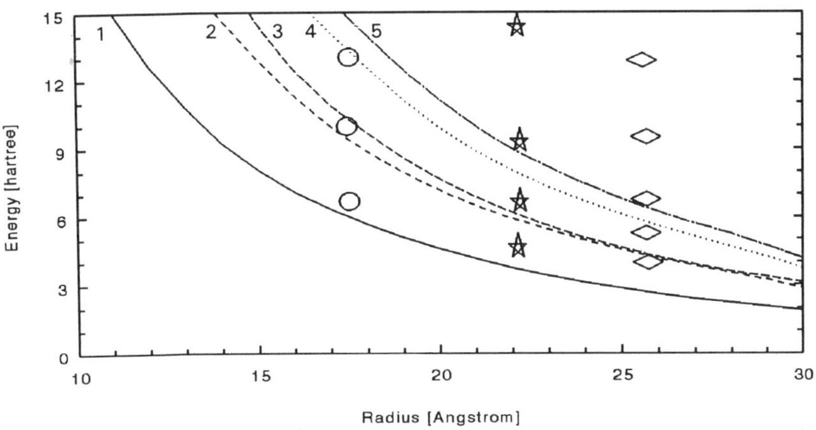

Fig. 2 Five lowest quantum levels of quantum dots of CdS obtained by first-order approximation of equation (6) and Table 1, L=0. Curve 1: 1s1s, curve 2: 1s2s, curve 3: 2p2p, curve 4: 1s3s, curve 5: 2p3p.These are compared with energy levels obtained from excitation spectra of thin films: Circles-low concentration of cadmium sulfide 0.051 mol/mol, stars-high concentration of cadmium sulfide 0.117 mol/mol, diamonds-GLYMO 0.127 mol/mol.

the excitation spectra on Fig. 3. The levels fit fairly well the experimental quantum levels of the CdS particles introduced into Glass films prepared by the sol gel method [18-22] where the CdS nanoparticles were formed in thin transparent films of either glass or Ormocer prepared by the sol gel method. The size dependence of the electronic levels in the glasses is analogous nanoparticles were formed in thin transparent films of either glass or Ormocer prepared by the sol gel method. The size dependence of the electronic levels in the glasses is

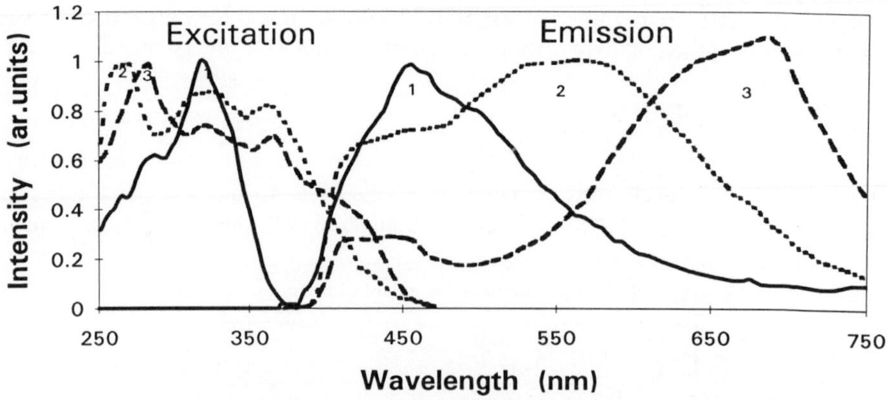

Fig. 3 Emission and excitation spectra of thin films containing CdS measured at 77 K. Curves 1: TEOS film containing initially cadmium nitrate 0.051 mol/mol TEOS. Curves 2: TEOS film containing initially cadmium nitrate 0.117 mol /mol TEOS. Curves 3: GLYMO film containing initially cadmium nitrate 0.127 mol/mol of GLYMO.

analogous to the behavior of small colloidal cadmium sulfide particles separated by electrophoresis [23]. Average radii of the CdS particles as indicated by the fit are 17, 22 and 26 Angstrom, in a good agreement with the measurements by X-ray diffraction and excitation maxima.

III. NONLINEAR OPTICAL GLASS MATERIALS

The nonlinear properties of glasses uniformly doped by semiconductor, metal or dye clusters are the mostly promising material for applications. This is due to their ability to be formed in desired shapes especially as wave guides and to their mechanical and optical qualities.

Glasses [24] and polymers [25, 26] have the advantage of being readily made into thin films for device application, but the disadvantage of having particle size distributions, which in present commercial samples typically vary by more than 10% [27]. The consequence is that the exciton resonances are broadly distributed in energy so that wavelength-selective NLO phenomena are difficult to achieve [28].

Since R.K. Jain and R.C. Lind [29] reported that some colored glass filters made of semiconductor fine particle-doped glass exhibited third-order optical nonlinearities of 10^{-9}–10^{-8} esu [29], nonlinear optical glass (NLOG) materials have been at the forefront of glass research. Main interests in the research are that NLOG has the potential to become a key material of all-optical (or photonic) switching devices which operate in picosecond or sub-picosecond response time, three orders of magnitude shorter than that of the limit attained by electronic switching devices [30]. Using all-optical switches, optical parallel data processing systems and optical time domain multiplexing transmission systems will be developed in the future.

The switching of these devices is achieved by the optical Kerr Effect, the change of refractive index of the material, n, with the intensity of the high power of input light, I. The effect is expressed by the simplified equation

$$n = n_0 + n_2 I \tag{8}$$

where n_0 and n_2 are the intensity-independent term and intensity-dependent coefficient respectively. The n_2, known as the nonlinear index, has a relationship to the third-order nonlinear susceptibility $\chi^{(3)}$, which is given by the equation

$$n_2 = 4\pi^2 \, \mathrm{Re}\left|\chi^{(3)}\right|(n_0{}^2 \varepsilon) \tag{9}$$

thus in the materials having a large value of $\chi^{(3)}$ the n value varies strongly with I. The main equation connecting the field induced polarization P, in the material with the susceptibility χ are

$$P = P_0 + \chi^{(1)} \cdot \overline{E} + \chi^{(2)} \cdot \overline{E} \cdot \overline{E} + \chi^{(3)} \overline{E} \cdot \overline{E} \cdot \overline{E} + \cdots \tag{10}$$

where P_0 is the static dipole in the absence of an electric field; $\chi^{(1)}$ is the first order or linear susceptibility and is the origin of the linear refractive index, n_0; $\chi^{(2)}$ and $\chi^{(3)}$ are second- and third-order nonlinear susceptibilities, respectively. Since E–$E_0\cos(\omega t)$, where E_0 is the field at maximum amplitude, Eq. (9) becomes

$$\begin{aligned} P &= P_o + \chi^{(1)} E_o \cos(\omega t) + \chi^{(2)} E_o{}^2 \cos^2(\omega t) + \chi^{(3)} E_o{}^3 \cos^3(\omega t) + \ldots \\ &= P_o + \chi^{(1)} E_o \cos(\omega t) + (1/2)\chi^{(2)}[1 + \cos(2\omega t)] E_o{}^2 \\ &+ (1/4)\chi^{(3)}[3\sin(\omega t) - \sin(3\omega t)] E_o{}^3 + \ldots. \end{aligned} \tag{11}$$

In order to have an estimate of the value of $\chi^{(3)}$ which is applicable for practical all-optical switches, we use as an example the case of the Mach-Zehnder interferometer switch. Assuming a laser beam of 50 mW intensity being launched into a wave guide having 50 μm^2 cross-sectional area, a $\chi^{(3)}$ of 10^{-9} esu is required in the material having the effective length of 10 mm to cause a phase change of π in the beam. The absorption coefficient, α, must be less than 1 cm^{-1}, so a figure of merit of the material F= $\chi^{(3)}/\alpha\tau$, is required to be greater than 10^3 for switching less than one picosecond in response time τ [31].

III.A. Potential Glasses and Discussions

There are two kinds of glass which have large F values; one is glass having high refractive index, the other are glasses doped with semiconductor or metal fineparticles. As shown in Table 2 , high-index glass such as lead-silicate, lead-bismuth-gallate, and chalcogenide glasses, have the relative high $\chi^{(3)}$ values of 10^{-13} to 10^{-12}esu [32, 33, 34]. These values are far less than the values required to compose the wave guide circuit devices, although these glasses have very low α values and small τ values due to the non

Table 2 Nonlinear properties of glass [31]

Particle Glass preparation method	$\chi^{(3)}$esu	α (cm^{-1})	τ (sec)	remarks
SiO_2	$1x10^{-14}$	10^{-5}	10^{-14}	(32)
$PbO-SiO_2$ (SF59)	$7.5x10^{-14}$	$6X10^{-3}$		(33), (34)
$PbO-Bi_2O_3-GaO_3$	$4.2x10^{-13}$	--	--	(34)
Chalcogenide (As_2S_3)	$1.7x10^{-12}$		--	(34)
$CdSSeK_2O-ZnO-SiO_2$ base filter Corning CS3-68	$1.3x10^{-8}$	~3	--	(29)
ibid Shott OG30	$1x10^{-8}$	----	50	(36)
CdSe $Na_2O-ZnO-B_2O_3$ SiO_2 base glass melt-heat treatment	----	~50	0.3-3	(39)
CdS Silica Sol-gel	$1.5x10^{-10}$	----	----	(41)
CuCl $Na_2O-B_2O_3-SiO_2$ base glass melt-heat treatment	$3x10^{-6}$	~600	----	(42)
CdSe $CdO-ZnO-P_2O_5$ base glass melt-heat treatment	$5x10^{-7}$	~1200	----	(43)
CdTe silica Laser evaporation	$6x10^{-7}$	8000	----	(44)
CdS $Na_2O-B_2O_3-SiO_2$ base glass. Sol-gel	$6.3x10^{-7}$	----	----	(46)
Au doped glass	$5x10^{-11}$	30	2	(47)

resonant non linearity effect of the material. Thus their figure of merit is calculated as 10^3-10^4. Considering the practical devices composed of these glasses, the devices need a very high input power or a very large effective length; however, some practical devices will be developed for specified systems in the near future. Trials of the preparation and evaluation of the devices have begun using double-core fibers or ion-exchanged wave guide circuits.

Various studies on the third-order non linearities of glass doped with fine particles have been reported. Table 2 shows a chronological order list of the studies on the glass given by the main papers [35-48]. In the initial stage since 1983 when non-linear optical properties were investigated using the commercially available colored filters, it was confirmed that the K_2O-ZnO-SiO_2 base glass doped with CdSSe fine particles had $\chi^{(3)}$ values of 10^{-9}-10^{-8}esu.

In 1987 the optical bistable switching effect was observed by Yumoto et al.[49] in the study of nonlinearities of colored filters. A response time of 25 picoseconds was confirmed in both on and off switchings. Borrelli et al. [50] in 1987 pointed out that the colored glass filters had a large spread of particle sizes and that their quantum size effect was insufficient. Then, as shown in Table 2, current researches have moved to studies on new fine particle materials as well as new preparation methods. CdSe, CdS, CuCl, CdTe, and Au are typical new fine particle materials, while sputtering, sol-gel, and laser evaporation techniques are novel preparation methods [31]. Very large $\chi^{(3)}$ values of 10^{-8}-10^{-6}esu were obtained in the current studies of semiconductor or metal fine particle-doped glasses.

Because of the resonant nonlinearity of the fine particle-doped glass, the materials have greater $\chi^{(3)}$, bigger α, and longer τ than materials showing non resonant non-linearity.

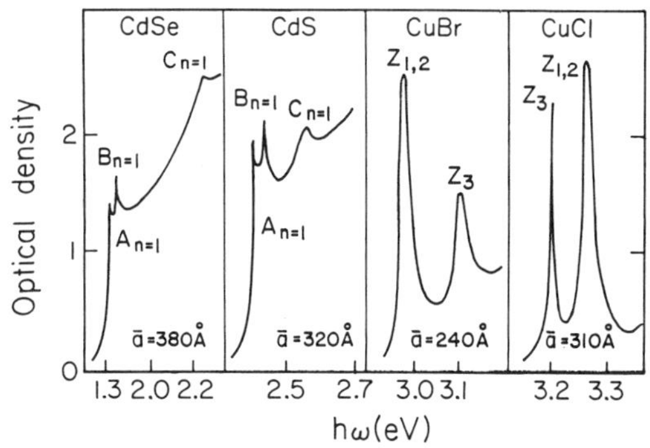

Fig. 4 Low temperature (T=4.2 K) absorption spectra of glasses with microcrystals: CdSe (a=380 Angstrom), CdS (a=320 Angstrom), and CuCl (a=310 Angstrom) [51].

The figure of merit of the fine particle-doped glasses are 10 -10^2 at the present time. There is still a gap in the third-order non linearity of the fine particle-doped glasses before the practical wave-guide circuit type switches can be composed. The present research on the materials has entered into theoretical studies in depth on the generation mechanism of the third-order non linearity of fine particles, the new preparation methods which have fine controllability in the particle size, the spread of particle sizes, and the concentration and crystal structure of particle materials. The laser evaporation method [31] is a novel approach to the preparation of the fine particle-doped silica glass. Due to the instant evaporation of the semiconductor materials by the pulsed laser heating in the inert gas atmosphere, high quality fine particles with the desired stoichiometry can be synthesized by this method. The size of particles and the concentration of doping materials in silica glass are controllable over a wide range by the choice of laser power, heating time, and gas pressure.

The optimum of third-order non linearity of the fine particle-doped glass will become clear in the coming two or three years. The theoretical studies, of the new materials such as metal-covered semiconductor particles, and the controllable preparation methods will provide a breakthrough of the glass for the development of practical all-optical switching devices. As an enhancement of photoluminescence was observed recently in the silicon fine particle-doped glass, the glass doped with fine particle materials hides unknown possibilities to itself. The glass is expected to grow into a key material for photonic applications in the early part of the next century.

The optical transparency of an oxide glass matrix makes it possible to apply all of the methods of optical spectroscopy to study such zero-dimensional system. Fig. 4 taken from ref. [51] shows the absorption spectra of glass samples containing CdSe, CdS, CuBr and CuCl microcrystallites. It is seen that spectra reveal the typical excitonic structure of near band-gap transitions. This structure is due to the spin-orbit splitting in cubic materials (CuBr, CuCl) and to the spin-orbit and crystal field splitting in hexagonal semiconductors (CdS, CdSe). The absorption spectra as well as luminescence and Raman spectra show that the semiconductor particles grown in glass matrix have the crystalline structure and sufficiently high spectroscopic quality. The X-ray scattering experiments were carried out recently for the investigation of structural modifications of CdS [52] and CdSe [53] microcrystals.

Fig. 5a [51] shows the absorption spectra of glasses doped with CuCl micro crystals of different sizes. As can be seen, the absorption is of excitonic nature down to smallest sizes. The decrease of the radius of microcrystals leads to large blue shift for both excitonic lines. This effect is due to the size quantization of an exciton as a whole, because the binding energy of the exciton in the material is very high ($E_{ex} \approx 0.2$eV) and the radius of exciton is rather small ($R_{ex} \approx 9$ Angstrom). The size dependence of spectral position of both excitonic lines are shown in Fig. 5b. For the exciton originating from the upper band Γ_7 the dependence is described by the simple expression

$$\hbar\omega_{Z_3} = E_g - E_{ex} + 0.67\frac{\hbar^2}{2M_s\bar{a}^2}\pi^2 \qquad (12)$$

where E_g is the band gap and E_{ex} is the exciton binding energy. The numerical factor results from the averaging of the Lifshits size distribution function. The slope of the curve is determined by the value of the excitonic translational mass M_s [51].

The special treatment is needed to describe the size dependence of spectral position for excitons originating from the degenerate Γ_8 valence subband . It was shown that this dependence is given by the following expression [54]

$$\hbar\omega_{Z_{1,2}} = E_g - E_{ex} + \Delta + 0.67\frac{\hbar^2\pi^2}{2M_h\bar{a}^2}[\Phi(M_l / M_h)]^2 \qquad (13)$$

where M_h and M_l are the "heavy" and "light" exciton masses, Δ is the value of spin-orbit splitting, $\Phi(M_l/M_h)$ is the root of a transcendental equation [38].

Comparison between experimental and theoretical results enabled [51] to determine the masses of exciton for all three exciton bands: $M_s=(1.9\pm0.2)m_0$, $M_h=(2.6\pm0.2)m_0$, and $M_l=(1.5\pm0.2)m_0$. One should to emphasize that the effective mass approximation gives the possibility to describe experimental results for the values of microcrystal radius down to about $\alpha_0 \approx 15$Angstrom.

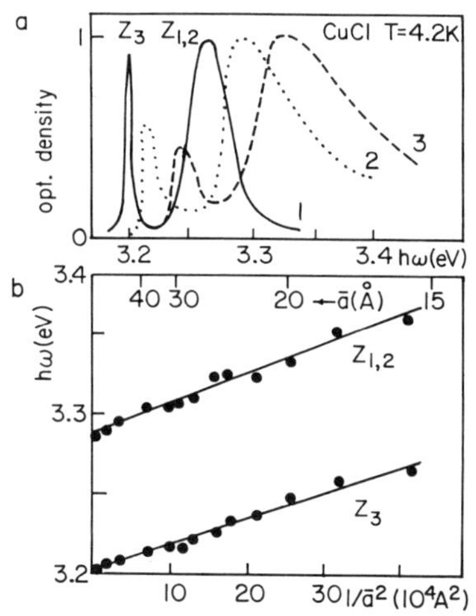

Fig. 5(a) Absorption spectra of glasses with CuCl microcrystals: (1) a=310 Angstrom, (2) a=29 Angstrom, (3) a=20 Angstrom. (b) Spectral position of oscillations vs. $1/a^2$ [51].

The absorption spectra of CdS microcrystals of different sizes are shown in Fig. 5a [51]. It is seen that for the semiconductor material with a low binding energy of the exciton ($E_{ex}\approx30$ meV for CdS) the decrease of microcrystal size leads to a large blue shift of the absorption band edge. Oscillations in absorption spectra of microcrystals with radius less than the exciton radius, $R_{ex}\approx30$Angstrom are due to quantum sublevels of the conduction band [55]. As is seen in Fig. 5b, the simple analytical expression

$$\hbar\omega_{l,n} = E_g + 0.71\frac{\hbar^2\pi^2}{2m_e a^2} = \Phi^2_{l,n} \tag{14}$$

with the electron effective mass $m_e = 0.2m_0$ gives the possibility to describe the size dependence of the blue shift only for the forbidden gap width. It is clear that a nonparabolicity of the electron dispersion law and the finite depth of the quantum well will have to be taken into account.

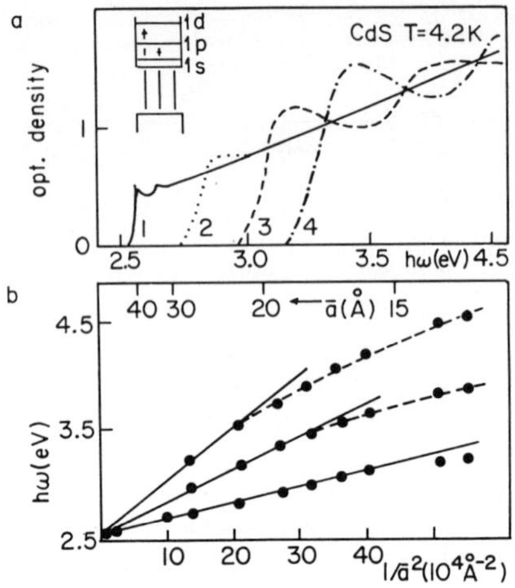

Fig. 6 (a) Absorption spectra of glasses with CdS microcrystals: (1) a=330 Angstrom, (2) a=23 Angstrom, (3) a=15 Angstrom, (4) a=12 Angstrom. (b) Spectral positions of oscillations vs. $1/a^2$ [51].

III.B. Semiconductor-Doped Glasses and Ormosils prepared by the sol gel method

Semiconductor microcrystallites in the 1-10 nm diameter range embedded in various matrices have shown that they possess high third order nonlinearities and fast switching speeds. As mentioned above oxide glasses provide highly transparent and very stable matrices for those semiconductor quantum dots and metal clusters. However, the high temperature processes of the conventional glass fusion methods normally limit the amounts of volatile dopant that can be incorporated into the glasses. Furthermore, the presence of impurities in the starting materials of glasses may cause some detrimental effects, such as the photodarkening phenomenon observed in commercial semiconductor doped filter glasses. For the nonlinear optical applications. it is therefore highly desirable to seek innovative processing methods to prepare the superfine particle-glass composites at lower temperature and free from impurities. The sol gel process is the most active method for this purpose [56-59].

The sol-gel technique offers a low temperature method for synthesizing amorphous materials which are either totally inorganic in nature or composed of inorganics and organics. The process is based on hydrolysis and condensation reactions of organometallic compounds in alcoholic solutions. The most widely investigated system involves silica-based glasses which are prepared by polymerization of a silicon alkoxide, $Si(OR)_4$ [60-62]. The process is based on inorganic polymerization reactions, starts from molecular precursors, and a macromolecular oxide network is obtained via hydroxylation condensation reactions [63, 64] which can be controlled by the chemical design of molecular precursors. The viscosity properties of sols can be adjusted allowing easy deposition of transparent coatings onto glass, ceramic, or polymeric substrates [65, 66]. Sol-gel chemistry is performed in solution at lower temperatures than conventional chemical methods. Homogeneous doping by mixing components at a molecular level, synthesis of metastable or amorphous phases allowing larger concentrations of chromophores, and synthesis of mixed organic-inorganic materials can then be performed.

The ability to synthesize inorganic polymers using sol gel processing with little or no heating makes it possible to dope these gels with a variety of organic and organometallic molecules [67-75] especially dyes having desired luminescent properties [76]. The emission properties of the molecules have been used to optically probe sol-gel chemistry [67-72] and structure [67, 74-76]. Another active area for organic doped sol-gel glasses is to use dopants to induce selected optical properties and to synthesize new optical materials [56, 78-80].

In the case of organosilicate glasses (organically modified ceramics or ormocers), the silicate network may be modified by organic substituents such as alkyl (e.g. methyl) groups or other functional groups (e.g. 3-glycidoxypropyl) [81] which may form organic copolymers that penetrate the silicate structure. In principle, Si may be substituted by Al, B, Ti, or Zr to yield ceramics of variable mechanical properties [82]. At present there is no consistent picture about the microstructure of these materials. In the investigation of glassy systems, several analytical methods [64, 83] have been used among which are neutron and light scattering.

The sol-gel method has therefore been adopted by a number of workers as an alternative way of fabricating these so-called "quantum dot" materials. Most published work report the use of sol-gel derived SiO_2 glass as the matrix material [84, 85]. Takada et al.[86] elected to use a Na_2-O-B_2O_3-SiO_2 gel derived glass as the matrix and a temperature of 580°C was sufficient to achieve complete densification. CdS crystallites ranging from 40 - 60 Angstrom were prepared with concentrations up to 8 wt %. The glass nanocomposites have excellent optical quality and high values of $\chi^{(3)}$ as shown in Table 3. When better control of particle size distribution is achieved, the $\chi^{(3)}$ values are expected to increase further, thus rendering these gel-derived nanocomposites as acceptable candidates for a variety of photonic applications.The Ormosils described above have also been used as matrix for the CdS semiconductor quantum dot materials in thin films [18-20]. As in the case of the borosilicate glass matrix materials, the Ormosil nanocomposites showed excellent optical quality and high values of $\chi^{(3)}$ comparable to those

Table 3 -Third-order susceptibility mesured by DFWM for CdS quantum dot borosilicate glasses [86].

	λ=450 nm		λ=460 nm	
CdS	Abs.Coeff.		Abs.Coeff	
(wt.%)	$\alpha(cm^{-1})$	$\chi^{(3)}$(esu)	$\alpha(cm^{-1})$	$\chi^{(3)}$(esu)
8	1833	$\sim 4.5x10^{-7}$	1715	$\sim 6.3x10^{-7}$
3.8	915	$\sim 3.1x10^{-7}$	849	$\sim 4.1x10^{-7}$
1.4	493	~2.9x10-7	417	$\sim 3.3x10^{-7}$
0.5	217	~2.9x10-7	184	$\sim 1.3x10^{-7}$

given in Table 3. The photo darkening shown by filter glasses made from fusion methods is greatly reduced. These then are also promising candidates for nonlinear optics.

A schematic representation of formation of oxide glass by the sol-gel method is:

Hydrolysis:

$$M(OR)_4+4H_2O \rightarrow M(OH)_4+4ROH \qquad (15)$$

Condensation:

$$M(OH)_4+M(OH)_4 \rightarrow (OH)_3\text{-}M\text{-}O\text{-}M\text{-}(OH)_3+H_2O \qquad (16)$$

where M can be Si, Ti and Zr, and R is an alkyl group. An example of Ormocer - Ormosil formation is polycondensation of polydimethyl siloxane (PDMS form) with silicon alkoxide. PDMS is

$$\begin{array}{ccccccccccccc} & O & & Me & & Me & & & O & \\ & | & & | & & | & & & | & \\ O\text{-} & Si & \text{-}O\text{-} & Si & \text{-}O\text{-} & Si & \text{-}O\text{-} & & Si & \text{-}O \\ & | & & | & & | & & & | & \\ & O & & Me & & Me & & & O & \end{array} \qquad \text{PDMS} \qquad (17)$$

and

Polycondensation Reaction of PDMS and TEOS

$$2Si(OH)_4 + HO - \left[\begin{array}{ccc} Me & & Me \\ | & & | \\ -Si & -O- & Si- \\ | & & | \\ Me & & Me \end{array} \right]_n - OH \rightarrow \qquad (18)$$

$$
\begin{array}{ccccccc}
OH & \lceil Me & & Me \rceil & & OH & \\
| & | & & | & & | & \\
OH\text{-}Si\text{-}O\ - & | \text{-}Si\text{-} & O\text{-} & Si\text{-} | & - O - & Si - OH + 2H_2O & \quad (19) \\
| & | & & | & & | & \\
OH & \lfloor Me & & Me \rfloor_n & & OH &
\end{array}
$$

<u>Polycondensation Reaction of Metal Alkoxide and the Copolymer (phenylmethylsiloxane) [46] mentioned above</u>

$$
2M(OH)_4 + H\text{-}[copolymer]\text{-}OH
$$

$$
\begin{array}{ccccc}
 & O & & O & \\
 & | & & | & \\
\rightarrow - & M - O - [copolymer] - O - & & M - + H_2O & \quad (20) \\
 & | & & | & \\
 & O & & O &
\end{array}
$$

The inorganic component provides superior mechanical and photostability and the organic chain endows the elasticity. Another means of controlling particle size is the control of the particle to environment interface by use of surface active ligands. In [186] it was shown, that only by variation of a complexing ligand the ZrO_2 particle size could be changed from 1000 to 30 Angstrom. Similar results could be obtained by Sanchez et al. [65]. It was shown, that complexing of Zr alkoxides with methacrylic acid (MAS) can be used to keep particle sizes of the hydrolyzed product below 50 Angstrom which is sufficient for optical applications. In this case, the ZrO_2 content can be used for adapting the refractive index of a system consisting of ZrO_2 linked to MAS and γ-methacryloxypropyl triethoxysilane (MPTS) as a system to be polymerized with MAS. It could be shown, that below a ratio of MAS to $Zr(OR)_4 \leq 0.8$ the particle size increases rapidly and below 0.75 precipitation takes place. But, independent on H_2O addition, the Zr to MAS bond remains unaffected.

Particle size tailoring can be carried out with other systems like semiconducting quantum dots, based on semiconducting oxides or sulfides (ZnO, CdS, PbS), or metals, becoming semiconducting, too, if particle sizes become small enough [46]. It could be shown, that by proper choice of ligands CdS can be stabilized in form of 2 nm sized colloids and subsequently incorporated into polymeric matrices.

$$
\begin{array}{l}
Cd^{2+} + S^{2-} \rightarrow CdS\downarrow + HS(CH_2)_3Si(OR)_3 \\
\qquad\qquad\qquad\qquad \downarrow \\
\qquad Si(OR)_3 \wedge\wedge SH \quad HS \wedge\wedge Si(OR)_3 \qquad\qquad (21) \\
\qquad\qquad\qquad\quad CdS \\
\qquad Si(OR)_3 \wedge\wedge SH \wedge\wedge SH \quad HS \wedge\wedge Si(OR)_3
\end{array}
$$

In a similar way CdS can be stabilized with amino group containing silanes. It was interesting to observe that under identical reaction conditions by variation of the ligand type linked to the silane the optical spectra defined by the band gap energy can be varied. As ligands $NH_2CH_2CH_2NH(CH_2)_3Si(OR)_3$, $NH_2(CH_2)_3Si(OR)_3$ and $HS(CH_2)_3Si(OR)_3$ were used. The results are shown in Fig. 3 of the ref. [186].

$\chi^{(3)}$ effects have been measured on PbS coated CdS quantum dots. These materials show $\chi(3)$ of about 10^{-9} esu. Due to the small particle the materials are highly transparent.. In a recent paper by the groups of Schmidt and Klingshirn [88] semiconductor quantum dots have been prepared by various sol-gel routes. Semiconductor CdS and sandwiched CdS-PbS, were prepared via multifunctional inorganic-organic sol-gel processing. As a sulfur source, hexamethyldisilylthiane was employed. In precursor sols, the metal sulfide clusters are carrying functionalized silanes acting as stabilizing centers as well as inorganic and organic network formers. Hydrolysis and condensation produces an inorganic network yielding viscous liquids useful to prepare optically transparent films or monoliths. The final organic cross linking at T<100°C results in materials of variable spectral response, thickness and optical density. In preliminary degenerate four-wave mixing experiments, self-diffraction from laser induced gratings was observed on unsupported 200 μm thick CdS-PbS doped monoliths. The maximum first order gratings efficiency, measured at different wavelengths between 490 and 520 nm, was $0.5. \times 10^{-3}$ and the corresponding calculated effective third order susceptibility was of the order of 10^{-9} esu .

The group of Mackenzie [89] used polydimethylsiloxane (PDMS) to organically modify the silica network by the sol-gel method. Compared to a purely inorganic gel, an organically modified gel (Ormosil) has enhanced crack resistance and film forming abilities without compromising optical transparency. Cadmium sulfide nanometer-size particles have been formed in the Ormosil (Ormocer) matrix. Up to 20 wt% of CdS microcrystallites have been homogeneously dispersed into Ormosils. The particle size of the CdS particles are well within the quantum confinement region. Cadmium sulfide doped Ormosil films with enough thickness for future wave guide fabrication have been prepared by spin or dip coating methods. The linear and nonlinear optical properties of these quantum dot materials are heavily dependent on the semiconductor particle size and size distribution

$\chi^{(3)}$ values measured by DFWM method at 532 nm using a 25 ps laser pulse are ranged from 10^{-11} to 10^{-13}. The on resonance pump-and-probe experiments were conducted by using a Nd^{3+}:YAG pumped dye laser with 3ns pulse width and 10 Hz repetition rate. The on resonance $\chi^{(3)}$ values of CdS-doped Ormosils are on the order of 10^{-8} esu. Photo darkening of the CdS-doped ORMOSIL sample reveal very little change in third order response while a CdS-doped glass samples made by glass fusion method totally lose the nonlinear response.

1. Semiconductor quantum dots in glass in general

Recent increase in research activities in semiconductor microcrystals is motivated by the potential application to nonlinear optical devices [104 - 107] in addition to the interest in the basic physics of the excited states in mesoscopic systems [108, 109]. It has been pointed out that highly enhanced optical nonlinearities will be attained in microcrystals due to the zero-dimensional modification of the energy spectrum and the wave function of the optical excited state, the so called quantum size effect. Recent synthesis of semiconductor clusters and particles had given us some understanding on quantum size effects of these entities [110]. The Q particles so far have been made by controlling of homogenous nucleation in fluid solvents [111] and glasses [112, 113]. One of the practical interests in these materials is the very fast and strong optical nonlinear response which has a potential application for optical switching and signal processing. The energy spectrum and the wave function of the optically excited states due to the Quantum Size Effect (QSE) [114] resulting from the confinement of the electron and the hole in a small volume, whereby the continuous energy band characteristic of bulk materials splits into discrete levels. The positions of these electronic levels which are shifted to higher energies in respect to the levels of bulk have been calculated by several authors [115, 16]. Size quantization effects were observed in the optical spectra of small crystallites of CdS, CuCl and CuBr in bulk silica glass, obtained by the sol-gel method [112, 117-122].

CuCl and CuBr crystallites of 2-10 nm size in conventional glasses have shown large nonlinearities by Nogami et.al. [123-125] and subsequently the preparation and quantum size effect in CuCl and CuBr microcrystals-doped glasses prepared by the sol gel process was reported [120, 123-125]. Since the effective Bohr radius of exciton in these microcrystals is small (0.68 and 1.25 nm for CuCl and CuBr, receptively) the translational motion of the excitons is three dimensionally confined by the deep potential of the glass matrix. Gels synthesized by the sol gel process were heated to precipitate the copper halide crystals, of which the crystal structure and the size were then determined. The quantum size effect of the exciton and the resonant nonlinearity of the obtained glasses were also measured.

To confirm the formation of copper halide microcrystals, the X-ray diffraction was reviewed measured for the glass heated for 2 h at 800°C. The XRD peaks on the background are caused by the amorphous phase [79, 80]. The position of these peaks closely resembles that of the cubic CuCl or CuBr crystal, which the position and relative intensity are taken from the JCPDS files. The precipitated crystals are thus identified with the cubic CuCl or CuBr crystal and the peaks in the patterns can be indexed as (111), (220), and (311) diffraction lines, respectively.

The absorption spectra were measured at 77 K, reveal two peaks clearly resolved. These two peaks, which are located at 3.27 and 3.35 eV for CuCl or 3.10 and 3.30 eV for CuBr, have been assigned to the excitons associated with the two spin-orbit valence band, $Z_{1,2}$ and Z_3, respectively, in the CuCl and CuBr microcrystals [126]. These values of exciton energies are higher than those of the bulk CuCl or CuBr crystal, which proves the quantum size effect of the exciton in the CuCl or CuBr microcrystals. The energy shift increases upward with increasing heat-treatment time and temperature.

2. CdS doped films

The incorporation of small semiconductor particles into thin films with wave guiding properties has potential applications in optoelectronics and nonlinear optics.The nonlinear properties of CdS incorporated in thin films prepared by the sol-gel method and in composite organic inorganic polymers was described by us recently [127, 128].

The preparation of the glasses doped by CdS was according to the general procedures of preparation of glass films described in references [76, 129-131]. The thickness of the films is a function of the speed of withdrawal of substrate glass from the precursor solution, pH, concentration of alkoxide in the solution and orientation of the sample in respect to the direction of withdrawal. Control of these conditions is necessary in order to guarantee reproducibility of the coatings.

The films are prepared from precursor solution having the typical molar ratios: Tetramethoxysilane (TMOS) or Tetraethoxysilane (TEOS): water: alcohol = 1:4(6):11, where the number in parentheses refers to TEOS. The molar ratio of Cd varied from 0.0077 to 0.195 (0.12) mol of Cd per 1 mol of TMOS (TEOS). ORMOCER (Organically Modified Ceramics) films are prepared according to the recipes given in reference [128,

129] and modified in our group for incorporation of dyes and colorants [133, 134]. The fresh films are then exposed to gaseous hydrogen sulfide for a few minutes, dried and heated to 250 ^{o}C The thickness of the films is determined by an interface method.

The average sizes of the CdS particles in the films are measured by two independent methods. The first is Transmission Electron Microscopy (TEM) of thin films containing CdS stretched on copper grids. The second method is based on X-ray diffraction on powders, where Scherrer's equation is used to find the correlation between the width (in the units of 2θ) of the diffraction peaks with the average size of the particles. Concentration of CdS particles in the films is of the order of 10^{21} molec./cm^3. However, number of molecules in a slab of 1cm^{2}x10^{-5}cm (which is the order of magnitude of thickness of the films) is about 10^{16} molecules. Such a small number of molecules requires very long scan times of 100-300 seconds/ 0.05-0.02^{o}.

The X-ray diffraction spectra reveal the existence of both hexagonal (wurtzite) and cubic zinc blende) forms of CdS. The TEM reveals average size of CdS particles of 20 -50 Angstrom although some particles as large as 200 Angstrom can be seen [135]. The inhomogeneous distribution of particle sizes in the glass is expected and is analogous to the inhomogeneous spectral broadening, where there is statistical distribution of crystallographic sites having slightly different energies. However, the dominant behavior (the maximum intensities of absorption and emission) in the spectroscopically active sites, the inflection point and nonlinear optical properties are due to a majority of particles having similar sizes.

Relation between the size of the particles and their absorption spectra is seen in Figure 7 which shows the positions of absorption inflection points for these and other samples as a function of the molar fraction of Cd (full circle). The data in TEOS samples are also given (open circles). As can be seen from the figure, the wavelengths of the inflection points rise monotonically with the molar fraction of Cd. At molar fractions higher than about 0.15 the wavelengths of the inflection points reach a plateau of about 510 nm. These inflection points can be correlated with the size of CdS particles shown by the measurements by TEM or X-ray diffraction (XRD). Plot of the energy of the inflection point vs. diameter of the CdS particles shows the correct trend, i.e. blue shift with decreasing of the diameter.

3. Nonlinear properties of the CdS doped films

Thin sol-gel glass and ORMOCER films doped by CdS nanoparticles produce strong third harmonics generation (THG) signals when irradiated at 1.06μm and observed at 0.355μm [18]. The strongest THG signals were about 1500-2000 times stronger than those generated by the glass substrate alone . This can be compared with the recent results of Ohashi et al. [136], where THG signals from CdS in a swollen thin polymer film were about 1.5 times higher than the THG signals of quartz when excited by 1.06μm laser radiation. They observed also that the ratio increased to about 11 with the change of the fundamental from 1.06μm to 1.45μm.

These values of $\chi^{(3)}$ are much lower than those of pure CdS, but still quite high considering the low volume fraction of CdS particles in the film. At the wavelength of 1.06 μm, $\chi^{(3)}$ presumably contains some enhancement due to a three-photon resonance. According to the work of Ohashi et al.[136], we can expect 10 fold increase of the THG signals by changing the fundamental wavelength from 1.06 μm to 1.45 μm.

4. Energy levels of CdS quantum dots in thin glass films

Thin glass films doped by cadmium sulfide emit perceptible fluorescence only at 77 K. Figure 3 presents excitation and emission spectra of two TEOS films, curve 1, concentration of 0.051 mol of Cd per mol of TEOS and curve 2, concentration of 0.127 mol of Cd per mole of TEOS. Curve 3 presents the emission and excitation spectra of cadmium sulfide (0.117 mol per mol) in GLYMO film. The excitation spectra are well structured and the wavelengths of maxima can be determined.

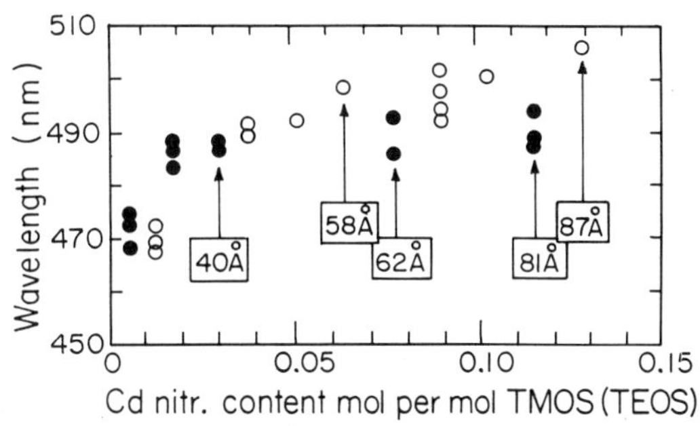

Fig. 7 Wavelength of inflection points of the absorption spectra as a function of concentration of Cd ions in the precursor solutions. Full circles: TMOS films; open circles: TEOS films. Number in boxes denote the diameters of CdS particles as determined from X-ray diffraction data [22].

Predicted energy levels are calculated from Table 1 and Equation (6) in the first-order approximation and drawn in Figure 2. These are compared with the energies calculated from the excitation spectra on Figure 3. The levels fit fairly well the experimental quantum levels of the TEOS samples. The GLYMO sample shows two such levels. The size dependence of the electronic levels in the glasses is analogous to the behavior of small colloidal cadmium sulfide particles separated by electrophoresis [137] Average radii of the CdS particles as indicated by the fit are 17 , 22 and 26 Angstrom, in good agreement with the measurements by X-ray diffraction.

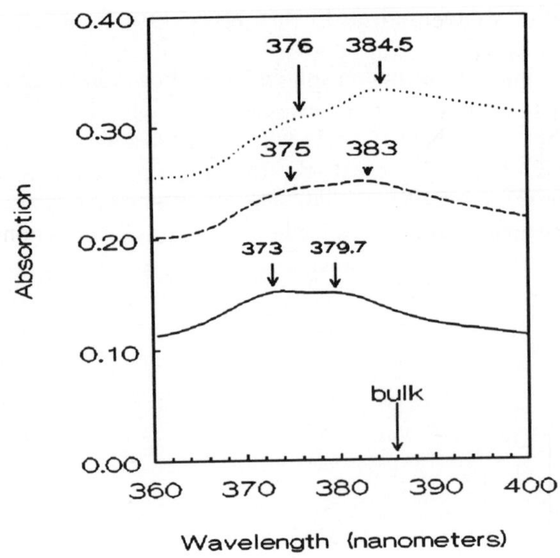

Fig. 8 Absorption spectra of CuCl particles in glass films measured at room temperature. Solid line: 0.061 mol, broken line: 0.091 mol, dotten line: 0.122 mol per mol of TMOS of Cu2O in the starting mixture. Position of the excitonic bands, Z1,2 and Z3 are shown by arrows. Red spectral shift with increasing concentration is evident [22].

5. CuCl quantum dots in thin films

Copper Chloride is a semiconducting material of which properties in nanometer-sized particles is a matter of increasing scientific interest. The reason for it is similar to that of the quantum dots of CdS, i.e. large optical nonlinearities, however, due to small Bohr radius of these particles (about 6.8 Angstrom as compared with the 19.7 Angstrom for CdS) leads to different size ranges for the optical nonlinearities. As in the case of CdS, the most desirable form of nonlinear semiconductor materials would be thin films, which are easy for miniaturization and can serve also as an nonlinear waveguide. This poses a challenge for a chemist of sol-gel glassess to prepare the thin glass films incorporating the quantum dots of CuCl possessing the desirable optical properties.

Preparation of thin films containing mainly monovalent copper presents an experimental difficulty since copper chloride tends easily to be oxidized to its divalent state. The best samples pertaining monovalent copper chloride in thin films are prepared by suspending Cu_2O in pure acetonitrile adding the suspension to the standard sol-gel precursor solution to which stoichiometric amount of HCl is added.

The precursor solution of the glass film consists of TMOS, Water and Methanol in the molar ratios, 1:9:5 acidified by 2.10^{-4} M/L of HCl. The particle size is controled by different concentration of Cu_2O in the precursor solution. The films so prepared show stability over a period of six weeks held in dark in vacuum. It should be noted that films prepared from CuCl exhibit only cationic Cu^+ rather than monocrystallites of CuCl as revealed from absorption spectrum. The absorption due to Z_3 excitons allows the calculation of the size of the nanoparticles, as was done for CuCl microcrystals by Masumoto et al. [145] and by Nogami et al. in silica glass [121]. When the crystal size is decreased from 77 to 25 Angstrom the energy of the Z_3 exciton is shifted from 3.234 eV to 3.275 eV, while the bulk value is 3.218 eV [146]. Fig. 8 presents the absorption spectra of CuCl microcrystallites prepared with the following initial concentrations of Cu_2O: 0.061,

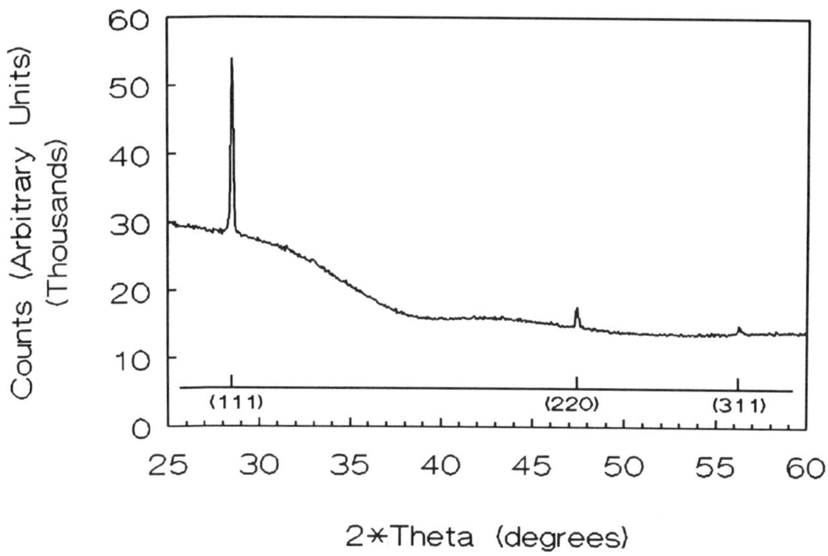

Fig. 9 Diffraction peaks and assignment of CuCl in thin films.

0.0091 and 0.122 mole per mol TMOS. The increase of the Crystallites with concentration is evident from the spectral shift to longer wavelengths.

The absorption spectra of each glass has two bands which result from the excitons associated with two spin-orbit valence subbands. The Z_3 exciton line originates from two-fold degenerate band Γ_7 and the line $Z_{1,2}$ line originates from four-fold $\Gamma_8$degenerate valence band Γ_8. Figure 9 present the results of the X-ray diffraction measurements for the quantum dots of CuCl in bulk sol-gel glasses. The cubic symmetry of the CuCl quantum dots in thin glass films is in accord with the reported data (146).

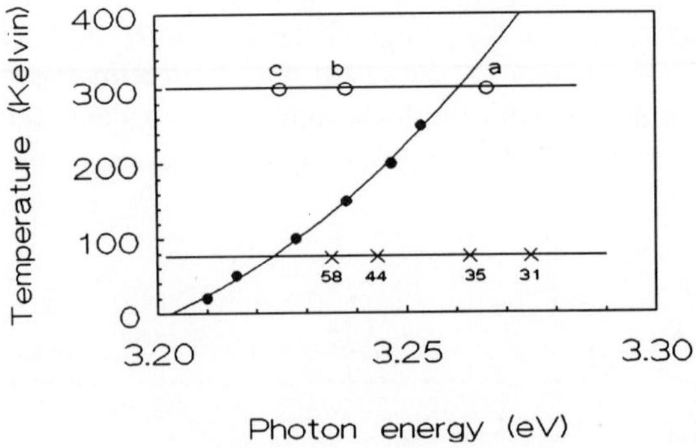

Fig. 10 Absorption maxima of CuCl in thin film as a function of concentration. Solid circles: 61 Angstrom diameter in NaCl [145]. Open circles: this work, a-0.061 mol, b-0.91 mol, and c-0.122 mol per mol of TMOS of Cu_2O in the starting mixture measured at the room temperature. Crosses: different sizes of CuCl particles at 77 K, according to [125].

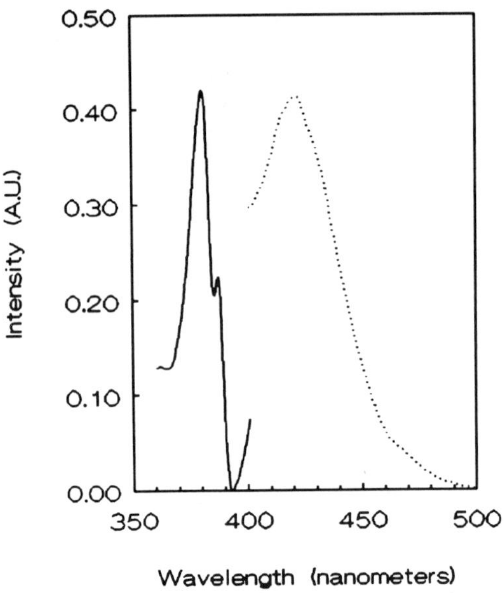

Fig. 11 Emission and excitation of CuCl in thin glass film.
Initial concentration of Cu_2O: 0.121 mol per mol TMOS [22].

Figure 10 presents the shift of the energies of the exciton Z_3 as a function of temperature and diameter of the microcrystallites of CuCl. The general trend is that the increase of temperature shifts the excitonic state to higher energy probably as a result of thermalization and the decrease of particle size results also in lifting up of the electronic states as an increase of quantum effect. By comparing our results with those obtained by other groups in crystals and bulk glass [145-164] we can see that the particle size of CuCl in films can be determined by the starting concentration of Cu_2O in the precursor solution.

Figure11 presents the excitation and emission spectrum of the CuCl particles in a glass film at 77 K. The exciton spectrum consists of two peaks which are assigned to the excitonic levels $Z_{1,2}$ (at high energy) and Z_3 (at lower energy). Although the emission takes place from the Z_3 level, it can be excited also via the $Z_{1,2}$ from which the energy realaxes with the assistance of lattice phonons to the luminescent level Z_3.

6. Preparation of thin glass films doped by quantum dots of CuBr

Similarly to the films doped by CuCl, CuBr doped films where obtained. The best CuBr doped film in which copper retained valency I where obtained from TMOS, Water and Methanol in the molar ratios 1:6:11 and 2.10^{-4}M HCl aqueous solution. Na_2SO_3, $CuBr_2$ and HBr were added after one hour hydrolysis.

Alternatively, we used our standard solution with molar ratios of TMOS/Water/Methanol = 1:6:11 to which Montmorillonite:TMOS of 1.7%w/w was added. After one hour of hydrolysis the following reactants: $Na_2SO_3/CuBr_2/HBr$ with the molar ratios of 0.041/0.0046/0.0015 in mol/mol TMOS units was added. The absorption spectra of two CuBr samples are presented in Fig.12.

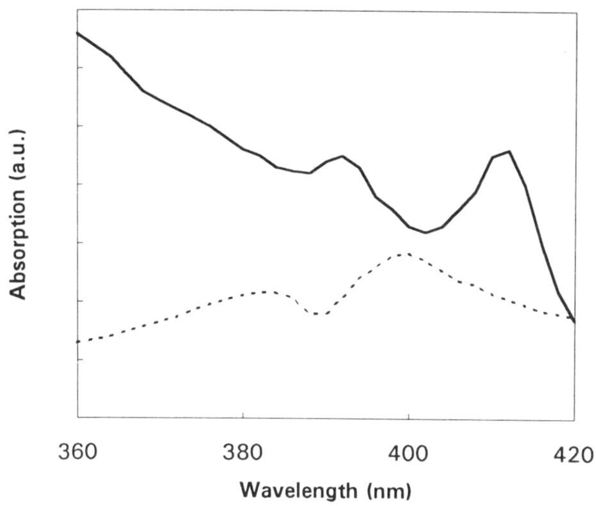

Fig. 12 CuBr films: full curve - TMOS:Water:Methanol=1:6:11 and $Na_2SO_3:CuBr_2:HBr$= 0.041:0.0046:0.0015, and dotted line - the same solution with Montmorillonite.

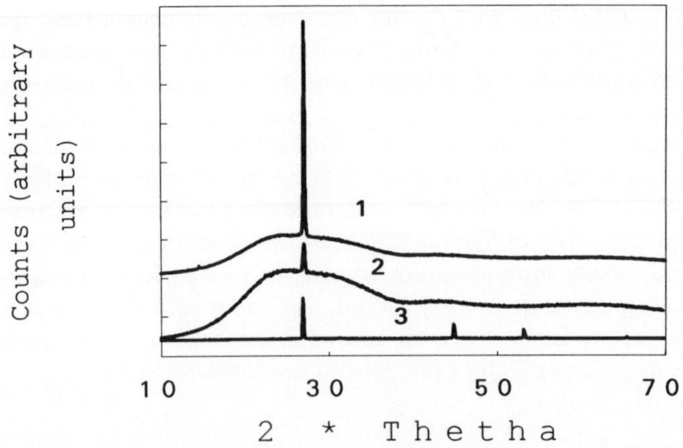

Fig. 13 XRD spectra of CuBr films. Curve 1 - CuBr film. Curve 2 - CuBr film with Montmorillonite. Curve 3 - Precipitate from Na_2SO_3 + $CuBr_2$ + HBr in aqueous solution.

XRD spectra were obtained in the 10°-70° region with a 10^{min} continuous scanning for the precipitates and, with 20^h step scanning (with the typical rate of $0^o06/72^{sec}$) for the sol-gel films. A Philips X-Ray Powder Diffractometer, Modell PW 1220 was used. The X-ray spectra are presented in Fig.13.

Using the Scherrer equation [165] it is possible to find the correlation between the halfwidth of the peaks ($\beta_{1/2}$) and the average size of the particle D_{hkl}, for the CuBr doped films:

$$D_{hkl} = \frac{0.9\,\lambda}{\beta_{1/2}\cos\theta} \qquad (22)$$

The average calculated diameter for the CuBr doped sol-gel films with Montmorillonite is 239 Angstrom as compared to the CuBr sol-gel films without Montmorillonite which range is from 482 to 519 Angstrom.

IV. CuCl-CuBr SOLID SOLUTIONS

The relative position of the $Z_{1,2}$ and Z_3 excitons discussed in the previous chapter were found by comparison with CuBr-CuCl solid solutions [166, 167]. The electron-hole exchange interaction plays an important role for the relative intensity of $Z_{1,2}$ and Z_3 exciton bands in CuCl-CuBr, which acquire relative opposite energies.

The Z_3 exciton in CuCl is an electron-hole composed of the Γ_7 valence band and the electron from the lowest conduction band of $\Gamma 6$ double-group symmetry, while $Z_{1,2}$ is an electron-hole pair composed of the Γ_8 valence band hole, and again an electron from the $\Gamma 6$ band (empty in the ground state).

Excitation by light close to $Z_{1,2}$ exciton energy produces a large quantity of $Z_{1,2}$ excitons, whereas an electron in Γ_7 is likely to move to Γ_8 states, leaving a hole on the Γ_7 band at slightly higher energy. This hole forms a Z_3 exciton in recombination with an electron in the upper , Cu 4s-like states. Such a Z_3 exciton can then annihilate to produce luminiscence. The $Z_{1,2}$ exciton shows a remarkable splitting of energies, when the microcrystallites are smaller. Hence, multiple excitations peaks appear in the excitation spectrum.

As the solid solution of the CuCl-CuBr system has been found to belong to the amalgamation type [167], the exciton band consists of two bands throughout the whole range of composition and their absorption and their absorption peaks shift continuously as the Br ion concentration is changed. Curves of shifts of the $Z_{1,2}$ and Z_3 band peaks with respect to the Br concentration, x, seem to crosseach other at x=0.23 and hence here the spin-orbit splitting becomes zero.

Takahashi and Goto [168] have found that the relative intensity of the two bands, measured photographically, varies a great deal and in a solid solution at x=0.23, the intensity of the exciton band on the low-energy side becomes very weak. Kato et al. [169] investigated absorption bands in evaporated films of $Cu(Br_xCl_{1-x})$ solid solutions in detail, photoelectrically. In CuCl the absorption area of the higher-energy band, $Z_{1,2}$, is larger than that of Z_3 and with increasing X the difference in absorption area of two bands becomes much larger. In CuBr the area of the lower-exciton band, $Z_{1,2}$, is larger than that of the Z_3 band but with increasing Cl concentration the area of Z_3 becomes larger than that of $Z_{1,2}$. That is to say with increasing separation between the two bands or with increasing spin-orbit splitting, the area of the higher-energy band becomes greater than that of the lower-energy one regardless of whether it is $Z_{1,2}$ or Z_3. In CuCl and CuBr the exciton band shapes are asymmetric Lorentzian as in the case of alkali halides. In the solid solutions of cuprous halides in which two exciton bands overlap, each band is also assumed to have the same asymmetric Lorentzian shape and the overall absorption curve is analysed into two bands. These bands were theoretically analyzed by Toyozawa [170].

V. SEMICONDUCTOR NANOPARTICLES IN ZEOLITES

Another interesting approach to size control has been the precipitation of cadmium sulfide in the very structured environment of zeolites [90, 110]. In this case the zeolite framework acts as a rigid macroligand for the growing CdS particle (Equation 23) which can control the cluster topography.

$$4Cd^{2+} + 4H_2S \xrightarrow{\text{zeolite}} [Cd_4S_4]\text{Zeolite} \quad (23)$$

Zeolite Y, a high-dielectric aluminosilicate host, has two types of cages available for cluster formation, the smaller 8 Angstrom sodalite units and the larger 43 Angstrom supercages. For the purposes of this discussion there are five sites (I, I', II, II', III) [28] available for cation siting within the sodalite and supercages. II-VI and I-VII quantum-confined clusters can be synthesized by well-understood ion-exchange methods, followed in the II-VI case by treatment with H_2S or H_2Se. It is important to note that the ion-exchange process can yield very different siting of cations depending on temperature, *p*H, solvent versus melt ion inclusion, other extra-framework ions, calcination, and loading levels. This process must be systematically controlled along with the conditions for treatment with H_2S or H_2Se to obtain materials that can be consistently reproduced and that contain monosize clusters.

For the synthesis of CdS in zeolite Y. careful control of the above ion-exchange parameters permits selective siting of Cd^{2+} at I' [91], which is within the sodalite cage. The zeolite is then treated with H_2S to form the CdS superlattice. At a low loading level of CdS in zeolite Y, isolated CdS molecular units are formed with an absorption peak around 290 nm and no emission even at 4.2 K. At higher loadings, x-ray single-crystal [28] and powder diffraction, EXAFS, and optical absorption data show that eight-atom $(CdX)_4$ (X=S or O) clusters can be precisely located within the sodalite cages [28, 91-101]. The discrete clusters within the small sodalite units of the CdS zeolite Y structure begin to interconnect electronically by quantum tunneling as the loading density within the Zeolite rises. The Cd atoms point toward each other through the double 6-rings linking the sodalite units with a Cd-Cd distance of 6.2 Angstrom. As this 3-D interconnection proceeds, the corresponding changes in optical properties indicate a progression toward a semiconductor super cluster with behavior intermediate between that of the discrete clusters and bulk semiconductor. The transition from clusters to aggregates upon increasing CdS loading is not continuous but rather abrupt, as judged from optical absorption and emission spectra. This suggest that the aggregation of the individual clusters inside the zeolite may be a percolative process.

When fully loaded, the quantum superlattice shows an exciton shoulder near 340 nm and emits at low temperature. Neither heating nor increasing CdS concentrations can further shift this exciton shoulder to the red. In contrast, when CdS is on the exterior surface of the zeolite crystallite, a continuous red shift of the absorption threshold toward the bulk value is observed, paralleling the behavior of CdS colloids.

Direct synthesis of CdS within the pore structure of zeolites leads to a novel super cluster with a structural geometry superimposed by the host framework [56]. Detailed X-ray powder diffraction and EXAFS analysis together with optical absorption data reveal discrete $(CdS, O)_4$ cubes located within the small sodalite units of the structure which begin to interconnect as the loading density within the zeolite rises. The discrete cube building blocks consist of interlocking tetrahedra of Cd and S with a CdS bond length of 2.47 Angstrom. At higher loadings these cubes begin to occupy adjacent sodalite units where the Cd atoms point toward each other through the double six-rings linking the sodalite moieties with a Cd-Cd distance of ~6 Angstrom. As this three-dimensional interconnection proceeds, the corresponding changes in optical properties indicate a progression toward a semiconductor super cluster with behavior intermediate between that of the discrete CdS cubes and bulk semiconductor. Semiconductor super clusters of this type represent a novel class of materials where the three-dimensional structure and electronic properties can be controlled by using different zeolites as the template. The unique stability of the semiconductor clusters inside the sodalite units is due to the coordination of Cd atoms with the framework oxygen atoms of the double six-ring windows. The stability of the super cluster comes from the interaction between clusters in the adjacent sodalite units. It is suggested that through-bound coupling is responsible for the interaction between clusters.

A final method of size control relies on a combination of the passivating properties of ligand additives and several real-time variables such as concentrations, pH, and reaction duration. Henglein & co-workers [102] have found that colloidal samples of semiconductor nanoclusters can be stabilized indefinitely by conducting the precipitation reaction in the presence of the inorganic polymer, sodium hexametaphosphate.

VI. METAL-CLUSTER-ORMOSIL NANOCOMPOSITES

In addition to semiconductor microcrystallites metal clusters of dimensions less than 100 Angstrom when in glass also exhibit high third order nonlinearity.

Studies on gold and silver colloid particles-doped glasses have shown that they possess high on resonance third-order optical nonlinearity comparable to those of semiconductor-doped glasses and a picosecond scale nonlinear response time. The high third order non linearity at surface plasma resonance was believed due to contributions from the electron quantum confinement effect and the hot photo excited electron. The magnitude of third order response of metal cluster-doped composites are dependent on the volume fraction of metal. The volume fraction of the gold colloid in glass prepared by the conventional glass fusion method is normally very low due to the low solubility of gold ion in glass melt. The maximum amount of a metal that can be incorporated into gel is determined by the solubility of the metal salt in sol-gel solution prior to gelation.

Metal-cluster-Ormosil nanocomposites [103] have been prepared by thermally or optically reducing the metal ions in Ormosils. Gold, silver, platinum, and copper clusters have been formed in Ormosils. Compared to other works on purely inorganic gel, metal clusters formation took place at very low temperature (from room temperature to 300°C). The low temperature process make it easier to study the mechanism of the nucleation and growth processes of these metal particles. The sizes of metal clusters formed are in the range of a few nanometers. Only crystalline particles with FCC structure are found in Ormosils. The optical absorption of these metal particles in Ormosils can be well explained by the free electron model. No third order response of the nanocomposites can be detected by nanosecond pump and probe experiments. This indicates that the metal clusters do not have any third order response shorter than 1 nanosecond. The fast response time of metal clusters is believed to be in a few picoseconds. Research on silver nanocomposites based on sol-gel glasses [104, 105] have been performed as well as nanocomposites based on Ormocer containing metal clusters of Ag, Au and Pt are now in progress [56].

Quantum dot formation can be obtained by use of amino grouping containing ligands from Ag or Au salts, too. In this case the formation of ammine complexes can be postulated according to eq.(24) [186].

$$AgNO_3 + 2H_2N(CH_2)_3Si(OR)_3 \rightarrow \begin{array}{c} (RO)_3Si(CH_2)_3NH_2 \\ \downarrow \\ Ag^+\, NO_3^- \\ \uparrow \\ H_2N(CH_2)_3Si(OR)_3 \end{array} \qquad (24)$$

As shown in [186], it is possible to form Ag colloids in the nano range by photoinitiated nucleation. The optical spectra of these colloids indicate a relatively narrow particle size distribution as colloidal solutions. The systems can be used to be incorporated into sol-gel materials and transformed into nano-scaled Ag halides, showing a reversible photochromic behavior.

Ormosils proved to be promising hosts for photo-active materials. Compared to the purely inorganic gel, an organically modified gel has enhanced mechanical properties, water resistance, and film forming abilities without compromising its optical transparency. The properties of Ormosils can be tuned by controlling organic/inorganic ratio, modifying molecular network, and controlling process parameters. A broad range of optically active materials such as semiconductor quantum dots, metal clusters, organics, as well as oxides and metal ions have been incorporated into Ormosils in our research. More research is needed to understand the fundamental aspects of Ormosils and the interaction between quest and host.

VII. THE NONLINEAR PROPERTIES OF DYES IN SOL-GEL GLASSES

The optical and nonlinear properties of dyes can be improved. when they are incorporated in inorganic glasses. These materials may be applicable for technological uses. The well-known technology of dyes in polymer matrices is not sufficient since both plastics and dyes undergo decomposition under prolonged irradiation and under harsh external conditions. On the other hand, one cannot use the current glass technology for incorporation of dyes because of the high temperature of the process at which the dyes decompose.

For this purpose we incorporated organic dyes into glasses of four types: sol-gel porous glasses, composite glasses, lead-tin fluoride glasses and boric acid matrices and characterized their linear and nonlinear optical properties [171-177, 179]· We studied fluorescein and its derivatives [171-173] in sol-gel glasses, in composite glasses and in boric acid glasses and determined triplet lifetimes and saturation intensities. Addition of a heavy ion, mercury, in the form of mercury acetate [172, 173], increases the $S_1 \rightarrow T_n$ transition rate followed by a decrease of saturation intensity I_s. Acridine-orange and acridine-yellow in composite glass and heavy glass have been studied as well [174-177, 179]. At low CW powers (typically up to 10 W/cm^2) the materials behave as saturable absorbers with typical triplet lifetimes of milliseconds to seconds and saturation intensities of the order of milliwatt/cm^2. Methyl-orange belongs to the class of materials which possess nonlinear optical properties as a result of a change in conformation. Time scale and nonlinear optical properties of methyl-orange in the milliseconds regime have been studied recently [175, 176].

Pump-and-probe experiments performed on these materials have shown nonlinear properties also on a nanosecond scale. In these experiments we irradiated the nonlinear materials by spectrally narrow nanosecond laser pulses, while monitoring the changes of transmission of the samples with a much weaker spectrally broad probe beam. We found in these experiments that acridine-orange in composite glass shows induced transmission, i.e. bleaching at laser powers below 6.3 MW/cm^2 and induced absorption at high powers [175]. Acridine-orange in heavy glass shows induced transmission at laser power of 1.7 MW/cm^2 and induced absorption at higher powers [176]. Both phenomena occur within the lifetime of the exciting pulse, their recovery time, i.e. relaxation to the state before the excitation, however is different: The induced absorption recovers on a time scale varying from minutes to hours, depending on the initial laser power, while the induced transmission conforms to the general scheme of saturable absorbers of this group of materials with recovery times of the order of tens of milliseconds.

Similar experiments done on methyl-orange are described in ref [175]. We observed induced absorption at both low and high excitation powers and a similar energy dependence of the induced absorption as in acridine. However, in contrast to the acridine, the relaxation of the absorption after switch-off of the laser is completed within less than a minute. At a continuous-wave (CW), low power excitation, methyl-orange undergoes induced bleaching due to different absorption coefficients of its cis and trans forms[178].

The nonlinear behaviour of Acridine Orange on a nanosecond scale due to singlet transitions at high intensity MW excitation was reported recently [178] in heavy glass, boric acid and in sol-gel glass.

Typical procedure for boric acid and sol-gel glass can be found in ref [76]. The acridine doped heavy hlass was prepared as follows: 52.2 mol% of SnF_2, 10.5 mol% of SnO, 5.1 mol% of PbF_2 and 32.1 mol% of [1/2P_2O_5) were mixed and heated to 400°C. The viscous liquid was cooled to 275°C and 1 mg of dye per each 10 gram of glass were added.

The glass was slowly cooled and then polished to 0.16 cm thickness. Final concentration of acridine orange in the sample is $3*10^{-3}$ M/L and extinction coefficient is $6*10^4$, which makes the number density of acridine molecules in this sample equal to about $1.8*10^{18}/\ cm^3$.

The composite glass doped by acridine orange was prepared by impregnation of sol-gel porous glass [181] by methylmethacrylate containing acridine. Concentration of the dye in the glass is also about 10^{-3} M/L. Boric acid samples were prepared by admixing the dye into ortoboric acid, heating to 250°C and sandwiching the viscous solution between two preheated microscopic slides. Concentration of the dye in such boric acid glass is about 10^{-3} M/L (which is difficult to determine exactly, due to sublimation of boric acid in the course of melting).

The spectral data of these materials were compared with the data measured on alcoholic solution of acridine orange in 0.1 cm glass cuvette having optical density similar to this of the samples.

Power - dependent measurements of lifetimes were performed using a 10 nsec. coumarin 485 nm dye laser pumped by an excimer laser in a single shot mode. The signals were captured by a streak camera and the luminescent signal was deconvoluved from the shape of the laser pulse.

Nonlinear absorption was investigated using a home-made two-beam laser spectrometer, consisting of two independently tunable dye lasers synchronously pumped by a second harmonic of Q-switched ruby laser (λ = 347 nm). Duration of both pump and probe pulses is 15-20 nsec, which allows to study of the dynamics of the singlet state in the quasi - stationary regime (typical lifetime of the excited singlet state of acridine orange in solids is about 2.5 nsec). The time resolution of the system is about 1 nsec. Details of the apparatus and the technique of pump-and-probe measurements can be found in references [182, 183].Table 1 summarizes the power dependence of luminescence lifetimes of acridine orange. The power dependence in the boric acid glass is linear, which shows that in this glass no nonlinear processes take place.

The power dependence for heavy glass can be rationalized by a competition between a process increasing the population of excited monomers and a process decreasing this population. Apparently the photoquenching has higher power threshold than the photodissociation of the dimers, which occurs between 1.6 and 6.3 MW/cm^2 for heavy glass and is above 25 MW/cm^2 in boric acid glass. It should be noted that in a pump-and-probe experiment done on the same sample in the power range of 1.5 MW/cm^2 to 30 MW/cm^2 we observed an onset of long-lived induced absorption at energy slightly higher than the lower limit [174-176]. It is due to photodissociation of dimers to monomers, which, after switching-off of the laser radiation, recombine back to dimers, by diffusion.

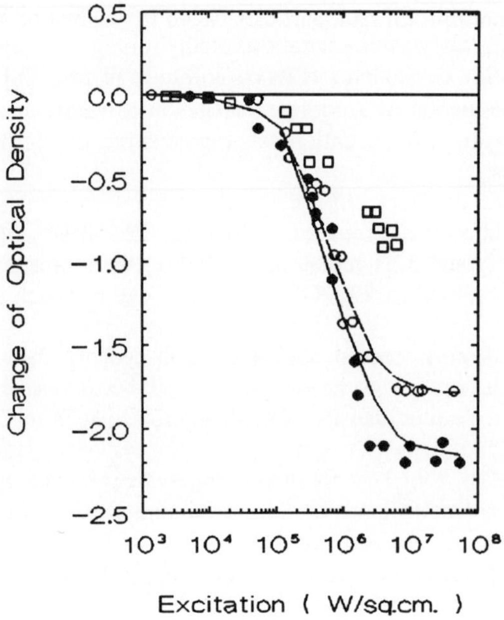

Fig. 14 Power dependence of acquired transparency of Acridine orange in heavy glass and in alcohol.Squares - 480 nm, cicles - 500 nm, and full dots - 515 nm [179].

The saturation experiments were using the pump-and-probe beam method at a few wavelengths as a function of intensity of the pump beam. The weak probe beam has approximately the same width and the same frequency. Figure 14 presents the results for acridine orange in heavy glass. Here we show how the transparency of the glass increases with the laser power. Since typical lifetime of acridine orange is about 2.5 nsec, we shall assume that the population of the singlet state of acridine reaches an equilibrium within the width of the pump pulse which is 15-20 nsec. In such a case, the results can be analyzed in terms of steady-state equations.

The behavior of the dyes incorporated in sol-gel glasses on a pico- and femto- second scale requires further studies which are in progress now.

VIII. CONCLUSIONS

In this work I tried to survey the state of art of nanoparticles in amorphous solids. It looks now that glasses or composite glass materials (Ormocers-Ormosils) prepared by the sol-gel method are the best hosts for semiconductors, metal or dye nanoparticles. The X ray spectra of these materials disclose the crystalline structure of the particles and their biggest sizes. The problem of uniformity of the sizes still remains to be solved. The smallest sizes can be detected by TEM or SEM as well as from excitation spectra. In our opinion the best solution would be incorporation of nanoparticles caged in zeolites or smectite clay minerals (e.g. montmorillonite) in glasses by the sol-gel method.

Such a situation was already tested with incorporation of CuBr into montmorillonite incorporated into glass films . The clay montmorillonite has a layer structure, each layer consists of an octahedral aluminate sheet sandwiched between two tetrahedral silicate sheets [185]. Between the layers nanoparticles can be incorporated.

Acknowledgments

The author is very grateful to Prof. C. K. Jϕrgensen for discussions and to Swiss National Science Foundation for the possibility (grant 20.-32127.91) to spend a month in Geneva where a part of this work was performed. To Prof. Georges Boulon for inviting me to France during which time a number of very useful discussions were held with him and his colleagues . To Dr H. Minti for helping with many of the experimental results and writing of this chapter. Dr. L. E. Brus has kindly given his permission to reproduce Fig. 1. And Professors Al. I. Efros and A. Ekimov for their kind permission to reproduce their figures: 4, 5, and 6. Finally I wish to thank Prof. Shmuel Yaariv and Dr. I. Lapides providing us with the relevant information about the clay minerals (Montmorillonite). We are grateful to the Israeli Ministry of Science and the European Community whose grant allowed to perform the experimental work at the Hebrew University. Finally, Professor FrÖlich is gratefully acknowledged for his kind help in reading the manuscript during the Erice School, providing highly useful remarks.

IX. REFERENCES

1. D. Z. Anderson, MRS Bull. 13, 30 (1988) and, A. M. Glass, Ed., "Photonic Materials", ibid., no.8.
2. Galen D. Stucky and James E. Mac Dougall, Science, 247, 669, 9 February (1990).
3. S. Schmitt-Rink, D. A. B., D. S. Chemla, Phys. Rev. B. 35, 8113 (1987); S. Scmitt-Rink, D.S. Chemla, D. A. B. Miller, Adv. Phys. 38, 89 (1989).
4. Y. Wang, N. Herron, W. Mahler, A. Suna, J. Opt. Soc. Am. B 6, 808 (1989).
5. MRS Bull. 14, 12, (1989).
6. M. L. Steigerwald et al., J. Am. Chem. Soc., 110, 3046 (1988).
7. J. G. Brennan et al., J. Am. Chem. Soc.,111, 4141 (1989).
8. A. I. Ekimov, Al. L. Efros, and A. A. Onuschchenko, Solid State Commun., 56, 921 (1985).
9. L. Brus, Appl. Phys. A 53, 465 (1991)
10. Y. Kayanama, Solid State Comm. 59, 405 (1986); S. V. Nair, S. Sinha, K. C. Rustagi, Phys. Rev. B 35, 4098 (1987); V. Mohan, J. B. Anderson, Chem. Phys. Lett., 156, 520 (1989).
11. Al. I. Efros, and A. L. Efros, Sov. Phys. Semicond., 16, 772 (1982).
12. L. E. Brus, J. Chem. Phys., 80, 4403 (1984).
13. A. Fojtik, H. Weller, U. Koch, and A. Henglein, Ber. Bunsenges. Physik Chem., 88, 649 (1984).

14. Michael L. Steigerwald, and Louis E. Brus, Annu. Rev. Mater. Sci., 19, 471 (1989).
15. L. E. Brus, New J. Chem. (France), 11, 23 (1987).
16. H. M. Schmidt, and H. Weller, Chem. Phys. Lett., 129, 615 (1986).
17. R. Reisfeld, SPIE 1758 Sol-Gel Optics II, 546 (1992).
18. H. Minti, M. Eyal, R. Reisfeld and G. Berkovic, "Quantum dots of cadmium sulfide thin glass films prepared by sol-gel technique", Chem. Phys. Lett., ***183***, 277-282 (1991).
19. R. Reisfeld, H. Minti and M. Eyal, "Active glasses prepared by the sol-gel method including islands of CdS or silver", SPIE Proc. ***1513*** , 360-67 (1991).
20. R. Reisfeld, " Semiconductor quantum dots in amorphous materials",A lecture given at the International School of Atomic and Molecular Spectroscopy, Erice, June 16-30 1991. To be published in " Optical Properties of Excited States in Solids". B. Di Bartolo editor, Plenum.
21. R. Reisfeld, V. Chernyak , M. Eyal and C.K. Jϕrgensen, " Photostable fluorescent glass films for Solar Concentrators", SPIE Proc. ***1727*** , 299-305 (1992)
22. Renata Reisfeld, Harry Minti, Marek Eyal and Valery Chernyak, Nonlinear properties of semiconductor quantum dots and organic molecules in glasses, prepared by the sol gel method. J. Nonlinear Optics, 1993.
23. L. Katsikas, A. Eichmuller, M. Gersig, and H. Weller, Chem. Phys. Lett., 172, 201 (1990).
24. D. Ricard, Physica A 157, 301 (1989).
25. Y. Wang, N. Herron, W. Mahler, A. Suna, J. Opt. Soc. Am. B 6, 808 (1989).
26. Y. Wang, and W. Mahler, Opt. Commun. 61, 233 (1987).
27. P. D. Persans, A. Tu, Y.-J. Wu, M. Lewis, J. Opt. Soc. Am., B., 6, 818 (1989).
28. Galen D. Stucky and James E. Mac Dougall, Science, Vol. 247, 669, 9 February (1990).
29. R. K. Jain, and R. C. Lind, J. Opt. Soc. Am., 73, 647 (1983).
30. P. W. Smith, Optical Computing and Nonlinear Materials, SPIE 881, 652 (1988).
31. Ken Koizumi, Int' Congress on Glass, V1, Bol. Soc. Ceram. VID 31-C, 45-59 (1992).
32. G. I. Stegeman, and R. H. Stolen, J. Opt. Soc. Am., B6, 652 (1989).
33. S. R. Friberg, and P. W. Smith, IEEE Quantum Electronics QE23, 2089 (1987).
34. D.W. Hall et al., Appl. Phys. Lett., 54, 1293 (1989).
35. S. S. Yao, et al., Appl. Phys. Lett., 46, 801 (1985).
36. P. Roussignol, et al., J. Opt. Soc. Am., B4, 5 (1987).
37. J. Yumoto, S. Fukushima, and K. Kubodera, Optics Letters, 12, 832 (1987).
38. M. Mitsunaga, H. Shinojima, and K. J. Kubodera, J. Opt. Soc. Am., B5, 1488 (1988).
39. N. Peyghambarian, et al., NATO ASI, Ser B 194, 191 (1989).
40. J. Yumoto, H. Shinojima, and N. Uesugi, Appl. Phys. Lett., 57, 2393 (1990).
41. M. Nogami, M. Watabe, and K. Nagasaka, Proc. SPIE Conf. on Sol-Gel Optics, 1328, 119 (1990).
42. N. Sugimoto, et al., "Science and Technology of New Glass 1991", S. Sakka, and N. Soga (eds), Ceram. Soc. of Japan, 394.
43. S. Omis, et al., ibid, 181.
44. S. Ohtsuka S., et al., to be published in The Proceedings of International Conference on Laser Advanced Materials (LAMP'92) in Nagaoka, Japan (1992).
45. T. Koyama, S. Ohtsuka, H. Nagata, and S. J. Tanaka, Crystal Growth, 117, 156 (1992).

46. J. D. Mackenzie, C. Y. Li, and X. Yuhuan, "Science and Technology of new glass 1991", S. Sakka, and N. Soga (eds), Ceram. Soc. of Japan, 1, 17.
47. F. Hache, D. Ricard, C. Flytzanis, and U. Krelibig, Appl. Phys., A47, 347 (1988).
48. H. Wakabayashi, H. Yamanaka, K. Kadono, T. Sakaguchi, and M. Miyata, "Science and Technology of New Glass 1991", Ceram. Soc. of Japan, 412.
49. J. Yumoto, S. Fukushima, and K. Kubodera, Optics Letters 12, 832 (1987).
50. N. F. Borrelli, D. W. Hall, H. J. Holland, and D. W. Smith, J. Appl. Phys., 61, 5399 (1987).
51. A. I. Ekimov, and Al. L. Efros, Acta Physica Polonica A, Vol. 79, No 1, 5 (1991).
52. C. T. Dameron, R. N. Reese, R. K. Mehra et al., Nature 338, 596 (1989).
53. M. G. Bowendi, A. R. Kortan, M. L. Steigerwald, L. E. Brus, J. Chem. Phys., 91, 7282 (1989).
54. A. I. Ekimov, A. A. Onushchenko, A. G. Plukhin, Al. L. Efros, Sov. Phys.-JETP 61, 891 (1985).
55. A. I. Ekimov, A. A. Onushchenko, JETP Lett., 40, 1136 (1984)
56. J. D. Mackenzie, J. Ceram. Soc. Japan, 101, 1 (1993).
57. C. J. Brinker, and G. W. Scherer, "Sol-Gel Science", Academic Press (1990).
58. L.C. Klein, "Sol-Gel Technology for Thin Films, Fibers, Preforms, Electronics and Speciality Shapes", Noyes Publications (1988).
59. J. D. Mackenzie, and D. R. Ulrich, John Wiley & Sons (1989), pp. 589-601.
60. J. D. Mackenzie, D. R. Ulrich (eds) (1988) Proceedings of the Third International Conference on Ultrastructure Processing. Wiley, New York.
61. C. J. Brinker, D. E. Clark, D. R. Ulrich (eds) Better ceramics through chemistry II [MRS Symposium Vol 73 (1986)] and Better ceramics through chemistry III [MRS Symposium Vol 121 (1988)] Materials Research Soc., Pittsburg.
62. H. Dislich , Thin films from the sol-gel process, in: L. C. Klein (ed), Sol-gel technology for thin films, fibers, preforms, electronics and speciality shapes, Noyes Publishers, Park Ridge NJ, pp 50-79 .(1988)
63. C. J. Brinker, G. Scherrer , Sol-gel science, the physics and chemistry of sol-gel processing, Academic Press, San Diego CA. (1989)
64. J. Livage, M. Henry, C. Sanchez , Prog in Solid State Chem., 18, 259 (1988).
65. C. Sanchez, J. Livage, New J. Chem., 14, 513 (1990).
66. Sol-Gel Science and Technology, Edited by M. A. Aegerter, M. Iafelicci, D. F. Souza, E. D. Zonotto, World Scientific , Singapoore (1989)
67. D. Avnir, D. Levy, R. Reisfeld, J. Phys. Chem., 88, 5956 (1984).
68. D. Avnir, V. R. Kaufman, R. Reisfeld, J. Non-Cryst. Solids, 74, 395 (1984).
69. R. Reisfeld, in: Sol-Gel Science and Technology, Edited by M. A. Aegerter, M. Jafelicci, D. F. Souza, E. D. Zanotto, World Scientific , Singapore (1989) p. 323.
70. R. Reisfeld, J. Non-Cryst. Solids, 121, 254 (1990).
71. R. Reisfeld, Theory and applications of spectroscopically active glasses prepared by the sol-gel method, Sol-Gel Optics, SPIE Proc 1328, 29 (1990).
72. R. Reisfeld, R. Zusman, M. Eyal, Chem. Phys. Lett., 147, 142 (1988).
73. V. R. Kaufman, D. Avnir, D. Pines-Rojanski, J. Huppert, J. Non-Cryst. Solids, 99, 379 (1988).
74. J. McKiernan, J. C. Pouxviel, B. Dunn, J. I. Zink, J. Phys. Chem., 93, 2129 (1989).
75. A. Makishima, T. Tani, J. Am. Ceram. Soc., 69, 4 (1986).
76. R. Reisfeld, Ch. K. Jϕrgensen, in: Structure and Bonding 77, 207 (1992), Reisfeld and Jϕrgensen editors, Springer-Verlag Berlin.

77. R. Reisfeld, V. Chernyak, M. Eyal, C. K. Jϕrgensen, Chem. Phys. Lett., 164, 307 (1989).
78. E. J. Pope, J. D. Mackenzie, MRS Bull, 12 29 (1987).
79. M. Nogami, Y.-Q. Zhu, Y. Tohyama, and K. Nagasaka, J. Non-Cryst. Solids, 134, 71 (1991).
80. M. Nogami, Y.-Q. Zhu, and K. Nagasaka, J. Non-Cryst. Solids, 134, 71 (1991).
81. H. Scholze, P. Strehlow, in: Tagungsblatt zum dritten Otto Schott-Kolloquium, Jena (1986).
82. J. Fricke, Phys. unserer Zeit, 17, 151 (1986).
83. J. Fricke, G. Reichenauer, J. Non-Cryst. Solids, 95, 1135 (1987).
84. E. J. A. Pope, J. D. Mackenzie, Mat. Sci. Res., Vol.20, 187 (1986), Plenum, NY.
85. C. Sanchez, SPIE Proc. 1328, 40 (1990).
86. T. Takada, T. Yano, A. Yasumori, M. Yamane, and J. D. Mackenzie, J. Non-Cryst. Solids, 147&148, 631 (1992).
87. D. Levi, R. Reisfeld, D. Avnir, Chem. Phys. Lett., 109, 593 (1984).
88. L. Spanhel, H. Schmidt, A. Uhrig, C. Klingshirn, Proc. MRS, Spring Meeting, San Francisco, April (1992),Mat. Res. Soc. Symp. Proc., in print.
89. C.-Y. Li, Y. J. Tseng, K. Morita, C. Lechner, Y. Hu, and J. D. Mackenzie, SPIE 1758, Sol-Gel Optics II, 410 (1992),
90. Y. Wang, N. Herron, J. Phys. Chem., 91, 257 (1987).
91. W. J. Mortier, E. Dendooven, and J. B. Uyttehoeven, in: Proc. Sixth Int. Zeolite Conf., D.
92. N. Herron et al., J. Am. Chem. Soc., 111, 530 (1989).
93. X. Liu, and J. K. Thomas, Langmuir 5, 58 (1989).
94. K. Moller, M. M. Eddy, G. D. Stucky, N. Herron, T. Bein, J. Am. Chem. Soc., 111, 2564 (1989).
95. J. E. MacDougall, et al., J. Am. Chem. Soc., 111, 8006 (1989).
96. R. D. Stramel, T. Nakamara, J. K. Thomas, J. Chem. Soc. Faraday Trans., 184, 1287
97. M. A. Fox, and T. L. Pettit, Langmuir, 5, 1056 (1989).
98. K. Moller, T. Bein, N. Herron, W. Mahler, Y. Wang, Inorg. Chem., 28, 2914 (1989).
99. D. H. Olson, R. J. Mikovsky, G. F. Shipman, E. Dempsey, J. Catal, 24, 161 (1972).
100. R. C. Burns, R. J. Gillespie, W. C. Luk, D. R. Slim, Inorg. Chem., 18, 3086 (1979).
101. T. Bein, D. B. Chase, R. D. Farlee, G. D. Stucky, Stud. Surf. Sci. Catal., 28, 311 (1986).
102. A. Fojtik, H. Weller, U. Koch, A. Henglein, Ber. Bunsenges, Phys. Chem., 88, 969 (1984).
103. J. Y. Tseng, C.-Y. Li, T. Takada, C. Lechner, and J. D. Mackenzie, SPIE Vol. 1758, Sol-Gel Optics II, 612 (1992).
104. D. Brusilovsky, M. Eyal and R. Reisfeld, Chem. Phys. Lett., ***153*** , 203-209.(1988).
105. R. Reisfeld, M. Eyal and D. Brusilovsky, Chem.Phys. Lett., ***153*** , 210-214.(1988).
106. S. Schmidt-Rink, D. A. B. Miller, D. S. Chemla, Phys. Rev., B35, 811 (1986).
107. E. Hanamura, Phys. Rev., B, 1228 (1988).
108. L. Efros, A. L. Efros, Sov. Phys. Semicond., 16, 772 (1982).
109. Y. Kayanuma, Phys. Rev., B38, 9797 (1988).
110. Y. Wang, and N. Heron, J. Phys. Chem., 95, 523 (1991).
111. L. Brus, Appl. Phys., A53, 465 (1991).
112. L. Spanhel, E. Arpac, H. Schmidt, J. Non-Cryst. Solids, J. Non-Cryst. Solids, 147&148, 657 (1992).

113. N. Tohge, M. Asuka, and T. Minami, SPIE Proc., **1328**, 125 (1990).
114. Y. Kayanuma, and K. Kuroda, Appl. Phys., A53, 475 (1991).
115. L. E. Brus, J. Chem. Phys., 80, 4403 (1984).
116. M. Nogami, M. Watabe, and K. Nagasaka, SPIE Proc., **1328,** Sol-Gel Optics, 119 (1990).
117. T. Rajh, M. I. Vucemilovici, N. M. Dimitrijevic, O. I. Micic, and A. J. Nozik, Chem. Phys. Lett., 143, 305 (1988).
118. S. Modes, and P. Lianos, Chem. Phys. Lett., 143, 305 (1988).
119. N. Tohge, M. Asuka, and T. Minami, Chem. Express., 5, 521 (1990).
120. M. Nogami, K. Nagasaki, and M. Takata, J. Non-Cryst. Solids, 122, 101 (1990).
121. M. Nogami, Y.-Q. Zhu, Y. Tohyama, and K. Nagasaka, J. Am. Ceram. Soc., 74, 238 (1991).
122. M. Nogami, Y.-Q. Zhu, and K. Nagasaka, J. Non-Cryst. Solids, 134, 71 (1991).
123. M. Nogami, K. Nagasaka, and E. Kato, J. Am. Ceram. Soc., 73, 2097 (1990).
124. M. Nogami, K. Nagasaka, and K. Kotani, J. Non-Cryst. Solids, 126, 87 (1990).
125. M. Nogami, Y.-Q. Zhu, Y. Tohyama, and K. Nagasaka, J. Am. Ceram. Soc., 74, 238 (1991).
126. T. Tamba, K. Hachisu, M. Ikezawa, J. Phys. Soc. Jpn, 50, 1579 (1981).
127. R. Reisfeld, Optical Materials on Dyes and Nanocrystallites in Amorphous Solids, in: Proc. US-France-Workshop on the Chemistry of Optical Materials, 28 Sept.-2 Oct., Maubuisson (1992).
128. R. Reisfeld, New Solid-State Tunable Lasers in the Visible. Proc. Second Israeli-German Joint Symposium on Light-Induced Processes in Molecular Assemblies, Jerusalem, March, 1419 (1993).
129. H. Schmidt, Sol-Gel Science and Technology, Edited by M. A. Aegerter, M. Iafelicci, D. F. Souza, E. D. Zonotto, World Scientific , Singapore (1989), p.432.
130. R. Reisfeld, ibid. p. 323.
131. R. Reisfeld, J. Non-Cryst. Solids, 121, 254 (1990).
132. H. Schmidt, and H. Wolter, J. Non-Cryst. Solids, 121, 428 (1990).
133. R. Reisfeld, D. Brusilovski, M. Eyal, E. Miron, Z. Burstein, and J. Ivri, Chem. Phys. Lett., 160, 4344 (1989).
134. R. Reisfeld, V. Chernyak, M. Eyal, and A. Weitz, SPIE Proceedings, 1016, 240-246 (1988).
135. Work in preparation with Reshef Tenne, and Lev Margulis at Weizmann Institute of Science, Israel.
136. Y. Ohashi, M. Ito, T. Hayashi, A. Nitta, H. Matsuda, S. Okada, H. Nakanishi, and M. Kato, in: "Nonlinear Optics of Organics and Semiconductors", Vol.36, T. Kobayashi (ed), Springer-Verlag, Berlin, Heidelberg, 81 (1989).
137. L. Katsikas, A. Eichmuller, M. Gersig, and H. Weller, Chem. Phys. Lett., 172, 201 (1990).
138. S. Graham, M. Eyal, M. Thoma, D. Brusilovsky, R. Reisfeld, and C. Klingshirn, J. Luminiscence, 48-49, 325 (1991).
139. Y. Kayanuma, Phys. Rev., B38, 9797 (1988).
140. S. Kono, T. Ishii, T. Sagawa, T. Kobayashi, Phys. Rev. B8, 795 (1973).
141. T. Takagahara, Phys. Rev., B39, 10206 (1989).
142. Al. L. Efros, and A. V. Rodina, Solid State Commun., 72, 645 (1989).
143. J. M. Luttinger, Phys. Rev., 102, 1030 (1956).
144. G. Fuxi, J. Non-Cryst. Solids, 129, 299 (1991).

145. Y. Masumoto, T. Wamura, A. Iwaki, Appl. Phys. Lett., 55, 2535 (1989).
146. T. Itoh, and M. Furumyia, J. of Luminiscence, 48&49, 704 (1991).
147. H.Kishishita , Phys. Stat. Sol(b), ***55***, 399 (1973).
148. C.Zaldo , and F.Jague, J. Phys. Chem. Solids, ***46***, 6, 689 (1985).
149. T.Mita ,K.Sotome , and M.Ueta, Solid State Com., ***33***, p1135, (1980).
150. M.Carderea , N.E.Christensen , and G.Fasol , Phys.Rev.Let., ***56***, 26, 2831, (1986).
151. Nakamura A., Yamada M., and Tokizaki T., Phys.Rev.B, ***40***, 585, (1989).
152. Kaifu Y., and Komatsu T., Phys.Stat.Sol.(b), ***48***, K125, (1971).
153. Ueta M., Ikazawa M., and Nagasaka S.-I., J.Phys.Soc.Japan, ***20,*** 1724, (1965).
154. Goto T., Yashiro T., and Ueta M., J.Phys.Soc.Japan, **20**, 2185, (1965).
155. Kratzig E., Timusk T., and Martienssen W., Phys.Stat.Sol., ***10***, 709, (1965).
156. Kaifu Y., Kawate Y., Nakanishi S., Niwa I., Nakazawa K., and Koba M., J.Phys.Soc.Japan, ***22***, 517, (1967).
157. Ringeissen J., Coret A., Nikitine S., in: "Proc. of the First Int.Conf.on localized excitons in solids", Irvine, Univ.of California, Sept. 18-22, 1967, p.297, ed.Wallis R.F., Plenum Press, NY, 1968,
158. Coret A., Ringeissen J., Nikitine S., in: "Proc. of the First Int.Conf.on localized excitons in solids", Irvine, Univ.of California, sept.18-22, 1967, p.306, ed.Wallis R.F., Plenum Press, NY, 1968,
159. Masayasu U., Goto T., J. Phys. Soc.Japan, ***20***, 401, (1965).
160. Kalm M., Uihlein Ch., Phys.Stat.Sol.(b), ***87,*** 575, (1978).
161. Itoh T., Iwabuchi Y., Kirihara T., Phys.Stat.Sol.(b), ***146***, 531, (1988).
162. Itoh T., Furumyia M., Ikehara T., Solid State Com., ***73***, 271, (1990).
163. Itoh T., Ikehara T., Iwabuchi Y., " J. Luminescence, ***45***, 29, (1990).
164. Ueta M., Goto T., J.Phys.Soc.Japan, ***20***, 401, (1965).
165. B. D. Cullity, Elements of X-ray diffraction, 2nd Ed., Addison-Wesley, New York, 102 (1978).
166. M. Ueta, H. Kanzaki, K. Kobayashi, Y. Toyozawa, and E. Hanamura (eds), "Excitonic Processes in Solids", Springer Series in Solid-State Science 60, Springer-Verlag, p.121 (1986)
167. M. Cardona, Phys. Rev., 129, 69 (1963).
168. T. Takahashi, T. Goto, J. Phys. Soc. Jpn., 25, 461 (1968).
169. Y. Kato, C. I. Yu, T. Goto, J. Phys. Soc. Jpn., 28, 104 (1970).
170. Y. Toyozawa, J. Phys. Chem. Solids, 25, 59 (1964).
171. R. Reisfeld, M. Eyal, R. Gvishi and C.K. Jorgensen, Proc. 7th Int'l Symp. on the Photochemistry and Photophysics of Coordination Compounds, Elmau, Germany, March 29 - April 2, 1987. Springer- Verlag, Heidelberg -New York, (1987) 313-316.
172. R. Reisfeld, M. Eyal and R. Gvishi, Chem. Phys. Lett., ***138*** (1987) 377-383.
173. M. Eyal and R. Reisfeld, High yield singlet-triplet transfer for efficient saturable absorbers. Proc. Int'l. Conf. on Luminescence (ICL'87), Beijing,China J. Luminescence, ***40-41*** (1988) 539-540.
174. S. Graham, R. Renner, C. Klingshirn, W. Schrepp, R. Reisfeld, D. Brusilovsky and M. Eyal, Cambridge, 1989. Inst. Phys. Conf. Ser. (Bristol, New York) No ***103***; Section 2.2, (1989) 157-162.
175. S. Graham, M. Eyal, M. Thoma, D. Brusilovsky, R. Reisfeld and C. Klingshirn, J. Lumines-cence, ***48-49***, 325-328 (1991).
176. S. Graham, M. Thoma, C. Klingshirn, M. Eyal, D. Brusilovsky, R. Reisfeld, S.V.Gaponenko V. Yu Lebed and L.G. Zimin, "Laser Induced Gratings and

Nonlinear Optics in Organic Materials", in Organic Materials for Nonlinear Optics II, Edited by R.A. Hann and Bloor, Royal Society of Chemistry, Thomas Graham House, Cambridge , 142-148 (1991).
177. S. Graham, C. Klingshirn, S. Gaponenko, I. Germanenko, V. Gribkovskii, L. Zimin, V. Lebedev, I. Malinovskii, A. Stupak, M. Eyal, D. Brusilovsky and R. Reisfeld, to be published in the NATO ASI Series, "Proc. Symp. Physics and Chemistry of Finite Systems: From Clusters to Crystals", Richmond, USA, October 8-12, 1991.
178. T. Todorov, L. Nikolova, N. Tomova and V. Dragostinova, IEEE J. Quant.Electron. ***QE-22*** (1986) 1262.
179. S. V. Gaponenko, V. P. Gribkovskii, L. G. Zimin, V. Yu. Lebed, I. E. Malinovskii, S. Graham, C. Klingshirn, M. Eyal, D. Brusilovsky, and R. Reisfeld, Optical Materials, 53-58 (1993).
180. C.K. Jϕrgensen and R. Reisfeld, SPIE Proceedings,. ***2288*** , 208-215.(1994).
181. R. Reisfeld, D. Brusilovsky, M. Eyal, E. Miron, Z. Burstein and J. Ivri, Chem. Phys. Lett. ***160,*** 43-44 (1989).
182. L.G. Zimin, S.V. Gaponenko, I.E. Malinovskii, V.U. Lebed, I.N. Germanenko, E.E. Podorova, V. A. Tsekhomskii, J. Modern Optics, ***37*** , 829-834 (1990).
183. L.G. Zimin, S.V. Gaponenko, I.E. Malinovskii, V.U. Lebed, I.N. Germanenko, J. Luminescence, ***46*** , 101-107 (1990).
184. S. Speiser and F.L. Chisena, Appl. Phys., ***B45*** , 137-144 (1988).
185. S. Yariv in International Reviews in Physical Chemistry, Vol.11, No.2, 345 (1992).
186. H. Schmidt, Bol. Soc. Esp. Ceram. VID, 31-C, 185 (1992).

UP-CONVERSION AND EXCITED STATE ABSORPTION IN LASER CRYSTALS AND GLASSES: ADVANTAGES AND DISADVANTAGES

F. Auzel

France Telecom
Centre National d'Etudes des Télécommunications, Paris
B - Laboratoire de Bagneux, B. P. 107
92225 Bagneux Cedex, France

ABSTRACT

After a brief history, the basic processes involved in up-conversion by energy transfer (Addition de Photons par Transferts d'Energie : APTE effect) are recalled. They are distinguished from other multiphoton effects which may take place in parallel, such as excited state absorption (ESA), cooperative sensitization, cooperative luminescence and harmonic generation. Relative orders of magnitude of these different up-conversion processes are given.

In usual materials, up-conversion by APTE is usually so dominant that cooperative effects are shown to be clearly observable only when APTE effects cannot take place due to either a large energy mismatch between single ion levels of adjacent ions or due to a lack of energy diffusion resulting from low ion concentrations.

ON the other hand, in materials with optical confinements such as optical fibers, excited state absorption may overtake other up-conversion processes. Such processes are shown to play active roles in some laser effects. In usual lasers they may be useful path or sometimes feed non-radiative processes. In anti-Stokes laser, where emission is at a shorter wavelength than pumping, up-conversion is used as the necessary pumping step.

An overview of recent results showing the positive and negative implications of up-conversion by APTE or ESA in laser materials shall finally be presented.

I. INTRODUCTION

When active ions are situated at sufficiently short distance for interactions between them to take place, two types of up-conversion processes may occur: summation of photon energy through energy transfers [1] (sometimes called APTE effect for Addition de Photons par Transferts d'Energie) or cooperative effects either by sensitization [2] emission [3] or absorption [4]. Both types are often mistaken one for the other because they present several similarities and may be simultaneously present in a given system for a given excitation.

Up to 1966 all energy transfers between rare-earth ions were of the types summarized in Fig. 1. That is the activator ion receiving the energy from a nearby sensitizer (s), was in its ground state. Then I proposed to consider cases where activators (A) were already in an excited state [1] as shown in Fig. 2.

This is evident afterwards because one has to realize that we can exchange energy differences between ions and not absolute energy.

Nonlinear Spectroscopy of Solids: Advances and Applications
Edited by B. Di Bartolo and B. Bowlby, Plenum Press, New York, 1994

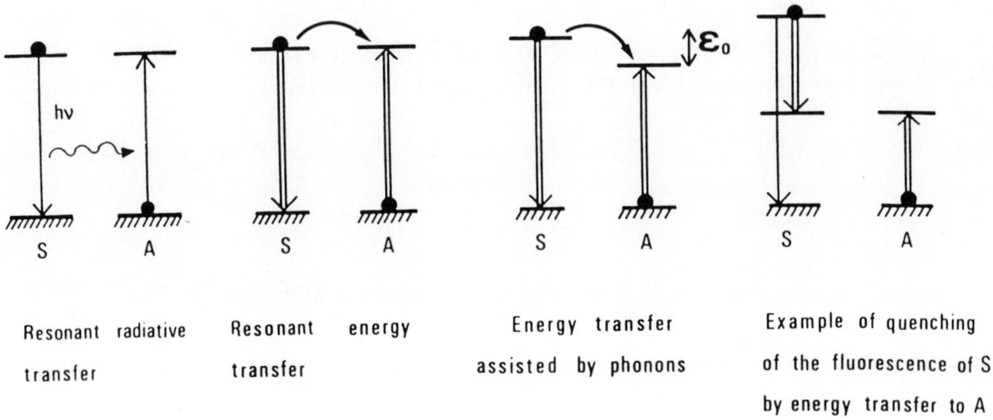

Fig 1 - Energy transfers from a sensitizer (S) to an activator (A) in ground state.

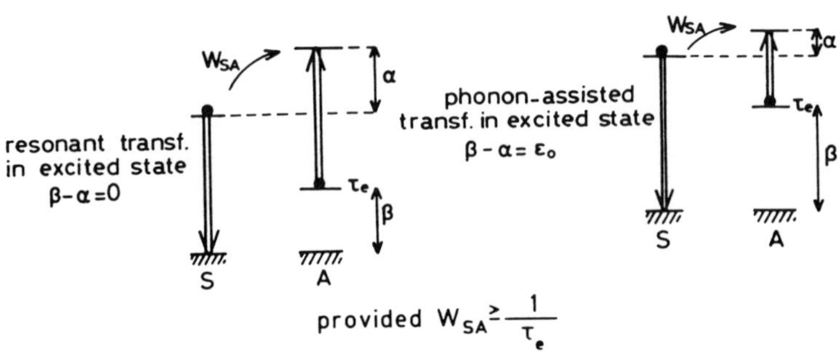

Fig. 2 - Energy transfer with the activator in metastable excited states.

The reason for proposing such up-going transfer was to point out that energy transfers then used [5] to improve the laser action of Er^{3+} by pumping Yb^{3+} in a glass matrix could also have the detrimental effect to increase reabsorption [6]. The simple proof of such effect was to look for and up-converted green emission (from $^4S_{3/2}$ of Er^{3+}) while pumping Yb^{3+} ($^2F_{7/2} \rightarrow {}^2F_{5/2}$), which we effectively observed [1] [6]. Of course the situation in Fig. 2 could repeat itself several times at the activator. This means that n-photon up-conversion by energy transfer was possible as demonstrated by the 3-photon up-conversion of 0.97 μm into blue light (0.475 μm) in the Yb^{3+} - Tm^{3+} couple [1]. Independently such IR → blue up-conversion was interpreted by Ovsyankin and Feofilov [2] as a 2-photon effect connected with a cooperative sensitization of Tm^{3+} by two Yb^{3+} ions, because of saturation in an intermediate step reducing intensity from a cubic to a quadratic law.

In order to make the terminology clearer a schematic comparison between APTE effect and other 2-photon up-conversion processes namely : 2-steps absorption, cooperative sensitization, cooperative luminescence, second harmonic generation (SHG) and 2-photons absorption excitation is presented on Fig. 3, together with typical efficiencies.

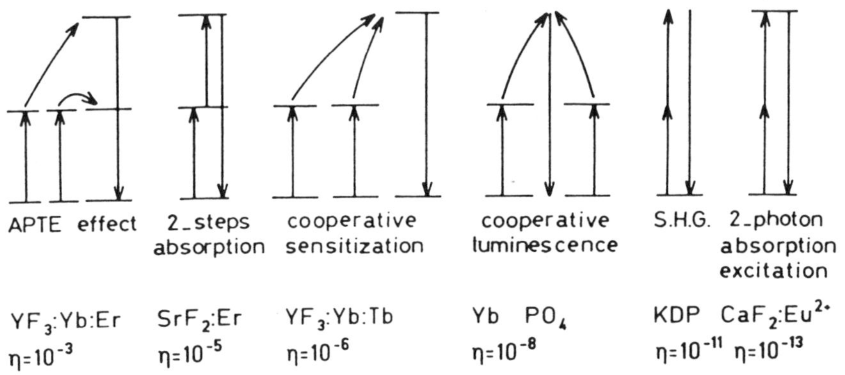

Fig.3 - Energy schemes for different 2-photon up-converting processes with quantum efficiency in cm^2/W for specified materials.

Since we are dealing with non-linear processes, usual efficiency, as defined in percent, has no meaning because it depends linearly on excitation intensity. Value are then normalized for incident flux and given in (cm^2/W) units.

A simple inspection of involved energy schemes shows that they differ at first sight by the resonances involved for in-and out-going photons : for highest efficiency, photons have to interact with the medium a longer time which is practically obtained by the existence of resonances. As shown, APTE effect is the most efficient because it is the process which approaches more the full resonance case.

However reality is some time not so simple and different up-conversion processes may exist simultaneously or their effects can be tentativelly reinforced reciprocally. For instance two-photons absorption and cooperative absorption have been theoretically investigated [7]. Also SHG and cooperative luminescence have been considered simultaneously in order to increase SHG by the partial resonance of cooperative luminescence [8] [9].

Many times in litterature up-conversion involving coupled ions are referred to as cooperative effects [10] [11] without demonstration, when in fact as can be guessed from their relative efficiencies, APTE effects are involved [12] [13]. The fact that APTE effect and cooperative ones are often mistaken one for the other is linked to a number of common properties. For instance, for 2-photon up-conversion, both processes shows a quadratic increases on excitation and on absorber concentrations ; both shows an emission lifetime equal to half the absorber lifetime. However as shown in next chapter the difference is more basic, though sometime difficult to establish experimentally except in special cases : when single ion resonances clearly do not exist or when diffusion between ions is prohibited by a too small concentration with still an interaction as in clusters.

II. PRIMARY PROCESSES

II.A. One Ion Processes

excited state absorption (ESA) as well as sensitizer excitation or activator emission are usually one ion, one phton processes involving in the general case electronic and phonon transitions. Non-radiative transition at the sensitizer or activator are also generally considered as a one-ion process generally involving many simultaneous phonons.

All these processes are the necessary buildings blocks involved in the simplest up-conversion phenomena.

In the following I shall present them in a simplified unifyied way.

Assuming validity of "Fermi Golden Rule", any transition probabilitiy can be written as :

$$W_{21} = \frac{2\pi}{\hbar} |< 2 | H | 1 >|^2 \rho_E \tag{1}$$

where H is the perturbation causing the transition from state 1 to state 2.

In the case of radiative transition H is usually the electric-dipole operator eR, the matrix element of which exists when state $|2>$ and $|1>$ are of opposite parity or have some traces of opposite parity.

In the case of H being a two-photon operator or a magnetic dipole one $|2>$ and $|1>$ may be of same parity.

In case of non-radiative transition, H is the " non Adiabatic Hamiltonian" [14].

In the general case of a vibronic transition $|2>$ and $|1>$ are the two vibronic states between which the transition takes place ; ρ is the density of final states which can be expressed as :

$$\rho = R_N \delta(E_2 - E_1 - E) \tag{2}$$

for a radiative transition at energy E between states at energy E_2 and E_1,

$$\rho = R_N \delta(E_2 - E_1) \tag{3}$$

for a non-radiative one.

R_N is usually given as a product of Franck-Condon Factors corresponding to the so-called "accepting modes" which "receive" the energy given up by the excited electronic state [14]. The matrix element $<|H|>$ can be considered as the "promoting" part of the transition [14].

An estimation of R_N at 0°K can be guessed by the following reasoning :

let p be the probability for one of a set of S equivalent modes to jump to the next state. p^N is the probability of having N simultaneous jumps in the same direction for N modes. But the number of different ways to choose the N simultaneous jumping modes among the total set of S is :

$$C_S^N = \frac{S(S-1)\dots(S-N+1)}{N!} \tag{4}$$

That is, the total probability p_t for N jumps among a set of S modes is :

$$p_t^{(N)} = \frac{p^N S(S-1)\dots(S-N+1)}{N!} \approx \frac{(pS)^N}{N!} \tag{5}$$

for S > N. The term pS represents the average number of modes which have been involved in a jump.

Following Toyosawa, So, the Huang-Rhuys coupling parameter, which is directly linked to the amount by which the normal coordinates are displaced by the electron-phonon coupling, can be viewed also as the number of phonons emitted or absorbed in the transition [15]. Then, it is possible to write : $pS = S_0$.

Then (5) can be written :

$$p_t^{(N)} = \frac{S_0^N}{N!} \tag{6}$$

Now the total probability for all N-processes is equal to unity and we have :

$$p_t = \Sigma_N k \frac{S_0^N}{N!} = k e^{S_0} = 1$$

that is

$$p(N) = \frac{e^{-S_0} S_0^N}{N!} \tag{7}$$

a Poisson distribution some time called the "Pekarian" function.

At temperature T, absorption and emission of phonons have to be simultaneously considered in such a way that :

$$N = k - l$$

where k is the number of phonons emitted and l the number of phonons absorbed. Then :

$$p_l(N,T) = e^{-S_0(\bar{n}+1)} \frac{S_0^k (\bar{n}+1)^k}{k!} e^{-S_0 \bar{n}} \frac{S_0^l (\bar{n})^l}{l!}$$

and in vibronic absorption for instance :

$$p_t(N,T) = e^{-S_0}(2\bar{n}+1) \frac{(\bar{n}+1)^{N/2}}{(\bar{n})^{N/2}} \Sigma_l \frac{S_0^{2} \bar{n} (\bar{n}+1)^{l+N/2}}{l!(N+l)!} \tag{8}$$

which is the R_N of Huang and Rhys [14].

If we consider eq. (8), as given by Huang and Rhys, the temperature dependence is rather complicated since it is contained in the phonon occupancy number,

$$\bar{n} = (\exp(h\omega_m/kT) - 1)^{-1}$$

one can write :

$$R_N = e^{-S_0} \frac{S_0^N}{N!} (\bar{n}+1)^N e^{-2\bar{n}S_0} \left[1 + \frac{S_0^2 \bar{n} - \bar{n} + 1)}{(n+1)} + \ldots \right] \tag{9}$$

For $S_0 \leq 1$, and not too high a temperature, we can retrain only the first term; that is, for a "Pekarian function" at 0°K we associate a temperature dependence of the form

$$(\bar{n}+1)^n e^{-2\bar{n}S_0} \tag{10}$$

In fact for small values of S_0, $\exp -2S_0\bar{n}$ can be neglected ($\approx 4\%$ for LaF_3 at 400°K) and one is left with the usual dependence $(\bar{n}+1)^N$ as experimentally verified, for instance by Riseberg and Moos [16].

From this, the small-coupling approximation of eq. (8) is equivalent even at temperature T to considering only phonon emission by $(n + 1)^N$ or only absorption by $(\bar{n})^N$. This possibility arises from the rapid decrease of P(N) with the number of phonons. This is shown in fig 4.

According to the value of S_0 and assuming that energy can be varied only by discrete equal values for N (one phonon frequency), P(N) is represented as in Fig 4. For a small coupling ($S_0 = 0.1$), the envelope of the distribution is nearly exponential. For medium (S = 1) and strong (S = 10) coupling the distribution becomes more symmetrical and approaches the usual Gaussian found for such cases [14].

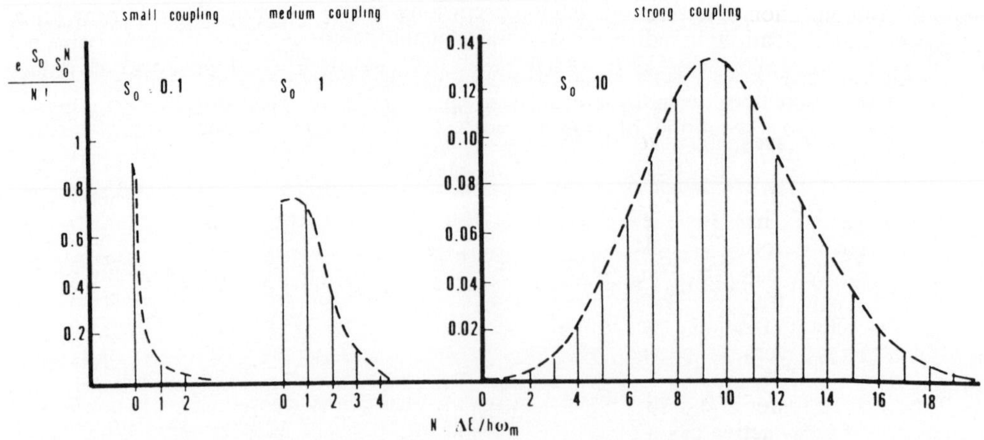

Fig 4. The "Pekarian function" for different coupling parameter (S_0) values.

Since this simple reasoning underlying "Pekarian" functions can be applied to any phenomenon in which multiphonon jumps are involved, one can now understand why when small coupling is considered, exponential gap laws are found whatever the process is. For strong coupling the Gaussian form of P(N) gives the Arrhenius limit form for non-radiative probability. For small coupling, N! is developped by Stirling's Formula giving for non-radiative transition :

$$W_{NR} = W_{NR}(0)\exp-\alpha(E_2 - E_1)$$

The precise knowledge of the phenomena comes as a modification to the statistics involved giving the gross features. Then such forms apply as well for non-radiative transition as for radiative ones. The N = 0 line is the so-called zero-phonon line the intensity of which decreases with coupling strength.

The electron-phonon coupling intensity depends essentially of the sensitivity of electrons in a given configuration to crystal field perturbation. Table I gives some typical orders of magnitude for S_0 found for different configurations.

Table I

Transition type	f - f	f - d	d - d	
active center type	rare-earth (trivalent)	rare-earth (bivalent)	transition metals	F - centers
S_0	10^{-2} - 10^{-1}	1 - 6	1.5 - 10	10 - 120

When the concentration of active ions is increased, long before the appearance of new lines due to pairs or modifications in radiative transition probabilities, a migration of energy between the centers is found. We are going to study this now, assuming that the multiphonon decay and the multiphonon radiative transitions remain one-center processes.

II.B. Basic Processes Involved at High Concentration

1. Energy Transfer. Having discussed single-ion processes, we shall deal now with multiions processes, namely energy transfers. Energy transfer occurs in a system when absorption and emission do not take place within the same center. It may occur without charge transport. Then one may distinguish between radiative and non-radiative, resonant and phonon-assisted energy transfer. Theoretical approaches start from a microscopic point of view with the interaction between two centers, whereas experimentally one is faced with a macroscopic result averaged over all the centers in the sample. Fig 5 shows the different energy transfer processes.

When the transfer is radiative, real photons are emitted by the sensitizer ions and are then absorbed by any activator ions within a photon travel distance. Because of this fact, such transfer depends on the shape of the sample.

Moreover, according to the degree of overlap between the emission spectrum of the sensitizer (S) and the absorption spectrum of the activator (A), the structure of the emission spectra of the sensitizer will change with activator concentration. Since photons are emitted anyway, the sensitizer lifetime is independent of the activator concentration. These three facts are the criteria used to distinguish between radiative and non-radiative resonant energy transfer.

Probability for such transfer between 2 ions at distance R is found to be [14] :

$$P_{SA}(R) = \frac{\tau_A}{4\pi R^2} \frac{1}{\tau_S} \int g_S(\nu) g_A(\nu) d\nu \qquad (11)$$

where τ_S is the sensitizer lifetime ; τ_A the absorption integrated cross section. The integral represents the spectral overlap between A and S. It should be noted that the distance dependence goes as R^{-2}. Such resonant radiative transfer may permit long-range energy diffusion between identical ions and gives rise to photon trapping effects. Trapping effects increase the apparent experimental lifetime and τ_S has to be measured on thin and lightly doped samples. These effects are particularly strong in Cr^{3+} and Yb^{3+} [16] [17].

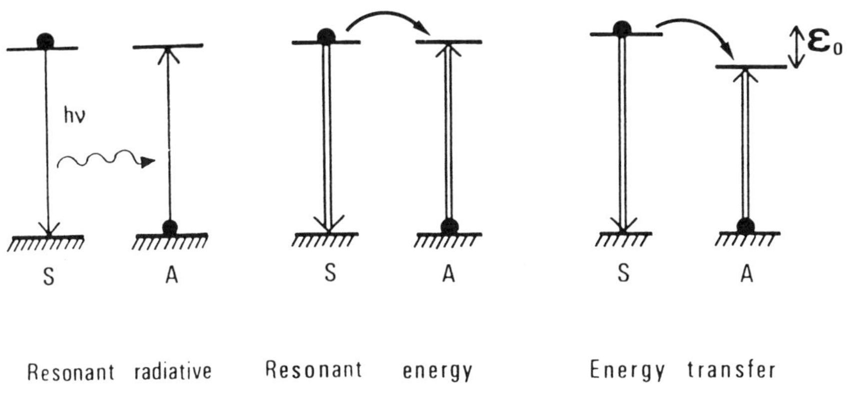

Fig 5. Different types of energy transfer to an activator ground state.

Let us consider the simple case of two ions, each with one excitable electronic state separated from its electronic ground state by nearly equal energy. With suitable interaction between the two electronic systems, the excitation will jump from one ion to the other before one is able to emit a quantum of fluorescence. The mutual interactions are the Coulomb interactions of the Van der Waals type between the two ions. Förster [18], who first treated such a case theoretically by quantum-mechanical theory, considered the dipole-dipole interactions. He assumed that the interaction is strongest if for both transitions electric-dipole transitions are permitted [19]. The interaction energy is then proportional to the inverse of the third power of the interionic distance and the transfer probability is given by :

$$P_{SA} = \frac{2\pi}{h} |< S^{*}A^{0} | H_{SA} | S^{0}A^{*} |^{2} \rho_{E} \qquad (12)$$

H_{SA} = Interaction Hamiltonian

ρ_E = Density of states provided by the vibrational motion contributing to the line-broadening of the transition, H_{SA} is proportional to the inverse sixth power of the separation. The wave-functions to be considered for the matrix element describe an initial state of the system with the sensitizer in its excited state and the activator in its ground state, the final state having the sensitizer in its ground state and the activator in its excited state.

Therefore, the transfer probability can be written as :

$$p_{SA} = \frac{1}{\tau_S}\left(\frac{R_0}{R}\right)^6 \qquad (13)$$

where τ_S is the actual lifetime of thge sensitizer excited state, including multiphonon radiative decay, and R_0 is the critical transfer distance for which excitation transfer and spontaneous deactivation of the sensitizer have equal probability.

However, Dexter pointed out [20] that this theory should be extended to include higher multipole and exchange interactions. In fact, for an isolated atom, one can consider the transition probability as decreasing as $(a_0/\lambda)^{2n}$ where a indice 0 is the Bohr radius, λ the wavelength, and n an integer.

Whereas, in an energy transfer process with a dependence on near-zone interactions, the transition probabilities drop off as $(a_0/\rho)^{2n}$ where $\rho$is the separation of the interacting ions. ρ can be as much as three orders of magnitude smaller than λ, so that the energy transfer effect tends to be more pronounced in systems with forbidden transitions [20]. This holds true for ions with transitions forbidden to first order, such as transition metal and lanthanide ions.

The energy transfer probability for electric multipolar interactions can be more generally written as [20] :

$$P_{SA} = \frac{(R_0/R)^S}{\tau_S} \qquad (14)$$

where S is a positive integer taking the following values :

S = 6 for dipole-dipole interactions

S = 8 for dipole-quadrupole interactions

S = 10 for quadrupole-quadrupole interactions.

It should be noted that, for dipole-dipole interactions, differences between radiative and non-radiative resonant transfer lay essentially in that for radiative transfer there is no critical R_0 depending only upon concentration. The variation goes as R^{-2} instead of R^{-6}, and the sensitizer lifetime does not depend upon the distance R.

Now, to be able to calculate effectively P_{SA} (R), eq. (14) is not very useful, because R_0 cannot be easily obtained theoretically. Applying Racah's tensorial methods straightaway at the beginning of the calculation of Dexter, eq. (12), permits one to develop calculations analogous to Judd's theory for radaitive transitions. The case of the multipolar interactions was ttreated in this way by Kushida [21] and calculations for magnetostatic, and exchange interactions were performed in our group [22].

The general form obtained is :

$$P_{SA}(R) = \frac{2\pi}{\hbar}\frac{I}{g_S^{*} g_A} \sum_{l_1 l_2} | C_{l_1 l_2} < S^{*} \left\| U^{(l_1)} \right\| S >< \mathring{A} \left\| U^{(l_2)} \right\| A^{*} > |^2 \qquad (15)$$

with $I = \int \gamma_S(E)\Gamma_A(E)dE$

where g_S^* is the degeneravcy of the S* (Å) level, $\gamma_S(E)$ ($\Gamma_A(E)$) is the normalized shape function of emission (absorption) spectrum. U(l_i) are the tensorial operators already seen for Judd's theory. $|C_{l_1 l_2}|^2$ can be considered as a parameter analogous to T_λ for oscillator strength.

This expression of the transfer probability has the following advantages :

- Radial and orbital parts have been separated
- Only a few reduced matrix elements need to be calculated. They are the same ones for the three interactions we consider (for any interaction leaving spins unchanged).
- Comparison between two different interactions can be made through comparison of the $C_{l_1 l_2}$ coefficients. They are independent of the states involved in the transfer and we call them $E_{l_1 l_2}$, $m_{L_1 l_2}$ and $X_{l_1 l_2}$ for electrostatic, magnetostatic and exchange interactions, respectively.
- Forced electric-dipole transitions, as calculated by Judd's method, can be included in eq. (15).
- This expression gives also a single mathematical form regardless of the interaction, which is a convenient result.

The somewhat complicated expressions for the different $C_{l_1 l_2}$ of 4f electrons are given in [22]. But we can note that :

- For electrostatic interaction ($E_{l_1 l_2}$) the $l_1 = 1$, $l_2 = 1$ term, corresponding to dipolar-dipolar interaction, being zero in first order, necessitates the introduction of Judd's T_λ parameters. The $E_{l_1 l_2}$values lay between $E_{22} \approx 30cm^{-1}$ for quadrupole-quadrupole intensities to $E_{66} \approx 3.10^{-1}cm^{-1}$, but all contain some dipole-dipole part due to the T_λ.
- For magnetostatic interactions ($M_{l_1 l_2}$) only terms with l_i = 1, 3, 5 are nonzero and have the following order of magnitude :

$$M_{11} \approx 1cm^{-1}, \quad M_{55} \approx 2.10^{-7}cm^{-1}$$

- For exchange interactions ($X_{l_1 l_2}$, we have $1 \le l_1 \le 6$, giving estimates of 1 to 10^{-1} cm^{-1} for the coefficients.

These results show that exchange or magnetostatic interactions can be found in cases of small dipole-dipole and quadrupole electrostatic interactions if the matrix elements allow them.

If now we consider two ions with excited states of different energies, the probability for energy transfer should drop to zero where the overlap integral $\int g_S(\nu)g_A(\nu)d\nu$ vanishes. However, it is found experimentally that energy transfer can take polace without phonon-broadened electronic overlap provided that the overall energy conservation is maintened by production or annihilation of phonons with energies approaching $k\Theta_d$, where Θ_d is the Debye temperature of the host matrix [23]. Then for small energy mismatch ($\approx 100cm^{-1}$ energy transfer assisted by one or two phonons can take place [24] [25]. However, for energy transfer between rare earths ions, energy mismatches as high as several thousand reciprocal centimeters are encountered. This is higher than the Debye cutoff frequency found in normally encountered hosts, so that multiphonon phenomena have to be considered here.

Miyakawa and Dexter showed that it is still legitimate to write the probability of energy transfer in the form of eq. (12), where $\rho(E)$ is taken as S_{SA} the overlap of the lineshape functions for emission in ion S and absorption of ion A, including the phonon sidebands in the lineshape. It is necessary to consider each partial overlap between the m-phonon emission line shape of ion S and the n-phonon absorption lineshape of ion A. This mathematical assumption has been given a physical meaning by our demonstration of the existence of multiphonon sidebands for trivalent rare-earth ions, even in such a case of very small electron-phonon coupling [26] even when they cannot be seen by usual spectroscopy.

Along the same lines as for vibronic-sidebands seen in II.A., S_{SA} can be expressed as follows [27] :

$$S_{SA} = \sum_N e^{-(S_{0S}+S_{0A})} \frac{(S_{0S}+S_{0A})^N}{N!} \alpha_{SA}(0,0;E)\delta(N, \Delta E/\hbar\omega), \qquad (16)$$

where S_{0S} and S_{0A} are the respective lattice coupling constants for the ions S and A, N the order of the multiphonon process with $N = \Delta E/\hbar\omega_m$, ΔE the energy mismatch between both ions, and $\hbar\omega_m$ the phonon cutoff frequency. $\sigma_{SA}(0,0;E)$ is the zero-phonon overlap integral between S and A. A "Pekar" function is found as for non-radiative decay.

The expression for S_{SA} with an energy mismatch of ΔE for small S_0 constant and for an occupation number $\bar{n} = (\exp(\hbar\omega.kT) - 1)^{-1}$ not exceeding unity at the operating temperature, can be approximated with Stirling's formula by :

$$S_{SA}(\Delta E) = S_{SA}(O)e^{-\beta\Delta E} \tag{17}$$

where $S_{SA}(O)$ is the zero-phonon overlap between S and A in the case where there is no energy mismatch between the two ions. β is given by :

$$\beta = (\hbar\omega)^{-1}(\log N/S_0(\bar{n}+1) - \log\left(1 + \frac{S_0}{S_{0S}}\right) \equiv \alpha_S - \gamma \approx \alpha_S - \log_{n^2} \tag{18}$$

involving α_S the non-radiative decay parameters and assuming identical electron-phonon coupling for ions A and S. This exponential dependence on energy mismatch agrees well with experiment [28].

Up to this point we have been dealing with the microscopic case of two ions interacting with one another. To discuss the case of real macroscopic samples with many ions and to obtain a link with experimental facts, a statistical analysis of the energy transfer is necessary.

We have then to think about the overlap integrals that, we have seen, arise in all transfer between two ions. In the microscopic case we are sure that the lineshapes involved can be only due to some homogeneous broadening even for transfer between two identical ions in different lattices sites.

In the macroscopic case, we can measure absorption and emission spectra taking into account all broadening processes averaged over the whole sample; for instance the inhomogeneous broadening process due to emission and absorption at centers in different lattice sites. Then the overlap integral measured experimentally from the usual spectra is a measure in excess of the real overlap since we take into account emission and absorption of centers at any distances, even those which cannot interract. The error is the largest for the processes at shortest interacting distance (exchange), and so is certainly negligible for radiative transfer, since photons can travel a much larger distance than the spread of the spatial disorder. The error is also smaller for systems with small inhomogeneous broadening having centers in only one type of lattice site, that is, without disorder.

Fluorescence line-narrowing techniques could give some idea about the homogeneous part of an emission line, but the statistical analysis for the whole sample should still be performed. Supposing only a sensitizer-activator interaction, an average transfer efficiency can be calculated [20]. This was studied in some detail by Inokuti and Hirayama [29]. They considered the number of activators situated at random in a sphere around a sensitizer such that the activator concentration is constant when the volume of the sphere and the number of activator ions considered goes to infinity. Then the average probability for transfer from one sensitizer to any acceptor is :

$$W_{SA} = N_A \int R^{\infty}_{\min} P_{SA}(R) 4\pi R^2 dR \tag{19}$$

Introducing eq. (19) into the expression for the intensity emitted by all sensitizers, each with different activator neighborhood, they obtained the following relation for the intensity decay of the emission of the sensitizer surrounded by many activators :

$$I(t) = \exp - \frac{t}{\tau_{SO}} - \Gamma\left(1 - \frac{3}{S}\right)\frac{C}{C_O}\left(\frac{t}{\tau_{SO}}\right)^{3/S} \tag{20}$$

where τ_{SO} is the decay constant of the sensitizer in the absence of activators ; C is the activator concentration ; C_0 is the critical activator concentration and S the parameter of the multipolar interaction.

The comparison between experimental decay and this theoretical expression has been widely used to determine the index of the multipolar interaction involved. This theory is valid only when there is no sensitizer-to-sensitizer transfer or activator-to-activator transfer. This formulation, therefore, has to be modified for high concentrations of sensitizers and activators. Then rapid energy migration between sensitizers or between activators is possible, because of the perfect resonance conditions. The general result is complicated [30] but, for large t, I(t) decays exponentially [31] :

$$I(t) = \exp\left(-\frac{t}{\tau_S} - \frac{t}{\tau_D}\right) \tag{21}$$

Then two cases can be distinguished :

i) In one case, spontaneous decay of excited sensitizers, diffusion among sensitizers and energy transfer between sensitizers and activators are about of the same order of magnitude.

For sufficiently long times and dipole-dipole interactions one has [31] :

$$\tau_D^{-1} = VN_S N_A \tag{22}$$

with $V = 8\pi C^{1/4} C_{SS}^{3/4}$

N_S is the sensitizer concentration ; N_A the activator concentration ; C is the sensitizer-activator energy transfer constant, such as $C = (R_0)^{s/\tau}$; C_{SS} is the sensitizer-sensitizer transfer constant.

ii) For high sensitizer concentration, the diffusion rate can be faster than spontaneous sensitizer decay or sensitizer-activator energy transfer. The limiting step is no longer diffusion and D appears to saturate with increasing donor concentration ; each activator experiences the same excited sensitizer neighbourhood. We take R as the minimum distance between sensitizer as permitted by the lattice ($R = R_{min}$). One has

$$\tau_D^{-1} = UN_A \tag{23}$$

with U a constant depending on the type of interaction as discussed earlier in this section through eq. (15).

Another approach to the macroscopic case is the use of the well-known rate equations which deal with the population of ions in a given state. This was used as a phenomenological approach in studies of lasers. The applicability of those equations in relation to the Inokuti and Hirayama statistics has been discussed by Grant [32].

The basic result of Grant is that the energy transfer probability is proportional to the activator concentration :

$$W_t = UN_A \tag{24}$$

this result is the same as obtained by fast-diffusion studies (eq.23).

The practical interest in considering diffusion is that the decays are again exponential, as when ions are not interacting. This validates the use of rate equations.

2. Cross-Relaxation. The cross-relaxation term usually refers to all types of energy transfer occuring between identical ions. In such a case the same kind of ion is both a sensitizer and an activator.

Cross-relaxation may give rise to the diffusion process already considered between sensitizers when the involved levels are identical or to self-quenching when they are different also to APTE effect when the energy of the activator is risen by energy transfer within one type of ions. In the first case there is no loss of energy whereas in the second case there is a change in the energy of the emitted photons either a loss or a gain.

Theoretically the same treatment is valid as for the more general case of energy transfer, however, experimentally it may be more difficult to distinguish between sensitizers and activators. So any of the microscopic processes discussed in 1) may happen with a maximum overlap when identical couple of levels are involved. From the macroscopic point of view, the diffusion-limited case predicts from eq. (22) :

$$\tau_D^{-1} = VN^2 \tag{25}$$

for $N_S = N_A = N$

and in the fast diffusion case :

$$\tau_D^{-1} = UN$$

3. Up-Conversion in Single Ion Level Description (APTE) and in Pair-Level One (Cooperative Effects). As seen in the introduction up-conversion by energy transfers is just a generalization of Dexter energy transfer [20] to the case of the activator being in a metastable state instead of being in its ground state ; this requires that the interaction between S and A (H_{SA}) be smaller than the vibronic interaction of S and A in order that both ions be described by single-ion levels coupled to the lattice. It is generally the case since for fully concentrated rare-earth crystals or for cluster, pairs level splitting is of the order of 0.5 cm^{-1} [33] ; in host with smaller concentration this interaction can even be weaker, whereas one-phonon or multi-phonon sidebands may modulate the levels positions by several 100 cm^{-1}. Further, up-conversion requires that the transfer probability for the second step (W_{SA}) be faster than radiative and non-radiative decay from the metastable level that is $W_{SA} > \tau^{-1}$ with τ the measured lifetime. W_{SA} is obtained from :

$$W_{SA} = \frac{2\pi}{\hbar} |< \psi_S^e \psi_A^o | H_{SA} | \psi_S^o \psi_A^e >|^2 \rho(E) \tag{26}$$

where the wave functions are simple products of single ion wave functions ; $\rho(E)$ describes, the dissipative density of states due to the coupling with the lattice.

All cooperative processes including up-conversion can be considered as transitions between pair level for both ions as a whole. A dipolar electric transition would be forbidden for such two-center transition and one needs product wave functions corrected to first order to account for interaction for electrons of different centers [34] :

$$\psi_{pair} = \psi^o(S)\psi^o(A) - \sum_{s''\neq 0} \sum_{a''\neq 0} \frac{< s'' a'' | H_{SA} | 00 >}{\delta_{s''} - 0 + \varepsilon_{a''} - 0} \psi_{s''}(S)\psi_{a''}(A) \tag{27}$$

for example for the ground state ; s'', a'' denote intermediate states for S and A ; $\delta_{s''}$, $\varepsilon_{a''}$ denote their corresponding energies. Then any one-photon transition in the cooperative description involves already four terms in the matrix element which cannot be reduced to (26) :

APTE up-conversion does not correspond to the same order of approximation as cooperative processes which have to be considered practically only when the first type cannot take place.

Such is the case when real levels do not exists to allows energy transfers ; it is the case for Yb^{3+} - Tb^{3+} up-conversion [33] or when concentration is to small to allows efficient transfer by energy diffusion between sensitizers. Then cooperative up-conversion is likely within clusters [35]. One may also look for crystal structures where the pairs clustering is built in [36].

In order to illustrate the difference between APTE and cooperative up-conversion we shall discuss an example of line-narrowing effect in n-photon summation as a mean to distinguish between both processes [37]. Irradiating Er^{3+} doped samples with IR radiation at 1.5 μm leads to various visible emissions :

Room temperature IR F-center laser excitation between 1.4 and 1.6 μm of 10 % Er^{3+} doped vitroceramic and of YF_3 : Er leads to emission bands from near IR to U.V. Such emission may be ascribed to multiphoton excitation respectively of order 1 to 5, either of the APTE or of the cooperative type as depicted respectively with energy levels of single ions (APTE) or with pair levels (Fig. 6.b and 6.a).

Successive absorptions in Fig. 6.a involve a combination of several J states. APTE effect, because of self matching by multiphonon processes, involves only J = 15/2 and J = 13/2 states.

Excitation spectra on Fig. 7 show a striking behaviour : each spectrum presents the same spectral structure with clearly an increasing narrowing with process order. The structure reproduces the Stark structure of the $^4I_{15/2} \rightarrow {}^4I_{13/2}$ first excited terms as can be obtained by a diffuse reflectance spectra. Then this is a direct proof of the validity of the APTE explanation, since a cooperative effect should show the convolution of all J states involved in the multiple absorption between pair levels [37].

The spectral narrowing can be understood by a rate equation treatment where higher excited population are neglected in front of the lower ones in order to obtained a tractable development (weak excitation assumption).

The emitted power from an n-photon summation is then given by :

$$P_n(\lambda) = \frac{W_n \ldots W_2}{(\omega_{n-1} \ldots \omega_2)} P_1^n(\lambda) \tag{28}$$

with symbols of Fig. 6 and $P_1(\lambda)$ the line shape of $^4I_{15/2} \rightarrow {}^4I_{13/2}$ absorption.

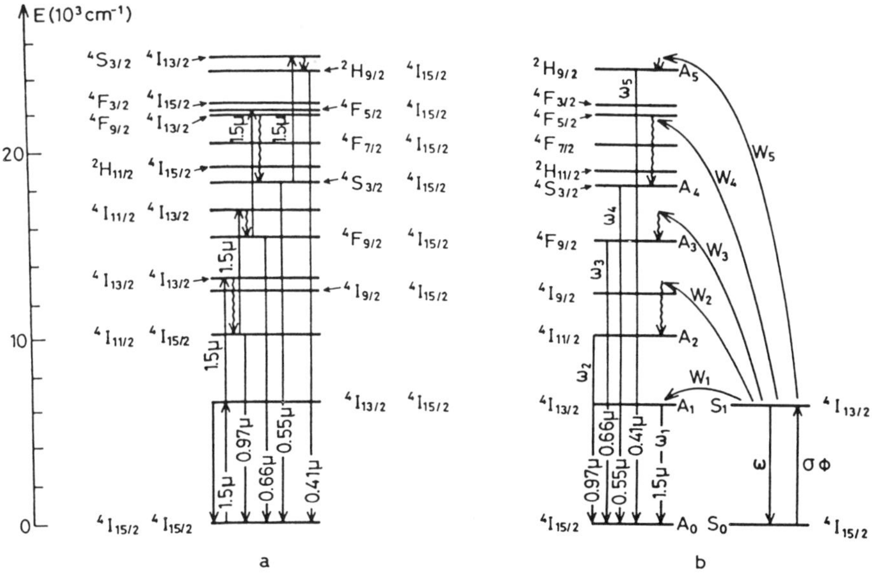

Fig 6. n-photon summation (n = 1 ...5) in Er^{3+} system by cooperative sequential pair absorption (a) and APTE effect (b) [12].

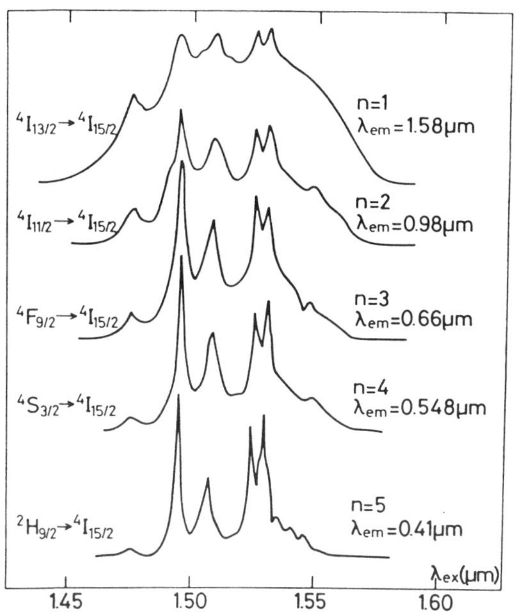

Fig 7. Excitation spectra for n-photon summation in $YF_3 : Er^{3+}$ [37].

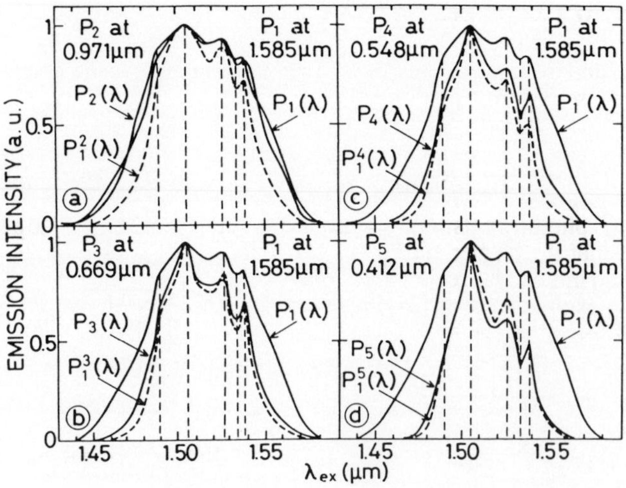

Fig 8. Comparison of calculated and experimental excitation spectra for Er^{3+} doped vitroceramics [12].

Fig. 8 shows that (28) is effectively verified except for the 2-photon case where the weak excitation assumption behind (28) may not be valid.

4. ESA and APTE Effect. The probability for ESA in a two-step absorption (W_{13}) connecting a state E_1 to E_3 by the intermediate state E_2, is just given by the product of the probabilities for each step :

$$W_{13} = W_{12}.W_{23} \tag{29}$$

For ESA by APTE effect we have also the product of two energy transfers probabilities.

Calculating the rate for the same ESA with APTE it comes :

$$N_A W_{13} = (N_A N_A^* W_{SA1}) N_S^* W_{SA2} \tag{30}$$

or

$$W_{13} = N_S^{*2} W_{SA1} W_{SA2} \tag{31}$$

were W_{SA} are the energy transfers probabilities for each step, and N_S^* the concentration of excited sensitizers which are given by :

$$N_S^* = N_S W_{12} \tag{32}$$

Assuming all W_{ij} to have same magnitude and W_{SA} also, as is typical for rare-earth ions, we have to compare

$$W_{13} = W_{12}^2 \quad \text{for single ion ESA} \tag{33}$$

and

$$W_{13} = W_{12}^2 N_S^2 W_{SA}^2 \quad \text{for ESA by APTE.} \tag{34}$$

Clearly the gain by APTE over one ion ESA can come from the product $N_S^2 W_{SA}^2$ which has to be > 1.

This points to and increase in sensitizer concentration (N_S) which leads to "fast diffusion" [38] and allows the use of rate equations in such multiions systems [39].

5. The Photon Avalanche Effect. While lookink for 2-steps ESA in Pr^{3+} doped $LaCl_3$ as a mean to detect an IR photon by its energy summation with a more energetic photon performing ESA, it was found that the higher energy photon alone could reduce the transmission of the sample above a given intensity threshold [40] see Fig 9. The effect was attributed to an increase of population on an excited state as resulting from a cross-relaxation process.

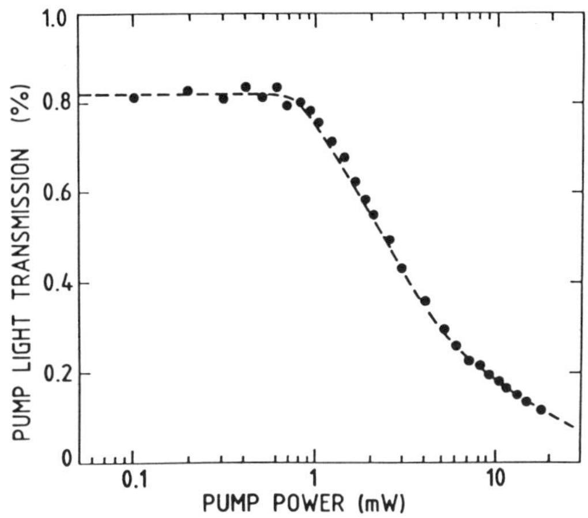

Fig 9. Decrease of transmission in a Pr^{3+} : $LaCl_3$ sample under $^3H_5$ - $^3P_1$ energy pumping. After [40]

The starting process is not completely determined [40]. In the Pr^{3+} case the $^3H_5 \rightarrow {}^3P_1$ absorption is initially very weak at room temperature because $^3H_5$ is about 2000 cm^{-1} above ground state (see Fig 10) ; however above about 1 mW of excitation this transition is increased ; the cross relaxation ($^3H_6$, $^3H_4$) $\rightarrow$ ($^3H_5$, $^3H_5$) increasing the $^3H_5$ population which reduce the transparancy of the sample at the ($^3P_1$ - $^3H_5$) energy. Because the more the ($^3P_1$ - $^3H_5$) energy is absorbed the more the $^3H_5$ population is increased, the process was termed "photon avalanche" [40] .

It is clearly a way to increase ESA in a sample.

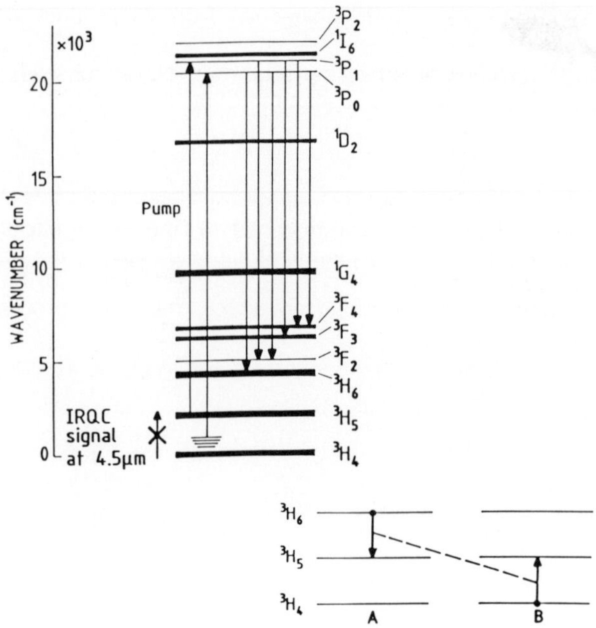

Fig 10. Pr^{3+} energy levels involved in the avalanche process. [40].

III. ROLES OF UP-CONVERSION PROCESSES IN LASER CRYSTALS AND GLASSES

III.A. Negative Aspects: ESA and ESA Enhanced by APTE

We have seen that APTE effect was found by its negative role in Yb^{3+} - Er^{3+} glass laser at 1.5 μm : being an anti-Stokes process its does induce reabsorption from excited states [6]. It has been shown to induce reabsorption in the high concentration Nd^{3+} stoechiometric laser materials [41] : up-conversion feeds non-radiative levels at the expense of the laser efficiency. In $Nd\ P_5\ 0_{14}$ materials such effect begins to be noticeable when population inversion reaches 10 %. The effect is described on Fig 11. Because of the nonlinearity of the effect the lifetime of $^4F_{3/2}$ is now pump intensity dependent.

We have observed up-conversion saturation in CW tunable laser with $MgF_2 : Ni^{2+}$ [42] ; however it seems the ions coupling plays an indirect rôle, up-conversion being due to ESA between single-ion levels.

Also we demonstrated the APTE effect to be a loss in the $LiYF_4 : Er^{3+}$ laser at 2.8 μm [43] : the experiment is described on Fig 12.

It is basically a 0.8 μm diode pumped $LiYF_4 : Er^{3+}$ laser at 2.8 μm. With a side fibre it is possible to monitor the emission spectra of the laser crystal at different values of the laser pumping power and also for the laser cavity being aligned or not. The results are shown on Fig 13.

A deterning of the laser cavity modifies only the 0.54 μm emission above laser threshold which demonstrates that the important APTE effect during laser action is not the one depopulating $^4I_{13/2}$ but the one depopulating $^4I_{11/2}$ (see Fig 14).

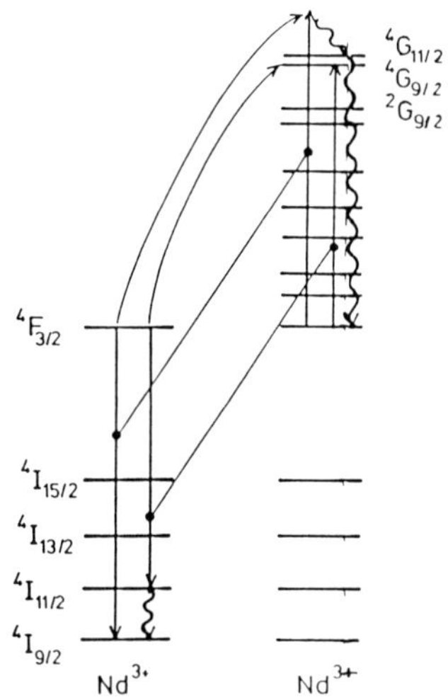

Fig 11. Intensity dependent $^4F_{3/2}$ lifetime of Nd^{3+} due to APTE quenching.[41]

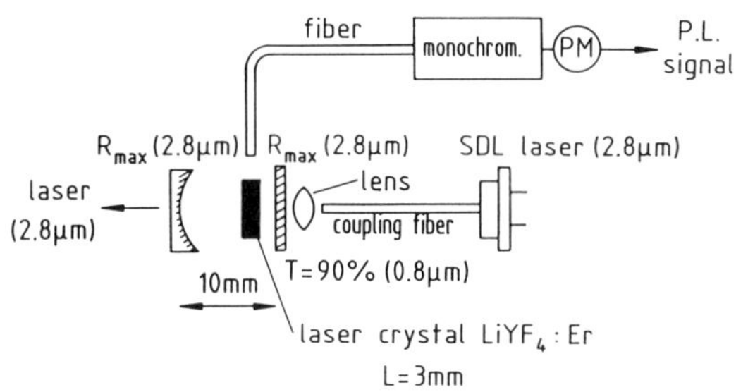

Fig 12. Up-conversion study of a LiYFI : Er^{3+} laser at 2.8 μm.

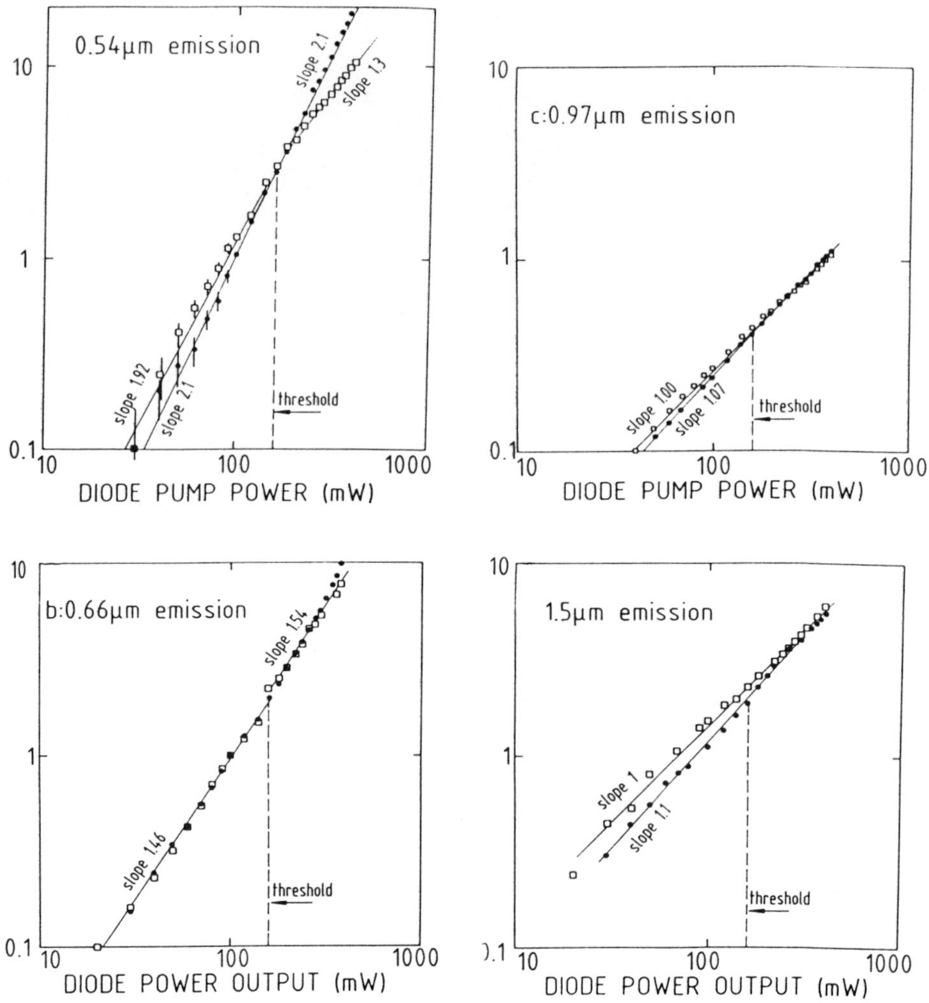

Fig 13. Up-conversion intensities at various pumping power of a 2.8 μm $LiYF_4 : Er^{3+}$ laser; (□) aligned laser cavity, (■) offset laser cavity.[43]

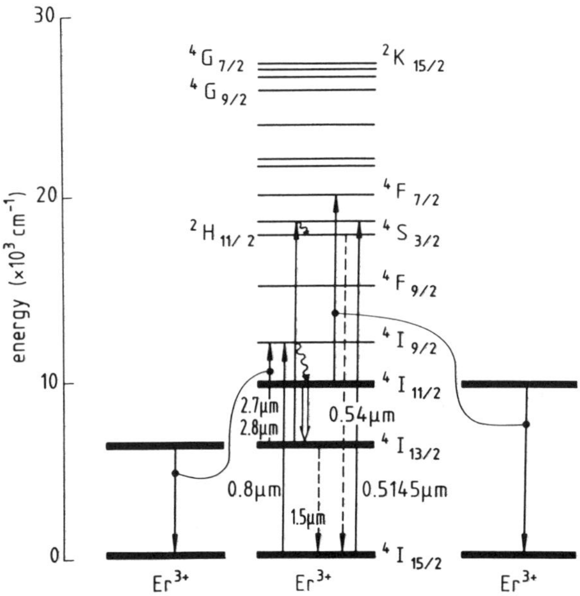

Fig 14. APTE effect taking place in $LiYF_4 : Er^{3+}$ 2.8 μm laser.

Then APTE is a drawback for such laser and not an advantage as sometimes believed [44] .

In fact the operating APTE effect depends on the excitation way : 1.5 μm excitation, on the opposite, enhances the APTE effect depopulating $^4I_{13/2}$ as demonstrated by the up-conversion pumping in $CaF_2 : Er^{3+}$ laser [45].

Also the same APTE process described by :

$(^4I_{13/2}, {}^4I_{15/2}) \rightarrow (^4I_{15/2}), {}^4I_{9/2})$

as been invoked to explain the gain saturation of Er^{3+} doped silica fibre amplifiers [46], when pumped at 1.48 μm even at doping concentration as low as 70 ppm. This points to the existence of Er^{3+} cluster in such glass fibres.

III.B. Positive Roles in Lasers: New Pumping Paths

Anti-Stokes Lasers in Crystals Because most of the luminescence processes are Stokes type in nature ($\lambda_{exc} < \lambda_{emission}$) lasers are also generally of the Stokes type : $\lambda_{pump} < \lambda_{emiss}$.

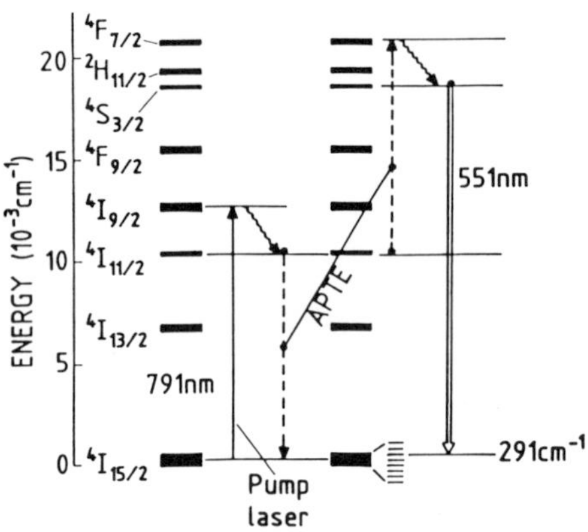

Fig 15. Energy scheme of the up-conversion diode pumped laser. After [48].

Because of the limitation in semiconductor laser diodes generally to the red or I.R. part of the spectrum, it appeared interesting to look for up-conversion pumped laser : $\lambda_{pump} > \lambda_{emiss}$. This would provide solid state laser operation in the visible using the now well developped IR semi-conductor pumping source; such anti-Stokes laser could be used for high density compact disc memories.

The first results in APTE anti-Stokes laser were obtained for the Yb^{3+} - Er^{3+} and Yb^{3+} - Ho^{3+} systems in BaY_2F_8 crystals [47] a few years after our finding of the APTE effect. Anti-Stokes lasers were then obtained under pulsed flash pumping. More recently CW up-conversion laser in the green has been obtained under IR laser diode pumping [48] of $LiYF_4 : Er^{3+}$ giving a renewed interest in APTE up-conversion. Fig 15.

However, such up-conversion lasers in crystals are still limited to low temperature [48] [49] [50] [51] [52] because the laser schemes are generally 3-level types.

Anti-Stokes Lasers in Glass Fibers Since the first demonstration in 1987 [53] of the feasibility of CW room-temperature 3-level rare-earth laser in Er^{3+} doped glass fibres, one could think that up-conversion pumped 3-level scheme could also be lased CW at room-temperature. This was demonstrated at CNET in 1990 [54] by Allain et al. The level scheme of the Ho^{3+} doped fluoride fiber laser is shown on Fig 16.

Because of the weak Ho^{3+} concentration (1200 ppm) it is believed that ESA within single ion levels system was taking place. However because clustering may some time takes place at much lower concentration there is still some doubt about the very pumping process as in many of the following up-conversion pumping fibre lasers :

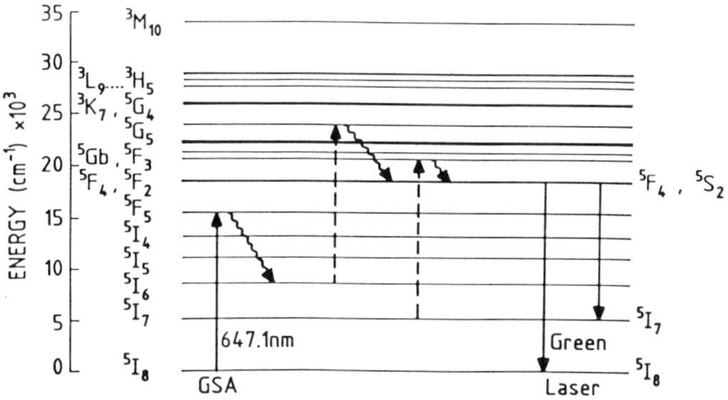

Fig 16. Energy scheme of the first CW, room temperature up-conversion laser [54].

Er^{3+} doped glass fibre have also shown CW-room temperature 3-level laser emission at 0.54 μm when pumped at 0.801 μm [55]. Because the pumping wavelength is in the laser diode range there is some hope that, compact fibre laser could be obtained.

Besides these 2-photon up-conversion pumped lasers even a 3-photon one has recently been obtained in a Tm^{3+} doped fibre [56]. Pumping at 1.12 μm, emission at 480 nm is obtained CW at room-temperature with a differential efficiency of 18%. Threshold being a rather low 30 mW.

Pr^{3+} doped fluoride fibre because of their low phonon energy have allowed anti-Stokes pumping lasers at blue, greens and red wavelength in the same fibre laser medium [57], also in CW room temperature conditions.

3. Up-Conversion Laser with Original Pumping Schemes. Besides pumping in the electronic rare-earth ions transitions, as shown above, up-conversion pumping was successful in multiphonon side band pumping with energy mismatch as large as 1000 cm^{-1}. It was the case for the Tm^{3+} fluoride fibre laser pumped at 1.06 μm in a 3-photon ESA process and lasing at 1.47 μm [58].

It was a 4-level scheme laser so providing, at room-temperature, CW oscillation with a differential efficiency of 27%.

Finally, it is worth mentionning the up-conversion avalanche laser in $LaCl_3 : Pr^{3+}$ single crystals [59]. A CW emission in a 4-level scheme is obtained at 644 nm through an up-conversion avalanche process as described in II.B. 5 under a 677 nm pumping which corresponds to an ESA pumping.

IV. CONCLUSION

Due to its efficiency and because of phonon assisted energy transfer, up-conversion by APTE is very general, though sometimes believed to be of the cooperative type. It is so much likely that considered matrices have smaller phonon energy giving long living metastable levels in rare-earth system. It has then a pervading nature as shown by the many studies that have been devoted to such effect in fluoride systems [43]-[45], [48]-[52] even in glass form [54] [60]. Unless precluded by lack of diffusion this very general effect appears at high concentration (10^{19} cm^{-3} for trivalent rare-earths ions) and more efficiently when excitation is concentrated in space, time and spectral domains because of its non-linear nature.

In fibre, because of the concentration of energy such non-linear processes are enhanced even the one with lower efficinecies such as ESA within single ions. Because the active ion concentration in fibre is rather low (70 ppm to 2000 ppm) ESA is believed to explain many of the observe up-conversion processes. However in the case of the existence of doping ion clusters such explanation should be revisited.

REFERENCES

1. F. Auzel, C.R. Acad. Sci. (Paris) 262, (1966) 1016 and 263, (1966) 819
2. V.V. Ovsyankin and P.P. Feofilov. Sov. Phys. JETP letters 4, (1966) 317
3. E. Nakazawa and S. Shionoya, Phys. Rev. Lett. 25, (1970) 1710
4. F. Varsanyi and G.H. Dieke 7, (1961) Phys. Rev. Lett. 442
5. E. Snitzer and R. Woodcock, Appl. Phys. Letters 6, (1965) 45
6. F. Auzel, Ann. Telecom. (Paris) 24, (1969) 363
7. J.R. Rios Leite and C.B. De Araujo, Chem. Phys. Letters 73, (1980), 71
8. R. Bonneville and F. Auzel, Opt. Com. 18, (1976) 51
9. V.V. Ovsyankin and A.A. Fedorov, Opt. and Spectrosc. 50, (1981), 565
10. J.P. Van der Ziel, L.G. Van Uitert, W.H. Grodkiewicz and R.M. Mikulyak, J. Appl. Phys. 60, (1986) 4262
11. G.J. Kintz, R. Allen and L. Esterowitz, Appl. Phys. Lett. 50, (1987) 1553
12. F. Auzel, J. of Lum. 31/32, (1984) 759
13. F. Auzel, S. Hubert and D. Meichenin, Appl. Phys. Lett. 54, (1989) 681
14. F. Auzel, in Radiationless Processes (B. Di Bartolo, ed.), Plenum, NYC (1980) p.213.
15. Y. Toyozawa, in Dynamical Processes in Solid State Optics, (R. Kubo, ed.) Benjamin (1967) p.90.
16. L.A. Riseberg and H.W. Moos, Phys. Rev. 174, 429 (1968).
17. F. Auzel and D. Pecile, C.R. Acad. Sci. Paris B277 155 (1973).
18. T. Förster, Ann. Phys. (Germany) 2 55 (1948).
19. T. Förster, Radiat. Res. (Supplement) 2 326 (1960).
20. D.L. Dexter, J. Chem. Phys. 21 836 (1953).
21. T. Kushida, J. Phys. Soc. Jpn 34 1318 (1973).

22. J.F. Pouradier and F. Auzel, J. de Phys. 39 825 (1978).
23. J.D. Axe and P.F. Weller, J. Chem. Phys. 40 3066 (1964).
24. R. Orbach, in Optical Properties of Ions in Crystals, (H.M. Crosswhite and H.W. Moos, eds.). Intersciences (1977) p.445.
25. R. Orbach, in Optical Properties of Ions in Solids, (ed. B. Di Bartolo) Plenum NYC (1975) p.355.
26. F. Auzel, Phys. Rev. B13 2809 (1976).
27. T. Miyakawa and D.L. Dexter, Phys. Rev. B1 2961 (1970).
28. R. Reisfeld and C.K. Jorgensen, in Laser and excited states of Rare Earths, Springer-Verlag, Berlin (1977).
29. M. Inokuti and F. Hirayama, J. Chem. Phys. 43 1978 (1965).
30. M. Yokota and O. Tanimoto, J. Phys. Soc. Jpn 22, 779 (1967).
31. M.J. Weber, Phys. Rev. B4, 2932 (1971).
32. W.J.C. Grant, Phys. Rev. 109, 648 (1958).
33. V.V. Ovsyankin, in "Spectroscopy of Solids containing Rare-earth ion" eds. A.A. Kaplyanskii and R.M. Mac Farlane, North-Holland Amsterdam (1987) p.405
34. M. Stavola and D.L. Dexter, Phys. Rev. B20, (1979) 1867
35. J.C. Vial, R. Buisson, F. Madeore and M. Poirier, J. Phys. 40, (1979) 913
36. N.J. Cockroft, G.D. Jones and R.W.G. Syme, J. of Lum. 43, (1989) 275
37. F. Auzel, in "Rare-Earth Spectroscopy", eds. B. Trzebiatowska, J. Legendziewicz and W. Strek, World Scientific, Singapore (1985), p.502
38. F. Auzel, in Optical Properties of excited States in Solids, (B. Di Bartolo, ed.) Plenum NYC (1992) p.305.
39. F. Auzel, in Spectroscopy of Solid State Laser Type Materials (B. Di Bartolo, ed.) Plenum NYC (1987) p.293.
40. A.W. Kueny, W.E. Case and M.E. Koch, J. Opt. Soc. Am. B6, 639 (1989).
41. M. Blätte, H.G. Danelmeyer and R. Verich, Appl. Phys. (Germany) 1 275 (1973).
42. R. Moncorgé, F. Auzel and J.M. Breteau, Phil. Mag. B51, 489 (1985).
43. F. Auzel, S. Hubert and D. Meichenin, Appl. Phys. Lett. 54, 681 (1989).
44. G.J. Kintz, R. Allen, and L. Esterowitz, Appl. Phys. Lett. 50, 1553 (1987).
45 S.A. Pollack and D.B. Chang, J. Appl. Phys. 64, 2885 (1988).
46. P. Blixt, J. Nilsson, J. Carlnäs, and B. Jaskorzynska, IEEE Trans. Photon. Techn. 3, 996 (1991).
47. H.G. Guggenheim and L.F. Johnson, Appl. Phys. Lett. 19 44 (1971).
48. F. Tong, W.P. Risk, R.M. Mac Farlane and W. Larth, Electron. Lett. 25, 1389 (1989).
49. T. Hebert, R. Wannemacher, W. Larth, and R.M. Mac Farlane, Appl. Phys. Lett. 57, 1727 (1990).
50. R.M. Mac Farlane, F. Tong, A.J. Silversmith, and W. Larth, Appl. Phys. Lett. J2 1300 (1988).
51. B.P. Scott, F. Zhao, R.S.F. Chang, and N. Djeu, Optics Lett. 18, 113 (1993).
52. R.R. Stephens and R.A. Mac Farlane, Optics Lett. 18, 34 (1993).
53. R.J. Mears et al. Electron. Lett. 22, 160 (1986).
54. J.Y. Allain, M. Monerie, and H. Poignant, Electron. Lett. 27 261 (1990).
55. T.J. Whitley, C.A. Millar, R. Wyatt, M.C. Brierley and D. Szebesta, electron. Lett. 27, 1785 (1991).
56. S.G. Grubb, K.B. Bennett, R.S. Cannon, and W.F. Humer, Electron.Lett. 28, 1243 (1992).
57. R.G. Smart, D.C. Hanna, A.C. Tropper, S.T. Davey, S.F. Carter and D. Szebesta, Electron. Lett. 27 1308 (1991).
58. T. Komukai, T. Yamamoto, T. Sugawa, and Y. Miyajima, Electron. Lett. 28, (1992).
59. M.E. Koch, A.W. Kueny, and W.E. Case, Appl. Phys. Lett. 56 1083 (1990).
60. D.C. Yeh, W.A. Sibley, M. Suscavage and D.G. Drexhage, J. Appl. Phys. 62 266 (1987).
61. J.M. Breteau, J.L. Ayral, F. Micheron and F. Auzel, J. Appl. Phys. 67 1102 (1990).

NON-LINEAR EFFECTS IN INSULATING LASER CRYSTALS

A. A. Kaminskii

Institute of Crystallography
Russian Academy of Sciences
Leninsky Prospekt. 59
117333 Moscow
Russia

ABSTRACT

The purpose of this article is to present our new ideas for using the non-linear transformation processes in insulating crystals doped with trivalent lanthanide ions, which possess second ($\chi^{(2)}$) and third ($\chi^{(3)}$) order non-linear optical susceptibilities, for the development and creation of new operating schemes and types of crystalline lasers.

I. SELF-FREQUENCY DOUBLED LASERS

A self-frequency doubled crystalline laser is a quantum generator in which the acentric activated crystals play two roles: 1) As the lasing medium and 2) as the non-linear element or frequency doubler of its own intrinsic stimulated emission (SE) [1]. Peculiarities of this type of laser explain Fig. 1 and Fig. 2. The list of known non-linear activated laser crystals used for the creation of such lasers is presented in Table 1.

II. SELF-PUMP-FREQUENCY DOUBLED LASERS

A self-pump-frequency doubled laser is a quantum generator in which the acentric activated crystal again plays two roles as a pump-frequency doubler and a lasing medium [2]. In the previous lasers, the non-linear transformation process (SHG) arises just after the SE of the Nd^{3+} ions. In the present case, the SHG occurs before the SE and serves only as an excitation channel of the activators (see Fig. 3). A sketch of the experimental set-up for obtaining SE from Nd^{3+} ions ($^4F_{3/2} \rightarrow {}^4I_{3/2}$ channel) in two non-linear crystals ($LaBGeO_5$ and $LiNbO_3$:MgO) is shown in Fig. 4. For pumping we used the focused emission (f = 20 cm.) of an external pulsed $LaF_3:Nd^{3+}$ crystal laser. At 300 K and a pump power density of about 0.1 GW/cm^2 the $LaBGeO_5:Nd^{3+}$ crystal (C_{Nd} = 1.4 at. %, $q_m \approx 54^o$, l = 12 and F = 4 mm) began to lase as π-polarized SE at 1.3141 μm.

Nonlinear Spectroscopy of Solids: Advances and Applications
Edited by B. Di Bartolo and B. Bowlby, Plenum Press, New York, 1994

Table 1. Acentric crystals doped with Ln^{3+} and Cr^{3+} ions for self-frequency doubled lasers.

1.	$LiNbO_3$-Tm^{3+}	
2.	$LiNbO_3$-Nd^{3+}	(*)
3.	$LiNbO_3$:Mg^{2+}-Nd^{3+}	(*)
4.	$YAl_3(BO_3)_4$-Nd^{3+}	
5.	$Ca_3Ga_2Ge_4O_{14}$-Nd^{3+}	(*)
6.	$Sr_3Ga_2Ge_4O_{14}$-Nd^{3+}	(*)
7.	$La_3Ga_5SiO_{14}$-Nd^{3+}	(*)
8.	$La_3Ga_5GeO_{14}$-Nd^{3+}	(*)
9.	$La_3Ga_{5.5}Nb_{0.5}O_{14}$-$Nd^{3+}$	(*)
10.	$La_3Ga_{5.5}Ta_{0.5}O_{14}$-$Nd^{3+}$	(*)
11.	$Nd_3Ga_5SiO_{14}$-Nd^{3+}	(*)
12.	$Ca_3Ga_2Ge_4O_{14}$-Cr^{3+}	(*)
13.	$Sr_3Ga_2Ge_4O_{14}$-Cr^{3+}	(*)
14.	$La_3Ga_5SiO_{14}$-Cr^{3+}	(*)
15.	$La_3Ga_5GeO_{14}$-Cr^{3+}	(*)
16.	$La_3Ga_{5.5}Nb_{0.5}O_{14}$-$Cr^{3+}$	(*)
17.	$La_3Ga_{5.5}Ta_{0.5}O_{14}$-$Cr^{3+}$	(*)
18.	$LaBGeO_5$-Nd^{3+}	(*)

(*) Development of our group

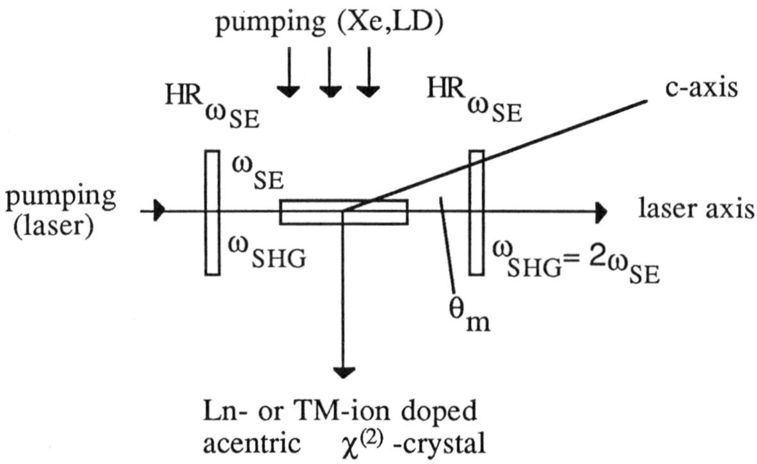

Figure 1. Sketch of self-frequency doubled crystalline laser. Denotation: Q_m is the phase matching angle for SHG, HR is the mirror of the laser resonator having high reflectivity at SE wavelength. Xe and LD are the pumping sources -- Xe flashlamp and laser diode respectively.

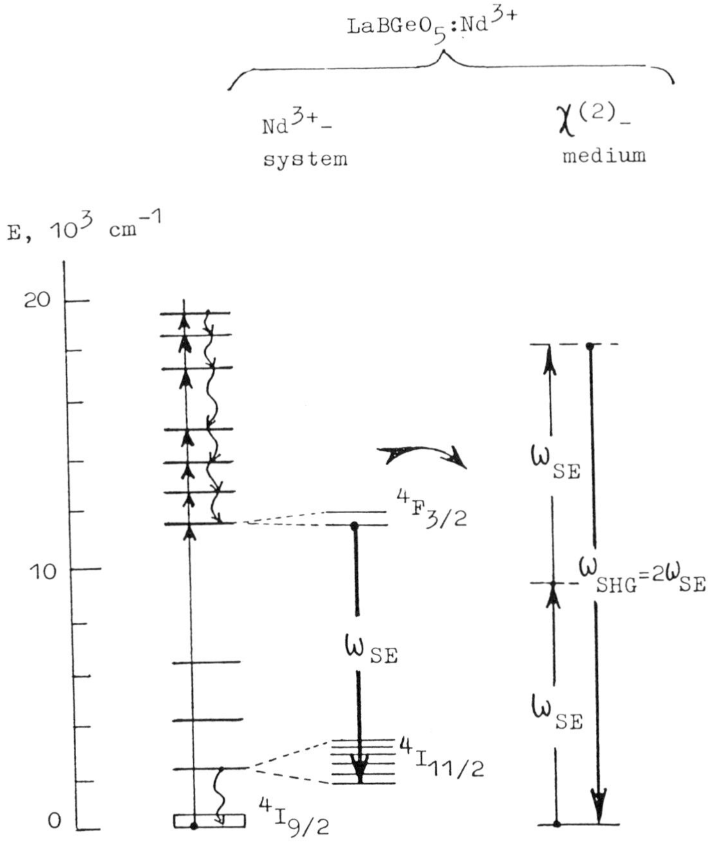

Figure 2. Simplified scheme of the energy pathways in a self-frequency doubled laser based on the acentric $LaBGeO_5:Nd^{3+}$ crystal. Ordinary and wavy arrows indicate pumping channels and non-radiative relaxation transitions.

III. SELF-PUMPED SRS CRYSTALLINE LASERS

A self-pumped stimulated Raman scattering (SRS) crystalline laser is a quantum generator in which the activated ion, possessing a high $\chi^{(3)}$-susceptibility, again plays two roles, first as a transformer (Raman shifter) and then as the lasing medium [2]. In these lasers the nonlinear process occurs before the SE acts in the activator ions (see Fig 5). A sketch of the experimental set-up for obtaining 3 μm SE from Er^{3+} ions in monoclinic crystals of $KY(WO_4)_2$ types ($KGd(WO_4)_2$, $KEr(WO_4)_2$, and $KLu(WO_4)_2$) is shown in Fig. 6. As the pumping source in this case we used red pulsed laser emission of $Al_2O_3:Ti^{3+}$ crystals. At 300 K all four of the above crystals doped with Er^{3+} ions (C_{Er}=20-100 at. %) started to lase at 3 μm ($^4I_{15/2} \rightarrow {}^4I_{13/2}$ channel) with a pump power density of about 0.3 GW/cm^2. The efficiency of SRS transformation in our experiments is 7-9%.

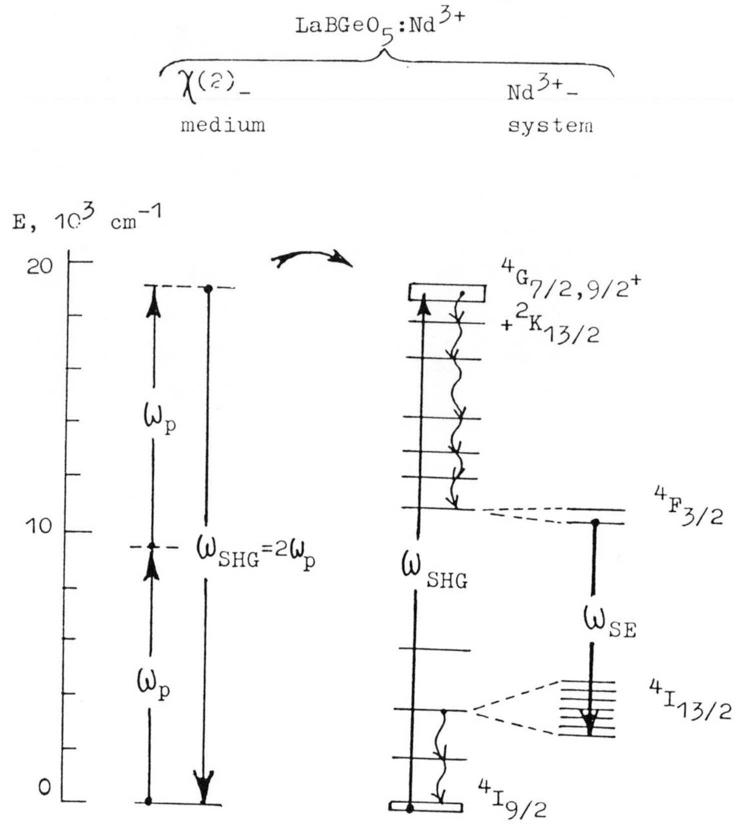

Figure 3. Simplified scheme of the energy pathways in a self-pump-frequency doubled laser based on the acentric $LaBGeO_5:Nd^{3+}$ crystal. Notation as in Fig. 2.

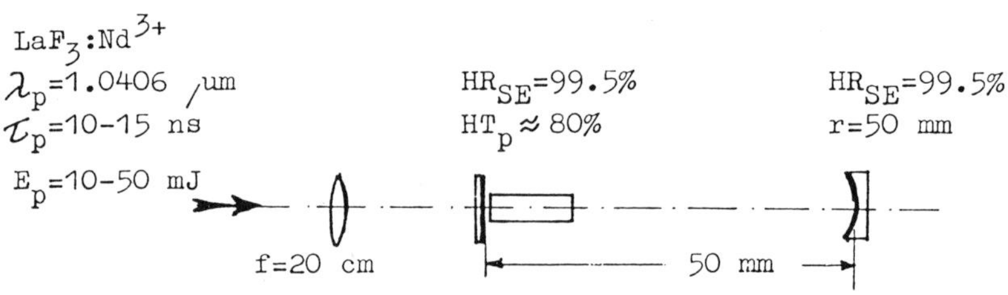

Figure 4. Sketch of a self-pump-frequency doubled laser based on the Nd^{3+} doped nonlinear crystals $LaBGeO_5$ and $LiNbO_3:MgO$. Denotation: HT is the flat mirror of semispherical laser resonator having high transmission of pumping wavelength. Other notation as in Fig. 1.

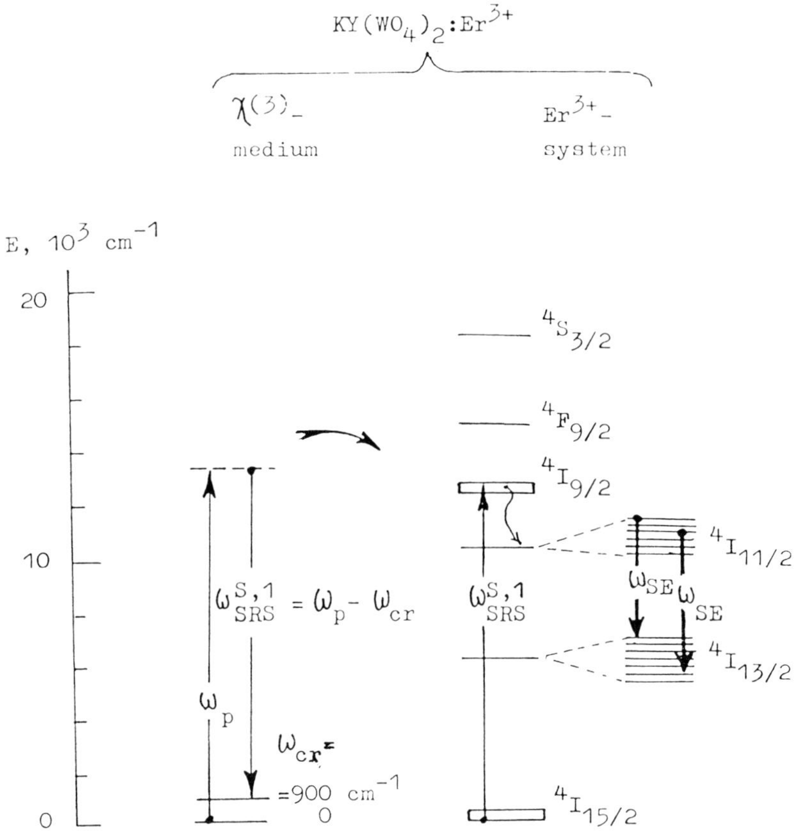

Figure 5. Simplified scheme of the energy pathways in self-pump SRS laser based on the monoclinic $KY(WO_4)_2:Er^{3+}$ crystal. Denotation: $W^{S,1}$ is the first Stokes frequency of pumping frequency (w_p), $w_{cr} = 900\ cm^{-1}$ is the frequency of the Raman-active vibration of the crystal-host tetrahedron $(WO_4)^{2-}$.

IV. CONCLUDING REMARKS

Using these new pumping principles based on SHG and SRS phenomena, new laser schemes of Ln^{3+} doped insulating crystals possessing second and third order non-linear optical susceptibilities have been proposed and realized. The non-linear properties of these insulating, activated crystals have been shown by our experiments to be a wide bridge-head for the creation and development of different types of laser pumping schemes based on optical parametric oscillation, as well as mixing and sum- or difference-frequency generation phenomena.

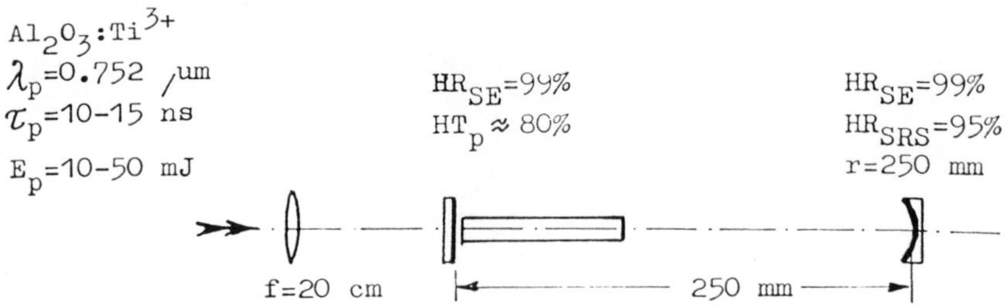

Figure 6. Sketch of a self-pumped SRS laser based on the $KY(WO_4)_2:Er^{3+}$ crystal. Notation as in Fig. 4.

REFERENCES

1. A. A. Kaminskii, Laser Crystals, Their Physics and Spectroscopy, Springer-Verlag, 1981 and 1990.
2. A. A. Kaminskii, Russian Journal of Quantum Electronics, 28, 6 (1993)

COMPUTER MODELING OF NONLINEARITIES

A.M. Buoncristiani [1], G. Armagan [2], A.A. Kaminskii [3]

[1] Christopher Newport University
Newport News, Virginia 23606, USA

[2] The College of William and Mary
Williamsburg, Virginia 23185, USA

[3] A.V.Shubnikov Institute of Crystallography
Academy of Sciences, Moscow 117333, Russia

ABSTRACT

Efforts to improve the performance of lasers used as spectroscopic probes or as measurement tools have resulted in significant refinements of the properties of the laser light (higher power, higher spectral brightness, longer coherence length and coherence time and shorter pulses). These efforts have also uncovered novel nonlinear responses in the laser itself and in its interaction with matter; these new phenomena challenge our understanding of laser dynamics and of the interaction of laser radiation with matter. This paper will describe some anomalous behavior arising from nonlinearities in laser processes and in material response. It will examine the nonlinearities that arise from inter-ionic processes and from stimulated emission and discuss the difficulties introduced by these nonlinearities in computations required to predict laser performance and the response of a nonlinear medium to laser excitation. Examples of these phenomena will be drawn from the YAG:Tm,Ho solid state laser system.

I. INTRODUCTION

The first generation of solid state lasers, principally the ruby and Nd:YAG lasers, operate on transitions between the electronic energy levels of a single type of ion substitutionally doped into a crystalline host. In these lasers active ions absorb the pump energy and transfer it, by internal electronic processes, to the upper laser level, where the collective dynamics of stimulated emission generate a coherent laser beam. In recent years there has been a strong research effort to develop lasers where excitation energy is transferred from one ion to another. This can occur in sensitized solid state lasers which require two or more types of ions to support lasing action or in single ion lasers relying on inter-ionic processes to transfer energy to the upper lasing level. In this type of laser, energy is absorbed by one ion, the sensitizer, and transferred to another ion, the activator, by a non-linear inter-ionic interaction; the dynamics of the sensitized laser depends critically upon the these nonlinear transfer processes.

The first reference to this type of energy transfer in lasers appeared early in the history of laser development. Johnson et al. suggested it as a means for improving the pumping efficiency of flash lamp lasers [1]. Subsequently, it was discussed by Kaminskii

[2] and by Auzel [3] who each pointed out a unique class of applications. Since the demonstration of energy transfer processes in solid state laser materials there have been many ingenious and innovative applications of these concepts leading to devices like the sensitized laser, the up-conversion laser and the avalanche laser. A good review of the early history of this research is given in reference [4]. It is important to recognize the significant difference in the *modus operandi* between the single ion laser and lasers using inter-ionic energy transfer. The dynamics of a laser depends on establishing a population inversion between the two states involved in the laser transition. In a single ion laser three or four electronic states of the same ion are involved in maintaining this population inversion and so single ion lasers are adequately described in terms of three level or four level models. In lasers utilizing inter-ionic processes more ions and more levels are involved; furthermore, the inter-ionic processes are non-linear in that their transition rates depend upon the current populations of both sensitizer and activator ions.

Successful design of sensitized lasers depends upon having a clear understanding of the various inter-ionic processes that can occur within the active medium. Some of these processes support the laser action and others lead to a loss of excitation. Characterization of these processes is often difficult because of the great number of inter-ionic processes that occur; identification of the various energy pathways that contribute to laser gain and the determination of numerical values for the related rate processes constitute convoluted tasks. There are many and varied means of describing the inter-ionic effects but not all of them can be easily incorporated into a laser model so that the challenge becomes to characterize this spectroscopic information in a way that can be used in a laser model.

In this paper we will concentrate on describing the nonlinearities that arise from two sources: non-radiative inter-ionic processes and stimulated emission; we shall not consider processes whose origin is the field-polarization interaction. In the following we will discuss a hierarchy of laser models, settling upon a suitable level of description for sensitized lasers; we will briefly discuss the numerical stability of laser models in general and these models in particular; we will present an example of energy transfer by considering a particularly clear example of the approach to equilibrium of the transfer of energy between the $^3F_4$ level of Tm and the $^5I_7$ level of Ho in Tm,Ho:YAG. Finally we will describe the incorporation of this spectroscopic information into a model for a 2 micron laser based on YAG: Tm,Ho.

II. LASER MODELS AND STABILITY

In this section we describe a model of laser action suitable for analysis of sensitized lasers. The objective of this model can be expressed by paraphrasing an often repeated remark by Albert Einstein, " A good laser model will be as simple as possible, but no simpler". We discuss only laser models which conform to the following technical assumptions: only paraxial rays and axial modes of the electromagnetic field are considered and the field variations are assumed to be slowly varying. Any laser model must describe the interaction between the propagating modes of the electro-magnetic field and the medium through which they pass. The semiclassical description treats three variables, the local electromagnetic field E(r,t), the material polarization P(r,t) and the population inversion N[r,t]; each of these variables has its own characteristic decay time and governing rate equation as shown in Table 1 [5, 6]. This system of equations has been extensively studied and shown to be equivalent to the Lorentz equations [7,8], which were the first equations to display the classical approach to chaos. They describe a system that is deterministic but not always predictable. Lasers used for measurement are operated far from these instabilities.

Table 1. Maxwell-Block-Haken Equations for Laser Dynamics

Dynamic Variable	Characteristic Decay Time	Differential Equation
Electric field, E	γ_C	$\dot{E} = \gamma_c P - \gamma_c E$
Material polarization, P	γ_P	$\dot{P} = \gamma_P E N - \gamma_P P$
Population inversion, N	γ_N	$\dot{N} = \gamma_N(\lambda + 1) E N - \gamma_N N - \gamma_N \lambda E P$

If, as we shall assume in the following, the polarization of the material relaxes faster than fluctuations in the field, that is if the polarization readjusts itself quickly after changes in excitation, the polarization of matter can be treated as if it is in an adiabatic equilibrium with the changing fields. The laser is then characterized by population variables which describe the occupancy of the various electronic states and the photon density and changes in these variables with time can be described by a set of rate equations.

As an illustration of the laser rate equations resulting from this adiabatic elimination, consider the four level model used to describe the Titanium dopedSapphire laser [9,10]. The energy level diagram for the Ti:Sapphire system is shown in Figure 1. Time evolution of the populations of the four energy levels is described by the following rate equations, developed as described above.

$$\frac{dn_3}{dt} = W_{03}\, n_0 - \frac{n_3}{\tau_3} \tag{1}$$

$$\frac{dn_2}{dt} = -\frac{n_2}{\tau_2} + \frac{n_3}{\tau_{32}} - \left(n_2 - \frac{g_1}{g_2} n_1\right) c\sigma\, \phi \tag{2}$$

$$\frac{dn_1}{dt} = -\frac{n_1}{\tau_1} + \frac{n_2}{\tau_{21}} + \frac{n_3}{\tau_{31}} + \left(n_2 - \frac{g_1}{g_2} n_1\right) c\sigma\, \phi \tag{3}$$

$$\frac{d\phi}{dt} = -\frac{\phi}{\tau_c} + \left(n_2 - \frac{g_1}{g_2} n_1\right) c\sigma\, \phi + S_p \tag{4}$$

The time dependence of the population density of each energy level is given by $n_i(t)$, $i = 0,1,2,3$. Since we consider that only the levels indicated in Figure 1 can be populated, these populations are constrained by the relation

$$N_{Ti} = n_0 + n_1 + n_2 + n_3\,, \tag{5}$$

where N_{Ti} is the concentration of Titanium dopants. The time dependence of the photon flux density in the laser cavity is given by $\phi(t)$. The rate at which the excitation is established by optical pumping is $W_{03}\, n_0$. Each of the excited levels decays electronically at a rate τ_i, $i = 1,2,3$, with the excitation branching to each of the lower levels so that the following constraint equations are satisfied

$$\frac{1}{\tau_i} = \sum_{j=1}^{i-1} \frac{1}{\tau_{ij}} \qquad i=1,2,3 \tag{6}$$

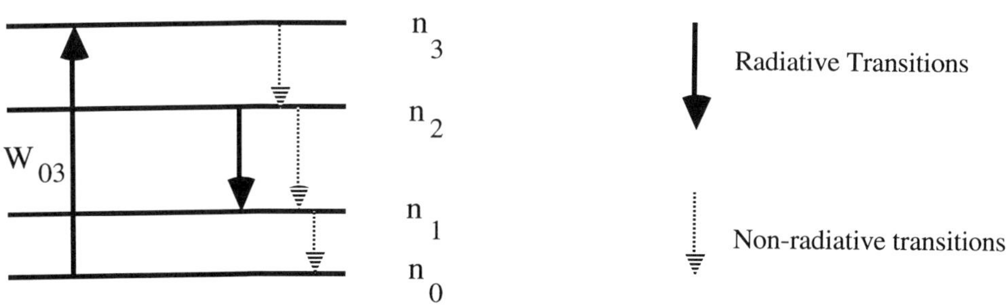

Figure 1. Schematic diagram of the low lying energy levels of Titanium doped Sapphire.

The laser cavity itself is assigned a decay time τ_c dependant upon the cavity losses. The terms bilinear in n_i and ϕ represent the stimulated emission (and absorption). Finally, the term Sp in the photon rate equation represents the spontaneous emission responsible for initiating the lasing process. This system of equations has been shown to be asymptotically stable [10]. In general, models of 3- and 4- level lasers with fast relaxation, such as the one above, are asymptotically stable, that is they may have only transient instabilities. However, they can be made more unstable by introducing additional phenomena for example, optical feedback, random fluctuations in the dynamical variable, or synchronous pulsed pumping. This leads to some striking instabilities in laser systems [11,12].

III. DESCRIPTION OF INTER-IONIC PROCESSES

The inter-ionic processes in sensitized lasers cause the non-radiative transfer of excitation from one ion to another. The ions are generally sufficiently far apart (from one to several nearest neighbor spacings) so that there is no significant overlap of their electronic wavefunctions. The primary interaction comes from one ion sensing the multi-polar field of another; the dipole-dipole interaction is the simplest case which we consider here. One of the interacting ions is identified as the donor of excitation to the other ion which is called the acceptor. Inter-ionic processes proceed at a rate that is proportional to the product of the occupation probabilities of the two initial states, that is, the transition rate is proportional to product of the population of the excited donor state, $n_D{}^*$ and the population of the unexcited acceptor state, n_A . In the case of a dipole-dipole interaction, the transition rate for the exchange of energy from a donor ion to an acceptor ion, separated by a distance R is proportional to the inverse sixth power of the separation [13,14],

$$W_{DA} = \frac{C}{R^6} = \frac{1}{\tau}\left(\frac{R_0}{R}\right)^6 \tag{7}$$

where C is called the microscopic interaction parameter; it is often expressed in terms of the donor decay time (in absence of the acceptor), τ, and a characteristic interaction distance, R_0. When the two ions are separated by R_0 the probability of the donor ion losing its excitation by means of an energy transfer reaction is equal to the probability of electronic decay. Dexter [15] has shown that the microscopic interaction parameter can be expressed as the product of three factors -- one is the radiative transition probability for the (isolated) donor ion, the second is the oscillator strength of the acceptor ion and the third is an overlap integral of the product of the donor emission spectrum and the acceptor absorption spectrum.

Several different types of processes can occur as described above. Energy transfer is the non-radiative transfer of excitation energy from one ion to another of the same type. Cross-relaxation is the transfer of a part of an ions excitation energy to a nearby ion; its inverse process is called up-conversion. Among the many factors that influence inter-ionic processes in a specific sample are the spatial distribution of ions in a sample or the distribution of the initial excitation.

As an illustration of an inter-ionic process we consider the energy transfer between the $^3F_4$ manifold of Thulium and the $^5I_7$ manifold of Holmium (refer to Fig. 4 for the energy level diagram for this system). Figure 2 shows the emission and absorption the emission and absorption spectra necessary to construct the overlap integral. The upper graph in Figure 2 gives the emission from Tm and the absorption from Ho related to the (forward) transfer from Tm to Ho while the lower graph shows the spectra related to transfer in the opposite direction. It is evident that there is a significant overlap for forward transfer but less for backward transfer. To demonstrate the transfer of energy we can examine the decay curves following a pulsed excitation populating the $^3F_4$ manifold of Tm. The two micron emissions from Tm and Ho are each relatively broad; they overlap in the central region but there is no overlap in the extremities. By observing the emission from either extremity (see Fig. 2) we can examine fluorescence coming from the Tm and Ho ions separately (observation at $\approx 1.7\ \mu$ gives the emission for Tm alone and at $\approx 2.1\ \mu$ gives the emission for Ho alone; sampling at $\approx 2\ \mu$ gives the combined emission from both types of ion).

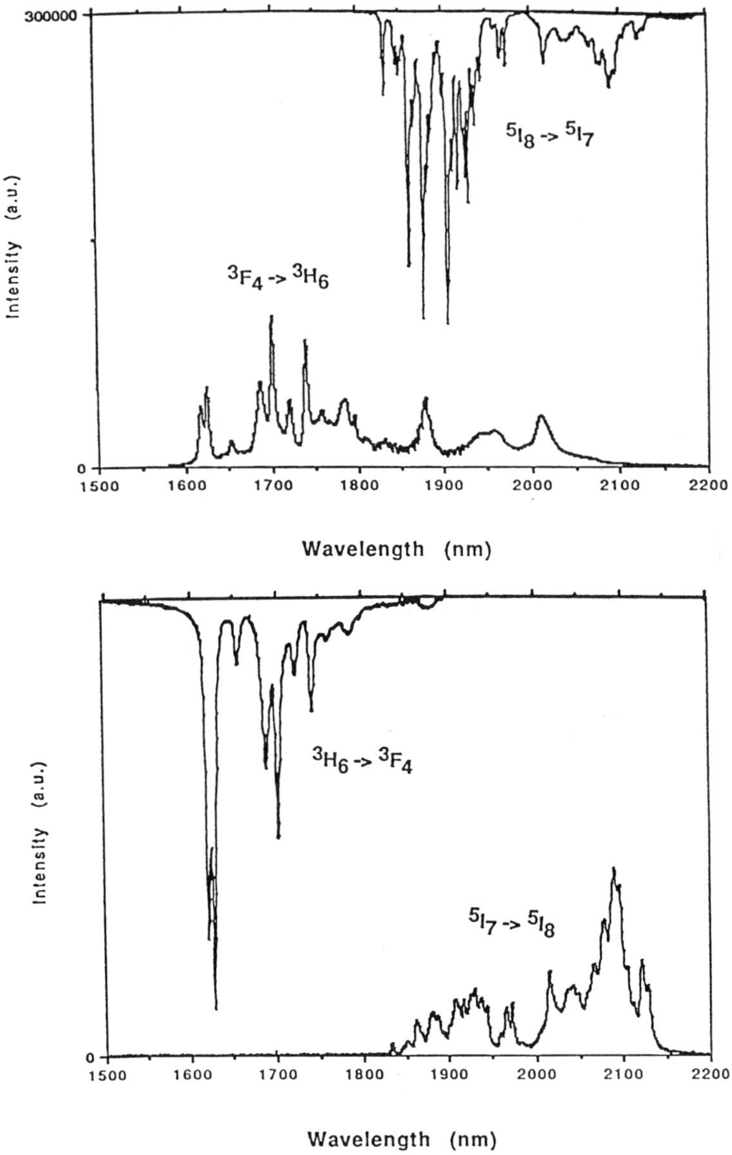

Figure 2. Absorption and emission curves contributing to the overlap integrals in the energy transfer between Tm and Ho at 2 microns. The upper graph shows the emission spectrum from Tm and the absorption spectrum from Ho and is related to the (forward) transfer from Tm to Ho. The lower graph shows the absorption spectrum from Tm and the emission spectrum from Ho and is related to transfer in the opposite direction.

Figure 3 shows the decay curves for Tm-Ho emission as observed through four band pass filters on a semi-log scale. The upper curve shows emission from only the Tm ions and the lower curve shows emission from only the Ho ions. Clearly the decay from Tm is faster than exponential with a corresponding rise in the emission from Ho. In the long term both emissions level off and proceed with the same exponential decay rate (a rate which is neither the decay rate for the individual ions alone). This is characteristic of the approach to equilibrium between the two excited states [17]. This clearly demonstrates the non-radiative exchange of energy between Tm and Ho ions.

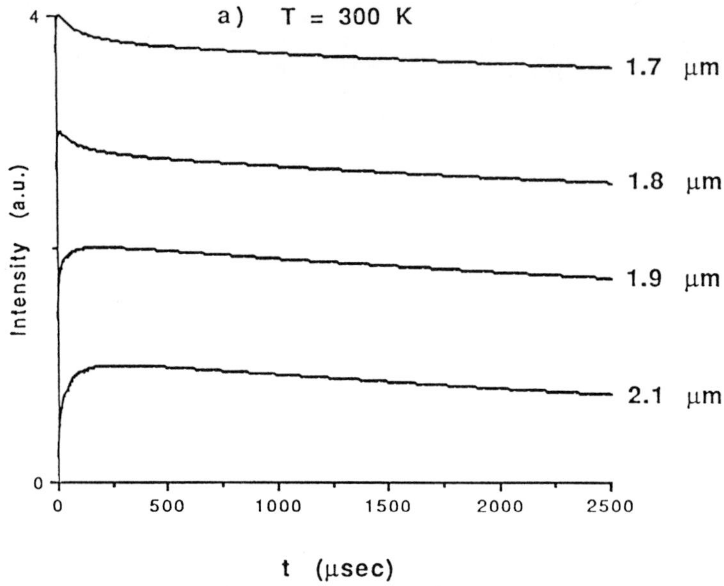

Figure 3. Decay curves showing the time dependence of the 2 micron emission from YAG:Tm(6%), Ho(0.5%) as veiwed through a series of bandpass filters. The emission measured through the filter centered ar 1.7 microns gives the emission from Tm; the emission measured through the filter centered ar 2.1 microns gives the emission from Ho.

IV. THE TM-HO LASER SYSTEM

We now focus attention on the specific case of a sensitized laser at 2 microns. There has been interest in the development of a 2 micron laser based on emission of the $^5I_7 \rightarrow {}^5I_8$ transition of the trivalent Ho ion. This transition has a long lifetime and a relatively large emission cross-section. Sensitization with Tm provides for absorption of optical energy and efficient transfer of this absorbed energy to the upper lasing level [18-20]. Furthermore Tm is easily pumped both by broad band sources (especially when Cr is used as a co-dopant) or directly by laser diodes.

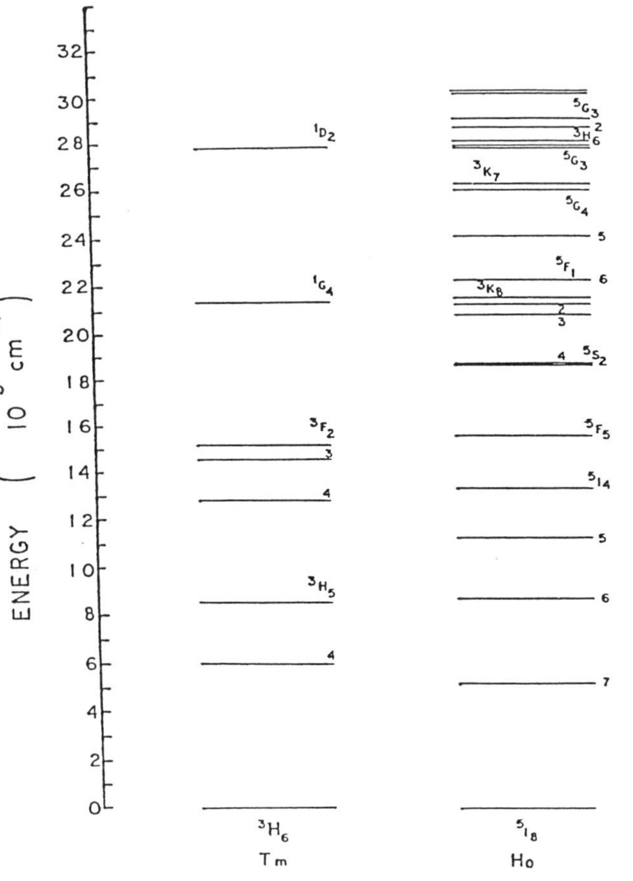

Figure 4. Energy level diagram for YAG: Tm,Ho

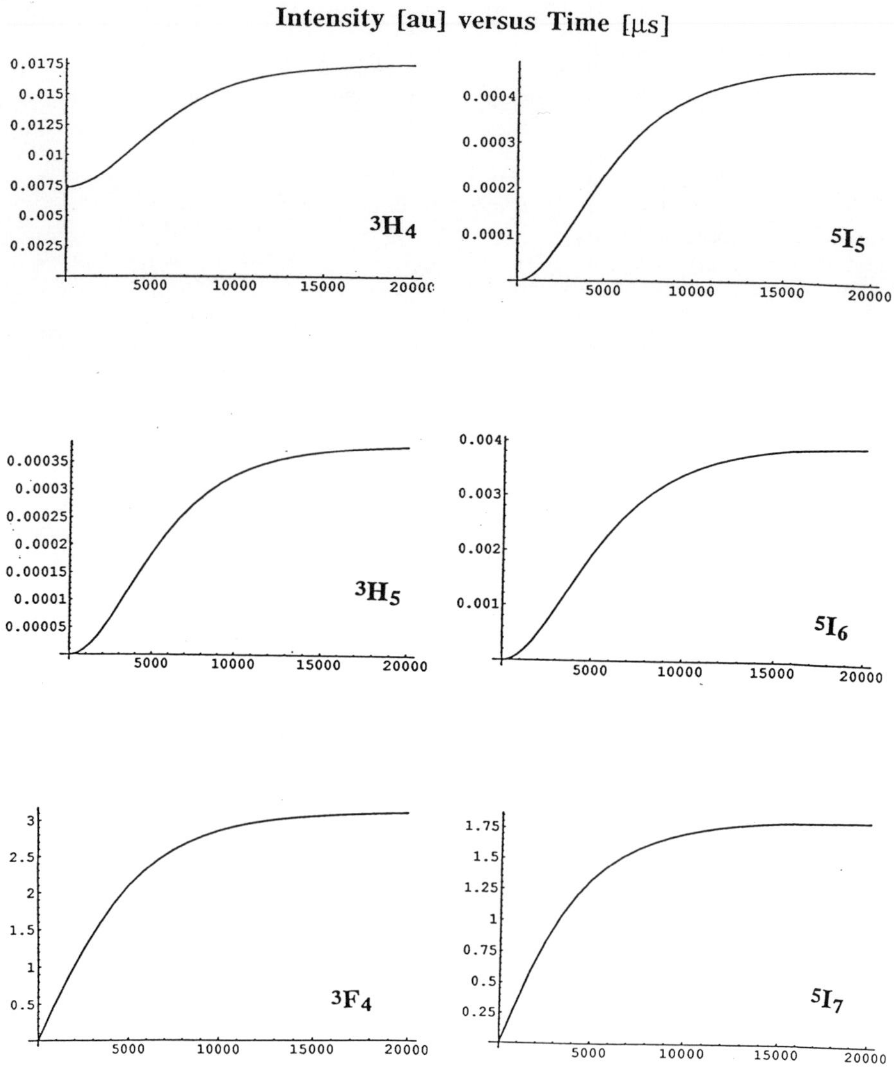

Figure 5. Time dependence of the emission from the radiating manifolds of the YAG:Tm,Ho solid state laser system as determined by numerical solution of the rate equations. Parameters in the rate equations are determined from steady state emission intensity at different pumping levels.

The energy level diagram of YAG: Tm, Ho is shown in Fig. 4. After the $^3H_4$ level of Tm is populated by excitation from a diode laser at $\approx$ 785 nm a sequence of inter-ionic processes take place which lead to emission at wavelengths about 2 μm : 1) the cross-relaxation process in Tm system ($^3H_4 \to {}^3F_4$, $^3H_6 \to {}^3F_4$), 2) the transfer of energy from the $^3F_4$ level of Tm to the $^5I_7$ level of Ho as discussed above, and 3) the thermalization of the $^3H_4$ level of Tm and the $^5I_7$ of Ho through a back transfer from Ho to Tm. In addition to processes which direct the pump energy to the upper laser level, there are also processes which draw energy away from the laser channel; for example, the up-conversion process{Tm: $^3H_4 \to {}^3H_6$; Ho: $^5I_7 \to {}^5I_5$} and the energy transfer between $^3H_4$ and $^5I_6$.

The processes participating in establishing Holmium lasing in this system involve four manifolds of the Tm ions and four manifolds of the Ho ions. The decay rates and branching ratios for the total decay (combined radiative and non-radiative decay) involves six independent parameters for each of these ions. In addition, there are two parameters for each nonlinear process. Treating the four nonlinear processes described in the previous paragraph leads to a total of twenty independent parameters. Some of these parameters can be obtained using transient spectroscopy and others inferred from emission and absorption spectroscopy. Alternatively, these parameters can be obtained from measurements of the steady state intensity of luminescence during optical pumping of different intensities. The population density of each of the six independent manifolds for this system at any time is given by $n_k(t)$ $k = 1, ..., 6$. The specific manifolds corresponding to the index k are, in increasing order: $^3H_4$, $^3H_5$, $^3F_4$, $^5I_5$, $^5I_6$, $^5I_7$. The rate equations for these population variables have the form

$$\dot{n}_k(t) = F_k(n(t); p) + R_k \qquad k= 1,..., 6 \tag{8}$$

R_k represents the intensity of the optical pumping of level k, n(t) represents the six population densities and p represents the 20 parameters. Furthermore, these equations are linear in the parameters

$$F_k(n(t); p) = \sum_{\alpha=1}^{20} F_{k\,\alpha}(n(t))\, p_\alpha \tag{9}$$

At steady state, $\dot{n}_k(t) = 0$, so that we can describe the luminescent intensity reached by each manifold under steady pumping the $^3H_4$ level by solving the above algebraic equations. We can solve for the individual parameters at each value of pump energy. Using these values of the parameters we can numerically solve the differential rate equations under the condition of fixed pump energy. The result of this solution of the rate equations gives the time dependent population of each of the manifolds. A typical time dependent population is shown in Figure 5.

REFERENCES

1. L.F.Johnson, L.G.von Uitert, J.J.Rubin and R.A.Thomas, Phys. Rev A, Vol. 133, p. 494 (1964).
2. A.A.Kaminskii, “Laser Crystals, Their Physics and Properties”, Springer, Heidelberg, 2nd edition (1990).
3. F.Auzel, C.R.Acad.Sci. (Paris), Vol. 262, p. 1016 (1966); Vol. 263, p. 819 (1966).
4. A.A.Kaminskii, Ann. Phys. Fr., Vol. 16, p. 639 (1991).
5. L.W. Casperson, in Proc. Intl. Meeting on Instabilities and Dynamics of Lasers and Nonlinear Optical Systems, R.W.Boyd, M.G.Raymer and L.N.Narducci, eds., p.58, Cambridge Univ. Press (1985).
6. C.O.Weiss, Opt. and Quantum Elect., Vol. 20, p. (1988).
7. H.Haken, Phys.Lett., Vol. 53A, p. 77 (1975).

8. E. Lorentz, J.Atmos.Sci., Vol. 10, p. 150 (1965).
9. P.Moulton, J.Opt.Soc.B, Vol.3, p. 125 (1986).
10. A.M.Buoncristiani, L Roberts and J.J.Swetits, Mathl. Comput. Modelling, Vol.12, p. 303 (1989).
11. F.T.Arecchi, R.Meucci and W.Gadomski, Phys.Rev. Lett., Vol. 58, p.2205 (1987).
12. J.J.Swetits and A.M.Buoncristiani, Phys.Rev A, Vol 38, p. 5430 (1988).
13. B. Di Bartolo, "Optical Interactions in Solids", John Wiley and Sons, Inc., New York (1968),
14. F. Auzel, in "Spectroscopy of Solid State Laser Type Materials", ed. B.Di Bartolo, Plenum Publishing Co. (9187).
15. D. L. Dexter, J. Chem. Phys. 21, 836 (1953).
16. G. Armagan, B. Di Bartolo and A. M. Buoncristiani, J. Luminescence 44 ,129 (1989). G. Armagan, B. Di Bartolo and A. M. Buoncristiani, J. Luminescence, 44 , 141 (1989).
17. G.Armagan, A.M.Buoncristiani and B.Di Bartolo, J. Luminescence, vol. 48 p. 171 (1991).
18. B.M.Antipenko, A.S.Glebov, T.I.Kiseleva, and V.A Pismennyi, Opt. Spect. (USSR) 63 (2) , 95 (1987).
19. T.Y.Fan, G.Huber, R.L.Byer, and P.Mitzscherlich, IEEE J. Quantum Electron. 24 , 924 (1988).
20. J.K.Tyminski, D.M.Franich, and M.Kokta, J. Appl. Phys. 65 (8) , 3181 (1989).

RATE EQUATION MODELING OF ENERGY TRANSFER PROCESSES

Guzin Armagan,[1] and Norman P. Barnes[2]

[1]The College of William and Mary
Department of Physics
Williamsburg, Virginia 23187

[2]NASA Langley Research Center, MS 474
Hampton, Virginia 23681

ABSTRACT

The dynamics of energy transfer processes in Tm-Ho laser materials are analyzed to determine the performance of 2.0 μm laser systems. These laser systems can utilize laser diode pumping and operate in the nominal eye-safe region of the spectrum. They are strong candidates for remote sensing applications such as the measurements of wind speed as well as CO_2 and H_2O concentrations. Here, the dynamics are analyzed using a rate equation model for two operating regimes, the weak excitation where most of the excitation is stored in the first excited manifolds of Tm and Ho and up-conversion is negligible and the strong excitation where up-conversion can no longer be ignored. In the first case closed form analytical solutions are obtained both in the steady state and the time dependent limits. In the second case, the steady state approach allows the expressions in closed form. These solutions are in terms of the population densities of all of the relevant manifolds rather than the first excited manifolds of Tm and Ho. A dynamical approach for the strong excitation case requires numerical techniques.

I. RATE EQUATION MODEL

The rate equations simply describe the time rate of change of population densities of energy manifolds due to various interaction processes that atoms/ions are added or removed from energy manifolds. The rate equations describing the energy transfer processes among ions can be derived from the density matrix equations [1]. In arriving at the rate equations from the density matrix elements, an approximation such that sensitizer ions must distribute energy among themselves in the same manifolds in a time short compared with the energy transfer to the activator ions was considered. Using a high concentration of sensitizer ions improves this approximation by increasing migration of energy among sensitizers.

I. A. Rate Equations for Tm and Ho

Energy level diagrams of Tm and Ho are shown in Fig. 1 [2,3]. Let us first consider a simplified case which the populations of Tm and Ho ions reside only in the ground and the

Nonlinear Spectroscopy of Solids: Advances and Applications
Edited by B. Di Bartolo and B. Bowlby, Plenum Press, New York, 1994

first excited manifolds. Consider also that Tm and Ho ions both play the role of sensitizer and activator. Assume that Tm ions are excited into the $^3F_4$ manifold with the excitation density rate R_2 [1/(m^3-sec)] and the energy transfer process $^3F_4 \rightarrow {}^3H_6$; $^5I_8 \rightarrow {}^5I_7$ and its inverse transfer process $^5I_7 \rightarrow {}^5I_8$; $^3H_6 \rightarrow {}^3F_4$ take place between Tm and Ho ions. The differential equations governing the dynamics of these 2 excited manifolds are given by,

$$dN_2(t)/dt = R_2 - p_{28}N_2(t)N_8(t) + p_{71}N_7(t)N_1(t) - N_2(t)/\tau_2 \quad (1)$$

$$dN_7(t)/dt = + p_{28}N_2(t)N_8(t) - p_{71}N_7(t)N_1(t) - N_7(t)/\tau_7 \quad (2)$$

with the ground manifold populations given by

$$N_1(t) = C_tN_s - N_2(t), \text{ and} \quad (3)$$

$$N_8(t) = C_hN_s - N_7(t), \quad (4)$$

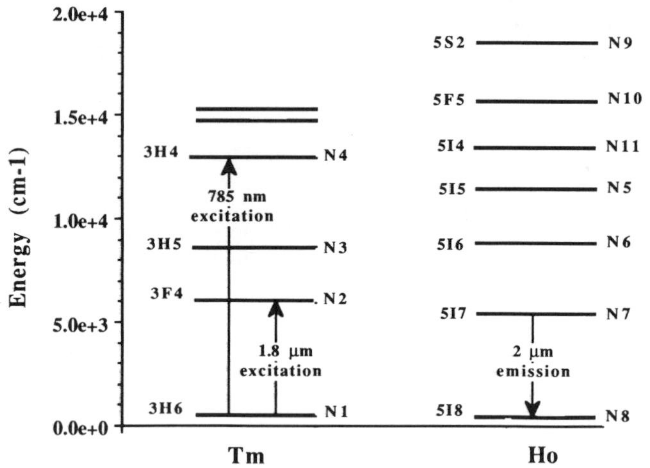

Fig.1. Energy manifold diagrams of Tm and Ho in YAG [2,3].

where p_{28}, in units [m^3/s], is the rate coefficient of the transfer process from Tm to Ho and p_{71} is the rate coefficient of inverse process, the transfer from Ho to Tm. τ_2 and τ_7 are the total lifetimes (including both radiative and non-radiative transitions) of $^3F_4$ and $^5I_7$ manifolds in the absence of an acceptor, and C_tN_s and C_hN_s [1/m^3] are the concentrations of Tm and Ho ions, respectively.

The excitation rate density is defined as the absorbed power per unit volume per unit energy and is given by,

$$R_2 = P_{abs}/(V_S h\upsilon_P) \quad (5)$$

where P_{abs} is the absorbed power, V_S is the volume of excitation, and $h\upsilon_P$ is the energy of the excitation photon.

I. B. Temperature Dependent Relation between Transfer Processes

The ratios of an energy transfer coefficient to its inverse coefficient can be approximated by a factor which can be calculated from a knowledge of the energy manifolds using Boltzmann statistics. Thus, for the energy transfer processes under consideration,

$$p_{71} = f_{71}\ p_{28}. \tag{6}$$

This is true if a single resonance transition in each type of atom contributes to the energy transfer process. If more than one transition of each type of atom contributes, the situation becomes more complicated [4]. An approximate formula for f_{71} can be written using Boltzmann relation for a single resonant transition in Tm $^3H_6$ - $^3F_4$ and Ho $^5I_8$ - $^5I_7$ as:

$$f_{71} = p_{71}/p_{28} = [(1/Z_7)\exp(-\Delta E_7/kT)]\,[(1/Z_1)\exp(-\Delta E_1/kT)]/ \\ [(1/Z_2)\exp(-\Delta E_2/kT)]\,[(1/Z_8)\exp(-\Delta E_8/kT)]. \tag{7}$$

Here $Z_i=\Sigma_n \exp(-E_n/kT)$, where the sum is over n Stark levels of manifold i, and ΔE_i is the energy gap between the lowest and the resonant Stark levels of the ith manifold.

II. SOLUTIONS OF RATE EQUATIONS

II. A. Steady State Solutions

Closed form solutions to the rate equations for the first excited manifolds of Tm and Ho are possible under the steady state excitation. Let us consider rate eqs. (1) and (2) under this condition. Since each of these equations can be solved for one rate coefficient when its inverse provided by the Boltzmann relation above, eqs.(1) and (2) give, respectively,

$$p_{28} = (R_2 - N_{20}/\tau_2)\ /\ [N_{20}\,(C_hN_s - N_{70}) - f_{71}N_{70}\,(C_tN_s - N_{20})]. \tag{8}$$

and

$$= (N_{70}/\tau_7)\ /\ [N_{20}\,(C_hN_s - N_{70}) - f_{71}N_{70}\,(C_tN_s - N_{20})]. \tag{9}$$

where N_{20} and N_{70} here are the steady state population densities of $^3F_4$ and $^5I_7$. Relating these two equations gives

$$R_2 = (N_{20}/\tau_2\ ) + (N_{70}/\tau_7). \tag{10}$$

Since the excitation rate density can, experimentally, be determined measuring the absorbed power and the excitation volume (see eq. (5)), this equation can be used in determining the geometrical proportionality factor, K, used in the expression for population density as discussed in Sec.III.

II.B. Time Dependent Solutions

In order to determine the time evolution of populations of Tm and Ho manifolds after exciting with a short pulse (time duration of pulse << lifetime), the differential equations (eqs.1-2) governing the dynamics of the first excited manifolds are rewritten as

$$dN_2(t)/dt = -\ aN_2(t) + bN_7(t) + cN_2(t)N_7(t) - N_2(t)/\tau_2 \tag{11}$$

$$dN_7(t)/dt = +\ aN_2(t) - bN_7(t) - cN_2(t)N_7(t) - N_7(t)/\tau_7 \tag{12}$$

where $a = p_{28}C_hN_s$, $b = p_{71}C_tN_s$ and $c = (p_{28} - p_{71})$. Note that eqs.(11) and (12) include a nonlinear population density term, $cN_2(t)N_7(t)$. In the case where the excitation density is weak, the nonlinear term, cN_2N_7, can be neglected; a solution of the form $N_2 = A_2 \exp(-\alpha t)$ and $N_7 = A_7 \exp(-\alpha t)$ can be tried. Substituting into the simplified differential equations, two solutions for α are obtained

$$\alpha_{a,b} = 1/2\,(\tau_2^{-1}+\tau_7^{-1}+a+b) \pm 1/2\,[(\tau_2^{-1}-\tau_7^{-1})^2 -2(a-b)(\tau_2^{-1}-\tau_7^{-1}) +(a+b)^2]^{1/2}. \quad (13)$$

If the energy exchange process is relatively fast compared with the fluorescent decay:

$$\alpha_a \simeq a + b \quad (14)$$

$$\alpha_b \simeq 1/2\,(b/\tau_2^{-1} + a/\tau_7^{-1}). \quad (15)$$

Since there are two possible exponents, the solution will be of the form

$$N_2(t) = A_2 \exp(-\alpha_a t) + B_2 \exp(-\alpha_b t) \quad (16)$$

$$N_7(t) = A_7 \exp(-\alpha_a t) + B_7 \exp(-\alpha_b t) \quad (17)$$

This closed form of equation should hold if either eq.(13) or eqs.(14) and (15) are used for solutions to α. If eqs.(14) and (15) are used, simple solutions exist for the differential equation. With $A_2 = \simeq -A_7$ and $aB_2 \simeq bB_7$, the solutions to the differential equations become

$$N_2(t) = A_2 \exp(-\alpha_a t) + B_2 \exp(-\alpha_b t) \quad (18)$$

$$N_7(t) = -A_2 \exp(-\alpha_a t) + (a/b)B_2 \exp(-\alpha_b t) \quad (19)$$

Common boundary conditions would be $N_2(t) = N_2(0)$ and $N_7(t) = N_2(0) = 0$ at t=0. In this case $A_2 = (a/b)\,B_2$ and $A_2 + B_2 = N_2(0)$, therefore $A_2 = N_2(0)\,a/(a+b)$, $B_2 = N_2(0)\,b/(a+b)$.

If the numerical values of the rate coefficients are obtained using steady state condition, by substituting these coefficients into the eqs.(18) and (19), the time evolution of Tm $^3F_4$ and Ho $^5I_7$ manifolds can be obtained.

III. MEASURING EXCITED STATE POPULATION DENSITY

III.A. Relation Between the Population Density and the Luminescence Spectrum

In order to determine the population densities using spectroscopic measurements, the relative luminescence signal has to be calibrated using a blackbody source. If the material is excited so that there are N_{i0} number atoms per unit volume in the state, the number of photons emitted per unit volume per unit time is equal to (N_{i0}/τ_{ij}) [photons/m^{3}s], where τ_{ij}^{-1} is the radiative transition probability from manifold i to manifold j.

If the spectroscopic measurement system intercepts a solid angle Ω, the fraction of the photons collected are $\Omega/4\pi$. Detector responsivity relates the photons incident on the spectroscopic system to a measurable parameter, the current in the detector. The measured detector current per wavelength interval, integrated over the spectral bandwidth of the luminescence emission of transition between ith and jth manifolds is

$$i_{ij} \simeq R_D\,(hc/\lambda_{ij})\,(\Omega_S/4\pi)\,V_S\,(N_{i0}/\tau_{ij}) \quad (20)$$

where R_D [A/W] is the responsivity of the detector, hc/λ_{ij} is the average energy of the emitted photons and V_S is the excited volume of the material. On the other hand, if the measurement system observes a blackbody radiation, it produces a current, i_B,

$$i_B \simeq R_D P_B (\Omega_B/4\pi) \quad (21)$$

where P_B [W] is the power emitted by blackbody radiation given by Planck distribution. Solving eqs.(20) and (21) for N_{i0}

$$N_{i0} \simeq K S_{ij} \tau_{ij} \quad (22)$$

where K [1/Joul-m] is the proportionality factor which depends on the geometry of the experimental set up and given by

$$K = (\Omega_B/\Omega_S)(1/hc) \quad (23)$$

and S_{ij} [W/m^2] is the total integrated intensity of the luminescence spectrum from manifold i to manifold j and given by

$$S_{ij} = (i_{ij}/i_B)\lambda_{ij} P_B/V_S. \quad (24)$$

III. B. Geometrical Proportionality Factor from Rate Equations

Substituting eq. (22) into eq. (10), the geometrical proportionality factor, K, can be obtained from the steady state solution of the rate equations and expressed as:

$$K = R_2 / [S_{21}(\tau_{21}/\tau_2) + S_{78}(\tau_{78}/\tau_7)]. \quad (25)$$

Using the spectroscopic measurements such as integrated intensities, radiative lifetimes and total lifetimes, and the excitation volume in the crystal, the values for K, N_2, N_7 and therefore p_{28} under steady state condition can be obtained. p_{28} (and p_{71} derived from p_{28} using Boltzmann statistics) can be substituted in the time dependent rate equations in order to determine the time rate of change of $N_2(t)$ and $N_7(t)$.

IV. STRONG EXCITATION CASE OF TM

IV. A. Steady State Excitation

If the excitation density of Tm is strong, the nonlinear processes play important role in the dynamics of Tm-Ho system. If the strong excitation of Tm into the $^3F_4$ manifold is considered, the up-conversion process,

$$^3F_4 \to {}^3H_6\,;\ {}^5I_7 \to {}^5I_5\ (p_{27}) \text{ and its inverse } {}^5I_5 \to {}^5I_7\,;\ {}^3H_6 \to {}^3F_4\ (p_{51}).$$

should be added to the rate equations. $^5I_5$ manifold of Ho is populated due to this up-conversion process. The transitions from $^5I_5$ to $^5I_6$ can proceed rapidly by nonradiative transitions in laser materials such as YAG. Therefore, the following energy transfer process involving $^5I_6$ manifold has also to be considered:

$$^5I_6 \to {}^5I_8\,;\ {}^3H_6 \to {}^3H_5\ (p_{61}) \text{ and its inverse } {}^3H_5 \to {}^3H_6;\ {}^5I_8 \to {}^5I_6\ (p_{38}).$$

This transfer populates $^3H_5$ manifold of Tm. Rate equations are then written for five manifolds, $^3H_5$ and $^3F_4$ of Tm , $^5I_5$, $^5I_6$ and $^5I_7$, of Ho. Under steady state condition, the equations for $^5I_5$, $^5I_6$ and $^5I_7$ yield the expressions for rate coefficients p_{27}, p_{61} and p_{28}, respectively. Equations for $^3H_5$ and $^3F_4$ also yield expressions for rate coefficients p_{61} and p_{28}, respectively. By relating the expressions for the same rate coefficients, additional information about population densities can be obtained as discussed in Secs. II. A and III.

If Tm is excited into the higher manifolds, more processes enter into the rate equations. For example, in the case of excitation into the $^3H_4$ manifold of Tm, the cross relaxation process:

$$^3H_4 \text{->} {}^3F_4 ;\ {}^3H_6 \text{->} {}^3F_4\ (p_{41}) \text{ and its inverse } {}^3F_4 \text{->} {}^3H_6 ;\ {}^3F_4 \text{->} {}^3H_4\ (p_{22})$$

and also if the excitation is strong, the up-conversion process:

$$^3H_4 \text{->} {}^3H_6 ;\ {}^5I_7 \text{->} {}^5S_2\ (p_{47}) \text{ and its inverse } {}^5S_2 \text{->} {}^5I_7 ;\ {}^3H_6 \text{->} {}^3H_4\ (p_{91})$$

have also to be included into the rate equations. It is also possible to obtain analytical expressions for all of the five rate coefficients, p_{27}, p_{61}, p_{28}, p_{41} and p_{47}, under steady state condition [5].

IV. B. Pulsed Excitation

Solutions of the rate equations for the time evolution of populations might be very complicated in the case of strong excitation of Tm. Even if only two excited manifolds are considered, the nonlinear term, $cN_2(t)N_7(t)$, in eqs. (11) and (12) can not anymore be neglected. As up-conversion, cross relaxation and other transfer processes are added into the rate equations, one would consider to find numerical solutions for the time evolution of population densities using computers. It is possible to solve these nonlinear equations, simultaneously, with a program called Mathematica [6]. Rise and curves of the Tm and Ho manifolds generated using this program can be compared with the experimental rise and decay curves [5].

ACKNOWLEDGEMENTS

This work was supported by NASA Grant NAS-19656-8.

REFERENCES

1. W. J. C. Grant, "Role of Rate Equations in the Theory of Luminescent Energy Transfer", Phys. Rev. B 4, 648 (1971).

2. J. B. Gruber et. al. "Spectra and Energy Manifolds of Tm^{3+}: $Y_3Al_5O_{12}$ ", Phys. Rev. B 40, 9464 (1989).

3. J. B. Gruber et. al. "Energy manifolds and Crystal Quantum States of Trivalent Holmium in Yttrium Aluminum Garnet", J. Appl. Phys. 69 (12), 8183 (1991).

4. N. P. Barnes et. al. "Ho:Tm Lasers I: Theoretical", presented at Laser Optics'93, St. Petersburg, Russia (June 1993)

5. G. Armagan et.al. "Determination of Tm-Ho Rate Coefficients from Spectroscopic Data", submitted to the OSA Proceedings of Advanced Solid State Lasers, Salt Lake City, Utah, (February 1994)

6. Wolfram Research, Inc. Mathematica, Version 2.1, 1992.

COLD FUSION FOUR YEARS LATER

Francesco Scaramuzzi

ENEA
Centro Ricerche Energia Frascati
Via Enrico Fermi, 27
00044 Frascati (Roma), Italy

In spite of the contradictory information forwarded by the mass-media, and in spite of the skepticism of the scientific community, a substantial research activity has been brought forward in this field in four years, since its starting in the Spring of 1989. Six international conferences and about 1000 papers witness the noticeable effort produced around the world, and particularly in Japan, the USA and in Italy.

The state of the art of cold fusion is presented, with particular reference to the experiments in which a noticeable amount of excess heat is produced (that cannot be explained in terms of chemical reactions), without the contemporary emission of nuclear energetic particles. Experiments are described in which the detection of ^4He, thought as an ash, seems to show that a nuclear reaction takes place in the lattice of palladium, when the deuterium concentration is sufficiently high.

SCIENCE AND THEOLOGY. SCIENTIFIC MENTALITY AND CHRISTIAN FAITH

Guiseppe Koch S. J.

Istituto Gonzaga
Palermo, Italy

In order to avoid the frequent confusion on the theme, pointing the attention to physics and cosmology among natural sciences, this seminar describes and tries to distinguish terms and realities such as science, technology, scientism, scientific mentality, faith, theology, religion, and religionism. What is said about faith and theology chiefly refers to Christian religion. Science and faith are both ways of coming to a knowledge of truth and reality.

In the realm of faith, theology is the most reflective intellectual activity practised among the community of believers. Among the successive phrases of the theological process, following Bernard Lonergan, the attention focuses on "intellectual conversion" and on the change of horizon -- roughly our way of thinking about the reality, the objectivity and the possibilities of human knowledge -- which it includes. Various levels of differences of horizons are described.

Possible conflicts between science and faith will be more or less easily overcome if the obstacles are rooted in reconcilable horizons. Otherwise it will hardly be possible to bring them into harmony.

Philosophy is suggested as an interface, not always well fitting, between science and theology. Big efforts are necessary to "fine-tune" this philosophical interface.

Three main channels of interaction between science and theology, scientific mentality and Christian faith are suggested:

1) Direct and reflective. Mostly among theories and doctrines of common interest.
2) Interaction through technology; largely unreflective but pervasive and powerful.
3) Interaction through science's impact on implicit philosophies, conceptual structures and various linguistic and cultural fields

The third level of interaction strongly challenges, with new stories and myths, the images and the symbols which, for a long time, have been at the basis of religious experience and language. Two possibilities are highlighted: to discover God, in a new and deeper way in these new stories and symbols, and to protect both science and religion from the dangers of idolatry.

We better understand today that religion is not founded on science, but on the ground of experienced revelation, and that, on the other hand, science is not to be conceived as an extension of religion.

Different focuses, grounds and methodologies of science and religion are discussed. The independence and autonomy of the two enterprises must be obviously acknowledged but we can not be satisfied with a sharp separation between the two. Main reasons for this statement are shown.

Continuous friendly dialogue is crucial to promote the perspective of a human knowledge really global and to overcome the gap between truth and values.

ENHANCED RED TO BLUE CONVERSION LUMINESCENCE OF Tm^{3+} IN Yb^{3+} DOPED FLUOROPHOSPHATE GLASSES

Gönül Özen

Centre National de la Recherche Scientifique
Laboratoire de Physicochimie des Matériaux (UPR-211)
1, Place Aristide Briand
92190 Meudon, France

The upconversion processes in fluorophosphate glasses doped with Tm^{3+} and codoped with Tm^{3+} and Yb^{3+} ions have been studied by exciting the $^3F_2$ and $^3F_3$ levels of Tm^{3+} ions. The absorption, upconversion emission, excitation and time resolved luminescence spectra were measured at room temperature.

An enhancement of blue emission corresponding to the $^1G_4 \rightarrow {}^3H_6$ transition was observed in the codoped sample when the excitation was into the $^3F_3$ level with 683 nm light. An upconversion mechanism due to the addition of photons by forward and back energy transfer processes between Tm^{3+} and Yb^{3+} ions is proposed to explain experimental results. Here, an excited Tm^{3+} ion relaxes to the ground state via $^3F_4(Tm) \rightarrow {}^2F_{5/2}(Yb)$ energy transfer. Then the excited Yb^{3+} ion gives its energy back to a Tm^{3+} ion in its $^3F_4$ level by $^2F_{5/2}(Yb) \rightarrow {}^1G_4(Tm)$ back energy transfer which enhances the blue emission originating from the $^1G_4$ level. This mechanism has been confirmed by measuring the excitation power dependence and the time resolved spectra of the upconverted emissions. According to the luminescence intensity versus excitation power measurements both emissions originating from the $^1D_2$ and $^1G_4$ levels are due to two photon processes. The time resolved spectrum of the blue emission intensities first show an increase and then decrease in time. The increase of the luminescence intensity confirms the back energy transfer from Yb^{3+} to Tm^{3+} ions in the codoped samples.

$^3F_4$ (Tm) emission intensity is found to decrease with Yb^{3+} concentration until 20 mole% rapidly and above this concentration decrease becomes slower. Intensity of Yb^{3+} and $^1G_4$ level of Tm^{3+} has the similar dependence on Tm^{3+} concentration in the range of 0.05 to 1 mole%.

CAPTURING A PHASE TRANSITION: THE DIELECTRIC CONSTANT OF GaAs FOLLOWING AN INTENSE ULTRASHORT OPTICAL EXCITATION

Eli Glezer

Gordon McKay Lab.
Harvard University
9 Oxford Str.
Cambridge, Massachusetts 02138

We have measured the time evolution of the real and imaginary parts of the dielectric constant of GaAs following an intense femtosecond optical excitation. We find, contrary to some previous assumptions, that the linear optical properties of the semiconductor change dramatically on a sub-picosecond time scale. The data show a collapse of the average optical gap, or average bonding anti-bonding energy level separation. The rate of collapse increases with pump fluence. The decrease in the gap indicates a structural transformation from a covalent, tetrahedrally coordinated crystal to a phase with metallic cohesive properties.

SECOND HARMONIC GENERATION IN MOVPE ZnTe WAVEGUIDES

Hans-Peter Wagner

Institut für Physik II - Festkörperphysik
Universität Regensburg
Universitätsstraße 31
8400 Regensburg, Germany

The semiconductor ZnTe with a direct gap energy of 2.27 eV at 300 K and its high nonlinear susceptibility $\chi^{(2)}$ is a promising material for second harmonic generation (SHG) of fundamental waves longer than 1.1 mm. We produced high quality single crystalline ZnTe / ZnSe / GaAs (001) waveguides by metalorganic-vapor-phase epitaxy (MOVPE). Using end-fine coupling we observed a strong non-phase matched second harmonic of the 1319 nm Nd:YAG fundamental (30 W peak power). Concepts of a phase matched SHG in waveguides is discussed.

The experimental support of H. Schmitze and H. Stauzl is kindly appreciated.

NEGATIVE DIFFERENTIAL MOBILITY OF SEMICONDUCTOR DEVICES

Emmanuel Anagnostakis

Postgraduate Faculty of Electronics and
Telecommunications of the Hellenic Army
22 Kalamakiou Ave.
174 55 Alimos
Athens, Greece

Experimental measurements for the evolution of Hall mobility versus incoming photon dose are recorded regarding a wide variety of differently prepared GaAs homodiodes. The emerging negative differential mobility feature is discussed as attributable to the onset of the occupancy of an excited conduction sub-level.

RARE-EARTH DOPED SEMICONDUCTORS FOR LINEAR AND NONLINEAR OPTICAL APPLICATIONS

Bruce W. Wessels
Northwestern University
Department of Material Sciences and Engineering
Materials and Life Sciences Building
2225 N. Campus Dr.
Evanston, Illinois 60208-3108

Rare-earth doped semiconductors are of interest for a number of linear and nonlinear optical applications. These include solid-state lasers and optically active waveguides. We have been investigating Er doped III-V semiconductor thin films. The optical properties of these materials have been examined for a range of alloy compositions. The mechanisms of rare-earth luminescence excitation and quenching have been determined and a model developed. The relationship between rare-earth luminescence efficiency and semiconductor host is discussed.

NONLINEAR SPECTROSCOPY IN CdSe: EXCITON COHERENCE AND DYNAMICS

John Erland

Fysisk Institut,
Odense Universitet
Campusvej 55
DK-5230 Odense M., Denmark

I will present the picosecond laser setup in Odense and discuss some recent degenerate four-wave mixing (DFWM) and laser induced grating (LIG) measurements.

At low temperatures the excitonic features of bulk CdSe were investigated. With DFWM, the dephasing time T_2 as well as the energy splitting between coherently excited levels were determined by observing quantum beats or, more precisely, polarisation interference. In LIG experiments the exciton lifetime T_1 and diffusion coefficient D were determined independently. From D the momentum relaxation t_p was calculated. The temperature dependence show that the excitons lose their momentum through interaction with acoustic phonons.

I would like to acknowledge Drs. K. H. Pantke, V. G. Lyssenko and B. S. Razbirin, as well as my supervisor Professor J. M. Hvam.

TWO PHOTON PROCESSES INVOLVING SIMULTANEOUS ELECTRIC AND MAGNETIC DIPOLE TRANSITION MOMENTS

Jaroslaw Sztucki

Institute for Low Temperature and Structure Research
Polish Academy of Sciences
P.O. Box 937
50-950 Wroclaw, Poland

A new mechanism of two photon transitions involving both electric and magnetic dipole operators has been proposed in order to explain an observation of odd parity vibronic sidebands of two photon spectra of Gd^{3+} ion in a centrosymmetric system. The expression for the transition intensity has been developed with the help of symmetry-adaptation techniques and Riecah-Wigner calculus.

It has been shown that a similar approach and expressions can be applied in all cases where the intermediate states belong to the ground configuration and the closure approximation cannot be used. The particular case of three photon transitions has been presented as an example.

LASER SPECTROSCOPY OF ATOMS AND IONS IN SUPERFLUID HELIUM

Harald Günther

Physikalisches Institut
University of Heidelberg
Philosophenweg 12
D-69120 Heidelberg, Germany

Laser spectroscopy of impurity ions and atoms immersed into superfluid Helium has been used for the investigation of both those point defects and the liquid Helium itself.

Foreign atoms and ions in liquid Helium form so called bubble and snowball structures. Like color center or matrix spectrscopy, the spectroscopy of trapped atoms and ions in the liquid Helium matrix opens insights into the nature of these imperfections.

Calculations were done starting from pair potentials describing the formation of those complex structures.

Recombination spectra of electrons with snowball type ions, like the alkaline and the transition metal ions (Cu, Ag, Au) reveal new and interesting effects. The non radiative transitions observed can be explained by the extension of multiphonon theory to quadrupole vibrations of the atomic bubbles. We observed and resolved the vibrational spectrum of a silver ion.

Additionally, we measured the mobility of various ions in liquid Helium by optical and electrical time of flight studies.

MIXED-MODE POLARITONS

Manfred Fiebig

Universität Dortmund
Lehrstuhl Exp. Physik II
44221 Dortmund, Germany

In non-cubic crystals longitudinal-transverse mixed states can be excited. They are denoted as mixed-mode polaritons. Mixed-mode polaritons in ZnO (wurtzite structure) are investigated with use of various two and three-photon techniques. The angular dependence of the longitudinal-transverse splitting can only be described quantitatively if one takes into account all three oscillators.

SYNTHETIC DIAMOND: THE OPTICAL BAND AT 1.883 eV

Luis Miguel Rino Cerveira da Silva

Departamento de Fisica
Universidade de Aveiro
Campus Universitário, Portugal

The 1.883 eV absorption band is the dominant feature in high pressure synthetic diamond grown using nickel containing a catalyst and without the presence of nitrogen getters. In absorption the optical band extends from 1.883 eV to about 2.2 eV and the dominant features are 3 sharp lines, of widths typically 2.5 meV at 1.883 eV, 1.905 eV, and 1.913 eV. The band is for the first time reported in photoluminescence and the results are consistent with the existence of 3 excited electronic states. We report the temperature dependence of the band in absorption and use it to calculate the band shape.

PICOSECOND RADIATIONLESS RELAXATION OF THE F-CENTER IN NaBr INDUCED BY HYDROGEN IMPURITIES

Emmanuel Gustin

Physics Department, UIA
1 Universiteitsplein
B-2610 Antwerpen-Wilrijk, Belgium

With a pump-probe laser transparency technique the ground-state recovery of the F-center in hydrogenated NaBr is measured. For a random distribution of F and substitutional hydrogen (H_S^-) centers a non-exponential F-center relaxation is observed, faster than the one in undoped crystals. The aggregation of F and H_S^- centers by subsequent optical bleachings is monitored by means of resonant Raman scattering. The corresponding relaxation measurements reveal three decay components, which are associated with the formation of $F_H(H_S^-)$ and $F_{H2}(H_S^-)$ centers and to the remaining F-centers. The relaxation time of the $F_H(H_S^-)$ center is established to be ≈ 290 ps, which is 20 times faster than the one of the pure F-center. An estimate of the decay rate assuming dipole-dipole coupling between the F-center electron and the H_S^- vibration shows that the interaction plays an important role in reducing the lifetime of the excited F-center. The frequency shift of the hydrogen vibration of the $F_H(H_S^-)$center with respect to the one of the H_S^-center cannot be explained by dipole-dipole coupling alone. This is possibly related to a displacement of the H_S^- equilibrium position of the $F_H(H_S^-)$ center away from the lattice site.

SILVER ATOMS ISOLATED IN RARE GAS CRYSTALS : TIME RESOLVED INVESTIGATION OF EXCITED STATE DYNAMICS

Ronald Kometer

Freie Universität Berlin
Fachbereich Physik
Arnimallee 14
1000 Berlin 33, Germany

The first excited state (5p ¨ 5s) of silver atoms in rare gas crystals shows a triplet structure in the excitation spectra. This structure, similar to alkali atoms in crystals, is attributed to a strong coupling of the excited p-state of the atom with the vibrational modes of the surroundings, called the Jahn - Teller Effect (JTE).

The aim of our experiments is to follow the excited state dynamics from the correlated vibration of the surroundings to the relaxed excited state, from which the emission comes.

SITE-SELECTION BY UP-CONVERSION IN $CsCdBr_3:Er^{3+}$

Philippe Goldner

Centre National de la Recherche Scientifique
Laboratoire de Physicochimie des Matériaux (UPR-211)
1, Place Aristide Briand
92190 Meudon, France

$CsCdBr_3$, which belongs to the AMX_3 hexagonal family, has a structure which consists of an array of parallel chains composed of $(CdBr_6)^{4-}$ octahedra that share faces (Cs^+ ions occupy positions between the chains). Besides its low dimensional structure, the main property of this matrix is due to the doping with trivalent ions. These impurities enter this host to form quasi-exclusively M^{3+}-Vacancy- M^{3+} complexes. This is observed even for very low concentration. For this reason, $CsCdBr_3$ allows one to study pair interactions and especially up-conversion processes. Indeed the strong coupling between the ions of one dimer leads to very efficient conversion for several rare earth ions under various excitations. Theoretically, intra-pair conversion dynamics can be treated by rate equations without fast diffusion assumptions.

Up-conversion is also a valuable tool for site-selection. A study of the different Er^{3+} centers in $CsCdBr_3$ has been carried out and gave the following results. Under direct excitation to the $^4S_{3/2}$ level three Er^{3+} centers are observed. One of them correspond to the isolated ions and the other to symmetric (M^{3+}-V-M^{3+}) and asymmetric (M^{3+}-M^{3+}-V) complexes. In the latter, an energy transfer occurs between the ions. In order to confirm the nature of the centers anti-stokes emissions have been studied for several excitations.

First, we compared emissions from the $^4S_{3/2}$ level obtained by excitation to the $^4F_{9/2}$, $^4I_{11/2}$, $^4F_{7/2}$ levels. Then, the excitation spectra of the $^4I_{15/2} \rightarrow {^4F_{7/2}}$ transition were analyzed by monitoring Stokes and anti-Stokes emissions.

The results show that conversion is due to an energy transfer in doubly excited dimers. Moreover, the nature of the different centers can be unambiguously determined by this method.

ELECTRON-HOLE DROPLET FORMATION IN INDIRECT $Al_xGa_{1-x}As$

Andreas Wörner

Fachbereich Physik
Universität Kaiserslautern
Erwin-Schrödinger-Straße
6750 Kaiserslautern, Germany

Analyzing the spectra of high excitation photoluminescence measurements, we could observe electron-hole droplet formation in indirect $Al_xGa_{1-x}As$ near the direct/indirect crossover.

The electron-hole droplet (EHD) state is a many particle effect due to the high excitation with the carrier density staying constant in time in difference to the electron-hole plasma. To get this EHD formation it is necessary to have a not too short carrier lifetime and some nucleation centers, as they are in $Al_xGa_{1-x}As$ because of its alloy nature.

We measured such an EHD formation in dependence of the excitation intensity, the lattice temperature and the aluminium concentration, which is responsible for the carrier lifetime, with the intention to get a contribution to the T/n-phase diagram for this material. As a further result we could estimate the importance of the sample quality for the EHD formation.

DYNAMICS OF CARRIER INDUCED SCREENING OF THE PIEZOELECTRIC FIELDS IN CdS/CdSe SUPERLATTICES

Wolfgang Werner Langbein

Fachbereich Physik
Universität Kaiserslautern
Erwin-Schrödinger-Straße
6750 Kaiserslautern, Germany

CdS/CdSe superlattices show very strong intrinsic piezoelectric fields due to hexagonal crystal structure and lattice mismatch of 3.9%. Time resolved luminescence experiments with high ps-excitation of about 10^{19} cm^{-3} reveal a shift of the luminescence energy of 400 meV and a logarithmic time dependence. This is described by a model calculating the band structure of these superlattices under high carrier densities.

OPTICAL SWITCHING IN CdS

Rüdiger Schmolke

Institut für Theor. Physik
TU-Berlin
Sekr. PN7-1
Hardenbergstraße 36
10623 Berlin, Germany

In previous experiments the change in transmissivity of a CdS crystal under excitation with 20 ns pulses of moderate intensity was investigated. The wavelength emitted by the exciting lasers corresponded to photon energies just below the band gap. The crystal was kept at about 2K. A switching of the crystal into a state of high transmissivity was found.

We analyzed this switching behavior with a three-level model taking into account acceptor and donor states as well as the conduction band. We could not confirm the conjecture that the switching is due to an optical bistability. The analysis of the spatio-temporal behavior of a reduced model showed the appearance of a switching front travelling through the crystal. This result was confirmed by new experiments.

PHOTOREFLECTANCE: A POOR MAN'S APPROACH TO NON-LINEAR SPECTROSCOPY

Dan Birkedal

Mikroelektronik Centret
DTH. building 345 East
DK-2800 Lyngby, Denmark

Photoreflectance (PR) have proven to be a valuable tool for characterizing of semiconductor structures. As a modulated measurement PR is an inherent non-linear technique, even though low excitation powers are used. In this presentation a description of the experimental technique is given and the fundamental modulation mechanism is discussed. We show how the Franz-Keldysh effect give rise for oscillations in the PR-spectrum. PR- spectra of ∂-doped GaAs structures are presented and the interpretation of these spectra is discussed.

THE EFFECT OF CRYSTAL ANISOTROPY ON THE INFRARED REFLECTIVITY OF 6H-SiC

Franz Engelbrecht

Universität Erlangen-Nürnberg
Lehrstuhl Angewandte Physik
Staudtstraße 7
D-8520 Erlangen, Germany

Polarized infrared reflectivity spectra of 6H-SiC single crystals under different angles of incidence are reported. A new sharp line in the reststrahl band was observed in the extraordinary ray, whose line shape is dependent on the angle of incidence. It was possible to show that crystal anisotropy of 6H-SiC is the reason for the new line in the reflectivity spectrum. A numerical calculation using a Lorentz oscillator model confirmed this result. Two weak one-phonon absorption lines in the reststrahl band region for E || c were observed. Furthermore, we found a new line at 859 cm^{-1} in the extraordinary ray and an unusual structure near w_{TO} in the reststrahl band of parallel c-cut 6H-SiC samples, which could not be explained in terms of the classical Lorentz oscillator model.

SECOND HARMONIC GENERATION IN DNA

Maryanne Large

Dept. of Pure & Applied Physics
Trinity College
University of Dublin
Dublin 2, Ireland

The chirality of the DNA molecule suggests that it should have a non zero $\chi^{(2)}$. Projected experiments have been designed to exploit this non-linear property by aligning the DNA molecules within a stretched polymer matrix. Preliminary experiments show a second harmonic signal.

INFRARED TO VISIBLE UP-CONVERSION IN $Cs_3Yb_2Cl_9$:Tm^{3+}

Toni Riedener

Institut für Anorganische Chemie
Universität Bern
Freiestraße 3
3012 Bern, Switzerland

Infrared to visible up-conversion by stepwise energy transfer is reported in Thulium doped $Cs_3Yb_2Cl_9$:Tm^3 (1-15% Tm). In this crystal structure the rare earth ions are arranged in M_2X_9 - dimers with a short M-M distance of 3.6Å.

The excitation of the Yb $^2F_{5/2}$ level (10670cm^{-1}) results in a series of emissions in near infrared ($^3H_4$ - $^3H_6$, 12,000 cm^{-1}), in the red ($^3F_3$ - $^3H_6$, 13,000 cm^{-1}), in the blue ($^1G_4$ - $^3H_6$ 21,000cm^{-1}) and ($^1D_2$ - $^3F_4$ 22,000cm^{-1}), and in the ultraviolet ($^1D_2$ - $^3H_6$, 27,000cm^{-1})

NONLINEAR BEHAVIOR OF LUMINESCENCE EMISSION INTENSITY OF RARE EARTH IONS WITH HIGH EXCITATION ENERGY

Guzin Armagan

Department of Physics
The College of William and Mary
Williamsburg, Virginia 23187 U.S.A.

The luminescence emission intensity of each transition of Tm and Ho in garnet and fluoride crystals has been measured as a function of absorbed power. The variation of the intensity with increasing absorbed power has been found to be linear, positive hyperbolic, and negative hyperbolic for the transitions originating from states near the excitation state, higher than the excitation state and lower than the excitation state respectively. By determining the relative intensity of each of each transition with respect to the others, it was possible to apply these measurements to a steady state rate equation model in order to understand the mechanism of energy transfer processes.

STUDY OF HEALING AND INTERDIFFUSION PROCESSES AT PARTICLE-PARTICLE JUNCTION DURING LATEX FILM FORMATION BY FLUORECENCE TECHNIQUE

Murat Canpolat

Istanbul Technical University
Faculty of Sciences and Letters
Department of Physics
Maslak-Istanbul 80626, Turkey

When amorphous polymers of same type are brought into contact at temperatures above the glass transition, the interface gradually heals and junction boundaries eventually disappear. Here we used steady state fluorescence technique to examine the healing and the interdiffusion of polymer molecules during annealing of latex film above glass transition. The film was prepared from a mixture of naphthalene (donor) and pyrene (acceptor) labeled poly (methylmethacrylate) latex particles. We report that during the annealing processes transparency of film changed considerably. Healing temperature and time were measured when latex film became completely transparent. Above healing temperature interdiffusion of polymer chains were observed by detecting the energy transfer from excited naphthalene to pyrene molecules. Healing and diffusion activation energies were calculated and found to be 9.84 kcal/molecule and 30 kcal/molecule respectively for the annealed latex film samples.

FAST SCINTILLATOR GLASSES WHICH ARE HARD TO NUCLEAR RADIATIONS

David B. Hollis

University of Paisley
Department of Physics
High Street
Paisley PA1 2BE
Scotland, U.K.

Hole reaction nuclear fission calorimetry experiments require a material which can trap and detect all emitted particles. Systems such as LEP and LHC will produce radiation showers, including gamma rays, neutrons, and protons.

Scintillator liquids and crystals are available, but too expensive for future experiments. A cheap, compact alternative is sought. Example: glass?

Requirements are:

Radiation hardness	-	high damage threshold
Short scintillation time	-	high count rates
Short stopping distance	-	compact device cheaper system fine spatial resolution
Physical and chemical	-	durability in service
Optical clarity	-	detection of scintillation
High yield	-	conversion of energy in particles to light photons

Glasses which scintillate, and are hard to nuclear radiations are already known. Example: NE110, a SiO_2, Li_2O, CaO, Al_2O_3 glass doped with three percent Ce_2O_3. However, the stopping for the more energetic nuclear particles is about 10 cm. We require 2 cm. Crystals with high scintillation yield are also well known, such as NaI, or BaF_2:Ce, however, these are expensive in large quantity;

We are developing heavy metal fluoride and oxide glasses at Paisley University, these are being tested with nuclear radiation by our colleagues Dr. R. Brown et al. at Rutherford, Appleton Lab, Oxford, and Dr. D. Imry et al. at Brunel University. We seek a glass hard to nuclear radiation, strongly scintillating, with good optical clarity and short stopping length. To date the fluoride glasses show promise, and the oxide glasses are now being tested.

TRANSIENT HOLD-BURNING AND EXCITED STATE DYNAMICS IN A CO-ORDINATION COMPOUND

Elmars Krausz

Research School of Chemistry
Australian National University
Canberra ACT 0200, Australia

Laser line narrowing and hole-burning in the charge transfer excited states of a metal/organic molecular ion in the solid state are described. These measurements establish the dipole moment of the excited state and quantify some intramolecular energy transfer processes. A remarkable linewidth of only 15 MHz is obtained with side structure, probably due to nuclear hyperfine interactions.

ELECTRODYNAMICS OF THIN FILM WAVEGUIDES

Walter Ebeling

Fachbereich Physik
Universität Bremen
W-2800 Bremen 33, Germany

The matrix formalism for the description of the electrodynamics of a multi-layer thin film waveguide was presented. This method allows an easy numerical determination of the radiation modes and the propagation constants of the waveguide. As an example, a $ZnSe/ZnS_xSe_{1-x}$ waveguide was analyzed.

TEMPERATURE DISTRIBUTION IN AN Er:YAG LASER ROD DURING FLASH LAMP PUMPING

Tanja Rupnik

Jozef Stefan Institute
Jamova 39
61111 Ljubjana, Slovenia

Crystals of Yttrium-Aluminum garnet ($Y_3Al_5O_{12}$), known as YAG represent the base of a large number of solid-state lasers. Interest in the Er-doped YAG crystal has been shown already in the 1970's for its possible use as a laser for medical applications. Since our bodies are composed mostly of water, the absorption of light by water is a fundamental consideration for medical lasers. The peak absorption of water is in the region of 3 μm. Since the Er:YAG laser emits a wavelength of 2.94 μm, it's very interesting for medicine and biology. Measurements of the transient radial temperature profile of the Er:YAG rod during flash lamp pumping are reported.

ELECTRONIC ENERGY LEVEL STRUCTURE OF Er^{3+} DOPED INTO 3 GARNET HOSTS

John R. Quagliano

Chemical and Laser Sciences Division
Los Alamos National Laboratory
Los Alamos, New Mexico 87545 U.S.A.

The $A_3B_2C_3O_{12}$ (space group Ia3d) family of host crystals, better known as the garnets, is the most technologically important class of solid-state laser host materials in existence. The electronic spectra of these crystalline hosts containing the Er^{3+} ion (or any other tripositive lanthanide ion) display numerous sharp, weak transitions, closely clustered in between large energy gaps over the UV to near-IR energy range. A comparison between the experimental energy levels and the computationally predicted levels was presented for four selected J-multiplets of the Er^{3+} ion when doped into the $Y_3Al_5O_{12}$ (YAG), $Y_3Sc_2Al_3O_{12}$ (YSAG), and $Y_3Sc_2Ga_3O_{12}$ (YSGG) hosts. The agreement between theory and experiment is considerably improved when two-particle correlation crystal field (CCF) effects are included in our model.

THRESHOLD BEHAVIOR OF MICROCAVITY LASERS

Deepak Boggavarapu

Optical Sciences Center
The University of Arizona
Tucson, Arizona 85721 U.S.A.

Recently the concept of threshold-less lasing has been proposed in regard to microcavity lasers (i.e. wavelength sized optical cavities). We study the quantum noise properties of microcavity lasers near threshold both theoretically and experimentally. Computations based on two-level quantum theory show that microlasers exhibit a high degree of second order coherence compared with conventional lasers as result of suppression of spontaneous emission into non-lasing modes, and there always exists a finite threshold for these lasers defined by the peak of the photon number correlation function corresponding to the spontaneous-stimulated transition.

CALCULATION OF THE THG, TWO AND THREE PHOTON ABSORPTION IN C_{60} AND C_{70} MOLECULES

Marianna Fanti

Università degli studi di Bologna
Dipartimento di Chimica «G. Ciamician»
Via F. Selmi, 2
40126 Bologna, Italy

The following is a brief history of the fullerene's discovery and synthesis, especially in regard to the main characteristics of C_{60} and other fullerenes. Partial results on the computation of the dispersion of third-order polarizability for C_{60} and C_{70} as a THG effect (third harmonic generation) are shown. In order to determine the resonances, it is important to calculate the two-and three-photon absorption spectra. All the results have been worked out using the semiempirical CNDO/S method followed by a Sum Over the virtual excited States (SOS) for the hyperpolarization computation.

LOW FREQUENCY RAMAN SCATTERING FROM THERMALLY TREATED OXIDE GLASSES

Krystyna E. Lipinska-Kalita

ENEA, CRE-Frascati, Dip. INN-SVIL
00044 - Frascati (Roma), C.P. 65 Italy

Many studies have been performed to understand the structure of disordered solids from light scattering investigations. The majority have concentrated on spectral features in the optical mode region of Raman scattering spectra. Recently, interest has developed in the low-frequency or acoustic Raman modes region. Part of this interest is a manifestation of theoretical works on the properties of light scattering from acoustic modes in disordered solids.

The region of very low-frequency inelastic scattering continuum has become of great interest in the last years and is also the subject of the work presented because it offers interesting possibilities for observing the nanoscale structures (heterogenities of amorphous matrices, clusters, and paracrystals) in the glassy network. Additionally, the Raman spectroscopy has proved to be also a powerful technique for determining the size of microcrystals in glassy matrices.

$BaLiF_3:Ni^{2+}$ - SPECTROSCOPIC ANALYSIS OF LASER CAPABILITIES

Evely Martins

IPEN/CNEN/SP
Department of Optoelectronics Materials
P.O. Box 11049
CEP 05422-970
São Paulo, Brazil

The properties of transition metal ions in insulating matrices have received renewed interest in recent years mainly as an outcome of their potential usefulness for construction of vibronic solid-state lasers. Ions with the configuration $3d^3$ are particularly appropriate candidates since for these ions the $^3T_2$ state approaches or falls below the $^3A_2$ state and gives rise to broad vibronic emission.

Most materials doped with Ni^{2+} in octahedral sites show a vibronically broadened emission band in the infrared around 1.5 µm which is useful for tunable solid-state lasers. Although Ni^{2+} in this symmetry has been investigated since 1963, CW-operation at room temperature has not yet been achieved. We have investigated spectroscopic properties of new Ni^{2+}-doped perovskites, $BaLiF_3$; we believe that they show promise as new laser materials.

STUDY OF A DIODE-PUMPED LASER YLF: Tm, Ho LASER AT 2μm

Alessandra Toncelli

Viale Mameli 31/c
57126 Livorno, Italy

Due to recent technological advances, diode lasers have become interesting pumping sources for solid-state lasers. Tm^{3+} ions display absorption lines near 800 nm and high power emitting diode-lasers are now available in this wavelength region. For this reason, we chose a YLiF4: 5% Tm, 2% Ho crystal as the active material for a laser. We measured the gain of the laser material, the slope efficiency, the threshold and the tuning range of the laser both at 77K and 210K. The experimental results are quite encouraging and we hope to improve the performance of the laser in order to obtain mono-mode emission and to have the laser working at higher temperatures.

DECAY BEHAVIOR OF EXCITED Nd IONS IN KYF_4 CRYSTALS

Chun-Lai Yang

Department of Physics
Boston College
Chestnut Hill, Massachusetts 02167 U.S.A.

The Nd doped KYF_4 crystal is a new fluoride laser material. This material is attractive owing to broader absorption linewidths and longer fluorescence lifetimes compared with Nd-doped YAG or YLF. Its general laser emission shifts to shorter wavelength (1.041 μm). The lifetimes of Nd^{3+} ions in $^4F_{3/2}$ excited states were measured at various temperatures from 20K to 300K for three Nd-doped KYF_4 samples with different concentrations. Other basic spectroscopic measurements were also presented. When the temperature is not too low (150-300K), the lifetimes change very little with temperature. But in the low temperature region, the lifetimes strongly depend on temperature, especially around 50K. There appeared a strange decay behavior, which was assumed to be due to the energy transfers between different Nd ion sites in KYF_4. The lifetime of Nd ions in KYF_4 crystal is long, which is favorable for diode-pumping. For these crystals there is strong concentration quenching. The relation of lifetime with concentration is linear at 300K, and it is nonlinear at low temperatures.

SUMMARY OF THE COURSE

Dr. Macfarlane began the summary of the meeting with an overview of nonlinear spectroscopy. First he gave the motivations for studying nonlinear spectroscopy:

1) Curiosity about the physical world
2) Intellectual beauty
3) Technological - being able to generate and control light.

He then reviewed some of the aspects of nonlinear spectroscopy. These include coherent phenomena, either in 2-level or multi-level systems, new selection rules, echo phenomena, sum and difference frequency generation, upconversion, multiphoton processes, excitons, etc. He then compared nonlinear spectroscopy and nonlinear optics.

Dr. Macfarlane then gave a review of the lectures presented at the school. These he broke into 4 main categories:

1) The tools of nonlinear spectroscopy
2) The lasers
3) Using nonlinear spectroscopy to study particular materials
4) Developing materials with large nonlinearities

Among the topics covered in the tools of nonlinear spectroscopy were the density matrix formulation, T_1 and T_2 processes, $\chi^{(2)}$ and $\chi^{(3)}$, quantum noise, coherence and various other definitions. These laid the groundwork for further lectures. This material was covered in lectures and seminars by Drs. Abram and Di Bartolo and by Mr. Bowlby. Also covered in the tools of nonlinear spectroscopy were the applications to basic phenomena. This included the echo, parametric amplification, solitons in fiber and the two-photon laser. These topics were covered by Drs. Hahn, Abram, Mollenauer and Gauthier respectively.

The second major topic of the school was the laser. Here such terms as Q switching, mode locking, frequency stabilization and cavity design are encountered. The lectures of Drs. Ferguson and Ferrario covered this material.

The third topic was the use of nonlinear spectroscopy to study particular materials. The coherent interaction of light with materials gives rise to echo effects, as pointed out by Drs. Hahn and Macfarlane. The "pure" multiphoton processes were examined by Dr. Frohlich. The properties of nanostructures were covered by Drs. Broido, Reisfeld and Woggon, while Dr. Baldacchini spoke of color centers. Also covered in this section was the behaviour of excitations at high densities. This material was covered by Drs. Klingshirn, Auzel, and Buoncristiani.

The final area of the course dealt with the development of materials with large nonlinearities. These included new sources of light and control of light, and the practical application of nonlinear phenomena. The latter was covered by Dr. Mollenauer in his seminar on solitons in optical fibers, while the former was covered in seminars by Drs. Auzel, Boulon, Buoncristiani, Kaminskii, and Macfarlane.

Dr. Macfarlane concluded the summary with a brief list of things to take away from Erice. First, a large pile of lecture notes. Second an appreciation for the power and beauty of nonlinear spectroscopy. Third, the motivation to follow up on the new ideas presented, and finally many new friends.

KEY TO THE PHOTOGRAPH

1. Erwin Hahn
2. Quirino Baldacchini
3. Bruce Wessels
4. Chun-Lai Yang
5. Alexander Kamınskıı
6. Joaquin Fernandez
7. Georges Boulon
8. Daniel Di Bartolo
9. Emmanuel Gustin
10. Emmanuel Anagnostakis
11. John Quagliano
12. Philippe Goldner
13. Bernd Henninger
14. Michel Mortier
15. Giuseppe Baldacchini0
16. Fabienne Pelé
17. Martin Buoncristiani
18. Paul Richerd Barber
19. Ronald Kometer
20. David Broido
21. Stefan Altner
22. Luis Miguel Rino Cerveira da Silva
23. Rolindes Balda
24. John Di Bartolo
25. Baldassre (Rino) Di Bartolo
26. Renata Reisfeld
27. Jean-Paul Faure
28. Harvey Tate
29. Robert Brown
30. Jørn Hvam
31. Henryk Lozykowski
32. Angela Siraco
33. Robert MacFarlane
34. Walter Ebeling
35. Brian Bowlby
36. Daniel Gauthier
37. Eli Glezer
38. Gönül Özen
39. Rüdiger Schmolke
40. Claus Klingshirn
41. Linn Mollenauer
42. Mrs. Mollenauer
43. Stamatios Kyrkos
44. Krystyna Lipinska-Kalita
45. Pinola Savalli
46. Jerry Pilarsky
47. Elmars Krausz
48. Roberto Simonelli
49. Tanja Rupnik
50. Alessandra Toncelli
51. Izo Abram
52. David Hollis
53. François Auzel
54. Manfred Fabig
55. Dan Birkedal
56. Deepak Boggavarapu
57. Franz Engelbrecht
58. Mustafa Kumru
59. Murat Canpolat
60. Maryanne Large
61. José Manuel Hernández
62. Wolfgang Werner Langbein
63. Ulrike Woggon
64. Reinhard Hannak
65. Andres Wörner
66. Evely Martins
67. Alfonso Lorenzo
68. Fernando Lahoz Zamarro
69. Miguel Angel Buñuel Magdalena
70. Hans-Peter Wagner
71. Alessandro Di Maio
72. Marianna Fanti
73. Guzin Armagan
74. Alberto Daidone, "Luce"

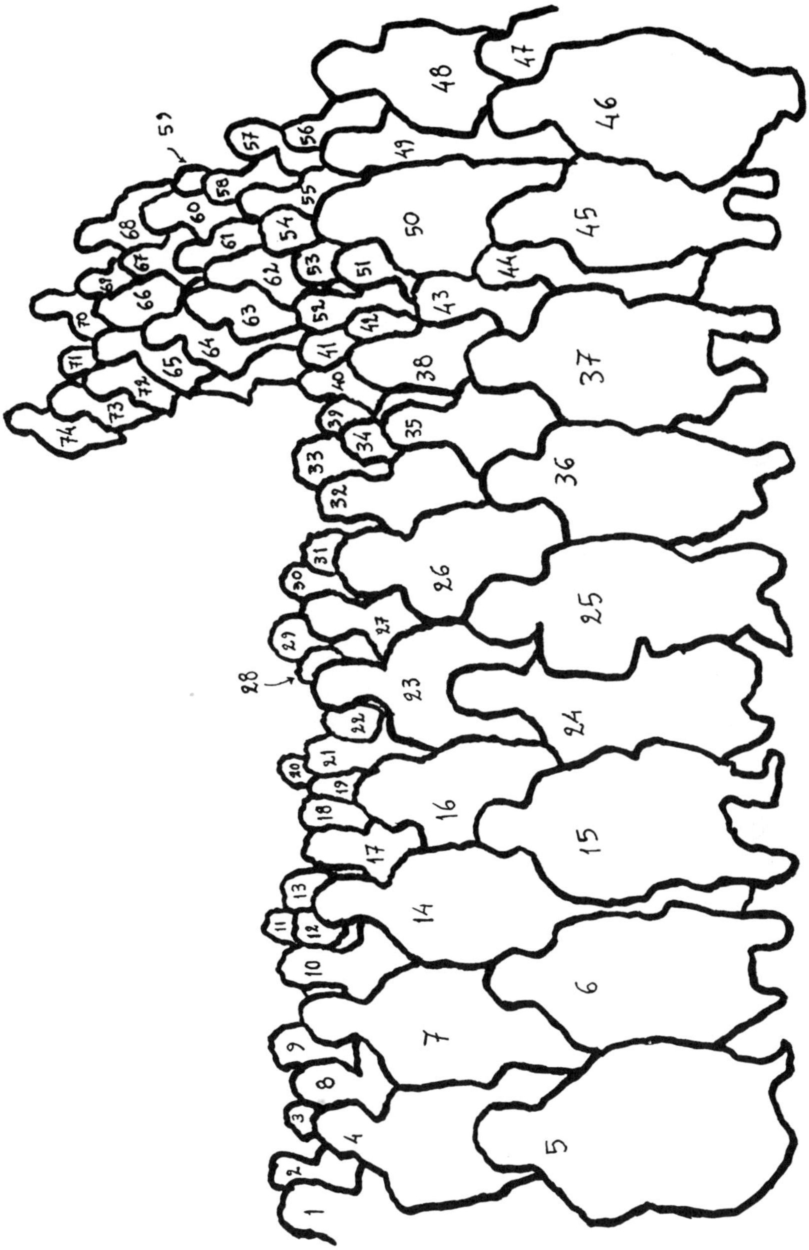

1
2
3
4
5
6
7
8
9
10
11
12
13
14
15
16
17
18
19
20
21
22
23
24
25
26
27
28
29
30
31
32
33
34
35
36
37
38
39
40
41
42
43
44
45
46
47
48
49
50
51
52
53
54
55
56
57
58
59
60
61
62
63
64
65
66
67
68
69
70
71
72
73
74

PARTICIPANTS

Participant	Fields of interest
Abram, Izo Centre National d'Études des Telecommunications 196, Ave. H. Ravera 92220 Bagneux, France phone:331-42317332 fax: 331-42534930 email: abram @bagneux.cnet.fr	• Quantum Optics • Nonlinear Optics
Altner, Stefan ETH - Zentrum Laboratorium für physikalische Chemie Universitätstraße 22 CH-8092 Zürich, Switzerland phone: 41-1-2564383 fax: 41-1-2523402 email: altner@ppc.lpc.ethz.ch	• Spectroscopy of inorganic and organic materials • Luminescence spectroscopy • Spectral holeburning • Applications of these techniques to optical computing
Anagnostakis, Emmanuel Postgraduate Faculty of Electronics and Telecommunications of the Hellenic Army 22 Kalamakiou Ave. 174 55 Alimos Athens, Greece phone: 30-1-9817810 fax:30-1-9843004	• Photoconductivity of semiconductor devices
Armagan, Guzin Department of Physics The College of William and Mary Williamsburg, Virginia 23187 U.S.A. phone: 804-8641570 email: g.armagan@larc.nasa.gov	• Spectroscopy of solid state laser materials
Auzel, François Centre National d'Études des Telecommunication 196, rue de Paris 92220 Bagneaux, France phone: 33-1-42317202	• Optical properties of solids doped with T.M. and R.E. ions, laser materials, amplifiers • Non radiative transitions • Up-conversion

Balda, Rolindes
Departamento de Física Aplicada I
Escuela Tecnica Superior de
Ingenieros Industriales y de Telecomunicacion
Alda, Urquijo, S/N
48013 Bilbao, Spain
phone: 34-4-4278055
fax: 34-4-4414041

- Optical properties of solid-state materials
- Time-resolved spectroscopy
- Photoacoustic spectroscopy

Baldacchini, Giuseppe
ENEA
Centro Ricerche Energia Frascati
Via E. Fermi, 27
00044 Frascati (Roma), Italy
phone: 39-06-9400-5365
fax: 39-06-9400-5400

- Optical and magneto-optical properties of solid state materials
- High resolution molecular spectroscopy

Barber, Paul Richard
Physics Department
University of Southampton
Highfield
Southampton SO9 5NH, U.K.
phone: 44703592066
fax: 44703585813
email: prb@mail.soton.ac.uk

- Rare earth doped fiber lasers
- Up-conversion

Birkedal, Dan
Mikroelektronik Centret
DTH. building 345 East
DK-2800 Lyngby, Denmark
phone: 45-45931222 ext. 5475
fax: 45-42887762
email: dan@mic.dth.dk

- Optical properties of low-dimensional semiconductor structures
- Quantum wells, superlattices

Boggavarapu, Deepak
Optical Sciences Center
The University of Arizona,
Tucson, Arizona 85721 U.S.A.
phone: 602-621-2452
fax: 602-621-4323
email: deepak@kilimanjaro.opt-sci.arizona.edu

- Nonlinear optics of low-dimensional semiconductor structures
- Semiconductor microcavity lasers

Boulon, Georges
Université Claude Bernard-Lyon 1
Laboratoire de Physico-Chimie
des Matériaux Luminescents
Unité de Recherche associée au CNRS
69622 Villeurbanne Cedex, France
phone:33-72431245
fax: 33-72431130

- Crystal growth and spectroscopy of solid-state laser-type materials

Bowlby, Brian †
Physics Department
Boston College
Chestnut Hill, Massachusetts 02167 U.S.A.
phone: 617-552-3577
fax:617-552-8478
email: bbowlby@csmvax.simmons.edu

• Spectroscopy of laser crystals with low symmetry

Broido, David
Physics Department
Boston College
Chestnut Hill, Massachusetts 02167, U.S.A.
phone: 617-552-3348
fax: 617-552-8478
email: broido@bcvms.bc.edu

• Theory of electronic and optical properties of lower-dimensional semiconductor systems

Brown, Robert S.
Physics Department
University of Southampton
Highfield
Southampton S095NH, U.K.
phone: 44-703-592194
fax: 44-703-585813
email: rsb@orc.soton.ac.uk

• Spectroscopy of bulk glasses and optical fibers doped with rare earth ions

Buñuel Magdalena, Miguel Angel
Física de la Materia Condensada
Facultad de Ciencias
Campus Universitario
50009 Zaragoza, Spain
phone:34-76-560741

• Time and site resolved spectroscopy of 3d and 4f ions in doped fluoride glasses

Buoncristiani, Martin
NASA Langley Research Center
MS468
Hampton, Virginia 23665-5225 U.S.A.
phone:804-864-1569
fax:804-594-7919
email: martinb@pcs.cnu.edu

• Solid state laser materials
• Laser Metrology (Lasers as sensors)

Canpolat, Murat
Istanbul Technical University
Faculty of Sciences and Letters
Department of Physics
Maslak-Istanbul 80626, Turkey
phone: 1-2766603

• Study of healing and long polymer chain using steady state fluorecence spectroscopy

---

† Secretary of the course

Di Bartolo, Baldassare (Rino)
Department of Physics
Boston College
Chestnut Hill, Massachusetts 02167 U.S.A.
phone: 617-552-3601
fax: 617-552-8478
email: dibartob@bcvms.bc.edu

- Luminescence spectroscopy
- Flash photolysis
- Photoacustic spectroscopy

Di Bartolo, Daniel L.[*]
Department of Psychology
Boston College
Chestnut Hill, Massachusetts 02167 U.S.A.

- Psychology

Di Bartolo, S. John
Department of Physics
Boston College
Chestnut Hill, Massachusetts 02167 U.S.A.

- Theoretical physics

Ebeling, Walter
Fachbereich Physik
Universität Bremen
W-2800 Bremen 33, Germany
phone: 421-218-4428
fax: 421-218-3601
email: ebeling@theO.physik.uni-bremen.de

- Nonlinear spectroscopy of II-IV compounds

Engelbrecht, Franz
Universität Erlangen-Nürnberg
Lehrstuhl Angewandte Physik
Staudtstraße 7
D-8520 Erlangen, Germany
phone:0049-9131-858420
fax: 0049-9131-858423

- Physical properties of SiC
- Infrared impuritiy absorption of SiC:N
- Infrared reflectivity of SiC
- Raman scattering in SiC
- Avalange breakdownin SiC
- SiC device tachnology: RIE

Erland, John
Fysisk Institut, Odense Universitet
Campusvej 55
DK-5230 Odense M., Denmark
phone: 45-66158600
fax: 45-66158760
email: je@dou.dk

- Ultrafast nonlinear spectroscopy of semiconductors

Fanti, Marianna
Università degli studi di Bologna
Dipartimento di Chimica «G. Ciamician»
Via F. Selmi, 2
40126 Bologna, Italy
phone: 39-51-259473
fax: 39-51-259456
email: marianna@treebeard.ciam.unibo.it

- Nonlinear optics and two- and three-photon absorption of the fullerenes and the other single molecules

---

[*] Administrative Assistant

Faure, Jean-Paul
C.E.A.
D.P.E./S.P.L.
Section des Composants Lasers et Optique
Centre d'Etude de Saclay
Bat 467
91191 Gif/ Yvette, France
phone: 33-1-69086856
fax: 33-1-69088691

• Nonlinear optics (third order effects in fibers)

Ferguson, Allister I.
Department of Physics and Applied Physics
University of Strathclyde
Glasgow G4 ONG
Scotland, U.K.
phone: 41-552-4400 ext. 3359
fax: 41-552-2891

• Laser physics
• Diode pumped solid state lasers
• Mode-locked solid state lasers
• Nonlinear frequency conversion

Fernandez, Joaquin
Departamento de Física Aplicada I
Escuela Tecnica Superior de
Ingenieros Industriales y de Telecomunicacion
Alda, Urquijo, S/N
48013 Bilbao, Spain
phone: 34-44278055
fax: 34-44414041

• Optical properties of solid state materials TR spectroscopy
• Photoacustic spectroscopy

Ferrario, Angelo
Quanta Systems
Via Venezia Giulia, 18
20157 Milano, Italy
phone: 39-2-33200239
fax: 39-2-3575140

• Turnable laser sources
• PS and FS lasers

Fiebig, Manfred
Universität Dortmund
Lehrstuhl Exp. Physik II
44221 Dortmund, Germany
phone: 49-231-7553535
fax: 49-231-7553569

• Nonlinear spectroscopy of solids
• Electronic properties of insulators and semiconductors

Flytzanis, Christos
Laboratoire d'Optique Quantique
École Polytechnique
F-91128 Palaiseau, Cedex, France
phone: 69-33169334124
fax: 69-33169333017

• Nonlinear optics (spectroscopy materials)
• Surfaces

Fröhlich, Dietmar
Universität Dortmund, Lehrstuhl
für Exp. Physik II
44221 Dortmund, Germany
phone: 49-231-7553530
fax: 49-231-7553569

• Nonlinear spectroscopy of solids
• Electronic properties of insulators and semiconductors

Gauthier, Daniel
Department of Physics
Duke University P.O. Box 90305
Durham, North Carolina 27708 U.S.A.
phone:919-660-2511
fax: 919-660-2525
email: gauthier@phy.duke.edu

- Quantum optics
- Nonlinear optics

Glezer, Eli
Gordon McKay Lab.
Harvard University
9 Oxford Str.
Cambridge, Massachusetts 02138 U.S.A.
phone:617-783-7144
fax:617-495-9827
email: eli_glezer@ lucifer.harvard.edu

- Linear and nonlinear spectroscopy of solids
- Ultrafast electron and lattice dynamics in semiconductors

Goldner, Philippe
Centre National de la
Recherche Scientifique
Laboratoire de Physicochimie
des Matériaux (UPR-211)
1, Place Aristide Briand
92190 Meudon, France
phone: 33-1-45075595
fax: 33-1-45075107
email: lize@bellevue.cnrs-bellevue.fr

- Spectroscopy of rare earth ions
- Up-conversion processes

Günther, Harald
Physikalisches Institut
University of Heidelberg
Philosophenweg 12
D-69120 Heidelberg, Germany
phone: 06221-569359
fax: 06221-475733
email: guenther@cosmo.physi.uni-heidelberg.de

- Atomic/low temperature physics
- Spectroscopy of atoms and ions in liquid helium

Gustin, Emmanuel
Physics Department, UIA
1 Universiteitsplein
B-2610 Antwerpen-Wilrijk, Belgium
phone:03-8202481
email: gustin@uia.ac.be

- Spectroscopy of solids
- Decay of F-centers in alkali halides and energy transfer processes

Hahn, Erwin
Department of Physics
University of California at Berkeley
Berkeley, California 94720 U.S.A.
phone:510-642-2305
fax: 510-643-8497

- Pulsed NMR, optics (laser)
- Optical pumping in GaAs (currently)

Heindl, Rudolf
Centre National de la
Recherche Scientifique
Laboratoires de Bellevue
1, Place Aristide Briand
92190 Meudon
France

- Radiothermoluminescence (materials-measurements-theory)PSL
- Optical and magnetic properties of trapped electrons
- Ionizing radiation detection at high intesities
- properties of F and V_k centers in halides

Henninger, Bernd
Technische Universität Berlin
Sekr. PN 4-1
Hardenbergstraße 36
10623 Berlin , Germany
phone:49-30-31423971
fax: 49-30-31422064

- Study of ps-spectroscopy (decay-time measurement) of CdTe, $Cd_{1-x}Mn_xTe$ QW structures

Hernández A., José Manuel
Instituto de Fisica, UNAM
Solid State Department
P.O. Box 20-364
01000, México D.F., Mexico
phone: 525-6225115
fax: 525-6161535
email: manuel@Ifunam.ifisicacu.unam.mx

- Photoluminescence and dielectric properties of impurities in inorganic crystals

Hollis, David B.
University of Paisley
Department of Physics
High Street
Paisley PA1 2BE
Scotland, U.K.
phone: 041-848-3000

- Glasses for active and passive infrared componenets
- Nonlinear optical devices e.g. photobleaching optical switching, frequency shifting

Hvam, Jørn M
Mikroelektronik Centret
Technical University of Denmark
Building 345 East
DK-2800 Lyngby, Denmark
phone: 45-45934610
fax: 45-45887762
email: hvam@mic.dth.dk

- Ultrafast nonlinear spectroscopy of semiconductor quantum structures

Kaminskii, Alexander
Institute of Crystallography
Russian Academy of Sciences
Leninsky pr. 59
Moscow 117333, Russia
phone: 7-095-1352210
fax: 7-095-1351011

- Physics of insulating laser crystals: search and characterization (spectroscopy and properties)
- Physics of crystalline lasers

Klingshirn, Claus
Fachbereich Physik
Universität Kaiserslautern
Erwin-Schrödinger-Straße
6750 Kaiserslautern, Germany
phone: 49-631-2052319
fax:49-631-2053300

- Semiconductor physics: linear and nonlinear optical roperties of SC
- Epitaxial growth of SC

Kometer, Ronald
AG Schwentner
Freie Universität Berlin
Fachbereich Physik
Arnimallee 14
1000 Berlin 33, Germany
phone:030-8384294
fax: 030-8383050

- Matrix isolation spectroscopy: investigation of dynamics of silver atoms in rare gas crystals by ps/fs pump-probe technique

Krausz, Elmars
Research School of Chemistry
Australian National University
Canberra ACT 0200, Australia
phone:61-6-2493577
fax: 61-6-2490750
email: krausz@rsc.anu.edu.au

- High resolution spectroscopy of solids : hole burning, line narrowing, optical energy transfer

Kumru, Mustafa
Firat University
Physics Department
23169 Elazig, Turkey
phone: 811-28500 /3365
fax: 811-22717
email: fizo8@trfirat.bitnet

- Physical properties of hydrogenated amorphous germanium carbide
- Optical and magnetic properties of transition metal complexes

Kyrkos, Stamatios*
Department of Physics
Boston College
Chestnut Hill, Massachusetts 02167 U.S.A.
phone:617-552-4588
fax: 617-552-0478
email: kyrkos@bcvms.bc.edu

- Theortical solid state physics

Lahoz Zamarro, Fernando
Física de la Materia Condensada
Facultad de Ciencias
Campus Universitario
50009 Zaragoza, Spain
phone: 34-76-560741

- Luminescence, EPR and raman spectroscopy of doped mixed crystals

---

* Secretary of the course

Langbein, Wolfgang Werner
Arbeitgruppe Klingshirn
Fachbereich Physik
Universität Kaiserslautern
Erwin-Schrödinger-Straße
6750 Kaiserslautern, Germany
phone:0631-2053112
fax: 0631-2053300
email: Langbein@sun.rhrk.uni-kl.de

- Linear and nonlinear spectroscopy of III-V and II-VI semiconductor structures

Large, Maryanne
Dept. of Pure & Applied Physics
Trinity College
University of Dublin
Dublin 2, Ireland
phone: 353-1-702-1469
fax: 353-1-711-759
email: mlarge@vax1.tcd.ie

- Nonlinear optics of DNA

Lipinska-Kalita, Krystyna E.
ENEA, CRE-Frascati, Dip. INN-SVIL
00044 - Frascati (Roma), C.P. 65 Italy
phone:39-6-94005207
fax: 39-6-94005400
email: froman@irmenea

- Microstructure of oxide glasses by very low frequency Raman scattering spectroscopy
- Glasses doped by nanocrystals

Lorenzo, Alfonso
Dpto Física de Materiales C-IV
Universidad Autónoma de Madrid
Cantoblanco 28049 Madrid, Spain
phone: 34-1-3974772
fax: 34-1-3978579

- Optical spectroscopy of rare earth doped $LiNbO_3$
- Time-resolved spectroscopy
- Site-selective spectroscopy

Lozykowski, Henryk J.
Dept. of Electrical & Computer Engineering
Ohio University
Athens, Ohio 45701-2979 U.S.A.
phone: 614-593-5072
fax: 614-593-0007

- The optical properties of bulk semiconductors, quantum wells and superlattices, rare earth doped II-VI, III-V semiconductors

Macfarlane, Roger M.
IBM Almaden Research Center K34/802
650 Harry Road
San Jose, California 95120 U.S.A.
phone: 408-927-2428
fax: 408-927-2100
email: macfarla@almaden.ibm.com

- Optical coherent transients
- Spectral hole-burning
- Up-conversion of light
- Upconversion lasers
- Rare earth spectroscopy

Martins, Evely
IPEN/CNEN/SP
Department of Optoelectronics Materials
P.O. Box 11049
CEP 05422-970
São Paulo, Brazil
phone: 55-11-2116011 ext. 1803
fax: 55-11-2123546

- ESA-Spectroscopy
- Solid-state lasers

Mollenauer, Linn
AT&T Bell Laboratories, Rm 4C-306
101 Crawfords Rd., P.O. Box 3030
Holmdel, New Jersey 07733 U.S.A.
phone: 908-949-5766
fax: 908-949-5784
email: linn@spin.att.com

- Solitons and other nonlinear effects in optical fibers
- Lasers
- Fiber optic technology in general

Mortier, Michel
Équipe de Physique de l'État Condensé
Université du Maine - Faculté des Sciences
Route de Laval
B.P. 535, 72017 Le Mans Cedex, France
phone: 43-833268
fax: 43-833366
email: dg@univ.lemans.fr

- Lattice dynamics of fluoride crystals and transitions ions spectroscopy
- Vibronic studies

Özen, Gönül
Centre National de la Recherche Scientifique
Laboratoire de Physicochimie des Matériaux
(UPR-211)
1, Place Aristide Briand
92190 Meudon, Bellevue, France
phone: 331-45075594
fax: 331-45075107
email: lize@bellevue.cnrs_bellevue.fr

- Up-conversion processes in glasses doped with rare earth ions
- Spectroscopy of solid-state laser materials

Pellé, Fabienne
Centre National de la Recherche Scientifique
Laboratoire de Physicochimie des Matériaux
(UPR-211)
1, Place Aristide Briand
92190 Meudon, Bellevue, France
phone: 033-145075593
fax: 033-145075107
email: lize@bellevue.cnrs-bellevue.fr

- Rare earth spectroscopy
- Up-conversion processes
- NLO inorganics (EFISH)

Quagliano, John R.
CLS-6 MS E535
Chemical and Laser Sciences Division
Los Alamos National Laboratory
Los Alamos, New Mexico 87545 U.S.A.
phone: 505-667-7732
fax: 505-665-4637
email: (Internet) jrq@beta.lanl.gov

- Semi-empirical modelling of metal ion energy states
- Solid-state lanthanide spectroscopy
- Characterization of chemical defects in solids

Reisfeld, Renata
Department of Inorganic
and Analytical Chemistry
The Hebrew University of Jerusalem
Jerusalem, Israel
phone: 972-2-585323
fax: 972-2-585319
email: (Bitnet) renata@vms.huji.ac.il

- Spectroscopy of inorganic ions in glasses
- Radiative and nonradiative transitions
- Solid-state lasers tunable in the visible
- Nonlinear optics
- Sensor- Sol Gel Glasses

Riedener, Toni
Institut für Anorganische Chemie
Universität Bern
Freiestraße 3
3012 Bern, Switzerland
phone: 41-31-654254
email: riedener@iac.unibe.ch

- Infrared to visible up-conversion in Tm-Yb-systems

Rino Cerveira da Silva, Luis Miguel
Departamento de Fisica
Universidade de Aveiro
Campus Universitário
3800 Aveiro, Portugal
phone: 351-034-25085
fax: 351-034-24965

- Optical spectroscopy of synthetic diamond and diamond films
- Time-resolved spectroscopy and uniaxial stress measurements

Runser, Claude
Institut de Physique & de Chimie
des Matériaux de Strasbourg
Unité Mixte 380046 - CNRS -
ULP - EHICS
Groupe d'Optique Nonlinéaire et
d'Optoélectronique
5, rue de l' Université
67084 Strasbourg Cedex, France

- Second order nonlinear properties of organic molecules with intra-molecular charge-transfer
- EFISHG studies
- Calculations

Rupnik, Tanja
Jozef Stefan Institute
Jamova 39
61111 Ljubljana, Slovenia
phone:38-61-159199
fax: 38-61-161029
email: tanja.rupnik@ijs.si

- Laser spectroscopy of solid-state laser materials (Er:YAG crystal)

Scaramuzzi, Francesco
ENEA
Centro Ricerche Energia Frascati
Via Enrico Fermi, 27
00044 Frascati (Roma), Italy
phone: 6-9400-5300
fax: 6-9400-5400

- Cryogenics
- Low temperature physics
- Cold fusion

Schmolke, Rüdiger
Inst. für Theor. Physik
TU-Berlin
Sekr. PN7-1
Hardenbergstraße 36
10623 Berlin , Germany
phone: 49-30-31424443
fax: 49-30-31421130
email: schmolke@itpz.physik.tn-berlin.de

- Nonlinear dynamics (theory)
- Semiconductor optics

Simonelli, Roberto
Via XXV Aprile , 16
27036 Mortara (PV), Italy
phone:39-384-93233
fax: 39-384-90585
email:simonelli@vaxpv.pv.infin.it

- Non linear optics
- Quantum optics
- MQW
- PNA and PND

Siraco, Angela◊
10 Spruce Circle
Andover, MA 01810 U.S.A.
phone: 508-475-2051

- Italian literature

Sztucki, Jaroslaw
Institute for Low Temperature
and Structure Research
Polish Academy of Sciences
50-950 Wroclaw
P.O. Box 937, Poland
phone: 48-71-35021
fax: 48-71-441029
email: widmo@plwrtu11

- Chemical physics
- Theoretical spectroscopy (esp. two photon) of rare earths and transition metals in solids
- Quantum groups

Tate, Harvey J.
Physics Department
University of Southampton
Highfield
Southampton SO9 5NH U.K.
phone:44-703-592194
fax:44-703-585813
email: hjt@orc.soton.ac.uk

- Spectroscopy of bulk and fiber glasses doped with rare earth ions
- Single ion spectroscopy

Toncelli, Alessandra
Viale Mameli 31/c
57126 Livorno, Italy
phone: 0586-809018

- Solid-state lasers

◊ Administrarive Assistant

Wagner, Hans-Peter
Institut für Physik II - Festkörperphysik
Universität Regensburg
Universitätsstraße 31
8400 Regensburg, Germany
phone: 49-941-943-2068
fax: 49-941-943-4226

- MOVPE growth of wide-gap II-VI semiconductors
- Linear and nonlinear optical properties of II-VI and second harmonic generation in II-VI waveguides

Wessels, Bruce W.
Northwestern University
Department of Material Sciences and Engineering
Materials and Life Sciences Building
2225 N. Campus Dr.
Evanston, Illinois 60208-3108 U.S.A.
phone: 708-491-3219
fax: 708-491-7820
email: b-wessels@nwu.edu

- MOVPE growth of III-V and oxide thin films
- Rare earth doped semiconductors
- Nonlinear optical properties of oxides

Woggon, Ulrike
Fachbereich Physik
Universität Kaiserslautern
Erwin-Schrödinger-Straße
6750 Kaiserslautern, Germany
phone: 49-631-2053112
fax: 49-631-2053300
email: Langbein@sun.rhrk.uni-kl.de

- Linear and nonlinear optical properties of quantum dots

Wörner, Andreas
Arbeitgruppe Klingshirn
Fachbereich Physik
Universität Kaiserslautern
Erwin-Schrödinger-Straße
6750 Kaiserslautern, Germany
phone: 49-631-2053168
fax: 49-631-2053300

- Nonlinear optical spectroscopy of $Al_xGa_{1-x}As$ at the crossover from direct to indirect semiconductor

Yang, Chun-Lai
Department of Physics
Boston College
Chestnut Hill, Massachusetts 02167 U.S.A.
phone: 617-552-3291
fax: 617-552-8478

- Luminescence spectroscopy of solid state laser materials

INDEX